D1220280

EQUILIBRIUM ANI
STATI

EQUILIBRIUM AND NONEQUILIBRIUM STATISTICAL MECHANICS

RADU BALESCU

Faculté des Sciences
Université Libre de Bruxelles
Brussels, Belgium

A Wiley-Interscience Publication
JOHN WILEY & SONS
New York · London · Sydney · Toronto

Library of Congress Cataloging in Publication Data:

Balescu, R.
Equilibrium and nonequilibrium statistical mechanics.
"A Wiley-Interscience publication."
Includes bibliographies.
1. Statistical mechanics. I. Title.
QC174.8.B34 1975 530.1'3 74-20907
ISBN 0-471-04600-0

Printed in the United States of America

10 9 8 7 6 5 4 3 2 1

PREFACE

In recent years many new and excellent books on statistical mechanics have appeared. Yet it was my impression for a long time that a certain type of book was still lacking on the present market. Being in this mood, I was enormously pleased—and honored—when Professor S. Rice suggested that I should write a book on this subject. I accepted his proposition gratefully and enthusiastically and started immediately on the project. The present book is the result of four years' work. In these four years the manuscript underwent numerous and serious transformations that reflected the changes in my own mental attitudes toward the subject. These transformations will remain, however, in a private domain. Only my close friends, co-workers, and students were exposed to them: I remain very grateful for their moral support and encouragement in all these years.

The decision to draw a final full stop at the end of a manuscript is always a very difficult one. Just as in research one never feels fully satisfied with one's own work, one feels tempted to continue the refinement process *ad infinitum.* But other factors work in the opposite direction. An active field like statistical mechanics is very far from being in a steady state. While one works on refining, and possibly contracting some chapter, new and often important results are flowing in from another side, and these in turn bring along the compelling need of a revision of some other chapter. It was my repeated experience that the result of such phenomena is an ever-expanding volume of the manuscript, the contractions being always overcompensated by unavoidable additions. For this reason, a compromise is necessary, and a pretty arbitrary decision is needed to terminate the manuscript.

The structure of this book is based on a few guiding ideas. It appears to me that in a faithful representation of present-day statistical mechanics the equilibrium theory and the nonequilibrium theory must be given equal weight. The lack of balance between these two aspects is, in my opinion, the major defect in most books existing on the market. Perhaps the only books in which this equilibrium is achieved are Gibbs' and Tolman's classics. As, however, these books were published in 1902 and in 1938, respectively, an updating was quite necessary. In the present book, equilibrium theory and nonequilibrium theory occupy roughly comparable space. In the presentation I have attempted to stress, as much as possible, the features that are similar in the two fields, in order to underline the structural unity of statistical mechanics.

A second general feature of this book is the lack of a rigid separation between classical and quantum statistical mechanics. This again proceeds from my search for a unified presentation. On purpose, in some chapters I go back and forth from the quantum to the classical language, while in other chapters I use a general symbolism that can be translated at will into one or the other. My view on this point is that statistical mechanics is something like a "transfer mechanics"* whose role is to transmit information from the microscopic to the macroscopic level. As such, it has developed a formalism of its own, which is well adapted to this function and which, basically, does not depend on the type of description of the underlying molecular level.

One of the most difficult decisions with which I was faced in writing this book was connected with the selection of the material. It is of course impossible, within a reasonable volume, to discuss or even list all the matters in a field of science whose size is quickly approaching the "thermodynamic limit." I therefore proceeded by sampling a few definite problems, which I discuss in rather great detail. There are, of course, important problems that are practically not touched upon, such as solid-state physics, low-temperature physics, superconductivity, relativistic statistical physics, let alone economical or sociological problems. I think, however, that the reader who assimilates the matters, methods, and ideas discussed in this book will have no difficulty in understanding the current literature in any other specific field.

In this connection, I wish to make a special remark. In the last few years, a series of important new results was developed in our group in Brussels by I. Prigogine and his co-workers, mainly in connection with the "causal dynamics," the "physical particle representation," and the generalized H theorem. Some readers may be surprised to see that these matters are not treated here. There are two simple reasons for this. First, the thorough discussion of these problems requires mathematical concepts and techniques beyond the average level of this book. These questions are more appropriate for a monograph than for a general textbook like this one. Second, and more important, such a monograph is presently being prepared by I. Prigogine and his co-workers, and I do not wish to compete with their work. I therefore limited myself to providing the reader with the necessary references to the relevant articles.

Among the subjects that *are* treated, I tried to establish a balance

* In the same sense as the "transfer RNA" of molecular biology.

between "classical" matters and new developments. I think that a nonnegligible amount of the latter is compulsory in any course of lectures in which one tries to avoid the dryness of established science and to offer a vista on actual scientific research. There is, of course, a risk involved in such a selection. First of all, recent (but also less recent) matters are often the subject of wild controversy. From that point of view I tried to remain neutral (insofar as this is possible). I did not try systematically to compare parallel theories of a single subject, but in my presentation I attempted a synthesis of the ideas of the various approaches. A more serious risk is related to the future fate of the new concepts selected for presentation. I think, however, that every physicist should at some time have the courage of making such bets on the future.

As for the general style and presentation, it should be realized that this book has essentially a pedagogical purpose. It is addressed primarily to physicists and to chemists. The mathematicians will certainly feel unhappy. I deliberately avoided mathematical rigor in favor of physical ideas. But lack of rigor, to me, does not necessarily imply sloppiness. My endeavor was, therefore, to present the matters as clearly as possible without too much sophistication, but in such a manner as to "pave the way" toward a rigorous treatment. At the other extreme, I also deliberately avoided purely philosophical discussions on the great questions such as the picture of the universe in statistical mechanics, the arrow of time, the origin of irreversibility, and the like. Like everybody else, I may have my personal, private opinions on these matters: they are not necessarily definitive. I therefore preferred to provide the reader with a physical and mathematical background and let him draw his own metaphysical conclusions.

As prerequisites for reading this book, a working knowledge of classical and quantum mechanics and of thermodynamics is necessary. The mathematical background is no more elaborate than the one required for quantum mechanics.

I wish to express here my deep gratitude to Professor Ilya Prigogine, with whom I have had the privilege of working for nearly twenty years. Since the first days of my apprenticeship I felt the effect of the passionate discussions we had together, of the rapid flow of ideas, and of his never failing enthusiasm. Without his stimulating contact over the years, this book would never have been written. He created around himself a most extraordinary group, often called the "Brussels school," in which professors, research physicists and chemists, and graduate students, working all together in strong interaction and in an atmosphere of mutual friendship, contributed to the advancement of statistical mechanics. To all the members of the group, I express my appreciation.

In particular, I wish to acknowledge the numerous discussions we have had over many years with my old friends and colleagues, Professors

Claude George, Françoise Henin, Gregor Nicolis, George Severne, and especially Pierre Résibois, who critically read some of the chapters.

My most sincere gratitude is due to my friend and co-worker Dr. Irina Paiva-Veretennicoff. She reviewed and checked most of the matters of this book, and "tested" many of them with her own students. Her permanent and enthusiastic optimism was an appreciable moral support during the elaboration of this book.

After the completion of the manuscript I received detailed comments from Professors I. Prigogine, S. Rice, and J. Yvon. These comments were valuable for the preparation of the final version. I wish to thank them sincerely for their help.

I also want to express my appreciation for very useful discussions with Acad. N. Bogoliubov, Professors R. Brout, Iu. Klimontovich, G. Prosperi, L. Rosenfeld, W. Schieve, V. Silin, L. van Hove, N. van Kampen, and H. Wergeland. Among the many colleagues and co-workers with whom I had most fruitful discussions, in Brussels and abroad, let me express my appreciation to Drs. J. Aderca, P. Allen, M. Baus, L. Brenig, P. Clavin, M. De Leener, Y. Dramaix, K. Haubold, K. Ichikawa, W. Kazimirowski, K. Kitahara, T. Kotera, A. Kuszell, L. Lanz, L. Lugiato, P. Mandel, A. Mangeney, J. Misguich, K. Nishikawa, J. Piasecki, E. Piña, M. Poulain, Professor A. Pytte, Drs. L. Reichl, A. Senatorski, R. Sergysels, Y. Soulet, J. Turner, J. Vardalas, and J. Wallenborn.

I wish to take this opportunity to express my gratitude to my alma mater, the Université Libre de Bruxelles, whose authorities never failed to support and encourage my own and my co-workers' activities.

Part of this work was done within the framework of the association of our laboratory with Euratom, in the field of statistical mechanics of plasmas. I gratefully acknowledge the support of this great European organization.

Part of the book was written during my stay as a Visiting Professor at the University of Texas in Austin in 1970–1971. I wish to thank Professors F. de Wette, W. Schieve, and R. Schechter for their hospitality and for many discussions.

I thank my wife and my daughters for their support and patience in undergoing the hardships that this enterprise imposed on our family life.

Last, but not least, I wish to thank Miss N. Galand, Mrs. L. Thomaes, Mrs. S. Deraumaux, and Miss M. Simon for their excellent secretarial work and Mr. P. Kinet and Mr. E. Kerckx for their technical assistance.

R. BALESCU

Brussels, Belgium
April 1974

CONTENTS

EQUILIBRIUM AND NONEQUILIBRIUM STATISTICAL MECHANICS

PART 1

GENERAL CONCEPTS OF STATISTICAL MECHANICS

CHAPTER 1
REVIEW OF HAMILTONIAN DYNAMICS

1.1. INDIVIDUAL AND GLOBAL BEHAVIOR

Statistical mechanics is the mechanics of large assemblies of (relatively) simple systems such as molecules in a gas, atoms in a crystal, photons in a laser beam, stars in a galaxy, cars on a highway, people in a social group, and so on. The main purpose of this science is to understand the behavior of the assembly as a whole in terms of the behavior of its constituents. Clearly, the whole does not behave as a simple superposition of its parts. The behavior of each constituent is modified by the mere presence in its neighborhood of another partner. A car driver is hindered by the presence of another car in front of his own, which he cannot pass. He must therefore change his own way of driving. This is the essence of the interaction process. Because of the cumulative effect of the interactions, the assembly as a whole can be completely and qualitatively different from its individual constituents. The most elementary symmetries of the motion of the individual particles can be broken by considering large assemblies of these. One of the most striking of these broken symmetries is the invariance under time inversion. We shall not dwell at length on a listing of problems on the first page of this book. This discussion is only an *apéritif* to be taken before starting to work. As we go on, the problems will appear, one after the other in a natural order.

The laws of motion of the individual parts are considered to be *given:* their derivation is not the object of statistical mechanics. A knowledge of these laws is, however, the underlying basis of this science. It is, of course, impossible in the framework of a single book to treat all the various kinds of systems listed above. We shall restrict our attention to a particular class of these. However, the general methods and ideas developed in this book can be adapted more or less successfully to the study of quite different problems.

The class of systems selected for study in this book (and actually in all existing books on statistical mechanics) is the class of systems governed by *Hamiltonian dynamics.* All the systems described individually, at least to a good approximation, by the laws of classical or of quantum mechanics belong to this class. Hence, we have at our disposal a very wide spectrum of different systems. (Of course, car drivers do not belong to this class, because the constituent units as such are already quite complex assemblies of molecules, whose laws of motion and of interaction are very complicated and not yet well understood.)

Hamiltonian dynamics is essentially characterized by a *structure:* here lies its beauty as well as its usefulness. This structure is common to classical and quantum systems. In this chapter we shall review classical and quantum dynamics by stressing particularly these structural aspects. It is important for the understanding of statistical mechanics to have a good feeling of this structure in order to keep track of where and how the Hamiltonian character is lost in the final stage of the description.

1.2. HAMILTONIAN DESCRIPTION OF CLASSICAL MECHANICS

In Hamiltonian dynamics a system is characterized, at any fixed instant of time, by a set of $2N$ numbers $q_1, \ldots, q_N, p_1, \ldots, p_N$. The q_i's are called *generalized coordinates,* and the p_i's are *generalized momenta* conjugate to q_i. The q_i's may represent positions of molecules in space, but they may also represent more abstract quantities, such as the amplitude of a wave, or some numbers characterizing the internal degrees of freedom of a molecule. The generalized momenta p_i are associated with the q_i's in a precise way, which is well known from mechanics. We will recall later the condition to be satisfied by p_i, q_i in order for them to be a conjugate pair. The number N of pairs q_i, p_i necessary to characterize completely the dynamical system is called the *number of degrees of freedom.*

To avoid too heavy notations, we will frequently use abreviations, whenever there is no risk of confusion. The following notations will be used interchangeably to denote the set of $2N$ q's and p's:

$$(q_1, \ldots, q_N, p_1, \ldots, p_N) \equiv (q, p) \tag{1.2.1}$$

We may think of this description in geometrical terms. The dynamical system under consideration can be represented by a *point* in a $2N$-dimensional space, spanned by a Cartesian reference frame of $2N$ mutually orthogonal axes, corresponding to the variables $(q_1, \ldots, p_N)$. This space is called the *phase space* (sometimes also the Γ space). It plays a fundamental role, being the natural framework of dynamics and statistical mechanics. Although it is impossible to represent such concepts properly on paper whenever $N > 1$, we may help fixing the ideas by using diagrams drawn in an (inappropriate) three-dimensional space, such as Fig. 1.2.1.

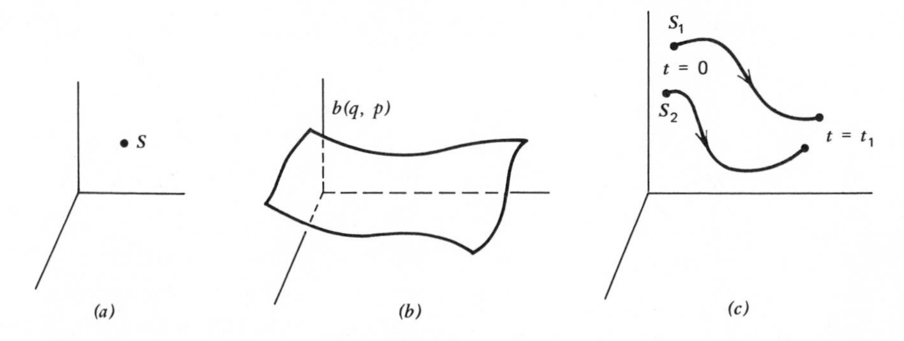

Figure 1.2.1. Various objects in phase space: (*a*) the point *S* represents the state of a system; (*b*) the graph of a smooth dynamical function $b(q, p)$; (*c*) trajectories of dynamical systems.

We will very often be interested in the value of certain quantities characterizing the system and that, in principle, could be measured. Examples of such quantities are the energy, the momentum, and the angular momentum. These quantities have a definite value for each state of the system (q, p). In other words, they can be characterized as the set of all real functions of the $2N$ variables $(q_1, \ldots, q_N, p_1, \ldots, p_N)$. They will be called *dynamical functions* and denoted collectively by $b(q, p)$: they describe all conceivable properties of the system. One could restrict oneself to certain types of functions. For instance, one could use all analytical functions of q, p, in which case every $b(q, p)$ is of the form:

$$b(q, p) = \sum_{n_1=0}^{\infty} \cdots \sum_{n_N=0}^{\infty} \sum_{m_1=0}^{\infty} \cdots \sum_{m_N=0}^{\infty} \bar{\beta}_{n_1 \cdots m_N} q_1^{n_1} \cdots q_N^{n_N} p_1^{m_1} \cdots p_N^{m_N} \quad (1.2.2)$$

where $\bar{\beta}_{n_1 \cdots m_N}$ are arbitrary real constants. We may also consider $b(q, p)$ as belonging to the class of functions that can be represented as Fourier series or integrals:

$$b(q, p) = \int dk_1 \cdots dk_N dj_1 \cdots dj_N \beta_{k_1 \cdots j_N} \exp\left[i \sum_{n=1}^{N} (k_n q_n + j_n p_n)\right] \quad (1.2.3)$$

where $\beta_{k_1 \cdots j_N}$ are arbitrary functions of k and j.*

* The coefficients β_{kj} may be singular "functions" of their arguments. For instance, the particular dynamical function $b(q, p) = q_1$ is represented by

$$q_1 \Rightarrow \beta_{k_1 \cdots j_N} = i\delta'(k_1) \prod_{r=2}^{N} \delta(k_r) \prod_{n=1}^{N} \delta(j_n)$$

where $\delta(k)$ is Dirac's singular delta function and $\delta'(k) = d\delta(k)/dk$.

The preceding description represents completely the state of the system at a given instant of time, say $t=0$. The main object of dynamics, however, is to study the evolution of the system in time. In the Hamiltonian description of dynamics, the motion is completely determined if we prescribe a given, privileged, dynamical function $H(q, p)$ called the *Hamiltonian.** This function completely characterizes the dynamical nature of the system. Physically, it is known that in most cases (but not always) $H(q, p)$ represents the total energy of the system.

Under the motion, the representative point of the system moves along a trajectory in phase space (see Fig. 1.2.1c). This trajectory can be represented by a set of $2N$ functions of time $q_i(t)$, $p_i(t)$, which are determined by solving *Hamilton's equations:*

$$\dot{q}_i = \frac{\partial H(q, p)}{\partial p_i}$$
$$\dot{p}_i = -\frac{\partial H(q, p)}{\partial q_i} \tag{1.2.4}$$

These $2N$ equations determine uniquely the value of q_i, p_i at all times, given their initial values $q_i(0)=q_i^0$, $p_i(0)=p_i^0$. Through each point of the phase space, passes one and only one trajectory satisfying Eq. (1.2.4).

Consider now an arbitrary dynamical function $b(q, p)$. As a result of the motion, its value will also change in time. Its rate of change will be given by

$$\begin{aligned} \dot{b}(q, p) &= \sum_{n=1}^{N} \left(\frac{\partial b}{\partial q_n} \dot{q}_n + \frac{\partial b}{\partial p_n} \dot{p}_n \right) \\ &= \sum_{n=1}^{N} \left(\frac{\partial b}{\partial q_n} \frac{\partial H}{\partial p_n} - \frac{\partial b}{\partial p_n} \frac{\partial H}{\partial q_n} \right) \end{aligned} \tag{1.2.5}$$

The second equality is obtained by using Hamilton's equations. The expression on the right-hand side plays such an important role in the theory that a special symbol is used to represent it:

$$[b, c]_P = \sum_{n=1}^{N} \left(\frac{\partial b}{\partial q_n} \frac{\partial c}{\partial p_n} - \frac{\partial b}{\partial p_n} \frac{\partial c}{\partial q_n} \right) \tag{1.2.6}$$

This expression is called the *Poisson bracket* of the two dynamical functions $b(q, p)$, $c(q, p)$. We may therefore write

$$\dot{b} = [b, H]_P \tag{1.2.7}$$

This is the most fundamental equation of Hamiltonian dynamics. It includes Eqs. (1.2.4) as special cases.

* We will usually assume that the Hamiltonian is independent of time.

The Poisson bracket can be regarded as a special algebraic operation on dynamical functions that associates with every pair b, c of these a new dynamical function $v=[b, c]_P$. One easily checks the following properties:

$$[b, c]_P = -[c, b]_P \tag{1.2.8}$$

$$[b, [c, d]_P]_P + [c, [d, b]_P]_P + [d, [b, c]_P]_P = 0 \tag{1.2.9}$$

$$[b, \alpha]_P = 0 \tag{1.2.10}$$

where α is a *scalar*, that is, a parameter independent of q and p. Any abstract operation having the properties (1.2.8)–(1.2.10) is called in algebra a *Lie bracket*. The Poisson bracket (1.2.6) is a particular "realization" of an abstract Lie bracket. These properties are very different from the rules of more usual operations such as addition or multiplication. For instance, Eq. (1.2.8) shows that the bracket is not commutative. It implies that

$$[b, b]_P = 0 \tag{1.2.11}$$

Equation (1.2.9) shows that the bracket is not associative either. This property is called the *Jacobi relation.*

We can now verify that the following rules connect the Poisson bracket operation with other algebraic operations:

$$[(b+c), d]_P = [b, d]_P + [c, d]_P \tag{1.2.12}$$

$$[\alpha b, c]_P = \alpha[b, c]_P \tag{1.2.13}$$

$$[bc, d]_P = b[c, d]_P + [b, d]_P c \tag{1.2.14}$$

The action of the operator $[\cdots, d]_P$ on a dynamical function is, therefore, similar to the action of a first-order differential operator. This is not surprising in the case of the Poisson bracket (1.2.6), but will remain true also in the more general cases we will meet later. Using these rules, we may calculate the Poisson bracket of any two elements, provided we know the "bracket multiplication table" of the fundamental elements q_r, p_r. This table is easily found from (1.2.6):

$$\begin{aligned} [q_r, q_s]_P &= 0 \\ [p_r, p_s]_P &= 0 \\ [q_r, p_s]_P &= \delta_{rs} \end{aligned} \tag{1.2.15}$$

where the Kronecker δ_{rs} is defined as usual:

$$\begin{aligned} \delta_{rs} &= 0 \quad \text{for} \quad r \neq s \\ &= 1 \quad \text{for} \quad r = s \end{aligned} \tag{1.2.16}$$

These relations define a *basic set of canonically conjugate variables.*

The set of all the dynamical functions will be called the *dynamical algebra* $\mathcal{D}$. The reason for this name is that one may combine any elements of $\mathcal{D}$ by means of the operations of addition, multiplication, and the Poisson bracket without leaving the set. A set in which these three operations are defined is called technically an algebra, or more precisely, a *Lie algebra.*

The Poisson bracket concept can be used for the definition of a class of *operators* acting on the dynamical functions. Let a be a fixed dynamical function belonging to $\mathcal{D}$. We define an operator $[a]$ acting on an arbitrary element b of the dynamical algebra $\mathcal{D}$ through

$$[a]b \equiv [b, a]_P \tag{1.2.17}$$

These operators $[a]$ play a fundamental role in Hamiltonian dynamics and in statistical mechanics.

Let us now come back to the law of motion and, before continuing, make the following remark. We start with an arbitrary dynamical function, represented at time zero by Eq. (1.2.2). By the motion, it will be transformed at time t into a new dynamical function $b(q, p; t)$, which can be looked at in two ways. It can be considered as the *same* function of the transformed variables:

$$b(q, p; t) = \sum_{n,m} \bar{\beta}_{nm} q_1^{n_1}(t) \cdots p_N^{m_N}(t) \tag{1.2.18}$$

This point of view corresponds to the first equation in (1.2.5). The coefficients $\bar{\beta}_{nm}$ are constants in this image. Alternatively, we may solve explicitly Hamilton's equations for $q_i(t)$, $p_i(t)$ as functions of time and of the initial values q_i, p_i. Substituting these functions into (1.2.18) we obtain a new function of q_i, p_i, that is, a new set of coefficients $\bar{\beta}_{nm}(t) \neq \bar{\beta}_{nm}$:

$$b(q, p; t) = \sum_{n,m} \bar{\beta}_{nm}(t) q_1^{n_1} \cdots p_N^{m_N} \tag{1.2.19}$$

In this picture, the variables $q_1 \cdots p_N$ are constant in time. To sum up, $b(q, p; t)$ may be considered either as the old function of new variables or as a new function of the old variables. The equivalence of these two points of view is an important property that will be discussed again at the end of this section.

In statistical mechanics, the second point of view is more natural. Indeed, the set (q, p) defines a phase space, once and for all. We want to inscribe all the phenomena in this unique phase space, hence we do not want to consider q and p as changing objects.

The central problem of dynamics can now be formulated as follows:

Given a dynamical function $b(q, p)$ at time $t=0$, what will be the dynamical function $b(q, p; t)$ corresponding to b at time t under the law of motion (1.2.7)?

As the correspondence is assumed to be continuous, we can solve this problem by expanding the solution in powers of t. Hence

$$b(t)=b+t\dot{b}+\tfrac{1}{2}t^2\ddot{b}+\cdots \tag{1.2.20}$$

where for simplicity we did not write the argument $t=0$ for dynamical functions evaluated at the initial time. We also omit writing explicitly the arguments q and p of the function b. The law of dynamics (1.2.7) gives the value

$$\dot{b}=[b, H]_P=[H]b \tag{1.2.21}$$

As however $\dot{b}$ is also a dynamical function, its own time derivative is also given by (1.2.7):

$$\ddot{b}=[\dot{b}, H]_P=[[b, H]_P, H]_P=[H]^2 b \tag{1.2.22}$$

Quite generally, the nth time derivative $b^{(n)}$ is given by

$$b^{(n)}=[H]^n b \tag{1.2.23}$$

Substituting these results into Eq. (1.2.20), we see that the result has the form of an exponential series. We therefore define formally the operator $\bar{U}(t)=\exp\{t[H]\}$ by the equation:

$$\begin{aligned} b(t)&=\sum_{r=0}^{\infty}(r!)^{-1}t^r[H]^r b \\ &=e^{t[H]}b\equiv\bar{U}(t)b \end{aligned} \tag{1.2.24}$$

This equation provides us with the formal solution to the initial value problem of Eq. (1.2.7). (It is still only formal, because at this stage we have not studied any question of convergence, integrability, etc.) The operator $\bar{U}(t)$ defines a transformation from the initial value b to the value of the function at time $b(t)$; it is sometimes called the *Green operator* or *propagator*.

Before going further, let us remark that *all scalars α remain unaffected* under the transformation $\bar{U}(t)$. Indeed because of Eq. (1.2.10):

$$e^{t[H]}\alpha=\alpha \tag{1.2.25}$$

It is very easy to verify that the set of all transformations $\bar{U}(t)$, corresponding to all possible values of the parameter t, has the structure of a *group*. Indeed, if we perform successively the transformations $\bar{U}(t_1)$, $\bar{U}(t_2)$, the result is the same as if we performed the single transformation

$\bar{U}(t_1+t_2)$:

$$e^{t_1[H]}\{e^{t_2[H]}b\}=e^{(t_1+t_2)[H]}b \tag{1.2.26}$$

The operation is clearly associative, and there exists a *neutral element* $\bar{U}(0)=1$:

$$e^{0[H]}b=b \tag{1.2.27}$$

and an *inverse element:*

$$e^{-t[H]}\{e^{t[H]}b\}=b \tag{1.2.28}$$

Technically, $\bar{U}(t)$ is called a *one-parameter continuous group* (or Lie group) of *transformations.* The privileged element H is called the *generator* of the group.

An important concept is the notion of an *infinitesimal transformation* of a continuous group: it is the limiting transformation $\bar{U}(\delta t)$ as $\delta t \to 0$; clearly:

$$b(\delta t)=b+\delta t\,[b, H]_P \tag{1.2.29}$$

As follows from the previous discussion, the transformation $\bar{U}(t)$ associates with every element b of $\mathcal{D}$ another element $b(t)$ of $\mathcal{D}$: it defines a mapping of $\mathcal{D}$ onto itself. But this mapping has a very special property that makes it important: *It preserves the algebraic structure of* $\mathcal{D}$. A mapping having this strong property is called an *automorphism.* More explicitly, this means that if $x, y, \ldots, v$ are elements of $\mathcal{D}$ interrelated by the operations of the algebra:

$$z=x+y \tag{1.2.30a}$$

$$w=\alpha x \tag{1.2.30b}$$

$$u=xy \tag{1.2.30c}$$

$$v=[x, y]_P \tag{1.2.30d}$$

then the same relations hold between transformed elements:

$$z_t=x_t+y_t \tag{1.2.31a}$$

$$w_t=\alpha x_t \tag{1.2.31b}$$

$$u_t=x_t y_t \tag{1.2.31c}$$

$$v_t=[x_t, y_t]_P \tag{1.2.31d}$$

It is sufficient to prove these relations for the infinitesimal transformation (1.2.29). Starting from the right-hand sides we easily obtain, for instance:

$$\begin{aligned} x_t y_t &= (x+t[x, H]_P)(y+t[y, H]_P) \\ &= xy+t([x, H]_P y + x[y, H]_P)+O(t^2) \\ &= xy+t[xy, H]_P = u_t \end{aligned}$$

The property needed here is the *derivation* property of the bracket (1.2.14). The reader can easily prove the other equations (1.2.31) in a similar way, using only the algebraic properties (1.2.8)–(1.2.14) of the Lie bracket.

The fact that the law of motion generates an automorphism is extremely important. It implies in particular that if the coordinates:

$$q_1, \ldots, q_N, p_1, \ldots, p_N$$

are taken as a basis in phase space, the transformed quantities

$$q_1(t), \ldots, q_N(t), p_1(t), \ldots, p_N(t)$$

form a perfectly equivalent basis. The quantities at time t obey the same equations of motion as those at time zero (because all these equations involve only the four fundamental operations of the algebra). This fact expresses the *invariance of the laws of mechanics under the motion.*

We may now ask ourselves if the group generated by the Hamiltonian H is the only invariance group of mechanics. The answer is clearly no. There is nothing special about the Hamiltonian. Actually, every element G of $\mathscr{D}$ generates a one-parameter group of automorphisms of the dynamical algebra:

$$x \to e^{\alpha[G]}x, \qquad \alpha \text{ a real parameter} \tag{1.2.32}$$

We have thus obtained an infinite family of transformations: they are called the *group of canonical transformations.* Each of these has the same properties as the special case $\bar{U}(t)$. Hence we can formulate a fundamental property: *Hamiltonian dynamics is invariant with respect to every canonical transformation.*

The invariance of the algebra $\mathscr{D}$ under canonical transformations has a remarkable consequence, which can be expressed as follows.

Let the basic variables q and p be transformed into $q(\alpha)$, $p(\alpha)$ under some canonical transformation generated by an element G:

$$e^{\alpha[G]}q = q(\alpha), \qquad e^{\alpha[G]}p = p(\alpha)$$

Then we have quite generally the following relation, valid for any dynamical function:

$$e^{\alpha[G]}b(q, p) = b(q(\alpha), p(\alpha)) \tag{1.2.33}$$

In other words, under a canonical transformation, an arbitrary dynamical function goes over into the *same* function of the transformed variables. Indeed, for any analytic function of the form (1.2.2) the quantity $b(q, p)$ is built up from (q, p) by a combination of operations of addition and multiplication. As the sum and product operations are preserved by canonical transformations [see Eqs. (1.2.30) and (1.2.31)], the result (1.2.33) follows immediately.

A special case of this result is the equivalence of the two forms (1.2.18) and (1.2.19).

1.3. HAMILTONIAN DESCRIPTION OF QUANTUM MECHANICS

We now consider the important problem of transcribing the results obtained in classical mechanics and ensuring a smooth transition to quantum mechanics. In doing this we assume that the reader is already familiar with the basic concepts of quantum mechanics, and with their physical justification.

The fundamental difference between quantum and classical mechanics is expressed in a nutshell by *Heisenberg's uncertainty principle:* the position and the momentum of a particle cannot be determined simultaneously with arbitrary precision. The smaller the error committed in the measurement of the momentum, the larger the uncertainty of the position (and vice versa): the product of the errors Δp, Δq is bounded below by a constant:

$$\Delta p \Delta q \geqslant ah \tag{1.3.1}$$

where h is the fundamental Planck constant, and a is a numerical factor of order unity, whose exact value is irrelevant. The constant h has the dimensions of an action, that is, (energy) · (time) or (position) · (momentum); its value is $h = 6.6256 \times 10^{-27}$ erg-sec. Planck's constant frequently occurs in the combination $h/2\pi$ for which one uses a special notation:

$$\hbar = \frac{h}{2\pi} = 1.0545 \times 10^{-27} \text{ erg-sec} \tag{1.3.2}$$

Clearly, the formalism of Section 1.2 cannot take account of Relation (1.3.1): if the observables are described by numerical functions, representing the actual values they can take in an experiment, nothing prevents us from specifying these functions with arbitrary precision. We need another kind of object in order to describe the concept of an "observable" in quantum mechanics.

It is well known that the genial idea of Heisenberg and Schrödinger was to note that all physical phenomena covered by quantum mechanics could be explained, if one associated with q and p two abstract linear operators $\hat{q}$, $\hat{p}$ acting on an abstract Hilbert space. These operators are assumed to be noncommuting; more precisely, it is required that

$$[\hat{q}, \hat{p}]_- \equiv \hat{q}\hat{p} - \hat{p}\hat{q} = i\hbar\,\hat{1} \tag{1.3.3}$$

where $\hat{1}$ is the identity operator (i.e., $i\hbar\,\hat{1}$ is an ordinary number, or a c-number as it is frequently called in quantum-mechanical literature). We usually omit writing the operator $\hat{1}$ explicitly.

We shall construct a realization of the quantum Hamiltonian dynamics along the same lines as in Section 1.2. The state of the system will be realized by an element of the Hilbert space (i.e., a wave function) instead of a point in phase space. The dynamical functions will be represented as operators acting on the Hilbert space instead of as functions on the phase space. We postpone the more detailed discussion of the states to the next two sections and proceed with the construction of the algebra of dynamical operators $\mathcal{D}_Q$.

We first need a prescription establishing a link between the abstract operators and the actual physical measurements of their values. The prescription is based on the following property of Hilbert spaces: For each operator, there (usually) exists a certain number of elements (or states) of the Hilbert space that are "almost" invariant under the action of that operator. More precisely, the operator $\hat{b}$ acting on such a state,* denoted by $|m\rangle$, transforms the latter into itself, up to a certain numerical factor b_m:

$$\hat{b}\,|m\rangle = b_m\,|m\rangle \tag{1.3.4}$$

The set of all possible *eigenvalues* b_m (corresponding to all possible *eigenstates* $|m\rangle$) is interpreted as the set of values that the observable associated with $\hat{b}$ can assume in any experiment. In general, this set of values is discrete: here lies a definite difference between classical and quantum mechanics. In a quantum-mechanical system the dynamical variables (such as the energy) can only take certain well-defined values: this is the essence of quantization. A further important remark is the following: As the eigenvalues b_m must represent observable numerical values of the dynamical functions, they must necessarily be real numbers. This implies that the operators $\hat{b}$ representing observables must necessarily be Hermitian, that is:

$$\hat{b}^\dagger = \hat{b} \tag{1.3.5}$$

where $\hat{b}^\dagger$ is the Hermitian conjugate of $\hat{b}$. (If $\hat{b}$ is represented by a

* We assume the reader to be familiar with Dirac's system of notation. An arbitrary element of the Hilbert space (or an arbitrary state) is denoted by a *ket* $|m\rangle$, m being a number or a set of numbers characterizing the state. The conjugate state (an element of the dual Hilbert space) is denoted by a *bra* $\langle m|$. The *bracket* $\langle m \mid m\rangle$ denotes the norm of the state $|m\rangle$.

complex matrix b_{mn}, the Hermitian conjugate is obtained by transposition, followed by complex conjugation of the matrix $b^\dagger_{mn} = (b_{nm})^*$.)

We now try to fit these ideas into the algebraic structure of Section 1.2. From our previous remarks, the basic building stones will be a set of Hermitian operators $\hat{q}_r$, $\hat{p}_r$. Comparing (1.3.3) to (1.2.15), it is very natural to translate the classical Poisson-bracket operation by the *commutator* of two operators (divided by $i\hbar$):

$$[b, c]_P \to (i\hbar)^{-1}[\hat{b}, \hat{c}]_- \equiv (i\hbar)^{-1}(\hat{b}\hat{c} - \hat{c}\hat{b}) \tag{1.3.6}$$

It is an easy exercise to verify that the identification has all the required properties of a Lie bracket (1.2.8)–(1.2.14). Note also that $(i\hbar)^{-1}$ times the commutator of two Hermitian operators is also a Hermitian operator.

The multiplication poses a problem that must be discussed now. In general, a product of noncommuting Hermitian operators is not Hermitian. Consider, for example:

$$(\hat{b}^2\hat{c})^\dagger = \hat{c}^\dagger\hat{b}^\dagger\hat{b}^\dagger = \hat{c}\hat{b}^2 \neq \hat{b}^2\hat{c}$$

One can always construct linear combinations of products of operators that are Hermitian. However, even this symmetrization procedure is not unique. Indeed, in the example above, we can consider two such symmetrized products:

$$\tfrac{1}{2}(\hat{b}^2\hat{c} + \hat{c}\hat{b}^2); \qquad \hat{b}\hat{c}\hat{b}$$

Both of these (or any linear combination) are Hermitian and could be associated with the classical dynamical function b^2c. We need, therefore, a unique rule that specifies how the elements of the algebra are constructed. It should be clearly realized that such a prescription is a postulate, hence there is no point in trying to prove anything about it. (This does not always seem clear in the literature!) A formulation that is quite generally used is due to H. Weyl. It can be stated as follows:

The dynamical functions of quantum mechanics are all operators that can be represented in the form*:

$$\hat{b}(\hat{q}, \hat{p}) = \int dk\, dj\, \beta(k, j) \exp(ik\hat{q} + ij\hat{p}) \tag{1.3.7}$$

where $\beta(k, j)$ is a numerical function (possibly singular) of the numbers k, j. (Clearly, k stands for the set $k_1, \ldots, k_N$; $dk \equiv dk_1 \cdots dk_N$; moreover $k\hat{q} = \sum_r k_r\hat{q}_r$.) In addition we require

$$\beta(-k, -j) = \beta^*(k, j) \tag{1.3.8}$$

* We restrict ourselves here to systems of particles having no internal degrees of freedom. The more general case is discussed at the end of this section.

(The asterisk denoting complex conjugation), in order to ensure the Hermitian character of the operator $\hat{b}$. It is clear that Weyl's rule automatically gives Hermitian operators. Moreover, it is a *correspondence rule* in the sense that it defines the quantum-dynamical operator corresponding to a given classical-dynamical function $b(q, p)$. If the classical function is expanded [as in Eq. (1.2.3)] in the form:

$$b(q, p) = \int dk\, dj\, \beta(k, j) \exp(ikq + ijp) \tag{1.3.9}$$

then Weyl's rule states that the corresponding operator is (1.3.7) with the *same* function $\beta(k, j)$.

Considering again our former example, the classical function q^2p can be represented as

$$q^2p = \int dk\, dj [i^2\delta''(k)][i\delta'(j)] \exp(ikq + ijp)$$

where $\delta'(k)$ denotes the derivative of Dirac's delta function: $\delta'(k) \equiv d\delta(k)/dk$. The corresponding operator will be

$$\int dk\, dj [i^2\delta''(k)][i\delta'(j)] \exp(ik\hat{q} + ij\hat{p}) = \tfrac{1}{3}(\hat{q}^2\hat{p} + \hat{q}\hat{p}\hat{q} + \hat{p}\hat{q}^2)$$

The latter result is obtained by expanding the exponential under the integral. In handling exponentials of operators great care is always required in taking account of the noncommutation. In particular, the exponential of a sum of operators is *not* equal to the product of the individual exponentials. In the particular case (valid for the Weyl operator) that the commutator of the operators $\hat{b}$ and $\hat{c}$ is a c-number [see Eq. (1.3.3)], the following formula holds:

$$\exp(\hat{b} + \hat{c}) = \exp(\hat{b}) \exp(\hat{c}) \exp(-\tfrac{1}{2}[\hat{b}, \hat{c}]_-) \tag{1.3.10}$$

This formula can be verified by series expansion. A useful consequence is

$$\exp(\hat{b}) \exp(\hat{c}) = \exp(\hat{c}) \exp(\hat{b}) \exp(-[\hat{c}, \hat{b}]_-) \tag{1.3.11}$$

We now possess all the elements necessary for the formal realization of the quantum dynamical algebra $\mathcal{D}_Q$. Restricting ourselves again to particles without internal degrees of freedom, the basic building stones are the elements of a set of $2N$ Hermitian operators:

$$\hat{q}_r, \hat{p}_r, \qquad r = 1, 2, \ldots, N$$

satisfying Eq. (1.3.3). The elements of the dynamical algebra $\mathcal{D}_Q$ are all the operators $\hat{b}$ constructed from $\hat{q}_r$ and $\hat{p}_r$ by means of the following four algebraic operations:

$$
\begin{array}{ll}
\alpha\hat{b} & \text{multiplication of an operator by a number } \alpha \\
\hat{b}+\hat{c} & \text{sum of operators} \\
\hat{b}\hat{c} & \text{product of operators} \\
(i\hbar)^{-1}[\hat{b},\hat{c}]_- & \text{Lie bracket} = (i\hbar)^{-1}\times\text{commutator}
\end{array}
\tag{1.3.12}
$$

In other words, if α is any complex number and $\hat{b}$, $\hat{c}$ are operators belonging to $\mathcal{D}_Q$, we state that the four operators defined in Eq. (1.3.12) also belong to the algebra.

We may note that the algebra defined in this way contains Hermitian as well as non-Hermitian operators, because the product of two Hermitian operators is not necessarily Hermitian, as explained above. One may try to use, instead of the ordinary product, the Weyl-symmetrized product for the construction of the algebra. This, however, is not a good idea, because this operation is not associative, as can easily be verified. It is much better to construct a larger set having the same structure as the classical set $\mathcal{D}$.* We simply add the constraint that only the Hermitian operators of $\mathcal{D}_Q$ represent observable quantities.

Having defined this realization of the dynamical algebra we can now translate all our results of Section 1.2 without any formal change: only the explicit meaning given to the symbols is changed according to (1.3.12).

Let us discuss briefly, in particular, the law of motion in quantum mechanics. We consider now the operators of $\mathcal{D}_Q$ as functions of the parameter t, the time (which is, of course, *not* an operator). The quantum operators can therefore be represented as follows:

$$\hat{b}(t) = \int dk\, dj\, \beta(k, j; t) \exp(ik\hat{q} + ij\hat{p}) \tag{1.3.13}$$

Again, a privileged role is played by a particular operator $\hat{H}$, the Hamiltonian, which generates the motion according to an equation corresponding to (1.2.7):

$$\dot{\hat{b}} = (i\hbar)^{-1}[\hat{b}, \hat{H}]_- \tag{1.3.14}$$

The formal solution of this equation can be represented as

$$
\begin{aligned}
\hat{b}(t) &= \exp(t[\hat{H}]_-)\hat{b} \\
&\equiv \hat{b} + \frac{t}{i\hbar}[\hat{b}, \hat{H}]_- + \frac{1}{2!}\frac{t^2}{(i\hbar)^2}[[\hat{b}, \hat{H}]_-, \hat{H}]_- + \cdots \\
&= \hat{\hat{U}}(t)\hat{b}
\end{aligned}
\tag{1.3.15}
$$

* Clearly, the set $\mathcal{D}_Q$ constructed by means of (1.3.12) contains in particular the symmetrized products of any operators, because these are obtained from the ordinary products by means of a linear combination, that is, an admissible algebraic operation.

Again, the *Green operator** $\hat{\hat{U}}(t)$ defines a transformation from the initial value $\hat{b}$ to the value of the operator at time t, $\hat{b}(t)$.

The transformations $\hat{\hat{U}}(t)$ form a group of *canonical transformations*, under which the algebra of quantum operators is invariant. We need, of course, no separate proof for these statements: the proof of Section 1.2 was based only on the algebraic properties of the operations, and hence goes over immediately in the quantum case.

It is sometimes useful to look at the canonical transformation discussed above from a slightly different point of view. Indeed, one will have no difficulty in proving, by series expansion, that

$$\hat{b}(t) \equiv \exp(t[\hat{H}]_{-})\hat{b} = \exp\left(\frac{it\hat{H}}{\hbar}\right)\hat{b}\exp\left(-\frac{it\hat{H}}{\hbar}\right) \tag{1.3.16}$$

Now it is well known from functional analysis (or from elementary quantum mechanics) that if $\hat{H}$ is a Hermitian operator, then $\hat{u}(t) \equiv \exp(it\hat{H}/\hbar)$ is a *unitary transformation* of the Hilbert space.† The right-hand side of Eq. (1.3.16) represents the image of the operator $\hat{b}$ under the unitary transformation $\hat{u}(t)$:

$$\hat{b}(t) = \hat{u}(t)\hat{b}\hat{u}^{-1}(t) \tag{1.3.17}$$

In this way, we have made contact with familiar concepts of quantum mechanics.

An important warning is necessary at this point. We have developed the structure of quantum mechanics by starting from classical mechanics and applying a correspondence rule. With every classical-dynamical function we associated a unique quantum operator. The correspondence may, however, not be inverted. There are quantum operators that have no classical counterpart. Such operators are needed for the complete description of systems of particles having *internal degrees of freedom*. A very well-known case is the *spin*. The description of a single electron (say) by means of all dynamical operators constructed on the basis of $\hat{q}$ and $\hat{p}$ does not explain all the properties of this system. We must add to

* A little word of caution might be in order here. The operator $\hat{\hat{U}}(t)$ is not of the same nature as the operators $\hat{b}$. In particular, it does not act on the basic Hilbert space of kets $|\rangle$ that is the substrate of the dynamical operators $\hat{b}$ of $\mathcal{D}_Q$. Rather, $\hat{\hat{U}}(t)$ acts on the operators of $\mathcal{D}_Q$ (which also have the structure of a vector space, and could be made into a Hilbert space by adding some additional axioms). Such operators acting on other operators are sometimes called "superoperators." Keeping this fact in mind, we shall, however, not burden the terminology by insisting on a rather unessential distinction.

† Note that the unitary operator $\hat{u}(t)$ is *not* a superoperator like $\hat{\hat{U}}(t)$! The right-hand side of (1.3.16) is a product of three operators belonging to $\mathcal{D}_Q$.

the description three new basic operators $\hat{s}_x$, $\hat{s}_y$, $\hat{s}_z$ connected to the spin, obeying the commutation rules:

$$[\hat{s}_i, \hat{s}_j]_- = i\varepsilon_{ijk}\hat{s}_k$$
$$[\hat{s}_i, \hat{q}]_- = 0 \tag{1.3.18}$$
$$[\hat{s}_i, \hat{p}]_- = 0$$

and having eigenvalues $\pm\frac{1}{2}$. (ε_{ijk} is the well-known antisymmetric Levi-Civitta symbol.) In other, more complex cases, other internal operators may have to be added to the description: isospin, strangeness, and so on. None of these observables have any classical correspondence. However, they can also be incorporated in the dynamical algebra. The knowledge of the commutation rules (1.3.18) allows us to construct algebraically any dynamical operator $\hat{b}$ depending not only on $\hat{q}$ and $\hat{p}$ but also on $\hat{s}_i$. Hence, the dynamical algebra $\mathscr{D}_Q$ of quantum operators is usually larger than the classical one $\mathscr{D}$, but its structure is exactly the same.

1.4. PURE STATES IN QUANTUM MECHANICS: BOSONS AND FERMIONS

In classical mechanics the description of the state of a system poses no serious problem. A system is completely defined at a given time by specifying the positions and momenta of all its particles. This information is equivalent to specifying the coordinates of a single point in phase space.

In quantum mechanics, such a specification is impossible. Because of Heisenberg's principle, the momentum and the position of a particle cannot be determined simultaneously with arbitrary precision. The best one can do in quantum mechanics is to consider the configuration space, that is, a space whose coordinates $(x_1, \ldots, x_N)$ are the positions of all the particles in the system (equal to "one-half" the phase space). The state of the system is described by specifying a complex *wave function* $\Psi(x_1, \ldots, x_N) \equiv \Psi(x)$.

In the case of systems having internal degrees of freedom, the wave function also depends on a set of quantum numbers specifying these variables. Typically, in the case of a set of electrons, the complete wave function is $\Psi_{\sigma_1 \cdots \sigma_N}(x_1, \ldots, x_N)$, where $\sigma_j = \pm\frac{1}{2}$. In order not to burden the notation we will usually not write these indices, except when they are of direct importance in a given problem.

The wave function is interpreted physically as a *probability amplitude.* In other words, its absolute square $|\Psi(x)|^2$ represents the probability density of finding the system at point x of configuration space. According to the laws of quantum mechanics, the wave function represents the maximum amount of information we can have about the system. Whenever the wave function is known precisely, the system is said to be in a *pure state.* This statistical aspect of quantum mechanics will be discussed again in Section 2.3. Let us briefly review here some of the most important rules of the quantum-mechanical formalism.

Let $\hat{b}$ be the operator associated with some observable (such as energy or momentum). The numerical value of $\hat{b}$ observed in some experiment is not definite in general; the only quantity we can specify is the average value $\bar{b}$ of the corresponding dynamical quantity in that state. This number represents the average outcome of a large number of experiments performed under identical conditions on the system in the state $\Psi(x)$. The expression of the average $\bar{b}$ is

$$\bar{b} = \int dx\ \Psi^*(x)\hat{b}\Psi(x) \tag{1.4.1}$$

The wave function is assumed to be normalized:

$$\int dx\ \Psi^*(x)\Psi(x) = 1 \tag{1.4.2}$$

These formulas can also be expressed in terms of the more general and more compact notation of Dirac. If the state $\Psi(x)$ corresponds to the ket $|\ \rangle$, Eqs. (1.4.1) and (1.4.2) can be transcribed as

$$\begin{aligned} \bar{b} &= \langle|\ \hat{b}\ |\rangle \\ 1 &= \langle|\ \hat{1}\ |\rangle = \langle\ |\ \rangle \end{aligned} \tag{1.4.3}$$

There are, however, some special cases in which a definite prediction can be made. This happens when the system is in an eigenstate $|n\rangle$ of $\hat{b}$: then the outcome of a measurement is the corresponding eigenvalue b_n with probability 1. Indeed, using (1.3.4) and the normalization of the wave function:

$$\bar{b} = \langle n|\ \hat{b}\ |n\rangle = b_n\langle n\ |\ n\rangle = b_n \tag{1.4.4}$$

In general, it is useful to expand the state of the system in a series of orthonormal basis functions $\varphi_m(x)$ or $|m\rangle$ spanning the Hilbert space (these functions could, but need not necessarily, be chosen as the eigenfunctions of some particular observable such as the energy):

$$\begin{aligned} \Psi(x) &= \sum_m c_m\varphi_m(x) \\ |\ \rangle &= \sum_m c_m\ |m\rangle \end{aligned} \tag{1.4.5}$$

The completeness of the basis is expressed by the property

$$\sum_n \varphi_n^*(x)\varphi_n(y) = \delta(x-y) \tag{1.4.6}$$

and the orthogonality condition is

$$\langle n \mid m \rangle \equiv \int dx\, \varphi_n^*(x)\varphi_m(x) = \delta_{n,m} \tag{1.4.7}$$

From this equation follows the expression of the expansion coefficients:

$$c_n = \langle n| \quad \rangle \equiv \int dx\, \varphi_n^*(x)\Psi(x) \tag{1.4.8}$$

Moreover, the normalization condition (1.4.2) implies

$$\sum_n c_n^* c_n \equiv \sum_n \langle n| \quad \rangle\langle \quad |n\rangle = 1 \tag{1.4.9}$$

Then Eq. (1.4.1) becomes

$$\bar{b} = \sum_m \sum_n c_m^* c_n \int dx\, \varphi_m^*(x)\hat{b}\varphi_n(x) \tag{1.4.10}$$

The *matrix element* b_{mn} of the observable $\hat{b}$ between the states φ_m and φ_n is defined as the number:

$$b_{mn} = \int dx\, \varphi_m^*(x)\hat{b}\varphi_n(x) \equiv \langle m|\, \hat{b}\, |n\rangle \tag{1.4.11}$$

Then the average value becomes

$$\bar{b} = \sum_m \sum_n c_m^* c_n b_{mn} = \sum_m \sum_n c_m^* c_n \langle m|\, \hat{b}\, |n\rangle \tag{1.4.12}$$

The typical systems of statistical mechanics are many-body systems. It is well known that in dealing with such systems in quantum mechanics, one needs a postulate that was not yet used. This is Pauli's *principle of undistinguishability of identical particles.* It can be formulated as follows.

The wave function of an assembly of identical particles can only be symmetric or antisymmetric under a permutation of two particles. This principle classifies all existing natural particles into *bosons* (characterized by symmetric wave functions) and *fermions* (characterized by antisymmetric wave functions). If the wave function is written as $\Psi(\ldots, x_k, \ldots, x_l, \ldots)$, where $x_1, \ldots, x_N$ are positions of the N particles, and if we fix our attention on the two particles k and l, leaving all others unchanged, Pauli's principle states that for every pair k, l:

$$\Psi(\ldots, x_k, \ldots, x_l, \ldots) = \theta\Psi(\ldots, x_l, \ldots, x_k, \ldots) \tag{1.4.13}$$

where

$$\theta = +1 \qquad \text{for bosons}$$
$$\theta = -1 \qquad \text{for fermions} \tag{1.4.14}$$

In both cases

$$|\Psi(\ldots, x_k, \ldots, x_l, \ldots)|^2 = |\Psi(\ldots, x_l, \ldots, x_k, \ldots)|^2 \tag{1.4.15}$$

which indeed means that the probability density of a given configuration does not change when two particles are interchanged. In other words, particles cannot be labeled.*

Because of the major importance of these ideas in statistical mechanics we shall review here the methods of dealing properly with the two types of particles. Let us assume that we have constructed a complete basis of functions for a one-particle system (these could be plane waves for point particles, or hydrogenic functions for atoms, etc.): We denote these states by $\varphi_{m_i}(x_i)$, where m_i is a set of quantum numbers, briefly called the *level*, characterizing the state of the single particle (for instance, the three components of the momentum, and the spin). We can now write a general N-particle state as a linear combination of products of φ_m's:

$$\Psi(x_1, \ldots, x_N) = \sum_{m_1} \cdots \sum_{m_N} c(m_1, \ldots, m_N)\varphi_{m_1}(x_1)\varphi_{m_2}(x_2) \cdots \varphi_{m_N}(x_N) \tag{1.4.16}$$

This is simply Eq. (1.4.5) written more explicitly. The basis functions $\varphi_m(x)$ for the N-body system are realized as direct products of N one-particle basis functions $\varphi_{m_i}(x_i)$. In general, however, such a function does not satisfy Eq. (1.4.12). It only does so if, for every pair k, l:

$$c(\ldots, m_k, \ldots, m_l, \ldots) = \theta c(\ldots, m_l, \ldots, m_k, \ldots) \tag{1.4.17}$$

We note in particular that two fermions cannot be found in the same level:

$$c(\ldots, m, \ldots, m, \ldots) = 0 \qquad \text{(fermions)} \tag{1.4.18}$$

This corresponds to the most familiar form of *Pauli's exclusion principle.*

The existence of the relations (1.4.17) clearly shows that a description of the form (1.4.16) is redundant. Equation (1.4.16) answers the question: "In which level is particle 1, in which level is particle 2, and so on?" But,

* Let us stress the fact that the property applies only to identical particles. For instance, the probability of finding an electron at point x and a proton at point y is different from the probability of the proton being at x and the electron at y. But we cannot distinguish a configuration with electron "number one" at x and electron "number two" at y from a configuration with electron "number two" at x and electron "number one" at y, a distinction that is assumed quite legitimate in a classical description.

because of the undistinguishability principle, such a question is meaningless. We must instead ask: "How many particles are in level m_1, how many particles are in level m_2, and so on?" Hence, the natural variables in a quantum many-body problem are the *occupation numbers* n_m, that is, the number of particles occupying the level m (e.g., number of particles of momentum $\hbar\mathbf{k}$). It is easy to find the connection between the two representations. Indeed, the probability density $|c(n_1, n_2, \ldots)|^2$ of finding n_1 particles in level 1, n_2 in level 2, and so on, equals the combined probability of all the permutations of a given configuration of particles $1, 2, \ldots, N$, compatible with our requirement:

$$|c(n_1, \ldots, n_i, \ldots)|^2 = \sum |c(m_1, \ldots, m_N)|^2$$

But because of (1.4.17), all the terms on the right-hand side are equal; on the other hand, there are $N!/\prod(n_i!)$ ways of distributing N particles among the occupied levels, hence:

$$|c(n_1, \ldots, n_i, \ldots)|^2 = \left(\frac{N!}{\prod_i (n_i!)}\right) |c(m_1, \ldots, m_N)|^2 \qquad (1.4.19)$$

This relation, valid for both bosons and fermions, allows us to transform the old expressions to the occupation number representation.

The occupation number representation is also called the *second quantization* representation. We note that in all interesting cases it involves an infinite number of variables because for each single particle there is an infinite number of allowed levels m (e.g., allowed values of the momentum, or atomic states, etc.). However, in every state of the N-body system only a finite number of n_m's are different from zero since clearly:

$$\sum_m n_m = N \qquad (1.4.20)$$

1.5. THE SECOND QUANTIZATION FORMALISM

We shall now develop a formalism that is particularly well adapted to the description of large quantum systems. It has the advantage of incorporating in a very simple form all the "quantum-statistical" effects arising from the Bose or Fermi symmetry requirements. The two cases will be treated separately.

A. Bosons

The explicit transformation of the Schrödinger equation to the occupation-number representation can be performed directly by using Eq. (1.4.19). We leave this as an exercise for the reader. We shall rather use a more intuitive method here.

We start with the Schrödinger equation:

$$i\hbar\partial_t \Psi(x_1, \ldots, x_N; t) = \hat{H}\Psi(x_1, \ldots, x_N; t) \tag{1.5.1}$$

To simplify the discussion, we assume that $\hat{H}$ is a sum of single-particle Hamiltonians:

$$\hat{H} = \sum_{j=1}^{N} \hat{h}^{(j)} \tag{1.5.2}$$

where $\hat{h}^{(j)}$ is an operator acting only on x_j. We now expand the states of the many-boson system according to Eq. (1.4.16). The matrix elements of the particular Hamiltonian (1.5.2) have the following obvious property:

$$\langle m_1, \ldots, m_N | \hat{H} | m_1', \ldots, m_N' \rangle = \sum_{j=1}^{N} \langle m_j | \hat{h}^{(j)} | m_j' \rangle \prod_{r=1 (r\neq j)}^{N}{}' \delta_{m_r', m_r} \tag{1.5.3}$$

By standard methods the Schrödinger equation (1.5.1) is transformed into an equation for the coefficients $c(m_1, \ldots, m_N)$:

$$i\hbar\partial_t c(m_1, \ldots, m_N; t) = \sum_{j=1}^{N} \sum_{m_j'} \langle m_j | \hat{h}^{(j)} | m_j' \rangle c(m_1, \ldots, m_j', \ldots, m_N; t) \tag{1.5.4}$$

The interpretation of the right-hand side is well known. The (one-particle) Hamiltonian causes transitions of particle j from level m_j' to level m_j: The evolution in time is determined by the sum of all possible transitions of this type. The probability amplitude of each transition is measured by the matrix element of the Hamiltonian.

We can now express the same idea in the language of occupation numbers (see Fig. 1.5.1). Before the transition there were $n'_{m_j'}$ particles in level m_j' and n'_{m_j} particles in level m_j. After the transition there are $n'_{m_j'} - 1$ in level m_j' and $n'_{m_j} + 1$ in level m_j. The process can be pictured as the *destruction* of a particle in level m_j' and the *creation* of a particle in level m_j.

It is easily seen that every quantum-mechanical process can be described in terms of destructions and creations of particles. We, therefore, understand that the fundamental operators of the second quantization formalism perform precisely these operations. Let us define them formally.

We first introduce a complete orthonormal set of wave functions characterized by definite values of the occupation numbers $\varphi(\ldots, n_m, \ldots)$, and such that:

$$(\varphi(\ldots, n_m, \ldots), \varphi(\ldots, n_m', \ldots)) = \prod_m \delta_{n_m, n_m'} \tag{1.5.5}$$

where (φ, φ') is an appropriately defined scalar product. The *destruction*

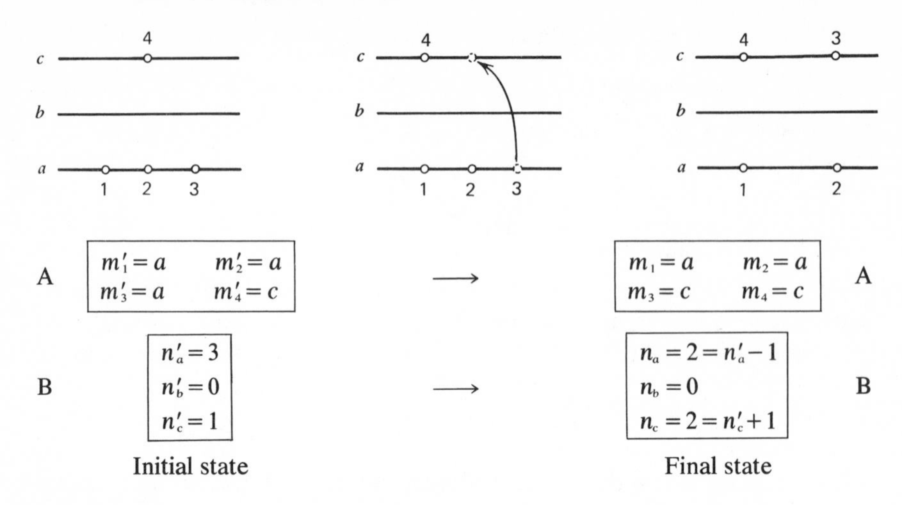

Figure 1.5.1. Description of a transition in ordinary quantum mechanics (A) and in the second quantization formalism (B).

operator a_m is then defined as*

$$\mathsf{a}_m \varphi(\ldots, n_m, \ldots) = (n_m)^{1/2} \varphi(\ldots, n_m - 1, \ldots) \qquad \text{(B)} \qquad (1.5.6)$$

Similarly we introduce the *creation operator* $\mathsf{a}_m^\dagger$

$$\mathsf{a}_m^\dagger \varphi(\ldots, n_m, \ldots) = (n_m + 1)^{1/2} \varphi(\ldots, n_m + 1, \ldots) \qquad \text{(B)} \qquad (1.5.7)$$

There is, of course, a pair of such operators for each level. It is easily verified that these operators are each other's Hermitian conjugate:

$$(\varphi(\ldots, n_m, \ldots), \mathsf{a}_m \varphi(\ldots, n_m + 1, \ldots)) \\ = (\mathsf{a}_m^\dagger \varphi(\ldots, n_m, \ldots), \varphi(\ldots, n_m + 1, \ldots)) \qquad (1.5.8)$$

It is now easily seen that, according to our discussion, the Hamiltonian operator must be of the form

$$\mathsf{H} = \sum_{m,m'} \langle m | \hat{h} | m' \rangle \mathsf{a}_m^\dagger \mathsf{a}_{m'} \qquad (1.5.9)$$

This operator performs the desired transitions, with the correct probability amplitude. Clearly, the particle index j must now be deleted, since the particles are no longer labeled by their name. Operators in the second quantization representation will be printed in sans serif type (a, b, etc.).

The operators a_m, $\mathsf{a}_m^\dagger$ are clearly not Hermitian: hence they do not

* In this section, all equations followed by the symbols (B) or (F) are valid *exclusively* for bosons, resp. fermions. Equations not followed by these symbols are valid in both cases (see below).

represent observable dynamical functions. One can, however, construct Hermitian operators by taking special combinations as we just did for the Hamiltonian. A fundamental operator of this type is the product $\mathsf{a}_m^\dagger \mathsf{a}_m$. From Eqs. (1.5.6) and (1.5.7) follows the property:

$$\mathsf{a}_m^\dagger \mathsf{a}_m \varphi(\ldots, n_m, \ldots) = n_m \varphi(\ldots, n_m, \ldots) \tag{1.5.10}$$

Hence the operator $\mathsf{a}_m^\dagger \mathsf{a}_m$ can be interpreted as the *number operator;* the states with definite numbers of particles are its eigenfunctions, and its eigenvalues are the occupation numbers, that is, the infinite set of nonnegative integers n_m.

Let us note that, in a system with a given finite number of particles N, the *sum* over all levels m of the occupation-number operators is a scalar:

$$\sum_m \mathsf{a}_m^\dagger \mathsf{a}_m = N \tag{1.5.11}$$

This equation expresses the constancy of the total number of particles and is equivalent to Eq. (1.4.20).

We also derive from the definition (1.5.6), (1.5.7)

$$\mathsf{a}_m \mathsf{a}_m^\dagger \varphi(\ldots, n_m, \ldots) = (n_m + 1)\varphi(\ldots, n_m, \ldots) \tag{1.5.12}$$

Hence, the operators a_m, $\mathsf{a}_m^\dagger$ do *not* commute with each other. It is easy to work out the following fundamental relations:

$$\begin{aligned} &[\mathsf{a}_m, \mathsf{a}_{m'}]_- = 0 \\ &[\mathsf{a}_m^\dagger, \mathsf{a}_{m'}^\dagger]_- = 0 \qquad \text{(B)} \\ &[\mathsf{a}_m, \mathsf{a}_{m'}^\dagger]_- = \delta_{m,m'} \end{aligned} \tag{1.5.13}$$

We recall that the bracket denotes the commutator: $[a, b]_- = ab - ba$.

We see from Eq. (1.5.6) that, starting from an eigenstate $\varphi(\ldots, n_m, \ldots)$, we can generate the eigenstate $\varphi(\ldots, n_m - 1, \ldots)$ by applying to the former the destruction operator. It is clear that this process cannot be continued indefinitely, otherwise we would obtain states with negative occupation numbers. Hence there exists a *vacuum state* $\varphi(0, \ldots, 0, \ldots) \equiv \varphi_0$ in which all levels are empty. Applying any a_m to this state, according to (1.5.6) we obtain an eigenvalue 0:

$$\mathsf{a}_m \varphi_0 = 0 \qquad \text{for all } m \tag{1.5.14}$$

Proceeding now in the opposite direction, we can systematically construct the basis $\varphi(\ldots, n_m, \ldots)$, by successively applying creation operators to φ_0. Thus

$$\mathsf{a}_m^\dagger \varphi_0$$

is a one-particle eigenstate, properly normalized. Similarly

$$\mathbf{a}_m^\dagger \mathbf{a}_{m'}^\dagger \varphi_0$$

is a two-particle eigenstate. It follows from the commutation relations (1.5.13) that this state is automatically symmetric under the permutation of m and m'. It is also clear from the same relations that nothing prevents us from putting as many particles as we want in a single level m. For instance,

$$(\sqrt{2})^{-1}\mathbf{a}_m^\dagger \mathbf{a}_m^\dagger \varphi_0$$

is a state with two particles in the level m. The factor $(\sqrt{2})^{-1}$ ensures its correct normalization. Hence, in constructing the many-body states in this way, we automatically obtain correct boson wave functions. The most general wave function of this type is

$$\Psi = c_0\varphi_0 + \sum_m c_m \mathbf{a}_m^\dagger \varphi_0 + \sum_m \sum_{m'} (n_m!n_{m'}!)^{-1/2} c_{mm'} \mathbf{a}_m^\dagger \mathbf{a}_{m'}^\dagger \varphi_0 + \cdots \quad (1.5.15)$$

This representation of the wave function is called the *Fock representation* after the name of its discoverer.

In a state like (1.5.15), the total number of particles is not definite, but subject to quantum uncertainty, that is, to fluctuations. This representation is, therefore, particularly well suited for problems in which the particles can be created or destroyed as a result of the interactions. This is the case in particular for particles associated with fields, such as photons, phonons, and the like. When we study material particles involved in processes that do not destroy their individuality (i.e., excluding the weak or strong nuclear interactions, or excluding chemical reactions), the total number of particles is conserved during the interactions. The allowed states are then of the form:

$$\Psi = \sum_{m_1} \cdots \sum_{m_N} \left[\prod_i n_{m_i}!\right]^{-1/2} c_{m_1,\ldots,m_N} \mathbf{a}_{m_1}^\dagger \mathbf{a}_{m_2}^\dagger \cdots \mathbf{a}_{m_N}^\dagger \varphi_0 \quad (1.5.16)$$

Of course, some of the m_i's can be identical (i.e., $n_{m_i} \neq 0, 1$).

Let us now consider the operators representing dynamical functions in the present formalism. The example of the simple Hamiltonian we have presented above is by no means exceptional. Every operator has the effect of causing a certain number of particles to make transitions from one level to another. In other words, it will destroy a certain number of particles in certain levels, and recreate them in other levels; the probability amplitude of this process is measured by the matrix element of the operator between the corresponding states. Hence the general form of such an operator, involving n particles, is

$$\mathsf{b}^{(n)} = (n!)^{-1} \sum_{m_1} \cdots \sum_{m_n} \sum_{m'_1} \cdots \sum_{m'_n} \langle m_1, \ldots, m_n | \hat{b}^{(n)} | m'_1, \ldots, m'_n \rangle$$

$$\times \mathsf{a}^{\dagger}_{m_n} \cdots \mathsf{a}^{\dagger}_{m_1} \mathsf{a}_{m'_1} \cdots \mathsf{a}_{m'_N} \quad (1.5.17)$$

Here $\langle m | \hat{b}^{(n)} | m' \rangle$ is the matrix element of the operator $\hat{b}^{(n)}$ in the ordinary representation (i.e., the one discussed in Section 1.3). The factor $(n!)^{-1}$ provides a convenient normalization of the expressions.

A necessary remark concerns the order in which the (noncommuting) operators are written in this kind of expression. We shall make the convention that all creation and destruction operators must be ordered in *normal product* form.* The latter is defined as follows:

In a normal product all creation operators $\mathsf{a}^{\dagger}$ stand to the left of all destruction operators a. Moreover, the creation operators are ordered (from left to right) in the reverse sequence as the corresponding destruction operators.† The normal product is denoted by the notation $(:\prod_j \mathsf{a}^{\dagger}_j \mathsf{a}_j :)$.

A typical example of normal ordering is

$$:\prod_{j=1}^{3} \mathsf{a}^{\dagger}_j \mathsf{a}_j : = \mathsf{a}^{\dagger}_3 \mathsf{a}^{\dagger}_2 \mathsf{a}^{\dagger}_1 \mathsf{a}_1 \mathsf{a}_2 \mathsf{a}_3$$

It is clear that *any* product of a, $\mathsf{a}^{\dagger}$ can be represented as a linear combination of normal products by using the commutation relations (1.5.13). For instance:

$$\mathsf{a}^{\dagger}_1 \mathsf{a}_1 \mathsf{a}_2 \mathsf{a}^{\dagger}_2 \mathsf{a}^{\dagger}_1 \mathsf{a}_1 = \mathsf{a}^{\dagger}_2 \mathsf{a}^{\dagger}_1 \mathsf{a}^{\dagger}_1 \mathsf{a}_1 \mathsf{a}_1 \mathsf{a}_2 + \mathsf{a}^{\dagger}_2 \mathsf{a}^{\dagger}_1 \mathsf{a}_1 \mathsf{a}_2 + \mathsf{a}^{\dagger}_1 \mathsf{a}^{\dagger}_1 \mathsf{a}_1 \mathsf{a}_1 + \mathsf{a}^{\dagger}_1 \mathsf{a}_1 \quad (B)$$

We now introduce another representation of the second quantization formalism that is useful in some problems. It may happen that we choose the quantum numbers m_i characterizing the "levels" of the particles to be simply their positions in space $\mathbf{x}_i$. We therefore go over to the *position representation* by introducing the "quantized wave functions":

$$\begin{aligned} \boldsymbol{\Psi}(\mathbf{x}) &= \sum_m \mathsf{a}_m \varphi_m(\mathbf{x}) \\ \boldsymbol{\Psi}^{\dagger}(\mathbf{x}) &= \sum_m \mathsf{a}^{\dagger}_m \varphi^*_m(\mathbf{x}) \end{aligned} \quad (1.5.18)$$

* Let us note that the Weyl rule cannot be applied here. Indeed $\mathsf{a}^{\dagger}$, a cannot play the role of $\hat{q}$, $\hat{p}$: They are not Hermitian and their commutation rules are different from (1.3.3).

† The second prescription is irrelevant in the case of bosons because the $\mathsf{a}^{\dagger}$'s commute among themselves, and so do the a's. However, this rule is important for fermions (see below).

These objects are *operators* in occupation-number space. Their interpretation is clear from their definition: $\mathbf{\Psi}(\mathbf{x})$ represents the destruction of a particle at point $\mathbf{x}$ (whatever its level m); similarly $\mathbf{\Psi}^\dagger(\mathbf{x})$ represents the creation of a particle at point $\mathbf{x}$.

We can invert Eq. (1.5.18) by using the orthogonality property of the basis functions $\varphi_m(\mathbf{x})$ [Eq. (1.4.7)]:

$$\begin{aligned} \mathsf{a}_m &= \int d\mathbf{x}\, \mathbf{\Psi}(\mathbf{x})\varphi_m^*(\mathbf{x}) \\ \mathsf{a}_m^\dagger &= \int d\mathbf{x}\, \mathbf{\Psi}^\dagger(\mathbf{x})\varphi_m(\mathbf{x}) \end{aligned} \tag{1.5.19}$$

It is easily seen that the commutation relations (1.5.13) imply similar relations for the $\mathbf{\Psi}$ and $\mathbf{\Psi}^\dagger$ operators. For instance:

$$\begin{aligned} [\mathbf{\Psi}(\mathbf{x}), \mathbf{\Psi}^\dagger(\mathbf{x}')]_- &= \sum_m \sum_{m'} [\mathsf{a}_m, \mathsf{a}_{m'}^\dagger]_- \varphi_m(\mathbf{x})\varphi_{m'}^*(\mathbf{x}') \\ &= \sum_m \sum_{m'} \delta_{m,m'}\varphi_m(\mathbf{x})\varphi_{m'}^*(\mathbf{x}') \\ &= \delta(\mathbf{x}-\mathbf{x}') \end{aligned}$$

where we used the completeness relation (1.4.6). We thus obtain:

$$\begin{aligned} &[\mathbf{\Psi}(\mathbf{x}), \mathbf{\Psi}(\mathbf{x}')]_- = 0 \\ &[\mathbf{\Psi}^\dagger(\mathbf{x}), \mathbf{\Psi}^\dagger(\mathbf{x}')]_- = 0 \qquad (\mathrm{B}) \\ &[\mathbf{\Psi}(\mathbf{x}), \mathbf{\Psi}^\dagger(\mathbf{x}')]_- = \delta(\mathbf{x}-\mathbf{x}') \end{aligned} \tag{1.5.20}$$

Note that the Kronecker delta symbol of Eq. (1.5.13) is replaced here by a Dirac delta function, because $\mathbf{x}$ is a continuous variable.

The operator $\mathbf{\Psi}^\dagger(\mathbf{x})\mathbf{\Psi}(\mathbf{x})$ can again be interpreted as the number of particles at point $\mathbf{x}$, and the total number N of particles is expressed similarly to (1.5.11):

$$\int d\mathbf{x}\, \mathbf{\Psi}^\dagger(\mathbf{x})\mathbf{\Psi}(\mathbf{x}) = N \tag{1.5.21}$$

We now substitute Eqs. (1.5.19) as well as (1.4.11) into (1.5.17):

$$\begin{aligned} \mathsf{b}^{(n)} = (n!)^{-1} \sum_{m_1}\cdots\sum_{m_n}\sum_{m_1'}\cdots\sum_{m_n'} \int d\mathbf{x}_1 \cdots \int d\mathbf{x}_n \int d\mathbf{x}_1' \cdots \int d\mathbf{x}_n' \int d\xi_1 \cdots \int d\xi_n \\ \times [\varphi_{m_1}^*(\xi_1)\cdots\varphi_{m_n}^*(\xi_n)\hat{b}_{(\xi)}^{(n)}\varphi_{m_1'}(\xi_1)\cdots\varphi_{m_n'}(\xi_n)] \\ \times [\varphi_{m_1}(\mathbf{x}_1)\cdots\varphi_{m_n}(\mathbf{x}_n)\mathbf{\Psi}^\dagger(\mathbf{x}_1)\cdots\mathbf{\Psi}^\dagger(\mathbf{x}_n)] \\ \times [\varphi_{m_n'}^*(\mathbf{x}_n')\cdots\varphi_{m_1'}^*(\mathbf{x}')\mathbf{\Psi}(\mathbf{x}_n')\cdots\mathbf{\Psi}(\mathbf{x}_1')] \end{aligned}$$

(The notation $\hat{b}_{(\xi)}^{(n)}$ is intended to stress that the operator $\hat{b}^{(n)}$ acts on the variables ξ.)

Using the completeness relations (1.4.6) we can easily perform the summations over the m_i and m'_i, and then use the resulting δ functions for doing the integrations over ξ_i and $\mathbf{x}'_i$ with the result:

$$\mathsf{b}^{(n)} = (n!)^{-1} \int d\mathbf{x}_1 \cdots d\mathbf{x}_n \boldsymbol{\Psi}^\dagger(\mathbf{x}_1) \cdots \boldsymbol{\Psi}^\dagger(\mathbf{x}_n) \hat{b}^{(n)} \boldsymbol{\Psi}(\mathbf{x}_n) \cdots \boldsymbol{\Psi}(\mathbf{x}_1) \quad (1.5.22)$$

where the "ordinary" operator $\hat{b}^{(n)}$ acts on the $\mathbf{x}$ variables lying to its right. We have here a nice and compact form for any n-particle operator $\mathsf{b}^{(n)}$.

An important particular case of the relationship between the operators $\boldsymbol{\Psi}^\dagger$ and $\mathsf{a}^\dagger$ is the case in which the basis functions $\varphi_m(\mathbf{x})$ are plane waves, labeled by a continuous wave vector $\boldsymbol{\kappa}$:

$$\varphi_{\kappa}(\mathbf{x}) = (8\pi^3)^{-1/2} e^{i\boldsymbol{\kappa}\cdot\mathbf{x}} \quad (1.5.23)$$

This basis corresponds to the *momentum representation* for a system of particles in an infinitely large volume. In view of its importance we collect here the main formulas appropriate to this case. The only slight difference with the previous ones comes from the continuous character of the wave vector $\boldsymbol{\kappa}$ replacing the index m. Hence Eq. (1.5.18) is replaced by

$$\begin{aligned} \boldsymbol{\Psi}(\mathbf{x}) &= (8\pi^3)^{-1/2} \int d\boldsymbol{\kappa}\ \mathsf{a}(\hbar\boldsymbol{\kappa}) e^{i\boldsymbol{\kappa}\cdot\mathbf{x}} \\ \boldsymbol{\Psi}^\dagger(\mathbf{x}) &= (8\pi^3)^{-1/2} \int d\boldsymbol{\kappa}\ \mathsf{a}^\dagger(\hbar\boldsymbol{\kappa}) e^{-i\boldsymbol{\kappa}\cdot\mathbf{x}} \end{aligned} \quad (1.5.24)$$

We write the argument of the creation and destruction operators $\mathsf{a}^\dagger$, a as $\hbar\boldsymbol{\kappa}$ for the sake of convenience; $\hbar\boldsymbol{\kappa}$ has the dimensions of momentum. The commutation relations obeyed by $\mathsf{a}^\dagger$, a are slightly different from (1.5.13): the Kronecker delta symbol is replaced by a Dirac delta function:

$$\begin{aligned} &[\mathsf{a}(\hbar\boldsymbol{\kappa}), \mathsf{a}(\hbar\boldsymbol{\kappa}')]_- = 0 \\ &[\mathsf{a}^\dagger(\hbar\boldsymbol{\kappa}), \mathsf{a}^\dagger(\hbar\boldsymbol{\kappa}')]_- = 0 \qquad \text{(B)} \\ &[\mathsf{a}(\hbar\boldsymbol{\kappa}), \mathsf{a}^\dagger(\hbar\boldsymbol{\kappa}')]_- = \delta(\boldsymbol{\kappa}-\boldsymbol{\kappa}') \end{aligned} \quad (1.5.25)$$

B. Fermions

It is tempting to formulate also the many-fermion problem in the occupation-number representation. It is clear that the preceding formalism cannot work as such, because it has built into it the structure of the boson symmetry. However, an apparently minor modification saves the situation. Let us tentatively consider Eqs. (1.5.13) and change all the minus signs into plus signs:

$$\begin{aligned} &[\mathsf{a}_m, \mathsf{a}_{m'}]_+ = 0 \\ &[\mathsf{a}^\dagger_m, \mathsf{a}^\dagger_{m'}]_+ = 0 \qquad \text{(F)} \\ &[\mathsf{a}_m, \mathsf{a}^\dagger_{m'}]_+ = \delta_{m,m'} \end{aligned} \quad (1.5.26)$$

where $[a, b]_+ \equiv ab + ba$ is the *anticommutator* of a and b. If we retain the interpretation in terms of creation and destruction operators we will see that these objects have strange algebraic properties, but which are precisely those needed to describe fermions.

First of all, we see that

$$\mathsf{a}_m^\dagger \mathsf{a}_m^\dagger \varphi(\ldots n_m \ldots) = \mathsf{a}_m \mathsf{a}_m \varphi(\ldots n_m \ldots) = 0 \qquad \text{(F)} \qquad (1.5.27)$$

Hence we can neither create nor destroy two particles in the same level. Therefore the occupation numbers of every level can only take the values 0 or 1. This can also be seen by noting the following property of the number operator $\mathsf{N}_m = \mathsf{a}_m^\dagger \mathsf{a}_m$:

$$\begin{aligned} \mathsf{N}_m^2 &= \mathsf{a}_m^\dagger \mathsf{a}_m \mathsf{a}_m^\dagger \mathsf{a}_m = -\mathsf{a}_m^\dagger \mathsf{a}_m^\dagger \mathsf{a}_m \mathsf{a}_m + \mathsf{a}_m^\dagger \mathsf{a}_m \\ &= \mathsf{a}_m^\dagger \mathsf{a}_m = \mathsf{N}_m \qquad \text{(F)} \end{aligned} \qquad (1.5.28)$$

The equality $\mathsf{N}_m^2 = \mathsf{N}_m$ can indeed only be satisfied by the operators $\mathsf{0}$ and $\mathsf{1}$.

We can again define a vacuum state φ_0 by Eq. (1.5.14) and a complete orthonormal set of eigenfunctions of the occupation number operators, by (1.5.15) or (1.5.16). But now clearly the permutation of two levels results in a change of sign. It follows that the explicit representations (1.5.6) and (1.5.7) are not valid: they do not satisfy Eq. (1.5.26). In order to obtain a similar representation, we must first make a convention of ordering the single-particle levels in some natural (but otherwise arbitrary) order:

$$m_1 < m_2 < \cdots < m_n < \cdots \qquad (1.5.29)$$

It is only by adopting such a convention that the *sign* of the wave function is made definite. We now define the *destruction operator* by

$$\mathsf{a}_m \varphi(\ldots, n_m, \ldots) = (-1)^{s_m} n_m \varphi(\ldots, n_m - 1, \ldots) \qquad \text{(F)} \qquad (1.5.30)$$

and the *creation operator* by

$$\mathsf{a}_m^\dagger \varphi(\ldots, n_m, \ldots) = (-1)^{s_m} (1 - n_m) \varphi(\ldots, n_m + 1, \ldots) \qquad \text{(F)} \qquad (1.5.31)$$

In these equations, s_m represents the number of occupied levels below the level m:

$$s_m = \sum_{j=m_1}^{m} n_j \qquad (1.5.32)$$

a number that is made definite by the convention (1.5.29).

We now clearly see that a particle can only be removed from an occupied level and be introduced into an empty level. The rule of the sign [which is necessary in order to ensure that Eqs. (1.5.26) are satisfied] is a complication of the fermion case. It means that the operators $a_m^\dagger$, $a_n^\dagger$, for $m \neq n$, are *not* independent as in the boson case (in which these operators commute). The result of their action on a given state depends on the occupation numbers of all the levels below them. However, in practice, we will see that this complication is not serious. Indeed, the *explicit* representations (1.5.6), (1.5.7) and (1.5.30), (1.5.31) are usually avoided; it is much simpler to calculate directly with the abstract operators $a^\dagger$, a using their commutation or anticommutation rules, rather than with their explicit matrix elements in the occupation-number representation.

Most formulas (all those that do not have an explicit label "B" or "F") can now be taken over for both fermions and bosons. Equations (1.5.13) and (1.5.26) can be rewritten in the unified form:

$$
\begin{aligned}
[a_m, a_{m'}]_\theta &\equiv a_m a_{m'} - \theta a_{m'} a_m = 0 \\
[a_m^\dagger, a_{m'}^\dagger]_\theta &= 0 \\
[a_m, a_{m'}^\dagger]_\theta &= \delta_{m,m'}
\end{aligned}
\tag{1.5.33}
$$

where θ has been defined in Eq. (1.4.14). In the momentum representation, with a continuous wave vector $\boldsymbol{\kappa}$, Eq. (1.5.25) is generalized as

$$
\begin{aligned}
[a(\hbar\boldsymbol{\kappa}), a(\hbar\boldsymbol{\kappa}')]_\theta &= 0 \\
[a^\dagger(\hbar\boldsymbol{\kappa}), a^\dagger(\hbar\boldsymbol{\kappa}')]_\theta &= 0 \\
[a(\hbar\boldsymbol{\kappa}), a^\dagger(\hbar\boldsymbol{\kappa}')]_\theta &= \delta(\boldsymbol{\kappa} - \boldsymbol{\kappa}')
\end{aligned}
\tag{1.5.34}
$$

Similarly, Eq. (1.5.20) can be written as

$$
\begin{aligned}
[\boldsymbol{\Psi}(\mathbf{x}), \boldsymbol{\Psi}(\mathbf{x}')]_\theta &= 0 \\
[\boldsymbol{\Psi}^\dagger(\mathbf{x}), \boldsymbol{\Psi}^\dagger(\mathbf{x}')]_\theta &= 0 \\
[\boldsymbol{\Psi}(\mathbf{x}), \boldsymbol{\Psi}^\dagger(\mathbf{x}')]_\theta &= \delta(\mathbf{x} - \mathbf{x}')
\end{aligned}
\tag{1.5.35}
$$

There is no difficulty whatever to integrate this formalism into the general framework worked out in Section 1.3. Indeed, Eq. (1.3.12) gives us the basic realization of the dynamical algebra $\mathcal{D}_Q$. Its translation into the second quantization language consists in just associating with each ordinary operator $\hat{b}$ an operator b defined by Eqs. (1.5.17) or (1.5.22). In particular, the characteristic Lie bracket of $\mathcal{D}_Q$ is still given by $(i\hbar)^{-1}$ times the commutator.*

* Let us warn the beginner that the dynamics is dictated by the *commutator*, both for bosons and for fermions. The anticommutator does not possess the required properties of a Lie bracket.

BIBLIOGRAPHICAL NOTES

1.2. The best and most modern exposition of classical mechanics and particularly of Hamiltonian dynamics remains H. Goldstein, *Classical Mechanics*, Addison-Wesley, Cambridge, Mass., 1953.

A remarkable exposition of classical mechanics from the point of view of its algebraic and group-theoretical structure is unfortunately published only as a report: E. C. G. Sudarshan, *Principles of Classical Dynamics*, Report NYO-10250, Department of Physics and Astronomy, University of Rochester, 1963.

A more easily accessible summary along the same lines appears in E. C. G. Sudarshan, in *Lectures in Theoretical Physics*, Brandeis Summer Institute (1961), Benjamin, New York, 1962.

A very important paper on Hamiltonian dynamics with an opening toward relativity is P. A. M. Dirac, *Rev. Mod. Phys.* **21**, 392 (1949).

1.3 There exist, of course, a large number of books on quantum mechanics. They cannot all be quoted here. The book that remains, after forty years, the best for a real study in depth of quantum mechanics is P. A. M. Dirac, *The Principles of Quantum Mechanics*, (1st ed., 1930), 4th ed., Clarendon, Oxford, 1958.

More recent very good and comprehensive books are:

L. I. Schiff, *Quantum Mechanics*, McGraw-Hill, New York, 1949.

A. Messiah, *Mécanique Quantique*, Dunod, Paris, 1959. (English translation: *Quantum Mechanics*, North Holland, Amsterdam, 1961.)

L. D. Landau and E. M. Lifshits, *Quantum Mechanics*, Pergamon, Oxford, 1960.

The Weyl correspondence rule was introduced in H. Weyl, *Gruppentheorie und Quantenmechanik*, 2nd. ed., Hirzel, Leipsig, 1931. (English translation: *The Theory of Groups and Quantum Mechanics*, Dover, New York, no date.)

1.4-1.5 A very clear, though old exposition of the second quantization formalism is V. Fock, *Z. Phys.* **75**, 622 (1932).

The method is also exposed in the books by Dirac and by Landau and Lifshits given above. An interesting exposition, oriented toward statistical physics is J. de Boer, "Construction Operator Formalism in Many-Particle Systems," in *Studies in Statistical Mechanics*, Vol. 3, edited by J. de Boer and G. E. Uhlenbeck, North Holland, Amsterdam, 1965.

CHAPTER 2
STATISTICAL ENSEMBLES

2.1. MACROSCOPIC PHYSICS AND MICROSCOPIC PHYSICS

In our everyday life we are surrounded by objects, the sizes of which are comparable to or larger than our own: a piece of wood, a river, a violin, a magnet, a combustion engine, and so on. Because of their easy access to our senses, these *macroscopic objects* were the first things man studied systematically. By the end of the eighteenth century these were the only objects of interest in physics, and their study, as such, continued and is still active today, after the revolution of ideas to be described below.

The results of centuries of experimental work are presently formalized into well-structured theories covering the field of macroscopic physics, with such subdivisions as: fluid mechanics, elasticity, thermodynamics, electromagnetism, and acoustics. The purpose of such theories is to inscribe all objects of our perception into the familiar framework of a four-dimensional space–time; we want to describe what happens in every point of space and at every instant of time. As an abstraction of our immediate perception, matter and energy are considered as a *continuum* in this framework. Therefore, the natural mathematical representation of physical quantities in this perspective is by means of *continuous or piecewise-continuous functions* of the space coordinates* $\mathbf{x}$ *and of the time* $t: B(\mathbf{x}, t)$. Such continuous functions describing physical quantities are called *fields*, and their behavior in space and time is governed by partial-differential equations or by integrodifferential equations.

This view of physics was profoundly altered by the advent and confirmation of the atomistic theory, through the nineteenth and twentieth centuries. As a result, it became clear that the continuity of matter and energy is but an illusion, and that, in the range of lengths of 10^{-7} cm and below, we would only see collections of a huge number of discrete particles, moving under the influence of their mutual interaction forces.

* A piecewise-continuous function is needed, say, to describe a fluid in a vessel. The boundaries are schematized as localized discontinuities, sharply separating regions of different nature.

Hence, the natural description of *microscopic physics* is in terms of dynamics of many bodies, which may be idealized as point particles or as small bodies having a few internal degrees of freedom. Fundamentally, their motion is governed by the laws of quantum mechanics, although classical mechanics is a very good approximation in many problems.

It is quite clear today that the laws of macroscopic physics give a very good description of macroscopic phenomena, and that the laws of microscopic physics adequately describe atomic and molecular phenomena. It is also clear that macroscopic phenomena are a manifestation, at our level, of the underlying microscopic world of atomic motions. Given the widely different character of the description of nature in the two perspectives, the need of an explanation of the laws of continuum physics as a consequence of the microscopic evolution of collections of discrete particles became stringent. This bridge between the two levels is provided by *statistical mechanics.*

This science necessarily deals with quite complex systems, consisting of a large number of particles. It was often said in the past that statistical mechanics must be constructed in order to cover our inability to solve the initial-value problem of dynamics in such a complicated case. This argument is, however, no longer very strong in our day. With present-day computers one can study in detail the molecular dynamics of systems made up of about 1000 particles and more. Such systems simulate already some of the features of real macroscopic bodies. It is important to note that even if an ideal computer existed, which could solve the initial-value problem for a system of 10^{23} particles, the solution as such would not help us answer the questions asked in macroscopic physics.

Consider the following simple but important example. A fundamental dynamical quantity is the *energy* of the system. From it we can derive the concept of *energy density* by considering the amount of energy per unit volume in a small element surrounding a point $\mathbf{x}$, and taking the limit of a vanishingly small volume element. The meaning of this concept is, however, quite different in microscopic and in macroscopic physics. In the first case, the energy density is a function of the positions and momenta of the individual particles. As a result of the motion of the particles, the energy density at a given point $\mathbf{x}$ is a violently and irregularly fluctuating function of time. In hydrodynamics, on the other hand, the concept of energy density is attached to a small element of fluid. It is a quite smooth, usually continuous function of the space coordinates. Its rate of change in time depends only on the specification of a few other macroscopic functions, such as the density and the velocity fields. The description is quite different; however, the underlying physical concept of energy density is the same in both cases. What we need, therefore, is a *correspondence rule* by which we associate in a unique way a macroscopic quantity to every possible microscopic dynamical function. Such a correspondence rule cannot be given by mechanics itself: a *postulate* is needed at this stage of the game.

Let us now consider a very simple macroscopic experiment. We take a metal rod prepared at time zero in such a way that its temperature varies linearly from one end to the other. We then let it evolve freely and measure the change of temperature as a function of time at a given point. It is a fact of experience that, if the experiment is repeated a large number of times under identical conditions, its result is quite reproducible. This fact implies that the specification of the initial temperature distribution is sufficient for the complete macroscopic specification of the problem.

On the other hand, if we think of the metal rod as a classical collection of atoms, it is obvious that the mere specification of a linear temperature distribution by no means fixes a unique mechanical initial condition. There is an enormous number of microscopic configurations that are all compatible with the given macroscopic constraint. If we repeat our experiment several times, as explained above, the overwhelming chances are that we start each time with our atoms in a different microscopic configuration. Nevertheless, experience shows that the phenomena observed on the macroscopic scale are insensitive to these differences. All the mechanical initial conditions compatible with the macroscopic constraint are, in a certain sense, equivalent and must be treated on an equal footing. One way of expressing this idea mathematically is to attach a certain *weight* to all the possible states of the system at time zero. For instance, we would attach a weight zero to all the states that are incompatible with the macroscopic constraint, and an equal weight to all states that are compatible. We are then tempted to *define* the macroscopic quantities (say, the energy) as the *average* of the microscopic quantity over *all* the states of the dynamical system, properly weighted as explained above. By this procedure, it is clear that, starting the system in any configuration compatible with the constraint results in equal values of the macroscopic quantities: we have an explanation of the reproducibility of macroscopic experiments.

We may now ask what is the predictive power of such a theory? Clearly, we cannot demand from it a detailed prediction of the outcome of any given experiment. The best we can hope is that *the formalism predicts the average result of a large number of experiments performed under identical conditions.* We cannot exclude the possibility of fluctuations away from this average. Such fluctuations cannot be described in detail by the theory.* It is, however, a fact of experience that—provided the system is large enough—the overwhelming majority of experiments described above yield a result that is extremely close to the average predicted by the theory. The validity of this statement is only limited by the size of the system: if we perform the experiment with a metallic film a few angstroms thick the fluctuations become quite important.

We thus arrived at the idea of a statistical description of the many-body system. The mathematical object representing the system is no longer a point in phase space, but a collection of points in phase space, each point being weighted by a certain number. Such a set of weighted points will be called an *ensemble*. The observable value of a dynamical function is identified with the ensemble average of the microscopic function. The value obtained by this prescription is interpreted as the average result of a large number of identical experiments.

One may be impressed at this point by the analogy of this interpretation with the interpretation of ordinary quantum-mechanical theory. In that case too, it is asserted that only average values of observables can be predicted by the theory. But the physical origin of the statistical nature of quantum theory is quite different. This important fact can be understood if we consider again our simple experiment of heat flow, but interpret it in quantum-mechanical terms. The piece of metal is now characterized microscopically by a certain wave function obeying the Schrödinger equation. In this particular state we may calculate the quantum-mechanical average of the energy and study its evolution in time. The wave function of a many-body system is, however, an extremely intricate object. By specifying at time zero only a macroscopic constraint (such as a given temperature gradient), we still have the choice among an enormous number of possible wave functions of the system, all compatible with the given constraint. To each of these states corresponds a well-defined quantum average of the energy, and these averages are usually different from each other. Hence, we are in the same position as in the classical case. Proceeding by analogy, we attribute a properly chosen weight to each of the possible states of the system. We then define the observable value of the energy as the ensemble average of the quantum averages of the microscopic energy. Clearly, *two* successive averaging processes are involved in a quantum-statistical system: the first one originating in Heisenberg's uncertainty principle, the second one coming from the uncertainty in the definition of the initial state of a many-body system.

* Note, however, that the formalism *can* predict the *statistical properties* of the fluctuations, such as the average square of the fluctuations, the statistical correlation between fluctuations of various quantities, and so on.

2.2. CLASSICAL ENSEMBLES: THE PHASE-SPACE DISTRIBUTION FUNCTIONS

In Section 2.1 we discussed intuitively the physical arguments underlying the idea of a statistical ensemble for the description of a many-body system. We now make these ideas quantitative.

We saw at the beginning of Section 2.1 that macroscopic quantities are described by continuous (or piecewise-continuous) functions of the physical space coordinates $\mathbf{x}$ and of the time $t : B(\mathbf{x}, t)$. They are *fields in physical space–time.* Microscopic dynamical quantities, on the other hand, are *functions of the phase-space coordinates* $(q_1, \ldots, q_N, p_1, \ldots, p_N) \equiv (q, p)$, which may, moreover, depend also on the *parameters* $\mathbf{x}$ and $t : b(q, p; \mathbf{x}; t)$.

The object of statistical mechanics is to provide a link between the two levels of description, whereby to each microscopic dynamical function $b(q, p; \mathbf{x}, t)$ there corresponds a unique macroscopic dynamical function:

$$b(q, p; \mathbf{x}, t) \rightarrow B(\mathbf{x}, t)$$

This correspondence defines a *mapping of the phase space into the physical space.* We must require a certain number of reasonable conditions in order to make the mapping definite.

Technically speaking, $B(\mathbf{x}, t)$ is called a *functional* of $b(q, p; \mathbf{x}, t)$: this means that, for every fixed value of the parameters $\mathbf{x}$ and t, to each choice of a function of $q, p : b(q, p)$, corresponds a number B. B is, therefore, a "function of a function." It will be sometimes denoted as follows:

$$B(\mathbf{x}, t) \equiv \langle b(q, p; \mathbf{x}, t) \rangle \equiv \langle b \rangle \tag{2.2.1}$$

It is quite natural to require that the functional be *linear:* given two dynamical functions b, c and two numbers β, γ, the following relation should hold:

$$\langle \beta b + \gamma c \rangle = \beta \langle b \rangle + \gamma \langle c \rangle \tag{2.2.2}$$

On the other hand, we require that a constant (independent of q and p) and, in particular, the unity, be not affected by the mapping; in other words

$$\langle 1 \rangle = 1 \tag{2.2.3}$$

It is pretty easy to construct an explicit operation, in terms of familiar concepts, that satisfies both Conditions (2.2.2) and (2.2.3). Consider indeed a function on the phase space $F(q, p)$ and construct the following expression:

$$B(\mathbf{x}, t) \equiv \langle b \rangle = \int dq \, dp \, b(q, p; \mathbf{x}, t) F(q, p) \tag{2.2.4}$$

the integration being over the entire phase space of the system.

Obviously, this expression satisfies the condition (2.2.2). In order that (2.2.3) be satisfied, we have to require

$$\int dq\, dp\, F(q, p) = 1 \tag{2.2.5}$$

Any function F satisfying Condition (2.2.5) is acceptable for the construction of the functional (2.2.4). The set of these functions is a subset of the set of dynamical functions: They are called *phase-space distribution functions*, or briefly *distribution functions*, or even *distributions*. We may now formulate the *basic postulate of statistical mechanics:*

> *The "state" of a system is completely specified at a given time by a certain distribution function $F(q, p)$, satisfying Eq. (2.2.5). The observable value $B(\mathbf{x}, t)$ of a dynamical function $b(q, p; \mathbf{x}, t)$ in such a system is postulated to be given by Eq. (2.2.4).*

It is clear that Eq. (2.2.4) and (2.2.5) strongly suggest the formalism of probability theory. Indeed, if we add the extra condition

$$F(q, p) \geqslant 0 \tag{2.2.6}$$

then $F(q, p)$ could be interpreted as *the probability density for finding the system at the point q, p in phase space.* Equation (2.2.6) ensures that this probability is a positive number, as it should be, and Eq. (2.2.5) means that the system is certainly "somewhere" in the phase space. Equation (2.2.4) is, then, just the usual prescription for constructing the average (or expectation value) of the random variable b. Hence, in the present picture the concept of "state of the system" is deeply modified as compared to the ordinary dynamical description of Section 1.2. The state is no longer specified at a given time t by a fixed set of values for q, p, that is, by a point in phase space. Rather, at time t, *every point* of phase space represents a possible state of the system. Each point, however, is *weighted* by the value $F(q, p)$ of the distribution function at that point. Given the distribution function, the value of all possible macroscopic observables can be calculated. Hence, as a consequence of the basic postulate, *the state of a system at a given time is completely defined by the specification of the distribution function $F(q, p)$.* We thus reached precisely the concept of an *ensemble* as defined qualitatively in Section 2.1.

We insist that the reader not be misled by the connotations of the word "probability." We show later that the fundamental laws (in particular, the laws of motion) of statistical mechanics are by no means different from those of ordinary exact mechanics: There is in particular no need of a probabilistic modification of the laws of motion (as many people still thought twenty years ago) in order to explain macroscopic physics. The definition (2.2.4) is the only extramechanical "statistical" assumption entering the theory. It postulates the definition of the quantities of interest $B(\mathbf{x}, t)$ in terms of the corresponding microscopic quantities.

Of course, Eq. (2.2.6) implies the choice of a distribution function $F(q, p)$ at a given time $(t = 0)$; this is a *nonmechanical problem.* We have to choose an initial distribution function that represents in the best way all that we know about the system. The way in which one proceeds for this choice was discussed qualitatively in Section 2.1 and will be considered again at length throughout the book. However, once this choice has been made, the subsequent evolution is strictly determined by the laws of exact mechanics. Let us develop this point further.

Suppose a distribution $F(q, p)$ was chosen to describe the initial state of the system. We want to determine the time dependence of the observable $B(\mathbf{x}, t)$, average of $b(q, p; \mathbf{x}, t)$. Let us assume that initially:

$$b(q, p; \mathbf{x}, t = 0) = b(q, p; \mathbf{x}) \tag{2.2.7}$$

where the rhs is a given dynamical function. Then

$$B(\mathbf{x}, 0) = \int dq\, dp\, b(q, \mathbf{p}; \mathbf{x}) F(q, p) \tag{2.2.8}$$

As the system moves according to the Hamiltonian equations of motion, the quantity $b(q, p; \mathbf{x})$ is transformed into $b(q, p; \mathbf{x}, t)$ according to Eq. (1.2.24); as a consequence, $B(\mathbf{x}, 0)$ is transformed into $B(\mathbf{x}, t)$ defined by

$$B(\mathbf{x}, t) = \int dq\, dp [e^{[H]t} b(q, p; \mathbf{x})] F(q, p) \tag{2.2.9}$$

This equation gives *the law of motion in physical space, induced by the Hamiltonian law in phase space.* It suggests an analogy with the "Heisenberg picture" of quantum mechanics: the state of the system is given once for all, and the evolution is described by the change in time of the dynamical functions.

There is an unpleasant difficulty in this description. We see from Eq. (2.2.9) that we have to solve a separate initial-value problem for every dynamical function $b(q, p; \mathbf{x}, t)$ whose macroscopic average $B(\mathbf{x}, t)$ is sought. We now show that the procedure can be enormously simplified, at least in principle. Indeed, it is sufficient to solve *one* single initial-value problem for a partial-differential equation: its solution provides us with all macroscopic functions $B(\mathbf{x}, t)$ for all times, through "simple" quadratures. We first prove a lemma.

Lemma. *If the same canonical transformation* $e^{[G]r}$ *(where* G *is an arbitrary dynamical function and* r *is a parameter) is applied to* ***both***

factors in the integrand of Eq. (2.2.4), the value of the integral remains unchanged:

$$\int dq\, dp\,[e^{[G]r}b(q, p; \mathbf{x}, t)][e^{[G]r}F(q, p)] = \int dq\, dp\, b(q, p; \mathbf{x}, t)F(q, p) \tag{2.2.10}$$

The proof is very simple; indeed by using Eq. (1.2.33):

$$\int dq\, dp\,[e^{[G]r}b(q, p; \mathbf{x}, t)][e^{[G]r}F(q, p)] = \int dq\, dp\, b(q_r, p_r; \mathbf{x}, t)F(q_r, p_r)$$

where

$$q_r = e^{[G]r}q$$
$$p_r = e^{[G]r}p$$

In the rhs we may make a change of integration variables, $q \to q_r$, $p \to p_r$:

$$\int dq\, dp\, b(q_r, p_r; \mathbf{x}, t)F(q_r, p_r) = \int dq_r\, dp_r\, |J|\, b(q_r, p_r; \mathbf{x}, t)F(q_r, p_r)$$

where J is the Jacobian of the transformation. But we know that the Jacobian of any canonical transformation is unity. Hence the result equals the rhs of Eq. (2.2.10).

Let us start from Eq. (2.2.9) and apply to both factors under the integral the canonical transformation $\exp(-[H]t)$:

$$B(\mathbf{x}, t) = \int dq\, dp\,[e^{-[H]t}e^{[H]t}b(q, p; \mathbf{x})][e^{-[H]t}F(q, p)] \tag{2.2.11}$$

Clearly, the two successive canonical transformations acting on b cancel each other. The effect of the procedure is to transfer the time dependence from the dynamical functions to the distribution function. Indeed, it is natural to introduce a time-dependent distribution function $F(q, p; t)$:

$$F(q, p; t) = e^{-[H]t}F(q, p) \tag{2.2.12}$$

In terms of this function, the observable at time t is expressed as

$$B(\mathbf{x}, t) = \int dq\, dp\, b(q, p; \mathbf{x})F(q, p; t) \tag{2.2.13}$$

and this clearly reduces to Eq. (2.2.8) at time 0. We have, therefore, a description in which the whole time dependence comes from the evolution of the state, whereas the dynamical functions are given and are constant in time. It could be called the "Schrödinger picture" of statistical mechanics. Although equilivalent to the "Heisenberg picture" (2.2.4), it appears more natural. In particular, the initial-value problem is much simpler in principle: The initial specification of a distribution function determines the subsequent evolution of all averages. Instead of a separate study for every observable, the problem of statistical mechanics is reduced to the study of the time dependence of a single function, $F(q, p; t)$.

It is very convenient to write Eq. (2.2.12) in differential form:

$$\frac{\partial F(q, p; t)}{\partial t} \equiv \partial_t F(q, p; t) = -[e^{-[H]t} F(q, p), H]_P$$

or else

$$\partial_t F(q, p; t) = [H(q, p), F(q, p; t)]_P \tag{2.2.14}$$

This is called the *Liouville equation* and is, beyond any doubt, the most important equation of statistical mechanics, just as the Schrödinger equation is the central equation of quantum mechanics.

The most important feature of the Liouville equation is its *linearity*. This property introduces an important touch of simplicity into an otherwise very complex theory and should be exploited as thoroughly as possible. To stress this feature, it is often convenient to write Eq. (2.2.14) in a slightly different form*:

$$\partial_t F(t) = LF(t) \tag{2.2.15}$$

where we introduced the linear operator L naturally defined as follows:

$$\begin{aligned} LF &\equiv [H, F]_P \\ &= \sum_{n=1}^{N} \left\{ \frac{\partial H}{\partial q_n} \frac{\partial F}{\partial p_n} - \frac{\partial H}{\partial p_n} \frac{\partial F}{\partial q_n} \right\} \end{aligned} \tag{2.2.16}$$

The fundamental operator L will be called the *Liouvillian* of the system. In the language of statistical mechanics, the Liouvillian plays exactly the same role that the Hamiltonian $H(q, p)$ plays in Section 1.2. In ordinary mechanics, the Hamiltonian fixes the law of evolution of the state of the system, that is, the motion of any point in phase space. In statistical mechanics, where the state is represented by a distribution function, the specification of the Liouvillian again completely fixes the law of evolution of the state, that is, of the function F. More abstractly, it can be stated

* One sometimes finds the equation (2.2.15) written in the form

$$i\,\partial_t F = L'F$$

Clearly

$$L' = iL$$

The reason for multiplying both sides of (2.2.14) by i is to make the analogy with the Schrödinger equation more obvious. As, however, H and F are real functions by definition, we think that the introduction of the imaginary unit is somewhat artificial.

that the Hamiltonian and the Liouvillian are two different realizations of the *generator of the infinitesimal time translations.* They reflect the same structure, that is, the same group of transformations. Only the objects to be transformed are different in the two cases: points in phase space in the ordinary dynamics, elements of the space of functions F in statistical dynamics.

For all Hamiltonian systems the two pictures of the evolution are equivalent: Given a Hamiltonian H, the Liouvillian is completely determined by Eq. (2.2.16). The difference between the two descriptions lies only in the definition of what we call the state of the system, *not* in its law of evolution. We may mention, however (although such problems are not treated in this book), that the Liouvillian is a good starting point for generalizations. One can conceive of problems for which a Hamiltonian does not exist, but for which a Liouvillian can be constructed. In other words, Eq. (2.2.15) can be taken as the basis of a theory even when L is not defined by Eq. (2.2.16) but by some appropriate generalization.

In terms of the Liouvillian, the formal solution of the initial-value problem (2.2.12) is written as follows:

$$F(q, p; t) = e^{Lt} F(q, p) \tag{2.2.17}$$

Although this equation represents the solution of the problem of statistical mechanics, it is still much too formal. We do not really know what precise meaning to attribute to the exponential of the operator L. The problem of transforming Eq. (2.2.17) into a precise, meaningful, and operational expression is, in a nutshell, the object of nonequilibrium statistical mechanics.

Equation (2.2.17) is the solution of the initial-value problem of the Liouville equation. It is not difficult to show that if the initial $F(q, p)$ is an acceptable distribution, it is acceptable at all later times. In particular

$$F(q, p; t) \geqslant 0, \qquad \text{all } t \tag{2.2.18}$$

$$\int dq\, dp\, F(q, p; t) = 1 \qquad \text{all } t \tag{2.2.19}$$

These properties can be formulated as follows;

> The sign and the normalization of the distribution function are invariants of the motion in phase space.
>
> The transformations (2.2.17) leave the set of distribution functions invariant.

It is easily seen that these properties are automatically satisfied. Indeed, consider Eq. (2.2.19):

$$\int dq\, dp\, F(q, p; t) = \int dq\, dp\, 1[e^{-[H]t}F(q, p)]$$

$$= \int dq\, dp (e^{-[H]t}1)[e^{-[H]t}F(q, p)]$$

$$= \int dq\, dp\, 1F(q, p) = 1$$

where we first used the fact that the neutral element 1 is unaffected by the canonical transformation (1.2.25) and then invoked the lemma (2.2.10) and the initial condition (2.2.5).

The proof of Eq. (2.2.18) is also very simple and is left as an exercise for the reader.

We now close this section with an important warning. We have defined in this section a mapping of the phase space into the physical space:

$$b(q, p) \to B$$

This means precisely: to every $b(q, p)$ there corresponds one and only one B. It is tempting to reverse the statement and say hastily "To every B there corresponds one and only one $b(q, p)$." *This statement is wrong.* We will soon see that there exist macroscopic quantities that cannot be expressed in the form (2.2.4) of an average of a dynamical function weighted with the distribution function F. Such quantities will actually play a very important role in thermodynamics. A typical example is entropy. The set of macroscopic quantities can, therefore, be subdivided into two classes: the *mechanical quantities*, which are of the form (2.2.4), and the *thermal quantities*, which are not of this form.

Although these problems will be discussed later in detail, we may already ask the question "Does the existence of nonmechanical macroscopic quantities imply a failure of the formalism?" The answer is definitely no. On the contrary, the existence of observable thermal quantities reveals that the distribution function is not a mathematical fiction, but a true physical object, whose properties can be observed and measured. Let us explain this in two examples in which we anticipate forthcoming results.

The *temperature* of a macroscopic body only has an unambiguous meaning in thermal equilibrium. Therefore, it is a concept attached not to the dynamics of a particular molecule or small set of molecules, but rather to the state of the macroscopic system as a whole. Hence, we cannot define the temperature as an average of a microscopic function taken with an arbitrary distribution function [as implied by Eq. (2.2.4)]: It rather appears as *a parameter characterizing the particular distribution function describing the system in thermal equilibrium.**

Another important example is *entropy*. This quantity is defined both in equilibrium and out of equilibrium. But again, entropy is not a property of a single particle. It describes the overall "state of disorder" of the system. The entropy can (in certain cases) apparently be written in the form (2.2.4). However, the quantity $b(q, p)$ is not a given, fixed function, but depends on the distribution function. *The entropy is therefore not a linear functional of* $F(q, p; t)$. As the system evolves in time, *both* b and F change, contrary to the situation of Eq. (2.2.13).

We only mention these important facts at this point. Later we will see how they come quite naturally into the theory.

2.3. QUANTUM ENSEMBLES: THE von NEUMANN DENSITY OPERATOR

We now extend the ideas developed in the previous section to the treatment of quantum systems. It was pointed out already in Section 2.1 that a quite new feature appears in this case. Quantum mechanics has a fundamental statistical character due to the Heisenberg uncertainty relations: Even if we possess the maximum possible information about the state of the system, we can only make statistical predictions about the values of the observables.

More precisely, let the state of the system be defined by a wave function $\Psi(x)$, where x stands for the collection of position coordinates of N particles. We expand this wave function in a series of orthonormal basis functions, as in Eqs. (1.4.5) or (1.4.16):

$$\Psi(x) = \sum_r c_r \varphi_r(x) \tag{2.3.1}$$

(We know from Section 1.4 that the coefficients c_r must be symmetric or antisymmetric with respect to permutations of any two indices in the set r.) The average value of an arbitrary operator $\hat{b}$ in the state $\Psi(x)$ was

* Let us clarify this statement. One sometimes finds the following definition: "The temperature is the average kinetic energy of an ideal gas in equilibrium (multiplied by a suitable constant)." This statement is correct, but is not a definition of the same kind as (2.2.4). It is only a consequence of the fact that the equilibrium distribution function depends in a special way on the parameter T. Actually, the average of *any* microscopic dynamical function in equilibrium depends on temperature (because it is calculated with a T-dependent distribution function). Conversely, the average of the kinetic energy (or of any other dynamical function) calculated for a nonequilibrium state bears no relation to temperature.

given in Eq. (1.4.12):

$$\begin{aligned}\bar{b} &= \sum_r \sum_s c_r^* c_s b_{rs} \\ &= \sum_r \sum_s c_r^* c_s \langle r|\, \hat{b}\, |s\rangle\end{aligned} \tag{2.3.2}$$

All the results discussed above are valid when the state of the system $\Psi(x)$ is definitely known, that is, when the system is in a *pure state.* This information is, however, not always available. Very often, and practically always when dealing with many-body systems, we only have a partial knowledge of the microscopic state of the system. All we can define is the probability γ_i of finding the system in some state $\Psi^{(i)}(x)$ among many possible states. This step is precisely the indeterminacy corresponding to the classical problem of statistical mechanics (Section 2.2). It implies the introduction of a *statistical ensemble* in quantum mechanics. Indeed, the single system is replaced by an ensemble of identical systems (same Hamiltonian) starting at time zero in different states $\Psi^{(i)}$.

We now ask what is the average value of some observable when the system is specified in this statistical way. In order to calculate this value, we expand each of the possible states in terms of the orthonormal basis $\varphi(x)$:

$$\Psi^{(i)}(x) = \sum_r c_r^{(i)} \varphi_r(x) \tag{2.3.3}$$

The average value of $\hat{b}$ in the state $\Psi^{(i)}(x)$ is given by

$$\bar{b}^{(i)} = \sum_r \sum_s c_r^{(i)*} c_s^{(i)} b_{rs} \tag{2.3.4}$$

This is the usual quantum average of Eq. (2.3.2). But we only know that the system has probability γ_i of being in state $\Psi^{(i)}(x)$. Hence we must perform a *second* averaging process in order to determine the overall average value $\langle b \rangle$ measurable in our statistically specified system:

$$\langle b \rangle = \sum_i \gamma_i \bar{b}^{(i)} = \sum_i \gamma_i \sum_r \sum_s c_r^{(i)*} c_s^{(i)} b_{rs} \tag{2.3.5}$$

where it must be assumed that the probability law γ_i satisfies the conditions:

$$\begin{aligned}\gamma_i &\geqslant 0 \\ \sum_i \gamma_i &= 1\end{aligned} \tag{2.3.6}$$

Let us now introduce a matrix ρ_{sr} defined as follows:

$$\rho_{sr} = \sum_i \gamma_i c_s^{(i)} c_r^{(i)*} \tag{2.3.7}$$

(Note the order of the indices!) The average (2.3.5) can be expressed as

$$\langle b \rangle = \sum_r \sum_s b_{rs} \rho_{sr} \tag{2.3.8}$$

Let us now denote by $\hat{\rho}$ the operator whose matrix elements are identified with ρ_{sr}:

$$\rho_{sr} = \int dx \; \varphi_s^*(x) \hat{\rho} \varphi_r(x) \equiv \langle s | \, \hat{\rho} \, | r \rangle \tag{2.3.9}$$

$\hat{\rho}$ is called *von Neumann's density matrix* (or *operator*). One then clearly sees a matrix product appearing in Eq. (2.3.8), which can be written as

$$\langle b \rangle = \sum_r \langle r | \, \hat{b} \hat{\rho} \, | r \rangle \tag{2.3.10}$$

Hence the average of the observable $\hat{b}$ is expressed as the sum of the diagonal elements, or the *trace*, of the matrix $\hat{b}\hat{\rho}$. But the trace of a matrix (or operator) is well known to be an intrinsic characteristic, independent of the representation chosen. Hence, we can write

$$B \equiv \langle b \rangle = \text{Tr} \; \hat{b} \hat{\rho} = \text{Tr} \; \hat{\rho} \hat{b} \tag{2.3.11}$$

where the second equality results from the invariance of the trace of a product of operators under cyclical permutation of the factors.

We have now reached the point where we can make contact with our general formalism of Section 2.2. Equation (2.3.11) expresses the macroscopic value $\langle b \rangle$ of the observable $\hat{b}$ as a *linear functional* of this observable, obeying Eq. (2.2.2). It clearly also obeys the condition (2.2.3). Indeed, using Eq. (1.4.9) we have

$$\text{Tr} \; \hat{\rho} = \sum_i \sum_r \gamma_i c_r^{(i)*} c_r^{(i)} = \sum_i \gamma_i$$

Using, moreover, Eq. (2.3.6) we obtain

$$\text{Tr} \; \hat{\rho} = 1 \tag{2.3.12}$$

Hence Eqs. (2.3.11) and (2.3.12) are to be considered as the quantum analogs of the classical equations (2.2.4) and (2.2.5).

We can now forget the detailed way in which we arrived at our results and take a more global view. Define the set of density operators as a subset of the quantum-dynamical algebra, whose members are characterized by the condition (2.3.12). Any member of this subset is an acceptable density operator, and we may formulate the quantum version of the *basic postulate* of statistical mechanics as follows:

The "state" of a system in statistical mechanics is completely specified, at a given time, by a certain density operator $\hat{\rho}$ satisfying Eq. (2.3.12). The observable value $B \equiv \langle b \rangle$ of a dynamical function $\hat{b}$ in such a system is given by Eq. (2.3.11).

Let us now discuss the problems related with the positive character of $\hat{\rho}$. Contrary to the classical equation (2.2.4), Eq. (2.3.11) does not suggest a simple probabilistic interpretation. In order to understand the situation more clearly, we write Eq. (2.3.8) by separating the contributions of the diagonal and of the nondiagonal terms of the matrices:

$$\langle b \rangle = \sum_r b_{rr}\rho_{rr} + \sum_{r \neq s}\sum b_{rs}\rho_{rs} \tag{2.3.13}$$

The first term of the right-hand side can be understood in terms of probabilities. Indeed, the diagonal matrix elements of the density operator are given by Eq. (2.3.7) as

$$\rho_{rr} = \sum_i \gamma_i c_r^{(i)} c_r^{(i)*} = \sum_i \gamma_i \left| c_r^{(i)} \right|^2$$

Hence, using also Eqs. (2.3.6) and (2.3.12):

$$\begin{gathered} \rho_{rr} \geqslant 0 \\ \sum_r \rho_{rr} = 1 \end{gathered} \tag{2.3.14}$$

From these properties it follows that ρ_{rr} can be interpreted as the probability of finding the system in state r. If, therefore, the matrix ρ_{rs} happened to be diagonal in the basis $\{\varphi_r\}$ chosen here, that is, $\rho_{rs} = 0$ for $r \neq s$, the definition of the observable value B would be identical with a weighted average of the same kind as in classical mechanics.

However, the matrix ρ_{rs} is in general not diagonal; moreover, even if it were diagonal in a given basis it would no longer be so in another representation: the diagonal character of a matrix is not an intrinsic property. The nondiagonal terms have no definite sign: they cannot be interpreted as probabilities. These terms are linked to the *interference effects* that are well-known quantum-mechanical features without a classical counterpart. (These are responsible for such effects as the diffraction of electrons and the tunnel effect.) They are consequences of the wavelike properties of matter. The nondiagonal coefficients ρ_{rs} can also be related to the phase correlation between the states r and s.

Hence, although the basic postulate of quantum-statistical mechanics defines the observable value B as a linear functional of the microscopic operator $\hat{b}$, this functional is not interpretable in purely probabilistic terms, because of the existence of nondiagonal interference terms. For simplicity, however, we call this functional a "quantum average."*

One can easily derive a stronger property from Eq. (2.3.7). The following property holds for arbitrary complex numbers α_r:

$$\sum_r \sum_s \rho_{rs}\alpha_r\alpha_s^* \geq o \tag{2.3.15}$$

In other words, the quadratic form associated with the matrix ρ_{rs} is nonnegative. This property can be considered as the closest analog of the classical equation (2.2.6). It must be added as an extra condition restricting the class of operators acceptable as density operators.

Let us now consider the law of motion in this formalism. Just as in Section 2.2, we want to determine the time dependence of the macroscopic observable $B(\mathbf{x}, t)$, average of the operator $\hat{b}(\mathbf{x}, t)$ (which can possibly depend on the position $\mathbf{x}$ in physical space as well as on the time t). If the initial condition is

$$\hat{b}(\mathbf{x}, t=0) = \hat{b}(\mathbf{x}) \tag{2.3.16}$$

then clearly

$$B(\mathbf{x}, t) = \text{Tr}\{\exp([\hat{H}]t)\hat{b}(\mathbf{x})\}\hat{\rho} \tag{2.3.17}$$

This equation gives the law of motion of a macroscopic observable, induced by the microscopic Hamiltonian description in the "Heisenberg" picture.

We can again transfer the evolution operator from $\hat{b}$ to the density matrix†; indeed, the lemma (2.2.10) can be generalized as follows:

Lemma. *If the same canonical transformation* $\exp\{[\hat{G}]r\}$ *(where* $\hat{G}$ *is an arbitrary dynamical operator and r is a parameter) is applied to both factors under the trace in Eq. (2.3.11), the value of the average is not changed:*

$$\text{Tr}\{\exp([\hat{G}]r)\hat{b}\}\{\exp([\hat{G}]r)\hat{\rho}\} = \text{Tr}\ \hat{b}\hat{\rho} \tag{2.3.18}$$

The proof is again very simple. Using Eq. (1.3.16) we can write

$$\begin{aligned}
&\text{Tr}\{\exp([\hat{G}]r)\hat{b}\}\{\exp([\hat{G}]r)\hat{\rho}\} \\
&\qquad = \text{Tr}\left[\exp\left(\frac{ir\hat{G}}{\hbar}\right)\hat{b}\exp\left(-\frac{ir\hat{G}}{\hbar}\right)\right]\left[\exp\left(\frac{ir\hat{G}}{\hbar}\right)\hat{\rho}\exp\left(-\frac{ir\hat{G}}{\hbar}\right)\right] \\
&\qquad = \text{Tr}\left[\exp\left(\frac{ir\hat{G}}{\hbar}\right)\hat{b}\hat{\rho}\exp\left(-\frac{ir\hat{G}}{\hbar}\right)\right]
\end{aligned}$$

* Clearly, there also exist observables that are not quantum averages of microscopic dynamic operators. The discussion of Section 2.2 about these observables applies without any change to the quantum case as well.

† For brevity, we omit writing the argument $\mathbf{x}$ on which $\hat{b}$ or B may depend.

The lemma results from the fact that the trace of an operator (here, $\hat{b}\hat{\rho}$) is invariant under unitary transformations.

Using this lemma we can prove immediately, just as in Section 2.2, that Eq. (2.3.17) is equivalent to

$$B(\mathbf{x}, t) = \mathrm{Tr}\, \hat{b}\hat{\rho}(t) \tag{2.3.19}$$

where

$$\hat{\rho}(t) = \exp(-[\hat{H}]t)\hat{\rho} \tag{2.3.20}$$

Alternatively, following Eq. (1.3.16), $\hat{\rho}(t)$ can be written in the form:

$$\hat{\rho}(t) = \exp\left(-\frac{i\hat{H}t}{\hbar}\right)\hat{\rho}\, \exp\left(\frac{i\hat{H}t}{\hbar}\right) \tag{2.3.21}$$

in which it is clear that $\hat{\rho}(t)$ is obtained from $\hat{\rho}$ via a unitary transformation generated by $\hat{H}$.

We have again obtained a description in which the whole time dependence is contained in the density matrix rather than in the dynamical operators. This can be called the "Schrödinger picture" of statistical mechanics.

By taking the time derivative of Eq. (2.3.20) we find the differential equation obeyed by the density matrix:

$$i\hbar\, \partial_t \hat{\rho}(t) = [\hat{H}, \hat{\rho}(t)]_- \tag{2.3.22}$$

This is the fundamental equation of quantum-statistical mechanics, called the *von Neumann equation.* It plays the same basic role as the classical Liouville equation. Quantum-statistical mechanics is essentially the study of the solutions of Eq. (2.3.22).

Let us finally note that the von Neumann equation, as written in Eq. (2.3.22), is independent of the representation chosen for the operators. In particular, if the second quantization formalism of Section 1.5 is used, the von Neumann equation keeps its form:

$$i\hbar\, \partial_t \boldsymbol{\rho}(t) = [\mathsf{H}, \boldsymbol{\rho}(t)]_- \tag{2.3.23}$$

where H and $\boldsymbol{\rho}$ are now to be considered as operators in the occupation number space.

2.4. HAMILTONIAN AND LIOUVILLIAN OF A SYSTEM OF INTERACTING PARTICLES

We now illustrate the concepts discussed so far on a general level by applying them to a specific case. The class of systems to be considered here is, however, so important that most of the topics studied in this book actually relate to its particular cases. Hence the results derived here will be used throughout the book.

Consider first a *classical* picture. The system under study is defined as a set of N identical point particles, each of mass m. The system is enclosed in a box of volume $\mathcal{V}$. The phase space of the system is then spanned by N position vectors $\mathbf{q}_1 \ldots \mathbf{q}_N$ and N momentum vectors $\mathbf{p}_1 \ldots \mathbf{p}_N$. We may also assume that an external field is present. The Hamiltonian of the system can then be written quite naturally as a sum of three terms:

$$H = H^0 + H' + H^F \tag{2.4.1}$$

The term H^0 represents the *free motion* of noninteracting particles in the absence of an external field. It is simply the kinetic energy. It has the important property of being a sum of N terms, each depending only on the momentum of a single particle:

$$H^0 = \sum_{j=1}^{N} H_j^0 \tag{2.4.2}$$

with

$$H_j^0 = \frac{p_j^2}{2m} \tag{2.4.3}$$

The term H' represents *interactions* between the particles: it is the potential of the intermolecular forces. It cannot, of course, be represented as a sum of terms, each depending on a single particle: Indeed, its main role is to provide a coupling between the particles. The simplest assumption consists of writing H' as a sum of terms, each depending (nonadditively) on the canonical variables of two particles*:

$$H' = \sum_{j<n=1}^{N}\sum V_{jn} \tag{2.4.4}$$

This assumption implies that the force exerted on a particle n by a particle j is the same whether the two particles are isolated or are surrounded by the other particles of the system. This is not always true, but it represents a quite good approximation in a very large number of important physical systems.

In many cases, we can further restrict the form of the interaction potential. The real function V_{jn} usually depends only on the positions $\mathbf{q}_j$, $\mathbf{q}_n$, not on the momenta (this assumption excludes magnetic interactions). Moreover, it actually depends only on the distance $\mathbf{r}_{jn} = \mathbf{q}_j - \mathbf{q}_n$ of the two particles, and more specifically on the absolute value of this vector $|\mathbf{r}_{jn}| \equiv r_{jn}$:

$$V_{jn} = V(|\mathbf{q}_j - \mathbf{q}_n|) = V(r_{jn}) \tag{2.4.5}$$

* The summation prescription in Eq. (2.4.4) will be discussed in detail in Section 3.1.

As the particles are identical, each term V_{jn} in (2.4.4) represents the *same* function of the variable r_{jn} pertaining to the couple *jn*. The form of this function depends, of course, on the nature of the particles: For charged particles it is the Coulomb potential r_{jn}^{-1}, for neutral molecules it is the Lennard-Jones potential, and so on. Various particular cases will be discussed at length in this book. The form (2.4.5) has the obvious symmetry consequence:

$$V_{jn} = V_{nj} \tag{2.4.6}$$

Moreover, we note the important property:

$$\frac{\partial V_{jn}}{\partial \mathbf{q}_j} = -\frac{\partial V_{jn}}{\partial \mathbf{q}_n} \tag{2.4.7}$$

This equality is the expression of Newton's third law (action = reaction).

In some cases, it is useful to work with the Fourier transform of the interaction potential. This function is defined as usual by the two mutually inverse equations:

$$V(r) = \int d\mathbf{k}\ \tilde{V}_k e^{i\mathbf{k}\cdot\mathbf{r}} \tag{2.4.8}$$

$$\tilde{V}_k = (8\pi^3)^{-1} \int d\mathbf{r}\ V(r) e^{-i\mathbf{k}\cdot\mathbf{r}} \tag{2.4.9}$$

It is easily checked that Eqs (2.4.5) and (2.4.6) imply that $\tilde{V}_k$ is a real function of the absolute value k of the wave vector $\mathbf{k}$:

$$\tilde{V}_{\mathbf{k}} = \tilde{V}_{-\mathbf{k}} = \tilde{V}_k \tag{2.4.10}$$

The term H^F in Eq. (2.4.1) describes the action of an external field deriving from a potential V^F. This field acts separately on each particle; therefore, it is again a sum of terms, each depending on the canonical coordinates of a single particle, just like H^0:

$$H^F = \sum_{j=1}^{N} V_j^F \tag{2.4.11}$$

In the case of an electrical or of a gravitational field, V_j^F only depends on the position $\mathbf{q}_j$ of particle j; in the case of a magnetic field it also depends on $\mathbf{p}_j$. Moreover, V_j^F may also depend explicitly on time:

$$V_j^F = V^F(\mathbf{q}_j, \mathbf{p}_j; t) \tag{2.4.12}$$

We now derive the form of the Liouvillian of this system. It is convenient to introduce the following notations for the differential operators occurring very frequently in this book:

$$\begin{aligned}\partial_t &= \frac{\partial}{\partial t}\\ \boldsymbol{\nabla}_j &= \frac{\partial}{\partial \mathbf{q}_j}\\ \partial_j &= \frac{\partial}{\partial \mathbf{p}_j}\\ \partial_{jn} &= \frac{\partial}{\partial \mathbf{p}_j} - \frac{\partial}{\partial \mathbf{p}_n}\end{aligned} \tag{2.4.13}$$

The Liouvillian (2.2.16) can now be written as follows:

$$L = \sum_{n=1}^{N} [(\boldsymbol{\nabla}_n H) \cdot \partial_n - (\partial_n H) \cdot \boldsymbol{\nabla}_n] \tag{2.4.14}$$

As the Liouvillian depends linearly on H, it follows that the natural decomposition (2.4.1) of the Hamiltonian implies a similar decomposition of the Liouvillian:

$$L = L^0 + L' + L^F \tag{2.4.15}$$

The free-particle Liouvillian is, in turn, a sum of terms, each acting on the canonical coordinates of a single particle:

$$L^0 = \sum_{j=1}^{N} L_j^0 \tag{2.4.16}$$

where L_j^0 is easily evaluated by using Eqs. (2.4.14):

$$L_j^0 = [H_j^0, \ldots]_P = (\boldsymbol{\nabla}_j H_j^0) \cdot \partial_j - (\partial_j H_j^0) \cdot \boldsymbol{\nabla}_j$$

Hence, using (2.4.3),

$$L_j^0 = -\left(\frac{\mathbf{p}_j}{m}\right) \cdot \boldsymbol{\nabla}_j = -\mathbf{v}_j \cdot \boldsymbol{\nabla}_j \tag{2.4.17}$$

where we introduced the *velocity* $\mathbf{v}_j$:

$$\mathbf{v}_j = \frac{\mathbf{p}_j}{m} \tag{2.4.18}$$

Consider now the interaction term. Clearly:

$$L' = \sum_{j<n=1}^{N} \sum L'_{jn} \tag{2.4.19}$$

Taking account of Eq. (2.4.5) we have

$$L'_{jn} = [V_{jn}, \ldots]_P = (\boldsymbol{\nabla}_j V_{jn}) \cdot \partial_j + (\boldsymbol{\nabla}_n V_{jn}) \cdot \partial_n$$

Using, moreover, Eq. (2.4.7) we get*

$$L'_{jn} = (\nabla_j V_{jn}) \cdot \partial_{jn} \tag{2.4.20}$$

Finally, by a quite similar calculation we obtain the external-field Liouvillian:

$$L^F = \sum_{j=1}^{N} L_j^F \tag{2.4.21}$$

with

$$L_j^F = (\nabla_j V_j^F) \cdot \partial_j - (\partial_j V_j^F) \cdot \nabla_j \tag{2.4.22}$$

The Liouville equation for our system can now be written quite explicitly as follows:

$$\partial_t F = -\sum_{j=1}^{N} \mathbf{v}_j \cdot \nabla_j F + \sum_{j<n=1}^{N} \sum (\nabla_j V_{jn}) \cdot \partial_{jn} F + \sum_{j=1}^{N} [(\nabla_j V_j^F) \cdot \partial_j F - (\partial_j V_j^F) \cdot \nabla_j F] \tag{2.4.23}$$

We now consider the *quantum-mechanical description* of the same system of N interacting particles. For simplicitly we only consider systems in absence of external field, $V^F = 0$. In the ordinary position representation, the Hamiltonian is again of the form (2.4.1), but is now an operator. In particular, the kinetic energy is of the form (2.4.2) with

$$\hat{H}_j^0 = -\left(\frac{\hbar^2}{2m}\right) \nabla_j^2 \tag{2.4.24}$$

The interaction energy is given by Eqs. (2.4.4) and (2.4.5); this operator is simply a multiplicative operator.

We now consider the Hamiltonian in the second quantization formalism. By using the general prescription (1.5.22) we obtain the Hamiltonian in the form:

$$\mathsf{H} = \int d\mathbf{x}\, \Psi^\dagger(\mathbf{x}) \left[-\left(\frac{\hbar^2}{2m}\right) \nabla^2 \right] \Psi(\mathbf{x}) + \frac{1}{2} \int d\mathbf{x} \int d\mathbf{x}'\, \Psi^\dagger(\mathbf{x}) \Psi^\dagger(\mathbf{x}') V(\mathbf{x}-\mathbf{x}') \Psi(\mathbf{x}') \Psi(\mathbf{x}) \tag{2.4.25}$$

This form automatically incorporates the Fermi–Dirac or the Bose–Einstein symmetry, according to the nature of the operators $\Psi^\dagger$, Ψ. Very often it is useful to go over to the momentum representation. We, therefore, use Eq. (1.5.23) in (2.4.25) and introduce also the Fourier

* Note that when the operator L'_{jn} is applied to an arbitrary function $f(x_j, x_n)$, the gradient ∇_j acts only on the function V_{jn}, not on $f(x_j, x_n)$. [Here, x_j denotes the set (q_j, p_j).]

transform of the potential (2.4.8):

$$
\begin{aligned}
\mathsf{H} = & -\frac{\hbar^2}{2m}\frac{1}{8\pi^3}\int d\boldsymbol{\kappa}\, d\boldsymbol{\kappa}'\int d\mathbf{x}\; e^{i(\boldsymbol{\kappa}'-\boldsymbol{\kappa})\cdot\mathbf{x}}\mathsf{a}^\dagger(\hbar\boldsymbol{\kappa})(i\boldsymbol{\kappa}')^2\mathsf{a}(\hbar\boldsymbol{\kappa}') \\
& +\frac{1}{2}\frac{1}{(8\pi^3)^2}\int d\boldsymbol{\kappa}_1\, d\boldsymbol{\kappa}_2\, d\boldsymbol{\kappa}_1'\, d\boldsymbol{\kappa}_2'\, d\mathbf{l}\int d\mathbf{x}\, d\mathbf{x}' \\
& \times \exp[-i\boldsymbol{\kappa}_1\cdot\mathbf{x} - i\boldsymbol{\kappa}_2\cdot\mathbf{x}' + i\mathbf{l}\cdot(\mathbf{x}-\mathbf{x}') + i\boldsymbol{\kappa}_2'\cdot\mathbf{x}' + i\boldsymbol{\kappa}_1'\cdot\mathbf{x}] \\
& \times \mathsf{a}^\dagger(\hbar\boldsymbol{\kappa}_1)\mathsf{a}^\dagger(\hbar\boldsymbol{\kappa}_2)\tilde{V}_1\mathsf{a}(\hbar\boldsymbol{\kappa}_2')\mathsf{a}(\hbar\boldsymbol{\kappa}_1') \\
= & \frac{\hbar^2}{2m}\int d\boldsymbol{\kappa}\, d\boldsymbol{\kappa}'\; \delta(\boldsymbol{\kappa}-\boldsymbol{\kappa}')\mathsf{a}^\dagger(\hbar\boldsymbol{\kappa})\kappa'^2\mathsf{a}(\hbar\boldsymbol{\kappa}') \\
& +\frac{1}{2}\int d\boldsymbol{\kappa}_1\, d\boldsymbol{\kappa}_2\, d\boldsymbol{\kappa}_1'\, d\boldsymbol{\kappa}_2'\, d\mathbf{l}\; \delta(-\boldsymbol{\kappa}_1+\mathbf{l}+\boldsymbol{\kappa}_1')\; \delta(-\boldsymbol{\kappa}_2-\mathbf{l}+\boldsymbol{\kappa}_2') \\
& \times \mathsf{a}^\dagger(\hbar\boldsymbol{\kappa}_1)\mathsf{a}^\dagger(\hbar\boldsymbol{\kappa}_2)\tilde{V}_1\mathsf{a}(\hbar\boldsymbol{\kappa}_2')\mathsf{a}(\hbar\boldsymbol{\kappa}_1')
\end{aligned}
$$

and finally:

$$\mathsf{H} = \mathsf{H}^0 + \mathsf{H}' \tag{2.4.26}$$

$$,\mathsf{H}^0 = \frac{\hbar^2}{2m}\int d\boldsymbol{\kappa}\; \kappa^2\mathsf{a}^\dagger(\hbar\boldsymbol{\kappa})\mathsf{a}(\hbar\boldsymbol{\kappa}) \tag{2.4.27}$$

$$\mathsf{H}' = \frac{1}{2}\int d\boldsymbol{\kappa}_1\, d\boldsymbol{\kappa}_2\, d\mathbf{l}\; \tilde{V}_1\mathsf{a}^\dagger(\hbar\boldsymbol{\kappa}_1)\mathsf{a}^\dagger(\hbar\boldsymbol{\kappa}_2)\mathsf{a}(\hbar\boldsymbol{\kappa}_2+\hbar\mathbf{l})\mathsf{a}(\hbar\boldsymbol{\kappa}_1-\hbar\mathbf{l}) \tag{2.4.28}$$

BIBLIOGRAPHICAL NOTES

The concept of a classical ensemble in phase space was introduced by J. W. Gibbs, *Elementary Principles in Statistical Mechanics*, Yale Univ. Press, New Haven, 1902 (reprinted by Dover, New York, 1960).

The density matrix was introduced in quantum mechanics by von Neumann and by Dirac:

J. von Neumann, *Mathematische Grundlagen der Quantenmechanik*, Berlin, 1932.
P. A. M. Dirac, *The Principles of Quantum Mechanics*, 4th ed. (1st ed., 1930), Clarendon, Oxford, 1958.

In most textbooks, the ensembles are discussed mainly for the equilibrium case. The presentation given here appears to us to be clearer, especially when dealing with nonequilibrium ensembles. It naturally opens the way for a study of the time evolution. Moreover, it lays the stress on the algebraic structure of the theory. Our presentation is very close to the one adopted in D. G. Currie, T. F. Jordan, and E. C. G. Sudarshan, *Rev. Mod. Phys.* **35,** 350 (1963).

An advantage of the algebraic and group-theoretical viewpoint is that it opens the door very naturally to a clear, covariant special-relativistic formulation of statistical mechanics. This was shown in the paper quoted above and was very much developed since. See, for instance:

R. Balescu and T. Kotera, *Physica* **33,** 558 (1967).

R. Balescu, T. Kotera, and E. Piña, *Physica* **33,** 581 (1967).

CHAPTER 3
REDUCED DISTRIBUTION FUNCTIONS

3.1. THE CLASSICAL DISTRIBUTION VECTOR

The formalism developed in Chapter 2 is quite general, and requires no assumption whatever concerning the form of the Hamiltonian or of the other dynamical functions. However, a little reflection will show us that the dynamical functions playing a role in the problems of real interest have simple features that we shall presently bring out. It is, therefore, desirable to exploit these features and develop a formalism well adapted to this structure. As a bonus, it will turn out that this formalism provides an especially clear and elegant description of systems with a very large number of degrees of freedom, that is, precisely those that are of particular importance in statistical mechanics.

In the present section we discuss classical systems. In order to fix the ideas we discuss the system of mutually interacting particles described in Section 2.4, although a similar discussion can be made for other types of systems. As an orientation we first review some particular important dynamical functions.

Consider the Hamiltonian (2.4.1). For convenience we introduce here the abbreviation x_j for the set of canonical variables pertaining to particle j:

$$x_j \equiv (\mathbf{q}_j, \mathbf{p}_j) \tag{3.1.1}$$

The Hamiltonian is a function of all the canonical variables $x_1, \ldots, x_N$. It has, however, a very special functional form: it can be written as

$$H(x_1, \ldots, x_N) = \sum_{j=1}^{N} H_1(x_j) + \sum_{j<n=1}^{N}\sum H_2(x_j x_n) \tag{3.1.2}$$

where $H_1(x_j) \equiv H_j^0 + V_j^F$ and $H_2(x_j x_n) = V_{jn}$. Hence H is a sum of one-particle terms plus a sum of two-particle terms. There are no contributions depending nonadditively on more than two particles. This type of functional dependence is very characteristic for most important dynamical functions.

A very important class of observables are the *local densities* of macroscopic quantities at a given point $\mathbf{x}$ of physical space. A simple example is provided by the *number density* of particles. To obtain this macroscopic quantity we must average a highly discontinuous dynamical function. If we calculate the density at point $\mathbf{x}$ (in physical space), $n(\mathbf{x})$, a particle contributes a term $\delta(\mathbf{x}-\mathbf{q}_j)$: indeed, either particle j is not at point $\mathbf{x}$—then it does not contribute to $n(\mathbf{x})$—or it is at $\mathbf{x}$, and contributes an infinite amount (because the volume of a point particle is assumed to vanish). The total density is given by a superposition of such terms for all particles; the average is

$$n(\mathbf{x}) = \int dq\, dp \sum_{j=1}^{N} \delta(\mathbf{q}_j - \mathbf{x}) F(q, p) \tag{3.1.3}$$

Hence the microscopic dynamical function corresponding to the number density is

$$n(x_1, \ldots, x_N; \mathbf{x}) = \sum_{j=1}^{N} \delta(\mathbf{q}_j - \mathbf{x}) \tag{3.1.4}$$

It is a sum of one-particle functions. Other densities can be constructed in the same way.

A subtler type of properties is provided by the *correlations*. These measure the influence of the phenomena occurring at a given point on the phenomena occurring at another point. For instance, we may be interested in a simultaneous measure of the density at two points $\mathbf{x}_1$, $\mathbf{x}_2$. Macroscopically, we define this two-point number density $n_2(\mathbf{x}_1, \mathbf{x}_2)$ as follows: Consider two volume elements of size $\mathcal{V}_{x_1}$, $\mathcal{V}_{x_2}$ centered around the points $\mathbf{x}_1$ and $\mathbf{x}_2$ respectively. Let N_{x_1} and N_{x_2} be the number of particles in each of these cells. We then define

$$n_2(\mathbf{x}_1, \mathbf{x}_2) = \lim_{\substack{\mathcal{V}_{x1}\to 0\\ \mathcal{V}_{x2}\to 0}} \frac{N_{x_1} N_{x_2}}{\mathcal{V}_{x_1} \mathcal{V}_{x_2}}$$

We may clearly define s-point densities in a similar way. The microscopic dynamical function corresponding to the two-point number density can be obtained by a straightforward extension of the previous argument:

$$n_2(x_1, \ldots, x_N; \mathbf{x}_1, \mathbf{x}_2) = \sum_{j=1}^{N} \sum_{n=1}^{N} \delta(\mathbf{q}_j - \mathbf{x}_1)\delta(\mathbf{q}_n - \mathbf{x}_2) \tag{3.1.5}$$

This is a sum of terms depending on one particle (for $j = n$) plus a sum of two-particle terms. Its average is

$$n_2(\mathbf{x}_1, \mathbf{x}_2) = \int dq\, dp \sum_{j} \sum_{n} \delta(\mathbf{q}_j - \mathbf{x}_1)\delta(\mathbf{q}_n - \mathbf{x}_2) F(q, p) \tag{3.1.6}$$

In general, it is clear that the average of the product of microscopic densities is *not* equal to the product of the averages. In real systems it is usually found that the function $n_2(\mathbf{x}_1, \mathbf{x}_2)$ differs from $n(\mathbf{x}_1)n(\mathbf{x}_2)$ when the distance $\mathbf{x}_1 - \mathbf{x}_2$ is sufficiently small. This deviation, which is (usually) due to the intermolecular interactions, is called the *two-particle correlation*. These correlations can be measured experimentally and also affect the calculation of many important macroscopic properties such as the pressure, the energy density, and the like. These problems will be amply discussed later.

We may imagine subtler types of correlations, for example, three- or four-body correlations (irreducible to combinations of two-body correlations). However, these soon become undetectable in practice, and they do not affect the calculation of physically interesting quantities.

After this orientation, we consider this matter more systematically. The dynamical functions characterizing our system are functions of the canonical variables $x_1, \ldots, x_N : b(x_1, \ldots, x_N)$. As the particles in the system are identical, we may restrict our interest to those functions in which all particles play the same role: these are the only ones that represent physically relevant quantities. Hence we assume that the acceptable functions are symmetric under permutation of any two variables:

$$b(x_1, \ldots, x_j, \ldots, x_n, \ldots, x_N) = b(x_1, \ldots, x_n, \ldots, x_j, \ldots, x_N) \quad (3.1.7)$$

We now note that any function of this type can be represented in the following form:

$$b(x_1, \ldots, x_N) = b_0 + \sum_{j=1}^{N} b_1(x_j) + \sum_{j<n=1}^{N}\sum b_2(x_j x_n) + \sum_{j<n<m=1}^{N}\sum\sum b_3(x_j x_n x_m) + \cdots + b_N(x_1, \ldots, x_N) \quad (3.1.8)$$

In this formula, $b_s(x_{j_1}, \ldots, x_{j_s})$ is a function of s variables only, which cannot be decomposed into a sum of functions depending on less variables. In order to agree with Eq. (3.1.7), we must assume that all the terms $b_1(x_j)$ are the *same* function of the various variables x_j, $j = 1, \ldots, N$; similarly for $b_2, b_3, \ldots$. Moreover, all the functions $b_s(x_{j_1}, \ldots, x_{j_s})$ must be symmetric under the permutation of their arguments, and b_0 is a constant.

Let us explain in some detail the prescription of summation in Eq. (3.1.8). Clearly, in the term $b_2(x_j x_n)$, the particles j and n must be different, otherwise the contribution would be of type b_1. Moreover, as $b_2(x_j x_n) = b_2(x_n x_j)$, the simple statement that $j \neq n$ would result in counting every distinct term twice. Hence we require that in the double summation we only include terms such that $j<n$. Thus, for $N=3$:

$$\sum_{j<n=1}^{3}\sum b_2(x_jx_n)=b_2(x_1x_2)+b_2(x_1x_3)+b_2(x_2x_3)$$

Equivalently, we may prescribe that in each term involving $b_s(x_1,\ldots,x_s)$ we sum over all values of the indices $1,\ldots,s$, with the only restriction that they are all distinct. In that case, we correct the overcounting by dividing the sum by $s!$ Hence, Eq. (3.1.8) can also be written as

$$b(x_1,\ldots,x_N)=b_0+\sum_{j=1}^{N} b_1(x_j)+(2!)^{-1}\sum_{j\neq n}\sum b_2(x_jx_n)$$
$$+\cdots+(s!)^{-1}\sum_{j_1\neq\cdots\neq j_s}\cdots\sum b_s(x_{j1}x_{j2},\ldots,x_{js})+\cdots+b_N(x_1,\ldots,x_N) \quad (3.1.9)$$

To check the correctness of this prescription, we consider again $N=3$:

$$\frac{1}{2!}\sum_{j\neq n}\sum b_2(x_jx_n)=\tfrac{1}{2}[b_2(x_1x_2)+b_2(x_1x_3)+b_2(x_2x_1)$$
$$+b_2(x_2x_3)+b_2(x_3x_1)+b_2(x_3x_2)]$$
$$=\tfrac{1}{2}[b_2(x_1x_2)+b_2(x_2x_1)]+\tfrac{1}{2}[b_2(x_1x_3)+b_2(x_3x_1)]$$
$$+\tfrac{1}{2}[b_2(x_2x_3)+b_2(x_3x_2)]$$

which, because of the symmetry of b_2, is identical to our previous result.

From the previous discussion it appears that the *dynamical functions of real physical interest involve only a finite number of irreducible functions* $b_0, b_1, \ldots, b_S$, say $S=2$ or 3. In other words, for these dynamical functions $b_s\equiv 0$ for $s>S$. This shows that in classifying the dynamical functions according to Eq. (3.1.8) we bring out a feature of great simplicity, which is very helpful, especially in dealing with systems of many particles.*

We now consider the process of averaging of these functions. The state of the system (at a given time) is defined by a distribution function $F(q,p)\equiv F(x_1,\ldots,x_N)$. Since all the particles are identical, we assume that this function is symmetric under permutations:

$$F(x_1,\ldots,x_j,\ldots,x_n,\ldots,x_N)=F(x_1,\ldots,x_n,\ldots,x_j,\ldots,x_N) \quad (3.1.10)$$

Next we use Eq. (2.2.4) to calculate the average of the dynamical function b of Eq. (3.1.8) or, better, (3.1.9). The average of the constant b_0 is just b_0, as follows from the postulate (2.2.5). The contribution of the

* This simplicity is, however, illusory, as will be presently shown.

one-particle terms can be written as

$$\int dx_1 \cdots dx_N \left[\sum_{j=1}^{N} b_1(x_j)\right] F(x_1, \ldots, x_N)$$
$$= N \int dx_1\, dx_2 \cdots dx_N\, b_1(x_1) F(x_1, \ldots, x_N) \quad (3.1.11)$$

Indeed, because of the symmetry properties described above, all terms in the integral of the sum on the left-hand side are equal and the expression equals N times the contribution of an arbitrary single term corresponding, say, to particle 1. We now note that all integrations except the one on x_1 involve only the distribution function F. Hence, we may introduce a new function, called a *reduced one-particle distribution function:*

$$f_1(x_1) = N \int dx_2 \cdots dx_N\, F(x_1, x_2, \ldots, x_N) \quad (3.1.12)$$

In terms of this function, the average in Eq. (3.1.11) becomes simply

$$\int dx_1 \cdots dx_N \left[\sum_{j=1}^{N} b_1(x_j)\right] F(x_1, \ldots, x_N) = \int dx_1\, b_1(x_1) f_1(x_1)$$

We now consider the general term in Eq. (3.1.9). Using again the symmetry argument, we note that the average of the s-fold sum consists of $[N!/(N-s)!\, s!]$ equal terms, this number being the number of ways of choosing s particles (regardless of their order) among N particles:

$$\int dx_1 \cdots dx_N \left[\frac{1}{s!} \sum \sum \cdots \sum_{j_1 \neq j_2 \neq \cdots \neq j_s} b_s(x_{j_1}, \ldots, x_{j_s})\right] F(x_1, \ldots, x_N)$$
$$= \frac{N!}{(N-s)!\, s!} \int dx_1 \cdots dx_s \cdots dx_N b_s(x_1, \ldots, x_s) F(x_1, \ldots, x_s, \ldots, x_N) \quad (3.1.13)$$

We now define the *reduced s-particle distribution function* $f_s(x_1 \cdots x_s)$, $s \leq N$, as

$$f_s = \frac{N!}{(N-s)!} \int dx_{s+1} \cdots dx_N\, F(x_1, \ldots, x_s, x_{s+1}, \ldots, x_N) \quad (3.1.14)$$

Clearly, this is again a symmetric function under permutation of the particle labels

$$f_s(x_1, \ldots, x_j, \ldots, x_n, \ldots, x_s) = f_s(x_1, \ldots, x_n, \ldots, x_j, \ldots, x_s) \quad (3.1.15)$$

Note in particular that, as a consequence of the normalization condition (2.2.5) we obtain:

$$f_0 \equiv 1 \quad (3.1.16)$$

and also the normalization of the reduced distributions:

$$\int dx_1 \cdots dx_s \, f_s(x_1, \ldots, x_s) = \frac{N!}{(N-s)!} \tag{3.1.17}$$

The rhs of Eq. (3.1.13) now reduces to

$$\frac{1}{s!} \int dx_1 \cdots dx_s \, b_s(x_1, \ldots, x_s) f_s(x_1, \ldots, x_s)$$

Hence, we may express the average of the dynamical function b completely in terms of reduced distribution functions:

$$\langle b \rangle = \sum_{s=0}^{N} (s!)^{-1} \int dx_1 \cdots dx_s \, b_s(x_1, \ldots, x_s) f_s(x_1, \ldots, x_s) \tag{3.1.18}$$

In formulas of this kind, the convention is made that the term corresponding to $s=0$ involves no integration.

The reduced distribution functions can also be defined by a different set of equations, which in some cases is quite useful. The one-particle function $f_1(y_1)$ can be written as

$$f_1(y_1) = \int dx_1 \cdots dx_N \left\{ \sum_{j=1}^{N} \delta(y_1 - x_j) \right\} F(x_1, \ldots, x_N) \tag{3.1.19}$$

where the integration is now over *all* the N variables $x_1, \ldots, x_N$ of the system. The effect of the δ function is to "cancel" the effect of one of these integrations. It is easily seen that, because of the symmetry of $F(x_1, \ldots, x_N)$ with respect to permutations, the sum involves N identical terms, and that the result is identical to Eq. (3.1.12), in which $x_1 \to y_1$. More generally,

$$f_s(y_1, \ldots, y_s) = \int dx_1 \cdots dx_N \left[\sum\sum_{j_1 \neq j_2 \neq \cdots \neq j_s} \cdots \sum \delta(y_1 - x_{j_1}) \right.$$
$$\left. \times \, \delta(y_2 - x_{j_2}) \cdots \delta(y_s - x_{j_s}) \right] F(x_1, \ldots, x_N) \tag{3.1.20}$$

(Note carefully the summation prescription $j_1 \neq j_2 \neq \cdots \neq j_s$!)

It is immediately checked that the normalization (3.1.17) is satisfied:

$$\int dy_1 \cdots dy_s \, f_s(y_1, \ldots, y_s)$$
$$= \int dx_1 \cdots dx_N \sum_{j_1 \neq \cdots \neq j_s} \cdots \sum \int dy_1 \cdots dy_s \, \delta(y_1 - x_{j_1}) \cdots \delta(y_s - x_{j_s})$$
$$\times F(x_1, \ldots, x_N)$$
$$= \sum_{j_1 \neq \cdots \neq j_s} \cdots \sum \int dx_1 \cdots dx_N \, F(x_1, \ldots, x_N)$$
$$= \sum_{j_1 \neq \cdots \neq j_s} \cdots \sum 1 = N(N-1) \cdots (N-s+1)$$

In this perspective, the reduced distribution functions appear as averages of the *singular* dynamical functions

$$\sum_{j_1 \neq \cdots \neq j_s} \cdots \sum \delta(y_1 - x_{j_1}) \cdots \delta(y_s - x_{j_s}),$$

which depend on the set of parameters $y_1, \ldots, y_s$. Comparing Eqs. (3.1.19), (3.1.20) to Eqs. (3.1.4), (3.1.5) we see that the reduced s-particle function can be interpreted as the s-point density in phase space.

The results obtained in this section suggest a very compact and elegant formulation of statistical mechanics. We may group together all the reduced distribution functions into a set :

$$\mathfrak{f} \equiv \{f_0, f_1(x_1), f_2(x_1 x_2), \ldots, f_N(x_1, \ldots, x_N)\} \tag{3.1.21}$$

The knowledge of all these functions enables us to calculate all averages of dynamical functions; hence it completely specifies the state of the system. The set $\mathfrak{f}$ can be looked at as a formal N-dimensional "vector": it is therefore called the *distribution vector* of the system.

Similarly, any symmetric dynamical function $b(x_1, \ldots, x_N)$ is fully characterized by specifying the form of the functions b_0, $b_1(x_1)$, $b_2(x_1 x_2), \ldots$ in the representation (3.1.9). These functions can also be grouped into a formal vector $\mathfrak{b}$:

$$\mathfrak{b} \equiv \{b_0, b_1(x_1), b_2(x_1 x_2), \ldots, b_N(x_1, \ldots, x_N)\} \tag{3.1.22}$$

The average of any dynamical function now appears as a formal *scalar product* of the two vectors $\mathfrak{f}$ and $\mathfrak{b}$, defined as

$$\langle b \rangle \equiv (\mathfrak{b}, \mathfrak{f}) = \sum_{s=0}^{N} (s!)^{-1} \int dx_1 \cdots dx_s \, b_s(x_1, \ldots, x_s) f_s(x_1, \ldots, x_s) \tag{3.1.23}$$

We note that the components of the distribution vector are not entirely independent. Indeed, as the various f_s are generated, through Eq. (3.1.14), from a single function F, the following relation holds:

$$f_r(x_1, \ldots, x_r) = \frac{(N-s)!}{(N-r)!} \int dx_{r+1} \cdots dx_s \, f_s(x_1, \ldots, x_s), \qquad r < s \leq N \tag{3.1.24}$$

Of course, the relation cannot be inverted: the s-particle function contains more information than f_r if $r < s$.

The new representation that we developed is, of course, entirely equivalent to the representation of Chapter 2. It seems even, at first glance, to possess an extraordinary advantage. We saw, indeed, that most physically relevant dynamical functions have the property that $b_s(x_1 \cdots x_s) \equiv 0$ for s larger than 2 or 3. Equation (3.1.18) then implies that the knowledge of the first two or three reduced distributions would be sufficient for the solution of most practical problems. This advantage is, however, illusory. It will be seen in Section 3.4 that the reduced distributions do not obey closed equations [whereas $F(x_1, \ldots, x_N)$ obeys the closed Liouville equation]. Rather, the exact determination of a given f_s requires the knowledge of the *whole set* of reduced distributions, $f_1, f_2, \ldots, f_N$. Approximation schemes can, however, be developed consistently, by which the set of equation for f_s can be cut down at some level or other. The search for such approximation schemes is one of the objects of statistical mechanics.

The advantage of the description in terms of reduced distributions, therefore, does not lie here: The methods used in the study of the equations for the f_s are not any simpler or any more complex than those involved in the study of the Liouville equation for F. The main advantage is more subtle and will be discussed qualitatively in the next sections.

3.2. THE PRINCIPLE OF MACROSCOPIC EQUIVALENCE

Statistical mechanics deals with systems consisting of a large number N of particles enclosed in volumes, $\mathcal{V}$, which are large compared to molecular dimensions. Its ultimate purpose is the understanding of macroscopic physics as a manifestation of the microscopic N-body problem. In order to get a feeling of the role played by these large parameters in the typical problems of statistical mechanics, we must first analyze some basic concepts of macroscopic physics.

It is a fact of experience that the mechanical quantities describing a *macroscopic* system such as a piece of metal or a sample of fluid are of two kinds. A first class includes quantities the value of which is proportional to the size of the system: these are called *extensive quantities.* Typical examples are the volume $\mathcal{V}$, the mass M, and the total energy E. A second class consists of quantities having at every point inside the system a well-defined value, which remains practically constant when the size of the system is increased: these are called *intensive quantities.* Typical examples are the mass density ρ, the pressure P, the temperature T, the specific heat at constant volume per unit mass c_v, and the viscosity η. The two- or three-point densities defined in Section 3.1 are also of this kind.

Dividing any extensive quantity by an arbitrarily chosen extensive quantity, such as the volume, results in an intensive quantity. It follows that the macroscopic system can be entirely described in terms of *one* extensive variable $\mathcal{V}$, and a set of intensive variables.

The following statement is a common experimental result of macroscopic physics: *The values of the intensive variables at any given point in the system are determined by the local environment. If the system is not in a steady state, the rate of change of the intensive quantities is determined again by the local environment, at least for times that are not too long.* These properties are characteristically described in mathematical language by a *field theory*. We already stated before that macroscopic physics is precisely a field theory appropriate to the view of matter as a continuum. It is of outstanding importance for understanding the essence of statistical mechanics to have a very clear feeling of these statements. We therefore illustrate them by a few simple examples.

As a first example, we consider *time-independent states*, described by the Laplace equation in two dimensions:

$$\nabla^2 U(x, y) = 0 \tag{3.2.1}$$

U may represent a temperature distribution within a solid body, or a distribution of concentrations within a mixture of fluids, or an electric potential, and so on. In order to describe a concrete system we must formulate a boundary-value problem.

A. We first take a trivial but important case. We assume that U is given and constant on a circle of radius L. (This problem may represent the distribution of temperature inside a cylindrical rod of infinite height immersed in a thermostat.) It is then well known that the only solution of the Laplace equation inside the circle is a constant. Hence, this problem corresponds to a *spatially homogeneous* system: the intensive quantity U is the same at all points within the system. If we consider a sequence of systems of increasing radius L and assume that U has the same value at their boundaries, then the value of U will be the same in all points of each of these systems.

B. We now consider a less trivial problem. For the same equation (3.2.1) we consider that, on a circle of radius L, the quantity U has the constant value $u_1 L$ in the first quadrant ($0 \leq \varphi \leq \pi/2$), 0 in the second, $-u_1 L$ in the third, and 0 in the fourth (see Fig. 3.2.1). This problem can be solved exactly: the value of U inside the circle is given by*

$$U(x, y) = \frac{u_1 L}{\pi}\left(\tan^{-1}\frac{2Lx}{L^2 - r^2} + \tan^{-1}\frac{2Ly}{L^2 - r^2}\right) \tag{3.2.2}$$

* This problem is taken from W. R. Smythe, *Static and Dynamic Electricity*, McGraw-Hill, New York, 1950, Chapter 4, Problem 8.

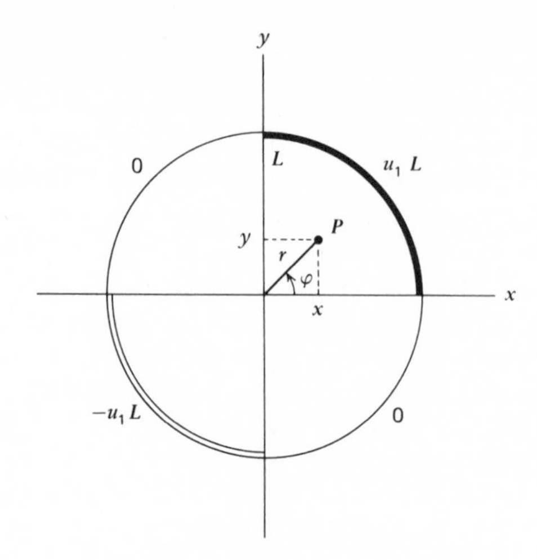

Figure 3.2.1. A boundary-value problem for the Laplace equation.

We have here a typical *spatially inhomogeneous state.* The value of U is different in each point of the system. We now consider again a sequence of systems of increasing L, concentric with the first. For all these systems we keep the constant u_1 fixed. As a result, the value of U at the boundaries in the first and third quadrants increases proportionally to the size of the system. This way of scaling up the systems guarantees the constancy of the *gradient* of U within the system, that is, the constancy of the local environment. We fix our attention on a point P, at given distance r from the origin, in each of these systems of our sequence. As we consider larger and larger systems, the value of U at P is obtained approximately by expanding the right-hand side of Eq. (3.2.2) in a series of powers of r/L:

$$U(x, y) \approx \frac{u_1}{\pi} 2(x+y)\left(1+\frac{r^2}{L^2}+\cdots\right) \tag{3.2.3}$$

Hence, at all points located well inside the system, the intensive quantity U has a value practically *independent of the size of the system.* The approximation is the better, the larger the system. Hence, this value can just as well be computed by assuming that the system has infinite radius.*

* Had we scaled up the systems by keeping $U = \pm U_1$ constant at the boundaries (in the first and third quadrants), it is easily seen that $U \to 0$ everywhere as $L \to \infty$. This is due to the progressive flattening of the internal gradients in the scaling process: the *local* conditions would not have been kept constant.

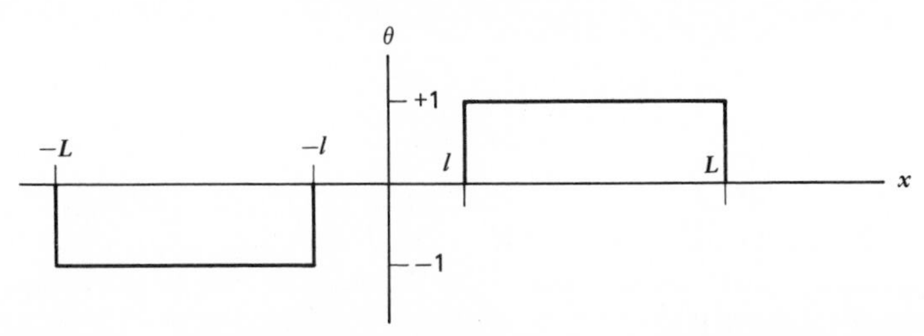

Figure 3.2.2. An initial-value problem for the heat-conduction equation.

C. We now consider a problem that is both *inhomogeneous in space* and *nonstationary in time.* More specifically, we consider a one-dimensional heat-conduction problem. Denoting by T_0 the temperature at the origin and by θ the deviation from that value ($\theta = T - T_0$), we have

$$\frac{\partial\theta(x,t)}{\partial t} = \kappa\frac{\partial^2\theta(x,t)}{\partial x^2} \tag{3.2.4}$$

We consider a system in which, at time zero, we prescribe the following temperature distribution (see Fig. 3.2.2):

$$\begin{aligned} &\theta = 0, && -\infty < x < -L \\ &\theta = -1, && -L < x < -l \\ &\theta = 0, && -l < x < l \qquad t = 0 \\ &\theta = +1, && l < x < L \\ &\theta = 0, && L < x < \infty \end{aligned} \tag{3.2.5}$$

This problem is a representation of the following experiment. At time zero we have two thermostats of size $(L-l)$ and at temperatures $T_0 - 1$ at left and $T_0 + 1$ at right. We place a system of length $2l$ and temperature T_0 in contact with the first thermostat at the left end and with the second one at the right end. We then switch off the temperature-regulation system of the thermostats and measure the temperature distribution at subsequent times $t > 0$. The solution is given by a well-known formula*:

$$\theta(x;t) = \frac{1}{2(\pi\kappa t)^{1/2}}\int_{-\infty}^{\infty} dx' \exp\left\{-\frac{(x-x')^2}{4\kappa t}\right\}\theta(x';0) \tag{3.2.6}$$

This formula is very easily applied to our particular initial condition (3.2.5) with the result:

$$\theta(x;t) = \frac{1}{2}\left\{\Phi\left(\frac{l+x}{2(\kappa t)^{1/2}}\right) - \Phi\left(\frac{l-x}{2(\kappa t)^{1/2}}\right) + \Phi\left(\frac{L-x}{2(\kappa t)^{1/2}}\right) - \Phi\left(\frac{L+x}{2(\kappa t)^{1/2}}\right)\right\} \tag{3.2.7}$$

* See, for example, H. Margenau and G. M. Murphy, *The Mathematics of Physics and Chemistry*, Van Nostrand, New York, 1943.

where $\Phi(\xi)$ is the error function:

$$\Phi(\xi) = \left(\frac{2}{\pi^{1/2}}\right) \int_0^{\xi} dt\, e^{-t^2} \tag{3.2.8}$$

We consider again a class of systems of increasing total length $2L$, keeping l fixed. We shall be mainly interested in the evolution in time of the temperature distribution within the "small" system of size $2l$, that is, in values of x such that

$$|x| < l \tag{3.2.9}$$

Moreover, the characteristic time in which a quasisteady, quasilinear temperature distribution is reached in the small system is $t_R = l^2/4\kappa$. In our experiment we shall be mainly interested in times t such that

$$t < \frac{l^2}{4\kappa} \tag{3.2.10}$$

If we consider systems in our sequence such that

$$L \gg l \tag{3.2.11}$$

then, for all values of x and t of order (3.2.9), (3.2.10), the last two bracketed error functions in Eq. (3.2.7) can be approximated by their asymptotic value:

$$\Phi\left(\frac{L+x}{2(\kappa t)^{1/2}}\right) \approx \frac{e^{-L^2/4\kappa t}}{L/2(\kappa t)^{1/2}}\left[1 + O\left(\frac{x^2}{L^2}\right)\right] \tag{3.2.12}$$

We thus see that the boundaries have a negligible influence on the distribution of temperatures and on its law of evolution in time inside the small system. Only for times of order $t \sim L^2/4\kappa$ do these terms become important. Hence, provided $L \gg l$ and $t \ll l^2/4\kappa$, the evolution inside the "small system" can be studied *as if* the total system were of infinite size. In Fig. 3.2.3 we show the temperature distribution inside the small system at various times, computed from Eq. (3.2.7) in the limit $L = \infty$. The temperature reaches an almost linear distribution as $\bar{t} \to 1$. At later times the temperature starts decaying slowly, mainly at the boundaries (we have imposed no fixed temperature constraint). In Fig. 3.2.4 we show the complete solution, *including* the boundary terms, at three different times. We see that for the longest time, $\bar{t} = 4$, a system with $L = 4l$, corresponding to a thermostat only 1.5 times the length of the system of interest, already gives a negligible correction. For $\bar{t} = 0.5$, even the case $L = 2l$, corresponding to a thermostat half the size of the system, gives a negligible correction. These pictures show vividly how rapid the convergence toward the limit value ($L = \infty$) is in practical problems.

We now sum up the conclusions to be drawn from the analysis of these simple examples. In most problems of macroscopic physics to which statistical mechanics is applied we are interested in the study of the *bulk properties* of matter, that is, the properties that are intrinsic to the system of interest and do not depend on specific external constraints. These properties are very weakly influenced by the boundary conditions imposed externally to the system.

We may take advantage of this situation in order to simplify the study of these problems by eliminating all spurious difficulties relating to the boundaries. We can do this clearly in the following way.

Suppose we are asked to study the intensive quantities characterizing some real system of volume $\mathcal{V}$. We associate with this problem a sequence of systems, labeled by an index k, differing from the original system by their volume $\mathcal{V}_k$. These systems are assumed to be properly scaled at their boundaries in order to possess the same symmetry as the original system, as was done in the simple examples above. Let μ_k be the value of some intensive quantity in the system labeled k. It is in general a

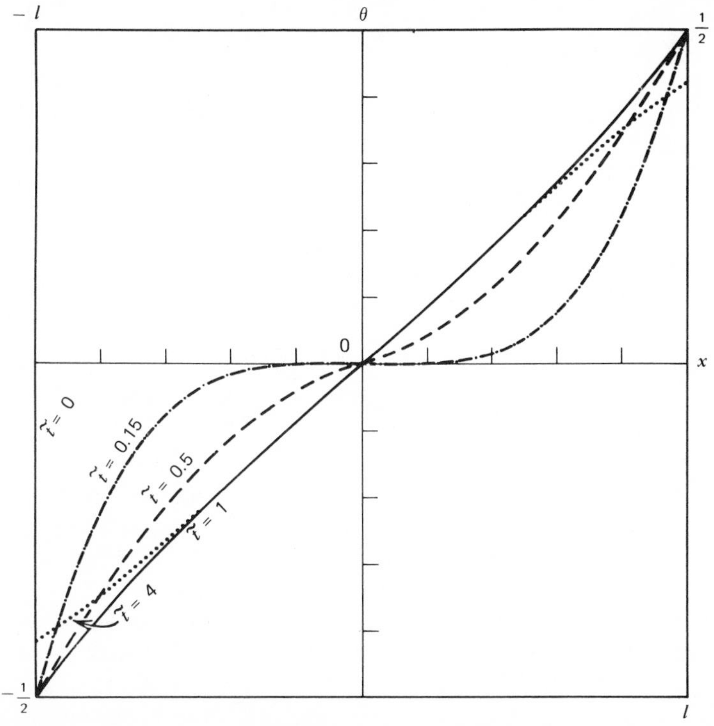

Figure 3.2.3. Temperature distribution in the small system at several values of the reduced time: $\tilde{t} = 4\kappa t/l^2$, in the limit $L \to \infty$.

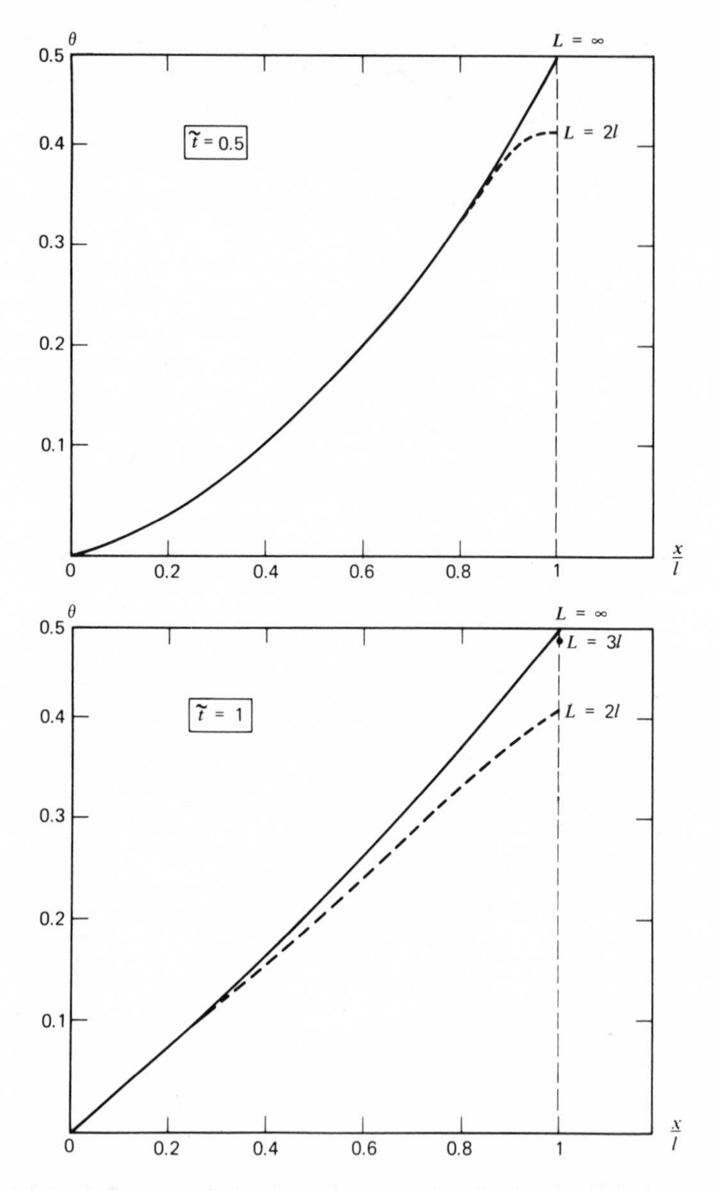

Figure 3.2.4. Influence of the boundary on the distribution of temperatures inside the small system.

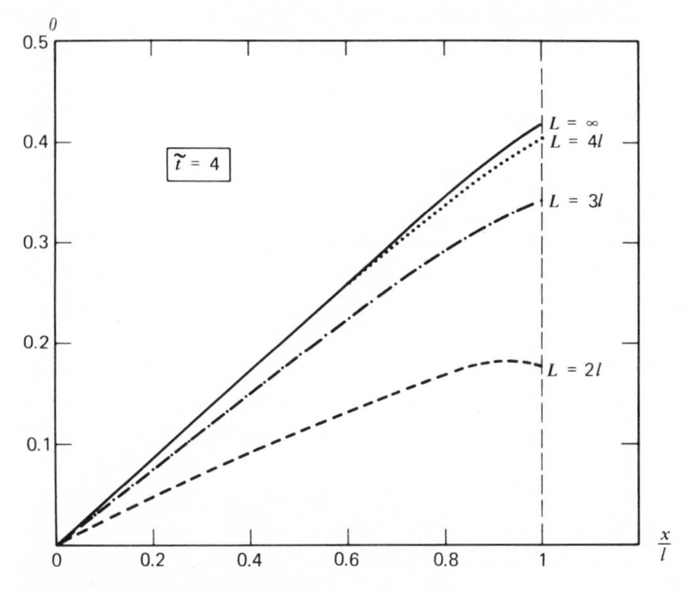

Figure 3.2.4. (Continued)

function of the position **x** and of time; moreover, it depends on the size of the system $\mathscr{V}_k$. We now formulate, as an experimental fact, a *principle of macroscopic equivalence*. This principle requires the intensive quantity μ_k to be of the form:

$$\mu_k = \mu + \mu_k'(\mathscr{V}_k) \tag{3.2.13}$$

where μ is a quantity independent of the size of the system, and μ_k' is a volume-dependent remainder such that

$$\mu_k'(\mathscr{V}_k) \to 0 \qquad \text{as} \qquad \mathscr{V}_k \to \infty \tag{3.2.14}$$

In a homogeneous system this property is assumed to hold uniformly throughout the system. In an inhomogeneous system the property is supposed to hold for all points that are sufficiently far from the boundary. For nonstationary systems it holds for times sufficiently short compared to a characteristic time T_k, which itself is an increasing function of the volume. As results from the simple examples discussed above, the convergence in Eq. (3.2.14) is usually quite rapid in practical problems.

From now on we shall restrict our interest to the study of the *bulk value* of the intensive quantities, that is, to the term μ in Eq. (3.2.13). With the reservations expressed above, this value is common to all the systems in the class defined above, provided their volume is sufficiently large. Such systems are called *macroscopically equivalent*. Because of this equivalence, it is not necessary to use the real system for the calculation of the quantity μ: We may use any other system that is macroscopically equivalent to it. *In particular, we may use the fictitious system corresponding to* $\mathcal{V}_k = \infty$. In this case the calculations are usually the simplest, because they are not contaminated by spurious boundary effects.

3.3. THE THERMODYNAMIC LIMIT

We now consider the microscopic picture and try to translate the analysis of Section 3.2 in molecular terms. We must first find a way of identifying extensive and intensive quantities. Proceeding as in Section 3.2, we attempt to construct a sequence of systems of increasing size and find conditions under which these systems can be considered as macroscopically equivalent. It is understood in all subsequent arguments that if the system is inhomogeneous, the scaling of the systems in the sequence is made in agreement with the overall symmetry, as described in Section 3.2.

We must first realize that, even in the macroscopic view, the process of increasing the size in the construction of our sequence is not a purely geometrical operation. If we double the size of the (real or fictitious) box in which the system is enclosed, we must fill the additional space with matter of the same kind as in the original system. We must therefore also double the amount of matter, that is, double the mass. If this is not done, the two systems are certainly not macroscopically equivalent: the mass density, an important intensive quantity, would have only one-half its original value in the new system. In the microscopic view, doubling the amount of matter means doubling the number N of molecules. Thus, when we construct our sequence of systems we must carefully specify that, when the volume is increased, the number of molecules is increased in the same ratio: the overall number density $N/\mathcal{V} = n$ is, therefore, maintained constant in all the systems of the sequence. In particular, when we consider the limiting system corresponding to $\mathcal{V} \to \infty$, we must also let $N \to \infty$, but not independently. We must require

$$\begin{aligned} &\mathcal{V} \to \infty \\ &N \to \infty \qquad\qquad\qquad\qquad \text{(T-limit)} \\ &\frac{N}{\mathcal{V}} = n: \qquad \text{finite constant} \end{aligned} \tag{3.3.1}$$

If the principle of macroscopic equivalence is valid, this particular limiting system describes the same local behavior as any other finite system of the sequence. The very important limiting process defined by (3.3.1) is given the name: *bulk limit*, or *thermodynamic limit*, or briefly *T limit*. It is a characteristic mathematical tool of statistical mechanics.

At this stage we are not yet in a position to state that all the systems in the sequence constructed above are macroscopically equivalent: It is clearly not sufficient that they have the same overall number density. We must discuss all the other, truly dynamical intensive quantities. The most typical of these are the densities, the two-point densities, and the like.* All these quantities are defined locally, at one point, two points, and so on. From the discussion at the beginning of Section 3.1 we can infer a general definition of the microscopic dynamical functions that correspond to local densities. Consider a one-particle dynamical function $\beta_1(\mathbf{q}_1, \mathbf{p}_1)$. In order to define the corresponding local density $b_1(\mathbf{q}_1, \mathbf{p}_1; \mathbf{x})$ at point $\mathbf{x}$, we argue that, in a discontinuous molecular description, there can be a contribution to this function only if a molecule happens to be located in $\mathbf{x}$, that is, if $\mathbf{q}_1 = \mathbf{x}$. Therefore

$$b_1(\mathbf{q}_1, \mathbf{p}_1; \mathbf{x}) = \delta(\mathbf{q}_1 - \mathbf{x})\beta_1(\mathbf{q}_1, \mathbf{p}_1)$$

More generally, an s-point density will be of the form:

$$b_s(\mathbf{q}_1, \mathbf{p}_1, \dots, \mathbf{q}_s, \mathbf{p}_s; \mathbf{x}_1, \dots, \mathbf{x}_s) = \delta(\mathbf{q}_1 - \mathbf{x}_1) \cdots \delta(\mathbf{q}_s - \mathbf{x}_s)\beta_s(\mathbf{q}_1, \mathbf{p}_1, \dots, \mathbf{q}_s, \mathbf{p}_s) \quad (3.3.2)$$

As stated in Section 3.1, the intensive quantities of physical interest are such that $b_s \equiv 0$ for $s > S$, where S is a finite number, much smaller than N, usually $S = 2$ or 3. The average of such quantities is given by Eq. (3.1.18); but by our restriction the upper limit in the sum is not really N, but the small, finite number S:

$$\langle b \rangle = \sum_{s=0}^{S} (s!)^{-1} \int dy_1 \cdots dy_s \, b_s(y_1, \dots, y_s; \mathbf{x}_1, \dots, \mathbf{x}_s) f_s(y_1, \dots, y_s) \quad (3.3.3)$$

We now calculate this expression for each one of the systems belonging to the sequence constructed at the beginning of this section. We thus regard it as a function of $\mathcal{V}$ and of N. The dependence on these parameters can only come from:

1. The limits of integration in Eq. (3.3.3)
2. The function b_s
3. The function f_s

Because of the particular form (3.3.2), we can perform the integrations over $\mathbf{q}$ immediately:

$$\int dy_1 \cdots dy_s \, b_s(y_1, \dots, y_s; \mathbf{x}_1, \dots, \mathbf{x}_s) f_s(y_1, \dots, y_s) = \int_{-\infty}^{\infty} d\mathbf{p}_1 \cdots d\mathbf{p}_s \, \beta_s(\mathbf{x}_1, \mathbf{p}_1, \dots, \mathbf{x}_s, \mathbf{p}_s) f_s(\mathbf{x}_1, \mathbf{p}_1, \dots, \mathbf{x}_s, \mathbf{p}_s)$$

* We restrict our discussion here to the macroscopic observables that can be expressed as averages of microscopic quantities. Observables that are not of this type (see end of Section 2.2) will be discussed later.

The natural limits of integration over the momenta are $\pm\infty$: they are independent of the volume of the box in which the system is enclosed. If β_s and f_s are sufficiently regular functions of the momenta, these integrals will be convergent, finite numbers for every given N and $\mathcal{V}$. On the other hand, the functions β_s, representing dynamical functions for a closed finite system of s particles ($s \leq S$) cannot depend on N nor on $\mathcal{V}$. Hence, the dependence on these parameters can only come from the reduced distribution functions $f_s(y_1 \cdots y_s)$.

These functions do, indeed, depend on the volume and on the number of particles, as can be seen from Eq. (3.1.14). This formula involves an integration over $N-s$ particles; the integrations over $\mathbf{q}_j$ are extended over the box of volume $\mathcal{V}$. Although the exact dependence on these parameters is quite complicated, a qualitative physical argument allows us to visualize the problem. From the definitions (3.1.19), and (3.1.20) there follows a possible interpretation of f_s as an s-point number density in phase space. We now note that the behavior of a molecule located at a point $\mathbf{q}$ is influenced by the nearby presence of another molecule, because of the mechanical interaction forces between these molecules. Therefore, the value of the function $f_s(\mathbf{q}_1\mathbf{p}_1, \ldots, \mathbf{q}_s\mathbf{p}_s)$ is determined physically by the mutual interactions of the s distinguished molecules, and also by the interactions of each one of these with all the other $(N-s)$ molecules of the system. The interactions of the second type are averaged over all the positions of the $(N-s)$ particles.* We also note that the interaction forces usually have a finite range, say r_c. Hence, the values and the shape of the function $f_s(x_1 \cdots x_s)$ is determined by the molecules located in a neighborhood of radius r_c around each of the s distinguished molecules.

We now come back to our sequence of increasingly large systems and study the influence of increasing N at constant density n on the value of $f_s(x_1, \ldots, x_s)$, for fixed s and fixed coordinates $x_1, \ldots, x_s$. First we note that when $N<s$, $f_s \equiv 0$, trivially. When $N=s$ the function jumps discontinuously to the finite value† $F(x_1, \ldots, x_s)$. The addition of an extra particle considerably disturbs the environment of the group $(1 \cdots s)$ and thus affects the value of f_s (see Fig. 3.3.1). As we add more and more particles, with the corresponding change of volume, the average environment of the distinguished particle gradually builds up to a practically invariant value. Therefore, when N becomes much larger than s (always at constant density), the value of f_s becomes less and less sensitive to the addition of extra particles. In conclusion, we may expect on physical

* The momenta are irrelevant in this discussion because (in a nonrelativistic theory) the interaction forces are independent of the momenta.

† Provided the coordinates $\mathbf{q}_1, \ldots, \mathbf{q}_s$ all lie within the volume $\mathcal{V} = N/n$.

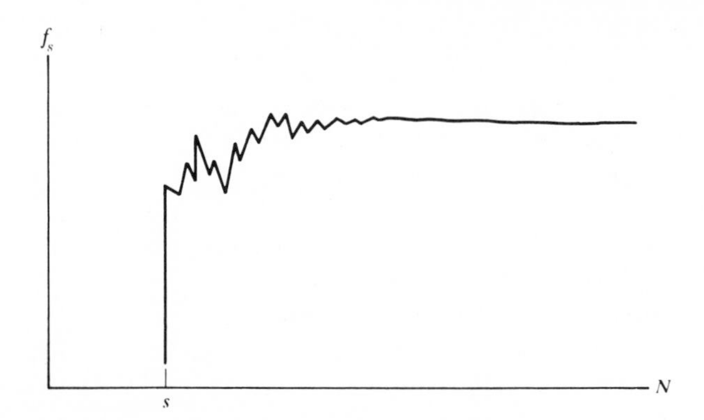

Figure 3.3.1. Dependence of $f_s(x_1, \ldots, x_s)$, for fixed $x_1, \ldots, x_s$, on the number of particles, as N increases at constant density (schematic).

grounds that $f_s(x_1, \ldots, x_s)$ is strongly N dependent for small values of N, but tends rapidly towards a constant asymptotic value for sufficiently large N. This asymptotic value will, of course, depend on the density n.

This discussion justifies our adoption of the following statement as a *postulate* restricting the systems to be considered in statistical mechanics to those that lead to well-behaved macroscopic behavior: *the reduced distribution functions $f_s(x_1 \cdots x_s)$, for any finite s, tend toward finite functions, independent of N in the thermodynamic limit (3.3.1)*. Thus, the reduced s-particle distribution function behaves in the characteristic way described at length in Section 3.2. Whenever our argument applies, the sequence of systems constructed here represents a class of macroscopically equivalent systems. All the observables defined by Eqs. (3.3.2) and (3.3.3) have the property expressed by Eqs. (3.2.13) and (3.2.14). Thus, the bulk value of these intensive quantities can be evaluated in any one of the systems of the class: the result will be the same. We may, in particular, use for this calculation the limiting system defined by Eq. (3.3.1).

The usual limitations of the principle of macroscopic equivalence, discussed in Section 3.2, apply here too. Even in a macroscopically homogeneous system, the local environment of the molecules located within a distance r_c of the boundary is different from the one in the bulk of the matter. The behavior of these molecules is not correctly described by going to the thermodynamic limit. However, these "surface films" do not influence the bulk properties of matter and will be disregarded here.

In time-dependent problems, some effects present in finite systems are also wiped out in the thermodynamic limit. The most celebrated of these is the so-called Poincaré recurrence. This effect is expressed in the following exact theorem of classical dynamics. Given a conservative mechanical system of N bodies enclosed in a finite region of space, starting in a given state at time zero, it will return arbitrarily close to the initial condition after a time T_p. Hence the motion of any finite mechanical system is quasiperiodic. Moreover, the period T_p tends toward infinity when $\mathcal{V} \to \infty$. Therefore, the results derived from a theory in the thermodynamic limit can only be valid for times much shorter than the Poincaré recurrence time. It turns out, however, that for all systems of interest in statistical mechanics the time T_p is so fantastically large that this is no limitation at all (for a cubic centimeter of gas T_p is of the order of billions of billions of years). Hence, we may safely postulate that the evolution in time of a sufficiently large finite system is the same as the evolution of the limiting system (3.3.1).*

We now briefly adapt the concepts and notations of Section 3.1 to the thermodynamic limit. As $N \to \infty$, the distribution vector $\mathfrak{f}$ has an infinite number of components:

$$\mathfrak{f} = \{f_0, f_1(x_1), f_2(x_1 x_2), \ldots\}$$

or simply

$$\mathfrak{f} = \{f_s(x_1, \ldots, x_s);\ s = 0, 1, 2, \ldots\} \tag{3.3.4}$$

Each component of this vector is supposed to be a *finite* function. Equation (3.1.23) for the average becomes, in the thermodynamic limit,

$$\langle b \rangle = (\mathfrak{b}, \mathfrak{f}) = \sum_{s=0}^{\infty} (s!)^{-1} \int dx_1 \cdots dx_s\, b_s(x_1, \ldots, x_s) f_s(x_1, \ldots, x_s) \tag{3.3.5}$$

For any intensive quantity, however, this equation is really the same as (3.1.23) because only a limited number S of terms is nonzero.

We note a few remarks on the "normalization" properties of the reduced distribution functions in the thermodynamic limit. Equation (3.1.17) clearly shows that the integral of all reduced distributions (except

* The reader is warned against a too direct analogy with the heat-conduction problem of Section 3.2. The Poincaré time is *not* the "long" time $L^2/4\kappa$ of that problem. The heat-conduction equation (3.2.4) is not a mechanical equation. It can, however, be derived from nonequilibrium statistical mechanics as an equation valid in the thermodynamic limit, thus for times much shorter than the Poincaré time T_p. (This will be discussed in Part 3 of this book.) In order to formulate our problem in the present framework, we consider a "very large" system of length 2Λ. Within it, we take a subsystem of length $2L$, with $L \ll \Lambda$: this is the total system described in Section 3.2 (Fig. 3.2.2). The "little" system of length $2l$, with $l \ll L$, is a subsystem of the subsystem. We are interested ultimately in the evolution of the little system. We then let $\Lambda \to \infty$ keeping L and l constant. In that limit, the heat-conduction equation appears as a valid description. We *then* let $L \to \infty$ (the order of limits implies the constraint $\Lambda/L \to \infty$) and obtain the solution shown in Fig. 3.2.3. We will often meet such successive limits again.

for the trivial f_0) tends to infinity in the T-limit.* The important point is, however, the order of infinity, that is, the dominant term in the limit. Clearly

$$\frac{N!}{(N-s)!} = N(N-1)(N-2)\cdots(N-s+1) \xrightarrow[\text{T-lim}]{} N^s \qquad (3.3.6)$$

The proper form of writing the normalization condition (3.1.17) in the T-limit is therefore:

$$\text{T-lim}\, N^{-s}\int dx_1 \cdots dx_s\, f_s(x_1, \ldots, x_s) = 1 \qquad (3.3.7)$$

Similarly, Eq. (3.1.14) becomes, in the T-limit:

$$f_r(x_1, \ldots, x_r) = \text{T-lim}\, N^{r-s}\int dx_{r+1} \cdots dx_s\, f_s(x_1, \ldots, x_r, x_{r+1}, \ldots, x_s), \qquad r<s \qquad (3.3.8)$$

We chose on purpose to give a lengthy discussion of the thermodynamic limit because this is one of the crucial concepts of statistical mechanics. The way we introduced it here shows, we hope, the vanity of an argument that appears stubbornly and periodically in the literature. Some authors indeed contend that the thermodynamic limit is a wrong concept, because real systems are finite. Therefore, they say, any theory based on the thermodynamic limit is bound to be unphysical. The fallacy in this argument is in the literal interpretation of Eq. (3.3.1) as describing a concretely infinite system. We showed in this section that the key for the physical interpretation lies in the *empirical* principle of macroscopic equivalence, which deals with *real, finite* macroscopic systems. This principle tells us that, *for the study of the local behavior of a macroscopic system, the volume is an irrelevant parameter.* We then applied this principle to the microscopic picture and were able to define a class of systems that are macroscopically equivalent. This being done, we chose to represent the real system by one particular member of this class, which we took to be the most convenient one for the subsequent mathematical treatment: This turns out to be the limiting member of the sequence, defined by Eq. (3.3.1). The reason why it is simpler is that all the complicated N-dependent contributions that burden the microscopic description but are macroscopically undetectable are smoothed out in the thermodynamic limit.

Are we entitled to assume that the T-limit describes all possible

* The mathematically minded reader will note that this integral is *not* related to a norm of $\mathfrak{f}$ or to an average consistent with the scalar product (3.3.5). It is just a special property of the individual components f_s.

systems? The answer is no. We may quote two typical examples to which the T-limit is not applicable. One is the case of a very thin metal film (of a thickness of a few angstroms): Such a system cannot be considered as macroscopic because in one of its dimensions it is of molecular size. Another example is provided by an ultradilute gas. In this case there are practically no collisions between the particles: the behavior is dictated by the collisions with the walls, that is, by boundary effects, which are experimentally well known from the study of the so-called "Knudsen regime." Such systems, although very interesting, will not be discussed in this book.

3.4. EVOLUTION IN TIME OF THE CLASSICAL DISTRIBUTION VECTOR

In the previous sections we have developed a formalism based on reduced distribution functions defining the state of the system at a given time. We now go on to the study of the time development of the system.

We consider again our classical system of Section 2.4, assuming first that the number of particles N and the volume are finite. The evolution of the system is governed by the basic Liouville equation (2.4.23), which can be written more briefly, with the notations of Section 2.4, as

$$\partial_t F = \sum_{j=1}^{N} (L_j^0 + L_j^F)F + \sum_{j<n}\sum L'_{jn} F \tag{3.4.1}$$

The equations of evolution of the reduced distributions will be generated by this equation.

We first derive two important auxiliary results. We note that the number of particles in the system is conserved. This implies that, for all times (and not only at time zero):

$$\int dx_1 \cdots dx_N \, F(x_1, \ldots, x_N; t) = \text{const} \qquad \text{all } t \tag{3.4.2}$$

Accordingly,

$$\partial_t \int dx_1 \cdots dx_N \, F(x_1, \ldots, x_N; t) = 0 \tag{3.4.3}$$

And, therefore, from Eq. (3.4.1)

$$\int dx_1 \cdots dx_N \left\{ \sum_{j=1}^{N} (L_j^0 + L_j^F) + \sum_{j<n}\sum L'_{jn} \right\} F = 0$$

As this relation must hold for every value of N, in presence as well as in absence of an external field, each term must vanish separately, and we can conclude that

$$\int dx_j\, L_j^0 F(x_1, \ldots, x_N) = 0$$

$$\int dx_j\, L_j^F F(x_1, \ldots, x_N) = 0 \qquad (3.4.4)$$

$$\int dx_j\, dx_n\, L'_{jn} F(x_1, \ldots, x_N) = 0$$

These relations can also be checked directly by using the explicit forms of the Liouvillians. Each of the terms L_j^0, L_j^F or L'_{jn} involves a derivative of F, either with respect to $\mathbf{q}$ or with respect to $\mathbf{p}$. We will always assume that F together with a sufficient number of its derivatives vanish at the boundaries of the system in configuration space and also for $\mathbf{p}_j = \pm\infty$.* This ensures the convergence of the integrals. It then follows from Green's theorem that

$$\int d\mathbf{q}_j\, \nabla_j F = 0$$

$$\int d\mathbf{p}_j\, \partial_j F = 0 \qquad (3.4.5)$$

and these results imply (3.4.4).

It is now easy to write an equation of evolution for the reduced distribution functions f_s. Using the definition (3.1.14) and Eq. (3.4.1), we obtain

$$\partial_t f_s(x_1, \ldots, x_s) = \partial_t \frac{N!}{(N-s)!} \int dx_{s+1} \cdots dx_N\, F(x_1, \ldots, x_s, x_{s+1}, \ldots, x_N)$$

$$= \frac{N!}{(N-s)!} \int dx_{s+1} \cdots dx_N \left(\sum_{j=1}^{N} L_j^1 F + \sum_{j<n=1}^{N} \sum L'_{jn} F \right) \qquad (3.4.6)$$

where we abbreviated $L_j^0 + L_j^F$ by L_j^1.

Consider the first bracketed term in this equation. We may write:

$$\sum_{j=1}^{N} L_j^1 = \sum_{j=1}^{s} L_j^1 + \sum_{j=s+1}^{N} L_j^1$$

If j belongs to the group $(1, \ldots, s)$, L_j^1 is not affected by the integrations and can be written in front of the integrals; if j belongs to the group $(s+1, \ldots, N)$, it follows from (3.4.4) that the corresponding terms in (3.4.6) vanish. Hence the first term in the right-hand side of that equation

* One may also consider other boundary conditions leading to the same results.

is simply

$$\sum_{j=1}^{s} L_j^1 \left\{ \frac{N!}{(N-s)!} \int dx_{s+1} \cdots dx_N \, F \right\} = \sum_{j=1}^{s} L_j^1 f_s(x_1, \ldots, x_s)$$

The second bracketed term in (3.4.6) is treated similarly. Three cases must be considered.

a. Both j and n belong to the group $(1, \ldots, s)$. The operator L'_{jn} can again be written in front of the integral with the result

$$\sum_{j<n=1}^{s}\sum L'_{jn} f_s(x_1, \ldots, x_s)$$

b. Both j and n belong to the group $(s+1, \ldots, N)$. The corresponding contributions vanish because of Eq. (3.4.4).

c. j belongs to the group $(1, \ldots, s)$ and n to the group $(s+1, \ldots, N)$. As x_n is now a dummy integration variable, and as F is symmetric in its arguments, we can write

$$\begin{aligned}
\frac{N!}{(N-s)!} & \int dx_{s+1} \cdots dx_N \sum_{j=1}^{s} \sum_{n=s+1}^{N} L'_{jn} F(x_1, \ldots, x_N) \\
&= \frac{N!}{(N-s)!} \int dx_{s+1} \cdots dx_N \sum_{j=1}^{s} (N-s) L'_{js+1} F(x_1, \ldots, x_N) \\
&= \frac{N!}{(N-s-1)!} \int dx_{s+1}\, dx_{s+2} \cdots dx_N \sum_{j=1}^{s} L'_{js+1} F(x_1, \ldots, x_N) \\
&= \sum_{j=1}^{s} \int dx_{s+1}\, L'_{js+1} \frac{N!}{(N-s-1)!} \int dx_{s+2} \cdots dx_N \, F(x_1, \ldots, x_N) \\
&= \sum_{j=1}^{s} \int dx_{s+1}\, L'_{js+1} f_{s+1}(x_1, \ldots, x_{s+1})
\end{aligned}$$

Collecting now all the partial results we obtain the final equation*:

$$\begin{aligned}
\partial_t f_s(x_1, \ldots, x_s) = & \sum_{j=1}^{s} L_j^0 f_s(x_1, \ldots, x_s) + \sum_{j=1}^{s} L_j^F f_s(x_1, \ldots, x_s) \\
& + \sum_{j<n=1}^{s}\sum L'_{jn} f_s(x_1, \ldots, x_s) \\
& + \sum_{j=1}^{s} \int dx_{s+1}\, L'_{js+1} f_{s+1}(x_1, \ldots, x_{s+1})
\end{aligned} \tag{3.4.7}$$

This is a system (or hierarchy) of equations determining the reduced

* Strictly speaking, this equation is valid for $s \geq 2$. In the case $s = 1$, the third term in the right side does not exist (one needs at least two particles for an interaction). In the case $s = 0$, $f_0 = 1$, hence the equation trivially reduces to $\partial_t f_0 = 0$.

distribution functions: it is called the *BBGKY hierarchy*, after the names of its authors (Bogoliubov–Born–Green–Kirkwood–Yvon). Contrary to the equation (3.4.1) for F, which is closed, we now have a set of N equations: the rate of change of f_s depends both on f_s and on the higher-order function f_{s+1}.

For the convenience of subsequent reference, we write down explicitly the first four equations of the hierarchy:

$$\partial_t f_0 = 0 \tag{3.4.8}$$

$$(\partial_t - L_1^0 - L_1^F) f_1(x_1) = \int dx_2 \, L'_{12} f_2(x_1 x_2) \tag{3.4.9}$$

$$(\partial_t - L_1^0 - L_2^0 - L_1^F - L_2^F) f_2(x_1 x_2) = L'_{12} f_2(x_1 x_2) + \int dx_3 \, (L'_{13} + L'_{23}) f_3(x_1 x_2 x_3) \tag{3.4.10}$$

$$\left(\partial_t - \sum_1^3 L_j^0 - \sum_1^3 L_j^F\right) f_3(x_1 x_2 x_3) = (L'_{12} + L'_{13} + L'_{23}) f_3(x_1 x_2 x_3) + \int dx_4 \, (L'_{14} + L'_{24} + L'_{34}) f_4(x_1 x_2 x_3 x_4) \tag{3.4.11}$$

We now note that Eqs. (3.4.7) do not depend explicitly on N. We are thus entitled to postulate that *the distribution functions f_s obey these same equations in the thermodynamic limit* (3.3.1). In that case, of course, the number of equations in the hierarchy is infinite. We may note here that the really rigorous proof of the existence of $f_s(t)$ in the thermodynamic limit poses a very difficult mathematical problem. The reason is that there always exist initial conditions leading to a catastrophic collapse of the system as time passes: there are even an infinite number of these in an infinite system. It must be shown then that such conditions are "pathological" or "exceptional" in some sense, provided a certain number of requirements are satisfied by the Hamiltonian. We cannot here go into the description of this work, which is not yet in a final stage at present: only very simple models have been studied in the literature.

As physicists, we rather take the point of view of assuming on physical grounds that the solutions to (3.4.7) exist and are sufficiently regular in the thermodynamic limit.

Equations (3.4.7) can be considered as components of a single equation for the distribution "vector" $\mathfrak{f}(t)$. The operators on the right-hand side of (3.4.7) are grouped in the form of a matrix $\mathscr{L}$ whose elements are operators determined in such a way as to reproduce the right-hand side of (3.4.7). Thus we write

$$\partial_t \mathfrak{f}(t) = \mathscr{L} \mathfrak{f}(t) \tag{3.4.12}$$

with

$$\mathscr{L} = \mathscr{L}^0 + \mathscr{L}^F + \mathscr{L}' \tag{3.4.13}$$

Equation (3.4.12), governing the time evolution of the distribution vector $\mathfrak{f}$, will be called *the (generalized) Liouville equation*, $\mathscr{L}$ being *the (generalized) Liouvillian*. The components of the vector equation (3.4.12) are written as follows:

$$\partial_t f_s(t) = \sum_{s'=0}^{\infty} \langle (s)| \mathscr{L}^0 + \mathscr{L}^F + \mathscr{L}' |(s')\rangle \, f_{s'}(t) \tag{3.4.14}$$

(Note that in a finite system, $f_{s'} \equiv 0$ for $s' > N$; hence the sum involves a finite number of terms. In the thermodynamic limit, however, the sum is truly infinite, because the vector $\mathfrak{f}$ has an infinite number of components.)

In order that Eq. (3.4.14) be equivalent to (3.4.7), the matrix elements $\langle (s)| \mathscr{L} |(s')\rangle$ must have the following values:

$$\langle (s)| \mathscr{L}^0 |(s')\rangle = \delta_{s',s}\left\{\sum_{j=1}^{s} L_j^0\right\}, \qquad s>0 \tag{3.4.15}$$

$$\langle (s)| \mathscr{L}^F |(s')\rangle = \delta_{s',s}\left\{\sum_{j=1}^{s} L_j^F\right\}, \qquad s>0 \tag{3.4.16}$$

$$\langle (s)| \mathscr{L}' |(s')\rangle = \delta_{s',s}\left\{\sum_{j<n=1}^{s}\sum L'_{jn}\right\} + \delta_{s',s+1}\int dx_{s+1}\left\{\sum_{j=1}^{s} L'_{js+1}\right\}, \qquad s>1 \tag{3.4.17}$$

$$\langle (1)| \mathscr{L}' |(s')\rangle = \delta_{s',2}\int dx_2 \, L'_{12} \tag{3.4.18}$$

$$\langle (0)| \mathscr{L}^0 |(s')\rangle = \langle (0)| \mathscr{L}^F |(s')\rangle = \langle (0)| \mathscr{L}' |(s')\rangle = 0 \tag{3.4.19}$$

There is no diagonal one-particle element of $\mathscr{L}'$ because a particle, of course, cannot interact with itself ($L'_{jj} = 0$). All elements contributing to f_0 vanish because f_0 is a constant, $\partial_t f_0 = 0$. The matrix $\mathscr{L}$ has, therefore, a peculiar structure, with nonzero elements only along two adjacent diagonals:

$$\begin{bmatrix}
0 & 0 & 0 & 0 & 0 & 0 & 0 & \cdots \\
0 & \blacksquare & \blacksquare & 0 & 0 & 0 & 0 & \\
0 & 0 & \blacksquare & \blacksquare & 0 & 0 & 0 & \\
0 & 0 & 0 & \blacksquare & \blacksquare & 0 & 0 & \\
0 & 0 & 0 & 0 & \blacksquare & \blacksquare & 0 & \\
0 & 0 & 0 & 0 & 0 & \blacksquare & \blacksquare & \\
\cdot & & & & & & \searrow & \searrow \\
\cdot & & & & & & & \\
\cdot & & & & & & &
\end{bmatrix}$$

At this point we introduce a graphical representation that will prove useful in our subsequent investigations. We represent a "state" $|(s)\rangle$

Table 3.4.1. Elementary Vertices in the Representation $\mathfrak{f} = \{f_s(x_1, \ldots, x_s)\}$

	Vertex	Operator
A	 j, n lines crossing at a vertex	L'_{jn}
B	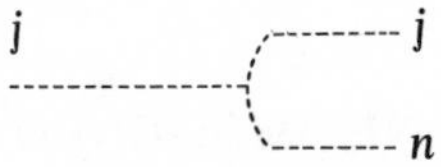 j line joined at a vertex to lines j, n	$\int dx_n \, L'_{jn}$

involving s particles by s dotted superposed lines. Each line bears the label of the corresponding particle. The operators $\mathscr{L}^0$ and $\mathscr{L}^F$ do not require a special symbol: they denote the propagation of independent particles. The interactions, on the other hand, seriously affect the state of the system by introducing correlations between the partners. We represent an interaction operator L'_{jn} by a vertex joining the lines j and n coming from the right. We now note, from Eq. (3.4.17), that there are two types of matrix elements involving L'_{jn}. One is diagonal in the number of particles (interactions within the group of particles considered): this will naturally be represented by the same two lines j, n coming out to the left of the vertex. The other type involves the interaction with an extra particle (a particle of the medium) followed by an integration over the

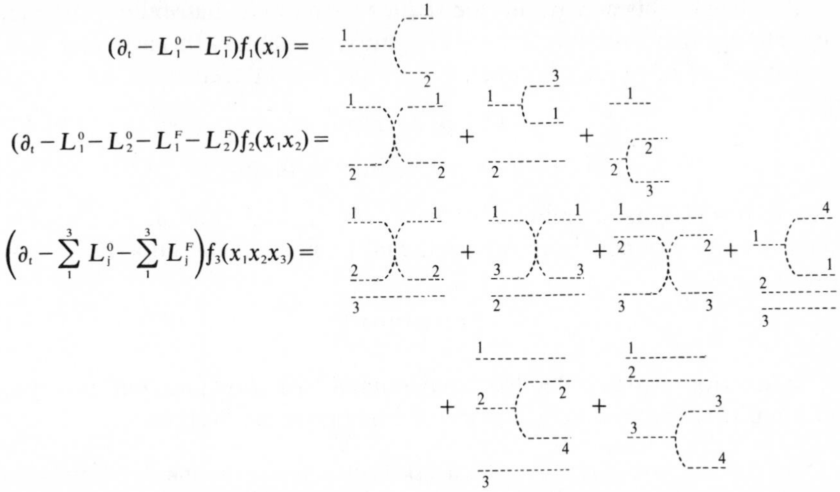

$$(\partial_t - L_1^0 - L_1^F) f_1(x_1) =$$

$$(\partial_t - L_1^0 - L_2^0 - L_1^F - L_2^F) f_2(x_1 x_2) =$$

$$\left(\partial_t - \sum_1^3 L_j^0 - \sum_1^3 L_j^F\right) f_3(x_1 x_2 x_3) =$$

Figure 3.4.1. Diagrams representing the three first equations of the BBGKY hierarchy [see Eqs. (3.4.9)–(3.4.11)].

coordinates of this extra particle. Such a process can be described as a transition from an $(s+1)$-particle state to a s-particle state. A single line comes out of the vertex to the left.

These two elementary diagrams are pictured in Table 3.4.1. With these rules in mind, the equations (3.4.9)–(3.4.11) are represented by the diagrams of Fig. 3.4.1.

At this stage, the diagram representation may seem utterly trivial. Its usefulness will, however, soon be realized.

3.5. HOMOGENEOUS AND INHOMOGENEOUS SYSTEMS: CORRELATIONS IN CLASSICAL SYSTEMS

We review in this section a few concepts that are of great importance in statistical mechanics and that are very clearly expressed in terms of reduced distribution functions.

We first note that among all possible states of a system, there exists a class of particularly simple states. A system is called *spatially homogeneous* (or simply homogeneous or uniform) when the local (intensive) physical properties are the same at all points in space. This property is expressed mathematically by the *translational invariance* of the reduced distribution functions:

$$f_s(\mathbf{q}_1+\mathbf{a}, \ldots, \mathbf{q}_s+\mathbf{a}, \mathbf{p}_1, \ldots, \mathbf{p}_s) = f_s(\mathbf{q}_1, \ldots, \mathbf{q}_s, \mathbf{p}_1, \ldots, \mathbf{p}_s) \quad (3.5.1)$$

where $\mathbf{a}$ is an arbitrary constant vector.*

Because of this constraint, the reduced s-particle distribution function effectively depends only on $s-1$ position variables. In particular, f_1 is independent of $\mathbf{q}$, and f_2 depends only on the difference $\mathbf{q}_1-\mathbf{q}_2$:

$$f_1(\mathbf{q}, \mathbf{p}) = n\varphi(\mathbf{p}) \quad (3.5.2)$$

$$f_2(\mathbf{q}_1, \mathbf{q}_2, \mathbf{p}_1, \mathbf{p}_2) = f_2(\mathbf{q}_1-\mathbf{q}_2, \mathbf{p}_1, \mathbf{p}_2) \quad (3.5.3)$$

where $n = N/\mathcal{V}$ is the number density (which is constant in a homogeneous system). For consistency with the normalization condition (3.1.17) we must have

$$\int d\mathbf{p}\, \varphi(\mathbf{p}) = 1 \quad (3.5.4)$$

The function $\varphi(\mathbf{p})$ is called the *momentum distribution function:* it plays a leading role in the kinetic theory of homogeneous systems.

* Note that the expression (3.5.1) for the translational invariance condition is only valid for a system of particles, described by canonical variables such that $\mathbf{q}_i$ represents the position in space of particle i. For any other choice of canonical variables, or for other types of systems, the condition will be expressed in a different form.

If we integrate the one-particle function over the momentum the result is the constant density n; but the integration of f_2 over the momenta $\mathbf{p}_1$, $\mathbf{p}_2$ results in a function depending on the distance between the particles, which can be represented as follows:

$$\int d\mathbf{p}_1\, d\mathbf{p}_2 f_2 = n_2'(\mathbf{q}_1 - \mathbf{q}_2) = n^2\{1 + \nu(\mathbf{q}_1 - \mathbf{q}_2)\} \tag{3.5.5}$$

The function $\nu(\mathbf{r})$ plays an important role in equilibrium statistical mechanics. It is called the *pair correlation function* or the *radial correlation function.*

None of the properties listed above applies to spatially *inhomogeneous systems:* in such systems the local properties vary from one point to another. We may also say that in such systems there exist *gradients* of local properties (density, velocity, etc.).

The distribution functions of more than one particle, whether homogeneous or not, introduce the fundamental concept of *correlations.* Consider an s-particle distribution at a fixed time. It may happen that this function is in the form of a product of one-particle distribution functions:

$$f_s^{\text{unc}}(x_1, \ldots, x_s) = \prod_{j=1}^{s} f_1(x_j) \tag{3.5.6}$$

If this happens, the system is said to be in an *uncorrelated* state, a name coming from the theory of probability. In such a state, the particles are statistically independent from each other; that is, the probability of finding a particle at some point is not influenced by the presence of any other particle. As a consequence, the average of the product of one-particle dynamical functions equals the product of the averages of the factors:

$$\int dx_1 \cdots dx_s \left[\prod_{j=1}^{s} b_1(x_j)\right] f_s^{\text{unc}}(x_1, \ldots, x_s) = \{\langle b_1 \rangle\}^s \tag{3.5.7}$$

where

$$\langle b_1 \rangle = \int dx\, b_1(x) f_1(x)$$

In general, however, Eq. (3.5.7) is not true. The physical reason therefore is the existence of mechanical interactions among the particles. Every particle influences the behavior of its neighboring particles, hence creating correlations. The range of correlations is at least equal to the range of the intermolecular forces; indirect correlations can actually have a much longer range. The existence of correlations can always be described by using the following representation of the distribution function:

$$f_s(x_1, \ldots, x_s) = \prod_{j=1}^{s} f_1(x_j) + g'_s(x_1, \ldots, x_s) \tag{3.5.8}$$

The function $g'_s(x_1, \ldots, x_s)$ provides a simple measure of the degree of correlation, that is, the deviation of the true s-particle distribution from a product. However, in many cases, it does not yield sufficient information. For $s=2$, $g'_2(x_1x_2)$ cannot be further analyzed. For $s=3$, however, we can imagine situations where particles 1 and 2 are correlated, the third one being independent, or particles 1, 2, and 3 could be mutually correlated. In neither case is the three-particle function a product of three one-particle functions, but the type of correlation is different in the two cases. In order to analyze all possible situations, we consider all possible ways in which the set $(1, \ldots, s)$ can be subdivided into disjoint subsets containing at least one particle. Interpreting each subset, or *cluster*, as representing a group of mutually correlated particles, statistically independent of the other subsets, we obtain the following representation of the reduced distribution functions, called the *cluster representation:*

$$f_2(x_1x_2) = f_1(x_1)f_1(x_2) + g_2(x_1x_2) \tag{3.5.9}$$

$$\begin{aligned} f_3(x_1x_2x_3) = {} & f_1(x_1)f_1(x_2)f_1(x_3) + f_1(x_1)g_2(x_2x_3) + f_1(x_2)g_2(x_1x_3) \\ & + f_1(x_3)g_2(x_1x_2) + g_3(x_1x_2x_3) \end{aligned} \tag{3.5.10}$$

$$\begin{aligned} f_4(x_1x_2x_3x_4) = {} & f_1(x_1)f_1(x_2)f_1(x_3)f_1(x_4) + f_1(x_1)f_1(x_2)g_2(x_3x_4) + f_1(x_1)f_1(x_3)g_2(x_2x_4) \\ & + f_1(x_1)f_1(x_4)g_2(x_2x_3) + f_1(x_2)f_1(x_3)g_2(x_1x_4) + f_1(x_2)f_1(x_4)g_2(x_1x_3) \\ & + f_1(x_3)f_1(x_4)g_2(x_1x_2) + f_1(x_1)g_3(x_2x_3x_4) + f_1(x_2)g_3(x_1x_3x_4) \\ & + f_1(x_3)g_3(x_1x_2x_4) + f_1(x_4)g_3(x_1x_2x_3) + g_2(x_1x_2)g_2(x_3x_4) \\ & + g_2(x_1x_3)g_2(x_2x_4) + g_2(x_1x_4)g_2(x_2x_3) + g_4(x_1x_2x_3x_4) \end{aligned} \tag{3.5.11}$$

and so on. Each term on the right-hand side of these equations will be called a *correlation pattern* [examples: $f_1(x_1)f_1(x_2)$; $g_3(x_1x_2x_3)$, $f_1(x_1)g_2(x_2x_4)g_4(x_3x_5x_6x_7)$]. The functions $g_s(x_1, \ldots, x_s)$ are called (irreducible) *s-particle correlation functions.* We note that, because of Eq. (3.1.15), the correlation functions must be symmetric under permutations of the particle labels:

$$g_s(x_1, \ldots, x_j, \ldots, x_n, \ldots, x_s) = g_s(x_1, \ldots, x_n, \ldots, x_j, \ldots, x_s) \tag{3.5.12}$$

It is useful to devise compact systems of notations for the correlation patterns in order to abbreviate formulas like (3.5.9)–(3.5.11). We will actually introduce *two* systems of notations and will use whichever of these is more suitable in the forthcoming expressions.

In a first system of notations, we agree to enumerate all the possible partitions of the set $(x_1, \ldots, x_s)$ $(s \geq 1)$, by attributing to each partition a fixed, conventional index Γ_s. We agree always to denote the completely uncorrelated state by $\Gamma_s = 0_s$ and the irreducible correlation function by $\Gamma_s = C_s$; the value of the indices attributed to the other partitions is not particularly relevant. A given correlation pattern is now characterized by the number (s), by the labels (say, $1, 2, \ldots, s$) of its constituent particles, and by the *correlation index* Γ_s; it is denoted by $\pi_s(x_1, \ldots, x_s; [\Gamma_s])$. In particular

$$\pi_s(x_1, \ldots, x_s; [0_s]) = \prod_{j=1}^{s} f_1(x_j) \tag{3.5.13}$$

$$\pi_s(x_1, \ldots, x_s; [C_s]) = g_s(x_1, \ldots, x_s) \tag{3.5.14}$$

The general cluster representation of the distribution functions will then be written as follows:

$$f_s(x_1, \ldots, x_s) = \sum_{\Gamma_s} \pi_s(x_1, \ldots, x_s; [\Gamma_s]) \tag{3.5.15}$$

Equations (3.5.9) and (3.5.10) are translated in this notation as follows:

$$\begin{aligned} f_1(x_1) &= \pi_1(x_1; [0_1]) \\ f_2(x_1x_2) &= \pi_2(x_1x_2; [0_2]) + \pi_2(x_1x_2; [C_2]) \\ f_3(x_1x_2x_3) &= \pi_3(x_1x_2x_3; [0_3]) + \pi_3(x_1x_2x_3; [1_3]) + \pi_3(x_1x_2x_3; [2_3]) \\ &\quad + \pi_3(x_1x_2x_3; [3_3]) + \pi_3(x_1x_2x_3; [C_3]) \end{aligned} \tag{3.5.16}$$

In some cases, however, we need more explicit formulas, exhibiting more vividly the particular type of correlation involved. In this case we materialize the partition by vertical bars and use the pretty obvious notation $\pi_s(i_1 \mid i_2 \mid i_3i_4i_5 \mid \cdots \mid \cdots i_s)$. For instance

$$\pi_9(1 \mid 4 \mid 25 \mid 36 \mid 789) \equiv f_1(x_1)f_1(x_4)g_2(x_2x_5)g_2(x_3x_6)g_3(x_7x_8x_9)$$

The correlation patterns have peculiar symmetry properties that are a direct consequence of the overall symmetry (3.1.15) of the reduced distribution functions. Any pattern is invariant under the following operations:

Permutation of labels within a subset
Permutation of complete subsets

For instance:

$$\pi_6(1 \mid 4 \mid 25 \mid 36) = \pi_6(1 \mid 4 \mid 52 \mid 36) = \pi_6(1 \mid 4 \mid 25 \mid 63)$$

$$\begin{aligned} \pi_6(1 \mid 4 \mid 25 \mid 36) &= \pi_6(4 \mid 1 \mid 25 \mid 36) = \pi_6(1 \mid 25 \mid 4 \mid 36) \\ &= \pi_6(1 \mid 4 \mid 36 \mid 25) = \cdots \end{aligned}$$

We may, therefore, agree on a "standard notation," in which the subsets of the partition are ordered in increasing size from left to right and the labels are written in increasing order in each subset. For instance, the standard notation for $\pi_8(57\,|\,623\,|\,4\,|\,18)$ is $\pi_8(4\,|\,18\,|\,57\,|\,236)$.

Considering again the first three reduced distributions, we write

$$\begin{aligned} f_1(x_1) &= \pi_1(1) \\ f_2(x_1x_2) &= \pi_2(1\,|\,2) + \pi_2(12) \\ f_3(x_1x_2x_3) &= \pi_3(1\,|\,2\,|\,3) + \pi_3(1\,|\,23) + \pi_3(2\,|\,13) + \pi_3(3\,|\,12) + \pi_3(123) \end{aligned} \tag{3.5.17}$$

We now briefly study the normalization properties of the correlation patterns. Consider Eq. (3.5.9), divide both sides by N^2, and integrate over the variables x_1, x_2:

$$N^{-2}\int dx_1\,dx_2\,f_2(x_1x_2) = \left\{N^{-1}\int dx_1\,f_1(x_1)\right\}\left\{N^{-1}\int dx_2\,f_1(x_2)\right\} + N^{-2}\int dx_1\,dx_2\,g_2(x_1x_2)$$

In the thermodynamic limit, the left-hand side equals 1 because of Eq. (3.3.7); by the same equation, the first term in the right-hand side also equals 1, hence*

$$\text{T-lim}\quad N^{-2}\int dx_1\,dx_2\,g_2(x_1x_2) = 0$$

This result is easily generalized:

$$\text{T-lim}\quad N^{-s}\int dx_1\cdots dx_s\,g_s(x_1,\ldots,x_s) = 0 \tag{3.5.18}$$

It then follows that, in the T-limit, the whole normalization of $f_s(x_1,\ldots,x_s)$ is contained in the uncorrelated term of the cluster representation: All other correlation patterns contribute zero to the normalization integral (3.3.7):

$$\text{T-lim}\quad N^{-s}\int dx_1\cdots dx_s\,\pi_s(x_1,\ldots,x_s;[0_s]) = 1 \tag{3.5.19}$$

$$\text{T-lim}\quad N^{-s}\int dx_1\cdots dx_s\,\pi_s(x_1,\ldots,x_s;[\Gamma_s]) = 0, \qquad \Gamma_s \neq 0_s \tag{3.5.20}$$

* For completeness, we note the following result, valid for a finite system [see Eq. (3.1.17)]:

$$\int dx_1\,dx_2\,g_2(x_1x_2) = -N$$

These formulas introduce an intrinsic and important difference between the correlation patterns $\pi_s([0_s])$ and all other patterns describing true correlation. This property will be used later.

3.6. QUANTUM DISTRIBUTION VECTOR: THE WIGNER FUNCTIONS

A comparison between Sections 1.2, 2.2, and 1.3, 2.3 leads to the justified impression that, in spite of the common Hamiltonian structure, there are considerable differences between the classical and quantum formalisms. Indeed, in the former case the basic set of dynamical variables is realized as an algebra of numerical functions, while in quantum mechanics it appears as an algebra of operators. We will see in the present section that there exists, however, a representation of quantum mechanics that is much closer to classical mechanics. This very remarkable fact was discovered by Wigner in 1932 and was more fully developed by Moyal in 1949 and by many other authors since.

For definiteness, we consider the system of N identical point particles of Section 2.4. We begin by noting that an arbitrary operator b of the quantum-dynamical algebra can always be represented in a form quite similar to Eq. (3.1.8). Indeed, using the second quantization formalism of Section 1.5, and in particular the representation of Eq. (1.5.21), we write

$$\begin{aligned}\mathsf{b} = b_0 &+ \int d\mathbf{x}_1\, \mathbf{\Psi}^\dagger(\mathbf{x}_1)\hat{b}_1\mathbf{\Psi}(\mathbf{x}_1) + \cdots \\ &+ (s!)^{-1}\int d\mathbf{x}_1 \cdots d\mathbf{x}_s\, \mathbf{\Psi}^\dagger(\mathbf{x}_1)\cdots \mathbf{\Psi}^\dagger(\mathbf{x}_s)\hat{b}_s\mathbf{\Psi}(\mathbf{x}_s)\cdots\mathbf{\Psi}(\mathbf{x}_1) \\ &+ \cdots \end{aligned} \tag{3.6.1}$$

The average value of such an operator is expressed in terms of the density matrix $\boldsymbol{\rho}$ as

$$\begin{aligned}\langle b\rangle &= \mathrm{Tr}\, \boldsymbol{\rho}\mathsf{b} \\ &= \sum_s (s!)^{-1}\int d\mathbf{x}_1\cdots d\mathbf{x}_s\, \mathrm{Tr}\{\boldsymbol{\rho}\mathbf{\Psi}^\dagger(\mathbf{x}_1)\cdots\mathbf{\Psi}^\dagger(\mathbf{x}_s)\hat{b}_s\mathbf{\Psi}(\mathbf{x}_s)\cdots\mathbf{\Psi}(\mathbf{x}_1)\}\end{aligned} \tag{3.6.2}$$

At this point we may note, as in Section 3.1, that the trace over the states of $(N-s)$ particles only involves the operator $\boldsymbol{\rho}$. Hence, we could introduce *reduced density matrices* defined as partial traces of $\boldsymbol{\rho}$, in analogy with the classical distribution functions. However, by means of a few more transformations, we will arrive at a still closer analogy with the equations of Section 3.1.

The key step in these transformations is provided by the *Weyl correspondence rule* explained in Section 1.3. In terms of this rule, to every quantum operator $\hat{b}_s$ represented as

$$\hat{b}_s = \int d\mathbf{k}_1\, d\mathbf{j}_1 \cdots d\mathbf{k}_s\, d\mathbf{j}_s\, \beta(\mathbf{k}_1\mathbf{j}_1 \cdots \mathbf{k}_s\mathbf{j}_s) \exp\left[i \sum_{n=1}^{s} (\mathbf{k}_n \hat{\mathbf{q}}_n + \mathbf{j}_n \cdot \hat{\mathbf{p}}_n)\right] \quad (3.6.3)$$

corresponds one and only one ordinary function b_s of the "classical" variables $(\mathbf{q}_1\mathbf{p}_1, \ldots, \mathbf{q}_s\mathbf{p}_s)$:

$$b_s(\mathbf{q}_1\mathbf{p}_1, \ldots, \mathbf{q}_s\mathbf{p}_s) = \int d\mathbf{k}_1\, d\mathbf{j}_1 \cdots d\mathbf{k}_s\, d\mathbf{j}_s\, \beta(\mathbf{k}_1\mathbf{j}_1, \ldots, \mathbf{k}_s\mathbf{j}_s)$$

$$\times \exp\left[i \sum_{n=1}^{s} (\mathbf{k}_n \cdot \mathbf{q}_n + \mathbf{j}_n \cdot \mathbf{p}_n)\right] \quad (3.6.4)$$

We first do the transformation in detail for $s = 1$. The corresponding term of Eq. (3.6.2), combined with (3.6.3), yields

$$\langle b_1 \rangle = \int d\mathbf{x} \int d\mathbf{k}\, d\mathbf{j}\, \beta_1(\mathbf{k}\mathbf{j}) \cdot \mathrm{Tr}\, \boldsymbol{\rho} \boldsymbol{\Psi}^\dagger(\mathbf{x}) \{\exp[i\mathbf{k} \cdot \mathbf{x} + i\mathbf{j} \cdot (-i\hbar\nabla)]\}\, \boldsymbol{\Psi}(\mathbf{x}) \quad (3.6.5)$$

where we introduced the explicit representation $\hat{\mathbf{q}} = \mathbf{x}$, $\hat{\mathbf{p}} = -i\hbar\nabla \equiv -i\hbar(\partial/\partial\mathbf{x})$. We now factorize the exponential appearing in the right-hand side, by using Eq. (1.3.10):

$$\exp(i\mathbf{k} \cdot \mathbf{x} + \hbar\mathbf{j} \cdot \nabla) = \exp(i\mathbf{k} \cdot \mathbf{x}) \exp(\hbar\mathbf{j} \cdot \nabla) \exp(-\tfrac{1}{2}[i\mathbf{k} \cdot \mathbf{x}, \hbar\mathbf{j} \cdot \nabla]_-)$$

$$= \exp(i\mathbf{k} \cdot \mathbf{x}) \exp(\hbar\mathbf{j} \cdot \nabla) \exp(\tfrac{1}{2}i\hbar\mathbf{k} \cdot \mathbf{j}) \quad (3.6.6)$$

Next, we note that the exponential of $\mathbf{a} \cdot \boldsymbol{\nabla}$ acting on a function $f(\mathbf{x})$ has the effect of shifting each component of $\mathbf{x}$ by the corresponding component of $\mathbf{a}$:

$$\{\exp(\mathbf{a} \cdot \boldsymbol{\nabla})\} f(\mathbf{x}) = f(\mathbf{x} + \mathbf{a}) \quad (3.6.7)$$

This identity is easily verified by noting that the series expansion of the exponential coincides with the Taylor expansion of the right-hand side. Hence, the exponential of a linear form in the gradient is simply a finite displacement operator.

Substituting (3.6.6) into (3.6.5), and using (3.6.7), we obtain

$$\langle b_1 \rangle = \int d\mathbf{x} \int d\mathbf{k}\, d\mathbf{j}\, \beta_1(\mathbf{k}, \mathbf{j})\, \mathrm{Tr}\{\boldsymbol{\rho} \boldsymbol{\Psi}^\dagger(\mathbf{x}) \exp[i\mathbf{k} \cdot (\mathbf{x} + \tfrac{1}{2}\hbar\mathbf{j})] \boldsymbol{\Psi}(\mathbf{x} + \hbar\mathbf{j})\} \quad (3.6.8)$$

We now note that the trace of the operator appearing under the integral is an ordinary ("c-number") function of the variables $\mathbf{x}$, $\mathbf{j}$, and $\mathbf{k}$. We introduce a function $f_1^W(\mathbf{q}, \mathbf{p})$ of the two variables $\mathbf{q}$, $\mathbf{p}$ by the following definition*:

$$\int d\mathbf{x} \exp[i\mathbf{k}\cdot(\mathbf{x}+\tfrac{1}{2}\hbar\mathbf{j})] \operatorname{Tr} \boldsymbol{\rho}\boldsymbol{\Psi}^{\dagger}(\mathbf{x})\boldsymbol{\Psi}(\mathbf{x}+\hbar\mathbf{j})$$
$$= \int d\mathbf{q}\, d\mathbf{p} \exp(i\mathbf{k}\cdot\mathbf{q}+i\mathbf{j}\cdot\mathbf{p}) f_1^W(\mathbf{q},\mathbf{p}) \quad (3.6.9)$$

This formula can be inverted by using the Fourier theorem:

$$f_1^W(\mathbf{q},\mathbf{p}) = (8\pi^3)^{-2} \int d\mathbf{k}\, d\mathbf{j} \int d\mathbf{x} \exp(-i\mathbf{k}\cdot\mathbf{q}-i\mathbf{j}\cdot\mathbf{p}) \exp[i\mathbf{k}\cdot(\mathbf{x}+\tfrac{1}{2}\hbar\mathbf{j})]$$
$$\cdot \operatorname{Tr} \boldsymbol{\rho}\boldsymbol{\Psi}^{\dagger}(\mathbf{x})\boldsymbol{\Psi}(\mathbf{x}+\hbar\mathbf{j})$$
$$= (8\pi^3)^{-1} \int d\mathbf{j} \int d\mathbf{x} \exp(-i\mathbf{j}\cdot\mathbf{p}) \delta(\mathbf{x}+\tfrac{1}{2}\hbar\mathbf{j}-\mathbf{q}) \operatorname{Tr} \boldsymbol{\rho}\boldsymbol{\Psi}^{\dagger}(\mathbf{x})\boldsymbol{\Psi}(\mathbf{x}+\hbar\mathbf{j})$$

Finally, we obtain after integration over $\mathbf{x}$:

$$f_1^W(\mathbf{q},\mathbf{p}) = (8\pi^3)^{-1} \int d\mathbf{j} \exp(-i\mathbf{j}\cdot\mathbf{p}) \operatorname{Tr} \boldsymbol{\rho}\boldsymbol{\Psi}^{\dagger}(\mathbf{q}-\tfrac{1}{2}\hbar\mathbf{j})\boldsymbol{\Psi}(\mathbf{q}+\tfrac{1}{2}\hbar\mathbf{j}) \quad (3.6.10)$$

This important function is called the *one-particle Wigner function.* Substituting (3.6.9) into (3.6.8) we get

$$\langle b_1 \rangle = \int d\mathbf{k}\, d\mathbf{j} \int d\mathbf{q}\, d\mathbf{p}\, \beta_1(\mathbf{k},\mathbf{j}) \exp(i\mathbf{k}\cdot\mathbf{q}+i\mathbf{j}\cdot\mathbf{p}) f_1^W(\mathbf{q},\mathbf{p})$$

Now there appears under the integral the phase-space dynamical function $b_1(\mathbf{q},\mathbf{p})$ corresponding to the quantum operator $\hat{b}_1$ through the Weyl correspondence rule (3.6.3), (3.6.4). Hence

$$\langle b_1 \rangle = \int d\mathbf{q}\, d\mathbf{p}\, b_1(\mathbf{q},\mathbf{p}) f_1^W(\mathbf{q},\mathbf{p})$$

This expression has the same form as a classical average, yet no approximation or departure from quantum mechanics entered the argument. The result is easily generalized. The *s-particle Wigner Function* $f_s^W(\mathbf{q}_1\mathbf{p}_1,\ldots,\mathbf{q}_s\mathbf{p}_s)$ is defined as

$$f_s^W(\mathbf{q}_1\mathbf{p}_1,\ldots,\mathbf{q}_s\mathbf{p}_s) = (8\pi^3)^{-s} \int d\mathbf{j}_1 \cdots d\mathbf{j}_s \exp\left(-i\sum_{n=1}^{s} \mathbf{j}_n\cdot\mathbf{p}_n\right)$$
$$\times \operatorname{Tr} \boldsymbol{\rho}\boldsymbol{\Psi}^{\dagger}(\mathbf{q}_1-\tfrac{1}{2}\hbar\mathbf{j}_1)\cdots\boldsymbol{\Psi}^{\dagger}(\mathbf{q}_s-\tfrac{1}{2}\hbar\mathbf{j}_s)\boldsymbol{\Psi}(\mathbf{q}_s+\tfrac{1}{2}\hbar\mathbf{j}_s)\cdots\boldsymbol{\Psi}(\mathbf{q}_1+\tfrac{1}{2}\hbar\mathbf{j}_1) \quad (3.6.11)$$

In terms of these functions the average of a general quantum observable

*Note that the scalar function $\exp(i\mathbf{k}\cdot\mathbf{x}+\tfrac{1}{2}\hbar i\mathbf{k}\cdot\mathbf{j})$ can be taken outside of the trace.

is written as

$$\langle b\rangle = \sum_{s=0}^{\infty} (s!)^{-1} \int d\mathbf{q}_1\, d\mathbf{p}_1 \cdots d\mathbf{q}_s\, d\mathbf{p}_s\, b_s(\mathbf{q}_1\mathbf{p}_1, \ldots, \mathbf{q}_s\mathbf{p}_s) f_s^W(\mathbf{q}_1\mathbf{p}_1, \ldots, \mathbf{q}_s\mathbf{p}_s) \tag{3.6.12}$$

This formula is the *same* as the classical formula (3.3.5).

The very existence of the Wigner functions is a quite unexpected feature of quantum mechanics. We know from our previous discussions that the phase space (q, p) of a system cannot have the same meaning in classical and in quantum mechanics. In the latter case we cannot represent a pure state of the system by a point in phase space, since we know from Heisenberg's principle that q and p cannot be known simultaneously with arbitrary precision. In spite of this fact, the *statistical* representation of a many-body system is possible in terms of a *distribution vector* $\mathfrak{f}^W$:

$$\mathfrak{f}^W \equiv \{f_s^W(\mathbf{q}_1\mathbf{p}_1, \ldots, \mathbf{q}_s\mathbf{p}_s);\quad s = 0, 1, 2, \ldots\} \tag{3.6.13}$$

whose components are s-particle Wigner functions. In terms of these functions, and in terms of the set of observables $b_s(\mathbf{q}_1\mathbf{p}_1, \ldots, \mathbf{q}_s\mathbf{p}_s)$, the formulation of quantum mechanics is identical to the classical one.

Of course, we must find some differences, at some point. In particular, the physical interpretation of the f_s^W cannot be the same as in classical mechanics. We shall come back to this problem in Section 3.7; let us now develop the features that are similar in both cases.

We first evaluate the normalization integral of $f_1^W(q, p)$:

$$\begin{aligned}
\int d\mathbf{q}\, d\mathbf{p}\, (8\pi^3)^{-1} \int d\mathbf{j}\, \exp(i\mathbf{j}\cdot\mathbf{p}) \operatorname{Tr} \boldsymbol{\rho} \boldsymbol{\Psi}^\dagger(\mathbf{q} - \tfrac{1}{2}\hbar\mathbf{j}) \boldsymbol{\Psi}(\mathbf{q} + \tfrac{1}{2}\hbar\mathbf{j}) \\
= \int d\mathbf{q} \int d\mathbf{j}\, \delta(\mathbf{j}) \operatorname{Tr} \boldsymbol{\rho} \boldsymbol{\Psi}^\dagger(\mathbf{q} - \tfrac{1}{2}\hbar\mathbf{j}) \boldsymbol{\Psi}(\mathbf{q} + \tfrac{1}{2}\hbar\mathbf{j}) \\
= \int d\mathbf{q} \operatorname{Tr} \boldsymbol{\rho} \boldsymbol{\Psi}^\dagger(\mathbf{q}) \boldsymbol{\Psi}(\mathbf{q}) = N \operatorname{Tr} \boldsymbol{\rho} = N
\end{aligned}$$

We used here Eqs. (1.5.21) and (2.3.12). A similar calculation provides the more general result:

$$\int d\mathbf{q}_1\, d\mathbf{p}_1 \cdots d\mathbf{q}_s\, d\mathbf{p}_s\, f_s^W(\mathbf{q}_1\mathbf{p}_1, \ldots, \mathbf{q}_s\mathbf{p}_s) = \frac{N!}{(N-s)!} \tag{3.6.14}$$

The normalization of the Wigner functions is thus identical to the normalization of the classical reduced distribution functions, Eq. (3.1.17).

We now introduce a slightly different representation of the Wigner functions, which is particularly convenient in practical calculations. It is obtained by going from the position representation to the momentum representation. To do so, we use Eq. (1.5.24), which we substitute into Eq. (3.6.10):

$$f_1^W(\mathbf{q}, \mathbf{p}) = (8\pi^3)^{-1} \int d\mathbf{j} \int d\boldsymbol{\kappa}\, d\boldsymbol{\kappa}' (8\pi^3)^{-1}$$
$$\times \exp[-i\mathbf{j}\cdot\mathbf{p} - i\boldsymbol{\kappa}\cdot(\mathbf{q}-\tfrac{1}{2}\hbar\mathbf{j}) + i\boldsymbol{\kappa}'\cdot(\mathbf{q}+\tfrac{1}{2}\hbar\mathbf{j})]$$
$$\times \mathrm{Tr}\, \boldsymbol{\rho}\mathbf{a}^\dagger(\hbar\boldsymbol{\kappa})\mathbf{a}(\hbar\boldsymbol{\kappa}')$$
$$= (8\pi^3)^{-1} \int d\boldsymbol{\kappa}\, d\boldsymbol{\kappa}'\, \delta(-\mathbf{p}+\tfrac{1}{2}\hbar\boldsymbol{\kappa}+\tfrac{1}{2}\hbar\boldsymbol{\kappa}')$$
$$\times \exp[i(\boldsymbol{\kappa}'-\boldsymbol{\kappa})\cdot\mathbf{q}]\, \mathrm{Tr}\, \boldsymbol{\rho}\mathbf{a}^\dagger(\hbar\boldsymbol{\kappa})\mathbf{a}(\hbar\boldsymbol{\kappa}')$$

With the change of variables

$$\boldsymbol{\kappa}' - \boldsymbol{\kappa} = \mathbf{k}, \qquad \boldsymbol{\kappa}' + \boldsymbol{\kappa} = 2\mathbf{K}$$

whose Jacobian is 1, we have

$$f_1^W(\mathbf{q}, \mathbf{p}) = (8\pi^3)^{-1} \int d\mathbf{k}\, d\mathbf{K}\, \delta(-\mathbf{p}+\hbar\mathbf{K})$$
$$\times \exp(i\mathbf{k}\cdot\mathbf{q})\, \mathrm{Tr}\, \boldsymbol{\rho}\mathbf{a}^\dagger(\hbar\mathbf{K}-\tfrac{1}{2}\hbar\mathbf{k})\mathbf{a}(\hbar\mathbf{K}+\tfrac{1}{2}\hbar\mathbf{k})$$
$$= (8\pi^3\hbar^3)^{-1} \int d\mathbf{k}\, \exp(i\mathbf{k}\cdot\mathbf{q})\, \mathrm{Tr}\, \boldsymbol{\rho}\mathbf{a}^\dagger(\mathbf{p}-\tfrac{1}{2}\hbar\mathbf{k})\mathbf{a}(\mathbf{p}+\tfrac{1}{2}\hbar\mathbf{k})$$

This calculation is easily extended to the s-particle function, which we write as

$$f_s^W(\mathbf{q}_1\mathbf{p}_1, \ldots, \mathbf{q}_s\mathbf{p}_s) = (8\pi^3)^{-s} \int d\mathbf{k}_1 \cdots d\mathbf{k}_s \exp\left(i\sum_{n=1}^{s} \mathbf{k}_n\cdot\mathbf{q}_n\right) f_s^W(\mathbf{k}_1\mathbf{p}_1, \ldots, \mathbf{k}_s\mathbf{p}_s) \tag{3.6.15}$$

where

$$f_s^W(\mathbf{k}_1\mathbf{p}_1, \ldots, \mathbf{k}_s\mathbf{p}_s) = \hbar^{-3s}\, \mathrm{Tr}\, \boldsymbol{\rho}\mathbf{a}^\dagger(\mathbf{p}_1-\tfrac{1}{2}\hbar\mathbf{k}_1)\cdots\mathbf{a}^\dagger(\mathbf{p}_s-\tfrac{1}{2}\hbar\mathbf{k}_s)$$
$$\times \mathbf{a}(\mathbf{p}_s+\tfrac{1}{2}\hbar\mathbf{k}_s)\cdots\mathbf{a}(\mathbf{p}_1+\tfrac{1}{2}\hbar\mathbf{k}_1) \tag{3.6.16}$$

The function $f_s^W(\mathbf{k}_1\mathbf{p}_1, \ldots, \mathbf{k}_s\mathbf{p}_s)$ is simply the Fourier transform of $f_s^W(\mathbf{q}_1\mathbf{p}_1 \cdots \mathbf{q}_s\mathbf{p}_s)$ with respect to the position variables $\mathbf{q}_1, \ldots, \mathbf{q}_s$. In order not to burden the notation, we do not use a different symbol for the Fourier transform: the nature of the arguments makes the distinction clear. We must note the following important relation between the arguments $\mathbf{k}_j$, $\mathbf{p}_j$ of the Wigner function on one hand and the argument $\mathbf{P}_j \equiv \mathbf{p}_j - \tfrac{1}{2}\hbar\mathbf{k}_j$ of the jth creation factor and the argument $\mathbf{P}'_j \equiv \mathbf{p}_j + \tfrac{1}{2}\hbar\mathbf{k}_j$ of the corresponding destruction factor of the normal product:

$$f^{W}(\cdots \mathbf{k}_j, \mathbf{p}_j \cdots) = \hbar^{-3s}\, \mathrm{Tr}\{\boldsymbol{\rho} \cdots \mathsf{a}^{\dagger}(\mathbf{P}_j) \cdots \mathsf{a}(\mathbf{P}'_j) \cdots\}$$

$$\mathbf{k}_j = \hbar^{-1}(\mathbf{P}'_j - \mathbf{P}_j) \tag{3.6.17}$$

$$\mathbf{p}_j = \tfrac{1}{2}(\mathbf{P}'_j + \mathbf{P}_j)$$

We must insist on the fact that the set $f_s^W(k, p)$ is as good a representation of the distribution vector $\mathfrak{f}^W$ as $f_s^W(q, p)$. Indeed, if the former is known, the latter can immediately be determined by using (3.6.15). But $f_s^W(k, p)$ *can also be used directly* for the calculation of averages. The dynamical functions $b_s(q, p)$ can be Fourier-transformed with respect to positions*:

$$b_s(\mathbf{q}_1\mathbf{p}_1, \ldots, \mathbf{q}_s\mathbf{p}_s) = \int d\mathbf{k}_1 \cdots d\mathbf{k}_s \exp\left\{i \sum_{n=1}^{s} \mathbf{k}_n \cdot \mathbf{q}_n\right\} b_s(\mathbf{k}_1\mathbf{p}_1, \ldots, \mathbf{k}_s\mathbf{p}_s) \tag{3.6.18}$$

Substituting this equation as well as (3.6.15) into Eq. (3.6.12) we easily obtain

$$\langle b \rangle = \sum_{s=0}^{\infty} (s!)^{-1} \int d\mathbf{k}_1\, d\mathbf{p}_1 \cdots d\mathbf{k}_s\, d\mathbf{p}_s\, b_s(-\mathbf{k}_1, \mathbf{p}_1, \ldots, -\mathbf{k}_s, \mathbf{p}_s) \times f_s^W(\mathbf{k}_1, \mathbf{p}_1, \ldots, \mathbf{k}_s, \mathbf{p}_s) \tag{3.6.19}$$

Of course, the use of the Fourier picture is not restricted to quantum mechanics. It can and has been used quite extensively in classical mechanics as well. However, it turns out that in quantum mechanics the Fourier picture leads to a formulation that is an order of magnitude simpler than the phase-space picture. Therefore, in practically all applications the Fourier transform of the Wigner function will be used.

We shall use an abbreviation similar to (3.1.1) in the Fourier picture. In order to make the distinction obvious we will use Greek symbols: $\xi_j, \eta_j, \ldots$ to denote the set of a wave vector and a momentum:

$$\xi_j = (\mathbf{k}_j, \mathbf{p}_j) \tag{3.6.20}$$

The distribution vector will now be considered as the set of Wigner functions in the Fourier picture:

$$\mathfrak{f}^W = \{f_s^W(\xi_1, \ldots, \xi_s);\ s = 0, 1, 2, \ldots\} \tag{3.6.21}$$

We now close this section with the study of the symmetry properties of the Wigner function. Consider, for simplicity, the two-particle function. If we permute the indices 1 and 2, it is clear that nothing changes:

$$\mathrm{Tr}\, \boldsymbol{\rho}\mathsf{a}^{\dagger}(\mathbf{p}_1 - \tfrac{1}{2}\hbar\mathbf{k}_1)\mathsf{a}^{\dagger}(\mathbf{p}_2 - \tfrac{1}{2}\hbar\mathbf{k}_2)\mathsf{a}(\mathbf{p}_2 + \tfrac{1}{2}\hbar\mathbf{k}_2)\mathsf{a}(\mathbf{p}_1 + \tfrac{1}{2}\hbar\mathbf{k}_1)$$
$$= \mathrm{Tr}\, \boldsymbol{\rho}\mathsf{a}^{\dagger}(\mathbf{p}_2 - \tfrac{1}{2}\hbar\mathbf{k}_2)\mathsf{a}^{\dagger}(\mathbf{p}_1 - \tfrac{1}{2}\hbar\mathbf{k}_1)\mathsf{a}(\mathbf{p}_1 + \tfrac{1}{2}\hbar\mathbf{k}_1)\mathsf{a}(\mathbf{p}_2 + \tfrac{1}{2}\hbar\mathbf{k}_2)$$

* Note that this corresponds to "half" of the transformation of Eq. (3.6.4).

Indeed, this operation is equivalent to a permutation of a pair $a^\dagger(1)$, $a^\dagger(2)$ and simultaneously of a pair $a(1)$, $a(2)$. Hence, both in the Fermi and in the Bose cases, this operation leaves the value of the normal product unchanged. Hence

$$f_2^W(\xi_1, \xi_2) = f_2^W(\xi_2, \xi_1) \tag{3.6.22}$$

This property is the same as in the classical case. However, quantum systems possess also a more subtle symmetry property. Indeed if we permute *only* $a(1)$ and $a(2)$, the value of the normal product either remains unchanged (for bosons) or changes sign (for fermions)*:

$$\begin{aligned}\operatorname{Tr} \rho a^\dagger(\mathbf{p}_1 - \tfrac{1}{2}\hbar\mathbf{k}_1)a^\dagger(\mathbf{p}_2 - \tfrac{1}{2}\hbar\mathbf{k}_2)a(\mathbf{p}_2 + \tfrac{1}{2}\hbar\mathbf{k}_2)a(\mathbf{p}_1 + \tfrac{1}{2}\hbar\mathbf{k}_1)& \\ = \theta \operatorname{Tr} \rho a^\dagger(\mathbf{p}_1 - \tfrac{1}{2}\hbar\mathbf{k}_1)a^\dagger(\mathbf{p}_2 - \tfrac{1}{2}\hbar\mathbf{k}_2)a(\mathbf{p}_1 + \tfrac{1}{2}\hbar\mathbf{k}_1)a(\mathbf{p}_2 + \tfrac{1}{2}\hbar\mathbf{k}_2)&\end{aligned}$$

But the expression in the right-hand side is a two-particle Wigner function of *different* arguments as in the left side. Indeed, by the rule of Eq. (3.6.17), this equation is translated as follows in terms of Wigner functions:

$$\begin{aligned}f_2^W(\mathbf{k}_1, \mathbf{p}_1; \mathbf{k}_2, \mathbf{p}_2) = \theta f_2^W[&\tfrac{1}{2}(\mathbf{k}_1 + \mathbf{k}_2) + \hbar^{-1}(\mathbf{p}_2 - \mathbf{p}_1), \tfrac{1}{2}(\mathbf{p}_1 + \mathbf{p}_2) + \tfrac{1}{4}\hbar(\mathbf{k}_2 - \mathbf{k}_1); \\ &\times \tfrac{1}{2}(\mathbf{k}_1 + \mathbf{k}_2) + \hbar^{-1}(\mathbf{p}_1 - \mathbf{p}_2), \tfrac{1}{2}(\mathbf{p}_1 + \mathbf{p}_2) + \tfrac{1}{4}\hbar(\mathbf{k}_1 - \mathbf{k}_2)]\end{aligned} \tag{3.6.23}$$

This is an important and characteristically quantum-statistical property. It clearly distinguishes bosons from fermions. For an s-particle function, a relation of the form (3.6.23) holds for every pair of variables that can be chosen among the s particles.

3.7. EVOLUTION IN TIME OF THE QUANTUM DISTRIBUTION VECTOR

We now generalize to quantum-mechanical systems the results of Section 3.4. We will again see that the structure of the equations of evolution of the Wigner functions is remarkably similar to their classical counterpart.

We derive these equations for a system described by the Hamiltonian (2.4.26)–(2.4.28). We perform the calculation in full detail for the function $f_2^W(\xi_1, \xi_2)$: This calculation contains all the ingredients appearing in the other cases. The time derivative of f_2^W is generated by the basic von Neumann equation for the density matrix (2.3.23). From Eq. (3.6.16) we

* It is easily checked that a permutation of $a^\dagger(1)$ and $a^\dagger(2)$ alone leads to the same result and must, therefore, not be considered separately.

therefore obtain

$$\begin{aligned}\partial_t f_2^W(\xi_1, \xi_2) &= \hbar^{-6}\mathrm{Tr}\,(\partial_t \rho)\,a^\dagger(\mathbf{p}_1 - \tfrac{1}{2}\hbar\mathbf{k}_1)a^\dagger(\mathbf{p}_2 - \tfrac{1}{2}\hbar\mathbf{k}_2)a(\mathbf{p}_2 + \tfrac{1}{2}\hbar\mathbf{k}_2)a(\mathbf{p}_1 + \tfrac{1}{2}\hbar\mathbf{k}_1)\\ &= \hbar^{-6}(i\hbar)^{-1}\mathrm{Tr}\,[H^0 + H', \rho]_- a^\dagger a^\dagger a a\\ &= \hbar^{-6}(i\hbar)^{-1}\,\mathrm{Tr}\,\rho[a^\dagger(\mathbf{p}_1 - \tfrac{1}{2}\hbar\mathbf{k}_1)a^\dagger(\mathbf{p}_2 - \tfrac{1}{2}\hbar\mathbf{k}_2)\\ &\qquad \times a(\mathbf{p}_2 + \tfrac{1}{2}\hbar\mathbf{k}_2)a(\mathbf{p}_1 + \tfrac{1}{2}\hbar\mathbf{k}_1),\, H^0 + H']_- \end{aligned} \tag{3.7.1}$$

The last step is based on a well-known identity about the trace of a product of operators:

$$\mathrm{Tr}\,ABC = \mathrm{Tr}\,BCA = \mathrm{Tr}\,CAB \tag{3.7.2}$$

The problem thus reduces to the evaluation of the commutator in the right-hand side of (3.7.1). This is a simple exercise, which is easily solved by using Eqs. (1.5.34) and the algebraic properties (1.2.12)–(1.2.14). The free-motion Hamiltonian H^0, Eq. (2.4.27), contributes:

$$\begin{aligned}&(i\hbar)^{-1}[a^\dagger a^\dagger a a, H^0]_-\\ &\quad = (i\hbar m)^{-1}\{(\mathbf{p}_1 + \tfrac{1}{2}\hbar\mathbf{k}_1)^2 - (\mathbf{p}_1 - \tfrac{1}{2}\hbar\mathbf{k}_1)^2 + (\mathbf{p}_2 + \tfrac{1}{2}\hbar\mathbf{k}_2)^2 - (\mathbf{p}_2 - \tfrac{1}{2}\hbar\mathbf{k}_2)^2\}a^\dagger a^\dagger a a\\ &\quad = -\left(\frac{i}{m}\right)(\mathbf{k}_1 \cdot \mathbf{p}_1 + \mathbf{k}_2 \cdot \mathbf{p}_2)a^\dagger(\mathbf{p}_1 - \tfrac{1}{2}\hbar\mathbf{k}_1)a^\dagger(\mathbf{p}_2 - \tfrac{1}{2}\hbar\mathbf{k}_2)\\ &\qquad \times a(\mathbf{p}_2 + \tfrac{1}{2}\hbar\mathbf{k}_2)a(\mathbf{p}_1 + \tfrac{1}{2}\hbar\mathbf{k}_1)\end{aligned} \tag{3.7.3}$$

The contribution of the interactions is somewhat more complicated. In a first step we find, from Eq. (2.4.28):

$$\begin{aligned}[a^\dagger a^\dagger a a, H']_- = \int d\mathbf{k}\, d\mathbf{l}\, \tilde{V}_l\{&a^\dagger(\mathbf{p}_1 - \tfrac{1}{2}\hbar\mathbf{k}_1)a^\dagger(\mathbf{p}_2 - \tfrac{1}{2}\hbar\mathbf{k}_2)a(\mathbf{p}_2 + \tfrac{1}{2}\hbar\mathbf{k}_2)a^\dagger(\hbar\mathbf{k})\\ &\times a(\hbar\mathbf{k} + \hbar\mathbf{l})a(\mathbf{p}_1 + \tfrac{1}{2}\hbar\mathbf{k}_1 - \hbar\mathbf{l}) + a^\dagger(\mathbf{p}_1 - \tfrac{1}{2}\hbar\mathbf{k}_1)a^\dagger(\mathbf{p}_2 - \tfrac{1}{2}\hbar\mathbf{k}_2)\\ &\times a^\dagger(\hbar\mathbf{k})a(\hbar\mathbf{k} + \hbar\mathbf{l})a(\mathbf{p}_2 + \tfrac{1}{2}\hbar\mathbf{k}_2 - \hbar\mathbf{l})a(\mathbf{p}_1 + \tfrac{1}{2}\hbar\mathbf{k}_1)\\ &- a^\dagger(\mathbf{p}_1 - \tfrac{1}{2}\hbar\mathbf{k}_1)a^\dagger(\hbar\mathbf{k})a^\dagger(\mathbf{p}_2 - \tfrac{1}{2}\hbar\mathbf{k}_2 - \hbar\mathbf{l})a(\hbar\mathbf{k} - \hbar\mathbf{l})\\ &\times a(\mathbf{p}_2 + \tfrac{1}{2}\hbar\mathbf{k}_2)a(\mathbf{p}_1 + \tfrac{1}{2}\hbar\mathbf{k}_1) - a^\dagger(\hbar\mathbf{k})a^\dagger(\mathbf{p}_1 - \tfrac{1}{2}\hbar\mathbf{k}_1 - \hbar\mathbf{l})\\ &\times a(\hbar\mathbf{k} - \hbar\mathbf{l})a^\dagger(\mathbf{p}_2 - \tfrac{1}{2}\hbar\mathbf{k}_2)a(\mathbf{p}_2 + \tfrac{1}{2}\hbar\mathbf{k}_2)a(\mathbf{p}_1 + \tfrac{1}{2}\hbar\mathbf{k}_1)\}\end{aligned}$$

This expression is not in normal product order: we must therefore bring it in such order by using again Eqs. (1.5.34) and making some pretty obvious changes in the names of the integration variables:

$$\begin{aligned}[a^\dagger a^\dagger a a, H']_- = \int d\mathbf{l}\, \tilde{V}_l\{&a^\dagger(\mathbf{p}_1 - \tfrac{1}{2}\hbar\mathbf{k}_1)a^\dagger(\mathbf{p}_2 - \tfrac{1}{2}\hbar\mathbf{k}_2)\\ &\times a(\mathbf{p}_2 + \tfrac{1}{2}\hbar\mathbf{k}_2 + \hbar\mathbf{l})a(\mathbf{p}_1 + \tfrac{1}{2}\hbar\mathbf{k}_1 - \hbar\mathbf{l}) - a^\dagger(\mathbf{p}_1 - \tfrac{1}{2}\hbar\mathbf{k}_1 + \hbar\mathbf{l})\\ &\times a^\dagger(\mathbf{p}_2 - \tfrac{1}{2}\hbar\mathbf{k}_2 - \hbar\mathbf{l})a(\mathbf{p}_2 + \tfrac{1}{2}\hbar\mathbf{k}_2)a(\mathbf{p}_1 + \tfrac{1}{2}\hbar\mathbf{k}_1)\}\end{aligned}$$

$$
\begin{aligned}
&+\hbar^{-3}\int d\mathbf{p}_3\, d\mathbf{l}\, \tilde{V}_l\{\mathrm{a}^\dagger(\mathbf{p}_1-\tfrac{1}{2}\hbar\mathbf{k}_1)\mathrm{a}^\dagger(\mathbf{p}_2-\tfrac{1}{2}\hbar\mathbf{k}_2)\mathrm{a}^\dagger(\mathbf{p}_3)\\
&\times \mathrm{a}(\mathbf{p}_3+\hbar\mathbf{l})\mathrm{a}(\mathbf{p}_2+\tfrac{1}{2}\hbar\mathbf{k}_2)\mathrm{a}(\mathbf{p}_1+\tfrac{1}{2}\hbar\mathbf{k}_1-\hbar\mathbf{l})\\
&+\mathrm{a}^\dagger(\mathbf{p}_1-\tfrac{1}{2}\hbar\mathbf{k}_1)\mathrm{a}^\dagger(\mathbf{p}_2-\tfrac{1}{2}\hbar\mathbf{k}_2)\mathrm{a}^\dagger(\mathbf{p}_3)\mathrm{a}(\mathbf{p}_3+\hbar\mathbf{l})\\
&\times \mathrm{a}(\mathbf{p}_2+\tfrac{1}{2}\hbar\mathbf{k}_2-\hbar\mathbf{l})\mathrm{a}(\mathbf{p}_1+\tfrac{1}{2}\hbar\mathbf{k}_1)-\mathrm{a}^\dagger(\mathbf{p}_1-\tfrac{1}{2}\hbar\mathbf{k}_1)\mathrm{a}^\dagger(\mathbf{p}_2-\tfrac{1}{2}\hbar\mathbf{k}_2+\hbar\mathbf{l})\\
&\times \mathrm{a}^\dagger(\mathbf{p}_3-\hbar\mathbf{l})\mathrm{a}(\mathbf{p}_3)\mathrm{a}(\mathbf{p}_2+\tfrac{1}{2}\hbar\mathbf{k}_2)\mathrm{a}(\mathbf{p}_1+\tfrac{1}{2}\hbar\mathbf{k}_1)\\
&-\mathrm{a}^\dagger(\mathbf{p}_1-\tfrac{1}{2}\hbar\mathbf{k}_1+\hbar\mathbf{l})\mathrm{a}^\dagger(\mathbf{p}_2-\tfrac{1}{2}\hbar\mathbf{k}_2)\mathrm{a}^\dagger(\mathbf{p}_3-\hbar\mathbf{l})\mathrm{a}(\mathbf{p}_3)\\
&\times \mathrm{a}(\mathbf{p}_2+\tfrac{1}{2}\hbar\mathbf{k}_2)\mathrm{a}(\mathbf{p}_1+\tfrac{1}{2}\hbar\mathbf{k}_1)\}
\end{aligned}
\tag{3.7.4}
$$

The partial results (3.7.3), (3.7.4) are now substituted into Eq. (3.7.1). We then note that, by using the definition (3.6.17), the right-hand side can be interpreted in terms of Wigner functions, as follows:

$$
\begin{aligned}
\partial_t f_2^W(\mathbf{k}_1,\mathbf{p}_1;\mathbf{k}_2,\mathbf{p}_2) &= -\left(\frac{i}{m}\right)(\mathbf{k}_1\cdot\mathbf{p}_1+\mathbf{k}_2\cdot\mathbf{p}_2)f_2^W(\mathbf{k}_1,\mathbf{p}_1;\mathbf{k}_2,\mathbf{p}_2)\\
&+(i\hbar)^{-1}\int d\mathbf{l}\,\tilde{V}_l\{f_2^W(\mathbf{k}_1-\mathbf{l},\mathbf{p}_1-\tfrac{1}{2}\hbar\mathbf{l};\mathbf{k}_2+\mathbf{l},\mathbf{p}_2+\tfrac{1}{2}\hbar\mathbf{l})\\
&-f_2^W(\mathbf{k}_1-\mathbf{l},\mathbf{p}_1+\tfrac{1}{2}\hbar\mathbf{l};\mathbf{k}_2+\mathbf{l},\mathbf{p}_2-\tfrac{1}{2}\hbar\mathbf{l})\}\\
&+(i\hbar)^{-1}\int d\mathbf{p}_3\int d\mathbf{l}\,\tilde{V}_l\{f_3^W(\mathbf{k}_1-\mathbf{l},\mathbf{p}_1-\tfrac{1}{2}\hbar\mathbf{l};\mathbf{k}_2,\mathbf{p}_2;\mathbf{l},\mathbf{p}_3+\tfrac{1}{2}\hbar\mathbf{l})\\
&-f_3^W(\mathbf{k}_1-\mathbf{l},\mathbf{p}_1+\tfrac{1}{2}\hbar\mathbf{l};\mathbf{k}_2,\mathbf{p}_2;\mathbf{l},\mathbf{p}_3-\tfrac{1}{2}\hbar\mathbf{l})\\
&+f_3^W(\mathbf{k}_1,\mathbf{p}_1;\mathbf{k}_2-\mathbf{l},\mathbf{p}_2-\tfrac{1}{2}\hbar\mathbf{l};\mathbf{l},\mathbf{p}_3+\tfrac{1}{2}\hbar\mathbf{l})\\
&-f_3^W(\mathbf{k}_1,\mathbf{p}_1;\mathbf{k}_2-\mathbf{l},\mathbf{p}_2+\tfrac{1}{2}\hbar\mathbf{l};\mathbf{l},\mathbf{p}_3-\tfrac{1}{2}\hbar\mathbf{l})\}
\end{aligned}
\tag{3.7.5}
$$

This equation can be written in a very simple and suggestive way. We introduce the abbreviations:

$$
\begin{aligned}
\partial_j &= \frac{\partial}{\partial\mathbf{p}_j}, &\qquad \partial_{ij} &= \frac{\partial}{\partial\mathbf{p}_i}-\frac{\partial}{\partial\mathbf{p}_j}\\
\boldsymbol{\delta}_j &= \frac{\partial}{\partial\mathbf{k}_j}, &\qquad \boldsymbol{\delta}_{ij} &= \frac{\partial}{\partial\mathbf{k}_i}-\frac{\partial}{\partial\mathbf{k}_j}
\end{aligned}
\tag{3.7.6}
$$

We moreover introduce the following operators:

$$
L_j^0 = -\left(\frac{i}{m}\right)\mathbf{k}_j\cdot\mathbf{p}_j \tag{3.7.7}
$$

$$
L'_{jn} = (i\hbar)^{-1}\int d\mathbf{l}\,\tilde{V}_l[\exp(-\tfrac{1}{2}\hbar\mathbf{l}\cdot\partial_{jn})-\exp(\tfrac{1}{2}\hbar\mathbf{l}\cdot\partial_{jn})]\exp(-\mathbf{l}\cdot\boldsymbol{\delta}_{jn}) \tag{3.7.8}
$$

[We recall that the exponential of a differential operator is a finite displacement operator, see Eq. (3.6.7).] It is now obvious that Eq. (3.7.5) can be rewritten in the form:

$$\partial_t f_2^W(\mathbf{k}_1, \mathbf{p}_1; \mathbf{k}_2, \mathbf{p}_2) = (L_1^0 + L_2^0) f_2^W(\mathbf{k}_1, \mathbf{p}_1; \mathbf{k}_2, \mathbf{p}_2) + L'_{12} f_2^W(\mathbf{k}_1, \mathbf{p}_1; \mathbf{k}_2, \mathbf{p}_2)$$
$$+ \int d\mathbf{p}_3 \, d\mathbf{k}_3 \, \delta(\mathbf{k}_3)(L'_{13} + L'_{23}) f_3^W(\mathbf{k}_1, \mathbf{p}_1; \mathbf{k}_2, \mathbf{p}_2; \mathbf{k}_3, \mathbf{p}_3) \quad (3.7.9)$$

We have thus been able to write the equation for the Wigner function f_2^W in *exactly the same form as the equation for the classical function* (3.4.10). The only difference lies in the meaning of the symbols L_j^0, L'_{jn}. The reader will easily convince himself that this is a quite general statement. In terms of the operators (3.7.7), (3.7.8), the Wigner functions f_s^W obey a system of equations that is the direct *quantum generalization of the BBGKY hierarchy:*

$$\partial_t f_s^W(\xi_1, \ldots, \xi_s) = \sum_{j=1}^{s} L_j^0 f_s^W(\xi_1, \ldots, \xi_s) + \sum_{j<n=1}^{s}\sum L'_{jn} f_s^W(\xi_1, \ldots, \xi_s)$$
$$+ \sum_{j=1}^{s} \int d\xi_{s+1} \, \delta(\mathbf{k}_{s+1}) \, L'_{js+1} f_{s+1}^W(\xi_1, \ldots, \xi_{s+1}) \quad (3.7.10)$$

The identity of structure between classical and quantum mechanics, when expressed in terms of Wigner functions, is an extremely remarkable feature. It will help us, particularly in nonequilibrium theory, in constructing a quite general and unified formalism, which can be translated at will into classical or quantum mechanics by simply inserting the corresponding definitions of the symbols.

The developments of Section 3.4 following Eq. (3.4.7) can be carried over, as they stand, into quantum mechanics. Let us only stress the fact that the complete dynamics of the system can be summed up in terms of the quantum distribution vector $\mathfrak{f}$ introduced in Eq. (3.6.21). This vector obeys a generalized *quantum Liouville equation* identical in structure to Eq. (3.4.12):

$$\partial_t \mathfrak{f}^W(t) = \mathscr{L} \mathfrak{f}^W(t) \quad (3.7.11)$$

The matrix elements of this operator are given precisely by (3.4.15)–(3.4.19) [interpreting dx_j as $d\mathbf{k}_j \, d\mathbf{p}_j \, \delta(\mathbf{k}_j)$]. Equation (3.7.10) thus provides a very convenient starting point for the study of quantum systems in the thermodynamic limit.

3.8. SPECIAL PROPERTIES OF WIGNER FUNCTIONS: QUANTUM CORRELATIONS

We found many analogies between the classical and the Wigner formalisms. There must, however, also be some differences! Many of the properties discussed in the present section are of this type.

First of all, it should be clear from the definitions (3.6.11) and (3.5.16) that the Wigner functions appear as average values (calculated with $\boldsymbol{\rho}$) of operators that are not positive definite. This means that *the Wigner functions are not everywhere positive* (or zero): they may have negative values. Consequently, *they cannot be interpreted as probability densities.* This is the price to be paid in order not to violate Heisenberg's uncertainty principle: The phase space cannot play the same role as in classical mechanics. A point can no longer be associated with the state of the system. It is, however, remarkable that the Wigner functions provide a perfectly self-consistent formalism for calculating averages, which is quite analogous to the probabilistic one, although the *interpretation* is different. This interpretation is, however, in many cases quite unimportant, because the phase-space distribution functions are not directly observable physical quantities.

In order clearly to show the link of this property with the Heisenberg principle, we note that, in virtue of this principle, the momentum of a particle can be exactly specified, if we renounce any statement about its position (and reciprocally). If our argument is correct, this would imply that the Wigner functions, integrated over the positions (or over the momenta) *can* be interpreted as probability densities. Consider indeed, for instance, the one-particle function, and integrate it over $\mathbf{q}$. Using Eq. (3.5.15) we obtain

$$\begin{aligned}\int d\mathbf{q}\, f_1(\mathbf{q}, \mathbf{p}) &= (8\pi^3\hbar^3)^{-1} \int d\mathbf{k}\, d\mathbf{q} \exp(i\mathbf{k}\cdot\mathbf{q}) \operatorname{Tr} \boldsymbol{\rho} \mathbf{a}^\dagger(\mathbf{p}-\tfrac{1}{2}\hbar\mathbf{k})\mathbf{a}(\mathbf{p}+\tfrac{1}{2}\hbar\mathbf{k}) \\ &= \hbar^{-3} \int d\mathbf{k}\, \delta(\mathbf{k}) \operatorname{Tr} \boldsymbol{\rho} \mathbf{a}^\dagger(\mathbf{p}-\tfrac{1}{2}\hbar\mathbf{k})\mathbf{a}(\mathbf{p}+\tfrac{1}{2}\hbar\mathbf{k}) \\ &= \hbar^{-3} \operatorname{Tr} \boldsymbol{\rho} \mathbf{a}^\dagger(\mathbf{p})\mathbf{a}(\mathbf{p}) \geqslant 0 \end{aligned} \tag{3.8.1}$$

We now clearly recognize the result as the average of the number of particles in state $\mathbf{p}$, an obviously nonnegative operator. A similar conclusion is reached for the integration over $\mathbf{p}$, starting conveniently from the representation (3.6.11).

A *spatially homogeneous system* is defined as in classical mechanics by the property of translational invariance of the Wigner functions, see Eq. (3.5.1). However, if we use the Fourier picture, this property is expressed in a slightly different way. It is seen from Eq. (3.6.15) that adding an arbitrary vector $\mathbf{a}$ to all coordinates $\mathbf{q}$ results in general in a change of $f_s^W(q, p)$, except for those values of $\mathbf{k}$ whose sum vanishes. Hence, only Fourier components with wave vectors adding up to zero can contribute to the Wigner function describing a homogeneous system:

$$f_s^W(\mathbf{k}_1, \mathbf{p}_1, \ldots, \mathbf{k}_s, \mathbf{p}_s) = \delta(\mathbf{k}_1 + \cdots + \mathbf{k}_s)\tilde{f}_s^W(\mathbf{k}_1, \mathbf{p}_1, \ldots, \mathbf{k}_s, \mathbf{p}_s)$$

$$\text{(homogeneous)} \tag{3.8.2}$$

In particular, the one-particle function is of the form:

$$f_1^W(\mathbf{k}, \mathbf{p}) = 8\pi^3 n\, \delta(\mathbf{k}) \varphi^W(\mathbf{p}) \tag{3.8.3}$$

where $\varphi^W(\mathbf{p})$ is the (true) momentum distribution function. The two-particle function depends effectively on a single wave vector:

$$f_2^W(\mathbf{k}_1, \mathbf{p}_1, \mathbf{k}_2, \mathbf{p}_2) = 8\pi^3\, \delta(\mathbf{k}_1 + \mathbf{k}_2) \tilde{f}_2^W(\mathbf{k}; \mathbf{p}_1, \mathbf{p}_2) \tag{3.8.4}$$

The radial distribution function is defined in a way similar to Eq. (3.4.5):

$$\int d\mathbf{p}_1\, d\mathbf{p}_2\, \tilde{f}_2^W(\mathbf{k}; \mathbf{p}_1, \mathbf{p}_2) = n_2'(\mathbf{k}) = n^2[1 + \tilde{\nu}(\mathbf{k})] \tag{3.8.5}$$

We now come to the discussion of *correlations* in quantum systems. Here we shall find serious differences with classical mechanics. Let us first discuss the two-particle function. If the particles were statistically independent, the function $f_2^W(\xi_1, \xi_2)$ would be factorized. However, in general, this is not true: hence we may write, as in Eq. (3.5.8):

$$f_2^W(\mathbf{k}_1, \mathbf{p}_1, \mathbf{k}_2, \mathbf{p}_2) = f_1^W(\mathbf{k}_1, \mathbf{p}_1) f_1^W(\mathbf{k}_2, \mathbf{p}_2) + g_2'(\mathbf{k}_1, \mathbf{p}_1, \mathbf{k}_2, \mathbf{p}_2) \tag{3.8.6}$$

This equation is certainly correct: it is merely a definition of the correlation function $g_2'(\xi_1, \xi_2)$. On the other hand, it has a formal defect that can be easily recognized. We know from Eq. (3.6.23) that the two-body function $f_2^W(\xi_1, \xi_2)$ possesses a characteristic symmetry property related to the fermion or boson nature of the particles. But the function $f_1^W(\xi_1) f_1^W(\xi_2)$ clearly does not possess this property, and therefore $g_2'(\xi_1, \xi_2)$ does not either. It follows that we can never set $g_2'(\xi_1, \xi_2) = 0$: the resulting state would violate Pauli's principle. This fact is physically clear. Because of the requirement of symmetry or antisymmetry of the global wave function of the many-body system, *the constituent particles cannot be independent.* Consider, for instance, a system of fermions: the presence of a particle of momentum $\mathbf{p}$ excludes the possibility for another particle of having momentum $\mathbf{p}$. In a classical system, the only source of correlations is the existence of interactions between the particles. In a quantum system we have a second source of correlations: the existence of quantum-statistical boson or fermion constraints. These are present even in an ideal gas of noninteracting particles. It would be useful to display these quantum-statistical correlations explicitly.

In order to obtain this result it suffices to write in the decomposition (3.8.6) together with $f_1^W(\xi_1) f_1^W(\xi_2)$ another term whose form is suggested by

Eq. (3.6.23); the new remainder is then called $g_2^W(\xi_1, \xi_2)$:

$$\begin{aligned} f_2^W(\mathbf{k}_1, \mathbf{p}_1; \mathbf{k}_2, \mathbf{p}_2) = & f_1^W(\mathbf{k}_1, \mathbf{p}_1) f_1^W(\mathbf{k}_2, \mathbf{p}_2) \\ & + \theta f_1^W[\tfrac{1}{2}(\mathbf{k}_1 + \mathbf{k}_2) + \hbar^{-1}(\mathbf{p}_2 - \mathbf{p}_1), \tfrac{1}{2}(\mathbf{p}_1 + \mathbf{p}_2) + \tfrac{1}{4}\hbar(\mathbf{k}_2 - \mathbf{k}_1)] \\ & \times f_1^W[\tfrac{1}{2}(\mathbf{k}_1 + \mathbf{k}_2) + \hbar^{-1}(\mathbf{p}_1 - \mathbf{p}_2), \tfrac{1}{2}(\mathbf{p}_1 + \mathbf{p}_2) + \tfrac{1}{4}\hbar(\mathbf{k}_1 - \mathbf{k}_2)] \\ & + g_2^W(\mathbf{k}_1, \mathbf{p}_1; \mathbf{k}_2, \mathbf{p}_2) \end{aligned} \tag{3.8.7}$$

The *sum* of the first two terms possesses the correct symmetry property (3.6.23); hence g_2^W does too. The second term represents physically a correlation: it is indeed factorized, but each of the two factors depends irreducibly on both particles 1 and 2. It represents *the two-body correlation resulting from quantum statistics.* On the other hand, it shares an important property with the first term in the right-hand side: *If the one-particle function $f_1^W(\mathbf{k}_1, \mathbf{p}_1)$ is known, this term is entirely determined* (it is a simple functional of the one-particle Wigner function). It is, therefore, often convenient to group this term with the first one. The sum can be called the *symmetrized (or antisymmetrized) correlation pattern* $\pi_2^Q(1 \mid 2)$ corresponding to the "classical" pattern $\pi_2(1 \mid 2)$.

We now express the symmetrized correlation pattern $\pi_2^Q(1 \mid 2)$ as the result of the action of a *symmetrization operator* $P(1 \mid 2)$ on the pattern $\pi_2(1 \mid 2)$:

$$\pi_2^Q(1 \mid 2) = P(1 \mid 2) \pi_2(1 \mid 2) \tag{3.8.8}$$

This formal equation can be made explicit as follows:

$$\pi_2^Q(\mathbf{k}_1\mathbf{p}_1 \mid \mathbf{k}_2\mathbf{p}_2) = \int d\mathbf{k}_1'\, d\mathbf{k}_2'\, \langle \mathbf{k}_1\mathbf{k}_2 |\, P(1 \mid 2)\, | \mathbf{k}_1'\mathbf{k}_2' \rangle f_1^W(\mathbf{k}_1'\mathbf{p}_1) f_1^W(\mathbf{k}_2'\mathbf{p}_2) \tag{3.8.9}$$

where the matrix element appearing in the right-hand side is an operator acting on the momenta $\mathbf{p}_1$, $\mathbf{p}_2$. It is easily checked that the following operator produces precisely the expected result:

$$\begin{aligned} & \langle \mathbf{k}_1\mathbf{k}_2 |\, P(1 \mid 2)\, | \mathbf{k}_1'\mathbf{k}_2' \rangle \\ & \quad = \{\delta[-\mathbf{k}_1' + \hbar^{-1}(\mathbf{p}_1 - \mathbf{p}_1) + \tfrac{1}{2}(\mathbf{k}_1 + \mathbf{k}_1)]\, \delta[-\mathbf{k}_2' + \hbar^{-1}(\mathbf{p}_2 - \mathbf{p}_2) + \tfrac{1}{2}(\mathbf{k}_2 + \mathbf{k}_2)] \\ & \qquad + \theta\delta[-\mathbf{k}_1' + \hbar^{-1}(\mathbf{p}_2 - \mathbf{p}_1) + \tfrac{1}{2}(\mathbf{k}_2 + \mathbf{k}_1)]\, \delta[-\mathbf{k}_2' + \hbar^{-1}(\mathbf{p}_1 - \mathbf{p}_2) + \tfrac{1}{2}(\mathbf{k}_1 + \mathbf{k}_2)]\} \\ & \qquad \times \exp\{\tfrac{1}{2}\hbar[(\mathbf{k}_1' - \mathbf{k}_1) \cdot \partial_1 + (\mathbf{k}_2' - \mathbf{k}_2) \cdot \partial_2]\} \\ & \quad = \sum_{(1|2)} (-)^P\, \delta[-\mathbf{k}_1' + \hbar^{-1}(\mathbf{p}_{i_1} - \mathbf{p}_1) + \tfrac{1}{2}(\mathbf{k}_{i_1} + \mathbf{k}_1)] \\ & \qquad \times \delta[-\mathbf{k}_2' + \hbar^{-1}(\mathbf{p}_{i_2} - \mathbf{p}_2) + \tfrac{1}{2}(\mathbf{k}_{i_2} + \mathbf{k}_2)] \\ & \qquad \times \exp\{\tfrac{1}{2}\hbar[(\mathbf{k}_1' - \mathbf{k}_1) \cdot \partial_1 + (\mathbf{k}_2' - \mathbf{k}_2) \cdot \partial_2]\} \end{aligned} \tag{3.8.10}$$

The summation runs over all permutations of the indices i_1, i_2, that is,

$(i_1=1, i_2=2)$ and $(i_1=2, i_2=1)$. The superscript P is defined as follows:

$$\begin{aligned} &\text{Bosons} && P\equiv 0 && \\ &\text{Fermions} && P=0 && \text{for all even permutations} \\ & && \;\;\,=1 && \text{for all odd permutations} \end{aligned} \tag{3.8.11}$$

This index fixes the correct sign of the terms in the Fermi case.

We note that the first term in (3.8.10), corresponding to the identity permutation, is simply the identity transformation $\delta(\mathbf{k}_1'-\mathbf{k}_1)\,\delta(\mathbf{k}_2'-\mathbf{k}_2)$; the second term has the effect of changing the variables to the value appearing in the second term of Eq. (3.8.7).

We now note that the third term in the right-hand side of (3.7.7) can also be written as the result of a symmetrization operator $P(12)$ acting on the pattern $g_2(12)$. Although this operator is simply the identity operator $\delta(-\mathbf{k}_1'+\mathbf{k}_1)\,\delta(-\mathbf{k}_2'+\mathbf{k}_2)$, it can also be suggestively written in the form (3.8.10):

$$\begin{aligned} P(12) = \sum_{(12)} (-)^P\, &\delta[-\mathbf{k}_1'+\hbar^{-1}(\mathbf{p}_{i_1}-\mathbf{p}_1)+\tfrac{1}{2}(\mathbf{k}_{i_1}+\mathbf{k}_1)] \\ &\times \delta[-\mathbf{k}_2'+\hbar^{-1}(\mathbf{p}_{i_2}-\mathbf{p}_2)+\tfrac{1}{2}(\mathbf{k}_{i_2}+\mathbf{k}_2)] \\ &\times \exp\{\tfrac{1}{2}\hbar[(\mathbf{k}_1'-\mathbf{k}_1)\cdot\partial_1+(\mathbf{k}_2'-\mathbf{k}_2)\cdot\partial_2]\} \end{aligned} \tag{3.8.12}$$

The only difference with (3.8.10) is in the summation prescription: Instead of summing over both permutations (12), (21) of the two particles, one keeps only the identical permutation (12). In other words, the permutation (other than the identical one) of the particles 12 belonging to the same subset is excluded, because $g_2(12)$ already possesses the correct symmetry.

The cluster decomposition (3.8.7), including quantum-statistical effects, can now be written very compactly in terms of the correlation patterns defined in Section 3.5:

$$f_2^W(1,2) = P(1\,|\,2)\pi_2(1\,|\,2) + P(12)\pi_2(12) \tag{3.8.13}$$

After this detailed discussion of the two-particle correlations, the generalization to the s-particle case is pretty obvious. We formulate it here without proof but urge the reader to work out the case $s=3$, which will convince him of the correctness of the argument. The quantum cluster representation of the s-particle Wigner function is written similarly to Eq. (3.5.15):

$$f_s^W(\xi_1,\ldots,\xi_s) = \sum_{\Gamma_s} P_s(\Gamma_s)\pi_s(\xi_1,\ldots,\xi_s;[\Gamma_s]) \tag{3.8.14}$$

The summation again runs over all partitions of the set $(1,\ldots,s)$, each partition being characterized by the index Γ_s. Each correlation pattern is now symmetrized or antisymmetrized by means of an operator $P_s(\Gamma_s)$ depending on the number s of particles and on the partition Γ_s. The action of this operator is defined explicitly as in (3.8.9):

$$(P\pi)(\mathbf{k}_1\mathbf{p}_1, \ldots, \mathbf{k}_s\mathbf{p}_s; [\Gamma_s]) = \int d\mathbf{k}'_1 \cdots d\mathbf{k}'_s$$

$$\times \langle \mathbf{k}_1, \ldots, \mathbf{k}_s | P_s(\Gamma_s) | \mathbf{k}'_1, \ldots, \mathbf{k}'_s \rangle \pi_s(\mathbf{k}'_1\mathbf{p}_1, \ldots, \mathbf{k}'_s\mathbf{p}_s; [\Gamma_s]) \quad (3.8.15)$$

with

$$P_s(\Gamma_s) = \sum_{(\Gamma_s)} (-)^P \delta[-\mathbf{k}'_1 + \hbar^{-1}(\mathbf{p}_{i_1} - \mathbf{p}_1) + \tfrac{1}{2}(\mathbf{k}_{i_1} + \mathbf{k}_1)] \cdots$$

$$\delta[-\mathbf{k}'_s + \hbar^{-1}(\mathbf{p}_{i_s} - \mathbf{p}_s) + \tfrac{1}{2}(\mathbf{k}_{i_s} + \mathbf{k}_s)]$$

$$\times \exp\left\{\tfrac{1}{2}\hbar \sum_{j=1}^{s} (\mathbf{k}'_j - \mathbf{k}_j) \cdot \partial_j\right\} \quad (3.8.16)$$

The only difference between the correlation patterns is in the prescription of the summation. The sum runs over all the permutations of the s particles, *excluding* (true) permutations of particles within the same subset. For instance, in the case $s = 3$, for the pattern $(1|2|3)$ all 3! permutations of the indices are included. But for the pattern $(1|23)$ there are only (3!/2!) terms in the sum, because permutations of the particles within the two-particle subset (23) must not be included. Finally, for the pattern (123) only the identity transformation appears in the sum. For further illustration of this rule, we have collected in Table 3.8.1 the various permutations to be retained in the symmetrization operators for all the patterns $P_3(\Gamma_3)$, as well as the sign of the corresponding term in the case of fermions.

An important remark can be made at this point. A quantum-statistical system can be completely characterized by the complete set of *unsymmetrical* correlation patterns $\pi_s(\xi_1, \ldots, \xi_s; [\Gamma_s])$ that are defined in the

Table 3.8.1. Values of the Indices i_1, i_2, i_3 in the Various Terms of Eq. (3.8.16), Contributing to the Symmetrizers of the Three-Particle Correlation Patterns. The Sign of the Corresponding Terms in the Case of Fermions Is Also Indicated

$P_3(1\|2\|3)$	$P_3(1\|23)$	$P_3(2\|13)$	$P_3(3\|12)$	$P_3(123)$
(+) 123	(+) 123	(+) 123	(+) 123	(+) 123
(−) 132		(−) 132	(−) 132	
(−) 213	(−) 213	(−) 213		
(+) 231				
(+) 312				
(−) 321	(−) 321		(−) 321	

same way as in classical mechanics. Indeed, if these are known, the s-particle Wigner function can be reconstructed by Formula (3.8.14). This remark is very important because, as will be seen in Section 14.3, the patterns $\pi_s(\Gamma_s)$ obey a closed (though infinite) hierarchy of equations, whereas the symmetrized patterns do not.

We may finally note that the symmetrization operator $P_s(\Gamma_s)$ has the simple form (3.8.17) only in the Fourier picture. Had we used the representation $\pi_s(\mathbf{q}_1\mathbf{p}_1, \ldots, \mathbf{q}_s\mathbf{p}_s; [\Gamma_s])$, the form of the symmetrization operator would have been excessively complex.

BIBLIOGRAPHICAL NOTES

The reduced distribution functions were introduced by several authors, but the first systematic treatment is probably due to Yvon: J. Yvon, *La Théorie Statistique des Fluides et l'Equation d'Etat*, Act. scient. et ind. No. 203, Hermann, Paris, 1935.

More recent general discussions of these functions can be found in the following books:

H. S. Green, *The Molecular Theory of Fluids*, North Holland, Amsterdam, 1952.

D. Massignon, *Mécanique Statistique des Fluides*, Dunod, Paris, 1957.

J. Yvon, *Les Corrélations et l'Entropie en Mécanique Statistique Classique*, Dunod, Paris, 1965.

A. Münster, *Statistical Thermodynamics*, Vol. 1, Springer, Berlin, and Academic Press, New York, 1969.

The initials BBGKY come from the following papers:

J. Yvon (1935): see above.

M. Born and H. S. Green, *Proc. Roy. Soc.* Lond. **A188,** 10 (1946).

J. G. Kirkwood, *J. Chem. Phys.* **14,** 180 (1946).

N. N. Bogoliubov, *J. Phys. USSR* **10,** 257, 265 (1946).

A rigorous mathematical formulation for equilibrium reduced distributions is given in D. Ruelle, *Statistical Mechanics, Rigorous Results*, Benjamin, New York, 1969.

A mathematical discussion of the BBGKY hierarchy is given in G. Galavotti, O. E. Lanford III, and J. L. Lebowitz, *J. Math. Phys.* **11,** 2898 (1970).

The Wigner functions were introduced in E. Wigner, *Phys. Rev.* **40,** 749 (1932).

The formalism was extensively developed in J. E. Moyal, *Proc. Cambridge Phil. Soc.* **45,** 99 (1949).

Among the many papers and books where Wigner functions were used, we quote: D. Massignon (see above).

I. L. Klimontovich and S. V. Temko, *Zh. Eksp. Teor. Fiz. USSR* **33,** 132 (1957).

H. Mori, I. Oppenheim, and J. Ross, in *Studies in Statistical Mechanics*, Vol. 1 (J. de Boer and G. E. Uhlenbeck, Eds.), North Holland, Amsterdam, 1962.

R. Balescu, *Statistical Mechanics of Charged Particles*, Wiley-Interscience, New York, 1963.

K. P. Gurov, *Osnovaniya Kineticheskoi Teorii*, Nauka, Moscow, 1966.

V. P. Silin, *Vvedenie v kineticheskuyu teoriyu gazov*, Nauka, Moscow, 1971.

PART 2

EQUILIBRIUM STATISTICAL MECHANICS

CHAPTER 4

EQUILIBRIUM ENSEMBLES AND THERMODYNAMICS

4.1. EQUILIBRIUM SOLUTIONS OF THE LIOUVILLE EQUATION

The simplest class of solutions of the classical Liouville equation (2.2.15) is the class of time-independent solutions, which obey the equation

$$LF^0 = 0 \tag{4.1.1}$$

or, explicitly

$$[H, F^0]_P = 0 \tag{4.1.2}$$

A set of solutions of this equation is immediately found: *any* function $F(H)$ of the Hamiltonian satisfies Eq. (4.1.2):

$$F^0(q, p) = F(H(q, p)) \tag{4.1.3}$$

The only restrictions on F are those imposed by Eqs. (2.2.5) and (2.2.6), and the finiteness of the resulting reduced distribution functions according to our postulate of Section 3.3.

Quite similarly, in quantum mechanics, any density operator that can be expressed as a function of the Hamiltonian operator $\hat{H}$:

$$\hat{\rho}^0 = \hat{R}(\hat{H}) \tag{4.1.4}$$

and satisfying Eqs. (2.3.12), (2.3.14) and (2.3.15) is a stationary solution of the von Neumann equation (2.3.22):

$$[\hat{H}, \hat{\rho}^0]_- = 0 \tag{4.1.5}$$

The functions (4.1.3) are not the only solutions of Eq. (4.1.2). Any function of the constants (or integrals) of the motion of the system also satisfies (4.1.2). Indeed, a constant of the motion α_n is defined by the relation [see Eq. (1.2.7)]

$$[\alpha_n, H]_P = 0 \tag{4.1.6}$$

Hence, in principle, any function of the form $F(H, \alpha_1, \ldots, \alpha_r)$ is a stationary solution. However, a simple algebraic definition like (4.1.6) does not give enough insight into the mathematical nature of the constants of the motion α_n. Indeed, it appears from a more detailed analysis that the Hamiltonian H (and for some systems, the total momentum and the total angular momentum) is qualitatively different from all the other constants of the motion, and as a result we may limit our consideration to functions of H alone. These problems are connected with the celebrated *ergodic theory*. This is a chapter of dynamics dealing with the global nature and stability of trajectories of dynamical systems. At one time it was thought that the ergodic theory would provide a firm basis for the justification of the postulates of statistical mechanics. Nowadays it is clear that the connection between statistical mechanics and ergodic theory is very thin. Nevertheless, we include an Appendix treating ergodic theory. We do this not only because traditionally this subject is discussed in books on statistical mechanics, but also because in the past ten years this field has received a strong new impetus. A number of very beautiful and important results have been found that activated the interest in ergodic theory *per se*.

Coming back to statistical mechanics, we will approach the problem of constructing the equilibrium ensemble in a much more pragmatic way, along the line discussed qualitatively in Section 2.1. The basic idea is that among all solutions of Eq. (4.1.2) or (4.1.5), there is a class that is compatible with our macroscopic knowledge about the system, for instance, all distributions that correspond to a specified value of the total energy. This class still contains an enormous number of functions. In the absence of further information we have no *a priori* reason for favoring one of these more than any other. Hence we shall naturally construct the equilibrium function by assigning equal statistical weights to all the functions compatible with our requirements. This procedure—implicitly used by Gibbs—was clearly formulated by Tolman in 1938 and called the principle of *equal a priori probabilities*. It has the advantage of simplicity, clarity, and flexibility. Clearly, the principle of equal *a priori* probabilities is not a mechanical, but a statistical assumption. But mechanics alone cannot solve the problem uniquely, as we saw above.

In equilibrium theory the role of dynamics is trivial: we exhausted this subject on the first page or so. The remaining problem is essentially the statistical one. The relative importance of the two aspects will be reversed in nonequilibrium theory, where the main stress is on the laws of evolution in time of an arbitrary, rather loosely specified initial condition. We may say that equilibrium statistical mechanics is mainly statistical, whereas nonequilibrium statistical mechanics is mainly mechanical.

4.2. THE MICROCANONICAL ENSEMBLE

Our next task is the construction of a distribution function representing a given type of equilibrium state. We start from the quantum description, which is clearer and more fundamental in many ways.

The simplest system we can consider is an *isolated system.* Such a system, which does not interact with the outside world, is characterized by a constant energy. However, it should be realized that such a system is necessarily an idealization: In reality, we cannot remove completely the interactions with the outside world. On the other hand, the number of states per unit energy interval is extremely large ($\sim a^N$) for a system consisting of many degrees of freedom. Hence any small (but macroscopic) uncertainty on the total energy will cover an extremely large number of possible states of the system; it would be illusory, in a correct quantum description, to specify exactly the value of the energy.

We therefore define an "isolated system" as having a total energy comprised between the values E and $E+\Delta E$, with the condition

$$E \gg \Delta E \tag{4.2.1}$$

Moreover, we assume that the system is enclosed in a finite volume $\mathcal{V}$, which we assume very large compared to typical volumes on the molecular scale (i.e., $10^{-24}\ \text{cm}^3$). If the system consists of material particles, we assume their number N very large (typically 10^{23}).

Each of these particles is actually a small mechanical system by itself. As such, it is characterized by a certain number s of degrees of freedom. For instance, a point particle has three degrees of freedom; for a rigid diatomic molecule, $s=5$, and so on. Hence for an N-particle system the total number of degrees of freedom is sN. It is customary to call the degrees of freedom (per particle) other than those of the center of mass *internal degrees of freedom.* Hence, the number of internal degrees of freedom is $s_i = s-3$.

Let us now construct a density matrix to represent this system. We shall write the density matrix in a representation in which the Hamiltonian is diagonal. As the density matrix of a stationary system can only depend on the Hamiltonian (4.1.4), it follows that $\hat{\rho}$ will also be diagonal:

$$\rho_{mn} = \Omega^{-1} a_m \delta_{mn} \tag{4.2.2}$$

where m (or n) stands for a quantum number characterizing completely an eigenstate of the system.* The numbers $(\Omega^{-1}a_m)$ must be nonnegative [see Eq. (2.3.14)]: they represent the probability of finding the system in the state m.

* There is a hidden subtlety in this statement, which even in the best books is treated only in passing. It is well known from quantum mechanics that the state of a system must be

(Footnote continued overleaf)

At this stage we formulate the *basic postulate of equilibrium statistical mechanics.* It expresses the fact that we know very little about the microscopic state of the system: we only assume that its energy lies in a narrow interval $(E, E+\Delta E)$. But we said already that for large systems such as ours there is an enormous number of eigenstates having energy values in this interval. We know nothing that allows us to choose any of these eigenstates as representing our system better than any other: they are all equally compatible with what we know of the system. The only reasonable assumption we can make is, therefore, that each of these states is equally probable as a realization of the macroscopic state of our system. This is the famous principle of equal *a priori* probabilities. Mathematically, the principle is translated into the following formula:

$$\begin{aligned} a_m &= 1 \qquad \text{if } E \leqslant E_m \leqslant E+\Delta E \\ &= 0 \qquad \text{otherwise} \end{aligned} \tag{4.2.3}$$

where E_m is the energy eigenvalue corresponding to the state m. The only factor we must still determine is the factor appearing in (4.2.2). This we do by imposing the normalization condition (2.3.12):

$$\begin{aligned} \operatorname{Tr} \hat{\rho} &= \sum_m \Omega^{-1} a_m \\ &= \Omega^{-1} \sum_m{}' 1 \qquad (E \leqslant E_m \leqslant E+\Delta E) \\ &= 1 \end{aligned}$$

Hence

$$\Omega = \sum_m{}' 1 \qquad (E \leqslant E_m \leqslant E+\Delta E) \tag{4.2.4}$$

The normalization factor Ω is, therefore, simply the number of states whose energy eigenvalue lies in the interval $(E, E+\Delta E)$. We call this quantity more briefly the *number of accessible states.*

---

described by a "complete set of commuting observables" (in Dirac's terminology). Our statement implies that the energy is, *by itself,* such a complete set: we exclude any other observables commuting with the Hamiltonian. This is, of course, a disguised formulation of the ergodic hypothesis.

If this statement is not valid, the assumption that $\hat{\rho}$ is a function of $\hat{H}$ alone does not imply that $\hat{\rho}$ is diagonal in the energy representation. Indeed, each energy eigenvalue is then multiply degenerate, because the complete description of the state requires additional quantum numbers. In that case, our assumption only tells us that $\hat{\rho}$ is diagonal in the energy quantum number, not in the others. Supplementary assumptions are necessary for the construction of a unique density matrix.

It will be seen in Section 5.2 that in the case of an ideal gas, which is indeed a nonergodic system possessing N invariants of motion (the separate energies of its particles), a stronger assumption is (implicitly) made for the construction of the density matrix.

Let us remark at this point that Ω is a function of the energy E, of the energy width ΔE, and also depends parametrically on the volume $\mathcal{V}$ and on the number of particles N:

$$\Omega \equiv \Omega(E; \Delta E; N, \mathcal{V}) \equiv \Omega(E; \Delta E) \tag{4.2.5}$$

By the last equality we mean that the parameters N and $\mathcal{V}$ will often not be written explicitly.

The microcanonical ensemble is of prime theoretical importance because it leads to a clear and satisfactory construction of an equilibrium ensemble by the very direct application of the principle of equal *a priori* probabilities. However, in most nontrivial problems this ensemble proves to be a mathematically difficult and unflexible tool.

On the other hand, taking the microcanonical ensemble as a starting point, one can develop methods of investigation that are much more directly useful and can be clearly connected with macroscopic physics. Therefore we go on now to these methods, omitting to develop further the microcanonical ensemble.

4.3. THE CANONICAL ENSEMBLE

Isolated systems are usually not very interesting in physics. A much more relevant object of study is a system that interacts with its surroundings, either by exchanging various forms of energy or by exchanging matter. Such systems are of primary interest in macroscopic thermodynamics, which studies precisely the laws governing these exchanges and the conditions of equilibrium of the system with its surroundings.

Consider first a very large isolated system U (we might call it the "universe"): it is described statistically by a microcanonical ensemble corresponding to energy E_U. We now fix our attention on a small subsystem S of U, which interacts with the "external world" W (i.e., the complement of S) (see Fig. 4.3.1).

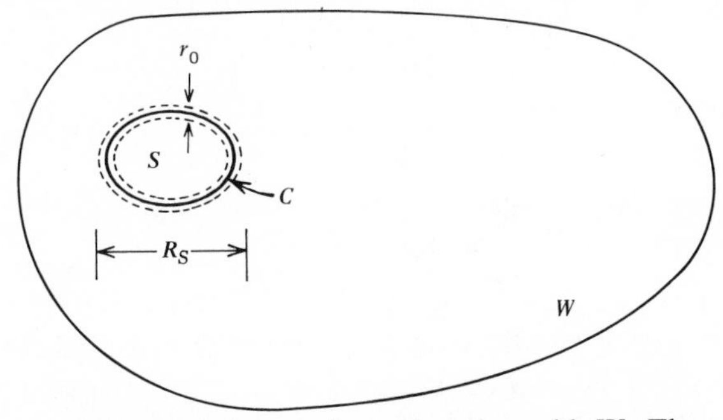

Figure 4.3.1. A subsystem S and the external world W. The corridor C is indicated by dotted lines.

The number of particles in S, N_S, is supposed to be much smaller than the number of particles N_W in the external world W; yet, we must assume that N_S is still an enormously large number, so that statistical mechanics can be validly applied to describe the system S. Hence:

$$1 \ll N_S \ll N_W \tag{4.3.1}$$

As the whole universe is in equilibrium, its density and all its local properties are uniform, except for fluctuations. We assume that by choosing the subsystem S we did not pick up a violently fluctuating region, and hence that the density in S is not very different from the density in W. This implies that the volumes $\mathcal{V}_S$, $\mathcal{V}_W$ of the system and of the external world should be very nearly in the same ratio to each other, as the corresponding numbers of particles

$$\frac{\mathcal{V}_S}{N_S} \approx \frac{\mathcal{V}_W}{N_W} \tag{4.3.2}$$

All the results to be derived presently are supposed to hold for very large systems. Recalling our considerations of Section 3.3, we must assume that N_S (and therefore $\mathcal{V}_S$) is sufficiently large in order that the principle of macroscopic equivalence be valid. In other words, we must be able to show that starting with S we can construct a sequence of systems with N_S increasing towards infinity, such that all the members of the set are macroscopically equivalent. This statement will be discussed in Section 4.7. Meantime we note that in this construction, the external world corresponding to the successive members of the sequence must also be of increasing size. Actually, Eq. (4.3.1) implies that N_W must increase faster than N_S. *Therefore, the thermodynamic limit in this case is defined by the following process*:

$$\begin{aligned} &N_S \to \infty, \qquad \mathcal{V}_S \to \infty \\ &N_W \to \infty, \qquad \mathcal{V}_W \to \infty \\ &\frac{N_S}{\mathcal{V}_S} = \frac{N_W}{\mathcal{V}_W} = n \\ &\frac{N_W}{N_S} \to \infty \end{aligned} \tag{4.3.3}$$

The energy of the universe U can be written quite generally as a sum of three terms:

$$E_U = E_S + E_W + H'_{SW} \tag{4.3.4}$$

where the terms on the right-hand side represent, respectively, the energy of S, the energy of W, and the interaction energy between S and W. The term H'_{SW} is the one that could introduce a real difficulty. We now show that, under realistic conditions, we may simply forget about it.

We assume that the molecules of the universe are described by a Hamiltonian of the form given by Eqs. (2.4.1), (2.4.4), with $\hat{H}^F \equiv 0$. In any realistic situation, the pair interaction potential $V(r)$ [Eq. (2.4.5)] has the shape shown schematically in Fig. 4.3.2. At short distances the force is repulsive as a result of the impenetrability of the molecules; at large distances the force is attractive, and falls off more or less rapidly as $r \to \infty$.

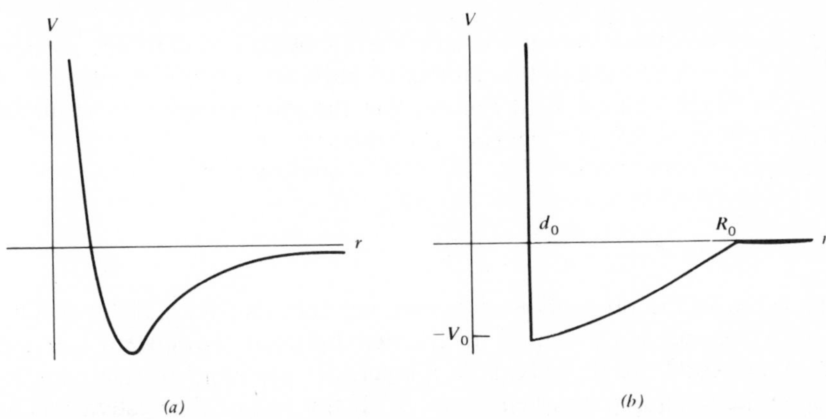

Figure 4.3.2. Intermolecular interaction potentials: (*a*) a realistic potential; (*b*) a model potential.

In order to simplify our discussion, we consider here a model potential that retains these characteristic features (Fig. 4.3.2*b*):

$$\begin{aligned} V(r) &= \infty & r &< d_0 \\ -V_0 \leq V(r) &\leq 0, & d_0 \leq r &\leq R_0 \\ V(r) &= 0, & r &> R_0 \end{aligned} \tag{4.3.5}$$

Because of the infinitely repulsive potential barrier, this model represents rigid spheres of diameter d_0 that attract each other provided their centers are separated by a distance smaller than R_0.

We now estimate the total potential energy of interaction $H'_{\mathcal{V}}$ of the molecules enclosed in an arbitrary volume $\mathcal{V}$. Clearly,

$$|H'_{\mathcal{V}}| \leq V_0 \times (\text{number of molecules in } \mathcal{V}) \times (\text{maximum number of molecules within the range of attractive force of a given molecule})$$

where V_0 is the absolute value of the lower bound on the attractive pair potential. If n is the number density, the number of molecules in $\mathcal{V}$ is $n\mathcal{V}$. The last factor is roughly given by the ratio between the volume of the sphere of interaction $\frac{4}{3}\pi R_0^3$ and the volume of the hard core region $\frac{4}{3}\pi d_0^3$.

Indeed, because of the presence of a rigid hard core, the maximum number of molecules in the interaction sphere equals the number of spheres of radius r_0 that fit into that domain; if $R_0 > d_0$, this number equals the ratio mentioned above times a geometrical factor α of order 1. Hence

$$|H'_V| \leqslant V_0 \times n\mathcal{V} \times \alpha\left(\frac{R_0^3}{d_0^3}\right) \tag{4.3.6}$$

On the other hand, the total kinetic energy, being a sum of one-particle terms, is proportional to the number of particles and hence, because of (4.3.2), to the volume $\mathcal{V}$. It follows that the total energy for the model Hamiltonian (4.3.5) is proportional to the volume.

We now come back to Eq. (4.3.4).It is obvious that

$$|E_W| \sim \mathcal{V}_W$$
$$|E_S| \sim \mathcal{V}_S$$

To estimate the interaction term H'_{SW} we note that the only molecules which can participate in an interaction between the system and the external world are those lying in a layer, or "corridor" of thickness R_0 along the boundary between S and W. If the size of the system S is of order R_S, its volume is approximately R_S^3 and the volume of the corridor is of order $R_S^2 R_0$. It then follows from (4.3.6) that

$$\frac{|H'_{SW}|}{|E_S|} \sim \frac{\mathcal{V}_{\text{corr}}}{\mathcal{V}_S} \sim \frac{R_S^2 R_0}{R_S^3} \sim \frac{R_0}{R_S} \sim R_0 \mathcal{V}_S^{-1/3} \tag{4.3.7}$$

Since R_0 is an intrinsic characteristic of the pair potential, it follows that the relative contribution of H'_{SW} can be made arbitrarily small by choosing the size of S sufficiently large compared to the range of the forces. In particular, in the thermodynamic limit (4.3.3) this ratio vanishes. We thus conclude that the interactions between the system and the external world, although physically very important in providing the necessary exchange of energy between the two subsystems, contribute numerically a negligibly small amount to the total amount of energy. The system and the external world can be considered as practically decoupled:

$$E_U \approx E_W + E_S \tag{4.3.8}$$

We also note that

$$E_S \ll E_W \tag{4.3.9}$$

Let us stress the fact that the nature of the Hamiltonian played a crucial role in the argument above. For more realistic potentials, the argument must be refined, and in some important cases it becomes quite involved.

In recent years these matters have become very well understood. They will be discussed again in Section 4.7.

We now come to our main problem:

Knowing that the whole universe is represented by a microcanonical ensemble, what is the probability P_m of finding the subsystem S in a given quantum state, characterized by the quantum numbers m and by the energy eigenvalue E_m?

We may remark first of all that the possibility of speaking of definite quantum states for the subsystem S alone arises from the extreme smallness of the interactions H'_{SW} in the Hamiltonian. Only in this case can we consider separate quasistationary states for S and W. (The Schrödinger equation for the composite system is separable and is therefore equivalent to a set of two independent equations for S and for W.) Clearly, the values of the *energy levels E_m depend on the volume $\mathcal{V}_S$ and on the number of particles N_S*. (To fix these ideas, think of the eigenvalues for one particle in a box of volume $\mathcal{V}$, and of the eigenvalues for N free particles in a box: in these elementary problems, the dependence of E_m on $\mathcal{V}$ can be determined explicitly.)

Within the framework of our assumptions, the main problem is very simple (see Fig. 4.3.3). Indeed, because of Eq. (4.3.8), if S has precisely the energy E_m, the external world can be in any one of the numerous states comprised between $E_W \equiv E_U - E_m$ and $E_W + \Delta E$. The number of such states is $\Omega_W(E_U - E_m; \Delta E)$, according to the results of Section 4.2. This is also the number of configurations of the whole universe, in which S is in the state m, corresponding to energy E_m, and *jointly* W has an energy within the appropriate range. This last statement follows from the fact that S and W are effectively decoupled and hence do not influence each other.

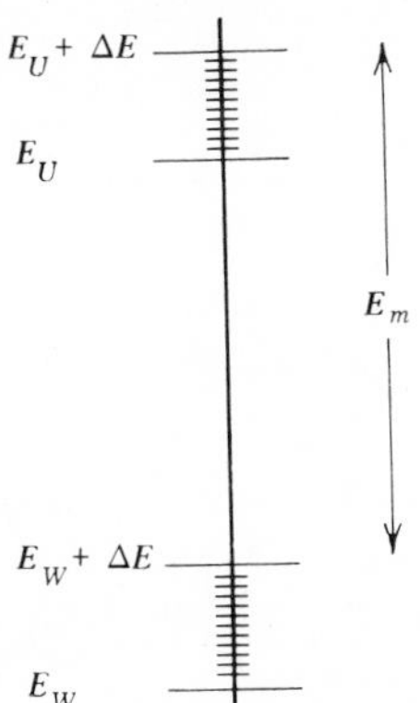

Figure 4.3.3. Scale of energies for the derivation of the canonical ensemble.

Because of our fundamental assumption of equal *a priori* probabilities, all the configurations counted in Ω_W have equal probability. Hence the probability p_m we are looking for is simply given by the ratio of the number of configurations of U in which S is precisely in the state m, that is, $\Omega_W(E_U - E_m; \Delta E)$, to the total number of accessible states of the universe, that is, $\Omega_U(E_U; \Delta E)$:

$$p_m = \frac{\Omega_W(E_U - E_m; \Delta E)}{\Omega_U(E_U; \Delta E)} \tag{4.3.10}$$

We may now use the assumption (4.3.9), which is *a fortiori* valid if we replace E_S by the single energy level E_m. This allows us to expand the numerator—or rather its logarithm, for better convergence—around E_U and retain only the linear term:

$$\ln \Omega_W(E_U - E_m; \Delta E) \approx \ln \Omega_W(E_U; \Delta E) - \beta E_m \tag{4.3.11}$$

where we introduce the parameter β by

$$\beta = \left(\frac{\partial \ln \Omega_W(E; \Delta E)}{\partial E}\right)_{E=E_U} \tag{4.3.12}$$

Hence

$$\Omega_W(E_U - E_m; \Delta E) \approx \Omega_W(E_U; \Delta E) e^{-\beta E_m} \tag{4.3.13}$$

and substituting into (4.3.10) we obtain

$$p_m = Z^{-1} e^{-\beta E_m} \tag{4.3.14}$$

This is our final result.* The probability p_m depends on two parameters that are independent of E_m: Z and β. The parameter Z can be expressed in an intrinsic way (i.e., in terms of quantities related to S alone). Indeed, the probability distribution (4.3.14) must be normalized to one:

$$\sum_m p_m = 1 \tag{4.3.15}$$

from which follows

$$Z = \sum_m e^{-\beta E_m} \tag{4.3.16}$$

the sum being over *all* states of the system S, without any restriction [as opposed to the summation prescription in the definition of the microcanonical Ω, see Eq. (4.2.4)].

The expression Z is called the *partition function.* It is one of the most important quantities of equilibrium statistical mechanics. Z clearly depends on the parameter β (which we have yet to interpret); it also

* An alternative derivation of this formula, more traditional and apparently simpler, is given in the Appendix to this chapter.

depends on the number of particles N and on the volume $\mathcal{V}$ of the system through the values of the energy levels E_m (we may now drop the subscripts S).

From the probability distribution p_m (4.3.14) we can construct a density matrix describing the system S (independently of the external world!):

$$\rho_{mn}^{(c)} = Z^{-1} e^{-\beta E_m} \delta_{mn} \tag{4.3.17}$$

These matrix elements are written in the representation in which the Hamiltonian (*of the system* S) is diagonal. The density matrix is then also diagonal, and hence it is a stationary solution of the von Neumann equation *for the system* S. Because of the simple analytical form (4.3.17) we can immediately translate the expression of the density matrix into any other representation, in which the energy is no longer diagonal:

$$\hat{\rho}^{(c)} = Z^{-1} \exp(-\beta \hat{H}) \tag{4.3.18}$$

and the partition function is expressed as

$$Z = \text{Tr} \exp(-\beta \hat{H}) \tag{4.3.19}$$

Equations (4.3.18) and (4.3.19) define a new equilibrium ensemble called the *canonical ensemble*. It was introduced for the first time by J. W. Gibbs (in the classical case) around 1900. It is very easy to define average values of operators $\hat{b}$ calculated in this ensemble:

$$\langle \hat{b} \rangle = \text{Tr}\, \hat{b} \hat{\rho}^{(c)} = Z^{-1}\, \text{Tr}\, \hat{b} \exp(-\beta \hat{H}) \tag{4.3.20}$$

It is very important, from the practical point of view, that we have been able to write the canonical density matrix in an *arbitrary* representation (this was not possible for the microcanonical ensemble because of its singular nature). Indeed, in order to evaluate ρ_{mn} in the form (4.3.17) or Z in the form (4.3.16) we need to know the eigenvalues of the Hamiltonian, that is, we must solve the Schrödinger equation for S, which is practically impossible for nontrivial systems. On the contrary, in the forms (4.3.18) and (4.3.19) we can use for the computations any convenient orthonormal set of functions as a basis, calculate the matrix elements of $\hat{H}$ in that basis (which is always feasible), and then devise some suitable approximation method for the calculation. One already appreciates the enormous flexibility of the canonical ensemble as compared to the microcanonical one.

From Eq. (4.3.18) it is very easy to proceed to the classical case. The *classical canonical distribution* is simply

$$F^{(c)}(q, p) = \bar{Z}^{-1} e^{-\beta H(q,p)} \tag{4.3.21}$$

where we denoted by $\bar{Z}$ the normalization constant defined as follows:

$$\bar{Z} = \int dq\, dp\, e^{-\beta H(q,p)} \tag{4.3.22}$$

The average value of a dynamical function is defined as

$$\langle b \rangle = \bar{Z}^{-1} \int dq\, dp\, b(q, p) e^{-\beta H(q,p)} \tag{4.3.23}$$

As usual, the integrations in (4.3.22), (4.3.23) cover the entire accessible phase space.

We denoted the normalization constant in Eq. (4.3.22) by $\bar{Z}$ rather than Z like the quantum partition function of Eq. (4.3.16) or (4.3.19). It is immediately seen that $\bar{Z}$ is *not* the exact analog of Z: indeed Z is a dimensionless number that results from a counting process, whereas $\bar{Z}$ has dimensions $[qp]^N$. The "states" of classical mechanics are continuously distributed and cannot, therefore, be counted. In order to find an analog of a "quantum state" in classical mechanics, we remember that a quantum state can only be defined, in the best case, within errors δp_i and δq_i of the phase-space coordinates satisfying the Heisenberg uncertainty relations*:

$$\delta q\, \delta p \sim h^{sN} \tag{4.3.24}$$

In other words, a point lying in a cell of volume h^{sN} in phase space cannot be distinguished from any other point lying in the same cell in the framework of quantum mechanics. We may expect, therefore, that a good classical analog of Z would be obtained by dividing $\bar{Z}$ by the volume of an elementary cell h^{sN}. However, this is not yet sufficient, as can be seen from the following argument. In quantum mechanics, the particles are *undistinguishable.* This means that a situation where particle 1 is in the cell a of phase space and particle 2 in cell b cannot be distinguished from a situation where particle 1 is in b and particle 2 in a. More generally, starting from a given configuration of N particles, we can perform $N!$ permutations of these particles; all the resulting situations are equivalent in quantum mechanics and are counted as a single state. On the contrary, there is no restriction on the integrations in Eq. (4.3.22): to every point $q_1 = q$, $q_2 = q'$, there corresponds a point $q_1 = q'$, $q_2 = q$ within the domain of integration. Hence, the result (4.3.22) even after division of h^{sN} would

* Remember the notation

$$\delta q\, \delta p = \prod_{j=1}^{N} \prod_{\sigma=1}^{s} \delta q_{j\sigma}\, \delta p_{j\sigma}$$

where j denotes the particle and σ the various degrees of freedom.

be $N!$ times larger than the result of the corresponding quantum-mechanical calculation, and must be corrected accordingly. Hence, the *"classical" definition of the partition function* is

$$Z = h^{-sN}(N!)^{-1}\int dq\, dp\, e^{-\beta H(q,p)} \tag{4.3.25}$$

Correspondingly, the canonical distribution (4.3.21) can also be expressed in terms of the partition function as

$$F^{(c)}(q, p) = h^{-sN}(N!)^{-1}Z^{-1}e^{-\beta H(q,p)} \tag{4.3.26}$$

The derivation given here for the "classical limit of the partition function" is admittedly heuristic. A rigorous derivation involves a systematic expansion of the quantum partition function in powers of $\hbar$. Such an expansion was done by Kirkwood: it yields Eq. (4.3.25) as the leading term when $\hbar \to 0$. This result gives the final justification of our choice (4.3.24) for the volume of a phase-space cell corresponding to a quantum state.

4.4. CONNECTION BETWEEN STATISTICAL MECHANICS AND THERMODYNAMICS

Classical thermodynamics starts from a few empirical observations, which are organized and axiomatized in the form of the "three laws." From these one can deduce an enormous number of relations between the various quantities characterizing the mechanical and thermal behavior of matter. Equilibrium thermodynamics remains a unique case in physics of a perfect logical structure. However, thermodynamics has its intrinsic weakness and incompleteness. First of all, it yields *relations* between the various quantities, but excludes any possibility of calculating their *absolute values.* For instance, thermodynamics determines a link between the specific heat at constant volume c_V and the specific heat at constant pressure c_P. If we know the value of c_P, we can predict c_V without further recourse to experiment; but we have no way of calculating c_P to start.

We will now see that statistical mechanics bridges this gap. The canonical ensemble provides us with a model of matter in thermal equilibrium and allows us to calculate all thermodynamical quantities in terms of quantities characterizing the microscopic properties of the molecules. The success and final justification of this model is amply supported by comparison with the experimental results. The interest of statistical mechanics is twofold: On one hand, it predicts the thermodynamical quantities from a knowledge of microscopic mechanics (e.g., the energy levels of molecules, as measured spectroscopically). In other cases, it allows one to deduce the microscopic properties (e.g., the nature of the intermolecular interactions) from measurements of macroscopic, thermodynamic properties. Last but not least, statistical mechanics opens the possibility of investigating the limitations of classical thermodynamics and of extending the domain of the study of matter to conditions where thermodynamics is no longer valid.

The quantities appearing in thermodynamics can be classified into three groups. The simplest are the *external parameters*, whose value is fixed by the external environment or by the experimenter in a precise way, without reference to the internal state of the system. Typical examples are the volume $\mathcal{V}$, the number of particles N, or the values of external fields (e.g., gravitational, electric, magnetic fields). To simplify the treatment, we assume that $\mathcal{V}$ and N are the only independent external parameters characterizing the systems considered in this section.

A second category consists of purely *mechanical quantities* such as the internal energy E or the pressure P. We know already how to deal with these from our discussions of Chapter 2: they are defined as averages of microscopic dynamical functions according to Eqs. (2.2.4) or (2.3.11).

There exists, however, a third distinct category, quite typical of thermodynamics.* Certain important quantities—such as temperature—have no microscopic meaning and can only be conceived at the macroscopic level. For instance, one can define the energy of a single molecule, but one cannot speak of the temperature of a single molecule. Because of this fact, we may expect that these *thermal quantities* (such as temperature, entropy, etc.) will be defined entirely in terms of properties of the ensemble distribution function, rather than as averages of single-particle properties. These are examples of *collective* properties whose value is determined by the overall distribution of all the particles in the system.

We now establish the desired link explicitly. We note first that, by definition, the external parameters need no statistical treatment, because their values are assumed to be precisely known.

Proceeding to the second category, it is very easy to identify the *internal energy* E as the average value of the Hamiltonian in the canonical ensemble:

$$E = \mathrm{Tr}\,\hat{H}\hat{\rho}^{(c)} = Z^{-1} \sum_m E_m e^{-\beta E_m} \tag{4.4.1}$$

In this way the energy is expressed as a function of the parameter β (characteristic of the canonical distribution) and of the external parameters $\mathcal{V}$ and N (entering implicitly through the eigenvalues E_m).

In order to find good definitions for the thermal quantities, we have to set up a picture of a thermodynamic process. We imagine again the large universe U of Section 4.3, but now we consider *two* subsystems S_1, S_2,

* This type of macroscopic observable was mentioned briefly at the end of Section 2.2.

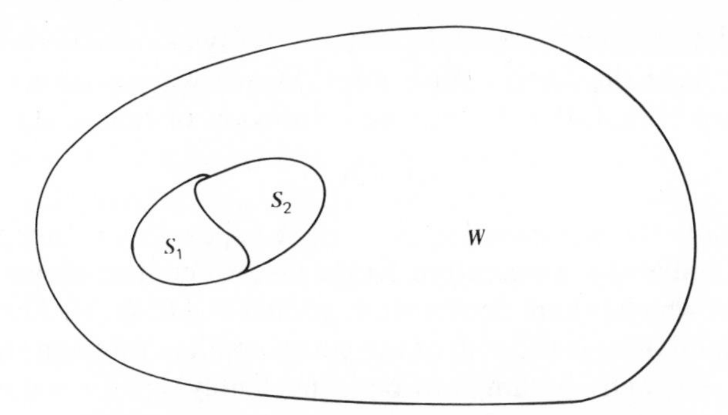

Figure 4.4.1. Two systems in thermal contact.

which we assume to be in mutual thermal contact, that is, they can exchange energy through a weak interaction (see Fig. 4.4.1). Again, if we assume both systems to be very large compared to the range of the intermolecular forces, their mutual interaction energy, although physically important for ensuring thermal contact, is nevertheless negligibly small compared to the energy content of S_1 and S_2 separately. A straightforward extension of the argument of Section 4.3 shows that the joint probability of finding system S_1 in state n (energy E_{1n}) and S_2 in state m (energy E_{2m}) is

$$p_{nm} = (Z_1^{-1} e^{-\beta_1 E_{1n}})(Z_2^{-1} e^{-\beta_2 E_{2m}}) \tag{4.4.2}$$

where

$$Z_i = \sum_r e^{-\beta_i E_{ir}}, \qquad i = 1, 2$$

In principle, β_1 is not necessarily equal to β_2. We now require that S_1 and S_2 be in thermal equilibrium with each other. In that case, the situation is equally well described by considering a single joint system S, with energy $E_{nm} = E_{1n} + E_{2m}$ immersed in the external world W. Its probability distribution is given by the canonical ensemble:

$$p_{nm} = Z^{-1} e^{-\beta(E_{1n}+E_{2m})} \tag{4.4.3}$$

with

$$Z = \sum_{n,m} e^{-\beta(E_{1n}+E_{2m})}$$

The condition of thermal equilibrium between S_1 and S_2 requires the identity of (4.4.2) and (4.4.3), which, in turn, requires

$$\beta_1 = \beta_2 = \beta \tag{4.4.4}$$

and jointly,

$$Z = Z_1 Z_2 \tag{4.4.5}$$

or equivalently

$$\ln Z = \ln Z_1 + \ln Z_2 \tag{4.4.6}$$

Equation (4.4.4) tells us that two canonical distributions representing systems in thermal equilibrium must have the same parameter β. On the other hand, thermodynamics tells us that two systems in thermal equilibrium must have the same temperature. Hence we are led to the provisional interpretation that β must be a function of temperature alone*:

$$\beta = \beta(T) \tag{4.4.7}$$

Consider now the partition function Z: it appears as a function of the volume $\mathcal{V}$ and of the temperature T (and also, of course, of the number of particles N, which is kept constant). Equation (4.4.6) shows that $\ln Z$ is an *additive function:* it is the sum of the corresponding functions $\ln Z_i$ of the parts of the system, assumed to be in negligibly small interaction. This property *suggests* that $\ln Z$ is an *extensive quantity*, that is, a quantity proportional to the number N of particles in the system.† Let us see whether we can identify $\ln Z$ with one of the known extensive thermodynamic functions. For clarity, let us represent Z in the form

$$Z = e^{-\beta A} \tag{4.4.8}$$

which introduces a function $A(\beta, \mathcal{V}, N)$. We now note the following relation:

$$\left[\frac{\partial}{\partial \beta}(\beta A)\right]_{V,N} = -\frac{\partial}{\partial \beta} \ln Z = -\frac{\partial}{\partial \beta}\left(\ln \sum_n e^{-\beta E_n}\right)$$
$$= Z^{-1} \sum_n E_n e^{-\beta E_n}$$

where the notation $[\ ]_{VN}$ denotes a partial derivative taken at constant volume and particle number. The right-hand side is nothing other than the internal energy, as defined by Eq. (4.4.1):

$$\left(\frac{\partial \beta A}{\partial \beta}\right)_{V,N} = E \tag{4.4.9}$$

This equation must be compared with the following well-known thermodynamic relation, connecting the free energy $A(T, \mathcal{V}, N)$ to the internal energy E:

$$\left(\frac{\partial (A/T)}{\partial (1/T)}\right)_{V,N} = E \tag{4.4.10}$$

We assumed that β is a function of temperature alone. The two equations above then suggest that the functional form is

$$\beta = \frac{1}{k_B T} \tag{4.4.11}$$

* Actually, thermodynamics tells us that all the intensive quantities are the same for S_1 and S_2. Hence β could depend on several other intensive quantities, such as, for instance, the pressure. The guess (4.4.7) will, however, be confirmed below.
† This is a suggestion, not a proof! See Section 4.7.

where k_B is a constant to be determined later. Moreover, we see that the function $A = -\beta^{-1} \ln Z$ or

$$A(T, \mathcal{V}, N) = -k_B T \ln Z(T, \mathcal{V}, N) \tag{4.4.12}$$

must be interpreted as the *free energy* of the system. Indeed, we found that it is connected to the known *mechanical function E* by the thermodynamic relation (4.4.10).

The only remaining question is the value of the constant k_B. One sees from Eq. (4.4.8) that it must have the dimensions of an entropy (i.e., energy per degree). It must be determined in such a way as to make all thermodynamic relations consistent. It is well known that the determination of a temperature scale in classical thermodynamics is connected (through the concept of the gas thermometer) to the laws of ideal gases. It will be shown in Chapter 5 (Section 5.2) that the equation of state of a classical ideal gas deduced from the canonical ensemble is

$$P\mathcal{V} = Nk_B T \tag{4.4.13}$$

This must be compared with the empirical equation of state

$$P\mathcal{V} = n_M RT \tag{4.4.14}$$

where R is the ideal gas constant ($R = 8.314 \times 10^7$ erg deg^{-1}-mole^{-1}) and n_M the number of moles in the system. Comparing the two equations, and remembering that there are N_A molecules in a mole (N_A being Avogadro's number, $N_A = 6.023 \times 10^{23}$), we find that *Boltzmann's constant* k_B has the value

$$k_B = \frac{R}{N_A} = 1.38054 \times 10^{-16} \text{ erg deg}^{-1} \tag{4.4.15}$$

Let us stress at this point that the temperature always appears naturally in the combination $k_B T$ in all statistical-mechanical expressions. Therefore the symbol β defined in Eq. (4.4.11) is quite widely used instead of, or together with T: it is one of the universally used notations in this field.

Equations (4.4.11), (4.4.12), and (4.4.15) give the final solution of our problem. Indeed, we have succeeded in calculating the free energy $A(T, \mathcal{V}, N)$ as a function of temperature, volume, and number of particles. We also obtained a definition of the temperature that agrees with the thermodynamic properties of this quantity. As was repeatedly stressed before, the temperature is not the average of a microscopic dynamical function. Rather, it appears from Eqs. (4.3.18) and (4.3.26) as a *parameter of the canonical distribution function.* The same remark holds for the free energy. These quantities are, therefore, characteristics of the equilibrium state of the system as a whole: There is no temperature or free energy of a single particle.

The free energy expressed as a function of $T, \mathcal{V}, N$ has the well-known property of being a *thermodynamic potential.* This means that all other thermodynamic functions can be calculated from it, by simple algebraic operations or by differentiations, some of which are recalled in Table 4.4.1. Because of this property, the *partition function appears as the central function of equilibrium statistical mechanics.* We need not make

Table 4.4.1. Expressions of the More Important Thermodynamic Functions in Terms of the Free Energy $A(T, \mathcal{V}, N)$ [or the Free Energy per Particle $a(T, n)$]

Name of Function	Symbol	Thermodynamic Definition
Extensive quantities		
Free energy	$A \equiv Na$	A
Free enthalpy	$G \equiv N\mu$	$A + P\mathcal{V}$
Internal energy	$E \equiv Ne$	$A + TS$
Enthalpy	$H \equiv Nh$	$A + TS + P\mathcal{V}$
Intensive quantities		
Pressure	P	$-\left(\frac{\partial A}{\partial \mathcal{V}}\right)_T = n^2\left(\frac{\partial a}{\partial n}\right)_T$
Entropy per particle	s	$-\left(\frac{\partial a}{\partial T}\right)_n$
Specific heat per particle at constant volume	c_V	$\left(\frac{\partial e}{\partial T}\right)_n$
Isothermal compressibility	χ_T	$-\mathcal{V}^{-1}\left(\frac{\partial \mathcal{V}}{\partial P}\right)_T = \left[n\left(\frac{\partial P}{\partial n}\right)_T\right]^{-1}$

separate calculations for every new thermodynamic function: $Z(T, \mathcal{V}, N)$ contains all the information about the thermodynamical equilibrium properties of the system. In this sense, one can state that the problem of equilibrium thermodynamics is completely solved in principle. Of course, the explicit calculation of the partition function is by no means simple: in general, it poses very difficult mathematical problems. These are, however, "technical" difficulties, not difficulties of principle, because we know in every case what we want to calculate.

To conclude, we note that we could have done all the calculations of this section in the classical case, starting from Eq. (4.3.25) instead of (4.3.16). All the results, and in particular the fundamental formula (4.4.12), are valid both in classical and in quantum mechanics.

4.5. THE GRAND CANONICAL ENSEMBLE

In deriving the canonical ensemble, we considered the distribution of a subsystem that is in thermal contact with a large isolated system, and hence is capable of exchanging energy with the external world. The number of particles in the subsystem was considered to be constant. We now go one step further by assuming that the subsystem can also exchange matter with the surrounding world; hence the number of particles is no longer constant.

It is not difficult to generalize the formal procedure of the previous section in order to treat this problem. We consider a large isolated system having energy E_U and containing N_U particles. We now ask what is the probability of finding the subsystem, containing N particles, in the quantum state m, having an energy E_{mN}.* The answer is similar to Eq. (4.3.10):

$$p_{mN} = \frac{\Omega_W(E_U - E_{mN}, N_U - N; \Delta E)}{\Omega_U(E_U, N_U; \Delta E)} \tag{4.5.1}$$

The only difference with (4.3.10) is that we now consider functions of two variables, E_{mN}, N. Again, if the subsystem is small, that is:

$$\begin{aligned} E_{mN} &\ll E_U \\ N &\ll N_U \end{aligned} \tag{4.5.2}$$

we may expand p_{mN} (or rather its logarithm) in powers of E_{mN}, N and keep only the linear terms:

$$\begin{aligned} \ln \Omega_W(E_U - E_{mN}, N_U - N) \approx \ln \Omega_W(E_U, N_U) &- \left(\frac{\partial \ln \Omega_W}{\partial E}\right)_{E=E_U, N=N_U} E_{mN} \\ &- \left(\frac{\partial \ln \Omega_W}{\partial N}\right)_{E=E_U, N=N_U} N \end{aligned} \tag{4.5.3}$$

The second term is identical to the one appearing in (4.3.11) and is related in the same way to $\beta = (k_B T)^{-1}$. For the third term we introduce a new symbol μ, to be interpreted later:

$$\left(\frac{\partial \ln \Omega_W}{\partial N}\right)_{E=E_U, N=N_U} = -\beta\mu \tag{4.5.4}$$

Going through calculations similar to those of Section 4.3, we find

$$p_{mN} = \Xi^{-1} e^{-\beta E_{mN}} e^{\beta \mu N} \tag{4.5.5}$$

The constant Ξ is again determined by a normalization condition, which

* Remember that the eigenvalue E_{mN} depends, of course, on the number of particles.

must, however, be generalized. As the number of particles is variable, we must now sum over all possible values of N in averaging:

$$\sum_{N=0}^{\infty} \sum_{m} p_{mN} = 1 \tag{4.5.6}$$

from which follows

$$\Xi \equiv \Xi(\beta, \mathscr{V}, \mu) = \sum_{N=0}^{\infty} \sum_{m} e^{-\beta E_{mN} + \beta\mu N} \tag{4.5.7}$$

The function Ξ, which depends on the parameters β, μ, and (implicitly) $\mathscr{V}$, is a generalization of the partition function that plays in the grand canonical ensemble the same role as Z plays in the canonical ensemble. It is called the *grand partition function.*

We now note that the probability coefficients p_{mN}, Eq. (4.5.5), are the diagonal matrix elements of the density matrix in a representation in which the Hamiltonian *and* the total particle-number operator are diagonal. It must indeed be clearly noted that N is now considered as an operator having eigenvalues equal to all nonnegative integers. The second quantization formalism is particularly well adapted to the problems studied in the grand canonical ensemble. The density matrix can easily be transformed into an expression valid in any arbitrary representation:

$$\boldsymbol{\rho}^{(g)} = \Xi^{-1} \exp\{-\beta \mathsf{H} + \beta\mu \mathsf{N}\} \tag{4.5.8}$$

where H is the Hamiltonian and N is the total number operator. This density matrix defines quite generally the *grand canonical ensemble.* The grand partition function now takes the intrinsic form:

$$\Xi(\beta, \mathscr{V}, \mu) = \mathscr{T}\!r \exp(-\beta \mathsf{H} + \beta\mu \mathsf{N}) \tag{4.5.9}$$

where the symbol $\mathscr{T}\!r$ means that the trace must also include a summation over the eigenvalues of the operator N.

The derivation of the classical limit is similar to the procedure of Section 4.3: we just give the results. The grand canonical distribution function $F_N^{(g)}(q, p)$ gives the probability density of finding N particles at the specified positions in the $2sN$-dimensional phase space corresponding to this system:

$$F_N^{(g)}(q, p) = h^{-sN}(N!)^{-1}\Xi^{-1} \exp[-\beta H_N(q, p) + \beta\mu N] \tag{4.5.10}$$

where $H_N(q, p)$ is the Hamiltonian of the N-particle system.

The classical grand partition function is

$$\Xi = \sum_{N=0}^{\infty} h^{-sN}(N!)^{-1} e^{\beta\mu N} \int dq\, dp \exp[-\beta H_N(q, p)] \tag{4.5.11}$$

One recognizes the characteristic factors $h^{-sN}(N!)^{-1}$ that ensure the correct transition from quantum to classical mechanics.

We now study the relations between the grand canonical ensemble and thermodynamics. The grand canonical ensemble represents an *open system*, in which the number of particles can vary. We recall that the equilibrium condition between two open systems requires, besides the equality of temperatures, the equality of *chemical potentials* μ. The latter is an intensive quantity defined as

$$\mu = \left(\frac{\partial A}{\partial \langle N \rangle}\right)_{T\mathcal{V}} \tag{4.5.12}$$

We denote now by $\langle N \rangle$ the *observable* number of particles, in order to distinguish it from the microscopic variable N.

The four best-known thermodynamic potentials, E, H, A, G correspond to a description in terms of the variables $(S, \mathcal{V}, \langle N \rangle)$, $(S, P, \langle N \rangle)$, $(T, \mathcal{V}, \langle N \rangle)$, and $(T, P, \langle N \rangle)$, respectively. Nothing prevents us from using a description in which one replaces the extensive variable $\langle N \rangle$ by the conjugate intensive variable μ. In order to construct a thermodynamic potential in the variables $(T, \mathcal{V}, \mu)$, we proceed in the same way as in the derivation, say, of the free enthalpy $G(T, P, \langle N \rangle)$ from the free energy $A(T, \mathcal{V}, \langle N \rangle)$, that is, by a "Legendre transformation." We recall that the total differential of $A(T, \mathcal{V}, \langle N \rangle)$ is expressed as

$$dA = -S\,dT - P\,d\mathcal{V} + \mu\,d\langle N \rangle \tag{4.5.13}$$

We introduce the new potential $J \equiv (-A + \mu\langle N \rangle)$. Its total differential is

$$dJ \equiv d(-A + \mu\langle N \rangle) = S\,dT + P\,d\mathcal{V} + \langle N \rangle\,d\mu \tag{4.5.14}$$

Hence J is the thermodynamic potential corresponding to the independent variables $(T, \mathcal{V}, \mu)$:

$$J(T, \mathcal{V}, \mu) = -A(T, \mathcal{V}, \mu) + \mu\,\langle N \rangle(T, \mathcal{V}, \mu) \tag{4.5.15}$$

(Note the important fact that the observable number of particles is considered now as a function of T, $\mathcal{V}$, μ.) From (4.5.14) follow the relations

$$\begin{aligned} \left(\frac{\partial J}{\partial T}\right)_{V,\mu} &= S \\ \left(\frac{\partial J}{\partial \mathcal{V}}\right)_{T,\mu} &= P \\ \left(\frac{\partial J}{\partial \mu}\right)_{T,V} &= \langle N \rangle \end{aligned} \tag{4.5.16}$$

We also note that $\mu\langle N \rangle$ is nothing other than the free enthalpy G:

$$\mu\langle N \rangle = G = A + P\mathcal{V} \tag{4.5.17}$$

Hence*

$$J(T, \mathcal{V}, \mu) = G - A = \mathcal{V}P(T, \mu) \tag{4.5.18}$$

The thermodynamic potential J has, therefore, a very simple physical meaning.

Let us now come back to the grand canonical ensemble. The "mechanical" quantities have an obvious definition. The internal energy is now

$$E(\beta, \mathcal{V}, \mu) = \sum_{N=0}^{\infty} \sum_{m} E_{mN} \Xi^{-1} e^{-\beta E_{mN} + \beta\mu N} \tag{4.5.19}$$

and similarly the observable number of particles $\langle N \rangle$:

$$\langle N \rangle(\beta, \mathcal{V}, \mu) = \sum_{N=0}^{\infty} \sum_{m} N \Xi^{-1} e^{-\beta E_{mN} + \beta\mu N} \tag{4.5.20}$$

The argument necessary for the definition of the "thermal" functions is a straightforward generalization of that of Section 4.4. We consider again the equilibrium of two subsystems S_1 and S_2 (which can exchange matter and energy) and find that not only $\beta_1 = \beta_2$ but also $\mu_1 = \mu_2$. Next, we find that $\ln \Xi$ is an extensive quantity, and hence is (probably) connected to the thermodynamic potential $J(T, \mathcal{V}, \mu)$. To make the link explicit we note that

$$\begin{aligned} \frac{\partial}{\partial \mu} (\beta^{-1} \ln \Xi) &= \frac{1}{\beta} \frac{\partial}{\partial \mu} \ln\left(\sum_N \sum_m e^{-\beta E_{mN} + \beta\mu N} \right) \\ &= \Xi^{-1} \sum_N \sum_m N e^{-\beta E_{mN} + \beta\mu N} = \langle N \rangle \end{aligned}$$

Comparing with the third equation (4.5.16) we see that μ can be identified with the chemical potential, and that the thermodynamic potential $J(T, \mathcal{V}, \mu)$ is given by

$$J(T, \mathcal{V}, \mu) \equiv \mathcal{V}P(T, \mu) = k_B T \ln \Xi(T, \mathcal{V}, \mu) \tag{4.5.21}$$

This is the basic formula of the grand canonical ensemble, analogous to Eq. (4.4.12). Its importance comes again from the fact that $J(T, \mathcal{V}, \mu)$ is a thermodynamic potential, from which all other thermodynamic quantities can be calculated very simply [see Table 4.5.1].

We note again that all the calculations could have been done classically, with the same results. Equation (4.5.21) is valid in both cases.

Finally, we may note that in some calculations it is convenient to use, instead of the chemical potential, a variable z, called the *fugacity*, which is simply related to μ:

$$z = e^{\beta\mu} \tag{4.5.22}$$

* Note that the pressure P, being an intensive quantity, cannot depend on the volume, which is extensive. It can only depend on the intensive variables T and μ.

Table 4.5.1. Expressions of the More Important Thermodynamic Functions in Terms of the Thermodynamic Potential $J(T, \mathcal{V}, \mu)$ or of the Pressure $P(T, \mu)$

Name of Function	Symbol	Thermodynamic Definition
Extensive quantities		
Thermodynamic potential $J(T, \mathcal{V}, \mu)$	$J = \mathcal{V}P$	J
Free energy	$A = \mathcal{V}\tilde{a}$	$\langle N\rangle\mu - J$
Free enthalpy	G	$J + A = \langle N\rangle\mu$
Internal energy	$E = \mathcal{V}\tilde{e}$	$\langle N\rangle\mu + TS - J$
Enthalpy	$H = \mathcal{V}\tilde{h}$	$\langle N\rangle\mu + TS$
Intensive quantities		
Number density	$n = \frac{\langle N\rangle}{\mathcal{V}}$	$\left(\frac{\partial P}{\partial \mu}\right)_T$
Pressure	P	$\frac{J}{\mathcal{V}}$
Entropy density	$\tilde{s} = \mathcal{V}^{-1}S$	$\left(\frac{\partial P}{\partial T}\right)_\mu$
Isothermal compressibility	χ_T	$-\mathcal{V}^{-1}\left(\frac{\partial \mathcal{V}}{\partial P}\right)_T = n^{-2}\left(\frac{\partial^2 P}{\partial \mu^2}\right)_T$

If Ξ is expressed in terms of $\mathfrak{z}$, one clearly sees an elegant connection between the grand partition function and the canonical partition function. Indeed, comparing (4.5.7) and (4.3.16) [or (4.5.11) and (4.3.26)] we can write

$$\Xi(T, \mathcal{V}, \mathfrak{z}) = \sum_{N=0}^{\infty} \mathfrak{z}^N Z(T, \mathcal{V}, N) \tag{4.5.23}$$

Hence, the grand partition function is a *generating function* for the canonical partition function. In other words, $N!\, Z(T, \mathcal{V}, N)$ is the coefficient of $\mathfrak{z}^N$ in the Taylor expansion of Ξ with respect to $\mathfrak{z}$.

4.6. EQUIVALENCE OF THE EQUILIBRIUM ENSEMBLES: FLUCTUATIONS

We constructed three different ensembles describing systems in thermal equilibrium. From their derivation, it appears that they correspond to different, well-specified conditions on the type of system they describe (given energy, given temperature, or given chemical potential). We show now that, actually, we may forget about these specifications, because the results obtained by calculating any thermodynamic quantity in the three formalisms are very nearly the same. This result is very important in practice. It allows us, in many cases, to use in a given problem interchangeably one or the other ensemble, the choice being motivated by practical convenience in the calculations.

The origin of this property can be understood qualitatively as follows. The microcanonical ensemble describes a system whose energy H has a fixed value (within limits that, classically, can be made arbitrarily narrow). In a system described by the canonical ensemble, the energy can take any value; only its average $\langle H\rangle$ is fixed. However, it will be shown that, if the system is sufficiently large, the probability of finding a member of the ensemble having a value of the energy significantly different from the average $\langle H\rangle$ is negligibly small. In other words, the overwhelming majority of the members of a canonical ensemble are clustered (in energy) near the average. Hence, the difference between a microcanonical and a canonical ensemble, as far as average values are concerned, fades out when the system is large. Similar arguments apply for the grand canonical ensemble.

To support this qualitative argument, it will be shown that the fluctuations of the energy, calculated in the canonical ensemble, are very small, and therefore the dispersion around the average is, indeed, negligible relatively to the average. Rigorous formulations of the equivalence problem have only been developed quite recently, and will be reviewed in Section 4.7.

We start the calculation of the energy fluctuation in a classical canonical ensemble from Eq. (4.3.26).* The normalization condition can be written [using Eq. (4.4.8)] as

$$h^{-sN}(N!)^{-1}\int dq\,dp\;e^{\beta A-\beta H(q,p)}=1 \tag{4.6.1}$$

from which we obtain, by differentiation with respect to β,

$$h^{-sN}(N!)^{-1}\int dq\,dp\;e^{\beta A-\beta H(q,p)}\left(\frac{\partial}{\partial\beta}(\beta A)-H(q,p)\right)=0 \tag{4.6.2}$$

Noting that

$$E(T,\mathcal{V},N)=h^{-sN}(N!)^{-1}\int dq\,dp\;H(q,p)\,e^{\beta A-\beta H(q,p)} \tag{4.6.3}$$

we see that Eq. (4.6.2) is equivalent to Eq. (4.4.9) giving the relation between internal energy and free energy. We now differentiate both sides

* For a change, the explicit calculations in this section will be done for classical systems. But the same results can be obtained in quantum statistics.

of (4.6.2) with respect to β:

$$h^{-sN}(N!)^{-1}\int dq\,dp\,e^{\beta A-\beta H}\left\{\left[\frac{\partial}{\partial\beta}(\beta A)-H\right]^2+\frac{\partial^2}{\partial\beta^2}(\beta A)\right\}=0 \quad (4.6.4)$$

Using (4.6.3) this gives the value of the mean-square deviation of the energy from the average:

$$\langle(H-E)^2\rangle \equiv \langle H^2\rangle - E^2 = -\frac{\partial^2}{\partial\beta^2}(\beta A) = -\frac{\partial}{\partial\beta}E$$

We note that the β derivative is taken at constant volume, and introduce the *heat capacity at constant volume* C_V:

$$C_V = \left(\frac{\partial E}{\partial T}\right)_V \quad (4.6.5)$$

We then obtain

$$\langle H^2\rangle - E^2 = k_B T^2 C_V \quad (4.6.6)$$

This is a fundamental formula. Its beauty lies in the fact that it relates the fluctuations of a microscopic quantity to an easily measurable macroscopic quantity, the *heat capacity*. It is intuitively reasonable that a medium with large heat capacity is favorable to local accumulations of energy in one region at the expense of the neighboring ones.

Equation (4.6.6) also contains the proof of our initial statement. Indeed, C_V is an extensive quantity, that is, proportional to N, whereas T is an intensive variable, independent of N. Hence,

$$\langle H^2\rangle - E^2 \sim N \quad (4.6.7)$$

The relative importance of the fluctuations is measured by the ratio of the square root of $\langle(\Delta H)^2\rangle$ to $\langle H\rangle$. As the internal energy is also an extensive variable, we obtain

$$\frac{\langle(H-E)^2\rangle^{1/2}}{E} \sim \frac{N^{1/2}}{N} = N^{-1/2} \quad (4.6.8)$$

This formula shows that the fluctuations, although very large in absolute value ($\sim N^{1/2}$) are nevertheless negligible compared to the much larger value of the mean energy ($\sim N$). In the thermodynamic limit this ratio vanishes. Hence, for all practical purposes, the probability of finding an energy value significantly different from E among the systems making up the canonical ensemble is null.

We now study the grand canonical ensemble. Its equivalence with the canonical ensemble will be established by showing that the fluctuation of the number of particles around the mean is negligibly small. The calculation is quite analogous to the previous one. Starting from the normalization condition

$$\sum_{N=0}^{\infty}\int dq\,dp\,F_N^{(g)}(q,p)=\sum_{N=0}^{\infty}h^{-sN}(N!)^{-1}\Xi^{-1}\int dq\,dp\,e^{-\beta H+\beta\mu N}=1 \quad (4.6.9)$$

and differentiating it twice with respect to μ we obtain the relation

$$\langle (N-\langle N\rangle)^2\rangle=\beta^{-1}\left(\frac{\partial^2 J}{\partial\mu^2}\right)_{VT}=k_B T\left(\frac{\partial\langle N\rangle}{\partial\mu}\right)_{VT} \quad (4.6.10)$$

where we made use of Eq. (4.5.16). From this formula, it is already apparent that $\langle (N-\langle N\rangle)^2\rangle$ is of order $\langle N\rangle$, and hence that the relative fluctuation of N is proportional to $\langle N\rangle^{-1/2}$ and can be neglected as $\langle N\rangle\to\infty$. This proves our statement.

We can, by a simple thermodynamical calculation, express this equation in terms of more familiar quantities.

Indeed,

$$\left(\frac{\partial\langle N\rangle}{\partial\mu}\right)_{VT}=\left(\frac{\partial\langle N\rangle}{\partial P}\right)_{VT}\left(\frac{\partial P}{\partial\mu}\right)_{VT}=\left(\frac{\partial\langle N\rangle}{\partial P}\right)_{VT}\left(\frac{\partial\langle N\rangle}{\partial\mathcal{V}}\right)_{T\mu}$$

the last step following from (4.5.14). We now note that the *extensive* variable $\langle N\rangle$, when expressed in terms of $\mathcal{V}$, T, μ can only have the form

$$\langle N\rangle=\mathcal{V}f(T,\mu)$$

where $f(T,\mu)$ is some function of the intensive variables T, μ. It then follows that $\partial\langle N\rangle/\partial\mathcal{V}=\langle N\rangle/\mathcal{V}$, hence

$$\left(\frac{\partial\langle N\rangle}{\partial\mu}\right)_{VT}=\frac{\langle N\rangle}{\mathcal{V}}\left(\frac{\partial\langle N\rangle}{\partial P}\right)_{VT}=\langle N\rangle\left(\frac{\partial}{\partial P}\frac{\langle N\rangle}{\mathcal{V}}\right)_{VT}$$

If we now consider $\langle N\rangle$ as a function of the intensive variables P, T it must have the form

$$\langle N\rangle=\mathcal{V}g(T,P)$$

Therefore, $\langle N\rangle/\mathcal{V}$ is independent of both $\mathcal{V}$ and $\langle N\rangle$; the derivative at constant $\mathcal{V}$ can be trivially replaced by a derivative at constant $\langle N\rangle$:

$$\left(\frac{\partial\langle N\rangle}{\partial\mu}\right)_{VT}=\langle N\rangle\left(\frac{\partial}{\partial P}\frac{\langle N\rangle}{\mathcal{V}}\right)_{\langle N\rangle,T}$$

$$=\langle N\rangle^2\left(\frac{\partial\mathcal{V}^{-1}}{\partial P}\right)_{\langle N\rangle T}=-\frac{\langle N\rangle^2}{\mathcal{V}^2}\left(\frac{\partial\mathcal{V}}{\partial P}\right)_{\langle N\rangle T}$$

We thus get the beautiful formula, analogous to (4.6.6) (see Table 4.5.1)

$$\frac{\langle N^2\rangle-\langle N\rangle^2}{\langle N\rangle}=-k_B T\frac{\langle N\rangle}{\mathcal{V}^2}\left(\frac{\partial\mathcal{V}}{\partial P}\right)_{\langle N\rangle T}=k_B T n\chi_T \quad (4.6.11)$$

The fluctuation of the number of particles is thus related to the *isothermal compressibility* of the system: a very compressible system (say, a dilute gas) is more favorable to large fluctuations than a solid, say.

We note that, for all stable systems, $(\partial \mathcal{V}/\partial P)_T$ is a negative quantity, which ensures a positive sign for the compressibility χ_T and for the mean-square deviation of $\langle N \rangle$. A very interesting situation occurs near the critical point of a gas–liquid phase transition. At this point $(\partial P/\partial \mathcal{V})_T$ vanishes. It follows that the density fluctuations become extremely large in the critical region. This fact manifests itself in the appearance of the well-known phenomenon of critical opalescence. We shall study the critical phenomena in more detail later (Chapters 9 and 10).

We conclude with a note of caution. It might seem that the problem of fluctuations is trivial because of the smallness of the phenomenon. We have actually proved here that the fluctuation of *total* energy or *total* number of particles is negligible: This result is fundamental for the justification of the methods of statistical mechanics. However, this by no means excludes the possibility of important *local* fluctuations, in *small* regions of the system of interest. These local fluctuations actually play a key role in the explanation of many important physical phenomena (scattering of light, plasma oscillations, etc.). We shall return to the discussion of these questions later.

*4.7. DYNAMICS AND THERMODYNAMICS: EXISTENCE OF THE THERMODYNAMIC LIMIT

We stressed repeatedly in this chapter the fact that the validity of the equilibrium ensembles as models of thermodynamics is not automatically granted, but depends crucially on the nature of the Hamiltonian. We now discuss this relation more closely, limiting ourselves to classical mechanics, and more particularly to the canonical ensemble. The key equation in this ensemble is Eq. (4.4.12). It was shown in Section 4.4 that the function $A(T, \mathcal{V}, N)$ has formal properties that allow us to identify it with the thermodynamic free energy, *provided it exists*! The fact that the problem of the existence arises is connected with our having repeatedly involved the thermodynamic limit for the empirical justification of many steps in the argument. The final rigorous justification of equilibrium statistical mechanics therefore rests on an *a posteriori* proof that the function $A(T, \mathcal{V}, N)$ exists in the T-limit. More precisely, we have to prove that $A(T, \mathcal{V}, N)$ is an extensive function or, equivalently, that the free-energy density $\tilde{a} = A/\mathcal{V}$ is finite in the T-limit (3.3.1), and therefore depends only on the density $n = N/\mathcal{V}$ (and on the temperature):

$$\lim_{\mathcal{V}\to\infty} \mathcal{V}^{-1} A(T, \mathcal{V}, n\mathcal{V}) = \lim_{\mathcal{V}\to\infty} [-\mathcal{V}^{-1} k_B T \ln Z(T, \mathcal{V}, n\mathcal{V})] = \tilde{a}(T, n) \tag{4.7.1}$$

Alternatively, one may consider the free energy per particle:

$$\lim_{N\to\infty} N^{-1} A\left(T, \frac{N}{n}, N\right) = -\lim_{N\to\infty} N^{-1} k_B T \ln Z\left(T, \frac{N}{n}, N\right)$$
$$= a(T, n) = n^{-1}\tilde{a}(T, n) \tag{4.7.2}$$

where again, $a(T, n)$ is a finite function. (In the subsequent equations, we omit writing explicitly the dependence of A on the temperature, which is irrelevant in this argument.)

In order to illustrate the arguments involved in such a proof, we consider again our system defined by Eqs. (2.4.1)–(2.4.5) with $H^F \equiv 0$. We first note that, because of Eqs. (2.4.2), (2.4.3), the contribution of the kinetic energy to Z is very simple. Indeed, in Eq. (4.3.25) the momentum integrations can be performed trivially. We may then write

$$Z(\mathcal{V}, N) = \Lambda^{-3N} Q'(\mathcal{V}, N) \tag{4.7.3}$$

where Λ is a number independent of $\mathcal{V}$ or N, defined as

$$\Lambda^{-1} = h^{-1} \int_{-\infty}^{\infty} dp \exp\left(-\left(\frac{\beta}{2m}\right) p^2\right) = \left(\frac{2\pi m}{\beta h^2}\right)^{1/2} \tag{4.7.4}$$

This number will be discussed in detail in Chapter 5. The function $Q'(\mathcal{V}, N)$, due to the interactions, is the truly nontrivial part of the partition function:

$$Q'(\mathcal{V}, N) = (N!)^{-1} \int_{\mathcal{V}} d\mathbf{q}_1 \cdots d\mathbf{q}_N \exp[-\beta H'(\mathbf{q}_1, \ldots, \mathbf{q}_N)] \tag{4.7.5}$$

It is obvious that $\tilde{a}(n)$ will exist only if the interaction energy H' satisfies restrictive conditions ensuring a "good behavior" of the integral (4.7.5). In order to understand these conditions we discuss again our model potential (4.3.5). In Eq. (4.3.6) we found an upper bound to the absolute value of the interaction energy of $N(= n\mathcal{V})$ particles in a volume $\mathcal{V}$. It is obvious from the argument that if the sign is taken into account, the interactions involved being attractive, the inequality (4.3.6) implies a lower bound on $H'_{\mathcal{V}}$:

$$H'(\mathbf{q}_1, \ldots, \mathbf{q}_N) \geqslant -N w_A \qquad \text{for all } \mathbf{q}_1, \ldots, \mathbf{q}_N \tag{4.7.6}$$

where, in our case, $w_A = \alpha V_0 R_0^3 / d_0^3$, that is, a finite number, independent of N or $\mathcal{V}$. This inequality is called *the stability condition on the interaction potential.* The important feature of this lower bound is its proportionality to the number N. Physically, the stability condition prevents the collapse of the system, that is, the possibility for the particles to "fall on each other" with the result $H' = -\infty$, which would clearly lead to a divergent integral Q'. In our model, the stability is brought about by the presence of a hard core in the pair potential.

In order to get a second condition, we consider that the N particles are divided in two groups of N_1 and N_2 particles, respectively, whose coordinates are spatially separated, and we define the *mutual potential energy* of the two groups $\Phi_{N_1 N_2}$ *as*

$$\begin{aligned}\Phi_{N_1N_2}&(\mathbf{q}_1, \ldots, \mathbf{q}_{N_1}; \mathbf{q}'_1, \ldots, \mathbf{q}'_{N_2}) \\ &= H'(\mathbf{q}_1, \ldots, \mathbf{q}_{N_1}; \mathbf{q}'_1, \ldots, \mathbf{q}'_{N_2}) - H'(\mathbf{q}_1, \ldots, \mathbf{q}_{N_1}) - H'(\mathbf{q}'_1, \ldots, \mathbf{q}'_{N_2})\end{aligned} \tag{4.7.7}$$

Returning to our model, we consider two groups of N_1 and N_2 particles, such that no two particles belonging to different groups are at a distance less than d_0. It is then obvious that the interaction between the two groups is purely attractive:

$$\Phi_{N_1N_2}(\mathbf{q}_1, \ldots, \mathbf{q}_{N_1}; \mathbf{q}'_1, \ldots, \mathbf{q}'_{N_2}) \leqslant 0 \tag{4.7.8}$$

for all $|\mathbf{q}_i - \mathbf{q}'_j| \geq R$, where R is a fixed number, $R \geq d_0$. Inequality (4.7.8) defines a *strongly tempered potential.* If it were not satisfied, because of the existence of sufficiently strong repulsive interactions at long distances, the system would "explode": the interactions would not be strong enough to keep the particles together, and the resulting free energy would diverge to $+\infty$. We now show that *the two conditions* (4.7.6), (4.7.8) *are sufficient for the existence of the free-energy density.*

We first note that if in the integral Q' we substitute for H' its lower bound from (4.7.7), we find

$$Z(\mathcal{V}, N) \leq \Lambda^{-3N}(N!)^{-1}\int_{\mathcal{V}} d\mathbf{q}_1 \cdots d\mathbf{q}_N \, e^{+\beta w_A N}$$

$$= \Lambda^{-3N}(N!)^{-1}\mathcal{V}^N e^{\beta w_A N} \tag{4.7.9}$$

We now take the logarithm and change the sign; we also note the following inequality (Stirling's theorem):

$$\ln N! \geq N \ln N - N$$

We then obtain

$$-\mathcal{V}^{-1}k_B T \ln Z(\mathcal{V}, N) \geq k_B T\mathcal{V}^{-1}(3N \ln \Lambda + N \ln N - N - N \ln \mathcal{V} - \beta N w_A)$$

which implies

$$\mathcal{V}^{-1}A(\mathcal{V}, n\mathcal{V}) \geq k_B T(\ln \Lambda^3 n - 1 - \beta w_A) \tag{4.7.10}$$

Hence the stability condition implies that the left-hand side is bounded below by a quantity independent of $\mathcal{V}$ and is therefore finite in the T-limit.

Next, we consider two subdomains Ω_1, Ω_2, of volumes $\mathcal{V}_1$, $\mathcal{V}_2$, separated by distances at least $R \geq d_0$.* If we reduce the volume of integration in (4.7.5) from $\mathcal{V}$ to $\mathcal{V}_1 + \mathcal{V}_2 < \mathcal{V}$, the result will be smaller than Z because the integrand is positive.

We now note that the partition function $Z(\mathcal{V}_1 + \mathcal{V}_2, N)$ is made up of a sum of contributions corresponding to: 0 particles in $\mathcal{V}_1$ and N in $\mathcal{V}_2$, 1 particle in $\mathcal{V}_1$ and $N-1$ in $\mathcal{V}_2, \ldots, N_1$ particles in $\mathcal{V}_1$ and $N - N_1$ in $\mathcal{V}_2, \ldots$, each of the corresponding integrals being multiplied by the number of ways one can choose N_1 particles among N:

$$Z(\mathcal{V}, N) \geq \frac{\Lambda^{-3N}}{N!} \sum_{N_1} \frac{N!}{N_1!\,(N-N_1)!} \int_{\mathcal{V}_1} d\mathbf{q}_1 \cdots \int_{\mathcal{V}_1} d\mathbf{q}_{N_1} \int_{\mathcal{V}_2} d\mathbf{q}_{N_1+1} \cdots \int_{\mathcal{V}_2} d\mathbf{q}_{N_1+N_2}$$

$$\times \exp[-\beta H'(\mathbf{q}_1, \ldots, \mathbf{q}_{N_1}) - \beta H'(\mathbf{q}_{N_1+1}, \ldots, \mathbf{q}_{N_1+N_2}) - \beta \Phi_{N_1 N_2}(\mathbf{q}_1, \ldots, \mathbf{q}_{N_1+N_2})] \tag{4.7.11}$$

Since all terms in the sum are positive, the inequality remains true if we keep only one term. We now make use of the strong tempering condition (4.7.8). It implies

* It is clear that $\mathcal{V}_1 + \mathcal{V}_2 < \mathcal{V}$ because of the presence of a corridor.

that $\exp(-\beta\Phi_{N_1N_2}) \geq 1$; hence, the inequality is strengthened as follows:

$$Z(\mathcal{V}, N) \geq \frac{\Lambda^{-3(N_1+N_2)}}{(N_1+N_2)!} \frac{(N_1+N_2)!}{N_1!\,N_2!} \int d\mathbf{q}_1 \cdots d\mathbf{q}_{N_1} \int d\mathbf{q}_{N_1+1} \cdots d\mathbf{q}_{N_1+N_2}$$

$$\times \exp[-\beta H'(\mathbf{q}_1, \ldots, \mathbf{q}_{N_1}) - \beta H'(\mathbf{q}_{N_1+1}, \ldots, \mathbf{q}_{N_1+N_2})]$$

But the integral in the right-hand side now factorizes into two factors that are simply the partition functions corresponding to N_1 particles in $\mathcal{V}_1$ and N_2 particles in $\mathcal{V}_2$:

$$Z(\mathcal{V}, N_1+N_2) \geq Z(\mathcal{V}_1, N_1) Z(\mathcal{V}_2, N_2) \tag{4.7.12}$$

This inequality is basic for the proof of the existence of the free-energy density. It implies

$$\frac{A(\mathcal{V}, n\mathcal{V})}{\mathcal{V}} \leq \frac{\mathcal{V}_1}{\mathcal{V}} \frac{A(\mathcal{V}_1, n_1\mathcal{V}_1)}{\mathcal{V}_1} + \frac{\mathcal{V}_2}{\mathcal{V}} \frac{A(\mathcal{V}_2, n_2\mathcal{V}_2)}{\mathcal{V}_2} \tag{4.7.13}$$

where

$$n = \frac{N}{\mathcal{V}}, \quad n_1 = \frac{N_1}{\mathcal{V}_1}, \quad n_2 = \frac{N_2}{\mathcal{V}_2}$$

so that

$$n\mathcal{V} = n_1\mathcal{V}_1 + n_2\mathcal{V}_2 \tag{4.7.14}$$

This inequality is readily extended to an arbitrary subdivision of the initial domain into m subdomains (with the appropriate corridors):

$$\frac{A(\mathcal{V}, n\mathcal{V})}{\mathcal{V}} \leq \sum_{j=1}^{m} \frac{\mathcal{V}_j}{\mathcal{V}} \frac{A(\mathcal{V}_j, n_j\mathcal{V}_j)}{\mathcal{V}_j} \tag{4.7.15}$$

with

$$n = \sum_{j=1}^{m} \frac{\mathcal{V}_j}{\mathcal{V}} n_j \tag{4.7.16}$$

Having established the two basic lemmas (4.7.10) and (4.7.15) we now go over to the proof of the existence of $\tilde{a}(n)$. To this purpose we construct a sequence of systems of increasing volumes, and prove that it is a class of macroscopically equivalent systems in the sense of Section 3.3. Successive systems in the sequence are enclosed in cubic boxes of side $(s_k + t)$. We assume that

$$s_{k+1} = 2s_k + t \tag{4.7.17}$$

where t is a fixed number, independent of k, and satisfying

$$t > d_0$$

In going from one system to the next, we put $2^3 = 8$ boxes of side s_k together, leaving corridors of width t between them (see Fig. 4.7.1).

We now assume that the $N_{k+1} = 8N_k$ particles of the box of side $s_{k+1} + t$ are equally divided among the 8 boxes of side s_k, with no particles in the corridors.

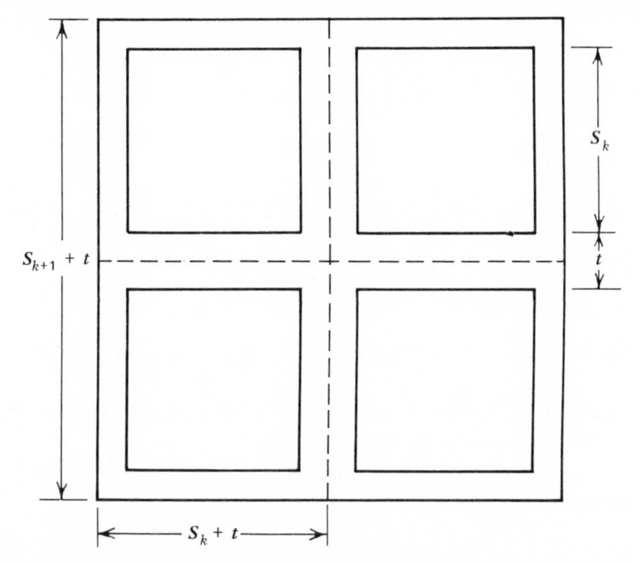

Figure 4.7.1. Geometrical construction for the proof of the existence of the free-energy density.

Then the basic inequality (4.7.15) tells us that

$$\frac{A[(s_{k+1}+t)^3, N_{k+1}]}{(s_{k+1}+t)^3} \leqslant \sum_{i=1}^{8} \frac{(s_k+t)^3}{(s_{k+1}+t)^3} \frac{A[(s_k+t)^3, N_k^{(i)}]}{(s_k+t)^3}$$

$$= 8\frac{(s_k+t)^3}{(2s_k+2t)^3} \frac{A[(s_k+t)^3, N_k]}{(s_k+t)^3} \qquad (4.7.18)$$

Hence

$$\frac{A(\mathscr{V}_{k+1}, n\mathscr{V}_{k+1})}{\mathscr{V}_{k+1}} \leqslant \frac{A(\mathscr{V}_k, n\mathscr{V}_k)}{\mathscr{V}_k} \qquad (4.7.19)$$

Thus, in going from one system to the next in the sequence, the free energy per unit volume can only decrease. But we know from Eq. (4.7.10) that this function has a lower bound, independent of the volume. Hence $A(\mathscr{V}, n\mathscr{V})/\mathscr{V}$ tends toward a finite limit as $\mathscr{V} \to \infty$: This limit is to be identified with the macroscopic free-energy density $\tilde{a}(n)$ and thus Eq. (4.7.1) is proved.

A simple modification of the argument gives us an extra bonus. Assuming that the particles are unevenly distributed among the cells, say N_{1k} particles, in half of the cells and N_{2k} in the other half, we easily derive, in the T-limit, the inequality

$$\tilde{a}(\tfrac{1}{2}n_1 + \tfrac{1}{2}n_2) \leqslant \tfrac{1}{2}\tilde{a}(n_1) + \tfrac{1}{2}\tilde{a}(n_2) \qquad (4.7.20)$$

or, more generally

$$\tilde{a}\left(\sum_i \omega_i n_i\right) \leqslant \sum_i \omega_i \tilde{a}(n_i) \qquad \text{with } \sum_i \omega_i = 1 \qquad (4.7.21)$$

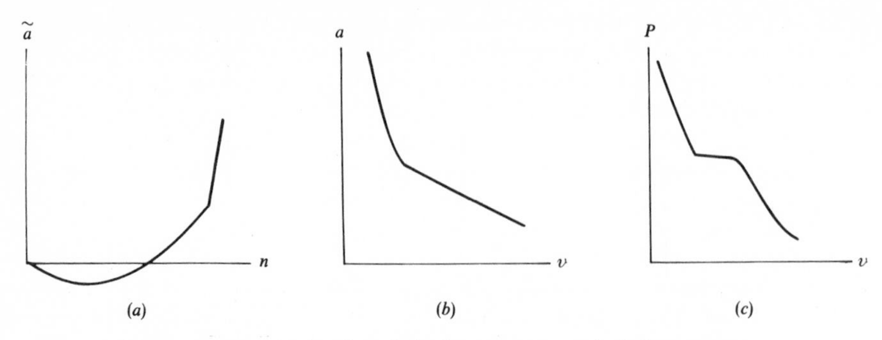

Figure 4.7.2. Behavior of the free energy and of the pressure.

Thus, the *free-energy density is a convex function of the density*. Its graph is therefore of the type shown in Fig. 4.7.2(a).

We now show that the convexity implies the *thermodynamic stability* of the system. It can be shown that

$$\lim_{n\to 0} \tilde{a}(n) = 0 \tag{4.7.22}$$

It then follows from (4.7.21) with $n_1 = 0$, $n_2 = n$, $\omega_2 = \omega \leqslant 1$:

$$\tilde{a}(\omega n) \leqslant \omega \tilde{a}(n) \tag{4.7.23}$$

We now consider the free energy *per particle: a* defined in Eq. (4.7.2), but expressed in terms of the *volume per particle:* $v = \mathcal{V}/N = n^{-1}$. Equation (4.7.23) then implies

$$a(v') \leqslant a(v) \qquad \text{for } v' = \frac{v}{\omega} \geqslant v \tag{4.7.24}$$

Hence the *free energy per particle is a monotonously decreasing function of the volume.* (Such a statement does not hold for $\tilde{a}$ as a function of n.) One can then prove rigorously the differentiability of $a(v)$, hence the pressure $P(v)$ exists and is everywhere positive:

$$P(v) = -\frac{\partial a(v)}{\partial v} \geqslant 0 \tag{4.7.25}$$

It is, moreover, monotonic nonincreasing in v, thus the inverse isothermal compressibility exists and is nonnegative:

$$\chi_T^{-1} = -v\left(\frac{\partial P}{\partial v}\right)_T \geqslant 0 \tag{4.7.26}$$

The stability of the potential (4.7.6) and the strong tempering condition (4.7.7) are sufficient for the proof of the existence of the free-energy density in the thermodynamic limit. The proof can be extended, and it turns out that all the concepts introduced in this chapter, such as the microcanonical ensemble, the grand canonical ensemble, and the equivalence of the three ensembles can be proved rigorously both in classical and in quantum-statistical mechanics. The first work in this direction is due to van Hove in 1949, who considered the classical canonical ensemble for a class of systems rather similar to the model systems of this chapter. Yang and Lee studied the classical grand canonical ensemble in 1952 but gave no proof of equivalence to the canonical ensemble. The main impetus in this field came with the works of Ruelle and Fisher in 1963–1964, followed by many others. Today equilibrium statistical mechanics is privileged as being one of the few domains of physics in which the basic concepts can be defined in full mathematical rigor.

An interesting question arises now: do the real systems satisfy the mathematical conditions of stability and strong tempering? Given a two-body (or many-body) potential, it is not always a trivial matter to verify that the two conditions are indeed satisfied. It can, however, be shown that the potentials describing the interactions of neutral molecules (such as the Lennard-Jones potential, to be amply discussed later) do satisfy these conditions. There is, however, one type of system to which these conditions do not apply: *systems of electrically charged particles.* Given the importance of these systems in nature, this exception appeared to be a serious flaw. How do these systems violate the conditions?

Considering first the stability condition, it is clear that a classical system of charged particles (or a plasma) with pure Coulomb interactions collapses [although Onsager proved already in 1939 that charged particles with (artificial) hard cores satisfy (4.7.6)]. However, it was only in 1967–1968 that Dyson and Lenard produced a rigorous proof that an electrically neutral assembly of charged particles is stable in the sense of Eq. (4.7.6), provided the negative particles and/or the positive particles are fermions.

The strong tempering condition (4.7.7) is the one that is badly violated: particles of the same charge repel each other at all distances. Fisher had shown in 1964 how the strong tempering condition can be relaxed by introducing the *weak tempering condition:*

$$\Phi_{N_1N_2}(\mathbf{q}_1, \ldots, \mathbf{q}_{N_1}; \mathbf{q}'_1, \ldots, \mathbf{q}'_{N_2}) \leqslant \frac{N_1N_2w_B}{R^{3+\varepsilon}} \tag{4.7.27}$$

whenever $|\mathbf{q}_i - \mathbf{q}_j| \geqslant R \geqslant R_0$, where R_0 and w_B are constants. He showed that all the previous results remain valid in this case, which allows for a repulsive tail, provided it decreases fast enough with distance. But the Coulomb forces, which decrease like R^{-1}, are not weakly tempered either. It took a great deal of ingenuity to Lebowitz and Lieb in 1969 to provide the final proof of the existence of the thermodynamic limit for charged particles. The physical clue to this statement is the screening of the interactions existing in a two-component plasma.

Now that the last big gap in the foundations of equilibrium theory is filled, we may proceed with confidence to the study of some of its developments.

APPENDIX: THE METHOD OF THE MOST PROBABLE DISTRIBUTION

We have purposely chosen to present in Section 4.3 a derivation of the canonical ensemble, that is not the simplest. Our object was to stress the fundamental fact that the existence of the canonical ensemble (hence of a thermodynamic description of matter) is subject to specific dynamical restrictions on the form of the Hamiltonian. Our presentation thus prepared the way for the modern rigorous theories sketched in Section 4.7.

As an alternative, we also present here the more traditional, shorter, but only *apparently* simpler derivation, based on the "method of the most probable distribution."

We consider again the isolated "universe" of Section 4.3, but regard it now as a collection of N systems S_i $(i = 1, 2, \ldots, N)$ (see Fig. 4.A.1). The systems S_i are

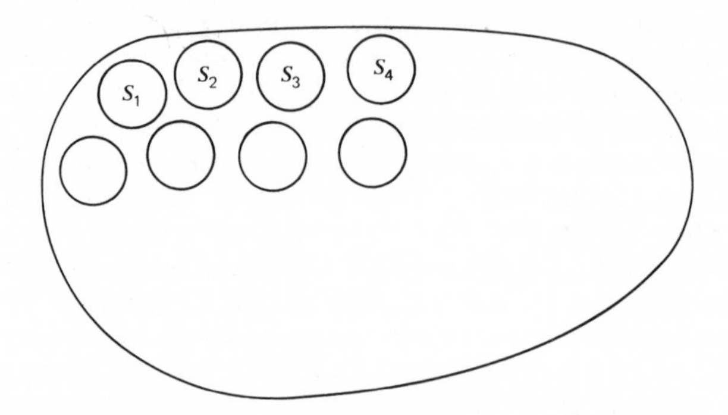

Figure 4.A.1. Representation of the "universe" in the method of the most probable distribution.

assumed dynamically identical: one may think of the universe as a concrete realization of a statistical ensemble according to the definition of Section 2.2. The systems S_i are allowed to interact, that is, to exchange energy, but only *very weakly*. Because of the weakness of the coupling, each system can be characterized by an individual energy, the total energy being the sum of these individual ones:

$$E = \sum_{i=1}^{N} E^{(i)} \tag{4.A.1}$$

More specifically, each system possesses its own spectrum of eigenvalues ε_l (the same spectrum for each of the systems in the set).*

The total energy E can be distributed in many ways among the constituent systems: each of these ways gives rise to a *configuration* described by a set of occupation numbers ν_l, giving the number of systems in the energy level ε_l. A given configuration can be realized in P ways by permutations of the systems:

$$P = \frac{N!}{\nu_1!\,\nu_2!\ldots\nu_l!\ldots} \tag{4.A.2}$$

* Clearly, one cannot escape at this stage the discussion of Section 4.3, in order to ascertain the validity of Eq. (4.A.1), which corresponds to the passage from (4.3.4) to (4.3.8).

The set of occupation numbers must obviously satisfy the conditions that the total number of systems is fixed and that the total energy is fixed (the universe is isolated):

$$\sum_l \nu_l = N \tag{4.A.3}$$

$$\sum_l \nu_l \varepsilon_l = E \tag{4.A.4}$$

the sums being over all the possible eigenstates of a system.

We now admit that the most probable configuration represents an almost certain configuration, in the limit of very large N. In other words, the function P—hence also $\ln P$—has an extremely sharp maximum. In order to find this maximum, subject to the conditions (4.A.3), (4.A.4), we use Lagrange's method, consisting of determining the free extremum of the function:

$$\ln P - \alpha \sum_l \nu_l - \beta \sum_l \nu_l \varepsilon_l \tag{4.A.5}$$

where α and β are Lagrange multipliers, to be determined later.

We introduce an asymptotic element by using throughout Stirling's approximation for the factorials:

$$\ln(n!) \approx n(\ln n - 1)$$

The variation of (4.A.5) then yields the condition

$$-\sum_l \ln \nu_l \, \delta\nu_l - \alpha \sum_l \delta\nu_l - \beta \sum_l \varepsilon_l \, \delta\nu_l = 0$$

from which we obtain

$$\ln \nu_l + \alpha + \beta\varepsilon_l = 0$$

Hence the most probable value for the occupation numbers is

$$\nu_l = e^{-\alpha - \beta\varepsilon_l}$$

The multiplier α can be immediately eliminated by using Condition (4.A.3):

$$\nu_l = \frac{N}{\sum_m e^{-\beta\varepsilon_m}} e^{-\beta\varepsilon_l} \tag{4.A.6}$$

Defining the probability p_l of finding a system (= *any* system) in state l as the relative number of systems in state l in the ensemble, we find

$$p_l = \frac{\nu_l}{N} = \frac{e^{-\beta\varepsilon_l}}{\sum_m e^{-\beta\varepsilon_m}} \tag{4.A.7}$$

which is just the canonical ensemble (4.3.14).

BIBLIOGRAPHICAL NOTES

The great ensembles of equilibrium statistical mechanics (microcanonical, canonical, grand canonical) were introduced by J. W. Gibbs, *Elementary Principles of Statistical Mechanics*, Yale Univ. Press, 1902 (reprinted by Dover, New York, 1960).

We give here a list of the principal general books on statistical mechanics. Most of these deal only, or mainly, with equilibrium statistical mechanics:

R. H. Fowler, *Statistical Mechanics*. Cambridge Univ. Press, Cambridge, 1936.

R. C. Tolman, *Statistical Mechanics*, Clarendon Press, Oxford, 1938.

J. Mayer and M. G. Mayer, *Statistical Mechanics*, Wiley, New York, 1940.

E. Schrödinger, *Statistical Thermodynamics*, Cambridge Univ. Press, Cambridge, 1948.

G. S. Rushbrooke, *Introduction to Statistical Mechanics*, Clarendon Press, Oxford, 1949.

A. I. Khinchine, *Mathematical Foundations of Statistical Mechanics*, Dover, New York, 1949.

L. D. Landau and E. M. Lifshits, *Statisticheskaya Fizika* (English translation: *Statistical Physics*, Pergamon, London, 1958).

A. Sommerfeld, *Vorlesungen über theoretische Physik*, Vol. 5; *Thermodynamik und Statistische Mechanik*, Dieterich, Wiesbaden, 1952 (English translation: *Lectures in Theoretical Physics*, Vol. 5, Academic Press, New York, 1964).

D. ter Haar, *Elements of Statistical Mechanics*, Rinehart, New York, 1954.

T. L. Hill, *Statistical Mechanics*, McGraw-Hill, New York, 1956.

K. Huang, *Statistical Mechanics*, Wiley, New York, 1963.

R. Kubo, *Statistical Mechanics*, an advanced course with problems and solutions, North Holland, Amsterdam, 1965. (This book contains the most extensive set of problems.)

F. Reif, *Statistical Physics*, Berkeley Physics Course, Vol. 5, McGraw-Hill, New York, 1967.

A. Münster, *Statistical Thermodynamics*, Vol. 1, Springer, Berlin, and Academic Press, New York, 1969.

A. Isihara, *Statistical Physics*, Academic Press, New York, 1971.

J. Kestin and J. R. Dorfman, *A Course in Statistical Thermodynamics*, Academic Press, New York, 1971 (with many problems).

R. P. Feynman, *Statistical Mechanics*, Benjamin, Reading, Mass., 1972.

Those who wish to revive their memory in thermodynamics are referred to the following books:

I. Prigogine and R. Defay, *Thermodynamique Chimique*, Desoer, Liège, 1950 (English translation: *Chemical Thermodynamics*, Longmans Green, London, 1954).

A. Sommerfeld (quoted above).

K. Huang (quoted above).

M. W. Zemansky, *Heat and Thermodynamics*, McGraw-Hill, New York, 1951.

P. T. Landsberg, *Thermodynamics with Quantum Statistical Illustrations*, Wiley-Interscience, New York, 1961.

P. H. Badger, *Equilibrium Thermodynamics*, Allyn and Bacon, Boston, 1967.

The semiclassical limit of the partition function was obtained by J. G. Kirkwood, *Phys. Rev.* **44,** 31 (1933).

Rigorous results on the thermodynamic limit (the history and the contents of these papers were discussed in Section 4.7):

L. van Hove, *Physica* **15,** 951 (1949).

C. N. Yang and T. D. Lee, *Phys. Rev.* **87,** 404 (1952).

D. Ruelle, *Helv. Phys. Acta* **36,** 183 (1963).

D. Ruelle, *Statistical Mechanics, Rigorous Results*, Benjamin, New York, 1969.

M. Fisher, *Arch. Rat. Mech. Anal.* **17,** 377 (1964).

L. Onsager, *J. Phys. Chem.* **43,** 189 (1939).

F. J. Dyson and A. Lenard, *J. Math. Phys.* **8,** 423 (1967).

J. Lebowitz and E. Lieb, *Phys. Rev. Lett.* **22,** 631 (1969).

These results are also reviewed by A. Münster (quoted above).

CHAPTER 5

EQUILIBRIUM PROPERTIES OF IDEAL SYSTEMS

5.1. DEFINITION OF IDEAL SYSTEMS

We now begin the application of the general theory developed in the first two chapters to the study of particular systems. Among these, a central importance belongs to a class of systems called *ideal systems*. These are defined by a Hamiltonian of the form

$$H^0 = \sum_{j=1}^{N} H_j^0 \tag{5.1.1}$$

where H_j^0 is a function involving only the coordinates and momenta of a finite (usually small) set of degrees of freedom, the set being denoted by the letter j. In quantum mechanics, H_j^0 is an operator acting only on a finite set j of coordinates of configuration space. A typical example is the case where j denotes a single atom, in this case the set j involves the three degrees of freedom of translation of the center of mass. For polyatomic molecules the degrees of freedom describing rotation and vibration must also be included. In some problems one must also take into account the intra-atomic electronic structure, particularly when excitation, ionization, or other chemical equilibria are studied. To simplify the terminology, we shall call the set j a *particle*, with the understanding that it may involve translational degrees of freedom as well as, possibly, internal ones. It must also be kept in mind that Hamiltonians of the form (5.1.1) may describe systems quite different from material particles (e.g., harmonic oscillators, phonons or excitons in crystals, etc.).

The main property of the Hamiltonian (5.1.1) is its *additive* form. It describes a set of *mutually independent particles*. The motion of each particle is not influenced by the motion of any other particle of the system. The dynamics of the N-body system splits up into N separate one-body dynamics.

The importance of these systems lies first in their simplicity. They are the models that can be treated the most thoroughly (and even in many cases exactly) in statistical mechanics.

On the other hand, it must be realized that ideal systems do not exist in nature. There is even a fundamental inconsistency in taking this concept too seriously. Indeed, we will see later (Chapter 13) that an ideal system, with the Hamiltonian (5.1.1), can never reach thermal equilibrium if it starts from an arbitrary state: the presence of interparticle interactions is indispensable for reaching equilibrium. If, however, the interactions are small in some sense—to be made precise later—then the end point of the time evolution of the system is an equilibrium state [say, described by Eq. (4.3.18)] whose properties are predominantly determined by the ideal part of the Hamiltonian (with small corrections due to the interactions). We therefore find the following fundamental dichotomy:

1. The *equilibrium properties of an ideal system* represent limiting values of the properties of certain real systems under well-defined conditions (to be studied in Chapter 6).
2. *An ideal system has no meaningful transport* (*nonequilibrium*) *properties.*

5.2. IDEAL SYSTEMS IN THE HIGH-TEMPERATURE LIMIT: THE BOLTZMANN GAS

The derivation of the thermodynamic properties of the most interesting natural systems, which can be assimilated to ideal ones, is essentially a quantum-mechanical problem. We will see, for instance, that in a dilute gas of molecules, the most interesting part of the thermodynamic properties comes from the internal degrees of freedom, which can only be treated quantum mechanically. However, in many cases the quantum effects will not manifest themselves. In particular, at sufficiently high temperatures the specific features introduced by the fermion or boson nature of the particles fade away. Consequently the calculations can be significantly simplified, as will be presently shown.

Let us, indeed, consider a system of N independent identical particles; its energy, in a given configuration, is

$$E = \sum_{j=1}^{N} \varepsilon^{(j)} \tag{5.2.1}$$

where $\varepsilon^{(j)}$ is the energy level occupied by particle j. We now wish to calculate the partition function (4.3.16); we are tempted to write*:

$$Z' = \sum_{m^{(1)}} \cdots \sum_{m^{(N)}} \exp\left(-\beta \sum_{j} \varepsilon^{(j)}\right) \tag{5.2.2}$$

Here $m^{(j)}$ denotes a complete set of quantum numbers characterizing the state (or "level") of the *single molecule* j. Typically, $m^{(j)}$ would consist of the components of the momentum of the center of mass, the vibrational and rotational quantum numbers, the spin, and so on. We have in Eq. (5.2.2) N independent summations over all states of each particle. This expression, however, is wrong, because it overcounts the states. Indeed, a given distribution of the particles among the various single-particle states m_k, characterized by the occupation numbers n_k, can be realized in $N!/n_k!\, n_l! \ldots$ ways by permuting the particles among themselves. Because of the undistinguishability principle of quantum mechanics, (Section 1.4) all these configurations are equivalent and have to be counted as a single one. Hence, the correct partition function is

$$Z = \sum_{m^{(1)}} \cdots \sum_{m^{(N)}} \frac{n_k!\, n_l! \ldots}{N!} \exp\left(-\beta \sum_{j} \varepsilon^{(j)}\right) \tag{5.2.3}$$

At low temperature, the dominant contribution to the sum comes from very low energy levels, because for energies much larger than $k_B T \equiv \beta^{-1}$ the exponential factor is negligibly small (see Fig. 5.2.1).

However, for *high temperatures*, the number of available energy levels (which give a significant contribution to the partition function) becomes very large: The N particles therefore tend to distribute themselves mostly on different levels, and the occupation numbers will be predominantly 0 or 1; hence $(n_k!) = 1$ for most configurations. Moreover, the average value of n_k will be much smaller than 1. In such cases, one can forget about the constraints of quantum statistics, because the configurations where the latter are violated are quite exceptional. Hence, we may approximate the partition function by the expression:

$$Z \approx (N!)^{-1} \sum_{m^{(1)}} \cdots \sum_{m^{(N)}} \exp\left(-\beta \sum_{j} \varepsilon^{(j)}\right) \tag{5.2.4}$$

* Note, in connection with the footnote of Section 4.2, p. 111, that in writing Eqs. (5.2.1) and (5.2.2) we use a representation in which not only the total energy of the gas, but also the individual energy of each particle, is diagonal. This is possible because in the ideal gas there are no interactions; hence the system is not ergodic. Each of the $\varepsilon^{(j)}$ is a constant of the motion, and their set is a complete set of commuting observables. In order to construct the equilibrium ensemble, it is assumed that $\hat{\rho}$ is totally diagonal, with respect to all energy quantum numbers, an assumption that is stronger than the one used in Section 4.2.

Each eigenvalue of the total energy is now multiply degenerate: a given total energy can be distributed in many ways among the N particles.

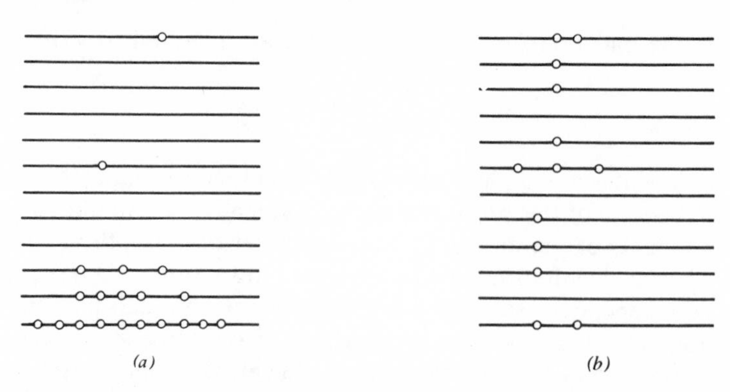

Figure 5.2.1. Typical distribution of bosons among single-particle energy levels: (*a*) low temperature; (*b*) high temperature.

The resulting, much simpler, picture of the gas is called the *Boltzmann approximation* and appears as a common limit of the Bose–Einstein and the Fermi–Dirac statistics, valid at high temperatures. More precise criteria of validity will be derived later. In this approximation, the states are counted as if the particles were distinguishable [i.e., as in classical mechanics or in Eq. (5.2.2)]; but the undistinguishability leaves a trace in the correcting factor $N!$ (see also Section 4.3).

We now note the most important property of the partition function for an ideal system in the Boltzmann approximation. As the summations in Eq. (5.2.4) are carried out independently, and as the exponential of a sum is the product of exponentials, we see that Z is the *product* of N factors:

$$Z = (N!)^{-1} \prod_{j=1}^{N} \left[\sum_{m^{(j)}} \exp(-\beta\varepsilon^{(j)}) \right] \tag{5.2.5}$$

But, as all particles have been assumed to be identical, the N factors are all equal and hence

$$Z = (N!)^{-1} Z_1^N \tag{5.2.6}$$

with

$$Z_1 = \sum_m e^{-\beta\varepsilon} \tag{5.2.7}$$

The summation is now carried out over all levels of a *single* particle.

As a consequence, the free energy of the ideal system is given by

$$\begin{aligned} A(T, \mathcal{V}, N) &= -k_B T \ln Z(T, \mathcal{V}, N) \\ &= k_B T \ln(N!) - N k_B T \ln Z_1(T, \mathcal{V}) \end{aligned} \tag{5.2.8}$$

As the number $N!$ is astronomically large, we can use for its evaluation a well-known asymptotic approximation, known as the *Stirling formula**:

$$\ln(N!) \sim N \ln\left(\frac{N}{e}\right) \tag{5.2.9}$$

Hence

$$A(T, \mathcal{V}, N) = -Nk_BT \ln[eN^{-1}Z_1(T, \mathcal{V})] \tag{5.2.10}$$

We may go one step further in the evaluation of Z_1. We note indeed that, for a molecule, the total energy can always be written as

$$\varepsilon = \varepsilon_{tr} + \varepsilon_i \tag{5.2.11}$$

where ε_{tr} is the energy associated with the center-of-mass motion (translation), and ε_i corresponds to the internal degrees of freedom (rotation, vibration, electronic excitation, etc.). These two terms depend on different variables (i.e., the corresponding Hamiltonians commute); hence the partition function Z_1 can again be factored:

$$Z_1 = Z_{1,tr}Z_i \tag{5.2.12}$$

If we consider a system in the absence of an external field, ε_{tr} is simply the kinetic energy of the center of mass:

$$\varepsilon_{tr} = \frac{p_x^2 + p_y^2 + p_z^2}{2m} \equiv \frac{p^2}{2m} \tag{5.2.13}$$

The allowed values of the momentum components depend on the size and the shape of the box containing the system, as well as on the boundary conditions. However, in the thermodynamic limit, the intensive quantities depend on none of these arguments. Therefore we may assume particularly simple boundary conditions, as they will be anyway irrelevant at the end of the calculation. We therefore choose to enclose the system in a cubic box of side L and volume $\mathcal{V} = L^3$ and assume that the wave function takes the same value on opposite faces of the cube (periodic boundary conditions). In that case it is known from elementary quantum mechanics that the eigenvalues of the momentum are

$$p_{i,n} = \hbar\left(\frac{2\pi}{L}\right)n_i, \qquad n_i = 0, \pm1, \pm2, \ldots \qquad i = x, y, z \tag{5.2.14}$$

The translational partition function is therefore

$$Z_{1,tr} = \sum_{n_x}\sum_{n_y}\sum_{n_z} \exp\left(-\beta\frac{h^2}{2mL^2}(n_x^2 + n_y^2 + n_z^2)\right) \tag{5.2.15}$$

* A simple heuristic derivation is given by Landau and Lifshitz: for large N, approximate the sum $\ln(N!) = \ln 1 + \ln 2 + \cdots + \ln N$ by the integral

$$\int_0^N dx \ln x = N \ln N - N$$

However, in any system of macroscopic size, the spacing between successive momentum levels is exceedingly small compared to the characteristic thermal momentum $(2mk_BT)^{1/2}$:

$$\frac{p_{i,n} - p_{i,n-1}}{(2mk_BT)^{1/2}} = \frac{h}{L(2mk_BT)^{1/2}} \ll 1 \tag{5.2.16}$$

Because of the largeness of L and the smallness of h, this criterion is very well satisfied even at the lowest accessible temperatures. The momentum is practically a continuous variable. The summations over the discrete variables n_i can therefore be replaced by integrations over the continuous variable p_i, provided we take account of the number $D(p_i)\, dp_i$ of momentum states in the interval dp_i. The prescription for this transition is

$$\begin{aligned}\sum_{p_i} f(p_i) &= \sum_{n_i} f(L^{-1}hn_i)\\ &\to \int dp_i\, D(p_i) f(p_i) = \int dp_i \frac{L}{h} f(p_i)\end{aligned} \tag{5.2.17}$$

The partition function (5.2.15) now becomes

$$Z_{1,\text{tr}} = \frac{L^3}{h^3}\int_{-\infty}^{\infty} dp_x\, dp_y\, dp_z \exp\left(-\frac{(p_x^2+p_y^2+p_z^2)}{2mk_BT}\right) \tag{5.2.18}$$

The resulting integrations are very simple, as the integrands are Gaussian functions:

$$Z_{1,\text{tr}} = \mathcal{V}(2\pi mk_BTh^{-2})^{3/2} \equiv \mathcal{V}\Lambda^{-3} \tag{5.2.19}$$

where the function $\Lambda = \Lambda(T)$ is defined by this equation.

Before going further, we note that this expression is in complete agreement with the result we would have obtained, had we used the *classical prescription* for the partition function. Indeed, the classical Hamiltonian describing this problem is independent of q:

$$H(q, p) = \frac{p^2}{2m}$$

Hence Eq. (4.3.25) becomes in this case

$$\begin{aligned}Z_{1,\text{tr}}^{\text{cl}} &= h^{-3}\int_{\mathcal{V}} d^3\mathbf{q}\int_{-\infty}^{\infty} d^3\mathbf{p}\exp\left(-\frac{p^2}{2mk_BT}\right)\\ &= h^{-3}\mathcal{V}(2\pi mk_BT)^{3/2}\end{aligned} \tag{5.2.20}$$

in complete agreement with (5.2.19). This result should not be surprising: considering p as a continuous variable wipes out the quantization. Specific quantum effects in the center-of-mass motion can only be detected if the particle is enclosed in a box of molecular size. In practically all cases of physical interest, the translational degrees of freedom of a macroscopic assembly of molecules can be treated classically.* The main interest of this result is in providing us with a direct and explicit check of the correctness of the correspondence rule between quantum and classical partition functions discussed at the end of Section 4.3.

Using now Eqs. (5.2.19) and (5.2.12) we can obtain a more explicit form of the free energy of a Boltzmann gas (in absence of external fields):

$$A(T, \mathcal{V}, N) = -Nk_BT \ln\left[e\frac{\mathcal{V}}{N}\left(\frac{2\pi mk_BT}{h^2}\right)^{3/2} Z_i\right] \tag{5.2.21}$$

with

$$Z_i = \sum_{\varepsilon_i} g_i \exp\left(-\frac{\varepsilon_i}{k_BT}\right) \tag{5.2.22}$$

where g_i is the degeneracy of the ith energy level.

This is as far as we can go without further specification of the nature of the molecules making up the gas. It is clear that the knowledge of the detailed structure is necessary for the computation of the internal energy spectrum ε_i and hence of Z_i. We shall discuss some more specific applications in Section 5.3. However, a number of important properties can already be deduced from (5.2.21).

We first note that the *internal partition function cannot depend on the volume* of the system. Indeed, by definition, the internal variables characterize an isolated molecule, for a given position of its center of mass. The corresponding variables (think, typically, of the amplitude of vibration of an oscillator characterizing the molecule) vary over a domain of the order of the size of the molecule, not the size of the system. Also, the energy levels corresponding to these subsystems cannot depend on the size of the system as do the translational levels [see (5.2.13), (5.2.14)]. Hence, Z_i *is a function of temperature alone.* We can therefore write

$$A(T, \mathcal{V}, N) = -N\left(k_BT \ln\frac{\mathcal{V}}{N} + k_BT + \tfrac{3}{2}k_BT \ln(2\pi mk_BTh^{-2}) - a_i(T)\right) \tag{5.2.23}$$

where

$$a_i(T) = -k_BT \ln Z_i(T) \tag{5.2.24}$$

In this form we have explicitly separated the volume and the temperature dependence. Moreover, if we consider structureless particles (say, electrons or monoatomic molecules at temperatures where electronic excitation can be neglected), $a_i(T) = 0$.

* A noteworthy exception is provided by a gas of bosons near the absolute zero temperature; see Section 5.6.

From Eq. (5.2.23) the existence of the free energy per particle in the thermodynamic limit is obvious. Hence Eq. (4.7.2) becomes

$$\lim_{N\to\infty} N^{-1} A\left(T, \frac{N}{n}, N\right) \equiv a(T, n)$$
$$= k_B T \ln n - k_B T - \tfrac{3}{2} k_B T \ln(2\pi m k_B T h^{-2}) + a_i(T) \tag{5.2.25}$$

Of course this result is especially simple. We saw in Section 4.7 that the main difficulty in the proof of the existence of thermodynamics comes from the interactions.

We may now very easily derive the thermodynamic properties of the ideal Boltzmann gas from Eq. (5.2.25). By using the relations of Table 4.4.1 the pressure is obtained from

$$P = n^2 \left(\frac{\partial a}{\partial n}\right)_T$$

which yields

$$P = n k_B T \tag{5.2.26}$$

This is the well-known equation of state of a perfect gas. One may note that its form is not influenced by the internal degrees of freedom. The entropy is given by

$$s = -\left(\frac{\partial a}{\partial T}\right)_n = k_B\{-\ln n + \tfrac{5}{2} + \tfrac{3}{2} \ln(2\pi m k_B T h^{-2}) - a_i'(T)\} \tag{5.2.27}$$

where $a_i'(T) = da_i(T)/dT$. The internal energy per particle is

$$e = a + Ts = \tfrac{3}{2} k_B T + a_i(T) - T a_i'(T) \tag{5.2.28}$$

We find another well-known law of the perfect gases: the energy is a function of temperature alone. For structureless particles $[a_i(T) = 0]$, the result can be interpreted as follows: every one of the 3 degrees of freedom per particle contributes $\frac{1}{2} k_B T$ to the energy. This is a special case of the so-called *equipartition principle*, valid only for classical systems. As will be seen in Section 5.3, if internal degrees of freedom are taken into account, the *equipartition principle* can be formulated as follows: In the classical limit, to each square term in the Hamiltonian corresponds a contribution of $\frac{1}{2} k_B T$ to the energy. Let us finally calculate the specific heat per particle, at constant volume:

$$c_V = \left(\frac{\partial e}{\partial T}\right)_n = \tfrac{3}{2} k_B - T a_i''(T) \tag{5.2.29}$$

For *structureless classical particles, the specific heat is constant;* only the internal degrees of freedom may introduce a temperature dependence.

5.3. MOLECULAR STRUCTURE AND THERMODYNAMICS

The main application of the formalism developed in Section 5.2 is the study of the thermodynamic properties of dilute gases, in the limit where the equation of state of a perfect gas (5.2.26) is a good approximation. If the function $a_i(T)$ of Eq. (5.2.25) could be neglected, we would have a very simple picture of the thermodynamic properties of the gas. The only molecular parameter entering these expressions is the *mass* of the molecules m, a readily accessible quantity.

It is obvious, however, that a molecule is a more or less complex assembly of atoms, the latter being in turn made up of nuclei and electrons.* This complex structure is characterized by a certain number of molecular parameters related to the various types of possible motions relative to the center of mass. For examples, a *principal moment of inertia* I_j is associated with every rotational degree of freedom, a *characteristic frequency* ν_k is associated with each vibrational degree of freedom, an *excitation energy* $\Delta\varepsilon_j$ is associated with each electronic state. All these internal degrees of freedom contribute to the Hamiltonian, and hence, via the partition function, to the thermodynamic properties. This is the reason why, for instance, the experimentally measured specific heats of real dilute gases are not simply constants equal to $\frac{3}{2}k_B$, but usually show a complicated temperature dependence.

In first approximation, the translational, rotational, and vibrational degrees of freedom can be considered as independent. Therefore, their contributions to the partition function are multiplicative and their contributions to the free energy are additive:

$$Z_i = Z_{rot} Z_{vib} \tag{5.3.1}$$

$$a_i = a_{rot} + a_{vib} \tag{5.3.2}$$

Before developing these expressions we note an important feature that is common to all internal degrees of freedom. The spacing between successive energy levels corresponding to the various dynamical processes depends, of course, on the characteristic parameters mentioned above. At temperatures such that $k_B T$ is much smaller than this spacing, the thermal motion cannot induce transitions to excited states. The molecule therefore remains in its ground state; the corresponding contribution to a_i is a constant and the contribution to c_V is zero. At such

* Under extreme conditions of pressure and temperature, arising in some systems of astrophysical interest (star interiors, neutron stars, etc.) one must dig further into the structure of matter, and take account of nuclear reactions, elementary particle transformations, and the like. We shall not deal with these problems here. [See, for instance, the recent paper by R. L. Bowers, J. A. Campbell, and R. L. Zimmerman, *Phys. Rev.* D **7**, 2289 (1973).]

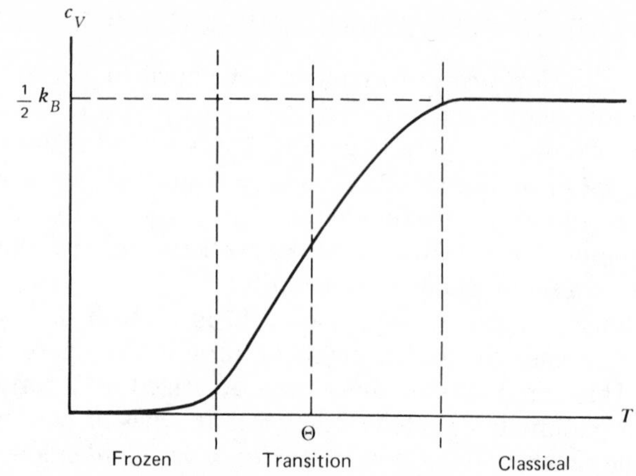

Figure 5.3.1. Typical contribution of an internal degree of freedom to the specific heat per particle, as a function of temperature. (Note that the behavior in the transition region is not necessarily monotonic.)

temperatures the corresponding degree of freedom is said to be *frozen.** On the contrary, at temperatures such that $k_B T$ is much larger than the characteristic separation, the corresponding energy can be treated as continuous, and the system behaves classically: The specific heat then equals the equipartition value $\frac{1}{2}k_B$. This state of affairs is conveniently described in terms of a *characteristic temperature* Θ_i for each internal degree of freedom. It is essentially the level spacing divided by k_B. The contribution of this degree of freedom to the specific heat is roughly sketched in Fig. 5.3.1, which summarizes the previous discussion. The total specific heat is a superposition of such sigmoid curves; as the characteristic temperatures Θ_i are usually very different for the various degrees of freedom, the resulting curve can be quite complicated. We now study some simple examples of internal degrees of freedom.

A. Rotation

We only consider the simple case of diatomic molecules. The energy eigenvalues of the rigid rotator are known from quantum mechanics:

$$\varepsilon_{\text{rot}} = \left(\frac{\hbar^2}{2I}\right) j(j+1), \qquad j = 0, 1, 2, \ldots \tag{5.3.3}$$

* As is well known, the spacing between the ground level and the first excited electronic state is an order of magnitude larger than the spacing between rotational or vibrational levels in a molecule. For this reason, at all usual temperatures the electronic excitation is frozen, and we did not consider it in Eqs. (5.3.1) and (5.3.2). Only in a few exceptional cases, when the ground state is a multiplet (for instance, in nitrogen oxide, NO) does electronic excitation play a role at usual temperatures.

where I is the moment of inertia of the rotator. Each level has a degeneracy $g = 2j + 1$. It is natural to define the characteristic temperature Θ_{rot} as

$$\Theta_{rot} = \frac{\hbar^2}{2Ik_B} \tag{5.3.4}$$

If the two atoms in the molecule are different (heteronuclear molecule) the partition function is*

$$Z_{rot} = \sum_{j=0}^{\infty} (2j+1) \exp\left[-\left(\frac{\Theta_{rot}}{T}\right) j(j+1)\right] \tag{5.3.5}$$

This sum cannot be evaluated in closed form. It is fortunate, however, that the characteristic temperatures Θ_{rot} are usually very small. Hence, at ordinary temperatures, the classical approximation, consisting in replacing the sum by an integral, is valid. Hence,

$$Z_{rot} \approx \int_0^{\infty} dj (2j+1) \exp\left[-\left(\frac{\Theta_{rot}}{T}\right) j(j+1)\right] \tag{5.3.6}$$

In the case of homonuclear molecules, we must take account of the fact that two orientations differing by an angle π are identical, and thus undistinguishable. The partition function must, therefore, be divided by 2 in order to correct for overcounting. More generally, we introduce a *symmetry number* σ equal to 1 for heteronuclear and equal to 2 for homonuclear molecules.† Then

$$Z_{rot} \approx \sigma^{-1} \int_0^{\infty} dj (2j+1) \exp\left[-\left(\frac{\Theta_{rot}}{T}\right) j(j+1)\right]$$

$$= \sigma^{-1} \int_0^{\infty} dy \exp\left[-\left(\frac{\Theta_{rot}}{T}\right) y\right] = \frac{T}{\sigma \Theta_{rot}}, \qquad T \gg \Theta_{rot} \tag{5.3.7}$$

* As the energy eigenvalue ε_j is $(2j+1)$-fold degenerate, one must count $(2j+1)$ states for each value of j in the partition function.

† In the low-temperature limit, the treatment of homonuclear molecules is not so simple. Z_{rot} is obtained from Eq. (5.3.5) by summing only over even or only over odd values of j. The choice depends on the total nuclear-spin state of the molecule: the total (nuclear × rotational) wave function must be antisymmetric. In the high-temperature limit, the "odd Z_{rot}" and the "even Z_{rot}" are practically equal to $\frac{1}{2}$ × the "total Z_{rot}." A very well-known problem in which these subleties play an important role is the case of hydrogen, H_2, which appears in two states (ortho and para) corresponding precisely to the two cases quoted above. This problem is discussed in most textbooks and will not be considered here.

The contribution of the rotation to the thermodynamic functions is therefore

$$
\begin{aligned}
a_{\mathrm{rot}} &= -k_B T \ln\left(\frac{T}{\sigma\Theta_{\mathrm{rot}}}\right) \\
s_{\mathrm{rot}} &= k_B \ln\left(\frac{T}{\sigma\Theta_{\mathrm{rot}}}\right) + k_B \\
e_{\mathrm{rot}} &= k_B T \\
c_{V,\mathrm{rot}} &= k_B
\end{aligned}
\tag{5.3.8}
$$

Clearly, in this classical approximation, the specific heat has the equipartition value $2\times\frac{1}{2}k_B$ (a rigid rotator corresponds to two square terms in the classical Hamiltonian).

B. Vibration

The vibrational motion of a polyatomic molecule can always* be described as a superposition of f independent harmonic oscillations, each of characteristic frequency ν_k. The number of normal modes f for an n-atomic molecule is $f=3n-5$ if the molecule is linear and $f=3n-6$ otherwise. The energy eigenvalues are then

$$\varepsilon_{\mathrm{vib}} = \sum_{k=1}^{f} \left(\tfrac{1}{2}+v\right)h\nu_k, \qquad v=0,1,2,\ldots \tag{5.3.9}$$

The partition function is†

$$Z_{vib} = \prod_{k=1}^{f} Z_k \tag{5.3.10}$$

and the free energy

$$a_{\mathrm{vib}} = \sum_{k=1}^{f} a_k \tag{5.3.11}$$

The natural characteristic vibration temperature of the normal mode k is

$$\Theta_k = \frac{h\nu_k}{k_B} \tag{5.3.12}$$

and the corresponding partition function can be evaluated exactly, being a

* Provided the amplitudes are small enough to avoid the contribution of anharmonic terms to the Hamiltonian.

† Note that in some cases two or more normal modes may have the same characteristic frequency (see Table 5.3.1). In those cases the contributions must be counted with the corresponding multiplicity. For instance, for CO_2 we have

$$Z_{\mathrm{vib}} = (Z_{960})^2 Z_{1900} Z_{3400}$$

Table 5.3.1. Characteristic Temperatures for Various Internal Degrees of Freedom. The Bracketed Numbers Denote the Number of Coincident Vibrational Frequencies[a]

	Θ_{rot}	Θ_{vib}
Hydrogen, H_2	85.4	6100
Hydrogen deuteride, HD	64.1	5300
Deuterium, D_2	42.7	4300
Hydrogen chloride, HCl	15.2	4140
Hydrogen iodide, HI	9.0	3200
Nitrogen, N_2	2.86	3340
Carbon dioxide, CO_2		960 (2) 1900 3400
Water, H_2O		2294 5180 5400
Carbon tetrachloride, CCl_4		313.7 (3) 451.8 (2) 659.0 1116.5 (3)

[a] Compiled from A. H. Wilson, *Thermodynamics and Statistical Mechanics*, Cambridge Univ. Press, London, 1957.

simple geometric series:

$$Z_k = \sum_{v=0}^{\infty} \exp\left\{-\left(\frac{\Theta_k}{T}\right)(v+\tfrac{1}{2})\right\} = \exp\left[-\frac{1}{2}\left(\frac{\Theta_k}{T}\right)\right]\sum_{v=0}^{\infty}\exp\left[-v\left(\frac{\Theta_k}{T}\right)\right]$$

$$= \frac{\exp[-\tfrac{1}{2}(\Theta_k/T)]}{1-\exp[-(\Theta_k/T)]} = \frac{1}{2\sinh(\Theta_k/2T)} \tag{5.3.13}$$

The thermodynamic functions can now be calculated by the usual procedure:

$$\begin{aligned} a_k &= \tfrac{1}{2}k_B\Theta_k + k_BT\ln\left[1-\exp\left(-\frac{\Theta_k}{T}\right)\right] \\ s_k &= -k_B\ln\left[1-\exp\left(-\frac{\Theta_k}{T}\right)\right] + k_B\left(\frac{\Theta_k}{T}\right)\left[\exp\left(\frac{\Theta_k}{T}\right)-1\right]^{-1} \\ e_k &= \tfrac{1}{2}k_B\Theta_k + k_B\Theta_k\left[\exp\left(\frac{\Theta_k}{T}\right)-1\right]^{-1} \\ c_{V,k} &= k_B\frac{(\Theta_k/T)^2\exp(\Theta_k/T)}{[\exp(\Theta_k/T)-1]^2} \equiv k_B\mathscr{E}\left(\frac{\Theta_k}{T}\right) \end{aligned} \tag{5.3.14}$$

The function $\mathscr{E}(x)$ appearing in the specific heat is called the *Einstein function.* It is well known and has been tabulated in detail.* It has the general shape of Fig. 5.3.1, tending toward $2\times\frac{1}{2}k_B$ as $T\to\infty$ (a normal mode contributes two squares to the classical Hamiltonian).

It is now clear, from these simple examples, that the knowledge of a few molecular parameters such as those of Table 5.3.1 enables us to predict in detail the thermodynamic functions of ideal gases. These parameters can easily be obtained, for instance, from spectroscopic data. We thus meet here with the first instance in which statistical mechanics provides us with a very direct connection between microscopic structure and macroscopic bulk properties. Needless to say, these statistical predictions have been thoroughly tested experimentally.

5.4. IDEAL SYSTEMS OF BOSONS OR FERMIONS

When we consider ideal systems at lower temperatures or at higher densities, we may no longer use the approximations of the previous subsection. The specific boson or fermion nature of their constituents now plays an important role in determining the thermodynamic properties. It is, therefore, natural to use in this study the second quantization formalism of Section 1.5, which automatically incorporates the effects of quantum statistics.

To be specific, we consider a system of N point particles in a volume $\mathscr{V}$. In order to cover physically interesting systems we assume that each particle has a nonzero total spin S.† Therefore, each level of the particle is labeled by the corresponding momentum eigenvalue $\mathbf{p}$ and by the value $\sigma\hbar$ of the z component of the spin. It is well known from elementary quantum mechanics that, for each value of $\mathbf{p}$, σ can assume only one among $2S+1$ integral or half-integral values: $\sigma=-S,\ -S+1,\ldots,S-1,S$. We also assume that, in absence of an external field, the energy eigenvalues $\varepsilon_{\mathbf{p},\sigma}$ depend only on $\mathbf{p}$, not on the spin:

$$\varepsilon_{\mathbf{p},\sigma}=\varepsilon_{\mathbf{p}}=\frac{p^2}{2m} \tag{5.4.1}$$

Hence, each of these eigenvalues is g-fold degenerate, where

$$g=2S+1 \tag{5.4.2}$$

* J. Hilsenroth and G. G. Ziegler, *Tables of Einstein Functions*, U.S. National Bureau of Standards Monograph 49, U.S. Govt. Printing Office, Washington, D.C., 1962.

† It is pointless to consider here vibrational or rotational degrees of freedom. These are not involved in permutations in space, and are therefore not affected by the Bose or Fermi statistics. Their treatment, even in quantum gases, is essentially the same as in Section 5.3. Even the inclusion of the spin is unessential from the point of view of quantum statistics. We only introduced it because of its importance in practical application.

Calling now $\mathsf{a}^\dagger_{\mathbf{p}\sigma}(\mathsf{a}_{\mathbf{p}\sigma})$ the creation (destruction) operator of a particle of momentum $\mathbf{p}$ and spin σ,* the Hamiltonian of the system can be written as

$$\mathsf{H}^0 = \sum_{\mathbf{p}} \sum_{\sigma} \varepsilon_{\mathbf{p}} \mathsf{a}^\dagger_{\mathbf{p}\sigma} \mathsf{a}_{\mathbf{p}\sigma} \tag{5.4.3}$$

The canonical partition function is now given by

$$Z = \mathrm{Tr}\, \exp(-\beta \mathsf{H}^0) = \sum_{\substack{\{n_{\mathbf{p}\sigma}\} \\ (\sum_{\mathbf{p}} \sum_{\sigma} n_{\mathbf{p}\sigma} = N)}} \exp\left(-\beta \sum_{\mathbf{p}} \sum_{\sigma} \varepsilon_{\mathbf{p}} n_{\mathbf{p}\sigma}\right) \tag{5.4.4}$$

This simple form is obtained because the Hamiltonian is diagonal in the occupation-number representation. The trace is calculated as follows: For each value of $\mathbf{p}$ and of σ, one sums over all possible values of the occupation number $n_{\mathbf{p}\sigma}$ of the corresponding level. The symbol $\sum_{(n_{\mathbf{p}\sigma})}$ therefore denotes an infinite number of such summations: one for each value of $\mathbf{p}$ and of σ.

We must, however, not forget that the *total number of particles N is fixed.* This apparently innocent condition introduces a serious complication in the evaluation of the right-hand side of (5.4.4). If it were absent, the partition function would simply factorize. Because of it, however, the summation can no longer be done. The various energy levels $\varepsilon_{\mathbf{p}}$ are not independent: There exist *correlations* among the particles, due to their indistinguishability. Thus we find again the quantum-statistical correlations discussed in Section 3.7.

In order to understand clearly the difference between the Boltzmann and the boson or fermion cases, we may note that in the previous subsection we did not use the second quantization formalism, which does not apply to distinguishable particles. If, however, we expressed the partition function in terms of occupation numbers, we should note that each configuration can be realized in $N!/\prod n_p!$ distinct ways for distinguishable particles (of spin $S=0$); hence [see Eq. (4.3.17)]

$$Z_B = \sum_{\substack{\{n_p\} \\ (\sum n_p = N)}} \frac{N!}{\prod n_p!} \prod_p e^{-\beta \varepsilon_p n_p}$$

We now note that the coefficient of the exponential is precisely the multinomial coefficient; hence the sum, *including the constraint*, is simply summed by Newton's multinomial formula:

$$Z_B = (e^{-\beta\varepsilon_1} + e^{-\beta\varepsilon_2} + \cdots)^N = \left(\sum_p e^{-\beta\varepsilon_p}\right)^N$$

which coincides with (5.2.6), except for the factor $N!$, which, as we know, cannot be found from a classical counting procedure. The difficulty in the Bose and the Fermi case lies in the absence of the multinomial coefficient.

* We briefly call "a particle of spin σ" a particle in a state such that the z component of its spin has the value $\sigma\hbar$.

We will now, for the first time, meet with the virtues of the grand canonical ensemble. Let us, indeed, try to derive the thermodynamical properties from the grand partition function (4.5.7), which, in the present case reduces to

$$\Xi=\sum_{N=0}^{\infty}\sum_{\substack{\{n_{\mathbf{p}\sigma}\}\\(\Sigma n_{\mathbf{p}\sigma}=N)}}\exp\Big(\beta\mu\sum_{\mathbf{p}}\sum_{\sigma}n_{\mathbf{p}\sigma}\Big)\exp\Big(-\beta\sum_{\mathbf{p}}\sum_{\sigma}\varepsilon_{\mathbf{p}}n_{\mathbf{p}\sigma}\Big)\tag{5.4.5}$$

The extra summation over N miraculously provides just the missing feature that was eliminated by the constraint. Indeed, we now rearrange the summations as a product of sums over n, each of these being *completely unconstrained:*

$$\Xi=\sum_{\{n_{\mathbf{p}\sigma}\}}\exp\Big[\beta\sum_{\mathbf{p}}\sum_{\sigma}(\mu-\varepsilon_{\mathbf{p}})n_{\mathbf{p}\sigma}\Big]=\prod_{\mathbf{p}}\prod_{\sigma}\sum_{n}\exp[\beta n(\mu-\varepsilon_{\mathbf{p}})]\tag{5.4.6}$$

In order to understand this transformation from (5.4.5) to (5.4.6) consider a simple system having only two possible energy levels and call $a=\exp[\beta(\mu-\varepsilon_1)]$, $b=\exp[\beta(\mu-\varepsilon_2)]$. The grand partition function is then

$$\Xi=\sum_{N=0}^{\infty}\sum_{\substack{n_1\quad n_2\\(n_1+n_2=N)}}\sum a^{n_1}b^{n_2}=\sum_{N=0}^{\infty}\sum_{n=0}^{N}a^{N-n}b^{n}$$

The summation prescription is therefore: fix N and sum over n from 0 to N; then add the partial results from $N=0$ to $N=\infty$, that is, add the horizontal rows in the following array:

$$\begin{array}{l}1\\ a+b\\ a^2+ab+b^2\\ a^3+a^2b+ab^2+b^3\\ \cdot\quad\cdot\quad\cdot\quad\cdot\end{array}$$

But we may also do the sum along diagonal rows parallel to the arrow. This means: fix the power of a and sum over all powers of b from 0 to ∞; then sum over all powers of a. The formula is

$$\Xi=\sum_{n_1=0}^{\infty}\sum_{n_2=0}^{\infty}a^{n_1}b^{n_2}=\Big(\sum_{n=0}^{\infty}a^{n}\Big)\Big(\sum_{m=0}^{\infty}b^{m}\Big)$$

which is the desired result.

We can now complete the calculation very simply.

In the case of *fermions*, the occupation numbers can only take the values $n=0$, $n=1$; hence the summation in (5.4.6) is trivial and yields the result

$$\Xi(T,\mathcal{V},\mu)=\prod_{\mathbf{p}}\prod_{\sigma}(1+e^{\beta(\mu-\varepsilon_{\mathbf{p}})})\qquad\text{(F)}\tag{5.4.7}$$

In the case of *bosons*, n can take all nonnegative values: we have to deal with the sum of a geometric series, which is also very simple:

$$\Xi(T, \mathcal{V}, \mu) = \prod_{\mathbf{p}} \prod_{\sigma} (1 - e^{\beta(\mu - \varepsilon_{\mathbf{p}})})^{-1} \qquad \text{(B)} \tag{5.4.8}$$

Before going further, we note that the validity of this summation presupposes the convergence of the geometric series, which implies that the exponential is less than 1. In order that this condition be satisfied—whatever the value of $\varepsilon_{\mathbf{p}}$, including $\varepsilon_{\mathbf{p}} = 0$—the chemical potential μ *must be negative:*

$$\mu < 0 \qquad \text{(B)} \tag{5.4.9}$$

There is no such limitation in the Fermi–Dirac case.

We may now treat both cases in parallel, by using the convenient symbol θ defined in Eq. (1.4.14), which equals $+1$ for bosons and -1 for fermions. Then

$$\Xi(T, \mathcal{V}, \mu) = \prod_{\mathbf{p}} \prod_{\sigma} (1 - \theta e^{\beta(\mu - \varepsilon_{\mathbf{p}})})^{-\theta} \tag{5.4.10}$$

Hence we see that the *grand partition function is again factorized* for an ideal quantum gas. However, the factors no longer refer to individual particles (as in the Boltzmann case), but to individual energy levels, a quite different concept: we have here an infinite product.

The grand potential $J(T, \mathcal{V}, \mu)$ is now readily obtained from Eq. (4.5.21):

$$J(T, \mathcal{V}, \mu) \equiv P\mathcal{V} = -\theta k_B T \sum_{\mathbf{p}} \sum_{\sigma} \ln(1 - \theta e^{(\mu - \varepsilon_{\mathbf{p}})/k_B T}) \tag{5.4.11}$$

From this equation, all thermodynamic relations can be derived by the formulas of Table 4.5.1. We may, however, further transform Eq. (5.4.11). First we note that, as $\varepsilon_{\mathbf{p}}$ is independent of σ, the summation over σ just gives a factor g, the degeneracy of the level $\varepsilon_{\mathbf{p}}$. We then note that, for a large system the allowed values of $\mathbf{p}$ are very closely spaced: hence the argument leading from Eq. (5.2.15) to Eq. (5.2.18) can again be used in transforming the summation into an integration* over $\mathbf{p}$. Using also Eq. (5.4.1) we obtain

$$P\mathcal{V} = -\theta g k_B T \mathcal{V} h^{-3} \int d\mathbf{p} \ln\left[1 - \theta \exp\left(\frac{\mu - (p^2/2m)}{k_B T}\right)\right] \tag{5.4.12}$$

Finally, we may evaluate the integral in polar coordinates, integrate over

* There is a notable exception to this procedure in the case of bosons at low temperatures; we shall discuss this case later (Section 5.6).

the angles, and substitute $\eta = p^2/2mk_BT$ for p in the final integration:

$$P = -2^{5/2}\pi\theta g m^{3/2}(k_BT)^{5/2}h^{-3}\int_0^\infty d\eta\,\sqrt{\eta}\,\ln\left\{1-\theta\exp\left[-\eta+\left(\frac{\mu}{k_BT}\right)\right]\right\} \tag{5.4.13}$$

Integrating by parts and introducing the important parameter Λ (which is a function of the temperature) [see also Eq. (5.2.19)]:

$$\Lambda = \left(\frac{h^2}{2\pi m k_BT}\right)^{1/2} \tag{5.4.14}$$

we obtain

$$P(T,\mu) = \tfrac{2}{3}k_BTg\Lambda^{-3}\left(\frac{2}{\pi^{1/2}}\right)\int_0^\infty d\eta\,\frac{\eta^{3/2}}{e^{\eta-(\mu/k_BT)}-\theta} \tag{5.4.15}$$

Hence we obtained an expression of the pressure, in terms of the temperature and of the chemical potential. The latter, however, is not a very convenient variable, as it is not easily measured. The usual way of expressing the pressure is in terms of the temperature and of the density, that is, in the form of an *equation of state.* In order to obtain it, we note that the density n can also be expressed in terms of the temperature and of the chemical potential, by means of a formula of Table 4.5.1:

$$\begin{aligned} n &= \left(\frac{\partial P}{\partial\mu}\right)_T \\ &= \tfrac{2}{3}\Lambda^{-3}\left(\frac{2}{\pi^{1/2}}\right)\int_0^\infty d\eta\,\frac{\eta^{3/2}e^{\eta-(\mu/k_BT)}}{[e^{\eta-(\mu/k_BT)}-\theta]^2}\end{aligned}$$

Performing an integration by parts, this formula can be written in the simpler form:

$$n = g\Lambda^{-3}\left(\frac{2}{\pi^{1/2}}\right)\int_0^\infty d\eta\,\frac{\eta^{1/2}}{e^{\eta-(\mu/k_BT)}-\theta} \tag{5.4.16}$$

The two equations (5.4.15) and (5.4.16) are a parametric representation of the *equation of state of an ideal quantum gas.* To obtain an explicit equation, we should be able to solve Eq. (5.4.16) for μ as a function of T and n, and substitute the result into Eq. (5.4.15). Clearly this is a difficult problem in general, to which we come back below.

Other thermodynamic functions are easily calculated from $P(T,\mu)$ by means of the relations of Table 4.5.1. As an example we consider the energy density:

$$\tilde{e}(T,\mu) \equiv \mathcal{V}^{-1}E = gk_BT\Lambda^{-3}\left(\frac{2}{\pi^{1/2}}\right)\int_0^\infty d\eta\,\frac{\eta^{3/2}}{e^{\eta-(\mu/k_BT)}-\theta} \tag{5.4.17}$$

The latter expression is remarkable. Indeed, comparing it with (5.4.15) we find that

$$P = \tfrac{2}{3}\tilde{e} \tag{5.4.18}$$

This relation holds for *all ideal systems:* Boltzmann, Bose, and Fermi. The fact is noteworthy, because we saw that individual thermodynamic functions have strikingly different expressions in the three cases. In particular, the equation of state of a boson or fermion ideal gas is quite different from the classical relation (5.2.21): Bose and Fermi gases are not "perfect gases" in the thermodynamic sense.

We now study more closely the equation of state of these systems. As mentioned above, the elimination of the chemical potential between the two equations (5.4.15) and (5.4.16) is a highly nontrivial problem. Although quantum ideal gases were studied for sixty years, an exact analytic form of the equation of state was only found very recently. In 1968 Leonard found a very elegant mathematical method for the inversion of the pressure and density formulas; this work was further developed by Nieto in 1970. It would lead us too far to go into the detailed exposition of Leonard's method here: we just quote the result. For a fermion gas the equation of state is

$$P = nk_BT + [2^{3/2}\pi g]^{-1}\Lambda^3 nP\int_1^\infty dt\, t^{-2}[\phi_n(t) - \phi_P(t)] \tag{5.4.19}$$

with

$$\phi_n(t) = \tan^{-1}\frac{[4(2\pi)^{1/2}g/n\Lambda^3](\ln t)^{1/2}}{1 + [4(2\pi)^{1/2}g/n\Lambda^3]t\int_1^\infty dy(\ln y)^{1/2}[y(y-t)]^{-1}} \tag{5.4.20}$$

$$\phi_P(t) = \tan^{-1}\frac{[8(2\pi)^{1/2}gk_BT/3P\Lambda^3](\ln t)^{3/2}}{1 + [8(2\pi)^{1/2}gk_BT/3P\Lambda^3]t\int_1^\infty dy(\ln y)^{3/2}[y(y-t)]^{-1}} \tag{5.4.21}$$

This equation is still in an implicit form (i.e., it is not solved for P in terms of n and T) and involves formidable integrals. To make it useful, a numerical work involving the tabulation of P is still necessary and has not been done as of yet.

For all practical purposes it is preferable to develop approximate equations, valid in some well-defined domain. In the present section we develop an approximation valid whenever the quantum-statistical effects are weak.

We assume that the fugacity, defined by Eq. (4.5.22), is small compared to 1:

$$\mathfrak{z} \equiv e^{\mu/k_BT} \ll 1$$

As the dependence of $\mathfrak{z}$ on T and n is not known at this stage, the meaning of this criterion is not yet clear. We will see, at the end of the calculation, under which condition the results are valid. The integral in (5.4.15) can be calculated by expanding the integrand:

$$\int_0^\infty d\eta \frac{\eta^{3/2}}{e^\eta \mathfrak{z}^{-1} - \theta} = \int_0^\infty d\eta\, \eta^{3/2} e^{-\eta} \mathfrak{z}(1 + \theta e^{-\eta} \mathfrak{z} + e^{-2\eta} \mathfrak{z}^2 + \cdots)$$

The integrations are elementary, and we find the result:

$$P = g k_B T \Lambda^{-3} \mathfrak{z}(1 + \theta 2^{-5/2} \mathfrak{z} + 3^{-5/2} \mathfrak{z}^2 + \cdots) \tag{5.4.22}$$

By the same method applied to Eq. (5.4.16) we find the expansion of the density:

$$n = g \Lambda^{-3} \mathfrak{z}(1 + \theta 2^{-3/2} \mathfrak{z} + 3^{-3/2} \mathfrak{z}^2 + \cdots) \tag{5.4.23}$$

This power series can be reversed by using Lagrange's theorem*:

$$\mathfrak{z} = g^{-1} \Lambda^3 n \left[1 - \theta 2^{-3/2} g^{-1} \Lambda^3 n + \left(\frac{1}{4} - \frac{1}{3^{3/2}}\right) g^{-2} \Lambda^6 n^2 + \cdots\right] \tag{5.4.24}$$

Finally, substituting the result into (5.4.22) we obtain

$$P = n k_B T \left[1 - \theta 2^{-5/2} g^{-1} \Lambda^3 n + \left(\frac{1}{8} - \frac{2}{3^{3/2}}\right) g^{-2} \Lambda^6 n^2 + \cdots\right] \tag{5.4.25}$$

or, with numerical values:

$$P = n k_B T [1 - 0.1768 \theta g^{-1} \Lambda^3 n - 0.0033 g^{-2} \Lambda^6 n^2 + \cdots] \tag{5.4.26}$$

We thus find an expression for the equation of state in the form of a series expansion in powers of the density. Such a form is called a *virial expansion* and is characteristic of imperfect gases. We stress again the fact that the deviation of the properties from the perfect-gas behavior are due to quantum statistics, not to real interactions. We note that the first correction from the perfect gas law is of opposite sign for fermions and for bosons. In the former case, at given density and temperature, the pressure is higher than for a perfect gas. The effect of Fermi statistics is similar to a repulsive force between particles. Clearly, this repulsive *pseudoforce* stems from the exclusion principle, which is unfavorable to configurations in which the particles are close together. An opposite situation appears for bosons, where the effect is similar to an attractive pseudoforce.

We can now examine more precisely the limits of validity of our

* The first few coefficients can be found in H. B. Dwight, *Tables of Integrals*, McMillan, New York, 1961, p. 15.

expansion, and hence the limits of validity of the Boltzmann approximation developed in Section 5.2. Clearly, the criterion of validity is dictated by a dimensionless number δ called the *degeneracy parameter:*

$$\delta = g^{-1}\Lambda^3 n = g^{-1}\left(\frac{h^2}{2\pi m k_B T}\right)^{3/2} n \ll 1 \tag{5.4.27}$$

We now clearly see that the Boltzmann approximation is, for given density, a *high-temperature* approximation; or, for given temperature, a *low-density* approximation. It is also important to note the dependence on the mass. Because of this dependence, the quantum-statistical effects are most important and are most easily detected in gases of light molecules, such as helium, at low temperatures. Hydrogen would in principle be a better candidate, but the intermolecular interactions are stronger (helium is a noble gas!): As a result the quantum-statistical deviations from the law of perfect gases are masked by effects due to nonideality (see Chapter 6).

5.5. THE BOSE–EINSTEIN AND THE FERMI–DIRAC DISTRIBUTIONS

For the more detailed study of the thermodynamic properties of quantum ideal systems, particularly in the high degeneracy region, $\delta \gg 1$, it is important to have a more precise picture of their microscopic structure. A key quantity in this respect is the *average occupation number* of the level $(\mathbf{p}, \sigma)$, denoted by $\langle n_{\mathbf{p}\sigma}\rangle$.

The overall system is described by the grand canonical density matrix:

$$\boldsymbol{\rho}^{(g)} = \Xi^{-1} \exp\left[\beta \sum_{\mathbf{p}} \sum_{\sigma} (\mu - \varepsilon_{\mathbf{p}})\, \mathbf{a}^{\dagger}_{\mathbf{p}\sigma}\mathbf{a}_{\mathbf{p}\sigma}\right] \tag{5.5.1}$$

Since the operator describing the number of particles in the level $(\mathbf{p}, \sigma)$ is $\mathbf{a}^{\dagger}_{\mathbf{p}\sigma}\mathbf{a}_{\mathbf{p}\sigma}$, the average of interest is

$$\langle n_{\mathbf{p}\sigma}\rangle = \mathcal{T}r\, \mathbf{a}^{\dagger}_{\mathbf{p}\sigma}\mathbf{a}_{\mathbf{p}\sigma}\boldsymbol{\rho}^{(g)} \tag{5.5.2}$$

It is clear that this quantity is quite directly related to the one-particle Wigner distribution function, as can be seen from Eqs. (3.8.1), (3.8.3). The reduced equilibrium distribution functions will be discussed in detail in Chapter 7. In the present section we limit ourselves to the derivation of a few results that are important for the discussions in Sections 5.6 and 5.7.

The average of Eq. (5.5.2) can be calculated by a straightforward extension of the previous arguments.

By reasoning in the same way as we did for the derivation of Eq. (5.4.6), we easily obtain

$$\langle n_{\mathbf{p}\sigma}\rangle = \frac{\left\{\sum_{n_{\mathbf{p}\sigma}} n_{\mathbf{p}\sigma} \exp[\beta(\mu-\varepsilon_{\mathbf{p}})n_{\mathbf{p}\sigma}]\right\} \prod_{\mathbf{p}'\sigma'(\neq\mathbf{p}\sigma)} \left\{\sum_{n_{\mathbf{p}'\sigma'}} \exp[\beta(\mu-\varepsilon_{\mathbf{p}'})n_{\mathbf{p}'\sigma'}]\right\}}{\prod_{\text{all }\mathbf{p}\sigma}\left\{\sum_{n_{\mathbf{p}\sigma}} \exp[\beta(\mu-\varepsilon_{\mathbf{p}})n_{\mathbf{p}\sigma}]\right\}}$$

$$= \frac{\sum_n n \exp[\beta(\mu-\varepsilon_{\mathbf{p}})n]}{\sum_n \exp[\beta(\mu-\varepsilon_{\mathbf{p}})n]}$$

The last expression can be reformulated as follows

$$\langle n_{\mathbf{p}\sigma}\rangle = \frac{\partial}{\partial(\beta\mu)} \ln\left\{\sum_n \exp[\beta(\mu-\varepsilon_{\mathbf{p}})n]\right\} \tag{5.5.3}$$

The sum appearing in the argument of the logarithm was already calculated [see Eqs. (5.4.7)–(5.4.10)], hence

$$\langle n_{\mathbf{p}\sigma}\rangle = \frac{\partial}{\partial(\beta\mu)} \ln\{1-\theta\exp[\beta(\mu-\varepsilon_{\mathbf{p}})]\}^{-\theta}$$

and finally

$$\langle n_{\mathbf{p}\sigma}\rangle = \frac{1}{e^{\beta(\varepsilon_{\mathbf{p}}-\mu)}-\theta} \tag{5.5.4}$$

The two functions written together in this equation are the celebrated *Bose–Einstein distribution:*

$$\langle n_{\mathbf{p}\sigma}\rangle_B = \frac{1}{e^{\beta(\varepsilon_{\mathbf{p}}-\mu)}-1} \tag{5.5.5}$$

and the *Fermi–Dirac distribution:*

$$\langle n_{\mathbf{p}\sigma}\rangle_F = \frac{1}{e^{\beta(\varepsilon_{\mathbf{p}}-\mu)}+1} \tag{5.5.6}$$

We may note that $\langle n_{\mathbf{p}\sigma}\rangle$ is independent of σ. Hence, there is an equal number of particles with each spin orientation in each level. The average total spin of the gas is therefore zero: the equilibrium state is "unpolarized."

Together with the function $\langle n_{\mathbf{p}\sigma}\rangle$, the function $1+\theta\langle n_{\mathbf{p}\sigma}\rangle$ plays an important role in quantum statistics. For fermions, this function has a clear interpretation. Indeed the microscopic variable $n_{\mathbf{p}\sigma}$ can only take the values 1 or 0. Hence a given level ($\mathbf{p}\sigma$) is either singly occupied or it is empty. This is also vividly expressed by saying that the level is occupied either by a particle or by a hole. The function $1-\langle n_{\mathbf{p}\sigma}\rangle$ thus represents the average number of holes in the level ($\mathbf{p}\sigma$). Such a simple picture is, of course, inadequate for the boson case. In both cases, we have

$$1+\theta\langle n_{\mathbf{p}\sigma}\rangle = \frac{e^{\beta(\varepsilon_{\mathbf{p}}-\mu)}}{e^{\beta(\varepsilon_{\mathbf{p}}-\mu)}-\theta} \tag{5.5.7}$$

5.6. HIGHLY DEGENERATE IDEAL FERMION GASES

We now go over to the study of quantum gases in the limit of high degeneracy,

$$\delta = \frac{\Lambda^3 n}{g} \geqslant 1 \tag{5.6.1}$$

In this case, the typical quantum effects are most clearly exhibited and, as a result, the fermion and boson gases behave quite differently and must be studied separately.

We begin with the fermion gases, which are simpler. We consider such a system, and, keeping its density constant, we lower the temperature. The particles then tend to distribute themselves in such a way as to minimize the energy. However, because of the exclusion principle, there can be no accumulation of particles on any level. As the temperature approaches zero, the most favorable energetic distribution consists in the occupation of the successive energy levels by one particle (or by g particles if the levels are degenerate), starting from the lowest level and ending when the particles are exhausted. The limiting level reached in this way is called the *Fermi energy* and is denoted by ε_F. Thus a fermion gas has a considerable "zero-point energy" at $T=0$. We will see that this is the most important difference with a boson gas.

The average occupation number $\langle n_{\mathbf{p}\sigma}\rangle$ at zero temperature is thus

$$\langle n_{\mathbf{p}\sigma}\rangle = \Theta(\varepsilon_F - \varepsilon_{\mathbf{p}}), \qquad T=0 \tag{5.6.2}$$

where $\Theta(x)$ is the Heaviside step function:

$$\begin{aligned} \Theta(x) &= 0, \qquad x<0 \\ &= 1, \qquad x>0 \end{aligned} \tag{5.6.3}$$

This function, as well as the complement $1-\langle n_{\mathbf{p}\sigma}\rangle$, is plotted in Fig. 5.6.1 as a function of the energy.

We now connect this result to the general form of the Fermi–Dirac distribution, Eq. (5.5.6). We want to study its limiting behavior as $T\to 0$,

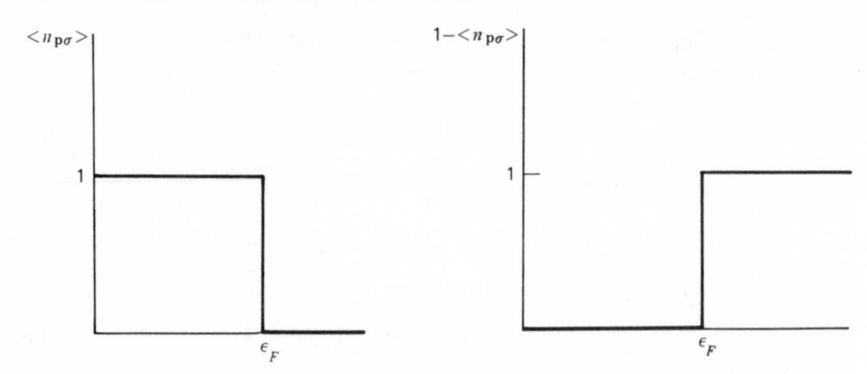

Figure 5.6.1. The Fermi–Dirac distribution function $\langle n_{\mathbf{p}\sigma}\rangle$ and its complement $1-\langle n_{\mathbf{p}\sigma}\rangle$ at zero temperature.

that is, as $\beta \to \infty$. The exponential then behaves as follows:

$$
\begin{aligned}
e^{\beta(\varepsilon-\mu)} &\to 0 \qquad \text{for } (\varepsilon-\mu)<0 \\
e^{\beta(\varepsilon-\mu)} &\to \infty \qquad \text{for } (\varepsilon-\mu)>0
\end{aligned}
\tag{5.6.4}
$$

It then follows that

$$
\begin{aligned}
\langle n_{\mathbf{p}\sigma}\rangle = \frac{1}{e^{\beta(\varepsilon-\mu)}+1} &\to 1, \qquad \varepsilon<\mu \\
&\to 0, \qquad \varepsilon>\mu
\end{aligned}
\tag{5.6.5}
$$

This equation is equivalent to (5.6.2), provided we make the identification (valid at $T=0$):

$$
\mu = \varepsilon_{\mathrm{F}}, \qquad T=0 \tag{5.6.6}
$$

It is very easy to determine the value of the Fermi energy. Indeed, from Eq. (5.4.16), fixing the number of particles, we immediately find, in the limit $T\to 0$ [taking $\varepsilon = k_{\mathrm{B}}T\eta$ as an integration variable and using Eqs. (5.6.2), (5.6.5)]

$$
\begin{aligned}
n &= 2^{5/2}\pi m^{3/2} h^{-3} g \int_0^\infty d\varepsilon\, \varepsilon^{1/2}\Theta(\varepsilon_{\mathrm{F}}-\varepsilon) \\
&= 2^{5/2}\pi m^{3/2} g (2\pi\hbar)^{-3}(\tfrac{2}{3}\varepsilon_{\mathrm{F}}^{3/2})
\end{aligned}
$$

from which follows the expression of ε_{F} as a function of the density n:

$$
\varepsilon_{\mathrm{F}} = \left(\frac{6\pi^2}{g}\right)^{2/3}\left(\frac{\hbar^2}{2m}\right) n^{2/3} \tag{5.6.7}
$$

To the Fermi energy corresponds a *Fermi momentum* p_{F} through the relation

$$
\varepsilon_{\mathrm{F}} = \frac{p_{\mathrm{F}}^2}{2m} \tag{5.6.8}
$$

This parameter depends on the density as follows:

$$p_F = \left(\frac{6\pi^2}{g}\right)^{1/3} \hbar n^{1/3} \tag{5.6.9}$$

The internal energy density is easily calculated from Eq. (5.4.17) with the result:

$$\tilde{e} = \tfrac{3}{5} n \varepsilon_F \tag{5.6.10}$$

and the pressure is obtained from (5.4.18)

$$P = \tfrac{2}{5} n \varepsilon_F = \frac{1}{5}\left(\frac{6\pi^2}{g}\right)^{2/3}\left(\frac{\hbar^2}{m}\right) n^{5/3} \tag{5.6.11}$$

We see that, because of its zero-point energy, the fermion gas has a considerable pressure at zero temperature: This is quite different from the boson gas where $P \to 0$ at zero temperature (the molecules in the ground state have no momentum).

We now calculate the values of the thermodynamic quantities at low temperatures, different from zero. In order to do so we need a systematic method (due to Sommerfeld) for calculating quantities of the type:

$$I = \int_0^\infty d\varepsilon \, g(\varepsilon) \frac{1}{e^{\beta(\varepsilon-\mu)}+1} \tag{5.6.12}$$

in the limit $k_B T \to 0$, that is, as $\beta \to \infty$. Here $g(\varepsilon)$ is an arbitrary function of ε such that $g(0)=0$. We also introduce

$$G(\varepsilon) = \int_0^\varepsilon d\eta \, g(\eta) \tag{5.6.13}$$

By an integration by parts, we transform (5.6.11) into

$$I = \beta \int_0^\infty d\varepsilon \, G(\varepsilon) \frac{e^{\beta(\varepsilon-\mu)}}{(e^{\beta(\varepsilon-\mu)}+1)^2} \tag{5.6.14}$$

We now note, remembering Eqs. (5.5.6) and (5.5.7), that

$$\frac{e^{\beta(\varepsilon-\mu)}}{(e^{\beta(\varepsilon-\mu)}+1)^2} = \langle n_{p\sigma}\rangle(1-\langle n_{p\sigma}\rangle) \tag{5.6.15}$$

At low temperatures $\langle n_{p\sigma}\rangle$ is not very different from the step function of Fig. 5.6.1: only the edges are somewhat rounded off, because some particles leave the Fermi sea (Fig. 5.6.2*a*). The function $1-\langle n_{p\sigma}\rangle$ is the average number of empty levels (or "holes"); its shape is complementary to $\langle n_{p\sigma}\rangle$ (Fig. 5.6.2*b*). It follows that the product $\langle n_{p\sigma}\rangle(1-\langle n_{p\sigma}\rangle)$ is different from zero only in a narrow domain near the Fermi level (Fig. 5.6.2*c*); the width of this range is of order $k_B T$.

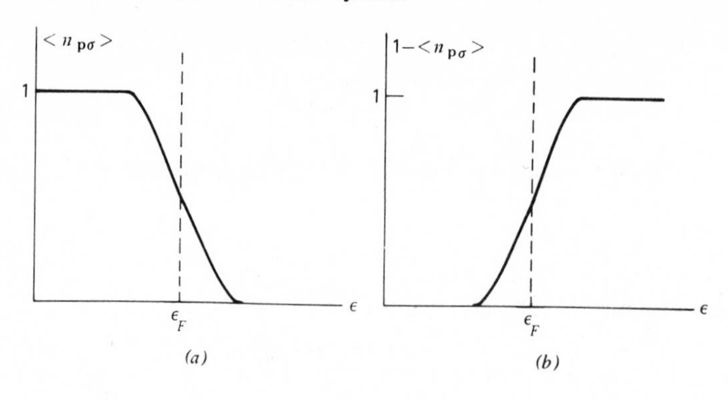

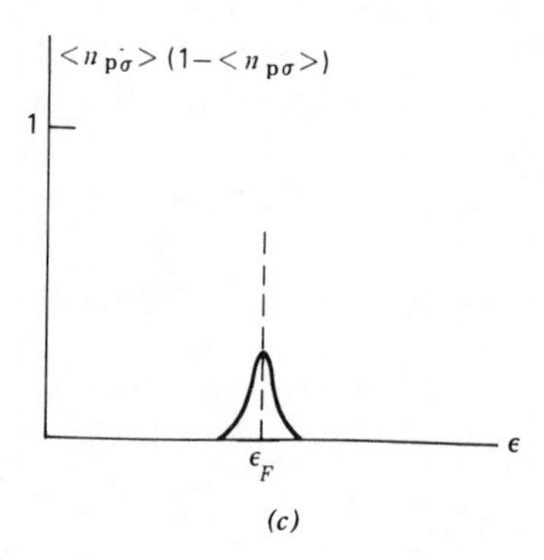

Figure 5.6.2. Fermi–Dirac distribution functions near zero temperature.

If we now assume that the function $G(\varepsilon)$ does not vary greatly over a range of energies $\delta\varepsilon \sim k_B T$, we may expand it around $\varepsilon = \mu$. After some elementary substitutions we find

$$I = \beta \int_{-\mu}^{\infty} dx\, G(x+\mu) \frac{e^{\beta x}}{(e^{\beta x}+1)^2}$$

$$\approx \int_{-\beta\mu}^{\infty} dy \sum_{n\geq 0} \frac{1}{n!} \frac{y^n}{\beta^n} G^{(n)}(\mu) \frac{e^y}{(e^y+1)^2}$$

where $G^{(n)}(\mu)$ is the nth derivative of $G(\varepsilon)$ evaluated at $\varepsilon = \mu$.

At low temperatures (large β) we may replace the lower limit of integration by $-\infty$ (the error is exponentially small). Note also that the ratio of exponentials in the integrand is an even function of y; hence only even powers of y contribute to the integral:

$$I \approx \left(\sum_{n\geq 0} \frac{1}{(2n)!} \frac{1}{\beta^{2n}} G^{(2n)}(\mu) \right) \left(2 \int_0^{\infty} dy\, y^{2n} \frac{e^y}{(e^y+1)^2} \right)$$

For the term $n = 0$ we have

$$\int_0^\infty dy \frac{e^y}{(e^y+1)^2} = \frac{1}{2}$$

For $n > 0$ we integrate by parts and find

$$\int_0^\infty dy\, y^{2n} \frac{e^y}{(e^y+1)^2} = 2n \int_0^\infty dy\, y^{2n-1}(e^y+1)^{-2}$$

$$= (1-2^{1-2n})(2n-1)!\,\zeta(2n) = (2^{2n-1}-1)\pi^{2n} B_n \qquad (5.6.16)$$

where B_n are the Bernoulli numbers, which are well known and tabulated.* We can now put together all these results and express them in terms of the original function $g(\varepsilon)$:

$$I \approx \int_0^\mu d\varepsilon\, g(\varepsilon) + \sum_{n\geq 1} \frac{2}{(2n)!}(2^{2n-1}-1)\pi^{2n} B_n (k_B T)^{2n} g^{(2n-1)}(\mu) \qquad (5.6.17)$$

Of course, only a finite number of terms in this asymptotic expansion are significant. The first few terms are, explicitly:

$$I = \int_0^\mu d\varepsilon\, g(\varepsilon) + \frac{\pi^2}{6}(k_B T)^2 g'(\mu) + \frac{7\pi^4}{360}(k_B T)^4 g'''(\mu) + \cdots \qquad (5.6.18)$$

Applying this result to Eq. (5.4.16) we find for the density [with $g(\varepsilon) = \varepsilon^{1/2}$]:

$$n = 2^{5/2}\pi m^{3/2} h^{-3} g\left\{\tfrac{2}{3}\mu^{3/2}\left[1 + \frac{\pi^2}{8}\left(\frac{k_B T}{\mu}\right)^2 + \cdots\right]\right\}$$

from which, together with (5.6.6), follows

$$\mu = \varepsilon_F\left[1 - \frac{\pi^2}{12}\left(\frac{k_B T}{\varepsilon_F}\right)^2 + \cdots\right] \qquad (5.6.19)$$

This equation gives the dependence of the chemical potential on temperature and on density (through ε_F) at low temperatures. For the energy density, we find similarly from Eqs. (5.4.17) and (5.6.10):

$$\tilde{e} = \tfrac{3}{5} n \varepsilon_F\left[1 + \frac{5\pi^2}{12}\left(\frac{k_B T}{\varepsilon_F}\right)^2 + \cdots\right] \qquad (5.6.20)$$

And therefore the specific heat per particle is

$$c_V = \frac{\pi^2}{2}\frac{k_B T}{\varepsilon_F} k_B \qquad (5.6.21)$$

* E. Jahnke and F. Emde, *Tables of Functions* Dover, New York, 1945; I. S. Gradshtein and I. M. Ryzhik, *Tables of Integrals, Sums, Series and Derivatives*, Academic Press, New York, 1965.

Table 5.6.1. Characteristic Parameters for the Electron Gas in Two Typical Metals

	ε_F (erg °K^{-1})	T_F (°K)
Li	6.34×10^{-12}	4.84×10^{4}
Na	4.32×10^{-12}	3.30×10^{4}

This result is characteristic of a fermion gas at low temperature; the specific heat is *linear* near $T=0$°K.

The most important system to which the concepts of the ideal fermion gas apply is the *electron gas in a metal.* It is well known that the outer electrons (valence electrons) of the metallic atoms are very weakly bound and therefore can be idealized as moving almost freely through the crystal lattice. If appropriate numerical values are used we get the typical orders of magnitude of Table 5.6.1. We see that the degeneracy parameter is so small, and the Fermi temperature so high, that the electron gas is a very strongly degenerate fermion gas. At ordinary temperatures this system can still be considered as being practically at the absolute zero ($T/T_F \sim 10^{-3}$).

One of the great early successes of quantum-statistical mechanics was Sommerfeld's great work of 1928, in which he demonstrated that the only way of explaining the then mysterious properties of the metals was to invoke the Fermi–Dirac statistics. When properly corrected for the effect of the periodic potential due to the ions in the lattice, this work became the foundation of the modern theory of metals. Actually, the theory works so well that it gives rise to a new puzzle. Why does the *ideal* gas model apply so well to the electrons, when we know that these interact with each other through Coulomb forces, and we expect the gas to be highly nonideal? The answer to this question was only given thirty years later. Qualitatively, it is precisely because of their interactions that the electrons build up around each other a polarization cloud that screens off the long-range Coulomb interaction. As a result, the entities in the metal are almost independent "quasiparticles." There also are phenomena that cannot be explained by the ideal gas model: they are related to collective excitations of the gas (plasma oscillations). However, these have such a high energy that they do not manifest themselves at ordinary temperatures. We shall not go into the details of these problems in this book, although the classical aspects of the Coulomb interactions will be studied in Section 6.5.

Let us briefly mention another type of degenerate fermion system, which has become a subject of active study in very recent years. When a small pellet of a mixture of solid deuterium and tritium is irradiated by a very intense and short pulse of laser light (say, 1000 J in 10^{-9} sec), a large amount of energy is deposited at the surface of the pellet. Among other processes, a succession of shock waves is set up, which are propagated towards the center. As a result, the system is very strongly compressed and heated: temperatures of 10^8 °K and densities as high as 10,000 times the solid densities ($n \approx 10^{26}\,\text{cm}^{-3}$) might be reached in this way. Under these conditions, thermonuclear fusion reactions can start in the center of the pellet. This *laser-driven fusion* process is now considered a very serious candidate for the realization of controlled thermonuclear fusion.

It is easily checked that under such conditions the electrons are very strongly degenerate. The compressed pellet is a true "microstar" produced in the laboratory. The detailed study of the behavior of matter in such extreme conditions is still an open problem of considerable interest.

5.7. HIGHLY DEGENERATE IDEAL BOSON GASES

We will now see that the ideal boson gas, as compared to the fermion gas, has a completely different behavior at low temperatures.

Consider the two basic equations (5.4.15), (5.4.16), which we rewrite as follows in terms of the fugacity $\mathfrak{z}$ [Eq. (4.5.22)] (we now take $g = 1$ for simplicity):

$$P(T, \mathfrak{z}) = k_B T \Lambda^{-3} Y(\mathfrak{z}) \tag{5.7.1}$$

$$n(T, \mathfrak{z}) = \Lambda^{-3} y(\mathfrak{z}) \tag{5.7.2}$$

where

$$Y(\mathfrak{z}) = \frac{4}{3\pi^{1/2}} \int_0^\infty d\eta \frac{\eta^{3/2}}{\mathfrak{z}^{-1} e^\eta - 1} \tag{5.7.3}$$

$$y(\mathfrak{z}) = \mathfrak{z} Y'(\mathfrak{z}) = \frac{2}{\pi^{1/2}} \int_0^\infty d\eta \frac{\eta^{1/2}}{\mathfrak{z}^{-1} e^\eta - 1} \tag{5.7.4}$$

Before starting the physical discussion it is important to have a good idea about the behavior of the functions $y(\mathfrak{z})$ and $Y(\mathfrak{z})$. We first note that only positive values of $\mathfrak{z}$ are allowed. ($\mathfrak{z}$ is defined as an exponential.) We derived in Eqs. (5.4.22), (5.4.23) the expansions

$$\begin{aligned} Y(\mathfrak{z}) &= \sum_{m=1}^\infty m^{-5/2} \mathfrak{z}^m \\ y(\mathfrak{z}) &= \sum_{m=1}^\infty m^{-3/2} \mathfrak{z}^m \end{aligned} \tag{5.7.5}$$

These representations are valid, provided the series converge. As both series have only positive terms, both functions start from zero [$Y(0) = 0$, $y(0) = 0$] and are monotonously increasing functions of $\mathfrak{z}$. A crucial

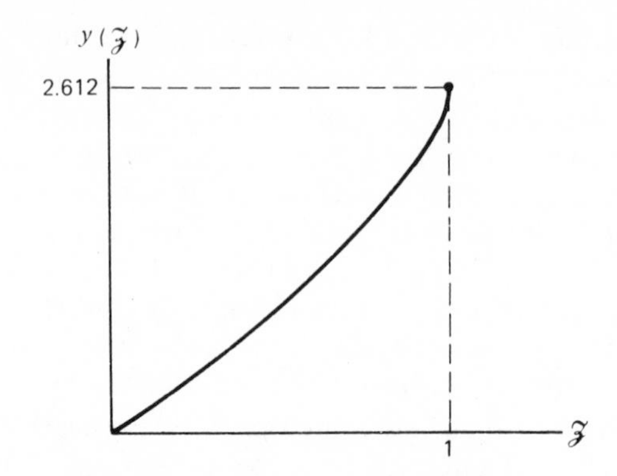

Figure 5.7.1. The function $y(ȥ)$.

change of behavior occurs at $ȥ = 1$, which turns out to be the radius of convergence of the two series (5.7.5). To understand why this is so, consider the equation

$$e^{\eta} - ȥ = 0 \tag{5.7.6}$$

This equation has a positive solution in η if and only if $ȥ > 1$. In that case there exists a value of η within the range $0 < \eta < \infty$ that annuls the denominator of the integrands (5.7.3), (5.7.4), causing the integrals to diverge. Hence the functions $Y(ȥ)$, $y(ȥ)$ simply do not exist for $ȥ > 1$. For $ȥ = 1$ they have a finite value, expressible in terms of the Riemann ζ function:

$$\begin{aligned} Y(1) &= \sum_{m=1}^{\infty} m^{-5/2} = \zeta(\tfrac{5}{2}) \approx 1.342 \\ y(1) &= \sum_{m=1}^{\infty} m^{-3/2} = \zeta(\tfrac{3}{2}) \approx 2.612 \end{aligned} \tag{5.7.7}$$

It can also be shown that the derivative of $y(ȥ)$ becomes infinite at $y = 1 : y'(1) = \infty$. The function $y(ȥ)$ is plotted in Fig. 5.7.1; $Y(ȥ)$ has a qualitatively similar shape.*

We now return to Eqs. (5.7.1) and (5.7.2). According to our general procedure we must eliminate the fugacity. We therefore consider Eq.

* It is interesting to compare this behavior with the corresponding Fermi functions $Y^F(ȥ)$, $y^F(ȥ)$. They also possess a series representation (5.4.22), (5.4.23), but now the terms alternate in sign. The radius of convergence of these series is also $ȥ = 1$, but this point is not a crucial singularity. The function represented by the series can be continued analytically beyond $ȥ = 1$. It is clear from the integral form analogous to (5.7.3), (5.7.4) that the denominator $(ȥ^{-1}e^{\eta} + 1)$ never vanishes for positive $ȥ$. Hence the functions exist for all $ȥ$ in the domain $0 \leqslant ȥ \leqslant \infty$.

(5.7.2), or

$$n\Lambda^3 = y(z)$$

At sufficiently high temperatures, the product $n\Lambda^3(\sim nT^{-3/2})$ will be small. By using the graph of Fig. 5.7.1 we determine a value of $z(<1)$, corresponding to a given value of the density and of the temperature (see Figs. 5.7.2a and 5.7.2a′). Keeping n constant we now consider lower temperatures: $n\Lambda^3$ then increases and so does z, until $z = 1$. This happens when $T = T_c$ defined as

$$n\Lambda_c^3 = y(1)$$

or

$$T_c = \left(\frac{h^2}{(2.612)^{2/3} 2\pi m k_B}\right) n^{2/3} \tag{5.7.8}$$

If the temperature is further lowered (still at the same density) Eq. (5.7.2) has no solution any more! Similarly, let us keep the temperature constant, but increase the density up to a critical (temperature-dependent) density n_c:

$$n_c\Lambda^3 = y(1)$$

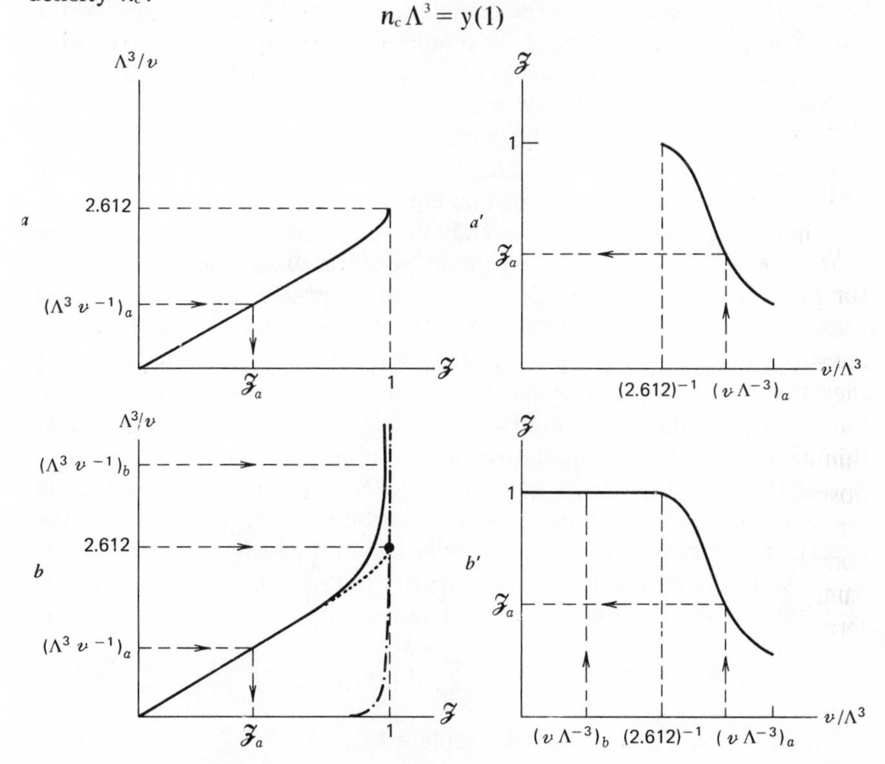

Figure 5.7.2. Graphical construction of the function $z(\mathcal{V}/\Lambda^3)$. a, a': Incorrect equation (5.7.2). b, b': Correct equation (5.7.13).

or

$$n_c^{-1} \equiv v_c = (2.612)^{-1} h^3 (2\pi m k_B)^{-3/2} T^{3/2} \tag{5.7.9}$$

(In this discussion it is sometimes useful to consider the specific volume v per particle instead of the density.) Again, for densities higher than n_c, there is no solution to Eq. (5.7.2). Everything happens as if, on trying to fit more particles into a given volume and trying to maintain equilibrium, the extra particles vanish mysteriously, falling into a trap and disappearing from the system. Clearly we have here an inconsistency: Our Eq. (5.7.2) must be incorrect.

It is the Bose statistics that gives us a clue in eliminating the paradox. We must not forget a general feature: When the temperature decreases, the equilibrium distribution of the particles tends towards the most favorable distribution in energy. In the case of bosons this distribution corresponds to an accumulation in the lowest energy state, which we (conventionally) denote by $\varepsilon_0 = 0$. Here lies the main difference with the fermion gas. There, such an accumulation is impossible because of the exclusion principle. The most favorable energetic distribution consisted in piling up the particles in the lowest levels, one per level. No level ever had an occupation number of macroscopic size. But we note that Eq. (5.7.2) does not account for the particles in the level $\varepsilon_0 = 0$: The integrand in that equation contains a factor $\eta^{1/2} = (\varepsilon / k_B T)^{1/2}$; hence particles with zero energy do not show up in this equation.

We therefore understand physically the nature of the "trap" accounting for the "missing particles." We must now formulate the problem in a satisfactory mathematical way. The flaw in the present reasoning is in the passage from Eq. (5.4.11) to Eq. (5.4.12), that is, in the evaluation of the thermodynamic limit. The passage from the sum to the integral is valid as long as the various terms in the sum are all finite, that is, the particles are thinly and smoothly distributed among the levels. However, in the case of bosons at very low temperature, there appears an accumulation of a sizable fraction of the N particles in the level $\varepsilon = 0$; hence the single term corresponding to $\varepsilon = 0$ in Eq. (5.4.11) contributes to $P\mathcal{V}$ an amount of the same order as the sum of all the other terms. In the limit $N \to \infty$, this single term would produce a deltalike singularity in (5.4.12). It must, therefore, be taken out of the sum and treated separately:

$$P\mathcal{V} = -k_B T \left\{ \ln(1 - z) + \sum_{p \neq 0} \ln\left[1 - z \exp\left(-\frac{\varepsilon_p}{k_B T}\right)\right]\right\} \tag{5.7.10}$$

Similarly, Eq. (5.4.16) must be replaced by

$$N = \frac{z}{1 - z} + \sum_{p \neq 0} \frac{1}{z^{-1} e^{-\beta \varepsilon_p} - 1} \tag{5.7.11}$$

Consider now a very large, but finite system. The sums, excluding $\mathbf{p}=0$, can be approximated by the corresponding integrals $Y(z)$ and $y(z)$. The first terms in the right-hand sides must, however, be discussed in more detail:

$$P(T, z) = k_B T[-\mathcal{V}^{-1}\ln(1-z) + \Lambda^{-3}Y(z)] \tag{5.7.12}$$

$$n(T, z) = \frac{1}{\mathcal{V}}\frac{z}{1-z} + \Lambda^{-3}y(z) \tag{5.7.13}$$

We now resume our previous discussion with Eq. (5.7.13). In the graphical construction of Fig. 5.7.2, we must add to $y(z)$ a term representing $\Lambda^3 z/\mathcal{V}(1-z)$. As long as z is significantly different from 1, this term is quite negligible because of the large factor $\mathcal{V}$ in its denominator. The value of z corresponding to a given $n\Lambda^3$ is not different from our former result. However, when z becomes "infinitesimally" close to 1, that is, $z = 1 - 0(N^{-1})$, the quantity $z/(1-z)$ becomes of order N and the first term in the rhs of (5.7.13) can no longer be neglected. The corresponding curve in the graph of Fig. 5.7.2*b* is a very steeply rising curve that goes to infinity as $z \to 1$ and that must be added to $y(z)$. This branch saves us: Indeed, we now find a solution to z for every value of $n\Lambda^3$. The paradox has disappeared. Moreover, we may now go to the thermodynamic limit. The extra contribution then becomes infinitely steep: it is a vertical line. Hence to all values of $n\Lambda^3$ below 2.612 [or $v\Lambda^{-3} > (2.612)^{-1}$] corresponds the formerly found solution, whereas to $n\Lambda^3 > 2.612$ [or $v\Lambda^{-3} < (2.612)^{-1}$] corresponds $z = 1$ (see Fig. 5.7.2b′):

$$\begin{aligned} z &= z_0(T, n), \qquad \Lambda^3 n \leqslant y(1) \\ z &= 1, \qquad \Lambda^3 n \geqslant y(1) \end{aligned} \tag{5.7.14}$$

where z_0 is the (temperature-and-density-dependent) root of the equation

$$y(z_0) = \Lambda^3 n \tag{5.7.15}$$

This root must be found by graphical or by numerical methods. Equation (5.7.14) is the correct equation replacing (5.7.2) in the thermodynamic limit. It takes due account of the macroscopic occupation of the ground state at temperatures below T_c.

Let us display more explicitly the behavior of this occupation number $\langle n_0 \rangle$ [see Eq. (5.5.5)]:

$$\langle n_0 \rangle = \frac{1}{e^{-\beta\mu} - 1} = \frac{z}{1-z} \tag{5.7.16}$$

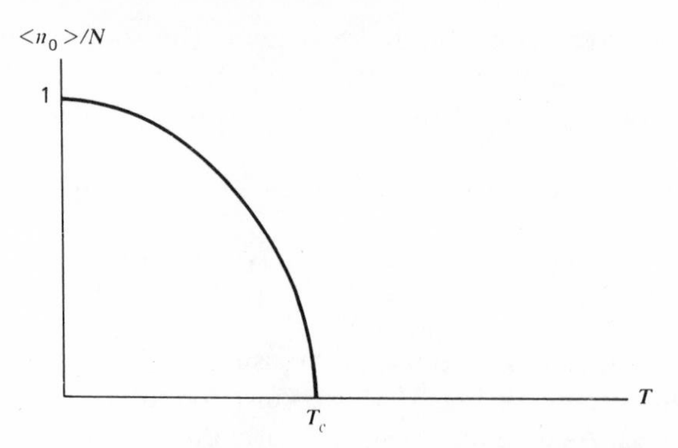

Figure 5.7.3. Fraction of particles in the ground state of a boson system at low temperature.

We note that Eq. (5.7.8) implies

$$\Lambda^{-3} y(1) = \Lambda^{-3} n \Lambda_c^3 = n\left(\frac{\Lambda_c}{\Lambda}\right)^3$$

$$= n\left(\frac{T}{T_c}\right)^{3/2}$$

Multiplying now both sides of Eq. (5.7.13) by $\mathcal{V}$, we find

$$N = \langle n_0 \rangle + N\left(\frac{T}{T_c}\right)^{3/2}, \qquad T < T_c$$

or

$$\langle n_0 \rangle = N\left[1 - \left(\frac{T}{T_c}\right)^{3/2}\right], \qquad T < T_c \tag{5.7.17}$$

This equation clearly shows the accumulation of particles in the ground state as $T \to 0$. This phenomenon is called the *Bose–Einstein condensation.* It is actually a condensation in momentum space, not in real space. The phenomenon has, however, common features with ordinary phase transitions and has been studied as a model of ordinary liquid–vapor transitions. Let us discuss some of these analogies.

Before going further we note that the fraction of particles in the ground state $\langle n_0 \rangle / N$ is a quantity that is nonzero below the critical point and vanishes abruptly at $T = T_c$ (Fig. 5.7.3). This behavior is characteristic of an *order parameter* associated with a phase transition (see Chapter 9).

We now consider the equation of state. We must again discuss the behavior of the first term in the rhs of (5.7.12). For z significantly different from 1, it is negligible because of the factor $\mathcal{V}^{-1}$. But even when $1 - z = 0(N^{-1})$, this term behaves like $\mathcal{V}^{-1} \ln N$, and also tends to zero in the T-limit. Hence the equation of state is

$$\begin{aligned} P(T, v) &= k_B T \Lambda^{-3} Y[z_0(T, v)], \qquad v > v_c \\ P(T, v) &= k_B T \Lambda^{-3} Y(1), \qquad v < v_c \end{aligned} \tag{5.7.18}$$

Hence the isotherms of an ideal Bose gas have a flat section for $0 < v < v_c$ (Fig. 5.7.4). This behavior is analogous to the region of phase coexistence in ordinary condensation (see Chapter 9). However, the analogy must not be pushed too far. If taken seriously it would describe the equilibrium of a "gaseous" phase of specific volume v_c and a "condensed" phase of zero specific volume (infinite density)! What the flat section really means is that the particles in the ground state, which have zero momentum, do not contribute to the pressure.

Among the many other thermodynamic properties one can calculate from (5.7.12) let us mention two important ones: The entropy per particle is given by

$$\begin{aligned} \frac{S}{k_B} &= \tfrac{5}{2}(\Lambda^3 n)^{-1} Y(z_0) - \ln z_0, \qquad T > T_c \\ &= \tfrac{5}{2}(\Lambda^3 n)^{-1} Y(1), \qquad T < T_c \end{aligned} \tag{5.7.19}$$

One sees that the entropy tends to zero (as $T^{3/2}$) for $T \to 0$. This is in agreement with the *third law of thermodynamics*. It represents a great success of quantum-statistical mechanics: Indeed, the Boltzmann expression (5.2.27) could not explain this property.

The specific heat per particle has a very interesting behavior (see Fig. 5.7.5):

$$\begin{aligned} \frac{c_V}{k_B} &= \tfrac{15}{4}(\Lambda^3 n)^{-1} Y(z_0) - \frac{9}{4} \frac{y(z_0)}{z_0 y'(z_0)}, \qquad T > T_c \\ &= \tfrac{15}{4}(\Lambda^3 n)^{-1} Y(1), \qquad T < T_c \end{aligned} \tag{5.7.20}$$

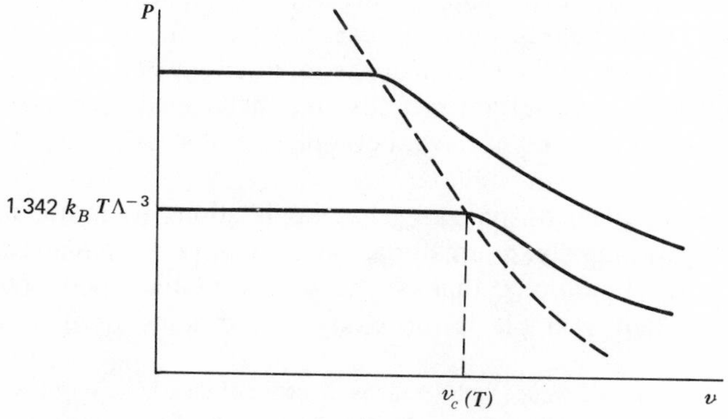

Figure 5.7.4. Isotherms of the ideal boson gas.

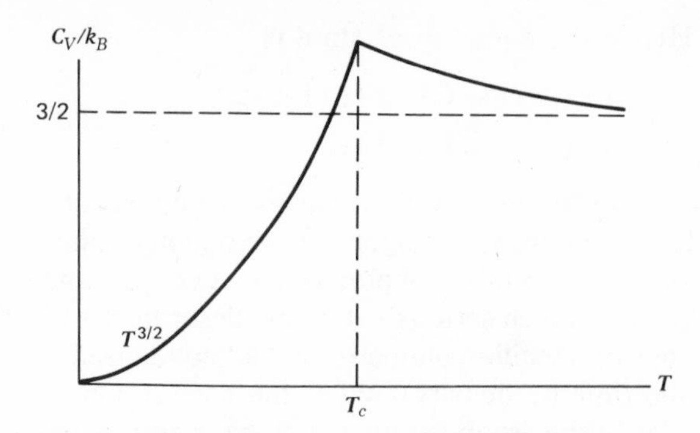

Figure 5.7.5. Specific heat of the ideal boson gas.

It starts at zero for $T=0$, growing as $T^{3/2}$. (Note the difference with a Fermion gas where $c_V \sim T$.) It remains continuous at T_c, where the curve switches to another shape, which tends to the classical value $\frac{3}{2}$ as $T \to \infty$ [see Eq. (5.2.29)]. At $T = T_c$, however, there is a cusp: The derivative dc_V/dT is discontinuous. This characteristic shape is at the origin of the name given to T_c: it is called a *lambda point*.

The problem of the Bose condensation is very interesting, but purely academic. No real physical system of molecules behaves at low temperatures as an *ideal* boson gas.* Helium is a good candidate, but it is a liquid at zero temperature, which means that the role of the intermolecular interactions certainly cannot be neglected. The behavior of liquid helium has some points in common with the behavior described above. There exists a critical temperature and a λ transition. Below the critical point, the liquid becomes *superfluid*, and this phenomenon is certainly connected to the Bose condensation of the particles in the ground state. However, the details of the behavior are very different from an ideal gas. A theory of liquid helium must necessarily be a theory of a *nonideal* boson system, in which the effects of the interactions and of the quantum statistics are combined. Much progress has been made recently in this field, although we do not yet have a complete and satisfactory theory of liquid helium.

The theory of ideal boson gases has applications to rather different kinds of systems, namely, nonmaterial systems. A very standard example is an assembly of *photons*, that is, the theory of blackbody radiation. Another important example is the study of harmonic oscillations of a

* We have here another difference with fermions. Because of their large zero-point energy, such systems may be considered at "zero temperature" (i.e., $T \ll T_F$) even at very high "real" temperatures.

crystal lattice, represented as an assembly of *phonons*. Compared to the material systems studied before, there are two main differences.

1. The energy spectrum is different: instead of being proportional to p^2, the energy levels are

$$\varepsilon_{\mathbf{p}} = cp$$

where c represents the speed of light or the speed of sound, respectively.

2. The number of particles is not fixed by a condition $\sum n_{\mathbf{p}} = N$, but, rather, is indefinite.

As a result of Condition 2 the theory of such systems becomes much simpler. The canonical partition function (5.4.4) can now be evaluated directly because it is factorized.

The reader now possesses all the tools for working out these problems by himself; hence we shall not continue their discussion here.

BIBLIOGRAPHICAL NOTES

The equilibrium properties of ideal systems are the most standard matter in statistical mechanics. They are treated in all the general textbooks quoted in Chapter 4. More special topics are the following.

The exact equation of state of boson and fermion gases in analytic form was obtained in

A. Leonard, *Phys. Rev.* **175,** 221 (1968).

M. M. Nieto, *J. Math. Phys.* **11,** 1346 (1970) (in this paper the relativistic equation of state is also treated).

The application of the Fermi statistics to the electron gas in a metal was done by

A. Sommerfeld, *Z. Phys.* **47,** 542 (1928).

This problem is treated in all textbooks on solid-state physics, for instance:

C. Kittel, *Introduction to Solid State Physics*, 3rd ed., Wiley, New York, 1968.

R. Peierls, *Quantum Theory of Solids*, Clarendon Press, Oxford, 1955.

J. M. Ziman, *Electrons and Phonons*, Clarendon Press, Oxford, 1960.

R. A. Smith, *Wave Mechanics of Crystalline Solids*, Wiley, New York, 1961.

These books also contain the application of Bose statistics to phonons, that is, quantized crystal vibrations.

The application of Bose statistics to photons can be found, for instance, in

M. Planck, *Wärmestrahlung*, Berlin, 1913 (English translation: *Theory of Heat Radiation*, reprinted by Dover, New York, 1959).

P. T. Landsberg, *Thermodynamics with Quantum Statistical Illustrations*, Wiley-Interscience, New York, 1961.

L. Landau and E. Lifshits (quoted in Chapter 4).

A brief, but thorough account of ideal degenerate gases, including relativistic effects (with an application to the theory of white dwarfs) is found in S. Chandrasekhar, *An Introduction to the Study of Stellar Structure*, Univ. Chicago Press, 1939 (reprinted, Dover, New York, 1958).

A simple and clear account of the problem of laser-driven fusion appears in J. Nuckolls, J. Emmett, and L. Wood, *Physics Today*, p. 46 (August 1973).

CHAPTER 6
SLIGHTLY NONIDEAL SYSTEMS IN EQUILIBRIUM

6.1. PERTURBATION EXPANSIONS IN STATISTICAL MECHANICS

The ideal systems considered in Chapter 5 are interesting as being the simplest conceivable systems that can be studied in statistical mechanics. From a theoretical point of view they are valuable because all the calculations can be done explicitly, and the main premises of statistical mechanics can be tested. For this reason, they are an important reference system to which other, more complicated systems can be compared.

On the other hand, ideal systems lack some of the most important features of truly nontrivial, realistic systems. In particular, we noted that such systems have no internal mechanism that could drive them to equilibrium. Also, from the practical point of view, any study of the real gases, liquids, or solids shows that their behavior is in general quite different from the ideal one. Their properties can only be understood when the *intermolecular interactions* are brought into the picture.

The study of assemblies of particles in mutual interaction thus appears as the central problem of statistical mechanics. But this problem is indeed a formidable one. Except for a very few simplified cases, such problems can no longer be solved exactly. One must, therefore, develop *approximation methods*, adapted to the various specific systems under study.

In order to fix the ideas, we consider in this chapter a classical system* of N point particles enclosed in a volume $\mathcal{V}$, in the absence of an external field. The corresponding Hamiltonian was given in Eq. (2.4.1) (with $H^F = 0$) and explained in Eqs. (2.4.2)–(2.4.10). We rewrite it as

$$H = H^0 + \lambda H' \tag{6.1.1}$$

We introduced here a dimensionless parameter λ, mainly in order to keep track of the order of magnitude of H' compared to H^0.

* The methods developed here can be extended to quantum systems as well, but they become much more involved because of the role of quantum statistics. The quantum nonideal gas will not be treated in this book. (See, however, the references at the end of this chapter.)

The standard method for the study of equilibrium properties is the evaluation of the partition function. For a classical system with the Hamiltonian (6.1.1), the partition function $Z(T, \mathcal{V}, N)$ is obtained from (4.3.26):

$$Z = h^{-3N}(N!)^{-1}\int d^N\mathbf{p}\, d^N\mathbf{q}\, e^{-\beta H^0}e^{-\beta\lambda H'} \tag{6.1.2}$$

We now note that the integrand is a product of a factor $\exp(-\beta H^0)$, which depends only on the momenta, and a factor $\exp(-\beta\lambda H')$, depending only on the positions. Hence Z factorizes as follows:

$$Z = Z_{tr}Q \tag{6.1.3}$$

where Z_{tr} is simply the ideal Boltzmann-gas partition function studied in detail in Section 5.2 [Eq. (5.2.20)]:

$$\begin{aligned} Z_{tr} &= h^{-3N}(N!)^{-1}\mathcal{V}^N\int d^N\mathbf{p}\exp(-\beta H^0) \\ &= (N!)^{-1}\mathcal{V}^N\Lambda^{3N} \end{aligned} \tag{6.1.4}$$

We are then left with the truly nontrivial factor Q, called the *configuration integral:*

$$Q(T, \mathcal{V}, N) = \mathcal{V}^{-N}\int d^N\mathbf{q}\exp(-\beta\lambda H') \tag{6.1.5}$$

The configuration integral is the central concept in the equilibrium statistical mechanics of interacting systems. From it, the deviations of the thermodynamic properties from ideal behavior can be obtained through the obvious relation:

$$A(T, \mathcal{V}, N) = A_{id}(T, \mathcal{V}, N) + A_c(T, \mathcal{V}, N) \tag{6.1.6}$$

where $A_{id}(T, \mathcal{V}, N)$ is the free energy of an ideal gas for the same values of $T, \mathcal{V}, N$ as the real gas, and the *configurational free energy* is given by

$$A_c(T, \mathcal{V}, N) = -k_BT\ln Q(T, \mathcal{V}, N) \tag{6.1.7}$$

Systems in which the contribution of the interaction Hamiltonian $\lambda H'$ to the thermodynamic properties is small compared to the contribution of H^0 are called *weakly coupled systems.* In this case we may formally expand the physical quantities in a power series in λ. If G is such a quantity (for instance, the configuration integral Q, or the free energy A, or the pressure P) we write

$$G(\lambda) = \sum_{p=0}^{\infty} G_p\lambda^p \tag{6.1.8}$$

The coefficient G_p is an expression involving p factors of H' [see, e.g.,

Eq. (6.2.1)]. We shall devise methods by which the coefficients G_p can be determined explicitly, that is, we set up a *perturbation theory* in statistical mechanics. This is a basic tool, used as well in equilibrium as in nonequilibrium problems.

As it stands, the series (6.1.8) is purely formal: nothing was said here about its convergence properties. It must, however, be kept in mind that, strictly speaking, there exist no weakly coupled gases in nature. Let us indeed discuss an important and quite typical example. The interaction Hamiltonian is a sum of pair potentials $V(r_{jn})$ as shown in Eqs. (2.4.4), (2.4.5). The shape of this potential is crucial in the determination of the properties of $Q(T, \mathcal{V}, N)$ (see also Section 4.7). A very widely used potential for the description of the mutual interaction of electrically neutral and nonpolar molecules is the *Lennard-Jones potential* (or 6–12 potential), which will be further discussed in forthcoming chapters (see Fig. 6.1.1). It is of the following form:

$$V(r) = 4\varepsilon\left[\left(\frac{\sigma}{r}\right)^{12} - \left(\frac{\sigma}{r}\right)^{6}\right] \tag{6.1.9}$$

It depends on two parameters: ε (having dimensions of an energy) and σ (having dimensions of a length). The r^{-6} term can be derived from

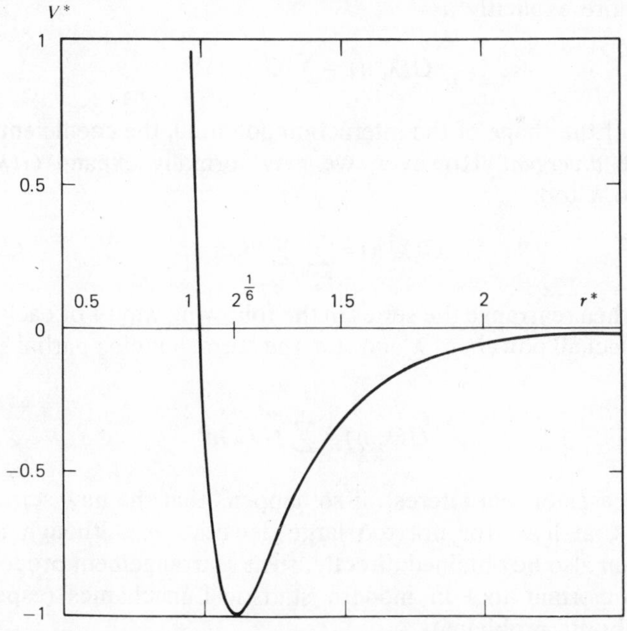

Figure 6.1.1. The Lennard-Jones (6–12) potential, Eq. (6.1.9), in reduced form: $V^* = V/\varepsilon$, $r^* = r/\sigma$.

quantum mechanics (London dispersion law). The r^{-12} term is not as firmly established; however, it is a good and simple approximation. The potential curve has a minimum, the coordinates of which are $(-\varepsilon)$ and $2^{1/6}\sigma$.* The potential is a superposition of two effects. A term proportional to $(-r^{-6})$ describes attractive forces. It dominates at large distances and tends to zero as $r \to \infty$. The second term, proportional to (r^{-12}), is repulsive. It dominates at small distances and gives rise to the steeply ascending branch of the curve: it tends quickly to infinity as $r \to 0$. This accounts for the impenetrability and the finite size of the particles. Clearly, even if ε is small compared to $k_B T$, the repulsive part of the potential cannot be treated as small. Thus, in all problems where the repulsive core plays an important role, the weak-coupling approximation breaks down.

There may, however, be other small parameters in the problem. If we study the properties of gases, for instance, the *density* n is obviously a small parameter. More generally, such nonideal systems are called *dilute systems*, and the possibility of an expansion in powers of n can be considered.

We will see that our basic λ expansion is useful in the present case too. Indeed, there exist now two parameters: λ and n; hence Eq. (6.1.8) can be written more explicitly as

$$G(\lambda, n) = \sum_{p=0}^{\infty} G_p(n)\lambda^p \tag{6.1.10}$$

Because of the shape of the interaction potential, the coefficients G_p turn out to be *divergent*. However, we may formally expand $G(\lambda, n)$ with respect to n too:

$$G(\lambda, n) = \sum_{p=0}^{\infty} \sum_{r=0}^{\infty} G_{pr} n^r \lambda^p \tag{6.1.11}$$

We may then rearrange the series in the following way: For each power of n, we collect all powers of λ and sum the corresponding partial series; the result is

$$G(\lambda, n) = \sum_{r=0}^{\infty} \bar{G}_r(\lambda) n^r \tag{6.1.12}$$

In many cases of real interest, it so happens that the new series (6.1.12) converges, at least for not too large densities n. Although the result (6.1.12) can also be obtained directly, such rearrangement procedures are a very important tool in modern statistical mechanics (especially in nonequilibrium problems).

* σ is the distance at which the potential vanishes: $V(\sigma) = 0$.

There exists a third standard case of slightly nonideal systems. A typical example is provided by the *Coulomb interactions* occurring in plasmas or in electrolyte solutions. Here the interactions are indeed weak,* but it still turns out that both the expansions in powers of λ (6.1.8) and of n (6.1.12) diverge. This is due to the fact that the interactions decrease too slowly at large distances. In other words, the difficulty lies in the *long range* of the interactions. It will be seen that in this case too, a rearrangement procedure will enable us to handle the problem. The idea here is, however, more subtle than in the previous case.

Before starting to develop these ideas, let us warn the reader about the method used in this chapter. We deliberately chose to expose the ideas in a very pedestrian, nonrigorous way. Many properties are stated without proofs. These are easily available in a number of standard textbooks. On the other hand, the machinery of combinatorial mathematics and of graph theory involved in the more rigorous proofs is pretty heavy. The main physical ideas and mathematical tools might be obscured by the façade of notations. Therefore, we hope that a simplified exposition such as the one chosen here can be considered as a thorough introduction to the subject and will facilitate the interested reader's further study.

6.2. THE λ EXPANSION OF THE CONFIGURATION INTEGRAL

We now assume that the gas is *weakly coupled.* In other words, we assume that, on the average, $\lambda H' \ll k_B T$. We may then formally expand the exponential in the integrand of Eq. (6.1.5) and obtain [using Eq. (2.4.4)]

$$Q = \mathcal{V}^{-N} \int d^N\mathbf{q} \sum_{b=0}^{\infty} (b!)^{-1}(-\lambda\beta)^b \left(\sum_{i<j}\sum V_{ij}\right)^b \tag{6.2.1}$$

On the other hand, we may expand each power of the potential by the multinomial formula:

$$\left(\sum_{i<j}\sum V_{ij}\right)^b = \underset{\sum_{s=1}^{B} m_s = b}{\sum_{m_1} \cdots \sum_{m_B}} \frac{b!}{m_1!\, m_2! \cdots m_B!} V_{kl}^{m_1} V_{mp}^{m_2} \cdots V_{tu}^{m_B} \tag{6.2.2}$$

where the number of factors in the product of V's in the right-hand side is $B = \frac{1}{2}N(N-1)$, corresponding to all possible pairs which can be made out of N particles.

The formulas (6.2.1), (6.2.2) provide in principle the desired expansion. However, their present form is still far from transparent and is certainly

* Assuming that we consider a domain where the hard core plays no significant role.

not suitable for numerical calculations. A deeper analysis will enable us to simplify considerably this expansion.

We first note that in Eq. (6.2.2), although we have in principle $\frac{1}{2}N(N-1)$ factors V_{jn}, most of the exponents m_s (in any low order b) are zero, because they are restricted by the condition $\sum m_s = b$. Hence we have *at most* b pairs contributing effectively in a term of order λ^b in Eq. (6.2.2). On the other hand, the nonvanishing exponents are distributed in all possible ways among the pairs; that is, together with, say, $V_{12}^2 V_{34}$ appear also the terms $V_{34}^2 V_{56}$, $V_{12}^2 V_{89}$, and so on. All these terms are affected by the same multinomial coefficient.

We now note that, although there can be no repeated indices on a given factor V_{ij} (e.g., V_{11} does not appear in the expansion), nothing prevents us from having repetitions of indices in different factors V_{ij} (e.g., $V_{12} V_{23} V_{24}^2$).

We next realize that many terms give identical contributions to the partition function. Indeed, when (6.2.2) is substituted into (6.2.1), all the indices become dummy integration variables. As a result, we have, for instance, in the fourth order:

$$\int d^N\mathbf{q}\ V_{12}^4 = \int d^N\mathbf{q}\ V_{13}^4 = \int d^N\mathbf{q}\ V_{78}^4 = \cdots = \mathcal{V}^{N-1} \int d\mathbf{r}\ V^4(r) \quad (6.2.3)$$

Hence, the only feature that determines the value of the integral is the fact that we have the fourth power of a factor V_{ij}; the name of the particles i, j is immaterial. Similarly

$$\int d^N\mathbf{q}\ V_{12} V_{23} V_{24} = \int d^N\mathbf{q}\ V_{13} V_{36} V_{39} = \cdots \quad (6.2.4)$$

The only property that counts here is the presence of three factors V_{ij}, having one particle label (and only one) in common. We note again that, in each group, the multinomial coefficients are the same for all "equivalent" terms.

After these preliminary remarks, we start developing a method by which the expansion (6.2.1), (6.2.2) can be considerably simplified. It is based on the following simple graphical representation. Consider a given term of order b in the right-hand side of Eq. (6.2.2). With every factor $V_{ij}^{m_s}$ for which $m_s \neq 0$, we associate a set of $m_s = m_{ij}$ lines (or "bonds") joining two points (or "vertices") labeled i and j. The resulting figure is called a *labeled graph*. Some examples are given in Fig. 6.2.1. To each graph correspond the following important numbers:

The number of bonds b
The multiplicity of each bond m_{ij}
The number of vertices p

$$V_{13}V_{35}V_{36}V_{56}$$
$$b=4$$
$$m_{ij}=1 \qquad \text{all } (ij)$$
$$p=4$$

Figure 6.2.1. Two labeled graphs, the corresponding products and their characteristics.

As mentioned above, when a given product (corresponding to a labeled graph) is integrated over all the q's, the result is a number that no longer depends on the names of the labels, but only on their mutual relationships. It is, therefore quite natural to group together all the labeled graphs that correspond to the same integral contributing to Q. The set of all b-bond labeled graphs corresponding to the same integral $\bar{J}(\Delta_b)$ will be called the *class* Δ_b. The evaluation of the partition function now amounts to the following operations:

A. Identify all the distinct classes Δ_b for given order b, and calculate the corresponding integral $\bar{J}(\Delta_b)$.

B. Determine the number $C(\Delta_b)$ of distinct labeled graphs belonging to the class Δ_b.

C. Determine the common multinomial coefficient $M(\Delta_b)$ corresponding to the graphs of class Δ_b.

D. The configuration integral is then expressed as

$$Q=\sum_{b=0}^{\infty}(b!)^{-1}(-\lambda\beta)^{b}\mu_{b} \tag{6.2.5}$$

with

$$\mu_{b}=\sum_{\Delta_b}M(\Delta_b)C(\Delta_b)\bar{J}(\Delta_b) \tag{6.2.6}$$

It turns out that in each order b, there is only a finite and (relatively) small number of classes. Hence if our program can be realized, it leads to a much simpler representation of Q compared to (6.2.1), (6.2.2). We now start carrying out our program.

Step A. As explained before, a class Δ_b is characterized by a certain type of relationship among the labels of a graph, irrespective of the names of the labels. Thus, among the terms of order b in Eq. (6.2.2), we must identify and group together in separate classes all terms having in common one of the following properties:

1. All the particle labels are distinct.
2. There is one label in common to two factors V_{ij}.
3. There is one label in common to three factors V_{ij}.
4. There are two labels in common to two factors V_{ij}.
5. There is one label in common to two factors V_{ij}, and another label in common to two other factors V_{ij}, and so on.

We now consider the graphical translation of this problem, which is shown in Fig. 6.2.2 for $b=4$. It is immediately clear that all labeled graphs that are of the form shown in Fig. 6.2.2*a* and differ from each other only by the labeling of their vertices satisfy Criterion 1. Moreover, they are the only ones that satisfy it. A similar statement holds for Criteria 2–5. Hence, we come to the conclusion that *a given class Δ_b is uniquely characterized by an unlabeled graph with b bonds.*

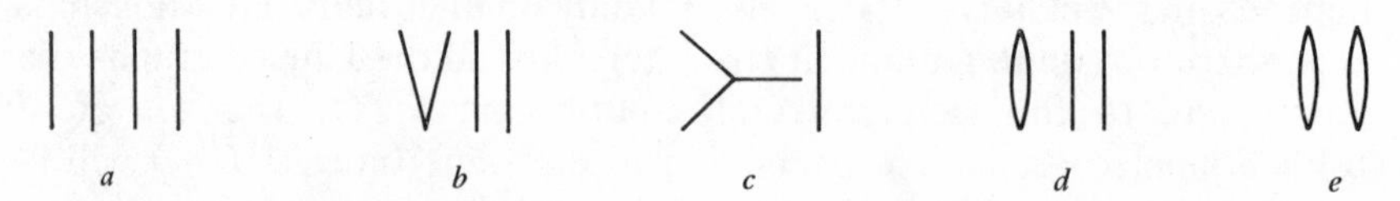

Figure 6.2.2. A few of the unlabeled graphs characterizing distinct classes for $b=4$.

To summarize this important point, we note that:

Each individual product in Eq. (6.2.2) corresponds to a labeled graph.

Each integral of an individual product corresponds to an unlabeled graph (see Fig. 6.2.3).

The problem of Step A can now be reduced to a geometrical one: *Identify all possible topologically distinct unlabeled graphs that can be constructed with b bonds.* Two graphs are defined as topologically equivalent if they have the same number of bonds and of vertices and the

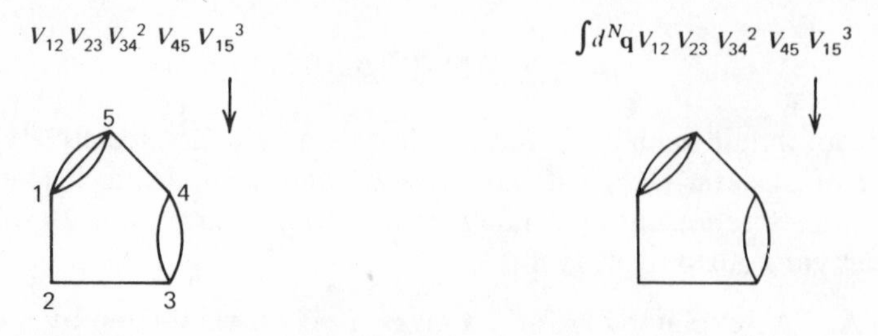

Figure 6.2.3. Labeled and unlabeled graphs.

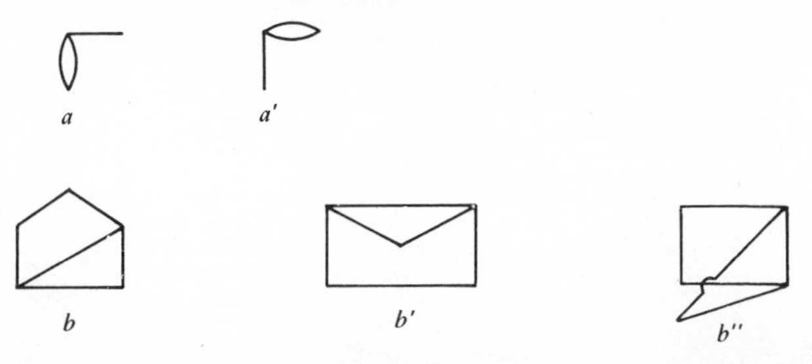

Figure 6.2.4. Examples of topologically equivalent unlabeled graphs.

same multiplicities, and if they can be superposed by letting their vertices move continuously in the plane, keeping the bonds (considered as perfectly extensible) always attached to the same vertices. In this process a vertex may cross a bond, but the new intersection thus created is not to be considered as a new vertex. Figure 6.2.4 shows some examples of topologically equivalent graphs.

It is easy to devise a systematic procedure for the construction of all the distinct classes. In first order ($b = 1$) there is obviously one single class, represented by a simple bond between two points. To obtain all possible distinct graphs to order $b = 2$ one adds to the graph of order 1 an additional line in all possible ways, that is:

(a) The line is added to the initial graph without any connection.
(b) The line is successively connected by one end to every vertex of the initial graph.
(c) The line is connected by its two ends to all possible pairs of vertices of the initial graph.

Among all the graphs obtained in this way one rejects all duplications, retaining every topologically distinct graph only once.

To construct all the graphs to order $b = 3$ one performs the operations (a), (b), (c) starting from every graph of order 2 (see Fig. 6.2.5). Continuing in this way all the distinct graphs are easily obtained.

Before going further we make some remarks about the nature of the graphs obtained in this way. We note that in a given order b we have graphs with a number of vertices ranging from 2 to $2b$. We can subdivide the various graphs in three categories. The *disconnected graphs* are those made up of two or more components having no particle in common (e.g., $2c$, $3f$, $3g$, $3h$). All the others are *connected*. These in turn are of two types. Graphs such as $2b$, $3c$, $3d$, $3e$ have at least one *articulation point* (these are denoted by A in Fig. 6.2.5), that is, a point such that if the bonds arriving at it are cut, the graph falls into disconnected pieces. Graphs with at least one articulation point will be called *reducibly connected*. Those

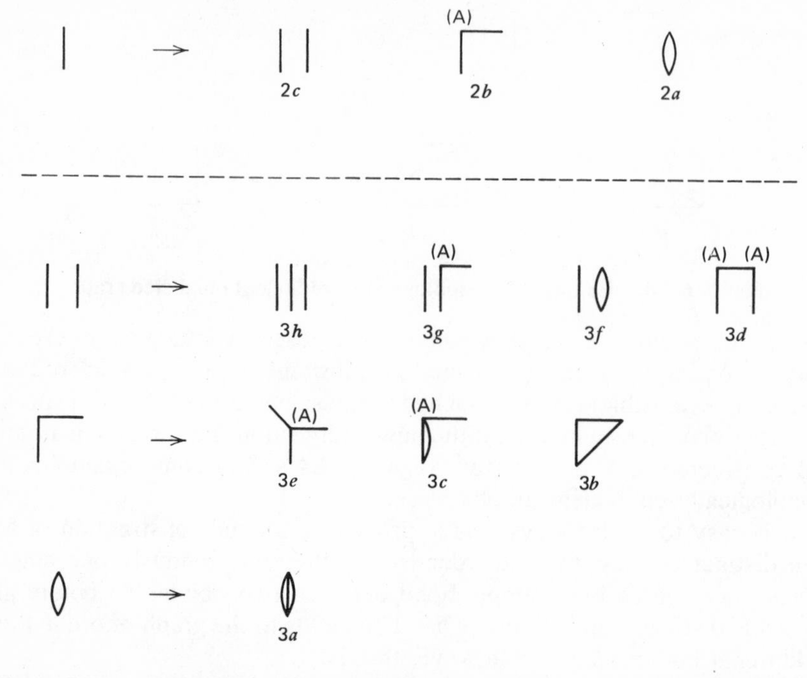

Figure 6.2.5. Generation of all the unlabeled graphs of order b from those of order $b-1$. Articulation points are denoted by (A). The numbering corresponds to Table 6.2.1.

that have no articulation point ($2a$, $3a$, $3b$) will be called *irreducibly connected* or simply *irreducible.*

By this method, we have identified all the classes in each order. The contribution of an individual member of a class to the configuration integral is very simply obtained. The vertices are first labeled by arbitrary numbers, for instance, the first natural numbers, in an arbitrary order. Then the graph is translated into an integral. For example, graph $3c$ contributes an integral:

$$\bar{J}_{3c} = \mathcal{V}^{-N} \int d^N\mathbf{q}\; V_{12}^2 V_{23} = \mathcal{V}^{-3} \int d\mathbf{q}_1\, d\mathbf{q}_2\, d\mathbf{q}_3\; V^2(\mathbf{q}_1 - \mathbf{q}_2) V(\mathbf{q}_2 - \mathbf{q}_3) \quad (6.2.7)$$

where we write $V^n(r)$ for $[V(r)]^n$. We may note now that the remaining integral in the right-hand side is proportional to the volume; indeed, as the V factors depend only on differences of $\mathbf{q}$'s we may make a change of variables to $\mathbf{q}_1$, $\mathbf{q}_2 - \mathbf{q}_1 = \mathbf{r}_{12}$, $\mathbf{q}_3 - \mathbf{q}_2 = \mathbf{r}_{32}$. We then have a free integration over $d\mathbf{q}_1$, which gives an extra factor $\mathcal{V}$. (The problem of the volume dependence will be discussed in more detail below.) Hence:

$$\bar{J}_{3c} = \mathcal{V}^{-2}\int d\mathbf{r}_{12}\, d\mathbf{r}_{32}\, V^2(r_{12})V(r_{32}) \tag{6.2.8}$$

Moreover, we note that the integral factorizes into two factors corresponding to the two components joining at the articulation point. This is a general property of the *reducibly connected* and of the *disconnected* graphs: they correspond to *factorized integrals.* This point is very important, as will be seen below.

It is also interesting for the subsequent considerations to write the integrals in the Fourier representation. This is easily done by substitution of Eq. (2.4.8) into Eq. (6.2.8):

$$\begin{aligned}
\bar{J}_{3c} &= \mathcal{V}^{-2}\int d\mathbf{r}_{12}\, d\mathbf{r}_{23}\int d\mathbf{k}\, d\mathbf{k}'\, d\mathbf{k}''\, \tilde{V}_{\mathbf{k}}\tilde{V}_{\mathbf{k}'}\tilde{V}_{\mathbf{k}''}e^{i\mathbf{k}\cdot\mathbf{r}_{12}}e^{i\mathbf{k}'\cdot\mathbf{r}_{12}}e^{i\mathbf{k}''\cdot\mathbf{r}_{23}} \\
&= \mathcal{V}^{-2}\int d\mathbf{k}\, d\mathbf{k}'\, d\mathbf{k}''\, \tilde{V}_{\mathbf{k}}\tilde{V}_{\mathbf{k}'}\tilde{V}_{\mathbf{k}''}8\pi^3\,\delta(\mathbf{k}+\mathbf{k}')8\pi^3\,\delta(\mathbf{k}'') \\
&= \mathcal{V}^{-2}(8\pi^3)^2\int d\mathbf{k}\, \tilde{V}_{\mathbf{k}}\tilde{V}_{-\mathbf{k}}\tilde{V}_0 = \mathcal{V}^{-2}(8\pi^3)^2\tilde{V}_0\int d\mathbf{k}\, \tilde{V}_{\mathbf{k}}^2
\end{aligned} \tag{6.2.9}$$

where we used the property (2.4.10). We see again very clearly that the contribution factorizes into $\tilde{V}_0$ (corresponding to the single bond) and the integral of $\tilde{V}_{\mathbf{k}}^2$ (corresponding to the double bond).

Step B. We now determine the coefficient $C(\Delta_b)$ in Eq. (6.2.6). This is done by counting how many labeled graphs correspond to the unlabeled graph characterizing the class Δ_b. In other words, we must count *in how many ways a given unlabeled graph can be labeled by means of labels chosen in the set 1, 2, 3, . . . , N.* From the illustration in Fig. 6.2.6 we get a

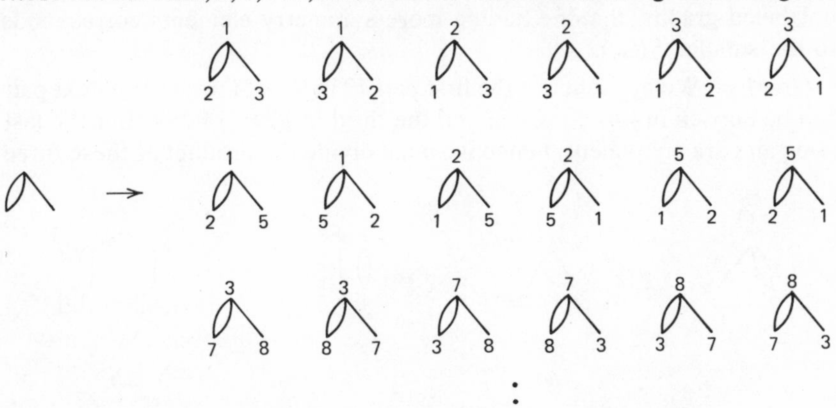

Figure 6.2.6. A few of the possible labelings of an unlabeled graph.

hint at the method of solution, which is decomposed into two substeps. Considering a diagram with p vertices, we first pick up a set of p labels among the N available. There are P possible choices, where

$$P(\Delta_b) = \frac{N!}{p!(N-p)!} \tag{6.2.10}$$

P is the number of rows in Fig. 6.2.6. In each row we drew *all the possible distinct labelings of a p-vertex graph by means of p given labels.* Let $S(\Delta_b)$ be the number of distinct labeled graphs obtained in this way, by permutations of the labels among the vertices. Since the names of the labels are immaterial in this argument, we may take them to be the first p natural numbers (first row of Fig. 6.2.6). Two labeled graphs are equivalent if they contain the same number of bonds between identically labeled vertices: two equivalent labeled graphs therefore correspond to the same product in Eq. (6.2.2). The determination of the number $S(\Delta_b)$ is a difficult problem: its value is not known for an arbitrary unlabeled graph Δ_b. $S(\Delta_b)$ strongly depends on the symmetry of the graph. Although there is no general method for its determination, it can be found by elementary reasoning in the case of small graphs. We illustrate these arguments in a few simple examples from Fig. 6.2.7.

Graph a. We fix the label 1 in the middle vertex: there are two possible ways left for labeling the two remaining vertices. Hence, $S(\Delta_a)$ equals twice the number of available labels (which can be fixed in the middle): $S(\Delta_a) = 2\times 3 = 6$ (see Fig. 6.2.6).

Graph b. We may argue as above, but now the labelings 213 and 312 are equivalent. Hence the number of distinct labelings is only one-half of the previous example: $S(\Delta_b) = 3$. This example shows that, among similar unlabeled graphs, the one having more symmetry elements corresponds to the smaller $S(\Delta_b)$.

Graph c. We may choose the first pair 12 in $\frac{1}{2}(6\times 5)$ ways. The next pair can be chosen in $\frac{1}{2}(4\times 3)$ ways, and the third in $\frac{1}{2}(2\times 1)$ ways. But the last two pairs are equivalent; hence we must divide the product of these three

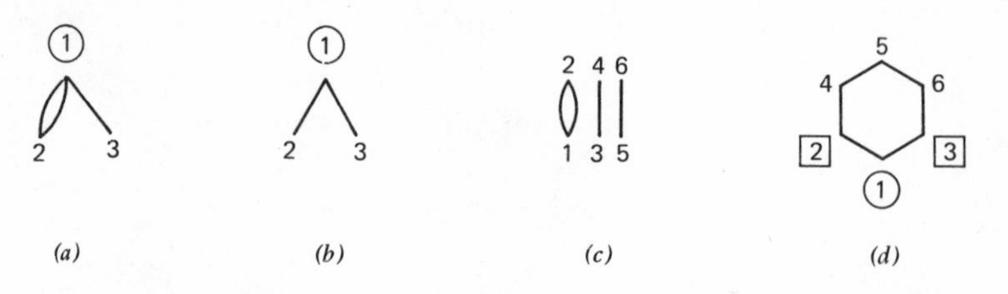

Figure 6.2.7. Examples for the computation of $S(\Delta_b)$.

Table 6.2.1. Contributions of Graphs to the Expansion of the Configuration Integral Q^a

No.	Graph	Type	$M(\Delta)$	$S(\Delta)$	$P(\Delta)$	Contribution	Value of integral
						Order 1	
1*a*		I	1	1	$\frac{N(N-1)}{2}$	$-\lambda \frac{\beta}{1!}\frac{1}{2}\frac{N(N-1)}{\mathcal{V}} J_{1a}$	$J_{1a} = \mathcal{V}^{-1}\int d\mathbf{q}_1\, d\mathbf{q}_2\, V_{12} = 8\pi^3 \tilde{V}_0$
						Order 2	
2*a*	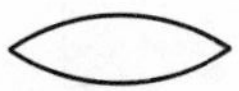	I	1	1	$\frac{N(N-1)}{2}$	$\lambda^2 \frac{\beta^2}{2!}\frac{1}{2}\frac{N(N-1)}{\mathcal{V}} J_{2a}$.	$J_{2a} = \mathcal{V}^{-1}\int d\mathbf{q}_1\, d\mathbf{q}_2\, V_{12}^2 = 8\pi^3 \int d\mathbf{k}\, \tilde{V}_{\mathbf{k}}^2$
2*b*		RC	2	3	$\frac{N!}{3!(N-3)!}$	$\lambda^2 \frac{\beta^2}{2!}\frac{N(N-1)(N-2)}{\mathcal{V}^2} J_{2b}$	$J_{2b} = J_{1a}^2$
2*c*		D	2	3	$\frac{N!}{4!(N-4)!}$	$\lambda^2 \frac{\beta^2}{2!}\frac{1}{4}\frac{N(N-1)(N-2)(N-3)}{\mathcal{V}^2} J_{2c}$	$J_{2c} = J_{1a}^2$
						Order 3	
3*a*	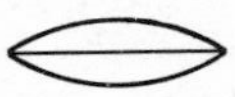	I	1	1	$\frac{N(N-1)}{2}$	$-\lambda^3 \frac{\beta^3}{3!}\frac{1}{2}\frac{N(N-1)}{\mathcal{V}} J_{3a}$	$J_{3a} = \mathcal{V}^{-1}\int d\mathbf{q}_1\, d\mathbf{q}_2\, V_{12}^3 = 8\pi^3 \int d\mathbf{k}\, d\mathbf{k}'\, \tilde{V}_{\mathbf{k}} \tilde{V}_{\mathbf{k}'} \tilde{V}_{\mathbf{k}+\mathbf{k}'}$

Table 6.2.1. *(Continued)*

No.	Graph	Type	$M(\Delta)$	$S(\Delta)$	$P(\Delta)$	Contribution	Value of integral
						Order 3 (*cont'd*)	
3*b*		I	6	1	$\frac{N!}{3!(N-3)!}$	$-\lambda^3 \frac{\beta^3}{3!} \frac{N(N-1)(N-2)}{\mathcal{V}^2} J_{3b}$	$J_{3b} = \mathcal{V}^{-1} \int d\mathbf{q}_1\, d\mathbf{q}_2\, d\mathbf{q}_3\, V_{12} V_{23} V_{31}$ $= (8\pi^3)^2 \int d\mathbf{k}\, \tilde{V}_k^3$
3*c*		RC	3	6	$\frac{N!}{3!(N-3)!}$	$-\lambda^3 \frac{\beta^3}{3!} 3 \frac{N(N-1)(N-2)}{\mathcal{V}^2} J_{3c}$	$J_{3c} = J_{2a} J_{1a}$
3*d*		RC	6	12	$\frac{N!}{4!(N-4)!}$	$-\lambda^3 \frac{\beta^3}{3!} 3 \frac{N(N-1)(N-2)}{\mathcal{V}^2} J_{3d}$	$J_{3d} = J_{1a}^3$
3*e*		RC	6	4	$\frac{N!}{4!(N-4)!}$	$-\lambda^3 \frac{\beta^3}{3!} \frac{N(N-1)(N-2)(N-3)}{\mathcal{V}^3} J_{3e}$	$J_{3e} = J_{1a}^3$
3*f*		D	3	6	$\frac{N!}{4!(N-4)!}$	$-\lambda^3 \frac{\beta^3}{3!} \frac{3}{4} \frac{N(N-1)(N-2)(N-3)}{\mathcal{V}^2} J_{3f}$	$J_{3f} = J_{2a} J_{1a}$
3*g*		D	6	30	$\frac{N!}{5!(N-5)!}$	$-\lambda^3 \frac{\beta^3}{3!} \frac{3}{2} \frac{N(N-1)\cdots(N-4)}{\mathcal{V}^3} J_{3g}$	$J_{3g} = J_{1a}^3$
3*h*		D	6	15	$\frac{N!}{6!(N-6)!}$	$-\lambda^3 \frac{\beta^3}{3!} \frac{1}{8} \frac{N(N-1)\cdots(N-5)}{\mathcal{V}^3} J_{3h}$	$J_{3h} = J_{1a}^3$

[a] Note that the volume dependence has been written explicitly; the integrals $J(\Delta)$ as defined here are volume independent, see Eq. (6.3.2).

numbers by the number of permutations of these pairs among themselves: 2! Hence

$$S(\Delta_c)=\frac{\frac{1}{2}(6\times 5)\frac{1}{2}(4\times 3)\frac{1}{2}(2\times 1)}{2!}$$

$$=\frac{6!}{2^4}$$

$$=45$$

Graph d. We fix label 1 at some vertex; then put labels 2 and 3 adjacent to it. The remaining labels can be distributed in 3! ways among the three vertices left. Permuting labels 2 and 3 does not give rise to any new graphs. Hence, the number of graphs with label 1 fixed equals 3! times the number of pairs made out of 5 particles. Starting with another particle in place of 1 does not give any new diagram. (It is equivalent with a rotation of the diagram by some angle.) Hence $S(\Delta_d)=\frac{1}{2}(5\times 4)\times 3!=\frac{1}{2}5!$. More generally, for every regular b-polygon (or "b-ring") we have $S_b=\frac{1}{2}(n-1)!$

Step C. The multinomial coefficient $M(\Delta_b)$, from Eq. (6.2.2), is very simply expressed in terms of the number b of bonds and the multiplicities m_b of the various bonds of the unlabeled graph Δ_b:

$$M(\Delta_b)=\frac{b!}{\prod_s m_s!} \tag{6.2.11}$$

Step D. We now collect our results by rewriting Eq. (6.2.6) as

$$\mu_b=\sum_{\Delta_b}\frac{b!}{\prod_s m_s!}\frac{N!}{p!(N-p)!}S(\Delta_b)\bar{J}(\Delta_b) \tag{6.2.12}$$

The summation is over *all distinct unlabeled graphs with b bonds.* This equation gives the solution to our problem of expansion of the configuration integral in powers of λ.

In Table 6.2.1 we collected the graphs and their contributions, complete through order $b=3$. The graphs are classified according to their irreducible, reducibly connected or disconnected type. We point out again the factorization of the reducibly connected and of the disconnected graphs: Its important role will appear clearly in the next section.

6.3. THE λ EXPANSION OF THE FREE ENERGY

An important problem that can be considered now is the behavior of the series derived in Section 6.2 in the thermodynamic limit. We therefore study the dependence of the terms in this series on the volume and on the number of particles.

First we note that the volume dependence comes in through the integrations over **q**. Consider thus again the integral corresponding to the graph 3*c*, Eqs. (6.2.7), (6.2.8):

$$\bar{J}_{3c} = \mathcal{V}^{-3}\int d\mathbf{q}_1\, d\mathbf{q}_2\, d\mathbf{q}_3\, V_{12}^2 V_{23}$$

$$= \mathcal{V}^{-3}\mathcal{V}\int d\mathbf{r}_{12}\, d\mathbf{r}_{23}\, V_{12}^2 V_{23} \tag{6.3.1}$$

In a first step we derived a trivial factor $\mathcal{V}^{-p}$ by integrating out all the particles not directly involved in the graph. We then noted that the remaining integral is proportional to the volume. Indeed, we may consider the three particles 123 as forming a real group, or *cluster.* We then transform the variables to those of particle 1, and to relative variables $\mathbf{r}_{12}$, $\mathbf{r}_{23}$. If the potential is sufficiently well behaved in order that the integral be convergent, the effective volume of integration over r_{ij} is not the whole volume, but a much smaller domain, the size of which is of the order of the range of the interactions r_0. Indeed, for $r > r_0$ the integrand is negligibly small. Actually, in the double integral the condition $r < r_0$ must be satisfied both for r_{12} and r_{23}, which further restricts the volume of integration. Particle 1 can move freely throughout the volume; therefore, the integration over $\mathbf{q}_1$ produces a factor $\mathcal{V}$. In this motion, however, the particle 1 "drags along" the two other particles, which, so to say, "stick" together. Hence the integral (6.3.1) depends on the volume as $\mathcal{V}^{-2} = \mathcal{V}^{-(p-1)}$.

On the other hand, if we consider a similar, but disconnected diagram, such as 3*f*:

$$\bar{J}_{3f} = \mathcal{V}^{-4}\int d\mathbf{q}_1\, d\mathbf{q}_2\, d\mathbf{q}_3\, d\mathbf{q}_4\, V_{12}^2 V_{34} = \mathcal{V}^{-4}\mathcal{V}\int d\mathbf{r}_{12}\, V_{12}^2 \mathcal{V}\int d\mathbf{r}_{34}\, V_{34}$$

The fourfold integral obtained in the first step is now proportional to $\mathcal{V}^2$ because the disconnected graph describes two clusters that can move freely through the volume. More generally, one can infer from this discussion that a graph, having p vertices and consisting of c connected components, depends on volume as follows*:

$$\bar{J}(\Delta_b) = \mathcal{V}^{-(p-c)} J(\Delta_b) \tag{6.3.2}$$

where $J(\Delta_b)$ is independent of the volume.

* It is clear from the argument that the statement is true only for large enough volumes, compared to the sphere of interaction, $\mathcal{V} \gg r_0^3$.

We now study the dependence on the number of particles N. It comes in through the factor $P(\Delta_b)$, Eq. (6.2.10). For large N ($N \gg p$):

$$P(\Delta_b) = \frac{N!}{p!(N-p)!} = N(N-1)\cdots(N-p+1) \approx N^p + O(N^{p-1})$$

Combining these two results, we see that a general graph contributes to Q a term of the following order in the T-limit*:

$$\text{T-lim } Q_\Delta \sim N^p \mathcal{V}^{-p+c} = N^c n^{p-c} \tag{6.3.3}$$

Hence the various terms in the λ expansion of Q behave differently in the thermodynamic limit: They all diverge proportionally to a power of N determined by their connectedness. Because of this lack of homogeneity one cannot give any meaning to Q in that limit.

This divergence of Q as a function of N is, however, not *a priori* a nuisance. Indeed, the quantity having a physical meaning is the free energy $A_c = -k_B T \ln Q$, not Q itself. Our next task is therefore to derive an expansion for A_c and check that all terms in *that* expansion are proportional to N, as one expects for an extensive thermodynamic quantity.

The derivation of a series expansion for A_c is not difficult if one notes an analogy with some basic concepts of mathematical statistics.

Let us introduce a formal averaging operation, denoted by a bar (in order to distinguish it from the usual statistical-mechanical ensemble average $\langle \cdots \rangle$):

$$\bar{g} = \mathcal{V}^{-N} \int d^N\mathbf{q} \; g(\mathbf{q}_1, \ldots, \mathbf{q}_N) \tag{6.3.4}$$

In particular, we may consider as typical "random" functions (in this sense) the various powers of the total potential energy H', Eq. (6.1.1), and define the *moments of* H' as

$$\mu_b \equiv \overline{H'^b} = \mathcal{V}^{-N} \int d^N\mathbf{q} \; H'^b, \qquad b = 0, 1, 2, \ldots \tag{6.3.5}$$

A well-known statistical method for evaluating the moments consists in introducing a moment generating function $\Phi(\alpha)$ defined as

$$\Phi(\alpha) = \overline{\exp(\alpha H')} = \mathcal{V}^{-N} \int d^N\mathbf{q} \; \exp(\alpha H') \tag{6.3.6}$$

Clearly, the configuration integral Q, Eq. (6.1.5), is precisely of this form, with $\alpha = -\beta\lambda$. $\Phi(\alpha)$ "generates" the moments in the sense that the

* Note that $p - c > 0$.

coefficient of $(-\beta\lambda)^b/b!$ in the power series expansion with respect to $(-\beta\lambda)$, Eq. (6.2.5), is nothing other than the bth moment μ_b of H'.

Together with the moment-generating function, one considers in statistics a *cumulant-generating function,* which is defined as $\ln \Phi(\alpha)$. The coefficient of $\alpha^b/b!$ in its series expansion is called the *cumulant of order* b, κ_b. Clearly, $\ln \Phi(-\beta\lambda)$ is nothing other than $(-\beta A_c)$, the configurational free energy. Hence we can immediately write down its series expansion in terms of the cumulants of H':

$$-\beta A_c = \ln \Phi(-\beta\lambda) = \sum_{b=1}^{\infty} \frac{(-\beta\lambda)^b}{b!} \kappa_b \tag{6.3.7}$$

The expansion (6.3.7) can be deduced from (6.2.5) in a quite straightforward way. Indeed, the cumulants κ_b are related to the moments μ_b in a well-known way.* We shall give here explicit expressions for the first four cumulants:

$$\begin{aligned} \kappa_1 &= \mu_1 \\ \kappa_2 &= \mu_2 - \mu_1^2 \\ \kappa_3 &= \mu_3 - 3\mu_2\mu_1 + 2\mu_1^3 \\ \kappa_4 &= \mu_4 - 4\mu_3\mu_1 - 3\mu_2^2 + 12\mu_2\mu_1^2 - 6\mu_1^4 \end{aligned} \tag{6.3.8}$$

We may now check, in the simplest case, a fundamental property. Let us calculate explicitly the cumulant κ_2. From the data of Table 6.2.1 we have

$$\begin{aligned} \kappa_2 &= \mu_2 - \mu_1^2 \\ &= [\tfrac{1}{2}N(N-1)\mathcal{V}^{-1}J_{2a} + N(N-1)(N-2)\mathcal{V}^{-2}(8\pi_3\tilde{V}_0)^2 \\ &\quad + \tfrac{1}{4}N(N-1)(N-2)(N-3)\mathcal{V}^{-2}(8\pi^3\tilde{V}_0)^2] \\ &\quad - \tfrac{1}{4}[N(N-1)]^2(\mathcal{V}^{-1}8\pi^3\tilde{V}_0)^2 \end{aligned}$$

Rearranging the terms we find the following coefficient for $\mathcal{V}^{-2}(8\pi^3\tilde{V}_0)^2$

$$(\tfrac{1}{4}-\tfrac{1}{4})N^4 + (1-\tfrac{6}{4}+\tfrac{2}{4})N^3 + (-3+\tfrac{11}{4}-\tfrac{1}{4})N^2 + \cdots$$

The coefficients of N^4 and of N^3 vanish identically. Hence, in the thermodynamic limit, J_{2a} has a coefficient of order N, whereas $\tilde{V}_0^2$ has a coefficient of order 1, that is, it is negligible compared to the former. Hence, the reducibly connected and the disconnected graphs are canceled in the cumulant by the counterterm μ_1^2 (up to terms negligible in the thermodynamic limit). Clearly, the factorization property of the reducible and disconnected graphs is essential in the argument. The property found

* The general expression of κ_b and explicit expressions for the first 10 cumulants can be found in M. G. Kendall, *Advanced Theory of Statistics,* Vol. 1, Griffin, London, 1952.

here in second order is quite general. We shall refrain from giving a formal proof, but the reader is urged to check the property at least in order 3. The result can be stated as follows.

The bth-order cumulant κ_b is given by the sum of all bth-order irreducibly connected graphs (up to terms of relative order N^{-1}):

$$\text{T-lim } \kappa_b = \sum_{\substack{\Delta \\ \text{(irred.)}}} \frac{b!}{\prod_s m_s!} N n^{p-1} S(\Delta_b) J(\Delta_b) \tag{6.3.9}$$

It follows that the configurational free energy per particle is expressed in terms of the following series, combining Eqs. (6.3.7) and (6.3.9):

$$-\beta a_c(T, n) = \sum_{b=1}^{\infty} \left(-\frac{\lambda}{k_B T}\right)^b C_b(n) \tag{6.3.10}$$

with

$$C_b(n) = \sum_{\substack{\Delta_b \\ \text{(irred.)}}} \frac{1}{\prod_s m_s!} n^{p-1} S(\Delta_b) \mathcal{V}^{-1} \int d\mathbf{q}_1 \cdots d\mathbf{q}_p \left(\prod V_{ij}\right)_{\Delta_b} \tag{6.3.11}$$

The summation is over *all irreducible graphs with b bonds;* $(\prod V_{ij})_{\Delta_b}$ is a product of factors V_{ij} connected according to the graph Δ_b.

Hence the series expansion of the free energy (6.3.10) is actually much simpler than the expansion (6.2.5) of the configuration integral: It is obtained from the latter by barring all reducibly connected and all disconnected graphs and dividing the result by N. As a result, *every term in the series becomes independent of* N. This important fact yields a check of the existence of the free energy per particle, *provided that:*

1. The integrals $J(\Delta_b)$ exist (in the thermodynamic limit).
2. The series converges.

These two conditions are by no means trivial, as will be seen in the next sections. Condition 1 is really a restriction on the potential functions V of the same kind as those discussed in Section 4.7; Condition 2 is a condition on temperature and density values* and is closely connected with the phase-transition problem.

For the purpose of reference, the contributions of the irreducible graphs are collected in Table 6.3.1. Here we have used the thermodynamic-limit approximation in replacing expressions such as $(N-1)(N-2)\mathcal{V}^{-2}$ by n^2.

Let us conclude with some comments on the thermodynamic functions

* Provided Condition 1 is satisfied!

Table 6.3.1. Contribution of Irreducible Graphs to the Free Energy per Particle, $-\beta a_c(n, T)$

Δ	Graph	$M(\Delta)$	$S(\Delta)$	Contribution	Value of integral J_Δ
					Order 1
$1a$		1	1	$-\lambda \frac{\beta}{1!} \frac{1}{2} n J_{1a}$	$\mathcal{V}^{-1} \int d\mathbf{q}_1\, d\mathbf{q}_2\, V_{12} = 8\pi^3 \tilde{V}_0$
					Order 2
$2a$		1	1	$\lambda^2 \frac{\beta^2}{2!} \frac{1}{2} n J_{2a}$	$\mathcal{V}^{-1} \int d\mathbf{q}_1\, d\mathbf{q}_2\, V_{12}^2 = 8\pi^3 \int d\mathbf{k}\, \tilde{V}_{\mathbf{k}}^2$
					Order 3
$3a$		1	1	$-\lambda^3 \frac{\beta^3}{3!} \frac{1}{2} n J_{3a}$	$\mathcal{V}^{-1} \int d\mathbf{q}_1\, d\mathbf{q}_2\, V_{12}^3 = 8\pi^3 \int d\mathbf{k}\, d\mathbf{k}'\, \tilde{V}_{\mathbf{k}} \tilde{V}_{\mathbf{k}'} \tilde{V}_{\mathbf{k}+\mathbf{k}'}$
$3b$	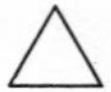	6	1	$-\lambda^3 \frac{\beta^3}{3!} n^2 J_{3b}$	$\mathcal{V}^{-1} \int d\mathbf{q}_1\, d\mathbf{q}_2\, d\mathbf{q}_3\, V_{12} V_{23} V_{31} = (8\pi^3)^2 \int d\mathbf{k}\, \tilde{V}_{\mathbf{k}}^3$
					Order 4
$4a$	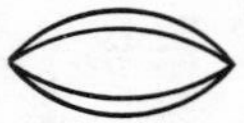	1	1	$\lambda^4 \frac{\beta^4}{4!} \frac{1}{2} n J_{4a}$	$\mathcal{V}^{-1} \int d\mathbf{q}_1\, d\mathbf{q}_2\, V_{12}^4 = 8\pi^3 \int d\mathbf{k}\, d\mathbf{k}'\, d\mathbf{k}''\, \tilde{V}_{\mathbf{k}} \tilde{V}_{\mathbf{k}'} \tilde{V}_{\mathbf{k}''} \tilde{V}_{\mathbf{k}+\mathbf{k}'+\mathbf{k}''}$

4*b*		12	3	$\lambda^4 \frac{\beta^4}{4!} 6n^2 J_{4b}$	$\mathscr{V}^{-1} \int d\mathbf{q}_1\, d\mathbf{q}_2\, d\mathbf{q}_3\, V_{12} V_{23} V_{13}^2 = (8\pi^3)^2 \int d\mathbf{k}\, d\mathbf{k}'\, \tilde{V}_{\mathbf{k}}^2 \tilde{V}_{\mathbf{k}'} \tilde{V}_{\mathbf{k}+\mathbf{k}'}$
4*c*		24	3	$\lambda^4 \frac{\beta^4}{4!} 3n^3 J_{4c}$	$\mathscr{V}^{-1} \int d\mathbf{q}_1 \cdots d\mathbf{q}_4\, V_{12} V_{23} V_{34} V_{14} = (8\pi^3)^3 \int d\mathbf{k}\, \tilde{V}_{\mathbf{k}}^4$
					Order 5
5*a*		1	1	$-\lambda^5 \frac{\beta^5}{5!} \frac{1}{2} n J_{5a}$	$\mathscr{V}^{-1} \int d\mathbf{q}_1\, d\mathbf{q}_2\, V_{12}^5 = 8\pi^3 \int d\mathbf{k}\, d\mathbf{k}'\, d\mathbf{k}''\, d\mathbf{k}'''\, \tilde{V}_{\mathbf{k}} \tilde{V}_{\mathbf{k}'} \tilde{V}_{\mathbf{k}''} \tilde{V}_{\mathbf{k}'''} \tilde{V}_{\mathbf{k}+\mathbf{k}'+\mathbf{k}''+\mathbf{k}'''}$
5*b*		20	3	$-\lambda^5 \frac{\beta^5}{5!} 10n^2 J_{5b}$	$\mathscr{V}^{-1} \int d\mathbf{q}_1\, d\mathbf{q}_2\, d\mathbf{q}_3\, V_{12} V_{23} V_{13}^3 = (8\pi^3)^2 \int d\mathbf{k}\, d\mathbf{k}'\, d\mathbf{k}''\, \tilde{V}_{\mathbf{k}}^2 \tilde{V}_{\mathbf{k}'} \tilde{V}_{\mathbf{k}''} \tilde{V}_{\mathbf{k}+\mathbf{k}'+\mathbf{k}''}$
5*c*		30	3	$-\lambda^5 \frac{\beta^5}{5!} 15n^2 J_{5c}$	$\mathscr{V}^{-1} \int d\mathbf{q}_1\, d\mathbf{q}_2\, d\mathbf{q}_3\, V_{12} V_{23}^2 V_{13}^2 = (8\pi^3)^2 \int d\mathbf{k}\, d\mathbf{k}'\, d\mathbf{k}''\, \tilde{V}_{\mathbf{k}} \tilde{V}_{\mathbf{k}'} \tilde{V}_{\mathbf{k}+\mathbf{k}'} \tilde{V}_{\mathbf{k}''} \tilde{V}_{\mathbf{k}+\mathbf{k}''}$
5*d*		60	12	$-\lambda^5 \frac{\beta^5}{5!} 30n^3 J_{5d}$	$\mathscr{V}^{-1} \int d\mathbf{q}_1 \cdots d\mathbf{q}_4\, V_{12} V_{23} V_{34} V_{14}^2 = (8\pi^3)^3 \int d\mathbf{k}\, d\mathbf{k}'\, \tilde{V}_{\mathbf{k}}^3 \tilde{V}_{\mathbf{k}'} \tilde{V}_{\mathbf{k}+\mathbf{k}'}$
5*e*		120	6	$-\lambda^5 \frac{\beta^5}{5!} 30n^3 J_{5e}$	$\mathscr{V}^{-1} \int d\mathbf{q}_1 \cdots d\mathbf{q}_4\, V_{12} V_{23} V_{34} V_{14} V_{13} = (8\pi^3)^3 \int d\mathbf{k}\, d\mathbf{k}'\, \tilde{V}_{\mathbf{k}}^2 \tilde{V}_{\mathbf{k}'}^2 \tilde{V}_{\mathbf{k}+\mathbf{k}'}$
5*f*		120	12	$-\lambda^5 \frac{\beta^5}{5!} 12n^4 J_{5f}$	$\mathscr{V}^{-1} \int d\mathbf{q}_1 \cdots d\mathbf{q}_5\, V_{12} V_{23} V_{34} V_{45} V_{15} = (8\pi^3)^4 \int d\mathbf{k}\, \tilde{V}_{\mathbf{k}}^5$

derived from Eq. (6.3.10). These can be calculated in a straightforward way. Indeed, the temperature dependence is quite explicit in the form $(-\beta\lambda)^b$; the density dependence also appears very simply, because each cumulant κ_b is made up of a finite number of terms proportional to powers of the density ranging from n to n^{b-1}. However, these expansions are not of much practical interest. If we regard (6.3.10) seriously as a λ expansion, we must realize that λ is not a parameter that we can vary arbitrarily: it is fixed once for all for a given system. Equation (6.3.10) may then allow us to compare the values of thermodynamic properties of different systems in the same conditions of temperature and density. But we have seen that weakly coupled systems do not really exist.

The main interest of the λ expansion lies in its being a valuable starting point for rearrangements adapted to a variety of problems.

6.4. THE DENSITY EXPANSION OF THE FREE ENERGY: THE VIRIAL COEFFICIENTS

We have already pointed out in Section 6.1 that the λ expansion is usually not a very realistic one and cannot be used as such in the applications. The reason is the finite size of the particles, which implies a very rapid rise in the potential energy at small distances (see Fig. 6.1.1). To obtain an expansion valid in presence of a repulsive core, we must look for another small parameter. If the system is sufficiently dilute, it is very easy to find such a parameter. Calling d_0 the measure of the distance of closest approach of two molecules (i.e., the size of the molecule), we construct the dimensionless parameter:

$$\gamma = d_0^3 n \tag{6.4.1}$$

The cubic root of this parameter is proportional to the ratio of the size d_0 of the molecules to the average distance between molecules. It is then natural to consider slightly nonideal systems defined by the condition:

$$\gamma \ll 1 \tag{6.4.2}$$

This condition clearly applies to *dilute gases*, but fails for liquids or solids. From Eq. (6.4.1) we see that an expansion in powers of γ coincides with a formal expansion in powers of the density n. According to the discussion of Section 6.1, we therefore collect from our basic λ expansion all terms proportional to $n, n^2, n^3, \ldots$. There is an infinite number of terms in each group: this clearly means that the interactions cannot be considered as small, but must be retained to all orders. Every partial series—corresponding to a given power of n—must then be summed.

In Table 6.4.1 we have collected the first few graphs (taken from Table

Table 6.4.1. Partial Summation of Graphs for the Density Expansion of the Free Energy

Graph	Contribution
	Order n
	$-\frac{\beta}{1!}\frac{1}{2}n\mathcal{V}^{-1}\int_{12} V_{12}$
	$\frac{\beta^2}{2!}\frac{1}{2}n\mathcal{V}^{-1}\int_{12} V_{12}^2$
	$\frac{\beta^3}{3!}\frac{1}{2}n\mathcal{V}^{-1}\int_{12} V_{12}^3$
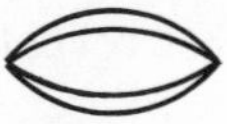	$\frac{\beta^4}{4!}\frac{1}{2}n\mathcal{V}^{-1}\int_{12} V_{12}^4$
	Order n^2
	$-\frac{\beta^3}{3!}n^2\mathcal{V}^{-1}\int_{123} V_{12}V_{23}V_{13}$
	$\frac{\beta^4}{4!}6n^2\mathcal{V}^{-1}\int_{123} V_{12}^2V_{23}V_{13}$
	$-\frac{\beta^5}{5!}15n^2\mathcal{V}^{-1}\int_{123} V_{12}^2V_{23}^2V_{13}$
	$\frac{\beta^6}{6!}15n^2\mathcal{V}^{-1}\int_{123} V_{12}^2V_{23}^2V_{13}^2$
	$-\frac{\beta^5}{5!}10n^2\mathcal{V}^{-1}\int_{123} V_{12}^3V_{23}V_{13}$
	$\frac{\beta^6}{6!}60n^2\mathcal{V}^{-1}\int_{123} V_{12}^3V_{23}^2V_{13}$
	Order n^3
	$\frac{\beta^4}{4!}3n^3\mathcal{V}^{-1}\int_{1234} V_{12}V_{23}V_{34}V_{14}$
	$-\frac{\beta^5}{5!}30n^3\mathcal{V}^{-1}\int_{1234} V_{12}^2V_{23}V_{34}V_{14}$
	$\frac{\beta^6}{6!}60n^3\mathcal{V}^{-1}\int_{1234} V_{12}^3V_{23}V_{34}V_{14}$
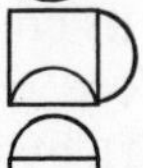	$\frac{\beta^6}{6!}90n^3\mathcal{V}^{-1}\int_{1234} V_{12}^2V_{23}^2V_{34}V_{14}$
	$\frac{\beta^6}{6!}45n^3\mathcal{V}^{-1}\int_{1234} V_{12}^2V_{23}V_{34}^2V_{14}$
	$-\frac{\beta^5}{5!}30n^3\mathcal{V}^{-1}\int_{1234} V_{12}V_{23}V_{34}V_{14}V_{13}$
	$\frac{\beta^6}{6!}90n^3\mathcal{V}^{-1}\int_{1234} V_{12}V_{23}V_{34}V_{14}V_{13}^2$

Table 6.4.1. (*Continued*)

Graph	Contribution
	Order n^3 (*cont'd*)
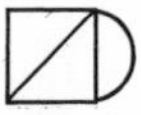 [graph]	$\frac{\beta^6}{6!}360n^3\mathcal{V}^{-1}\int_{1234} V_{12}^2V_{23}V_{34}V_{14}V_{13}$
 [graph]	$\frac{\beta^6}{6!}30n^3\mathcal{V}^{-1}\int_{1234} V_{12}V_{23}V_{34}V_{14}V_{13}V_{24}$
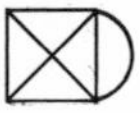 [graph]	$-\frac{\beta^7}{7!}630n^3\mathcal{V}^{-1}\int_{1234} V_{12}^2V_{23}V_{34}V_{14}V_{13}V_{24}$

6.3.1) corresponding to n, n^2, and n^3. A first look at these reveals a striking feature, which follows from (6.3.10):

The term of order n^p in the density expansion consists of all the irreducible graphs involving $p+1$ particles.

This link between the order in density and the number of particles involved in the corresponding process will also be found in the non-equilibrium theory. It motivates the traditional name of *cluster expansion* given to this series: The term of order n^p consists of the contributions of all "clusters" of $p+1$ particles.

We now proceed to the explicit summations. The terms in n are very simply summed:

$$-\beta a_c^{(1)}(n,T)=\tfrac{1}{2}n\mathcal{V}^{-1}\int d\mathbf{q}_1\,d\mathbf{q}_2\left[-\lambda\beta V_{12}+\frac{1}{2!}\lambda^2\beta^2V_{12}^2+\cdots+\frac{1}{s!}(-\lambda\beta V_{12})^s+\cdots\right]$$

$$=\tfrac{1}{2}n\mathcal{V}^{-1}\int d\mathbf{q}_1\,d\mathbf{q}_2\,f_{12} \tag{6.4.3}$$

where we introduce the *Mayer function* f_{ij}:

$$f_{ij}\equiv f(r_{ij})=\exp(-\lambda\beta V_{ij})-1 \tag{6.4.4}$$

We already note an important fact. The result of the summation is obtained by taking the simplest graph and replacing in its contribution the factor $(-\lambda\beta V_{12})$ by the factor f_{12}. We may, therefore, represent the complete sum (6.4.3) by the single graph $1a$ (Table 6.3.1), by simply giving a new meaning to the line. This process of reinterpretation of a graph by means of new rules, resulting from an infinite partial summation, will be called a *renormalization process.**

* This name stems from a (loose) analogy with the renormalization process of quantum field theory.

The Mayer function f_{ij} has several important properties. For $r_{ij} \to \infty$, $V_{ij} \to 0$ and $f_{ij} \to 0$ too. For $r_{ij} \to 0$, $V_{ij} \to \infty$ and $f_{ij} \to -1$. Hence, *f_{ij} is finite even when V_{ij} is infinite;* this is why a theory of strong repulsive interactions is possible in terms of f_{ij} and not of V_{ij}; this property ensures the convergence of the integrals in the region of the hard core.

The shape of the function $f(r)$ compared to $V(r)$ is shown schematically in Fig. 6.4.1. It is clearly seen that $f(r) \neq 0$ only when the particles are effectively correlated through interactions. If the forces are of sufficiently short range, the domain of integration in the integral of Eq. (6.4.3) is restricted to the range of interaction (i.e., to a volume much smaller than $\mathcal{V}$) for all variables but one. All integrals of type (6.4.3) [see Eq. (6.4.7)] are, therefore, exactly proportional to the volume in the case of short-range forces.

We now consider the terms of order n^2, that is, the clusters of three particles. We have arranged the diagrams in Table 6.4.1 in an order convenient for the summations. The first few terms will give us the hint for the complete series:

$$
\begin{aligned}
-\beta a_c^{(2)} &= n^2\mathcal{V}^{-1}\int d\mathbf{q}_1\, d\mathbf{q}_2\, d\mathbf{q}_3 \left(-\frac{1}{3!}\lambda^3\beta^3 V_{12}V_{23}V_{13} + \frac{6}{4!}\lambda^4\beta^4 V_{12}^2 V_{23}V_{13}\right. \\
&\quad \left. -\frac{15}{5!}\lambda^5\beta^5 V_{12}^2 V_{23}^2 V_{13} + \frac{15}{6!}\lambda^6\beta^6 V_{12}^2 V_{23}^2 V_{13}^2 + \cdots\right) \\
&= \frac{1}{3!} n^2\mathcal{V}^{-1}\int d\mathbf{q}_1\, d\mathbf{q}_2\, d\mathbf{q}_3 \left[(-\lambda\beta V_{12})(-\lambda\beta V_{23})(-\lambda\beta V_{13})\right. \\
&\quad + \frac{3}{2!}(-\lambda\beta V_{12})^2(-\lambda\beta V_{23})(-\lambda\beta V_{13}) \\
&\quad + \frac{3}{2!\,2!}(-\lambda\beta V_{12})^2(-\lambda\beta V_{23})^2(-\lambda\beta V_{13}) \\
&\quad \left. + \frac{1}{2!\,2!\,2!}(-\lambda\beta V_{12})^2(-\lambda\beta V_{23})^2(-\lambda\beta V_{13})^2 + \cdots\right]
\end{aligned}
$$

One now takes advantage of symmetry properties such as

$$
3\int d\mathbf{q}_1\, d\mathbf{q}_2\, d\mathbf{q}_3\, u_{12}^2 u_{23} u_{13} = \int d\mathbf{q}_1\, d\mathbf{q}_2\, d\mathbf{q}_3\, (u_{12}^2 u_{23} u_{13} + u_{12} u_{23}^2 u_{13} + u_{12} u_{23} u_{13}^2)
$$

and one finally finds

$$
\begin{aligned}
-\beta a_c^{(2)} &= \frac{1}{3!} n^2\mathcal{V}^{-1}\int d\mathbf{q}_1\, d\mathbf{q}_2\, d\mathbf{q}_3 \left[(-\lambda\beta V_{12}) + \frac{1}{2!}(-\lambda\beta V_{12})^2 + \cdots\right] \\
&\quad \times \left[(-\lambda\beta V_{23}) + \frac{1}{2!}(-\lambda\beta V_{23})^2 + \cdots\right]\left[(-\lambda\beta V_{13}) + \frac{1}{2!}(-\lambda\beta V_{13})^2 + \cdots\right] \\
&= \frac{1}{3!} n^2\mathcal{V}^{-1}\int d\mathbf{q}_1\, d\mathbf{q}_2\, d\mathbf{q}_3\, f_{12} f_{23} f_{13}
\end{aligned}
\tag{6.4.5}
$$

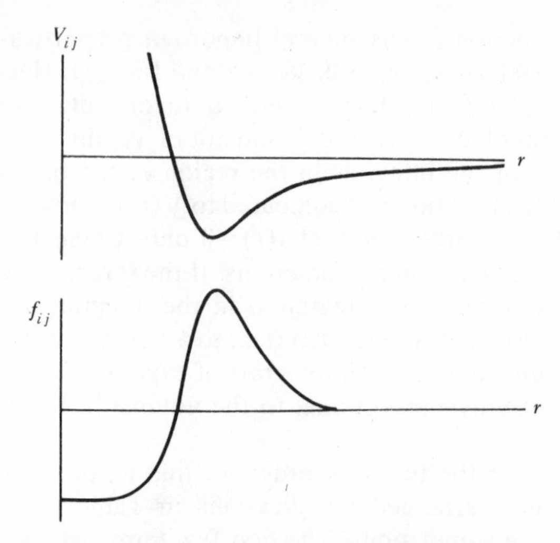

Figure 6.4.1. Comparison of $V(r)$ and $f(r)$.

We see that the renormalization process is again verified: the contribution to order n^2 derives from the simple triangle by reinterpreting the meaning of each line in terms of the Mayer functions f_{ij}.

It is easily checked that the same procedure works also in order n^3. The only (minor) difference is that the λ series corresponding to n^3 splits up into three independent subseries (visible in Table 6.4.1), each starting with one of the three different four-particle graphs having only single lines.

These examples are sufficient for suggesting the general result.* Define a *skeleton graph* as an irreducibly connected graph (no articulation points) with no multiple lines. The density expansion of the free energy per particle is then

$$-\beta a_c(n, T) = \sum_{p=1}^{\infty} n^p \bar{B}_{p+1}(T) \tag{6.4.6}$$

The coefficients $\bar{B}_p(T)$, which are functions of temperature, are defined as

$$\bar{B}_p(T) = (p!)^{-1} \sum_{\substack{\Delta^p \\ \text{(p-particle} \\ \text{skeleton graphs)}}} S(\Delta^p)\mathcal{V}^{-1} \int d\mathbf{q}_1 \cdots d\mathbf{q}_p \left(\prod f_{ij}\right)_{\Delta^p} \tag{6.4.7}$$

where $(\prod f_{ij})_{\Delta^p}$ is a product of f_{ij} factors connected according to the skeleton graph, and $S(\Delta^p)$ is the corresponding symmetry factor. The first few terms of the series (6.4.6) are given explicitly in Table 6.4.2.

* A formal proof can be found in R. Brout and P. Carruthers (reference at the end of this chapter).

Table 6.4.2. Renormalized Graphs in the Density Expansion of the Free Energy, Eq. (6.4.6)

Graph	$S(\Delta)$	$\bar{B}(\Delta)$
		Order n
	1	$\bar{B}_2 = \frac{1}{2}\mathcal{V}^{-1}\int_{12} f_{12}$
		Order n^2
	1	$\bar{B}_3 = \frac{1}{6}\mathcal{V}^{-1}\int_{123} f_{12}f_{23}f_{13}$
		Order n^3
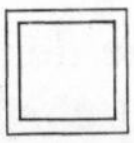	3	$\bar{B}_4^{(1)} = \frac{1}{8}\mathcal{V}^{-1}\int_{1234} f_{12}f_{23}f_{34}f_{14}$
	6	$\bar{B}_4^{(2)} = \frac{1}{4}\mathcal{V}^{-1}\int_{1234} f_{12}f_{23}f_{34}f_{14}f_{13}$
	1	$\bar{B}_4^{(3)} = \frac{1}{24}\mathcal{V}^{-1}\int_{1234} f_{12}f_{23}f_{34}f_{14}f_{13}f_{24}$
		Order n^4
	12	$\bar{B}_5^{(1)} = \frac{1}{5}\mathcal{V}^{-1}\int_{12345} f_{12}f_{23}f_{34}f_{45}f_{15}$
	60	$\bar{B}_5^{(2)} = \mathcal{V}^{-1}\int_{12345} f_{12}f_{23}f_{34}f_{45}f_{15}f_{13}$
	10	$\bar{B}_5^{(3)} = \frac{1}{6}\mathcal{V}^{-1}\int_{12345} f_{12}f_{23}f_{34}f_{45}f_{25}f_{14}$
	30	$\bar{B}_5^{(4)} = \frac{1}{2}\mathcal{V}^{-1}\int_{12345} f_{12}f_{23}f_{34}f_{45}f_{15}f_{14}f_{25}$
	10	$\bar{B}_5^{(5)} = \frac{1}{6}\mathcal{V}^{-1}\int_{12345} f_{12}f_{23}f_{34}f_{45}f_{25}f_{14}f_{24}$
	60	$\bar{B}_5^{(6)} = \mathcal{V}^{-1}\int_{12345} f_{12}f_{23}f_{34}f_{45}f_{15}f_{13}f_{35}$
	30	$\bar{B}_5^{(7)} = \frac{1}{2}\mathcal{V}^{-1}\int_{12345} f_{12}f_{23}f_{34}f_{45}f_{15}f_{14}f_{24}f_{25}$
	15	$\bar{B}_5^{(8)} = \frac{1}{4}\mathcal{V}^{-1}\int_{12345} f_{12}f_{23}f_{34}f_{45}f_{15}f_{13}f_{24}f_{35}$
	10	$\bar{B}_5^{(9)} = \frac{1}{6}\mathcal{V}^{-1}\int_{12345} f_{12}f_{23}f_{34}f_{45}f_{15}f_{13}f_{14}f_{24}f_{25}$
	1	$\bar{B}_5^{(10)} = \frac{1}{60}\mathcal{V}^{-1}\int_{12345} f_{12}f_{23}f_{34}f_{45}f_{15}f_{13}f_{14}f_{24}f_{25}f_{35}$

Having an expression for the free energy, we can calculate as usual all other thermodynamic functions. We are particularly interested in the pressure, that is, in the equation of state. Using the appropriate formula of Table 4.4.1 we write

$$\frac{P}{nk_BT}=1+n\frac{\partial}{\partial n}(\beta a_c) \tag{6.4.8}$$

We obtain from Eq. (6.4.6):

$$\frac{P}{nk_BT}=1+\sum_{p=1}^{\infty} n^p B_{p+1}(T) \tag{6.4.9}$$

where

$$B_p(T)=-(p-1)\bar{B}_p(T) \tag{6.4.10}$$

The equation of state expressed as a series in powers of the density is called the *virial expansion*, and $B_p(T)$ is called the *pth virial coefficient*. We see that it is expressed in terms of all the "irreducible cluster integrals" involving p particles.

Among the virial coefficients the best known, both theoretically and experimentally, is of course the simplest, that is, the *second virial coefficient* $B_2(T)$. Let us study two instructive examples.

Consider first the idealized system of *hard spheres* of diameter d_0 characterized by the following potential:

$$\begin{aligned} V^{HS}(r)&=\infty, \qquad r<d_0 \\ &=0, \qquad r>d_0 \end{aligned} \tag{6.4.11}$$

The Mayer function $f(r)$, Eq. (6.4.4), is

$$\begin{aligned} f^{HS}(r)&=-1, \qquad r<d_0 \\ &=0, \qquad r>d_0 \end{aligned} \tag{6.4.12}$$

The second virial coefficient is then easily calculated:

$$\begin{aligned} B_2^{HS}(T)&=-\tfrac{1}{2}\mathcal{V}^{-1}\int d\mathbf{q}_1\, d\mathbf{q}_2\, f(r_{12}) \\ &=-\frac{1}{2}\int_0^{d_0} dr(-4\pi r^2)=\tfrac{1}{2}(\tfrac{4}{3}\pi d_0^3) \end{aligned} \tag{6.4.13}$$

The second virial coefficient is simply four times the volume of the hard sphere. Two features are of interest here (see Fig. 6.4.2). $B_2(T)$ is *positive:* this is the case for all *repulsive interactions* whose effect is to increase the pressure as compared to the perfect gas. Second, we note that $B_2(T)$ is

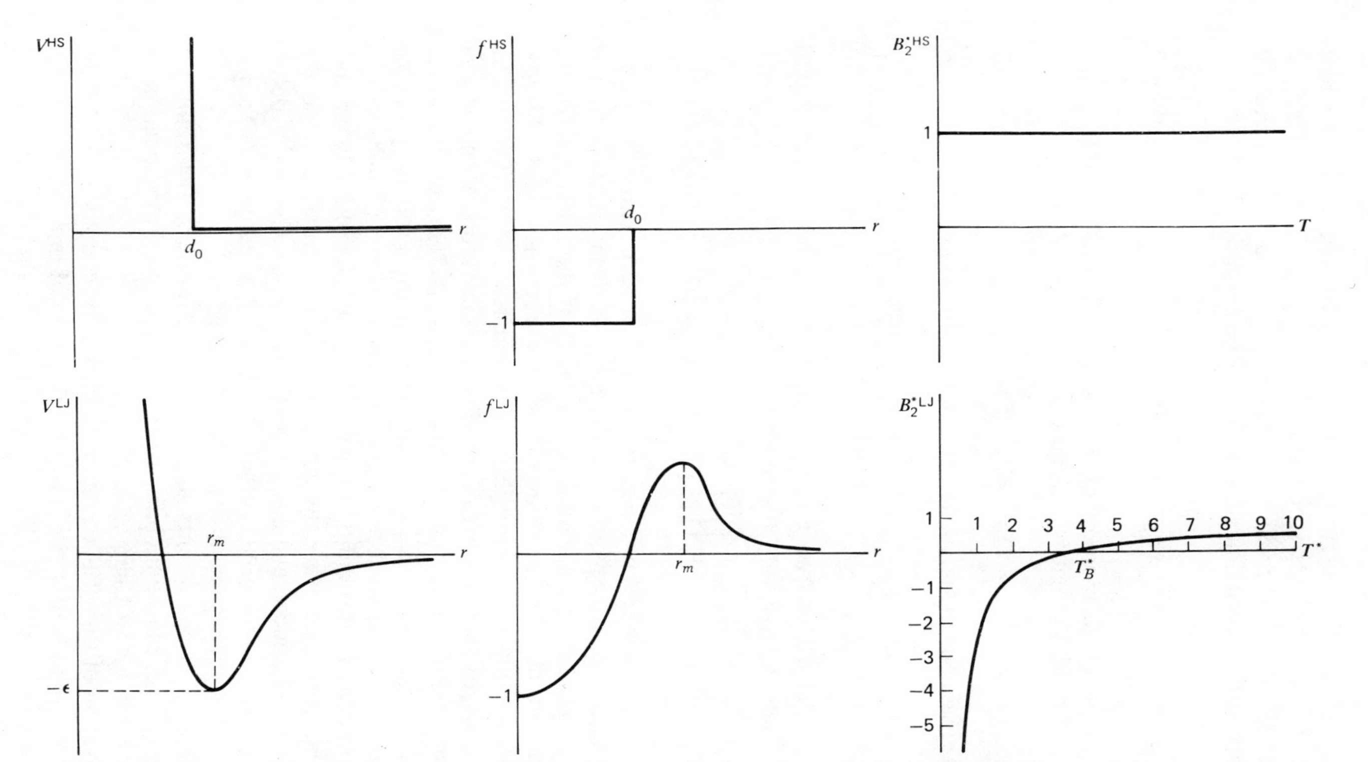

Figure 6.4.2. The intermolecular potential, the Mayer function, and the second virial coefficient in the case of hard spheres, and of the Lennard–Jones potential. For the hard-sphere gas, $B_2^{*\mathrm{HS}} = B_2/(4/6)\pi d_0^3$. For the Lennard–Jones gas, $B_2^{*-\mathrm{LJ}} = B_2/(2/3)\pi\sigma^3$ and $T^* = k_B T/4\varepsilon$. The graph for $B_2^{*\mathrm{LJ}}$ is based on data from J. A. Barker, P. J. Leonard and A. Pompe, *J. Chem. Phys.* **44,** 4206 (1966), where values of the third, fourth, and fifth virial coefficients are also tabulated.

independent of the temperature: This feature is specific to the hard spheres, and is easily understood from the singular form of $f(r)$, Eq. (6.4.12). Clearly, *all* virial coefficients are independent of the temperature, and the hard-sphere equation of state is of the form:

$$\left(\frac{P^{\mathrm{HS}}}{nk_BT}\right) = 1 + \psi(n) \qquad (6.4.14)$$

We now consider the more realistic *Lennard–Jones potential* (6.1.9). It is not difficult to show that for an interaction potential

$$V(r) = \alpha r^{-m}, \qquad m > 0 \qquad (6.4.15)$$

the second virial coefficient can be calculated analytically, with the result

$$B_2(T) = \frac{2\pi}{3}\left(\frac{\alpha}{k_BT}\right)^{3/m} \Gamma\left(\frac{m-3}{m}\right) \qquad (6.4.16)$$

where $\Gamma(x)$ is the well-known Euler gamma function. Hence, the second virial coefficient for the Lennard-Jones gas is

$$B_2^{\mathrm{LJ}}(T) = \frac{2\pi}{3}\sigma^3\left[\left(\frac{4\varepsilon}{k_BT}\right)^{1/4}\Gamma(\tfrac{3}{4}) - \left(\frac{4\varepsilon}{k_BT}\right)^{1/2}\Gamma(\tfrac{1}{2})\right] \qquad (6.4.17)$$

We now have a nontrivial function of the temperature, which no longer has a constant sign: it is plotted in Fig. 6.4.2. Its value results from a compromise between the negative contribution of the attractive part of $V^{\mathrm{LJ}}(r)$ and the positive contribution of the repulsive part. At low temperature the attractive contribution dominates and $B_2(T) < 0$. The coefficient changes sign at $T = T_B$ (the "Boyle temperature"), and at high temperatures it tends asymptotically to a constant. At these temperatures the molecules have such large kinetic energies that the attractive potential well is ignored and the molecules tend to behave like hard spheres.

The experimental data confirm very well this behavior of $B_2(T)$ for real, neutral molecules. Therefore, by fitting the experimental curve to the calculated one, the molecular parameters ε and σ can be easily determined. Thus, the second virial coefficient is a widely used quantity for the determination of molecular interaction parameters.

The higher virial coefficients are not so well known. We shall not discuss them here, but come back to the discussion of the equation of state of dense fluids in Chapter 8.

The derivation of the virial equation of state from statistical mechanics was one of the great landmarks in the development of this science. It was first obtained by Ursell in 1927. It was then further developed and simplified by Mayer in 1937 by the introduction of the graphs,* and was later studied by many authors (see the bibliography at the end of this chapter).

The traditional derivation of the virial expansion uses the method of the grand ensemble, in the spirit of our developments of Section 5.4. One derives in a first step an expansion of P in powers of the fugacity $\mathfrak{z}$ [analogous to Eq. (5.4.22)], and an auxiliary equation for n as a function of $\mathfrak{z}$ [see Eq. (5.4.23)]. The coefficients in these equations are related to cluster integrals corresponding to a certain class of graphs. In a second, rather complicated, step, one must eliminate the activities and get the virial expansion (6.4.9), in which the coefficients are cluster integrals corresponding to a different class of graphs (the one defined here).

The method of cumulant expansions was introduced for the first time by Kirkwood in 1938, for the problem of the Ising model (see Section 10.2). The application of these ideas to the classical virial expansion is due to Brout (1959). This method has many advantages. First of all, it is much simpler. The combinatorial problem appears only once, in the λ expansion, where it is very simple; the resummation procedure leading to the n expansion is pretty straightforward and avoids the complications appearing in the elimination of the fugacity. Although the λ expansion has not much physical meaning in itself for real gases, it provides a valuable "raw material" for all kinds of rearrangements (see, e.g., Section 6.5). Moreover, the λ expansions are the natural expansions in the nonequilibrium theory. Hence, this method provides a unifying line throughout all the fields of statistical mechanics.

6.5. CLASSICAL PLASMAS IN EQUILIBRIUM

We now consider a system of electrically charged particles. It will provide us with a typical (and most important) illustration of the difficulties encountered in statistical mechanics when the interaction forces have a very long range.

We construct a very simplified model of a plasma. In a real system, there must be at least two components present, say, electrons and positive ions, so that the total charge of the system vanishes. In order not to burden the notations, we rather consider a one-component system, say, an electron gas, in the presence of a homogeneous continuous background of positive charge. This background neutralizes the overall charge

* Mayer's theory of imperfect gases was historically one of the first theories using a graph technique systematically, and making full use of the correspondence between topological properties of graphs and analytical properties of integrals. This idea was later extensively used by Feynman in quantum field theory and has now become a routine tool of theoretical physics.

but has no dynamic role. We also consider that the particles have a finite size. We idealize this situation by assuming that the total interaction energy V_{ij}^{tot} is made up of two terms:

$$V_{ij}^{tot} = V_{ij}^{HS} + V_{ij} \tag{6.5.1}$$

where V_{ij}^{HS} is a hard-sphere potential, defined in Eq. (6.4.11), and V_{ij} is the Coulomb potential:

$$V_{ij} \equiv V(r_{ij}) = \frac{e^2}{r_{ij}} \tag{6.5.2}$$

Its Fourier transform, defined by Eq. (2.4.9), is

$$\tilde{V}_k = \frac{e^2}{2\pi^2 k^2} \tag{6.5.3}$$

The background is not retained in the Hamiltonian; however, we assume that the total average potential energy vanishes because of the overall electrical neutrality; hence

$$\tfrac{1}{2} \int d\mathbf{r}\, V(r) - \tfrac{1}{2}\tilde{V}_0 = 0 \tag{6.5.4}$$

where $\frac{1}{2}\tilde{V}_0$ is the potential energy due to the background, which is assumed to cancel exactly the contribution of the one-component plasma. This will be the only place in the theory where we take account of the background.*

The shape of the potential (6.5.1) is sketched in Fig. 6.5.1.

We now evaluate the free energy, starting from the series (6.3.10). Because of the form (6.5.1) of the potential, each graph can be decomposed into a set of graphs by replacing each original line in Table 6.3.1 by a contribution of V_{ij}^{HS} and a contribution of V_{ij}. We represent the former by a dotted line and the latter by a simple continuous line (see Fig. 6.5.2). We also know, from the discussion of Section 6.4, that the dotted lines give rise to divergent individual graphs. They must be grouped by the procedure exposed in that section. As a result, we may keep only diagrams without multiple dotted lines, provided we redefine the corresponding contribution in terms of f_{ij}^{HS} [replacing $(-\beta V_{ij}^{HS})$] (see Fig. 6.5.3).

If

$$d_0^3 n \ll 1 \tag{6.5.5}$$

we may retain in first approximation only the contribution of the graphs

* The generalization to a multicomponent plasma consisting of several species of charges $Z_\sigma e$ and concentrations n_σ is straightforward. The overall neutrality is then expressed by the condition $e \sum n_\sigma Z_\sigma = 0$. This condition ensures the vanishing of the average potential energy.

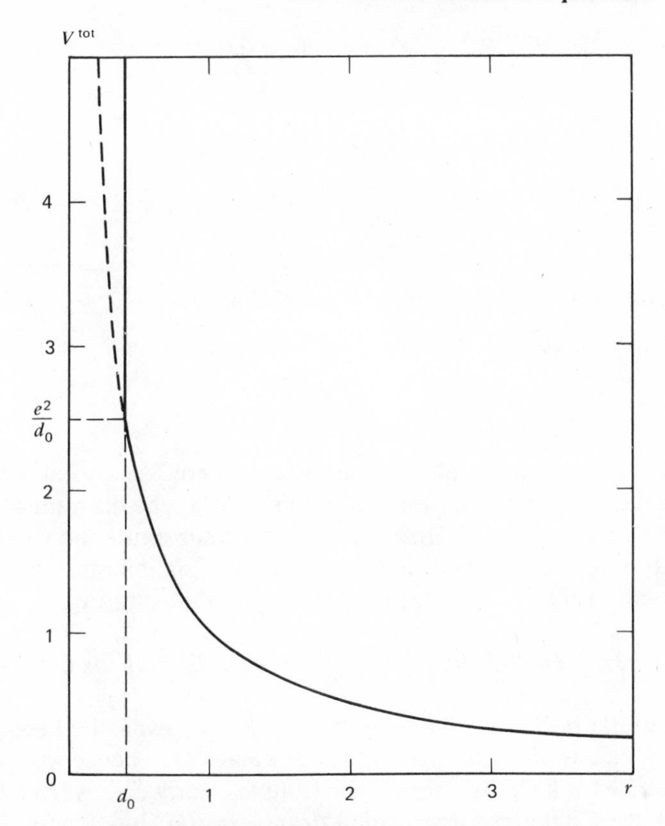

Figure 6.5.1. Interaction potential of charged particles with hard cores.

shown in the first line of Fig. 6.5.3, that is, the second virial coefficient, Eq. (6.4.13). In this approximation, we therefore neglect all other diagrams with dotted lines. This settles the contribution of the hard-sphere part of the potential.

We now come to the truly new part of our problem, that is, the treatment of the Coulomb potential. We may assume that the temperature

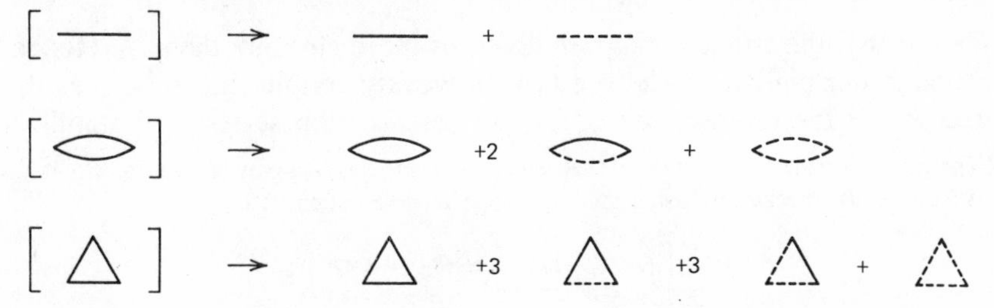

Figure 6.5.2. Splitting of the graphs corresponding to the potential (6.5.1).

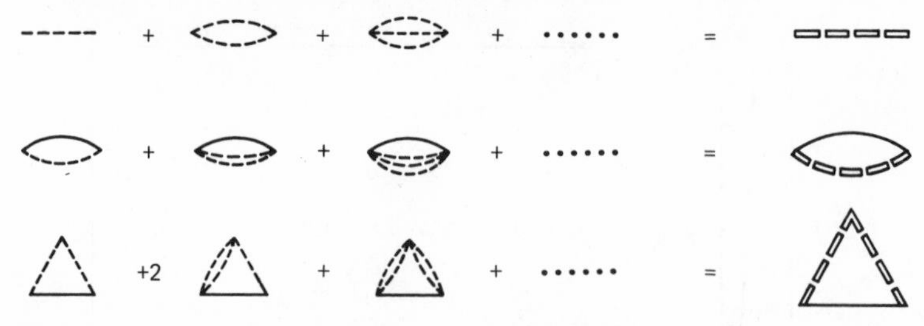

Figure 6.5.3. Renormalization of the V^{HS} lines.

is sufficiently high in order that

$$\frac{\beta e^2}{d_0} \ll 1 \tag{6.5.6}$$

Then (see Fig. 6.5.1) the plasma can be considered as a weakly coupled gas, and we could naturally think of retaining only the diagram of second order, $2a$, of Table 6.3.1.* However, when we substitute the value of the Coulomb potential, we find that the integral J_{2a} is infinite. This is most conveniently seen in the Fourier representation, using Eq. (6.5.3):

$$J_{2a} = 8\pi^3 \int d\mathbf{k}\, \tilde{V}_{\mathbf{k}}^2 = \frac{(8\pi^3)(4\pi)}{4\pi^4} \int_0^\infty dk\, k^2 k^{-4} = 8 \int_0^\infty dk\, k^{-2}$$

The integrand becomes infinite as $k \to 0$. It must always be kept in mind that small values of k correspond to large values of r; hence we have here a divergence for long distances.† The trouble clearly comes from the long range of the Coulomb force, which decreases too slowly as $r \to \infty$.

One might think that the situation could be cured by going to higher orders in λ $(= e^2)$; actually, the situation gets worse. The two diagrams in third order give

$$J_{3a} \sim \int d\mathbf{k}\, d\mathbf{k}'\, k^{-2} k'^{-2} |\mathbf{k} + \mathbf{k}'|^{-2}$$

$$J_{3b} \sim \int d\mathbf{k}\, k^2 k^{-6}$$

We see that the triangle diagram diverges more strongly than J_{2a}. Hence, although our plasma model is a typical weakly coupled gas (as far as the strength of the interactions goes), the perturbation series as it stands is

* The graph $1a$ vanishes, because of the cancellation with the background, see Eq. (6.5.4).

† This is easily checked by using the alternative representation of J_{2a}:

$$J_{2a} = \int d\mathbf{r}\, V_{12}^2 = 4\pi e^4 \int_0^\infty dr\, r^2 r^{-2} = 4\pi e^4 \int_0^\infty dr$$

This expression diverges at the *upper* limit.

meaningless, because the coefficients multiplying the small parameter are infinite. In order to make further progress we need a new idea.

The guiding line in the search for this idea is provided by a celebrated semiphenomenological approach to the problem: the *Debye–Hückel theory* of electrolyte solutions and plasmas, established in 1923, which was extremely successful. We briefly sketch its main idea:

We fix a reference particle in some point P, which we may take as the origin, and consider a "field particle" Q at a distance r from P in the plasma (see Fig. 6.5.4). We want to calculate the potential energy of Q.

In an ordinary gas, if the distance r is long enough, particle Q would not feel at all the influence of P; its potential energy would be determined by one or two nearest neighbors. *Because of the long range of the Coulomb forces*, the situation is quite different in a plasma. Even at a long distance, Q still feels a small, but nonnegligible influence of P. On the other hand, at such distances, there is a large number of particles present between P and Q, each one influencing Q. The potential energy of Q is, therefore, a cumulative effect of the weak interactions of Q with a very large number of particles. This potential energy is a *collective effect*, which clearly depends on the spatial distribution of the particles around any given particle. On the other hand, the spatial distribution depends on the interaction energy: if the interactions are repulsive, the local density of particles in the neighborhood of a given one will be less than the overall average number density n. Hence, the potential energy and the spatial distribution are intimately connected and must be determined together; this is a characteristic feature of a *self-consistent field.*

If there were no other particles than P and Q present, the probability density of finding Q at distance r from P, which is measured by the pair distribution function $n_2'(r)$ [Eq. (3.5.5)], divided by the average density of P, (i.e., n), would be given in equilibrium by the Boltzmann factor:

$$n^{-1}n_2'(r) = n\exp[-\beta V(r)] = n\exp\left(\frac{-\beta e^2}{r}\right) \qquad (6.5.7)$$

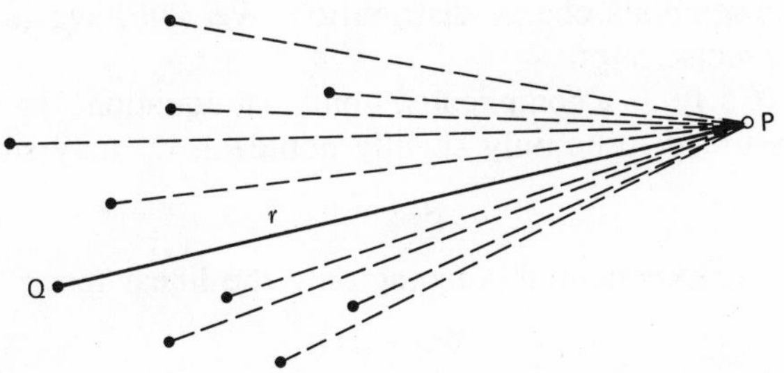

Figure 6.5.4. Phenomenological derivation of the Debye potential.

Because of the presence of the other particles, the true distribution is different; however, we may describe it in terms of an average effective potential energy $W(r)$ in the same form as (6.5.7). In a problem involving electrostatic interactions we may in turn express this effective potential energy in terms of an effective electrostatic potential $\varphi(r)$:

$$W(r) = e\varphi(r) \tag{6.5.8}$$

Hence

$$n^{-1}n_2'(r) = n \exp[-\beta e\varphi(r)] \tag{6.5.9}$$

In order to determine the potential $\varphi(r)$, Debye and Hückel assume that one may just write down the classical Poisson equation of electrostatics: $\nabla^2\varphi = -4\pi\rho_e$, substituting for the charge density ρ_e the spatial distribution of particles (multiplied by e) given by (6.5.9):

$$\nabla^2\varphi = -4\pi[en \exp(-\beta e\varphi) - en] \tag{6.5.10}$$

where the second term in the right-hand side takes account of the neutralizing background.

This equation—sometimes called the *Poisson–Boltzmann equation*—is the central point of the Debye–Hückel theory. It realizes the program of a self-consistent determination of the effective potential and of the pair distribution function. It also crystallizes the weakness of the theory from a fundamental point of view. Indeed, the Poisson equation is valid in macroscopic electrostatics of continuous media. Its application to a system of particles essentially means that we smooth out the individuality of the particles and replace them by a continuous charge distribution. Such a procedure requires theoretical justification. Its successful prediction of experimental data, however, suggests that there must be a deep basis in this picture. We can understand it qualitatively if we imagine that there is a very large number of particles within the effective range of the interactions. In that case (see Fig. 6.5.4) the field particle Q is influenced by so many other particles that the cumulative effect may be similar to the effect of a continuous charge distribution. We will have to make these ideas more precise later.

Equation (6.5.10) is a complicated nonlinear equation. However, as the system is assumed to be only slightly nonideal, we may suppose that

$$\beta e\varphi \ll 1 \tag{6.5.11}$$

and expand the exponential, keeping only the linear term:

$$\nabla^2\varphi = \kappa_D^2\varphi \tag{6.5.12}$$

where we introduced the fundamental parameter κ_D, called the *inverse*

Debye length, defined by

$$\kappa_D^2 = 4\pi e^2 n\beta \tag{6.5.13}$$

The spherically symmetrical solution of Eq. (6.5.12) is

$$\varphi(r) = e\,\frac{\exp(-\kappa_D r)}{r} \tag{6.5.14}$$

The factor e in front of the function has been chosen equal to the charge e, in order that $\varphi(r)$ reduce to e/r at small distances from the central particle, where the effect of other particles is not felt. Hence, the effective potential energy $W(r)$, Eq. (6.5.8), is

$$W(r) = e^2\,\frac{\exp(-\kappa_D r)}{r} = e^2\,\frac{\exp[-(4\pi\beta e^2 n)^{1/2} r]}{r} \tag{6.5.15}$$

We see that the main effect of the collective interactions is to provide a *screening* of the Coulomb interactions at large distances. The effective interaction $W(r)$ drops off exponentially at a distance κ_D^{-1}, which appears as the effective range of the interactions. Hence, because of the long range of the individual interactions, the particles adjust themselves in such a way as to screen out the long distance tail of the potential (Fig. 6.5.5).

We may now discuss more quantitatively the conditions of validity of the Debye approximation. We said already that the continuum model will be satisfactory if there are many particles within the range of the effective forces. This implies that the average distance between particles $n^{-1/3}$ should be much smaller than the Debye length κ_D^{-1}, or

$$\kappa_D^3 n^{-1} \ll 1 \tag{6.5.16}$$

Note that this condition is, in a sense, the reverse of condition (6.4.5) connecting the density to the hard core d_0. It would seem, therefore, that (6.5.16) is valid at high densities. This is, however, not so: The effective, self-consistently fixed, Debye length κ_D^{-1} depends on the density; on substituting (6.5.13) we therefore get (neglecting numerical factors)

$$e^3 n^{1/2} \beta^{3/2} \ll 1 \tag{6.5.17}$$

The validity condition therefore involves the smallness of the characteristic plasma parameter $e^3 n^{1/2} \beta^{3/2}$, which combines the charge, the density, and the temperature.

We now investigate whether it is possible to derive the results of the Debye theory from the fundamental cluster expansion (6.3.10). Clearly, an expression like (6.5.15) cannot originate from a single diagram, because the coupling parameter e^2 (i.e., λ) appears in an exponential. Hence we

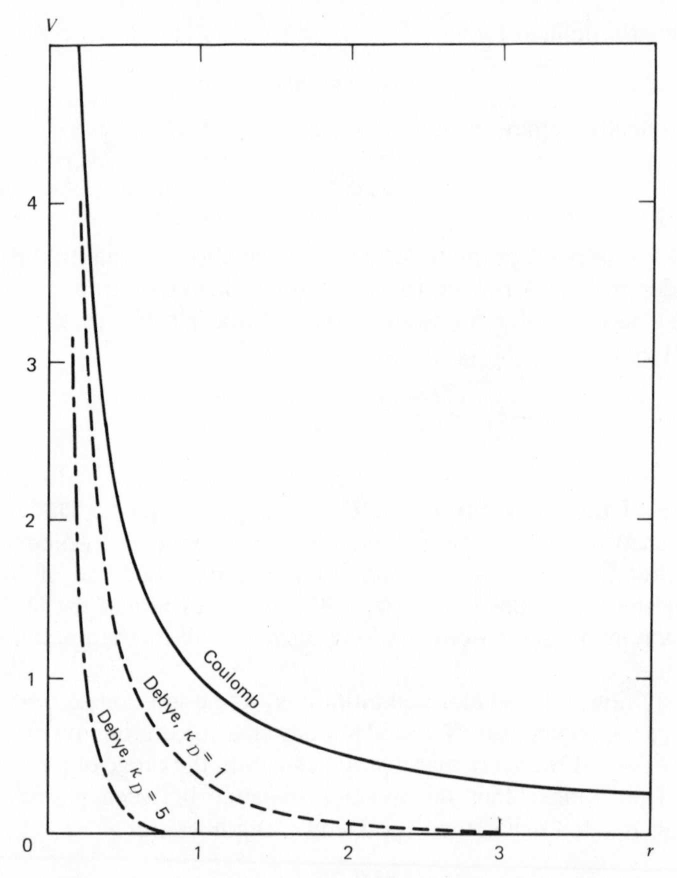

Figure 6.5.5. The Coulomb potential and the Debye potential.

must *sum* the contributions of a certain class of graphs, which we must identify. To do so, we note that Eq. (6.5.15) also involves the density n in the exponential, more precisely in the combination e^2n. Hence, an expression like (6.5.15), if expanded in powers of the density, contains terms of order:

$$e^2(e^2n)^p, \qquad p=0, 1, 2, \ldots \tag{6.5.18}$$

This remark is very useful in giving us a hint in the identification of the graphs. We tentatively adopt the following rule:

The dominant part of the free energy corresponding to the Debye approximation is obtained by summing all graphs depending on charge and density through the combination (6.5.18) for all values of p.

It will be verified *a posteriori* that these graphs give rise precisely to the leading terms when (6.5.16) is satisfied.

We now recall that an irreducible graph with b bonds and p verticles is proportional to $\lambda^b n^{p-1}$. Therefore, the terms of order (6.5.18) are represented by all the graphs with an equal number of bonds and of vertices. Such graphs are simple polygons or, as they are usually called, *rings*. In each order in λ (or e^2) there is just one ring (see also Table 6.3.1). Our problem is, therefore, to *sum the infinite series of ring graphs* shown in Fig. 6.5.6.

We first "regularize" the integrals by replacing the Coulomb potential (6.5.3) by the potential:

$$\tilde{V}_{\mathbf{k}}^{(\gamma)} = \frac{e^2}{2\pi^2(k^2+\gamma^2)} \tag{6.5.19}$$

With this potential, the cluster integrals are convergent. We then sum the relevant graphs, and at the end of the calculation we let $\gamma \to 0$.

Using the results of Table 6.3.1 we write

$$\begin{aligned}(-\beta a_c^{(\gamma)})_{\text{rings}} &= (8\pi^3)^{-1} \int d\mathbf{k} \left\{ \tfrac{1}{4}\beta^2 e^4 n [8\pi^3 \tilde{V}_{\mathbf{k}}^{(\gamma)}]^2 - \tfrac{1}{6}\beta^3 e^6 n^2 [8\pi^3 \tilde{V}_{\mathbf{k}}^{(\gamma)}]^3 \right. \\ &\quad \left. + \cdots + \frac{1}{2p}(-e^2\beta)^p n^{p-1} [8\pi^3 \tilde{V}_{\mathbf{k}}^{(\gamma)}]^p + \cdots \right\} \\ &= -\frac{1}{16\pi^3 n} \int d\mathbf{k} \left\{ \sum_{p=1}^{\infty} \frac{1}{p} (-1)^{p+1} [8\pi^3 \beta n e^2 \tilde{V}_{\mathbf{k}}^{(\gamma)}]^p - 8\pi^3 \beta n e^2 \tilde{V}_{\mathbf{k}}^{(\gamma)} \right\}\end{aligned} \tag{6.5.20}$$

The infinite series appearing here is just the expansion of the logarithmic function, hence [using also Eqs. (6.5.13) and (6.5.19)]

$$\begin{aligned}(-\beta a_c^{(\gamma)})_{\text{rings}} &= -(16\pi^3 n)^{-1} \int d\mathbf{k} \, \{\ln[1 + 8\pi^3 \beta n e^2 \tilde{V}_{\mathbf{k}}^{(\gamma)}] - 8\pi^3 \beta n e^2 \tilde{V}_{\mathbf{k}}^{(\gamma)}\} \\ &= -(16\pi^3 n)^{-1} 4\pi \int_0^\infty dk \, k^2 \left[\ln\left(1 + \frac{\gamma^2 + \kappa_D^2}{k^2}\right) - \frac{\gamma^2 + \kappa_D^2}{k^2} \right]\end{aligned} \tag{6.5.21}$$

We now let $\gamma \to 0$, obtaining

$$\begin{aligned}(-\beta a_c)_{\text{rings}} &= \lim_{\gamma \to 0} (-\beta a_c^{(\gamma)})_{\text{rings}} \\ &= -(4\pi^2 n)^{-1} \int_0^\infty dk \, k^2 \left[\ln\left(1 + \frac{\kappa_D^2}{k^2}\right) - \frac{\kappa_D^2}{k^2} \right]\end{aligned} \tag{6.5.22}$$

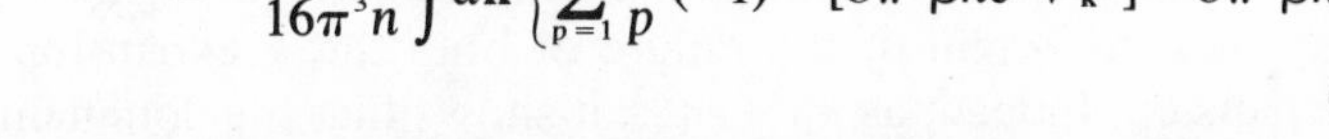

Figure 6.5.6. The ring graphs.

We thus obtain a remarkable result: Although every individual term in the series (6.5.20) gives rise to a divergent integral for $\gamma \to 0$, the integral of the sum is convergent. Moreover, the characteristic Debye parameter κ_D appears quite naturally in the result. The proof of this fundamental fact, that the long-range divergence can be lifted by a rearrangement and partial-summation method, was given for the first time by Mayer in a basic paper in 1950.

The evaluation of the remaining integral is elementary and gives the result (dropping now the subscript "rings"):

$$\beta a_c(n, T) = -\frac{\kappa_D^3}{12\pi n} \tag{6.5.23}$$

This result contains the *a posteriori* justification for our choice of graphs (6.5.18). Indeed, it shows that the relative correction to the ideal-gas free energy is proportional to the characteristic plasma parameter κ_D^3/n, and hence the result is appropriate to the condition (6.5.16).

Actually, Eq. (6.5.23) is precisely equal to the free energy calculated from the Debye theory on the basis of Eqs. (6.5.9), (6.5.14). Multiplying both sides of (6.5.23) by $k_B T$ we get another interesting form:

$$a_c(n, T) = -\frac{k_B T}{12\pi n}\kappa_D^3 = -\tfrac{1}{3}e^2\kappa_D \tag{6.5.24}$$

This result is illuminating from the mathematical point of view because it clearly shows the origin of the failure of both the λ expansion and the virial expansion. Indeed, as $\kappa_D \sim en^{1/2}$, it shows that the dominant correction to the free energy is of order $e^3n^{1/2} = \lambda^{3/2}n^{1/2}$. The occurrence of *half-integral powers* implies that the free energy of a plasma is *not an analytical function of* λ *or of* n. The origin, in both cases ($\lambda = 0$ or $n = 0$), is a branch point of the function $a_c(\lambda, n)$. Hence, this function does not possess a Taylor-series expansion near the origin, and this is indeed what we found.

Let us have a quick look at some of the other thermodynamic functions, derived by the formulas of Table 4.4.1:

$$\begin{aligned} e &= \tfrac{3}{2}k_B T - \tfrac{1}{2}e^2\kappa_D \\ c_V &= \tfrac{3}{2}k_B + \tfrac{1}{4}k_B\beta e^2\kappa_D \\ s &= s_{\text{ideal}} - \tfrac{1}{6}k_B\beta e^2\kappa_D \\ P &= nk_B T\{1 - \tfrac{1}{6}\beta e^2\kappa_D\} \end{aligned} \tag{6.5.25}$$

We can understand these corrections qualitatively. The internal energy is lower than the ideal value: This reflects a stabilization of the system by the interactions. The entropy is also lower: Of course, the collective

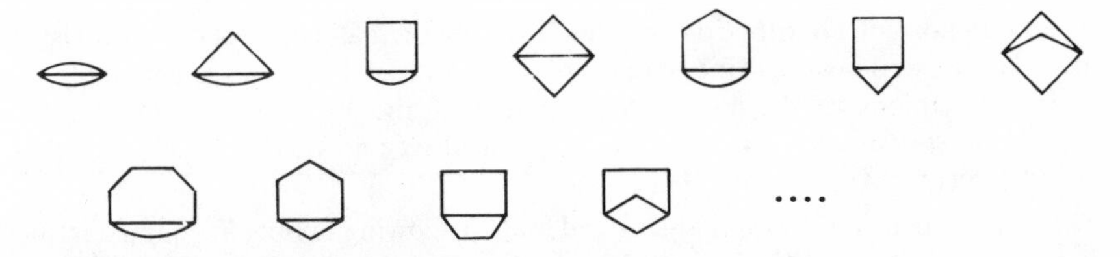

Figure 6.5.7. Graphs of order $e^4(e^2n)^p$ contributing to the free energy of a plasma beyond the Debye approximation.

effects introduce an ordered structure around each ion through the polarization cloud (6.5.9). If we supply energy to the plasma, this energy is used partly to increase the kinetic energy of the particles, and partly to destroy the polarization structure: whence the positive correction term to the specific heat.

In this way, we have solved completely the problem of the plasma in first approximation. The summation method can be continued, in principle, to higher approximations (whereas the Debye method cannot!). One could, for instance, envisage the next order as defined by the prescription $e^4(e^2n)^p$. This generates the graphs of Fig. 6.5.7. The rule is clear: Starting from the first graph, each original line is expanded into a chain (just as the rings were generated from the original graph in Fig. 6.5.6). However, it must be kept in mind that in higher orders the mixed contributions involving the hard core d_0 also begin to come into the picture. The explicit summations become rather complicated. The reader is referred to the literature for additional details.

BIBLIOGRAPHICAL NOTES

A very good reference for the discussion of the *molecular interaction potentials* is J. O. Hirschfelder, C. F. Curtiss, and R. B. Bird, *Molecular Theory of Gases and Liquids*, Wiley, New York, 1954.

The λ *expansion* and the *cumulant method* are due to R. Brout, *Phys. Rev.* **115,** 824 (1959).

See also R. Brout and P. Carruthers, *Lectures on the Many-Electron Problem*, Wiley-Interscience, New York, 1963.

The density expansion (*virial expansion*) was first introduced in H. D. Ursell, *Proc. Cambridge Phil. Soc.* **23,** 685 (1927).

It was developed by J. Mayer who introduced his famous graphs: J. Mayer, *J. Chem. Phys.* **5,** 67 (1937).

Among the many papers that followed, we quote

B. Kahn and G. E. Uhlenbeck, *Physica* **5,** 399 (1938).

M. Born and K. Fuchs, *Proc. Roy. Soc. (London)* **A166,** 391 (1938).

A very important contribution, in which the mathematical theory of graphs is used, is reviewed in G. E. Uhlenbeck and G. W. Ford, "Theory of Linear Graphs, with Applications to the Virial Development of the Properties of Gases," in *Studies in Statistical Mechanics,* Vol. 1 (J. de Boer and G. E. Uhlenbeck, eds.) North Holland, Amsterdam, 1962.

The virial expansion is throughly treated in the following textbooks, quoted in the notes to Chapter 4: Mayer and Mayer, Rushbrooke, ter Haar, Hill, Huang, Münster, and Hirschfelder et al. (see above).

The classical theory of *electrolytes* originated in P. Debye and E. Hückel, *Phys. Z.* **24,** 185 (1923).

It is thoroughly exposed in the following books:

H. Falkenhagen, *Elektrolyte,* S. Hirzel, Leipzig, 1953 (new edition, 1971).

H. L. Friedman, *Ionic Solution Theory,* Interscience, New York, 1962.

P. Résibois, *Electrolyte Theory, an Elementary Introduction to a Microscopic Approach,* Harper and Row, New York, 1968.

R. Balescu, *Statistical Mechanics of Charged Particles,* Wiley-Interscience, New York, 1963.

It is interesting to note that a very early attempt at a microscopic derivation of the Debye–Hückel theory is due to J. Yvon, *J. Phys. (Paris)* **7,** 93 (1936). His method was based on a (somewhat arbitrary) truncation of his hierarchy for the equilibrium pair distribution (see Section 7.4).

The statistical-mechanical derivation of the Debye theory by the summation of the ring graphs series is due to J. Mayer, *J. Chem. Phys.* **18,** 1426 (1950). See also the very clear exposition of E. E. Salpeter, *Ann. Phys.* (N.Y.) **5,** 183 (1958).

The extension of the *virial expansion methods to quantum systems* gave rise to an extensive literature, from which we quote only some of the more important papers:

T. Matsubara, *Prog. Theor. Phys.* **14,** 351 (1955).

N. M. Hugenholtz, *Physica* **23,** 533 (1957).

C. Bloch and C. de Dominicis, *Nucl. Phys.* **7,** 459 (1958); **10,** 181 (1959).

E. W. Montroll and J. C. Ward, *Phys. Fluids* **1,** 55 (1958).

A famous type of expansion, which was later of great use in nonequilibrium theory, is called the *binary collision expansion.* It is due to

A. J. F. Siegert and E. Teramoto, *Phys. Rev.* **110,** 1212 (1958).

T. D. Lee and C. N. Yang, *Phys. Rev.* **113,** 1165 (1959); **116,** 25 (1959); **117,** 12 (1960).

An excellent review of the whole field is C. Bloch, "Diagram Expansions in Quantum Statistical Mechanics," in *Studies in Statistical Mechanics,* Vol. 3 (J. de Boer and G. E. Uhlenbeck, eds.), North Holland, Amsterdam, 1965.

A more specialized topic is reviewed in K. Huang, "Imperfect Bose Gas," in *Studies in Statistical Mechanics,* Vol. 2 (J. de Boer and G. E. Uhlenbeck, eds.), North Holland, Amsterdam, 1964.

The *quantum plasmas* are important systems, with applications in solid-state physics. The treatment of their equilibrium properties can be found in

M. Gell-Mann and K. A. Brueckner, *Phys. Rev.* **106,** 364 (1957).

K. Sawada, *Phys. Rev.* **106,** 372 (1957).

R. Brout, *Phys. Rev.* **108,** 515 (1957).

J. Hubbard, *Proc. Roy. Soc.* (*London*) **A240,** 539 (1957); **A243,** 336 (1957).

R. Balescu (quoted above).

CHAPTER 7
REDUCED DISTRIBUTION FUNCTIONS IN EQUILIBRIUM

7.1. BASIC DEFINITIONS

In Chapters 4–6 we illustrated the main method of equilibrium statistical mechanics. The idea of this method may be summarized as follows. On the basis of the principle of equal *a priori* probabilities, a certain number of equilibrium ensembles can be constructed. The canonical and the grand canonical ensemble are the most important of these; moreover, they become equivalent in the thermodynamic limit. It is then shown that the normalization constants, that is, the *partition functions*, corresponding to these ensembles contain all the information necessary for the calculation of the thermodynamic quantities. Hence, the problem of equilibrium thermodynamics is reduced to the computation of the partition function.

We now consider an alternative formulation of the problem, involving the use of the *reduced distribution functions*, introduced in Chapter 3. In this method, the thermodynamic functions are expressed as averages of dynamical functions, evaluated with the equilibrium reduced distribution functions. This method is, therefore, much closer to the general philosophy of statistical mechanics, as discussed in Chapter 1. However, this philosophy must be implemented in a nontrivial way, in order to take account of those thermodynamical quantities that cannot be expressed as true averages of a dynamical function. These questions, which were mentioned in Chapter 1, will now be treated in detail.

One may ask, what is the advantage of developing another method of equilibrium statistical mechanics, since we know that the problem is completely solved (in principle) by the partition function? The reduced distributions can only give results equivalent to those provided by the partition function.

The answer to this objection is twofold. First, it must be realized that the equivalence is only true as far as the exact results are concerned. In most nontrivial realistic problems, however, only approximate forms of either the partition function or of the distribution functions can be calculated: the approximate results provided by the two methods are no longer necessarily equivalent in general. Under these circumstances, the introduction of new formulations—known to be equivalent in the "limit" of an exact treatment—is always beneficial. Indeed, different formulations suggest different approximation procedures, which may lead to better practical results.

Besides this pragmatic argument, there is another, much deeper justification for the development of the method of distribution functions. The method of the partition functions, in spite of its extreme elegance, is completely closed on itself. The expressions obtained in terms of the partition function are derived by using exclusively the particular functional form of the equilibrium ensemble. It is impossible to define anything like a nonequilibrium partition function. On the contrary, the set of reduced distribution functions is a concept valid as well in equilibrium as outside of equilibrium. This formulation is, therefore, the only universal one, providing a link between equilibrium and nonequilibrium theory. The development of such a universal theory is, and must be, the main aim of present-day statistical mechanics.

We first consider a classical system of interacting point particles of the same kind as described in Chapter 6. As we know from our general discussion of Chapter 3, the main interest lies with the one-body distribution $f_1(x_1)$ and the two-body distribution $f_2(x_1x_2)$. Taking into account the definition (3.1.12) and the form (4.3.26) of the canonical distribution function, together with Eqs. (6.1.2)–(6.1.5) (setting $\lambda = 1$) we find

$$\begin{aligned} f_1^0(\mathbf{q}_1, \mathbf{p}_1) &= Nh^{-3N}(N!)^{-1}Z^{-1}\int d\mathbf{p}_2 \cdots d\mathbf{p}_N \int d\mathbf{q}_2 \cdots d\mathbf{q}_N \exp(-\beta H^0 - \beta H') \\ &= (2\pi m k_B T)^{-3/2} n_1(\mathbf{q}_1) \exp\left(-\frac{\beta p_1^2}{2m}\right) \\ &\equiv n_1(\mathbf{q}_1)\varphi^0(p_1) \end{aligned} \tag{7.1.1}$$

with

$$n_1(\mathbf{q}_1) = N\mathcal{V}^{-N}Q^{-1}\int d\mathbf{q}_2 \cdots d\mathbf{q}_N\, e^{-\beta H'} \tag{7.1.2}$$

We introduced the notation $\varphi^0(p)$ for the famous *Maxwell distribution* of momenta:

$$\varphi^0(p) = (2\pi m k_B T)^{-3/2} \exp\left(-\frac{p^2}{2mk_B T}\right) \tag{7.1.3}$$

Q is the configuration integral defined by Eq. (6.1.5).

Whenever the system is homogeneous in space, the one-body distribution is independent of $\mathbf{q}_1$ (in the thermodynamic limit). The mathematical origin of this property is in the fact that the potential energy H' is translationally invariant. In all these situations, which are physically realized whenever the system is in a *fluid phase*, the configurational one-body distribution is simply equal to the constant number density:

$$n_1(\mathbf{q}_1) = n \qquad \text{(fluid)} \tag{7.1.4}$$

We know, however, that an equilibrium system can assume, under well-defined circumstances, an inhomogeneous distribution. The typical example is a *solid phase*, in which the atoms move predominantly by oscillating slightly around equilibrium positions forming an infinite regular lattice. In such situations, the one-body distribution becomes a periodic function of $\mathbf{q}_1$, exhibiting the symmetry of the crystal lattice. The statistical explanation of the transition to the new symmetry is by no means completely understood at present.

In a quite similar way, we find the two-body reduced distribution function in the form

$$f_2^0(\mathbf{q}_1, \mathbf{q}_2, \mathbf{p}_1, \mathbf{p}_2) = n_2'(\mathbf{q}_1, \mathbf{q}_2)\varphi^0(p_1)\varphi^0(p_2) \tag{7.1.5}$$

with

$$n_2'(\mathbf{q}_1, \mathbf{q}_2) = N(N-1)\mathscr{V}^{-N}Q^{-1}\int d\mathbf{q}_3 \cdots d\mathbf{q}_N\, e^{-\beta H'} \tag{7.1.6}$$

We see that, in equilibrium, the momentum dependence of the distribution functions is trivial. The truly interesting part is the configurational two-body distribution function $n_2'(\mathbf{q}_1, \mathbf{q}_2)$. In a homogeneous fluid phase, this function only depends on the relative distance between the particles. It is then convenient to introduce a slightly different function, differing from n_2' by the normalization:

$$n_2'(\mathbf{q}_1, \mathbf{q}_2) = n^2 n_2(|\mathbf{q}_1 - \mathbf{q}_2|) \equiv n^2 n_2(r_{12}) \qquad \text{(fluid)} \tag{7.1.7}$$

This function $n_2(r)$ will be called briefly the *pair distribution*. In an inhomogeneous system we can still define a pair distribution as follows: Express the function n_2' in terms of the variables $\mathbf{q}_1, \mathbf{q}_1 - \mathbf{q}_2 : n_2'(\mathbf{q}_1, \mathbf{q}_2) \equiv n_2''(\mathbf{q}_1, \mathbf{q}_1 - \mathbf{q}_2)$. We then introduce

$$n_2(\mathbf{q}_1 - \mathbf{q}_2) = (Nn)^{-1}\int d\mathbf{q}_1\, n_2''(\mathbf{q}_1, \mathbf{q}_1 - \mathbf{q}_2) \tag{7.1.8}$$

This definition clearly reduces to (7.1.7) in a homogeneous fluid.

Finally, we introduce the concept of correlation, which plays an important role. Introducing the cluster representation (3.4.9) and integrating over the momenta, we find, in a homogeneous system

$$n_2'(\mathbf{q}_1, \mathbf{q}_2) = n^2 + g_2'(r_{12})$$

It is again convenient to write the correlation function g_2' with an explicit n^2 factor; we therefore combine this definition with Eq. (7.1.7) and find [see also Eq. (3.4.5)]

$$n_2(r_{12}) = 1 + \nu_2(r_{12}) \qquad \text{(fluid)} \tag{7.1.9}$$

The function $\nu_2(r)$ will be called the *pair correlation*. This function plays a central role in statistical thermodynamics.

We may note that in a fluid, the correlation between two particles is expected to decrease very rapidly as the distance increases. Hence we may assume that

$$\lim_{r\to\infty} \nu_2(r) = 0 \qquad \text{(fluid)} \qquad \lim_{r\to\infty} n_2(r) = 1 \tag{7.1.10}$$

Also, one easily checks that Eq. (3.5.18) combined with (7.1.7), (7.1.9) implies the following normalization:

$$\text{T-lim}\, N^{-1} \int d\mathbf{r}\, \nu_2(r) = 0 \qquad \text{(fluid)} \tag{7.1.11}$$

It is now easy to generalize these concepts to an s-particle reduced distribution function. We have

$$f_s^0(\mathbf{q}_1, \mathbf{p}_1, \ldots, \mathbf{q}_s, \mathbf{p}_s) = n_s'(\mathbf{q}_1, \ldots, \mathbf{q}_s)\varphi^0(p_1) \cdots \varphi^0(p_s) \tag{7.1.12}$$

with

$$n_s'(\mathbf{q}_1, \ldots, \mathbf{q}_s) = \left[\frac{N!}{(N-s)!}\right] \mathcal{V}^{-N} Q^{-1} \int d\mathbf{q}_{s+1} \cdots d\mathbf{q}_N\, e^{-\beta H'} \tag{7.1.13}$$

In this way the configurational part n_s' of the reduced distribution is neatly separated: in equilibrium this is the only nontrivial part (for classical systems). We further define the functions $n_s(\mathbf{q}_1, \ldots, \mathbf{q}_s)$ by

$$n_s'(\mathbf{q}_1, \ldots, \mathbf{q}_s) = n^s n_s(\mathbf{q}_1, \ldots, \mathbf{q}_s) \tag{7.1.14}$$

s-body correlation functions analogous to $\nu_2(r)$ can also be defined by introducing the cluster representation (3.5.9); we shall, however, not continue this list of definitions.

The reader must be warned at this stage against the distressing variety of notations and names found in the literature in this field. Often n_2 is called a correlation function; in some papers our n_2 is denoted g_2, in others ν_2 is denoted by g_2, and so on. The reader is advised to check carefully the notation before reading any paper; the simplest way of being sure of the notation is to check Eqs. (7.1.10), which make a clear distinction between a distribution and a correlation.

7.2. THERMODYNAMIC QUANTITIES EXPRESSED IN TERMS OF REDUCED DISTRIBUTION FUNCTIONS

All the thermodynamic quantities that are defined originally as phase-space averages of dynamical functions can be very simply expressed in terms of reduced distribution functions. A typical example is the internal energy $E(T, \mathcal{V}, N)$, or better, the intensive *internal energy per particle* $e(T, n) = E/N$:

$$e(T, n) = N^{-1} \int d^N q\, d^N p\, H(q, p) F(q, p) \qquad (7.2.1)$$

where $F(q, p)$ is the phase-space distribution function, and $H(q, p)$ is the Hamiltonian. Assuming once more that the Hamiltonian is of the form defined in Section 2.4, we easily find by using Eqs. (3.1.12), (3.5.2), and (3.5.5):

$$e(T, n) = \int d\mathbf{p}\, H^0(\mathbf{p}) \varphi(\mathbf{p}) + \tfrac{1}{2} N^{-1} \int d\mathbf{q}_1\, d\mathbf{q}_2\, V(\mathbf{q}_1 - \mathbf{q}_2) n_2'(\mathbf{q}_1, \mathbf{q}_2) \qquad (7.2.2)$$

This definition is quite general, being valid both in equilibrium and out of equilibrium. We now specialize the definition by using the canonical distribution and assuming that the system is homogeneous.

The first term is then readily evaluated, using Eqs. (7.1.3):

$$\int d\mathbf{p}\, H^0(\mathbf{p}) \varphi^0(p) = (2\pi m k_B T)^{-3/2} 4\pi \int_0^\infty dp\, p^2 \left(\frac{p^2}{2m}\right) \exp\left(-\frac{p^2}{2mk_B T}\right)$$
$$= \tfrac{3}{2} k_B T$$

where we used the well-known integrals of Gaussian-like functions. Hence, introducing the pair distribution through Eq. (7.1.7) we obtain

$$e(T, n) = \tfrac{3}{2} k_B T + \tfrac{1}{2} n \int d\mathbf{r}\, V(r) n_2(r) \qquad \text{(fluid)} \qquad (7.2.3)$$

This equation provides a compact expression of the internal energy per particle in terms of the pair distribution $n_2(r)$.

The derivation of an expression for the *pressure* is somewhat more complicated. We follow here an elegant proof due to Bogoliubov.

Consider quite generally a function of the volume $\mathcal{V}$ that is represented as a multiple integral over the domain of volume $\mathcal{V}$, which, for simplicity, may be taken as a cube (the results, however, are general):

$$G(\mathcal{V}) = \int_{\mathcal{V}} \cdots \int_{\mathcal{V}} d^3\mathbf{q}_1 \cdots d^3\mathbf{q}_N\, g(\mathbf{q}_1, \ldots, \mathbf{q}_N) \qquad (7.2.4)$$

We now imagine that the volume is increased by a factor ξ^3. The new value of the integral can then be represented in two equivalent ways:

$$G(\xi^3 \mathcal{V}) = \int_{\xi^3 \mathcal{V}} \cdots \int_{\xi^3 \mathcal{V}} d^3\mathbf{q}_1 \cdots d^3\mathbf{q}_N\, g(\mathbf{q}_1, \ldots, \mathbf{q}_N)$$
$$= \int_{\mathcal{V}} \cdots \int_{\mathcal{V}} d^3(\xi\mathbf{q}_1) \cdots d^3(\xi\mathbf{q}_N)\, g(\xi\mathbf{q}_1, \ldots, \xi\mathbf{q}_N) \qquad (7.2.5)$$

In other words, we may either increase the volume in (7.2.4) or scale up the coordinates and integrate in the original domain. We easily establish the following relation:

$$\frac{\partial G(\xi^3\mathcal{V})}{\partial \mathcal{V}} = \frac{\xi}{3\mathcal{V}} \frac{\partial G(\xi^3\mathcal{V})}{\partial \xi}$$

As $\xi \to 1$, this relation becomes

$$\frac{\partial G(\mathcal{V})}{\partial \mathcal{V}} = \frac{1}{3\mathcal{V}}\left(\frac{\partial G(\xi^3\mathcal{V})}{\partial \xi}\right)_{\xi=1} \tag{7.2.6}$$

We now apply these formulas to the case of the pressure. The expression of this quantity in the classical canonical ensemble is taken from Table 4.4.1:

$$P = k_B T \frac{\partial}{\partial \mathcal{V}} \ln Z = \frac{k_B T}{Z} \frac{\partial Z}{\partial \mathcal{V}} \tag{7.2.7}$$

We evaluate the volume derivatives by first making a scale transformation in Z, according to Eq. (7.2.5)*:

$$Z(T, \xi^3\mathcal{V}, N) = \xi^{3N} \int d^N p \int_{\mathcal{V}} d^N q \, \exp\left[-\beta \sum_j H_j^0 - \beta \sum_{j<n}\sum V(\xi r_{jn})\right]$$

where $r_{jn} = |\mathbf{q}_j - \mathbf{q}_n|$. We now calculate the derivative with respect to ξ:

$$\frac{\partial Z(T, \xi^3\mathcal{V}, N)}{\partial \xi} = \frac{3N}{\xi} Z(T, \xi^3\mathcal{V}, N) + \xi^{3N} \int d^N p \, d^N q \left[-\beta \sum_{j<n}\sum V'(\xi r_{jn}) r_{jn}\right] e^{-\beta H}$$

where $V'(r) = dV(r)/dr$. Finally, applying Eq. (7.2.6) we get

$$P = \frac{k_B T}{3\mathcal{V}} \left(\frac{\partial \ln Z(T, \xi^3\mathcal{V}, N)}{\partial \xi}\right)_{\xi=1}$$

or

$$P = \frac{N}{\mathcal{V}} k_B T - \frac{1}{3\mathcal{V}} \tfrac{1}{2} N(N-1) \int d\mathbf{q}_1 \, d\mathbf{q}_2 \, V'(r_{12}) r_{12} \int d^N p \int d^{N-2} q \frac{e^{-\beta H}}{Z}$$

or, using (7.1.6) and (7.1.7),

$$P = n k_B T - \tfrac{1}{6} n^2 \int d\mathbf{r} \, r \left(\frac{dV(r)}{dr}\right) n_2(r) \qquad \text{(fluid)} \tag{7.2.8}$$

This is a quite general form for the equation of state of a fluid. The second term in the right-hand side accounts for all the deviations from the ideal gas law.

We now consider a related quantity, the *isothermal compressibility*. As we know from Section 4.6, this quantity is related to the fluctuations of

* We omit writing here the irrelevant factor $h^{-3N}(N!)^{-1}$.

the numbers of particles; we now wish to express it in terms of the canonical pair-correlation function. We consider a part of our total system, of volume Ω. As this partial system is not assumed to be closed, the situation is very much the same as in Section 4.5, where we introduced the grand ensemble. We now note that the average number of particles in the volume Ω is easily derived from Eq. (3.1.3) for the density [using also Eqs. (3.1.11), (3.1.12)]

$$\langle N_\Omega \rangle = \int_\Omega d\mathbf{x}\, n(\mathbf{x}) = \int_\Omega d\mathbf{x} \int d^N q\, d^N p \sum_{j=1}^{N} \delta(\mathbf{x}-\mathbf{q}_j) F(q,p)$$
$$= \int_\Omega d\mathbf{x} \int d\mathbf{q}_1\, d\mathbf{p}_1\, \delta(\mathbf{x}-\mathbf{q}_1) f_1(\mathbf{q}_1, \mathbf{p}_1) = \int_\Omega d\mathbf{q}_1\, n_1(\mathbf{q}_1) \qquad (7.2.9)$$

We may similarly find the average of the square of the number of particles in Ω:

$$\langle N_\Omega^2 \rangle = \int_\Omega d\mathbf{x} \int_\Omega d\mathbf{x}' \int d^N q\, d^N p \sum_{j=1}^{N} \sum_{n=1}^{N} \delta(\mathbf{x}-\mathbf{q}_j)\, \delta(\mathbf{x}'-\mathbf{q}_n) F(q,p)$$

Separating out the terms for which $j = n$, we may also write

$$\langle N_\Omega^2 \rangle = \int_\Omega d\mathbf{x} \int d^N q\, d^N p \sum_{j=1}^{N} \delta(\mathbf{x}-\mathbf{q}_j) F(q,p)$$
$$+ \int_\Omega d\mathbf{x} \int_\Omega d\mathbf{x}' \int d^N q\, d^N p \sum\sum_{j\neq n} \delta(\mathbf{x}-\mathbf{q}_j)\, \delta(\mathbf{x}'-\mathbf{q}_n) F(q,p)$$
$$= \int_\Omega d\mathbf{q}_1\, n_1(\mathbf{q}_1) + \int_\Omega d\mathbf{q}_1 \int_\Omega d\mathbf{q}_2\, n_2'(\mathbf{q}_1, \mathbf{q}_2) \qquad (7.2.10)$$

It then follows that

$$\langle N_\Omega^2 \rangle - \langle N_\Omega \rangle^2 = \int_\Omega d\mathbf{q}_1\, n_1(\mathbf{q}_1) + \int_\Omega d\mathbf{q}_1 \int_\Omega d\mathbf{q}_2 [n_2'(\mathbf{q}_1, \mathbf{q}_2) - n_1(\mathbf{q}_1) n_1(\mathbf{q}_2)]$$

If the subsystem is sufficiently large and is homogeneous, the density within the subsystem can be considered to be constant [see Eq. (7.1.4)] and we get simply

$$\langle N_\Omega \rangle = \int_\Omega d\mathbf{q}_1\, n = \frac{N}{\mathcal{V}} \Omega$$

Using also Eq. (7.1.7) we get

$$\frac{\langle N_\Omega^2 \rangle - \langle N_\Omega \rangle^2}{\langle N_\Omega \rangle} = \frac{N(\Omega/\mathcal{V}) \left\{ 1 + \int_{[\Omega]} d\mathbf{r} [n n_2(r) - n] \right\}}{N(\Omega/\mathcal{V})}$$

This expression has a well-defined limit when we *first* go to the thermodynamic limit ($N\rightarrow\infty$, $\mathcal{V}\rightarrow\infty$; $N/\mathcal{V}=n$) and *then* consider Ω to be arbitrarily large; we simply get

$$\frac{\langle N^2\rangle-\langle N\rangle^2}{\langle N\rangle}=1+n\int d\mathbf{r}\; \nu_2(r) \tag{7.2.11}$$

The integration is now carried out over an infinite volume. In order to evaluate the left-hand side we can make direct use of our result (4.6.11). The latter was derived in the grand canonical formalism; it is clear, however, that the device used here of singling out a subsystem Ω is quite equivalent to the grand canonical picture. Hence we may introduce the isothermal compressibility χ_T. We obtain the final result:

$$k_BT\left(\frac{\partial n}{\partial P}\right)_T\equiv nk_BT\chi_T=1+n\int d\mathbf{r}\; \nu_2(r) \tag{7.2.12}$$

This very beautiful formula, connecting the compressibility to the integral of the pair correlation function, plays a quite important role in many branches of statistical physics, particularly in the theory of critical phenomena, and in the theory of the equation of state of dense fluids.

We now come to the most delicate thermodynamic function, the *entropy*. It is defined, in the classical canonical ensemble, by the following formula taken from Table 4.4.1:

$$\begin{aligned} S&=k_B\ln Z+k_BT\frac{\partial}{\partial T}\ln Z=-k_B\beta A-k_B\beta\frac{\partial}{\partial\beta}\ln Z\\ &=\frac{k_B\int dq\,dp\,[-\beta A+\beta H(q,p)]\exp[-\beta H(q,p)]}{\int dq\,dp\,\exp[-\beta H(q,p)]}\\ &=k_B\int dq\,dp\,[-\beta A+\beta H(q,p)]F^{(c)}(q,p)\end{aligned}$$

At this stage, we can already see that the entropy is not the average of a microscopic dynamical function: Its definition involves the free energy A, a macroscopic quantity for which we do not know a microscopic dynamical function. The right-hand side can be written alternatively as

$$\begin{aligned} S&=-k_B\int dq\,dp\{\ln[h^{3N}N!\,F^{(c)}(q,p)]F^{(c)}(q,p)\}\\ &=-k_B\int dq\,dp\,[\ln\,F^{(c)}(q,p)]F^{(c)}(q,p)-k_B\ln(h^{3N}N!) \end{aligned} \tag{7.2.13}$$

This is a very celebrated formula, due to Gibbs. It expresses the entropy as a *nonlinear functional of the phase-space distribution function* $F^{(c)}(q,p)$. It is the "formal average" of the function $\ln F^{(c)}(q,p)$: the "dynamical" function, whose "average" is the entropy, depends on the state of the system. It measures the degree of disorder, as can be shown by using well-known elementary arguments. We already mentioned this peculiar behavior in Section 2.2

We now would like to express the entropy in terms of *reduced* distribution functions. In the case of an ideal gas this is a simple matter, as will be shown in Section 7.3. But in the general case this is a puzzling problem. Indeed, if we apply our usual argument of Section 3.1 and decompose the "dynamical function" $\ln F^{(c)}$ according to Eq. (3.1.8), we find terms depending irreducibly on an arbitrary number of particles.* We now show a way out of this dilemma.

Let us introduce a formal coupling constant λ, the same as in Chapter 6; that is, we replace the interaction Hamiltonian H' by $\lambda H'$. When $\lambda \to 0$, the system reduces to an ideal one, whereas for $\lambda \to 1$ we recover the real system. Clearly, the canonical distribution function becomes a function of λ:

$$F^{(c)}(q,p) \to F^{(c)}(q,p;\lambda)$$

All the thermodynamic functions are also functions of λ. Consider now the derivative with respect to λ of the entropy:

$$\begin{aligned}
\partial_\lambda S(\lambda) &= -k_B \int dq\, dp\, \partial_\lambda \{[\ln F^{(c)}(q,p;\lambda)] F^{(c)}(q,p;\lambda)\} \\
&= -k_B \int dq\, dp\, [1 + \ln F^{(c)}(q,p;\lambda)]\, \partial_\lambda F^{(c)}(q,p;\lambda) \\
&= -k_B \int dq\, dp\, [1 - \beta H^0(p) - \beta\lambda H'(q) - \ln Z(\lambda)]\, \partial_\lambda F^{(c)}(q,p;\lambda)
\end{aligned} \tag{7.2.14}$$

We have three types of terms in this formula, and we consider them separately. We first have the "average" of a constant:

$$\begin{aligned}
-k_B \int dq\, dp\, [1 - \ln Z(\lambda)]\, \partial_\lambda F^{(c)}(q,p;\lambda) \\
= -k_B [1 - \ln Z(\lambda)]\, \partial_\lambda \int dq\, dp\, F^{(c)}(q,p;\lambda) = 0
\end{aligned}$$

This result follows from the fact that the canonical distribution function is normalized for all values of the coupling parameter. We note that

* Or else, if we use the explicit form of the canonical distribution, we find an expression in terms of the free energy, which is unknown in terms of the distribution function.

precisely this step eliminates the troublesome partition function from the expression of S. We further get

$$\beta k_B \int dq\, dp\, H^0(p)\, \partial_\lambda F^{(c)}(q, p; \lambda) = \beta k_B \int dq\, dp \sum_{j=1}^{N} H^0(\mathbf{p}_j)\, \partial_\lambda F^{(c)}(q, p; \lambda)$$

$$= \beta k_B \int d\mathbf{p}_1\, H^0(\mathbf{p}_1)\, \partial_\lambda \varphi(\mathbf{p}_1; \lambda) = 0$$

This result follows from the general fact that the reduced momentum distribution of a classical system is simply the Maxwell distribution, whether the system is ideal or not [see Eqs. (7.1.1), (7.1.5)]. This function is therefore independent of λ. This fact is not true for quantum systems. We are finally left with the interaction term, which is transformed as follows:

$$\beta k_B \int dq\, dp\, \tfrac{1}{2} \sum_{j \neq n}\sum \lambda V_{jn}\, \partial_\lambda F^{(c)}(q, p; \lambda)$$

$$= \tfrac{1}{2} \beta k_B \int d\mathbf{q}_1\, d\mathbf{q}_2\, \lambda V(\mathbf{q}_1 - \mathbf{q}_2)\, \partial_\lambda n_2'(\mathbf{q}_1, \mathbf{q}_2; \lambda)$$

Finally, for a homogeneous fluid phase we obtain the following expression for the *entropy per particle:*

$$\partial_\lambda s(\lambda) = \tfrac{1}{2} \beta k_B n \int d\mathbf{r}\, \lambda V(r)\, \partial_\lambda n_2(r; \lambda) \qquad (7.2.15)$$

We now integrate both sides over λ from $\lambda = 0$ to $\lambda = 1$ and remember that $s(0)$ is the entropy of the corresponding ideal gas, $s(1) = s$ is the real entropy, and $n_2(r; 1) = n_2(r)$ is the real pair distribution:

$$\int_0^1 d\lambda\, \partial_\lambda s(\lambda) = s(1) - s(0)$$

$$= \tfrac{1}{2} \beta k_B n \int d\mathbf{r}\, V(r) \int_0^1 d\lambda\, \lambda\, \partial_\lambda n_2(r; \lambda)$$

A simple integration by parts then provides us with the final formula, valid for *classical systems:*

$$s = s(0) + \tfrac{1}{2} k_B \beta n \int d\mathbf{r}\, V(r) n_2(r) - \tfrac{1}{2} k_B \beta n \int d\mathbf{r}\, V(r) \int_0^1 d\lambda\, n_2(r; \lambda) \qquad (7.2.16)$$

Using also Eq. (7.1.9) we see that only the pair correlation contributes:

$$s = s(0) + \tfrac{1}{2} k_B \beta n \int d\mathbf{r}\, V(r) \left[\nu_2(r) - \int_0^1 d\lambda\, \nu_2(r; \lambda) \right] \qquad (7.2.17)$$

We thus obtained a very compact formula expressing the entropy density in terms of the *pair correlation function* alone. However, the expression is *not* a simple average, but involves the new operation of integration over the coupling constant. This operation is also called a *charging process*, because it involves a gradual switching on of the interactions. It clearly exhibits the nonmechanical character of the entropy.

7.3. REDUCED DISTRIBUTION FUNCTIONS OF IDEAL SYSTEMS IN EQUILIBRIUM

In Chapter 5 the thermodynamic properties of ideal systems were derived from the partition function. This method, however, gives us only an incomplete picture of the microscopic properties of these systems. We now illustrate the concepts discussed in Sections 7.1 and 7.2 in this particularly simple case.

We first consider a *classical (Boltzmann) gas* of point particles with the usual Hamiltonian:

$$H^0 = \sum_{j=1}^{N} \frac{p_j^2}{2m} \tag{7.3.1}$$

The canonical distribution function $F^c(q, p)$, Eq. (4.3.27), is now

$$F^{(c)}(q, p) = \mathcal{V}^{-N}(2\pi m k_B T)^{-3N/2} \exp\left[-\sum_{j=1}^{N} \frac{p_j^2}{2mk_B T}\right] \tag{7.3.2}$$

where we made use of Eq. (5.2.19). This distribution has trivially simple properties, which are important precisely because of their simplicity.

The one-particle reduced distribution is obtained form Eq. (7.1.1):

$$f_1^0(\mathbf{q}, \mathbf{p}) = n\varphi^0(p) \tag{7.3.3}$$

The one-particle distribution is thus independent of the positions and is simply proportional to the Maxwell distribution (7.1.3). The properties of this function are discussed in every elementary physics textbook.

We see that the N-particle phase-space distribution $F^{(c)}(q, p)$ is simply a product of one-particle functions:

$$F^{(c)}(q, p) = N^{-N} \prod_{j=1}^{N} f_1^0(\mathbf{q}, \mathbf{p}) \tag{7.3.4}$$

A fortiori, all reduced distribution functions $f_s(x_1, \ldots, x_s)$ of the gas are also factorized; for instance:

$$f_2^0(x_1, x_2) = f_1^0(x_1) f_1^0(x_2) \tag{7.3.5}$$

Comparing this form with Eq. (3.5.9) we conclude that

$$g_2^0(x_1 x_2) \equiv 0 \tag{7.3.6}$$

There are no correlations in a classical ideal gas at equilibrium. This is in agreement with the physical idea that the correlations in a classical system are due to interactions.

The configurational reduced distributions are also trivially simple:

$$\begin{aligned} n_2(r) &= 1 \\ \nu_2(r) &= 0 \end{aligned} \tag{7.3.7}$$

We may now check the more important thermodynamic functions. The internal energy per particle is obtained from Eq. (7.2.3):

$$e = \tfrac{3}{2}k_B T \tag{7.3.8}$$

The pressure is obtained from Eq. (7.2.8):

$$P = nk_B T \tag{7.3.9}$$

The isothermal compressibility follows from (7.2.12):

$$\chi_T = (nk_B T)^{-1} \tag{7.3.10}$$

All these results agree with the results obtained by the partition-function method in Section 5.2.

We now calculate the entropy per particle, by substituting Eq. (7.3.4) into the normalization condition (3.1.17):

$$\begin{aligned} s &= -k_B N^{-1} \int dq\, dp\, F^{(c)}(q, p) \ln[h^{3N} N!\, F^{(c)}(q, p)] \\ &= -k_B N^{-1} \int dq\, dp \Big[N^{-N} \prod_j f_1(\mathbf{q}_j, \mathbf{p}_j)\Big] \ln\Big[h^{3N}\Big(\frac{N}{e}\Big)^N N^{-N} \prod_{n=1}^{N} f_1(\mathbf{q}_n, \mathbf{p}_n)\Big] \\ &= -k_B \int d\mathbf{q}_j\, d\mathbf{p}_j [N^{-1} f_1(\mathbf{q}_j, \mathbf{p}_j)] \ln\Big[\Big(\frac{h^3}{e}\Big) f_1(\mathbf{q}_j, \mathbf{p}_j)\Big] \end{aligned}$$

Finally,

$$s = -k_B N^{-1} \int d\mathbf{q}\, d\mathbf{p}\, f_1(\mathbf{q}, \mathbf{p}) \ln\Big[\Big(\frac{h^3}{e}\Big) f_1(\mathbf{q}, \mathbf{p})\Big] \tag{7.3.11}$$

or alternatively

$$s = -k_B N^{-1} \int d\mathbf{q}\, d\mathbf{p}\, f_1(\mathbf{q}, \mathbf{p}) \ln f_1(\mathbf{q}, \mathbf{p}) + k_B \ln\Big(\frac{e}{h^3}\Big) \tag{7.3.12}$$

Equation (7.3.12) is a fundamental formula of statistical mechanics: it was first derived by *Boltzmann* in 1872. He only obtained the first term on the right-hand side. The second term (which is constant) is, however, important for several reasons. It ensures the correct dimensionality of the entropy: The latter must have dimensions identical with k_B, that is, energy per degree. This is evident from Eq. (7.3.11), because $h^3 f_1$ is dimensionless. The presence of this term shows the impossibility of deriving the thermodynamical properties of a gas from a purely classical theory. In such a theory one always misses two features, which are precisely recovered by the second term in Eq. (7.3.12): the Heisenberg uncertainty principle, which leads to the consideration of finite cells in phase space (factor h^3), and the undistinguishability (factor e). It is noteworthy that these "irreducible" quantum features show up in the entropy, not in the energy. We should also note that in thermodynamics one is usually not interested in the absolute value of the entropy (or of any other potential function), but only in *differences* of entropy between two states. In calculating such differences, the constant quantum correction term cancels out. Hence, *relative thermodynamic quantities can be calculated on a purely classical basis.* The reader will easily check that Eq. (7.3.12), together with (7.1.1), (7.1.4) agrees with our previous result (5.2.27).

Until now we only have considered classical systems. The *reduced distribution functions of quantum statistical systems* are considerably more complicated objects because, as we know, the correlations originate here not only from interactions but also from quantum statistics. Fortunately, for most fluids of interest (except, notably, helium-4), the quantum-statistical effects are not very important, and a classical approximation is sufficient.

We shall illustrate the typical quantum correlations in the simple case of *ideal boson or fermion systems.* For the same reasons as explained in Section 5.4, it is convenient to use the grand ensemble as the starting point for the calculations. (We know, however, that the result is equivalent, in the thermodynamic limit, to the results obtained from the canonical ensemble.) The *one-particle Wigner function* in equilibrium is then of the form (3.8.3):

$$f_1^0(\mathbf{k}, \mathbf{p}) = 8\pi^3 n\delta(\mathbf{k})\varphi_\theta^0(p) \tag{7.3.13}$$

with the momentum distribution $\varphi_\theta^0(p)$ defined by Eq. (3.6.1):

$$\varphi_\theta^0(p) = h^{-3}\, \mathcal{T}\!r\; \boldsymbol{\rho}^{(g)}\mathbf{a}^\dagger(\mathbf{p})\mathbf{a}(\mathbf{p}) \tag{7.3.14}$$

This quantity is, within the factor h^{-3}, equal to the average occupation number $\langle n_p\rangle$, which we calculated in Section 5.5 [see Eq. (5.5.2)]. Hence we obtain immediately for bosons

$$\varphi_+^0(p) = (h^3 n)^{-1}\frac{1}{e^{\beta(\varepsilon_\mathbf{p}-\mu)}-1} \tag{7.3.15}$$

and for fermions

$$\varphi_-^0(p) = (h^3 n)^{-1}\frac{1}{e^{\beta(\varepsilon_\mathbf{p}-\mu)}+1} \tag{7.3.16}$$

These functions play the same role in quantum statistics as the Maxwell distribution (7.1.3) in the classical case. They actually tend towards the Maxwell distribution whenever $\exp(\beta\mu) \ll 1$. In that case we obtain, using Eqs. (5.4.24) and (5.4.14),

$$\begin{aligned}\varphi_\theta^0(p) &= (h^3 n)^{-1}[e^{\beta(\varepsilon_p-\mu)} - \theta]^{-1} \\ &\approx (h^3 n)^{-1} e^{\beta\mu} e^{-\beta\varepsilon_p}(1 + \theta e^{\beta(\mu-\varepsilon_p)} + \cdots) \\ &\approx (h^3 n)^{-1} \Lambda^3 n e^{-\beta\varepsilon_p}(1 + \cdots) = (2\pi m k_B T)^{-3/2} e^{-\beta p^2/2m} \qquad (7.3.17)\end{aligned}$$

We saw in Section 5.5 that the "complementary" function $1 + \theta\langle n_p \rangle$ plays an important role together with $\varphi(\mathbf{p})$. In full generality we define this function as

$$\psi(\mathbf{p}) \equiv 1 + \theta h^3 n \varphi(\mathbf{p}) \qquad (7.3.18)$$

In equilibrium this function is

$$\psi_\theta^0(p) = \frac{e^{\beta(\varepsilon_p-\mu)}}{e^{\beta(\varepsilon_p-\mu)} - \theta} \qquad (7.3.19)$$

We note the following useful identity which is simply derived from Eqs. (7.3.15), (7.3.16), and (7.3.19):

$$\varphi_\theta^0(\mathbf{p})\psi_\theta^0(\mathbf{p}+\mathbf{p}') = \varphi_\theta^0(\mathbf{p}+\mathbf{p}')\psi_\theta^0(\mathbf{p}) \exp\left(\frac{\beta(\mathbf{p}\cdot\mathbf{p}' + \frac{1}{2}p'^2)}{m}\right) \qquad (7.3.20)$$

We now examine the *correlation properties in quantum systems.* As there are no interactions, we expect the correlation function $g_2^W(\mathbf{k}_1\mathbf{p}_1, \mathbf{k}_2\mathbf{p}_2)$ in Eq. (3.8.7) to vanish. This can indeed be checked by a direct calculation. However, we know that even in absence of interactions there are correlations due to quantum statistics. These can be easily evaluated by using the formalism of Section 3.8. The two-body distribution is given by Eq. (3.8.13) with $\pi_2(12) = 0$. Using Eqs. (3.8.10) and (7.3.13) we get

$$\begin{aligned}f_2^0(\mathbf{k}_1\mathbf{p}_1, \mathbf{k}_2\mathbf{p}_2) &= \int d\mathbf{k}_1'\, d\mathbf{k}_2'\, \langle \mathbf{k}_1\mathbf{k}_2 |\, P(1\,|\,2)\, | \mathbf{k}_1'\mathbf{k}_2' \rangle\, f_1^0(\mathbf{k}_1'\mathbf{p}_1) f_1^0(\mathbf{k}_2'\mathbf{p}_2) \\ &= \int d\mathbf{k}_1'\, d\mathbf{k}_2'\, \{\delta(-\mathbf{k}_1' + \mathbf{k}_1)\, \delta(-\mathbf{k}_2' + \mathbf{k}_2) \\ &\quad + \theta\delta[-\mathbf{k}_1' + \tfrac{1}{2}(\mathbf{k}_1 + \mathbf{k}_2) + \hbar^{-1}(\mathbf{p}_2 - \mathbf{p}_1)] \\ &\quad \times \delta[-\mathbf{k}_2' + \tfrac{1}{2}(\mathbf{k}_1 + \mathbf{k}_2) + \hbar^{-1}(\mathbf{p}_1 - \mathbf{p}_2)]\} \\ &\quad \times \exp\{\tfrac{1}{2}\hbar[(\mathbf{k}_1' - \mathbf{k}_1)\cdot\boldsymbol{\partial}_1 + (\mathbf{k}_2' - \mathbf{k}_2)\cdot\boldsymbol{\partial}_2]\} \\ &\quad \times (8\pi^3 n)^2\, \delta(\mathbf{k}_1')\, \delta(\mathbf{k}_2') \varphi_\theta^0(p_1)\varphi_\theta^0(p_2) \\ &= f_1^0(\mathbf{k}_1\mathbf{p}_1) f_1^0(\mathbf{k}_2\mathbf{p}_2) + G_2(\mathbf{k}_1\mathbf{p}_1, \mathbf{k}_2\mathbf{p}_2) \qquad (7.3.21)\end{aligned}$$

where G_2 is the quantum-statistical correlation we were looking for. We shall rather calculate the corresponding pair correlation defined as in (7.1.9) by an integration over the momenta:

$$\tilde{\nu}_2(\mathbf{k}_1\mathbf{k}_2) = n^{-2}\int d\mathbf{p}_1\, d\mathbf{p}_2\, G_2(\mathbf{k}_1\mathbf{p}_1, \mathbf{k}_2\mathbf{p}_2)$$

$$= n^{-2}(8\pi^3 n)^2\theta \int d\mathbf{k}_1'\, d\mathbf{k}_2'\, d\mathbf{p}_1\, d\mathbf{p}_2\, \delta[-\mathbf{k}_1' + \tfrac{1}{2}(\mathbf{k}_1 + \mathbf{k}_2) + \hbar^{-1}(\mathbf{p}_2 - \mathbf{p}_1)]$$

$$\times \delta[-\mathbf{k}_2' + \tfrac{1}{2}(\mathbf{k}_1 + \mathbf{k}_2) + \hbar^{-1}(\mathbf{p}_1 - \mathbf{p}_2)]\, \delta(\mathbf{k}_1')\, \delta(\mathbf{k}_2')$$

$$\times \varphi_\theta^0(\mathbf{p}_1 - \tfrac{1}{2}\hbar\mathbf{k}_1)\varphi_\theta^0(\mathbf{p}_2 - \tfrac{1}{2}\hbar\mathbf{k}_2)$$

The integrations are very simple and we come out with the result:

$$\tilde{\nu}_2(\mathbf{k}_1, \mathbf{k}_2) = 8\pi^3\theta h^3\, \delta(\mathbf{k}_1 + \mathbf{k}_2) \int d\mathbf{p}\, \varphi_\theta^0(\mathbf{p} - \tfrac{1}{2}\hbar\mathbf{k}_1)\varphi_\theta^0(\mathbf{p} + \tfrac{1}{2}\hbar\mathbf{k}_1) \quad (7.3.22)$$

This is a very nice explicit formula from which we can clearly see that the quantum correlations are explicit functionals of the one-particle distribution. We may now proceed to the distance-dependent pair correlation by means of Eq. (3.6.15), which implies

$$\nu_2(r) = \theta h^3 \int d\mathbf{k} \int d\mathbf{p}\, e^{i\mathbf{k}\cdot\mathbf{r}} \varphi_\theta^0(\mathbf{p} - \tfrac{1}{2}\hbar\mathbf{k})\varphi_\theta^0(\mathbf{p} + \tfrac{1}{2}\hbar\mathbf{k})$$

Taking now $\mathbf{p} + \frac{1}{2}\hbar\mathbf{k}$ and $\mathbf{p} - \frac{1}{2}\hbar\mathbf{k}$ as integration variables, we easily find

$$\nu_2(r) = \theta \left| \int d\mathbf{P} \exp\left(\frac{i\mathbf{P}\cdot\mathbf{r}}{\hbar}\right) \varphi_\theta^0(\mathbf{P}) \right|^2 \quad (7.3.23)$$

This very elegant formula is due to London and Placzek.

A distinctive feature of this phenomenon is that the *range* of the correlations is a function of temperature: it is actually measured by the parameter Λ of Eq. (5.4.14). We may, indeed, calculate the integral in the limit of small degeneracy by using Eq. (7.3.17). A simple calculation then yields

$$\nu_2(r) = \theta\{\exp(-\tfrac{1}{4}\rho^2) + \Lambda^3 n\theta[-2^{-1/2}\exp(-\tfrac{1}{4}\rho^2) + 2^{1/2}\exp(-\tfrac{3}{16}\rho^2)] + \cdots\},$$
$$\Lambda^3 n \ll 1 \quad (7.3.24)$$

where

$$\rho = (2\pi)^{1/2}\Lambda^{-1} r$$

We therefore see that the width of the function $\nu_2(r)$ is approximately measured by Λ. At high temperatures the correlation range tends to zero; when it reaches a length comparable to the size of the real particle, one can forget about the quantum correlations and recover the classical result. On the contrary, at low temperatures the range of correlation becomes quite considerable.

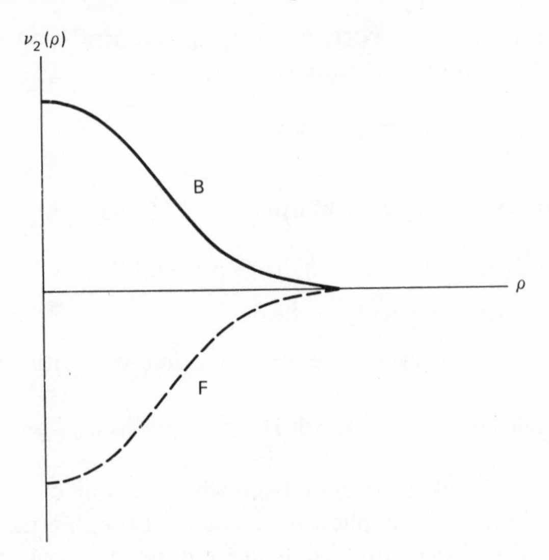

Figure 7.3.1. The pair-correlation function in equilibrium for an ideal quantum gas at moderately high temperatures.

We also note the sign of the correlation function. Clearly, from (7.3.23),

$$\begin{aligned} \nu_2(r) &> 0 \qquad \text{for bosons} \\ &< 0 \qquad \text{for fermions} \end{aligned} \tag{7.3.25}$$

This very clearly shows the effect of statistics acting as a pseudoforce, which is attractive in the boson case (enhanced probability of finding two particles near each other) and repulsive in the fermion case. The correlation functions are depicted in Fig. 7.3.1 (in the limit of small $\Lambda^3 n$).

Let us now briefly discuss the thermodynamical functions, expressed in terms of the reduced distribution functions. It is very simple to verify, by using Eqs. (7.3.15) and (7.3.16) and making a few changes of integration variables, that the definition:

$$e = N^{-1} \int d\mathbf{q}\, d\mathbf{p} \left(\frac{p^2}{2m}\right) f_1^0(\mathbf{q}, \mathbf{p}) \tag{7.3.26}$$

indeed yields the same result as (5.4.17). The case of the entropy is more complicated. The definition (7.3.12) is not valid in quantum statistics. It must be replaced by

$$s = -k_B N^{-1} \int d\mathbf{q}\, d\mathbf{p}\, \{f_1^0(\mathbf{q}, \mathbf{p})[\ln h^3 f_1^0(\mathbf{q}, \mathbf{p})] + (h^3 n)^{-1} \psi^0(\mathbf{q}, \mathbf{p}) \ln \psi^0(\mathbf{q}, \mathbf{p})\} \tag{7.3.27}$$

In the Fermi case, the interpretation of the extra term is pretty simple: It is a Boltzmann-type entropy formula for the holes, which must be added to the "classical" term in order to account for the correct entropy. In the Bose case the interpretation is less clear. It can be verified that this formula for the entropy gives the correct result, identical with the expression derived from the grand partition function.

The fact that we need to average a different function in classical and in quantum mechanics in order to obtain the entropy should not be surprising. It originates from the peculiar character of the entropy, which is not a true average of a dynamical function, but a nonlinear functional of the distribution function. For such expressions, the Wigner correspondence rule does not hold any more, and we have to construct the correct microscopic entropy by a direct comparison with the method of the partition function.

7.4. EQUILIBRIUM HIERARCHY FOR THE REDUCED DISTRIBUTION FUNCTIONS

The calculation of the s-particle distribution $n'_s(\mathbf{q}_1, \ldots, \mathbf{q}_s)$ by using the definition (7.1.13) involves $N-s$ integrations, a process that is complicated in general, especially when we consider the thermodynamic limit. We must try to obtain an alternative definition that would involve only operations on reduced functions. The idea is similar to the spirit of the BBGKY hierarchy of Section 3.4. If the functions defined by (7.1.12) are substituted in the hierarchy equations (3.4.7) one obtains a hierarchy for the functions n'_s; however, we shall derive a somewhat more convenient set of equations.

We start from Eq. (7.1.13) and differentiate both sides with respect to $\mathbf{q}_1$; using Eq. (2.4.4) we find

$$\nabla_1 n'_s(\mathbf{q}_1, \ldots, \mathbf{q}_s) = \left(\frac{N!}{(N-s)!}\right)\mathcal{V}^{-N} Q^{-1} \int d\mathbf{q}_{s+1} \cdots d\mathbf{q}_N \left(-\beta \sum_{n=2}^{N} \nabla_1 V_{1n}\right) e^{-\beta H'} \tag{7.4.1}$$

We now proceed as in Section 3.4, splitting the sum in the rhs:

$$\sum_{n=2}^{N} = \sum_{n=2}^{s} + \sum_{n=s+1}^{N}$$

The first part yields

$$\left(-\beta \sum_{n=2}^{s} \nabla_1 V_{1n}\right)\left(\frac{N!}{(N-s)!}\right)\mathcal{V}^{-N} Q^{-1} \int d\mathbf{q}_{s+1} \cdots d\mathbf{q}_N \, e^{-\beta H'}$$
$$= \left(-\beta \sum_{n=2}^{s} \nabla_1 V_{1n}\right) n'_s(\mathbf{q}_1, \ldots, \mathbf{q}_s)$$

In the second sum, $\mathbf{q}_n$ is a dummy integration variable; therefore the integral of the sum yields $(N-s)$ equal terms:

$$\left(\frac{N!}{(N-s)!}\right)\mathcal{V}^{-N}Q^{-1}\int d\mathbf{q}_{s+1}\cdots d\mathbf{q}_N\left(-\beta\sum_{n=s+1}^{N}\boldsymbol{\nabla}_1 V_{1n}\right)e^{-\beta H'}$$

$$=\left(\frac{N!}{(N-s)!}\right)\mathcal{V}^{-N}Q^{-1}(N-s)\int d\mathbf{q}_{s+1}(\beta\boldsymbol{\nabla}_1 V_{1\,s+1})\int d\mathbf{q}_{s+2}\cdots d\mathbf{q}_N\, e^{-\beta H'}$$

$$=\int d\mathbf{q}_{s+1}(-\beta\boldsymbol{\nabla}_1 V_{1\,s+1})n'_{s+1}(\mathbf{q}_1,\ldots,\mathbf{q}_{s+1})$$

Collecting now the results, multiplying by $k_B T$, and using Eq. (7.1.14), we obtain

$$-k_B T\,\boldsymbol{\nabla}_1 n_s(\mathbf{q}_1,\ldots,\mathbf{q}_s)$$
$$=\sum_{n=2}^{s}(\boldsymbol{\nabla}_1 V_{1n})n_s(\mathbf{q}_1,\ldots,\mathbf{q}_s)+n\int d\mathbf{q}_{s+1}(\boldsymbol{\nabla}_1 V_{1\,s+1})n_{s+1}(\mathbf{q}_1,\ldots,\mathbf{q}_{s+1}),\qquad s\geqslant 2 \tag{7.4.2}$$

For $s=1$, the first term on the right-hand side is absent:

$$-k_B T\,\boldsymbol{\nabla}_1 n_1(\mathbf{q}_1)=n\int d\mathbf{q}_2(\boldsymbol{\nabla}_1 V_{1\,s+1})n_2(\mathbf{q}_1,\mathbf{q}_2) \tag{7.4.3}$$

We thus obtain a hierarchy of equations determining the configurational distribution functions in equilibrium. This hierarchy, derived by *Yvon* in 1935, is quite analogous to (and of course, compatible with) the general BBGKY hierarchy (3.4.7) for the general, time-dependent functions. Of course, the same difficulty appears here as in the general case: In the T-limit we have an infinite set of linear equations. It can only be solved by making appropriate approximations, which always boil down to breaking off the hierarchy after a limiting value of s, usually $s=2$.

Before going further, we may rewrite Eqs. (7.4.2) in a very suggestive form. By using an argument similar to the Debye theory of Section 6.5, we define a *potential of average force*, $W_s(\mathbf{q}_1,\ldots,\mathbf{q}_s)$ through the following equation:

$$n_s(\mathbf{q}_1,\ldots,\mathbf{q}_s)=\exp[-\beta W_s(\mathbf{q}_1,\ldots,\mathbf{q}_s)] \tag{7.4.4}$$

Dividing both sides of Eq. (7.4.2) by n_s we obtain

$$-k_B T\,\boldsymbol{\nabla}_1\ln n_s(\mathbf{q}_1,\ldots,\mathbf{q}_s)$$
$$\equiv\boldsymbol{\nabla}_1 W_s(\mathbf{q}_1,\ldots,\mathbf{q}_s)$$
$$=\sum_{n=1}^{s}\boldsymbol{\nabla}_1 V_{1n}+n\int d\mathbf{q}_{s+1}(\boldsymbol{\nabla}_1 V_{1\,s+1})\frac{n_{s+1}(\mathbf{q}_1,\ldots,\mathbf{q}_{s+1})}{n_s(\mathbf{q}_1,\ldots,\mathbf{q}_s)} \tag{7.4.5}$$

We thus see that, if the s particles were isolated, the average force

$(-\nabla_1 W_s)$ acting on particle 1 would be simply the true mechanical force expressed by the first term on the right-hand side. But the particles are also interacting with the medium: This results in an extra term, representing the force on particle 1 due to a particle $s+1$, averaged with the weight function n_{s+1}/n_s, which is the conditional probability of finding a particle $s+1$ in $\mathbf{q}_{s+1}$, knowing that particles $1, \ldots, s$ are fixed at $\mathbf{q}_1, \ldots, \mathbf{q}_s$.

The traditional method of cutting the hierarchy is the use of the famous *superposition approximation* due to Kirkwood (1935). It consists in expressing the three-body configurational distribution in terms of the pair distribution in the following way:

$$n_3(\mathbf{q}_1, \mathbf{q}_2, \mathbf{q}_3) = n_2(\mathbf{q}_1, \mathbf{q}_2) n_2(\mathbf{q}_2, \mathbf{q}_3) n_2(\mathbf{q}_1, \mathbf{q}_3) \tag{7.4.6}$$

This is an *ad hoc* assumption. Although it has been extensively used and studied for 35 years, it is still not possible to justify it or assess its domain of validity. The only tests are either comparisons of its predictions with experimental data or tests of internal consistency. An interesting consequence of the superposition approximation is the following:

$$W_3(123) = W_2(12) + W_2(23) + W_2(13)$$

or, more generally:

$$W_s(1, \ldots, s) = \sum_{i<j=1}^{s}\sum W_2(ij) \tag{7.4.7}$$

In other words, the average potential appears to be pairwise additive, just like the mechanical potential [Eq. (2.4.4)]. There is clearly no compelling reason that it be so, because statistical collective effects are likely to play an important role in this concept. However, not much is known about the validity of (7.4.7).

When Eq. (7.4.5) is combined (for $n=2$) with the superposition assumption, one obtains a closed *nonlinear* equation:

$$-k_B T \, \nabla_1\{\ln n_2(r_{12})\} = \nabla_1 V(r_{12}) + n \int d\mathbf{q}_3 \{\nabla_1 V(r_{13})\} n_2(r_{13}) n_2(r_{23}) \tag{7.4.8}$$

This equation bears the name Born–Green–Yvon (*BGY equation*). It has been thoroughly studied. Besides this equation, there exists, however, a host of other approximate integral equations for the pair distribution. They represent generalizations or improved versions of the Born–Green–Yvon equations and were derived in the hope of obtaining reasonable approximations for the description of dense fluids. We cannot describe all these equations here, but shall come back to the subject in Section 8.3, where we discuss the most successful of these equations and the experimental data for dense fluids.

7.5. CONNECTION BETWEEN THE PARTITION FUNCTION AND THE REDUCED DISTRIBUTION FUNCTIONS

The two great methods of equilibrium statistical mechanics, based on the partition function and on the reduced distribution functions, respectively, are by no means unrelated, as can be expected from the identity of their results. The connection between these methods has been expressed in a very elegant fashion by Bogoliubov and further developed by Lebowitz and Percus. In addition to revealing an important structural feature of the theory, this method is suitable for applications, as will be seen in the next chapter.

We begin by recalling some elementary concepts of functional analysis.* A *functional* $F[\psi(y)]$ is a "function of a function" $\psi(y)$. In other words, with each function $\psi(y)$ (within a certain domain, for instance, all continuous functions of y in the interval $0 \leq y \leq 1$) the functional associates a real (or complex) number. The rules of ordinary calculus can be extended to provide rules of calculation with functionals. In order to understand this extension, we may think of a functional as a "limiting" object. Consider, indeed, an ordinary function of s variables: $f(x_1, \ldots, x_s)$. In order to specify the value of this object, we must fix the values of s numbers. We may use the following representation for this situation. We plot on a horizontal axis the value of the *index* y of the variable x_y, and on a vertical axis the *value* of the variable x_y. A set of values $(x_1, \ldots, x_s)$ corresponds to a set of s isolated points in this plane (see Fig. 7.5.1). With each such *set* of points, the function $f(\{x_y\})$ associates a well-defined number. The *set* $\{x_y\}$ can be regarded as a function of the discrete variable y. The function $f(\{x_y\})$ associates a number with every different function x_y of the discrete index y; that is, $f(\{x_y\})$ can be regarded as a functional of the function x_y. We may now conceive of functions of an increasing number of variables: this means that we need more and more points in our plane in order to specify the value of the function f. Finally, we can imagine a transition to a limiting situation where we need to specify the values of a continuously infinite set of variables in order to define the value of f. In other words, the index y becomes a continuous variable, and the set of variables x_y becomes a function $\psi(y)$ of the continuous index y. In this limit, the function $f(\{x_y\})$ becomes a functional $F[\psi(y)]$: with every curve, (or segment of a curve), the functional F associates a definite number.

Returning to the ordinary function $f(\{x_y\})$, we now change the value of

* The present exposition is purposely heuristic. The reader interested in a rigorous mathematical formalism is referred to specialized textbooks for full details.

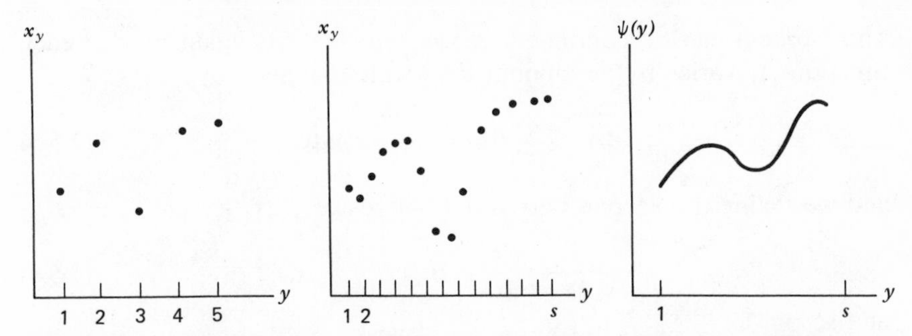

Figure 7.5.1. Passage from a function of a finite number of variables to a functional (see main text).

x_y, for every y, by a small amount dx_y (see Fig. 7.5.2). As a result, the function changes by an amount df which (in the limit $dx_y \to 0$) is a homogeneous linear combination of all the dx_y's:

$$df = \sum_{y=1}^{s} A_y(x_1, \ldots, x_s)\, dx_y \tag{7.5.1}$$

The coefficients A_y depend, in general, on the values $x_1, \ldots, x_s$ from which we start the variation: they are still functions of $\{x_y\}$. We then define the partial derivative of f with respect to x_y as the coefficient of dx_y in this linear form:

$$\frac{\partial f}{\partial x_y} = A_y(x_1, \ldots, x_s) \tag{7.5.2}$$

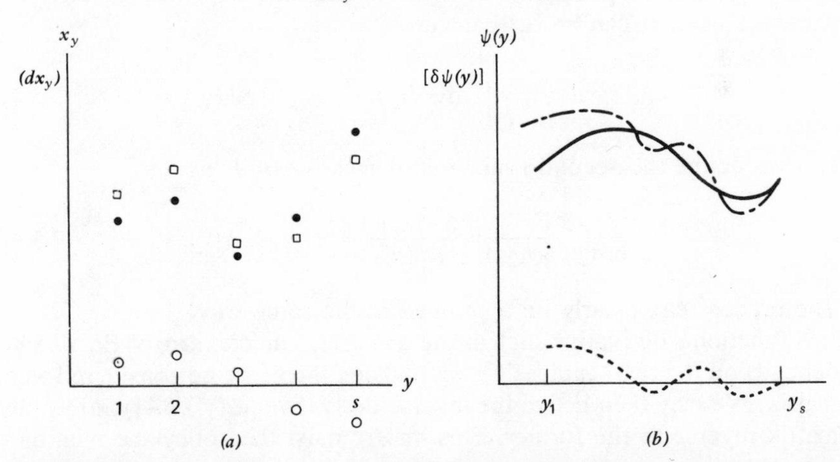

Figure 7.5.2. The concept of functional differential. In (*a*) the black dots represent the values of x_y; the open circles are the displacements dx_y, the squares are the new values $x_y + dx_y$. In (*b*) the continuous line is $\psi(y)$, the dotted line is $\delta\psi(y)$, and the dashed line is $\psi(y) + \delta\psi(y)$.

This process can be continued. If we vary the x_y's again by dx_y, each function A_y varies by an amount dA_y, which is given by

$$dA_y = \sum_{y'=1}^{s} B_{yy'}(x_1, \ldots, x_s)\, dx_{y'} \tag{7.5.3}$$

and we define the second partial derivatives of f as

$$\frac{\partial^2 f}{\partial x_y\, \partial x_{y'}} = \frac{\partial A_y}{\partial x_{y'}} = B_{yy'}(x_1, \ldots, x_s) \tag{7.5.4}$$

and so on.

Proceeding to the functionals, the definitions are easily generalized. The variation of the argument corresponds to changing the value of the function $\psi(y)$ by a certain amount $\delta\psi(y)$ in every point y (see Fig. 7.5.2). The functional F then changes by an amount that [in the limit $\delta\psi(y) \to 0$] is a linear form in $\delta\psi(y)$; the summation over discrete indices is, however, replaced by an integration over the continuous "index" y:

$$\delta F = \int_{y_1}^{y_s} dy\, A[y; \psi(y)]\, \delta\psi(y) \tag{7.5.5}$$

and we define quite naturally the *functional derivative* of F by

$$\frac{\delta F}{\delta\psi(y)} = A[y; \psi(y)] \tag{7.5.6}$$

$A[y; \psi(y)]$ is still a functional of $\psi(y)$ but may also depend separately on the variable y. It can be further varied:

$$\delta A[y; \psi(y)] = \int_{y_1}^{y_s} dy'\, B[y, y'; \psi(y')]\, \delta\psi(y') \tag{7.5.7}$$

and we define the second functional derivative of F as

$$\frac{\delta^2 F}{\delta\psi(y)\, \delta\psi(y')} = \frac{\delta A[y; \psi]}{\delta\psi(y')} = B[y, y'; \psi] \tag{7.5.8}$$

The process can clearly be continued in the same way.

A functional derivative such as the one in the middle term of Eq. (7.5.8) depends on the two "indices" y, y', and can therefore be considered as a matrix. We may then *define* the inverse derivative $\delta\psi(y')/\delta A[y; \psi]$ as the matrix inverse to the former. This matrix must then obey the relation:

$$\int dy'' \frac{\delta A[y; \psi]}{\delta\psi(y'')} \frac{\delta\psi(y'')}{\delta A[y'; \psi]} = \delta(y - y') \tag{7.5.9}$$

We now apply these concepts to the theory of distribution functions.

Consider a system whose Hamiltonian is of the form (2.4.1):

$$H^T = H^0 + H' + \Psi \equiv H + \Psi \tag{7.5.10}$$

where H, H^0, and H' have their usual meaning, and Ψ is the potential of an arbitrary external force, depending only on the coordinates of the particles [see Eq. (2.4.11)]:

$$\Psi = \sum_{j=1}^{N} \psi_j = \sum_{j=1}^{N} \psi(\mathbf{q}_j) \tag{7.5.11}$$

The partition function of this system can be written in the form (6.1.3):

$$Z[\psi] = (N!)^{-1} \Lambda^{3N} \mathcal{V}^N Q[\psi] \tag{7.5.12}$$

with the configuration integral defined by

$$Q[\psi] = \mathcal{V}^{-N} \int d\mathbf{q}_1 \cdots d\mathbf{q}_N \exp[-\beta H' - \beta \Psi] \tag{7.5.13}$$

We now look at the logarithm of the partition function and at the logarithm of the configuration integral, as functionals of the potential ψ. Calculating the variation with respect to ψ yields

$$\delta \ln Z[\psi] = \delta \ln Q[\psi]$$

$$= -\beta Q^{-1}[\psi] \mathcal{V}^{-N} \int d\mathbf{q}_1 \cdots d\mathbf{q}_N \Big(\sum_j \delta\psi_j\Big) \exp\Big[-\beta H' - \beta \sum_i \psi_i\Big]$$

Using the symmetry of the integrand, we note that all the terms in the sum over j give the same contribution, hence

$$\delta \ln Z[\psi] = -\beta Q^{-1}[\psi] \mathcal{V}^{-N} N \int d\mathbf{q}_1 \cdots d\mathbf{q}_N \, \delta\psi_1 \exp(-\beta H' - \beta \Psi)$$

$$= -\beta \int d\mathbf{q}_1 \, \delta\psi(\mathbf{q}_1) N \mathcal{V}^{-N} Q^{-1}[\psi] \int d\mathbf{q}_2 \cdots d\mathbf{q}_N \exp(-\beta H' - \beta \Psi)$$

Comparing this result with Eq. (7.1.2), we see that the factor of $\delta\psi(\mathbf{q}_1)$ in the integrand is simply the one-particle distribution, in presence of the external field ψ:

$$\delta \ln Z[\psi] = -\beta \int d\mathbf{q}_1 \, \delta\psi(\mathbf{q}_1) \, n_1(\mathbf{q}_1; \psi) \tag{7.5.14}$$

Using now Eqs. (7.5.5), (7.5.6), we obtain

$$-k_B T \frac{\delta \ln Z[\psi]}{\delta\psi(\mathbf{q})} = n_1(\mathbf{q}; \psi) \tag{7.5.15}$$

In particular, taking now $\psi = 0$, we find

$$-k_BT\left[\frac{\delta \ln Z[\psi]}{\delta\psi(\mathbf{q})}\right]_{\psi=0} = n_1(\mathbf{q}) \tag{7.5.16}$$

Hence, the *one-body distribution is the first functional derivative of the free energy with respect to the external field* ψ, in the limit of a vanishing field. This beautiful formula provides a very interesting interpretation of the equilibrium one-particle distribution. This function measures the response of the free energy (i.e., a macroscopic variable) to a small variation of the microscopic Hamiltonian.

We now calculate the second variation of ln Z. We must be careful in treating the variable $\mathbf{q}_1$, which is not integrated over. We find

$$\begin{aligned}
\delta^2 \ln Z[\psi] &= \delta n_1(\mathbf{q}_1; \psi)\\
&= N\delta\frac{\int d\mathbf{q}_2 \cdots d\mathbf{q}_N \exp(-\beta H' - \beta\sum \psi_i)}{\int d\mathbf{q}_1 \cdots d\mathbf{q}_N \exp(-\beta H' - \beta \sum \psi_i)}\\
&= -\beta N\mathcal{V}^{-N}Q^{-1}[\psi]\int d\mathbf{q}_2 \cdots d\mathbf{q}_N\, e^{-\beta H' - \beta\Psi}\sum_{j=2}^{N}\delta\psi_j\\
&\quad - \beta N\mathcal{V}^{-N}Q^{-1}[\psi]\,\delta\psi_1\int d\mathbf{q}_2 \cdots d\mathbf{q}_N\, e^{-\beta H'-\beta\Psi}\\
&\quad + \beta N\mathcal{V}^{-2N}Q^{-2}[\psi]\left\{\int d\mathbf{q}_2 \cdots d\mathbf{q}_N\, e^{-\beta H'-\beta\Psi}\right\}\\
&\quad \times\left\{\int d\mathbf{q}_1 \cdots d\mathbf{q}_N\, e^{-\beta H' - \beta\Psi}\sum_j \delta\psi_j\right\}\\
&= -\beta\int d\mathbf{q}_2\, \delta\psi(\mathbf{q}_2)N(N-1)\mathcal{V}^{-N}Q^{-1}[\psi]\int d\mathbf{q}_3 \cdots d\mathbf{q}_N\, e^{-\beta H'-\beta\Psi}\\
&\quad - \beta\,\delta\psi(\mathbf{q}_1)n_1(\mathbf{q}_1;\psi) + \beta n_1(\mathbf{q}_1;\psi)\int d\mathbf{q}_2\,\delta\psi(\mathbf{q}_2)N\mathcal{V}^{-N}Q^{-1}[\psi]\\
&\quad \times\int d\mathbf{q}_1\, d\mathbf{q}_3 \cdots d\mathbf{q}_N\, e^{-\beta H'-\beta\Psi}\\
&= -\beta\int d\mathbf{q}_2\,\delta\psi(\mathbf{q}_2)\{n_2'(\mathbf{q}_1, \mathbf{q}_2;\psi) + n_1(\mathbf{q}_1;\psi)\,\delta(\mathbf{q}_1 - \mathbf{q}_2)\\
&\quad - n_1(\mathbf{q}_1;\psi)n_1(\mathbf{q}_2;\psi)\}
\end{aligned}$$

where we used Eqs. (7.1.2), (7.1.6). Finally:

$$\begin{aligned}
(k_BT)^2\frac{\delta^2 \ln Z[\psi]}{\delta\psi(\mathbf{q}_1)\,\delta\psi(\mathbf{q}_2)} &= -k_BT\frac{\delta n_1(\mathbf{q}_1;\psi)}{\delta\psi(\mathbf{q}_2)}\\
&= n_2'(\mathbf{q}_1, \mathbf{q}_2;\psi) - n_1(\mathbf{q}_1;\psi)n_1(\mathbf{q}_2;\psi)\\
&\quad + n_1(\mathbf{q}_1;\psi)\,\delta(\mathbf{q}_1 - \mathbf{q}_2)
\end{aligned} \tag{7.5.17}$$

and for $\psi=0$

$$(k_B T)^2 \left(\frac{\delta^2 \ln Z[\psi]}{\delta\psi(\mathbf{q}_1)\, \delta\psi(\mathbf{q}_2)} \right)_{\psi=0} = n_2'(\mathbf{q}_1, \mathbf{q}_2) - n_1(\mathbf{q}_1) n_1(\mathbf{q}_2) + n_1(\mathbf{q}_1)\, \delta(\mathbf{q}_1 - \mathbf{q}_2) \tag{7.5.18}$$

Hence, the *pair distribution function is directly related to the second functional derivative of the free energy*, or equivalently to the first functional derivative of the one-body distribution. Obviously, this process can be continued for all the higher distribution functions.

An interesting and important relation is obtained from a consideration of the inverse functional derivative of the distribution function. Let us first introduce the following *notation:*

$$-\frac{1}{k_B T} \frac{\delta\psi(\mathbf{q}_2)}{\delta n_1(\mathbf{q}_1; \psi)} \equiv \frac{1}{n_1(\mathbf{q}_1; \psi)} \delta(\mathbf{q}_1 - \mathbf{q}_2) - C(\mathbf{q}_1, \mathbf{q}_2; \psi) \tag{7.5.19}$$

This relation *defines* the *direct correlation function* $C(\mathbf{q}_1, \mathbf{q}_2; \psi)$. Combining Eqs. (7.5.9), (7.5.17), and (7.5.19) we obtain the following equation:

$$\int d\mathbf{q}_3 \{ n_2'(\mathbf{q}_1, \mathbf{q}_3; \psi) - n_1(\mathbf{q}_1; \psi) n_1(\mathbf{q}_3; \psi) + n_1(\mathbf{q}_1; \psi)\, \delta(\mathbf{q}_1 - \mathbf{q}_3) \}$$
$$\times \{ [n_1(\mathbf{q}_2; \psi)]^{-1} \delta(\mathbf{q}_3 - \mathbf{q}_2) - C(\mathbf{q}_3, \mathbf{q}_2; \psi) \} = \delta(\mathbf{q}_1 - \mathbf{q}_2)$$

which reduces to

$$[n_1(\mathbf{q}_1; \psi)]^{-1} [n_2'(\mathbf{q}_1, \mathbf{q}_2; \psi) - n_1(\mathbf{q}_1; \psi) n_1(\mathbf{q}_2; \psi)]$$
$$= n_1(\mathbf{q}_1; \psi) C(\mathbf{q}_2, \mathbf{q}_1; \psi) + \int d\mathbf{q}_3 [n_2'(\mathbf{q}_1, \mathbf{q}_3; \psi) - n_1(\mathbf{q}_1; \psi) n_1(\mathbf{q}_3; \psi)] C(\mathbf{q}_2, \mathbf{q}_3; \psi) \tag{7.5.20}$$

We now set $\psi=0$ and assume that the system is spatially homogeneous; this means that Eqs. (7.1.4), (7.1.7), and (7.1.9) apply and reduce the equation to the following one:

$$\nu_2(\mathbf{q}_1 - \mathbf{q}_2) = C(\mathbf{q}_1 - \mathbf{q}_2) + n \int d\mathbf{q}_3\, \nu_2(\mathbf{q}_1 - \mathbf{q}_3) C(\mathbf{q}_3 - \mathbf{q}_2) \tag{7.5.21}$$

This is a very famous relation called after its authors the *Ornstein–Zernike equation* (briefly, O–Z equation). It provides a relation between the pair correlation function $\nu_2(r)$ and the direct correlation function $C(r)$. It is an exact relation; however, it is void as long as we cannot find a second independent equation to determine the two functions. Various approximate equations, of various degrees of sophistication, have been suggested in order to complete the O–Z equation. Some of them will be studied in subsequent chapters.

The O–Z equation takes a specially simple form in the Fourier representation. We introduce the functions $\tilde{\nu}_{\mathbf{k}}$ and $\tilde{C}_{\mathbf{k}}$ by

$$\tilde{\nu}_{\mathbf{k}} = \int d\mathbf{r}\, e^{-i\mathbf{k}\cdot\mathbf{r}} \nu_2(r) \tag{7.5.22}$$

$$\tilde{C}_{\mathbf{k}} = \int d\mathbf{r}\, e^{-i\mathbf{k}\cdot\mathbf{r}} C(\mathbf{r}) \tag{7.5.23}$$

We now note that the integral on the right-hand side of (7.5.21) is a convolution of $\nu_2(r)$ and $C(r)$; such an expression transforms into an ordinary product of the Fourier transforms, hence

$$\tilde{\nu}_{\mathbf{k}} = \tilde{C}_{\mathbf{k}} + n\tilde{\nu}_{\mathbf{k}}\tilde{C}_{\mathbf{k}}$$

or

$$\tilde{\nu}_{\mathbf{k}} = \frac{\tilde{C}_{\mathbf{k}}}{1 - n\tilde{C}_{\mathbf{k}}} \tag{7.5.24}$$

These equations become very interesting in the limit $k = 0$. Indeed, as is seen from (7.5.23), we have simply

$$\tilde{\nu}_0 = \int d\mathbf{r}\, \nu_2(r)$$
$$\tilde{C}_0 = \int d\mathbf{r}\, C(r) \tag{7.5.25}$$

Hence $\tilde{\nu}_0$ is directly related to the compressibility through Eq. (7.2.12):

$$k_B T \left(\frac{\partial n}{\partial P}\right)_T = 1 + n\tilde{\nu}_0 \tag{7.5.26}$$

Using now Eq. (7.5.24) we get the following, very useful relation:

$$(k_B T)^{-1} \left(\frac{\partial P}{\partial n}\right)_T = 1 - n\tilde{C}_0 = 1 - n\int d\mathbf{r}\, C(r) \tag{7.5.27}$$

This formula, which is, in a sense, complementary to Eq. (7.2.12), will be used in cases where the direct correlation function is simpler than the pair distribution, which is often so.

BIBLIOGRAPHICAL NOTES

The topics of this chapter are extensively treated in the general textbooks by Hill and by Münster (see Bibliographical Notes to Chapter 4).

An important, more specialized reference is S. Rice and P. Gray, *Statistical Mechanics of Simple Liquids*, Wiley-Interscience, New York, 1965.

Yvon's hierarchy appeared for the first time in his book of 1935, quoted in the Bibliographical Notes to Chapter 3.

The *superposition approximation* was introduced by J. G. Kirkwood, *J. Chem. Phys.* **3,** 300 (1935).

The *BGY equation* appears in M. Born and H. S. Green, *Proc. Roy. Soc.* (London) **A188,** 10 (1946).

Other integral equations appear in

J. G. Kirkwood (quoted above).

J. E. Mayer, *J. Chem. Phys.* **15,** 187 (1947).

J. G. Kirkwood and Z. W. Salsburg, *Disc. Faraday Soc.* **15,** 28 (1953).

These and other equations are extensively reviewed by Münster.

The *functional methods* were introduced systematically in statistical mechanics by N. N. Bogoliubov, *Problemy Dinamicheskoi Teorii v Statisticheskoi Fizike,* Gostekhizdat, Moscow, 1946 [English translation: Problems of a Dynamical Theory in Statistical Physics, in *Studies in Statistical Mechanics,* Vol. 1 (J. de Boer and G. E. Uhlenbeck, eds.). North Holland, Amsterdam, 1962].

It is, however, interesting to note that the first application of functional methods to statistical mechanics appeared in a not too well-known paper by Yvon: J. Yvon, *Revue Scientifique,* **1939,** 662. He applied these methods to a perturbation theory for the calculation of the pair distribution function.

A very clear paper on the functional methods, which we followed closely in Section 7.5, is J. L. Lebowitz and J. K. Percus, *J. Math. Phys.* **4,** 116 (1963).

The O–Z *equation* appeared in L. S. Ornstein and F. Zernike, *Proc. Acad. Sci. Amsterdam* **17,** 793 (1914).

CHAPTER 8

DENSE FLUIDS IN EQUILIBRIUM

8.1. PAIR CORRELATION FUNCTION AND SCATTERING PHENOMENA

The theory of dense fluids is the main field of application of the method of reduced distribution functions. There are several reasons for this state of affairs.

One reason is purely theoretical. The methods of investigation of the partition function introduced in Chapter 6 become quite inadequate in the present case. Indeed, they are based on the idea of a perturbation expansion in powers of the density (or of some other parameter). Such expansions converge very poorly when we consider dense gases or liquids. In the best cases we would need to retain an enormous number of graphs, whose contributions we would be unable to calculate exactly. In the worst cases, the series would no longer converge, so that they would no longer represent the partition function at all.

Series expansions can be devised for the pair distribution function also; but, for the same reason, they cannot be expected to be more successful. However, in the case of the pair distribution, it is possible to "short-circuit" the difficulties by making more or less sophisticated assumptions about the properties of reduced distribution functions. Typically, one would start from the Yvon hierarchy of equations of Section 7.4 and introduce an *a priori* assumption that truncates it at the level of the pair distribution. Or one would start from the formal series expansion, select a certain (infinite) class of graphs, and show that the corresponding approximate pair distribution obeys a closed equation. We must stress the fact that such procedures can *never* be justified on a completely rational basis: They always involve a guessing process, whose result may turn out to be more or less lucky. Nevertheless, in recent years some approximation procedures of this type provided remarkably good results; they will be discussed in the forthcoming sections.

There is a much more important reason for studying the pair distribution function of a dense fluid (rather than the partition function). The pair distribution function can actually be determined experimentally in full detail. This fact provides a much more precise and sensitive test of the theoretical models than the mere measurements of pressure, specific heat, or other thermodynamic quantities.

The crucial phenomenon in this respect is the *elastic scattering of electromagnetic waves by a fluid.* The experiment consists in irradiating a sample of fluid by a beam of radiation of wave vector $\mathbf{k}_i$ and circular frequency ω. If the energy $\hbar\omega$ of the quanta is much larger than the characteristic excitation energy of the molecules, the scattering occurs without any change of frequency; in other words, the wave vector $\mathbf{k}_f$ of the scattered wave is equal in magnitude to $|\mathbf{k}_i|$. Calling $\mathbf{k}$ the momentum transfer: $\mathbf{k}=\mathbf{k}_f-\mathbf{k}_i$, it is clear that the intensity of the scattered radiation depends only on this variable.

The wave scattered by an individual particle j in the direction $\mathbf{k}$ is assumed to have an amplitude $\alpha_j(\mathbf{k})$, and its intensity is $|\alpha_j(\mathbf{k})|^2$: this intensity is the same for all particles. The function $\alpha_j(\mathbf{k})$ depends on the detailed atomic or molecular structure of the particle: It can be calculated or measured, but its form is not relevant to us now.

We are interested, rather, in the intensity of the wave scattered by the set of N particles making up the fluid. If these intensities were simply additive, the result would be:

$$I_0(\mathbf{k})=N\,|\alpha_1(\mathbf{k})|^2 \tag{8.1.1}$$

However, we know that the intensities of the waves are not additive. Rather, we must add together the amplitudes and square the sum to get the total intensity. The waves scattered by two particles, say 1 and j differ in phase by an amount depending on the distance:

$$\alpha_j(\mathbf{k})=\alpha_1(\mathbf{k})\exp[-i\mathbf{k}\cdot(\mathbf{q}_j-\mathbf{q}_1)] \tag{8.1.2}$$

Hence, the total intensity is obtained by averaging the absolute square of the total amplitude over all positions of the particles:

$$\begin{aligned}
I(\mathbf{k}) &= \left\langle\left|\sum_{j=1}^{N}\alpha_j(\mathbf{k})\right|^2\right\rangle=\left\langle\left|\alpha_1(\mathbf{k})\sum_{j=1}^{N}e^{-i\mathbf{k}\cdot(\mathbf{q}_j-\mathbf{q}_1)}\right|^2\right\rangle\\
&= |\alpha_1(\mathbf{k})|^2\left\langle\sum_{j=1}^{N}\sum_{n=1}^{N}e^{-i\mathbf{k}\cdot(\mathbf{q}_j-\mathbf{q}_n)}\right\rangle\\
&= N^{-1}I_0(\mathbf{k})\int d\mathbf{x}\,d\mathbf{x}'\,e^{-i\mathbf{k}\cdot(\mathbf{x}-\mathbf{x}')}\left\langle\sum_{j=1}^{N}\sum_{n=1}^{N}\delta(\mathbf{x}-\mathbf{q}_j)\delta(\mathbf{x}'-\mathbf{q}_n)\right\rangle
\end{aligned}$$

Now using Eq. (7.2.10) we obtain

$$\begin{aligned}
I(\mathbf{k}) &= I_0(\mathbf{k})\left\{N^{-1}\int d\mathbf{x}\,n_1(\mathbf{x})+N^{-1}\int d\mathbf{x}\int d\mathbf{r}\,e^{-i\mathbf{k}\cdot\mathbf{r}}n_2'(\mathbf{r})\right\}\\
&= I_0(\mathbf{k})\left\{1+n^{-1}\int d\mathbf{r}\,e^{-i\mathbf{k}\cdot\mathbf{r}}n_2'(\mathbf{r})\right\}
\end{aligned} \tag{8.1.3}$$

In a fluid without correlations, $n_2'(r) = n^2$ and we would simply get

$$I^{(0)}(\mathbf{k}) = I_0(\mathbf{k})\{1 + n\delta(\mathbf{k})\} \tag{8.1.4}$$

In other words, there would simply be a large increase of the intensity in the forward direction, where it would be unobservable because it would be masked by the incident beam. The real cause of the scattering phenomenon (at nonzero angles) is the presence of a correlation, that is, of density fluctuations in the fluid.

It is convenient to subtract the uninteresting term $n\delta(\mathbf{k})$ from the total intensity, obtaining then

$$\begin{aligned} i(\mathbf{k}) &\equiv I(\mathbf{k}) - I_0(\mathbf{k})n\delta(\mathbf{k}) \\ &= I_0(\mathbf{k})\left\{1 + n^{-1}\int d\mathbf{r}\, e^{-i\mathbf{k}\cdot\mathbf{r}}[n_2'(r) - n^2]\right\} \end{aligned}$$

or, introducing the pair correlation (7.1.9) as well as its Fourier transform (7.5.22), we finally get

$$\begin{aligned} a_{\mathbf{k}} \equiv \frac{i(\mathbf{k})}{I_0(\mathbf{k})} &= 1 + n\int d\mathbf{r}\, e^{-i\mathbf{k}\cdot\mathbf{r}}\nu_2(r) \\ &= 1 + n\tilde{\nu}_{\mathbf{k}} \end{aligned} \tag{8.1.5}$$

The function $a_{\mathbf{k}}$ is called the *structure factor* of the fluid.

Hence, the experimentally observable scattering intensity is directly related to the Fourier transform of the pair correlation function. This is a quite fundamental result.

We note that the intensity pattern is actually a function of the scattering angle. Indeed, from the geometry of Fig. 8.1.1 we find that

$$k = 2k_i \sin\tfrac{1}{2}\theta = \left(\frac{2\omega}{c}\right)\sin\tfrac{1}{2}\theta \tag{8.1.6}$$

Hence, probing the fluid with radiation in a certain range of frequencies ω and observing the scattering pattern at various angles provides enough information for the determination of $\nu_2(r)$ through a Fourier analysis of the results.

The choice of frequencies is crucial in these experiments. Indeed, we

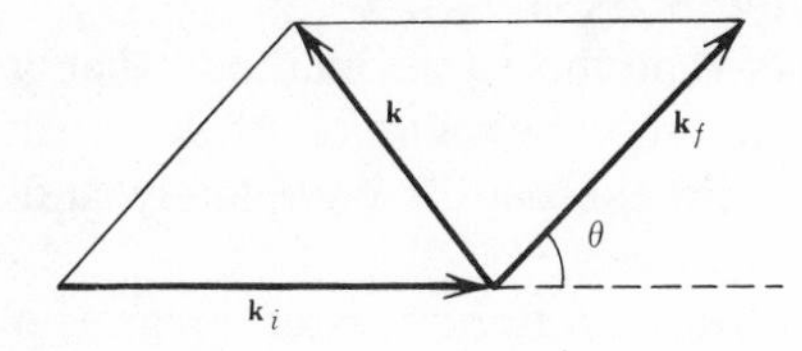

Figure 8.1.1. Geometry of the scattering.

need frequencies corresponding to wavelengths $\lambda \equiv 2\pi/k$, of the order of the range of the correlation function, that is, we need to do the experiment with x rays. If we use longer wavelengths, such as visible light, no structure will show up. To see this, we transform the middle side of Eq. (8.1.5) by integrating over the orientation of the vector **r**:

$$a_k = 1 + 4\pi n \int dr\, r^2 \left(\frac{\sin kr}{kr}\right) \nu_2(r) \tag{8.1.7}$$

If the wavelength is much longer than the range over which $\nu_2(r) \neq 0$, that is, the product kr is very small in the effective range of integration, then $(\sin kr/kr)$ can be effectively replaced by 1:

$$a_k \approx 1 + 4\pi n \int dr\, r^2 \nu_2(r) \tag{8.1.8}$$

and, using Eq. (7.2.12), we find

$$a_k = nk_B T \chi_T \tag{8.1.9}$$

Hence, the scattering in the optical domain is determined simply by an overall thermodynamic quantity, the isothermal compressibility.

Equation (8.1.9) provides us with the direct explanation of the phenomenon of *critical opalescence.* Indeed, as we lower the temperature of the fluid, approaching the critical temperature, the isotherms become flat near the critical volume. Hence χ_T becomes very large, and so does the amount of scattering.

8.2. DENSITY EXPANSION OF THE PAIR DISTRIBUTION FUNCTION

The pair distribution function can be represented as a formal series expansion in powers of the density:

$$n_2(r_{12}) = \exp[-\beta V(r_{12})]\left(1 + \sum_{p=3}^{\infty} \beta_{p;2}(r_{12}) n^{p-2}\right) \tag{8.2.1}$$

It is clear from the definitions (7.1.5) and (7.1.6) that the leading term in the series is simply the Boltzmann factor $\exp[-\beta V(r_{12})]$, which would be exactly equal to $n_2(r_{12})$ if the influence of the particles other than 1 and 2 could be neglected (see Section 7.4).

It is pretty obvious that the same methods that were developed in Chapter 6 can be applied to the evaluation of the coefficients $\beta_{p;2}(r)$. For this reason we shall skip the details completely and simply quote the results.

The coefficients $\beta_{p;2}$ turn out to be cluster integrals of the same kind as $\bar{B}_p$ in Eq. (6.4.7). The main difference is that $\beta_{p;2}$ still depends on the distance r_{12}, because one must not integrate over the positions $\mathbf{q}_1$, $\mathbf{q}_2$ in the cluster integral. As a result, the vertices 1, 2 play a special role in the representative graph: they are called *roots*. Because of the presence of roots, the combinatorial factors in the graphs are different. The final rule tells us that $\beta_{p;2}$ is of the form:

$$\beta_{p;2} = \frac{1}{(p-2)!}\int d\mathbf{q}_3 \cdots d\mathbf{q}_p \sum_{\Delta} S'_{\Delta}\left(\prod f_{ij}\right)_{\Delta} \tag{8.2.2}$$

The summation runs over all skeleton graphs of p points with two roots 1, 2, which become irreducibly connected when a line 1–2 is inserted, but which contain no such line. The coefficient S'_{Δ} is again the number of distinct labeled graphs obtained from a prototype by permutation of the labels (other than 1 and 2) among the vertices (other than the roots). The following remark is important in understanding the relation between the expansion (8.2.1) and the virial expansion (6.4.9) of the equation of state: The coefficient $\beta_{p;2}(r)$ corresponds to the virial coefficients B_p, as they both involve p-particle cluster integrals. Therefore, an expansion (8.2.1) truncated at $p = p_0$, that is, to order n^{p_0-2}, yields through Eq. (7.2.8) an equation of state correct to order n^{p_0-1}.

The graphs contributing to $\beta_{3;2}$ and $\beta_{4;2}$ according to this rule are shown in Table 8.2.1. We shall not further develop this subject, but shall come back to it in the next section.

Table 8.2.1. Graphs for the Density Expansion of the Pair Distribution Function, $n_2(r)\exp[\beta V(r)]$

Graph	S'_{Δ}	Contribution
	Order n	
	1	$\beta_{3;2} = \int_3 f_{13}f_{23}$
	Order n^2	
	2	$\beta^{(1)}_{4;2} = \int_{34} f_{13}f_{34}f_{24}$
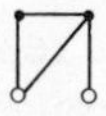	2	$\beta^{(2)}_{4;2} = \int_{34} f_{13}f_{34}f_{14}f_{24}$
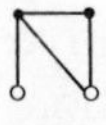	2	$\beta^{(3)}_{4;2} = \int_{34} f_{13}f_{34}f_{24}f_{23}$
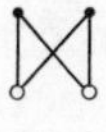	1	$\beta^{(4)}_{4;2} = \frac{1}{2}\int_{34} f_{13}f_{24}f_{14}f_{23}$
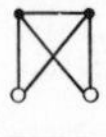	1	$\beta^{(5)}_{4;2} = \frac{1}{2}\int_{34} f_{13}f_{34}f_{24}f_{14}f_{23}$

8.3. THE PERCUS–YEVICK AND THE HYPERNETTED-CHAINS EQUATIONS

As was noted already in Section 8.1, the density expansion developed in Section 8.2 is not of direct use in the high-density domain. The only way in which it can be utilized is to select infinite summable subclasses of graphs and show that their sum obeys a closed integral equation, which represents an approximation to the exact equation (7.4.2) for n_2 in the Yvon hierarchy. The criteria for defining the class to be considered are purely heuristic and pragmatic. Very roughly speaking, one chooses whatever diagrams can be effectively summed in a sufficiently simple way, and one checks *a posteriori* whether the results compare favorably with the experimental data.

The same integral equations can be derived in a much shorter way by using a method due to Lebowitz and Percus and based on functional Taylor expansions. We will use this latter idea here in order to familiarize the reader with this very elegant method.

Consider, quite generally, two functionals $A(\mathbf{q}; \psi)$ and $B(\mathbf{q}; \psi)$ of a function $\psi(\mathbf{q})$ (see Section 7.5). $A(\mathbf{q}; \psi)$ will be called the generating functional and $B(\mathbf{q}; \psi)$ the independent functional. The functional A can be expanded in powers of the functional B by using a theorem that generalizes the Taylor–McLaurin expansion:

$$A(\mathbf{q}_1; \psi) = A(\mathbf{q}_1) + \int d\mathbf{q}_2\, [B(\mathbf{q}_2; \psi) - B(\mathbf{q}_2)]\left(\frac{\delta A(\mathbf{q}_1; \psi)}{\delta B(\mathbf{q}_2; \psi)}\right)_{\psi=0}$$
$$+\frac{1}{2!}\int d\mathbf{q}_2\, d\mathbf{q}_3\, [B(\mathbf{q}_2; \psi) - B(\mathbf{q}_2)][B(\mathbf{q}_3; \psi) - B(\mathbf{q}_3)]$$
$$\times\left[\frac{\delta^2 A(\mathbf{q}_1; \psi)}{\delta B(\mathbf{q}_2; \psi)\delta B(\mathbf{q}_3; \psi)}\right]_{\psi=0} + \cdots \tag{8.3.1}$$

We now keep in mind the results of Section 7.5. We choose the generating functional and the independent functional to be related to the one-particle distribution function, and we note that*

$$\frac{\delta A(\mathbf{q}_1; \psi)}{\delta B(\mathbf{q}_2; \psi)} = \int d\mathbf{q}_3\, \frac{\delta A(\mathbf{q}_1; \psi)}{\delta \psi(\mathbf{q}_3)}\, \frac{\delta \psi(\mathbf{q}_3)}{\delta B(\mathbf{q}_2; \psi)} \tag{8.3.2}$$

We then realize that the first-order term in (8.3.1) will be related to the pair distribution and to the direct correlation function through Eqs. (7.5.18) and (7.5.19), whereas the higher functional derivatives will involve higher-order distributions. If we want a closed equation involving only $n_2(r)$ and $C(r)$ we have to assume that the expansion can be truncated after the first term. No other argument can be claimed for the justification

* Note that Eq. (7.5.9) is a particular case of Eq. (8.3.2).

of the procedure. However, we must note that the functional formulation is so flexible that we still have an enormous number of possibilities resulting from the freedom in the choice of the functionals A and B and of the function ψ. Considerable ingenuity is required in this choice. We will consider two such choices that turned out particularly successful.

For the external potential ψ that modifies the Hamiltonian we take the interaction potential with all the particles of an extra particle (labeled 0), identical to the ones of the fluid, located at a fixed position q_0:

$$\psi = \sum_{i=1}^{N} \psi_i \equiv \sum_{i=1}^{N} \psi(\mathbf{q}_i; \mathbf{q}_0) = \sum_{i=1}^{N} V(\mathbf{q}_i - \mathbf{q}_0) \tag{8.3.3}$$

Hence the variation of ψ appears as a generalization of the "charging" process of the kind used in Section 7.2 for the definition of the entropy.

As an independent functional $B(\mathbf{q}; \psi)$ we choose the one-particle distribution in the presence of the field:

$$B(\mathbf{q}; \psi) = n_1(\mathbf{q}; \psi) \tag{8.3.4}$$

This choice is particularly clever, because all the functions occurring in (8.3.1) now have a simple physical meaning. Indeed:

$$n_1(\mathbf{q}; 0) = n_1(\mathbf{q})$$

which is the one-particle distribution of the unperturbed system. On the other hand, in the presence of the field we have

$$\begin{aligned} n_1(\mathbf{q}; \psi) &= N \frac{\int d\mathbf{q}_2 \cdots d\mathbf{q}_N \exp\left\{-\beta \sum\sum_{i<j=1}^{N} V_{ij} - \beta \sum_{j=1}^{N} V_{0j}\right\}}{\int d\mathbf{q}_1\, d\mathbf{q}_2 \cdots d\mathbf{q}_N \exp\left\{-\beta \sum\sum_{i<j=1}^{N} V_{ij} - \beta \sum_{j=1}^{N} V_{0j}\right\}} \\ &= \frac{(N+1)NQ_{N+1}^{-1} \int d\mathbf{q}_2 \cdots d\mathbf{q}_N \exp\left\{-\beta \sum\sum_{i<j=0}^{N} V_{ij}\right\}}{(N+1)Q_{N+1}^{-1} \int d\mathbf{q}_1 \cdots d\mathbf{q}_N \exp\left\{-\beta \sum\sum_{i<j=0}^{N} V_{ij}\right\}} \end{aligned}$$

Hence $n_1(\mathbf{q}; \psi)$ appears as the ratio of the pair distribution to the one-particle distribution in an $(N+1)$-particle system. In the thermodynamic limit we simply get

$$n_1(\mathbf{q}_1; \psi) = \frac{n_2'(\mathbf{q}_1; \mathbf{q}_0)}{n_1(\mathbf{q}_0)} \tag{8.3.5}$$

Finally we choose for the generating functional A:

$$A(\mathbf{q}_1; \psi) = n_1(\mathbf{q}_1; \psi) e^{\beta V_{01}} \tag{8.3.6}$$

With these choices, Eq. (8.3.1), truncated at the linear term and combined with (8.3.2), becomes

$$n_1(\mathbf{q}_1;\psi)e^{\beta V_{01}} = n_1(\mathbf{q}_1) + \int d\mathbf{q}_2\,[n_1(\mathbf{q}_2;\psi) - n_1(\mathbf{q}_2)]\int d\mathbf{q}_3 \times\left\{\frac{\delta n_1(\mathbf{q}_1;\psi)e^{\beta V_{01}}}{\delta[-\beta\psi(\mathbf{q}_3)]}\frac{\delta[-\beta\psi(\mathbf{q}_3)]}{\delta n_1(\mathbf{q}_2;\psi)}\right\}_0 \tag{8.3.7}$$

We now note, using Eq. (7.5.18):

$$\begin{aligned}\frac{\delta n_1(\mathbf{q}_1;\psi)e^{\beta V_{01}}}{\delta(-\beta V_{03})} &= e^{\beta V_{01}}\frac{\delta n_1(\mathbf{q}_1;\psi)}{\delta(-\beta V_{03})} + n_1(\mathbf{q}_1;\psi)\frac{\delta e^{\beta V_{01}}}{\delta(-\beta V_{03})}\\ &= e^{\beta V_{01}}\{n_2'(\mathbf{q}_1;\mathbf{q}_3;\psi) - n_1(\mathbf{q}_1;\psi)n_1(\mathbf{q}_3;\psi)\\ &\quad + n_1(\mathbf{q}_1;\psi)\delta(\mathbf{q}_1-\mathbf{q}_3)\} - n_1(\mathbf{q}_1;\psi)e^{\beta V_{01}}\delta(\mathbf{q}_1-\mathbf{q}_3)\end{aligned} \tag{8.3.8}$$

Also using Eq. (7.5.19), we obtain

$$\begin{aligned}&\int d\mathbf{q}_3\left\{\frac{\delta n_1(\mathbf{q}_1;\psi)e^{\beta V_{01}}}{\delta(-\beta\psi_3)}\frac{\delta(-\beta\psi_3)}{\delta n_1(\mathbf{q}_2;\psi)}\right\}_0\\ &= \int d\mathbf{q}_3\,\{e^{\beta V_{01}}[n_2'(\mathbf{q}_1,\mathbf{q}_3;\psi) - n_1(\mathbf{q}_1;\psi)n_1(\mathbf{q}_3;\psi)]\\ &\qquad\times[(n_1(\mathbf{q}_2;\psi))^{-1}\delta(\mathbf{q}_2-\mathbf{q}_3) - C(\mathbf{q}_2,\mathbf{q}_3;\psi)]\}_0\\ &= \int d\mathbf{q}_3\,[n_2'(\mathbf{q}_1,\mathbf{q}_3) - n_1(\mathbf{q}_1)n_1(\mathbf{q}_3)]\{[n_1(\mathbf{q}_2)]^{-1}\delta(\mathbf{q}_2-\mathbf{q}_3) - C(\mathbf{q}_2,\mathbf{q}_3)\}\\ &= n_1(\mathbf{q}_1)C(\mathbf{q}_2,\mathbf{q}_1)\end{aligned} \tag{8.3.9}$$

where the Ornstein–Zernike (O–Z) equation (7.5.20) was used in the last step. Equation (8.3.7), combined with (8.3.9) and (8.3.5) as well as (7.1.4), (7.1.7), now becomes

$$\begin{aligned}n_2(\mathbf{q}_0-\mathbf{q}_1)e^{\beta V_{01}} &= 1 + n\int d\mathbf{q}_2\,[n_2(\mathbf{q}_0-\mathbf{q}_2)-1]C(\mathbf{q}_1-\mathbf{q}_2)\\ &= n_2(\mathbf{q}_0-\mathbf{q}_1) - C(\mathbf{q}_0-\mathbf{q}_1)\end{aligned}$$

We have thus obtained a very simple relation between the direct correlation function and the pair distribution:

$$C(r) = [1 - e^{\beta V(r)}]n_2(r) \tag{8.3.10}$$

When this approximate relationship is combined with the Ornstein–Zernike equation (7.5.21), we obtain a *closed nonlinear integral equation for the pair distribution:*

$$n_2(r)e^{\beta V(r)} = 1 - n\int ds\,[e^{\beta V(s)}-1]n_2(s)[n_2(|\mathbf{r}-\mathbf{s}|)-1] \tag{8.3.11}$$

This important equation is called the *Percus–Yevick* (or, briefly, P–Y) *equation.*

Before discussing it in more detail we illustrate the flexibility of the functional formalism by showing how a different equation can be obtained by making a different choice for the generating functional. We now take

$$A(\mathbf{q}; \psi) = \ln[n_1(\mathbf{q}; \psi)e^{\beta V_{01}}] \tag{8.3.12}$$

and maintain the same definitions for B and ψ. Repeating the calculations, we now find a different form for the direct correlation function

$$C(r) = n_2(r) - 1 - \ln n_2(r) - \beta V(r) \tag{8.3.13}$$

and the corresponding integral equation is

$$\ln n_2(r) = -\beta V(r) + n \int d\mathbf{s}\, [n_2(s) - 1 - \ln n_2(s) - \beta V(s)][n_2(|\mathbf{r} - \mathbf{s}|) - 1] \tag{8.3.14}$$

This equation is called the *hypernetted-chains equation* (*HNC*). It was first derived by a graphical method: Its strange name refers to the kind of topology of the graphs retained in this approximation.

As a first orientation in the discussion of these equations, we consider the behavior of these approximations in the limit of low densities. As was said before, the P–Y and the HNC can both be derived by summing a certain class of graphs chosen among those that contribute to the density expansion (8.2.1). We show in Table 8.3.1 the graphs retained in the two equations and also the complete set of diagrams, up to order n^3.*

The first correction term $[O(n)]$ is identical in the three cases. This implies that the second and the third virial coefficient in the expansion of the pressure (6.4.9) based on the P–Y and the HNC equations are exact. Starting in order n^2, we see that only part of the diagrams are retained in the approximations. The striking conclusion is that more diagrams are retained in the HNC than in the P–Y equation. Hence in principle, the HNC equation is a better approximation than the P–Y equation.

However, an apparent paradox arises here. When actual calculations are performed, it appears that the P–Y equation gives far better numerical results than the HNC equations. This is true not only for the individual virial coefficients (up to the sixth for hard spheres), but consistently over the whole range of densities (for temperatures above the critical temperature). The numerical results will be discussed in detail in Section 8.6. At this point we only wish to stress this paradox as a warning. It clearly shows that in this type of expansions, *the actual degree of approximation*

* In order to simplify the picture, we here make no distinction between the root points 1 and 2. As a result, for instance, the second and third graphs in order n^2 in Table 8.2.1 are replaced by a single diagram with the combinatorial factor $2+2=4$.

Table 8.3.1. Graphs for the Density Expansion of the Pair Distribution $n_2(r)\exp[\beta V(r)]$ Retained in Various Approximations

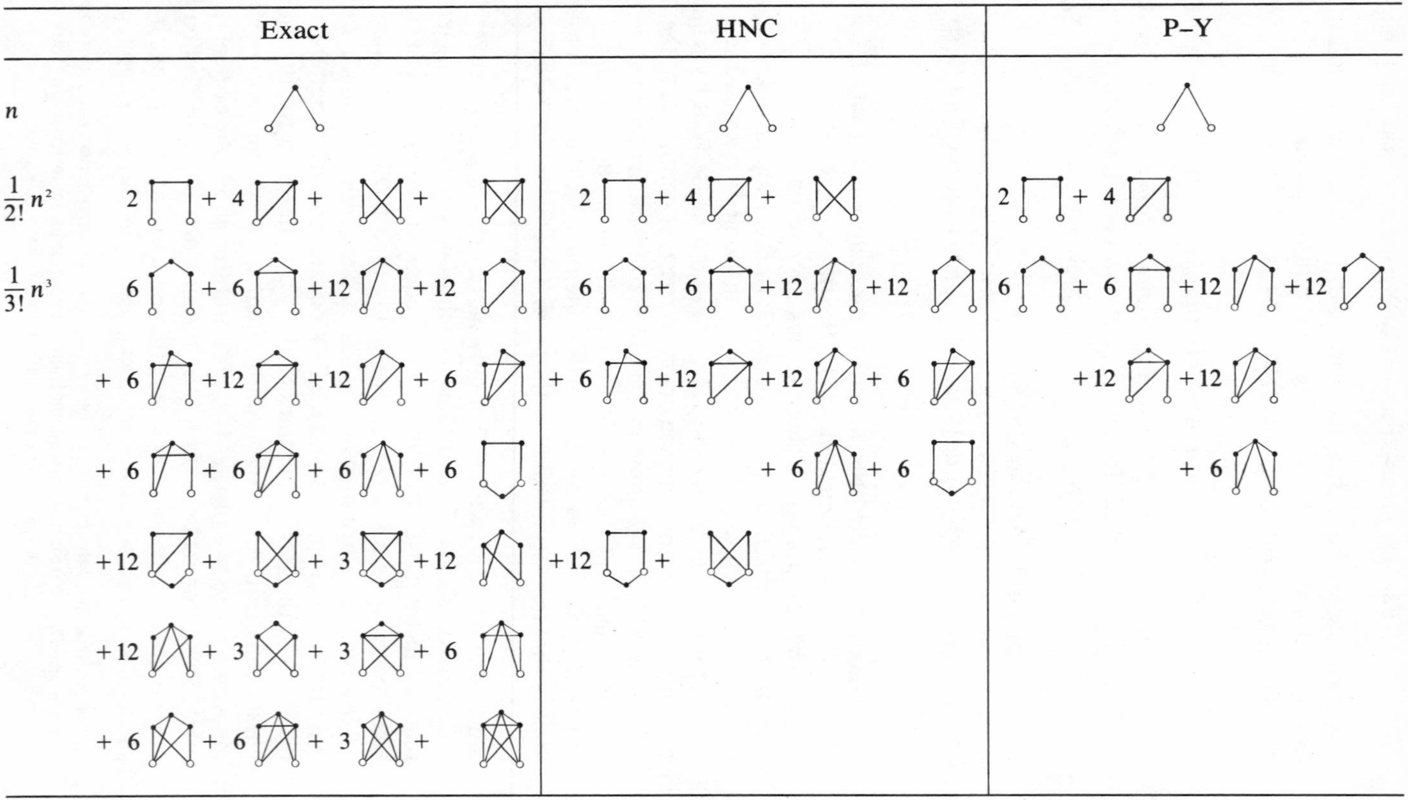

	Exact	HNC	P–Y
n			
$\frac{1}{2!}n^2$	2 + 4 + +	2 + 4 +	2 + 4
$\frac{1}{3!}n^3$	6 + 6 + 12 + 12	6 + 6 + 12 + 12	6 + 6 + 12 + 12
	+ 6 + 12 + 12 + 6	+ 6 + 12 + 12 + 6	+ 12 + 12
	+ 6 + 6 + 6 + 6	+ 6 + 6	+ 6
	+ 12 + + 3 + 12	+ 12 +	
	+ 12 + 3 + 3 + 6		
	+ 6 + 6 + 3 +		

is not necessarily improved by adding more graphs to the theory. This situation is typical of nonconvergent or semiconvergent series. It means that the diagrams neglected in the P–Y equation, which correspond to both positive and negative contributions, cancel each other to a reasonable degree, so that the result is a close approximation to the exact value. Adding a few more diagrams, as in the HNC approximation, may result in adding contributions of a predominant sign, so that the cancellation no longer occurs, and the result is worse.

We know from Sections 6.4 and 6.5 that a partial summation of a series means that the terms of the original expansion are reclassified according to their dependence on a new parameter. The summation of the subseries yields the coefficients of the expansion in powers of the new parameter. In Section 6.4 and 6.5 we knew what the new parameter should be (the density for slightly nonideal gases or the Debye parameter for plasmas), and we could therefore define precisely the criteria for the classification of the graphs. In the case of dense fluids we do not know what the significant parameter is, hence we are led to making guesses and checking the results *a posteriori*. It appears at present that the P–Y equation, or its modifications, is the best approximation known so far.

*8.4. SOLUTION OF THE PERCUS–YEVICK EQUATION FOR HARD SPHERES

Besides giving very good numerical results, the P–Y equation has a quite unique feature: It can be solved *exactly* in the case of a hard-sphere potential. This is the only case so far in which a nonlinear integral equation for $n_2(r)$ has been solved analytically. The solution is due to Wertheim and Thiele, who obtained it independently in 1963. Although the hard-sphere potential is rather unrealistic it will be seen in Section 8.6 that the Wertheim–Thiele solution is quite important even in practical cases.

The hard-sphere potential has the following, quite singular features:

$$\begin{aligned} V(r) &= +\infty, \qquad r<d_0 \\ &= 0, \qquad r>d_0 \end{aligned} \tag{8.4.1}$$

from which follows

$$\begin{aligned} e^{-\beta V(r)} = \Theta(r-d_0) &= 0, \qquad r<d_0 \\ &= 1, \qquad r>d_0 \end{aligned} \tag{8.4.2}$$

where $\Theta(x)$ is the Heaviside step function and d_0 is the diameter of the hard spheres. Writing now the typical P–Y relation (8.3.10) in the form:

$$e^{-\beta V(r)} C(r) = (e^{-\beta V(r)} - 1) n_2(r) \tag{8.4.3}$$

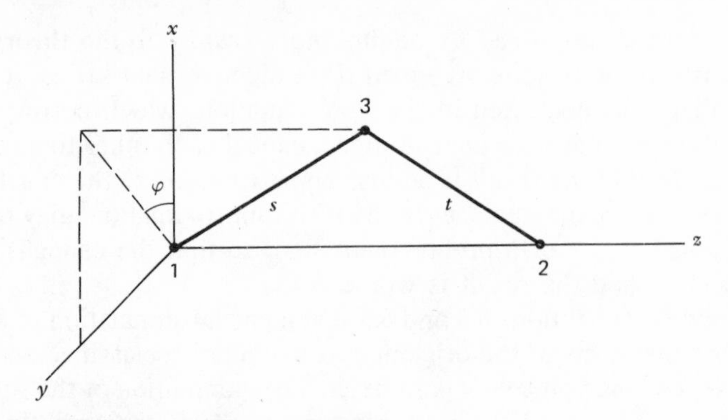

Figure 8.4.1. Bipolar coordinate system.

we see that for hard spheres we have

$$n_2(r) = 0, \qquad r < d_0 \tag{8.4.4}$$

$$C(r) = 0, \qquad r > d_0 \tag{8.4.5}$$

The relation (8.4.4) is an exact property of the hard-sphere pair distribution; (8.4.5) is, however, true only in the P–Y approximation.

In order to write the integral in Eq. (8.3.11) in the most explicit way, a system of *bipolar coordinates*, as shown in Fig. 8.4.1 is the most convenient: The position of the third particle is measured by its distances s and t to the first and second (fixed) particles and by the azimuthal angle φ. The volume element in these variables is $d^3\mathbf{s} \equiv (st/r)ds\, dt\, d\varphi$. Let us also introduce the independent function $\tau(r)$:

$$\tau(r) = e^{\beta V(r)} n_2(r) \tag{8.4.6}$$

From Eqs. (8.4.3) and (8.4.2), (8.4.4) we see that, for hard spheres:

$$\begin{aligned} \tau(r) &= -C(r), \qquad r < d_0 \\ \tau(r) &= n_2(r), \qquad r > d_0 \end{aligned} \tag{8.4.7}$$

The P–Y equation (8.3.11) can now be written as

$$r\tau(r) = r - 2\pi n \int_0^\infty ds \int_{|r-s|}^{r+s} dt\, [1 - e^{-\beta V(s)}] s\tau(s) [e^{-\beta V(t)} t\tau(t) - t] \tag{8.4.8}$$

Again using Eq. (8.4.2), this equation reduces to

$$r\tau(r) = r - 2\pi n \int_0^{d_0} ds\, s\tau(s) \int_{|r-s|}^{r+s} dt\, \Theta(t - d_0) t\tau(t) + 2\pi n \int_0^{d_0} ds\, s\tau(s) \int_{|r-s|}^{r+s} dt\, t$$

Introducing finally the auxiliary variable $h(r)$:

$$r\tau(r) = h(r) \tag{8.4.9}$$

we obtain the P–Y equation for hard spheres in its simplest form:

$$h(r) = Ar - 2\pi n \int_0^{d_0} ds\, h(s) \int_{|r-s|}^{r+s} dt\, \Theta(t - d_0) h(t) \tag{8.4.10}$$

straight-forward

where

$$A = 1 + 4\pi n \int_0^{d_0} ds\, sh(s) \tag{8.4.11}$$

A is a constant, independent of r; but it is also a functional of the unknown function $h(s)$.

We now Laplace-transform the P–Y equation:

$$\int_0^{\infty} dr\, e^{izr} h(r) = (-iz)^{-2} A - 2\pi n \int_0^{\infty} dr\, e^{izr} \int_0^{d_0} ds\, h(s) \int_{|r-s|}^{r+s} dt\, \Theta(t-d_0) h(t) \tag{8.4.12}$$

The triple integral in the right-hand side of this equation, which we call K, is most easily evaluated by doing the r integration in the first place:

$$\begin{aligned} K &= \int_0^{d_0} ds\, h(s) \int_s^{\infty} dt\, \Theta(t-d_0) h(t) \int_{t-s}^{t+s} dr\, e^{izr} \\ &= (iz)^{-1} \left\{ \int_0^{d_0} ds\, h(s) e^{izs} \int_{d_0}^{\infty} dt\, h(t) e^{izt} - \int_0^{d_0} ds\, h(s) e^{-izs} \int_{d_0}^{\infty} dt\, h(t) e^{izt} \right\} \end{aligned}$$

We now define two functions of the Laplace variable z, whose physical meaning is derived from Eqs. (8.4.4), (8.4.5), (8.4.7), and (8.4.9):

$$F(z) = \int_0^{d_0} ds\, h(s) e^{izs} = -\int_0^{\infty} ds\, sC(s) e^{izs} \tag{8.4.13}$$

$$G(z) = \int_{d_0}^{\infty} ds\, h(s) e^{izs} = \int_0^{\infty} ds\, sn_2(s) e^{izs} \tag{8.4.14}$$

Equation (8.4.12) is now transformed into a purely algebraic equation:

$$F(z) + G(z) = (iz)^{-2} A - 2\pi n (iz)^{-1} G(z) [F(z) - F(-z)] \tag{8.4.15}$$

from which follows

$$G(z) = \frac{(iz)^{-2} A - F(z)}{1 + 2\pi n (iz)^{-1} [F(z) - F(-z)]} \tag{8.4.16}$$

This formula does not yet provide us with an explicit solution, because both A and $F(z)$ still depend on the unknown function. It is, however, possible to go beyond it by making use of the very stringent analytical properties of $F(z)$ and $G(z)$. This very elegant argument, due to Wertheim, will be sketched now.

We first note that $F(z)$, being the Laplace transform of a regular function in a *finite* domain $(0, d_0)$, is an *entire function* of z, that is, has no singularities anywhere in the finite region of the plane. On the other hand, $G(z)$, the Laplace transform of a regular function in a half-infinite domain (d_0, ∞) is a function of z, *regular in the upper half-plane* S_+ (i.e., $\mathrm{Im}\, z > 0$), but it may have singularities in the lower half-plane or on the real axis. Our function $G(z)$, Eq. (8.4.16), actually has a double pole in $z = 0$.

These are quite general statements from the theory of Laplace transforms. The reader who would like to convince himself of these statements without a long proof may consider the simple example: $h(s)=s$. Then

$$\int_0^{d_0} ds\, se^{izs} = \frac{(izd_0-1)e^{izd_0}+1}{(iz)^2}$$

This function is regular in $z=0$ where its value is $\frac{1}{2}d_0^2$ (as can be seen by expanding the numerator). Hence, it has no singularities in any finite region of the plane; that is, it is entire. On the other hand,

$$\int_0^{\infty} ds\, se^{izs} = \frac{1}{(iz)^2} \qquad \text{for} \quad z \in S_+$$

This function is regular in the upper half-plane, but has a double pole in $z=0$.

We now construct the following function:

$$H(z) \equiv (iz)^4 G(z)\{(iz)^{-2}A - F(-z)\} \tag{8.4.17}$$

This function has no pole in $z=0$; hence it is regular in the whole upper half-plane and on the real axis. Besides, it is an *even* function of z; therefore, it is also regular in the lower half-plane, and hence $H(z)$ is an *entire* function. This result is important, because entire functions have a quite peculiar behavior: They are completely determined by their behavior at infinity. More precisely, a famous theorem of Rouché tells us that an entire function behaving like z^n (n, a nonnegative integer) for $z \to \infty$ can only be a polynomial of degree n. Let us, therefore, analyze the behavior of $H(z)$ at infinity.

The explicit from of $H(z)$ is

$$H(z) = (iz)^4 \frac{(iz)^{-4}A^2 - (iz)^{-2}A[F(z)+F(-z)] + F(z)F(-z)}{1+2\pi n(iz)^{-1}[F(z)-F(-z)]} \tag{8.4.18}$$

We must obtain estimates for the rate of growth of $F(z)$ and $F(-z)$, as $z \to \infty$, say in the lower half-plane S_-. If $h(s)$ is continuous, it is bounded in the domain $(0, d_0)$, hence

$$F(z) \leq \bar{h}\int_0^{d_0} ds\, e^{izs} = \frac{\bar{h}[e^{id_0z}-1]}{iz}$$

and therefore

$$F(z) \sim \frac{e^{id_0z}}{z} \tag{8.4.19}$$

$F(z)$ grows exponentially as $z \to \infty$ in S_-. For $F(-z)$, we do the following integration by parts:

$$F(-z) = -\int_0^{d_0} ds\, sC(s)e^{-isz}$$

$$= \frac{d_0C(d_0)e^{-id_0z}}{iz} + \frac{[d_0C'(d_0)+C(d_0)]e^{-id_0z} - C(0)}{(iz)^3} + O(z^{-3})$$

Hence, the exponentials die out very strongly and, as $z \to \infty$ (S_-):

$$F(-z) \sim -\frac{C(0)}{(iz)^2} = \frac{A}{(iz)^2} \tag{8.4.20}$$

The second equality follows from the relation

$$C(0) = -A \tag{8.4.21}$$

which is easily obtained by using the right-hand side of the P–Y equation (8.4.10) to define $C(0)$.

We now see that the strong exponential growth of $F(z)$ dominates as $z \to \infty$ (S_-) and, coming back to (8.4.18), we obtain

$$\begin{aligned} H(z) &\sim (iz)^4 \frac{\{-(iz)^{-2}A + F(-z)\}F(z)}{2\pi n(iz)^{-1}F(z)} \\ &\sim (iz)^5 \frac{-(iz)^{-2}A + (iz)^{-2}A + O(z^{-3})}{2\pi n} = O(z^2) \end{aligned} \tag{8.4.22}$$

Hence, $H(z)$ grows as z^2, as $z \to \infty$. The Rouché theorem, therefore, tells us that $H(z)$ is a polynomial of degree 2; as it is also an even function, it must be of the form

$$H(z) = \lambda_1 + \lambda_2 z^2 \tag{8.4.23}$$

The coefficients can be obtained by expanding the right side of Eq. (8.4.18) around $z = 0$, thus finding*

$$\begin{aligned} \lambda_1 &= A \\ \lambda_2 &= -2F(0) + \left(\frac{2\pi n}{3i}\right)F'''(0) \end{aligned} \tag{8.4.24}$$

where the strokes denote derivatives with respect to z. These equations, however, are not yet the final solution because $F(z)$ is still unknown.

To go further, we combine Eqs. (8.4.17) and (8.4.23) to get

$$G(z)\{(iz)^{-2}A - F(-z)\} = \lambda_1(iz)^{-4} + \lambda_2(iz)^{-2} \tag{8.4.25}$$

which is used to eliminate the term $G(z)F(-z)$ from Eq. (8.4.15):

$$F(z) = \lambda_1(iz)^{-2} - 2\pi n\lambda_2(iz)^{-3} - 2\pi n\lambda_1(iz)^{-5} + 2\pi nG(z)\{-(iz)^{-1}F(z) + (iz)^{-3}A\} \tag{8.4.26}$$

We now consider the inverse Laplace transform of this equation

$$\begin{aligned} (2\pi)^{-1}\int_C dz\, e^{-izr}F(z) &= (2\pi)^{-1}\int_C dz\, e^{-izr}\{\lambda_1(iz)^{-2} - 2\pi n[\lambda_2(iz)^{-3} + \lambda_1(iz)^{-5}]\} \\ &\quad + n\int_C dz\, e^{-izr}G(z)\{-(iz)^{-1}F(z) + (iz)^{-3}A\} \end{aligned} \tag{8.4.27}$$

where C is a contour parallel to the real axis and lying in the upper half-plane S_+. We consider this equation for $r < d_0$. From our previous discussion, we know that $G(z)$ is regular in the upper half-plane, and behaves like e^{izd_0}/z. Hence the

* The Taylor expansion is apparently an infinite series; but by (8.4.23) we know that the coefficients of $z^n (n > 2)$ must vanish identically as a result of internal cancellations.

"effective" exponential in the last term of Eq. (8.4.27) is $e^{-iz(r-d_0)}$. For $r<d_0$, the contour must therefore be closed by a semicircle of infinite radius in the *upper* half-plane. The integrand being regular in S_+ and behaving nicely at infinity, the corresponding integral vanishes. The remaining terms are easily evaluated by closing the contour in S_-:

$$(2\pi)^{-1}\int_C dz\, e^{-izr}F(z)=\lambda_1 r+\left(\frac{2\pi n}{2!}\right)\lambda_2 r^2+\left(\frac{2\pi n}{4!}\right)\lambda_1 r^4, \qquad r<d_0$$

or, using Eq. (8.4.13):

$$C(r)=-\lambda_1-\pi n\lambda_2 r-\left(\frac{\pi n}{12}\right)\lambda_1 r^3, \qquad r<d_0 \tag{8.4.28}$$

We may now combine this equation with Eqs. (8.4.24):

$$\begin{aligned}\lambda_1&=A=1-4\pi n\int_0^{d_0} ds\, s^2C(s)\\&=1+\tfrac{4}{3}\pi n d_0^3\lambda_1+\pi^2n^2d_0^4\lambda_2+\tfrac{1}{18}\pi^2n^2d_0^6\lambda_1\\\lambda_2&=-2F(0)-\tfrac{2}{3}\pi i n F'''(0)\\&=\int_0^{d_0} ds\,\{2sC(s)+\tfrac{2}{3}\pi i n(is)^3sC(s)\}\\&=-d_0^2\{1+\tfrac{1}{6}\pi n d_0^3+\tfrac{1}{144}\pi^2n^2d_0^6\}\lambda_1-\{\tfrac{2}{3}\pi n d_0^3+\tfrac{1}{9}\pi^2n^2d_0^6\}\lambda_2\end{aligned}$$

We thus have now a set of two linear algebraic equations for λ_1, λ_2 from which the *explicit* solution is readily found:

$$\begin{aligned}\lambda_1&=\frac{(1+2\eta)^2}{(1-\eta)^4}\\\lambda_2&=\frac{-d_0^2(1+\frac{1}{2}\eta)^2}{(1-\eta)^4}\end{aligned} \tag{8.4.29}$$

where we introduced the convenient dimensionless density:

$$\eta=\tfrac{1}{6}\pi n d_0^3 \tag{8.4.30}$$

Using also the dimensionless distance $r^*=r/d_0$, we obtain

$$C(r^*)=(1-\eta)^{-4}\{-(1+2\eta)^2+6\eta(1+\tfrac{1}{2}\eta)^2r^*-\tfrac{1}{2}\eta(1+2\eta)^2r^{*3}\} \tag{8.4.31}$$

This is the final solution for the direct correlation function.

From this solution, we can calculate the complete $F(z)$ by means of Eq. (8.4.13), and substitute in Eq. (8.4.26) to obtain an explicit expression for $G(z)$, the Laplace transform of the pair distribution. We do not quote here the rather heavy formula: It is sufficient to stress that the problem is completely solved.

We now derive the equation of state in the P–Y approximation. We may start from the general equation (7.2.8). In order to evaluate the quite singular "force" dV/dr, we note that Eq. (8.4.2) provides us with a step function, whose derivative is a delta function:

$$\frac{d}{dr}e^{-\beta V(r)} \equiv -\beta\frac{dV}{dr}e^{-\beta V(r)} = \delta(r-d_0)$$

hence we may consistently use

$$\frac{dV}{dr} = -k_B T\delta(r-d_0) \tag{8.4.32}$$

which, upon substitution in (7.2.8), and with the remark that $n_2(d_0) = -C(d_0)$, yields

$$P = nk_B T[1-\tfrac{2}{3}\pi n d_0^3 C(d_0)] \tag{8.4.33}$$

With our form (8.4.31) this equation becomes

$$\frac{P}{nk_B T} = \frac{1+2\eta+3\eta^2}{(1-\eta)^2} \tag{8.4.34}$$

Alternatively, one could start from the compressibility equation (7.5.27). We see from Eqs. (8.4.7), (8.4.9), (8.4.11), and (8.4.24) that

$$\lambda_1 \equiv 1 - 4\pi n\int_0^{d_0} dr\, r^2 C(r) = (k_B T)^{-1}\left(\frac{\partial P}{\partial n}\right)_T \tag{8.4.35}$$

Substituting the value (8.4.29) and integrating over n, we obtain

$$\frac{P_c}{nk_B T} = \frac{1+\eta+\eta^2}{(1-\eta)^3} \tag{8.4.36}$$

If the result were exact, the two equations (8.4.34) and (8.4.36) would coincide; since, however, the P–Y equation is only an approximation, they do not. We therefore denote the pressure obtained from the compressibility equation by a subscript c. The numerical results obtained from these equations will be discussed in Section 8.6.

8.5. THE MONTE CARLO AND THE MOLECULAR-DYNAMICS METHODS

It may seem that the Wertheim–Thiele solution of the P–Y equation is just an academic exericse that could never be checked experimentally, because the hard-sphere system is a quite unrealistic one. This, however, is not true, because since the advent of fast electronic computers, quite revolutionary methods of investigation of the many-body problem have evolved. These methods are, in a sense, half-way between theory and experiment and are frequently called *numerical experiments.* In these experiments the behavior of a many-body system is simulated by the motions of a relatively small set of particles (between 50 and 1000). It turns out that when the conditions of the experiment are chosen carefully these "small" systems behave in many respects as a real large system.

The thermodynamic quantities obtained from the simulation can be significantly compared to real experimental data. All the existing numerical experiments are variants of two basic methods that will be now briefly described.

The first of these is called the *Monte Carlo method.* The idea here (as suggested by its name) is to replace the exact dynamics by a "game," that is, a stochastic process. By means of this trick, it is possible to evaluate averages rather easily in the canonical ensemble, as will be shown presently. We first give the rules of the game. We start out with N point particles in some simple initial condition, say, with all the particles distributed at the vertices of a regular cubic lattice. The computer keeps the record of the coordinates of all these particles. We then move particle No. 1 by changing its coordinates as follows:

$$\begin{aligned} x &\to x + \alpha\xi_1 \\ y &\to y + \alpha\xi_2 \\ z &\to z + \alpha\xi_3 \end{aligned} \tag{8.5.1}$$

where α is the maximum displacement, and ξ_1, ξ_2, ξ_3 are *random numbers,* comprised between (-1) and 1. (Such random numbers are easily generated by a computer.) As a result, after the move, particle No. 1 is equally likely to be anywhere within a cube of side 2α.* At this stage the computer is asked to calculate the total potential energy of the system in the new configuration:

$$H' \equiv \sum_{i<j} V(r_{ij})$$

If it turns out that the energy increment $H'_{\text{final}} - H'_{\text{initial}} \equiv \Delta H'$ is negative, the move is "accepted" and particle 1 is put in the new position. If, however, $\Delta H'$ is positive, we accept the move only with probability $\exp(-\beta\Delta H')$. This is done by throwing a (somewhat sophisticated) die: we generate a new random number ξ_4 $(0 \leq \xi_4 \leq 1)$; if $\xi_4 < \exp(-\beta\Delta H')$ the move is accepted; otherwise the particle is put back to its initial position. Whatever the result (i.e., whether the move is accepted or not) we calculate the value of the relevant microscopic quantity A in this configuration and call this value A_1. This complicated procedure actually takes a relatively short time on a modern fast computer.

The game is then repeated with particles 2, 3, 4, . . . , up to N, thus completing a cycle; then it is started again with particle 1, and so on. It is clear that the initial ordered structure is rapidly destroyed, and the system approaches equilibrium (as will be shown below). After M moves the

* Periodic boundary conditions are assumed. Hence if the move drives the particle outside the cubic container of the system, it re-enters on the opposite side.

average value of A is evaluated simply as

$$\langle A \rangle = M^{-1} \sum_{j=1}^{M} A_j \tag{8.5.2}$$

Let us show that the rules of the game are equivalent to weighting each configuration by $\exp(-H'/k_B T)$. We consider a large ensemble of identical systems of N particles. Let ν_r be the number of systems of the ensemble in the configuration r. We now make a move in all the systems of the ensemble; its result is to bring a system from state r to configuration s with an *a priori* probability P_{rs}. [The *a priori* probability is the probability of the move before discriminating on the basis of $\exp(-\beta \Delta H')$.] By the rules of the game, $P_{rs} = P_{sr}$; indeed, if r and s differ from each other only by the position of a particle and if these positions are within each other's cube of side 2α defined in (8.5.1), the transitions $r \to s$ and $s \to r$ occur with equal probability, since all the positions within the cube are equally likely. Assume now that $H'_r > H'_s$; then the number of systems in the ensemble moving from r to s is simply $\nu_r P_{rs}$; but the number of systems moving from s to r is, by our rules, $\nu_s P_{sr} \exp[-\beta(H'_r - H'_s]$, since $H'_r - H'_s > 0$. Hence, the net number of systems moving from s to r is

$$P_{rs}\{\nu_s \exp[-\beta(H'_r - H'_s)] - \nu_r\}$$

Therefore, if

$$\frac{\nu_r}{\nu_s} > \frac{\exp(-\beta H'_r)}{\exp(-\beta H'_s)}$$

there will be, on the average, more systems moving from r to s. Clearly, after a sufficient number of moves, the ensemble reaches a stationary equilibrium state in which, on the average, there are no more changes in the configurations:

$$\nu_r \sim \exp(-\beta H'_r)$$

In other words, the system reaches a canonical equilibrium distribution. The average $\langle A \rangle$ first fluctuates irregularly, but, as M increases, it stabilizes at its equilibrium value. The number of moves it takes in practice to reach equilibrium depends on a judicious choice of the size of α.

There is no difficulty at all in applying this method to a gas of hard spheres: its application is even simpler because of Eq. (8.4.2). A move is simply rejected whenever the spheres overlap.

The Monte Carlo method and its variants are specifically designed to compute *equilibrium quantities.* It cannot, of course, tell us *how* the system reaches equilibrium, because the approach to this state is imposed by the computer in the form of a stochastic process: it does not result from the natural dynamics of the system.

Another method that is more powerful and can yield much more information is the *molecular-dynamics* method. Its principle is simpler than that of the Monte Carlo procedure. It consists in solving Newton's equations of motion for a many-body system. Typical and classical works are Alder and Wainwright's pioneering "experiments" on systems with 32 to 500 hard spheres (1959) and the monumental work of Rahman on 864 particles interacting through a realistic Lennard–Jones potential, and simulating argon atoms (1964).

As usual in computer work, Newton's equations:

$$\begin{aligned} \frac{dx_i}{dt} &= v_i \\ \frac{dv_i}{dt} &\equiv a_i = m^{-1} \sum_{j(\neq i)} F_{ij}(x_i, x_j) \end{aligned} \tag{8.5.3}$$

are replaced with finite difference equations, by chopping the time scale in finite intervals Δt. Assuming the positions of all the molecules given as $x_i^{(n-1)}$ at time t_{n-1} as well as the positions, velocities and accelerations, $x_i^{(n)}$, $v_i^{(n)}$, $a_i^{(n)}$ at time t_n, we make a first guess for the positions at time t_{n+1}:

$$\bar{x}_i^{(n+1)} = x_i^{(n-1)} + 2\Delta t v_i^{(n)} \tag{8.5.4}$$

With these predicted positions, we calculate predicted accelerations $\bar{a}_i^{(n+1)}$ by substituting (8.5.4) into (8.5.3). With these values we make a second approximation for the positions and velocities:

$$\begin{aligned} v_i^{(n+1)} &= v_i^{(n)} + \tfrac{1}{2}\Delta t(\bar{a}_i^{(n+1)} + \bar{a}_i^{(n)}) \\ x_i^{(n+1)} &= x_i^{(n)} + \tfrac{1}{2}\Delta t(v_i^{(n+1)} + v_i^{(n)}) \end{aligned} \tag{8.5.5}$$

This interpolation procedure can be repeated until the difference between predicted and corrected values is smaller than a prescribed number, after which the solution goes on to t_{n+2}. All the positions, velocities, and accelerations at all times are recorded on a magnetic tape. This tape, therefore, contains an enormous amount of information (using an excellent phrase of Rahman, the tape is really the "Maxwell demon" who knows everything about the motion of all molecules). This information can then be used in many ways by electronic processing. For instance, one can obtain the pair correlation function by counting the number of particles at a given distance from a prescribed particle at a given time, say, when the system has reached equilibrium. This is the kind of data of interest in our present problem. But, clearly, one can get much more out of the tape, in particular the detailed way in which the system approaches equilibrium. This kind of problem will be discussed in Part 3.

The numerical methods have become a quite important aspect of statistical physics in recent years, because they possess very peculiar advantages. On one hand, by comparing their results with "real" experiments one can get very precise information about the parameters of the interaction forces and other structural details of the molecules. On the other hand, they are of great value to the theorist. They enable him to do experiments with simplified models that can be handled theoretically but cannot be found in nature (such as hard spheres). Even in realistic situations, they allow him to keep a very tight control on the "experimental" conditions, to work in absence of foreign interfering influences and therefore to isolate the relevant phenomenon. Of course, the systems they can handle are very small compared to real ones. However, it appears that for many purposes a system of 1000 particles behaves already as if in the thermodynamic limit.

8.6. PRESENT STATE OF THE THEORY OF DENSE FLUIDS

We are now ready for a synthetic discussion of the results of the various approaches discussed in this chapter. We begin with the systems of *hard spheres*. These are the simplest fluids and have been studied very thoroughly. As will be seen, they, moreover, have an intrinsic importance, which became clearer in recent years.

In zeroth approximation, the pair distribution reduces to the Boltzmann factor $\exp(-\beta V)$, which is simply a step function (Fig. 8.6.1a). The first correction, corresponding to the second virial coefficient, already shows a very interesting feature (Fig. 8.6.1b). Its analytic form is

$$\begin{aligned} n_2(r^*) &= 0, && r^* < 1 \\ n_2(r^*) &= 1 + 8\eta(1 - \tfrac{3}{4}r^* + \tfrac{1}{16}r^{*3}), && 1 < r^* < 2 \\ n_2(r^*) &= 1, && r^* > 2 \end{aligned} \tag{8.6.1}$$

where $r^* = r/d_0$ and η was defined in (8.4.30). The interesting property of this function is the existence of a region (or "shell"), $1 < r^* < 2$ in which $n_2(r^*) > 1$, implying that the probability of finding a particle at such a distance from the central one is larger than the average. This implies the existence of an effective attraction between the two particles, in spite of the absence of any attractive interactions in the Hamiltonian. Just as in the case of the pseudoattraction existing in an ideal boson gas, we have here a collective effect involving the interplay of the many particles in the system. The physical origin is different here, of course, but can be easily understood as a shielding effect. When the second particle is at a distance

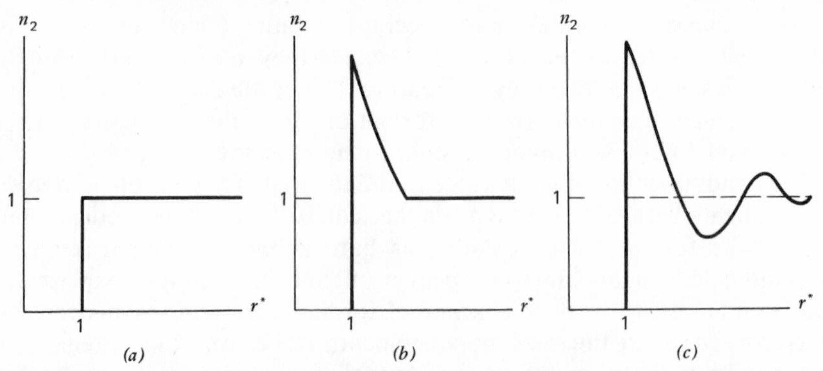

Figure 8.6.1. Qualitative behavior of the pair distribution for hard spheres. (*a*) zeroth approximation, (*b*) first approximation, (*c*) high density.

between one and two diameters from the central one, there is no possibility of inserting a third particle in between. As a result, the second particle suffers less collisions on the side facing the central one than on the opposite side: the effect is a net average attraction toward the center.

As the density increases, the structure in the fluid becomes more and more apparent (Fig. 8.6.1*c*). The first peak is followed by a dip, and then by a second peak of less height, around $r^* \approx 2$. Around each particle there appear, therefore, two, and maybe three "coordination shells" in which the probabilities of finding a second particle are maximal. This structure, called *short-range order*, is characteristic of a liquid. It "prepares," so to say, the structure of the crystal (in which the pair distribution is strictly periodic and undamped). The liquid–solid transition, however, is not yet clearly understood. The qualitative change of the radial distribution as the density increases is strikingly shown in Fig. 8.6.2, which is based on molecular-dynamics calculations (see Section 8.5). These data allow us to check the validity of the various approximate theories discussed in previous sections. Such a comparison is shown in Fig. 8.6.3 at a density of $\eta = 0.463$ (which is well within the liquid range). It is clear that the Born–Green–Yvon equation (BGY) (7.4.8) gives a result that deviates quite badly from the exact result.* The hypernetted-chain (HNC) approximation (8.3.4) gives a qualitatively fair curve, but the best approximation

* It must be mentioned, however, that Rice and Lekner extended the domain of validity of the Kirkwood superposition approximation (which is at the basis of the BGY equation) by assuming the following ansatz:

$$n_3(123) = n_2(12)n_2(13)n_2(23)\exp[\tau(123)]$$

The correction $\tau(123)$ is then expanded in cluster fashion, for a hard-sphere fluid. Using the two first terms of this expansion, together with a Padé approximant, these authors obtain a remarkably accurate equation of state for hard spheres. The basic reason for this success is not clear, however.

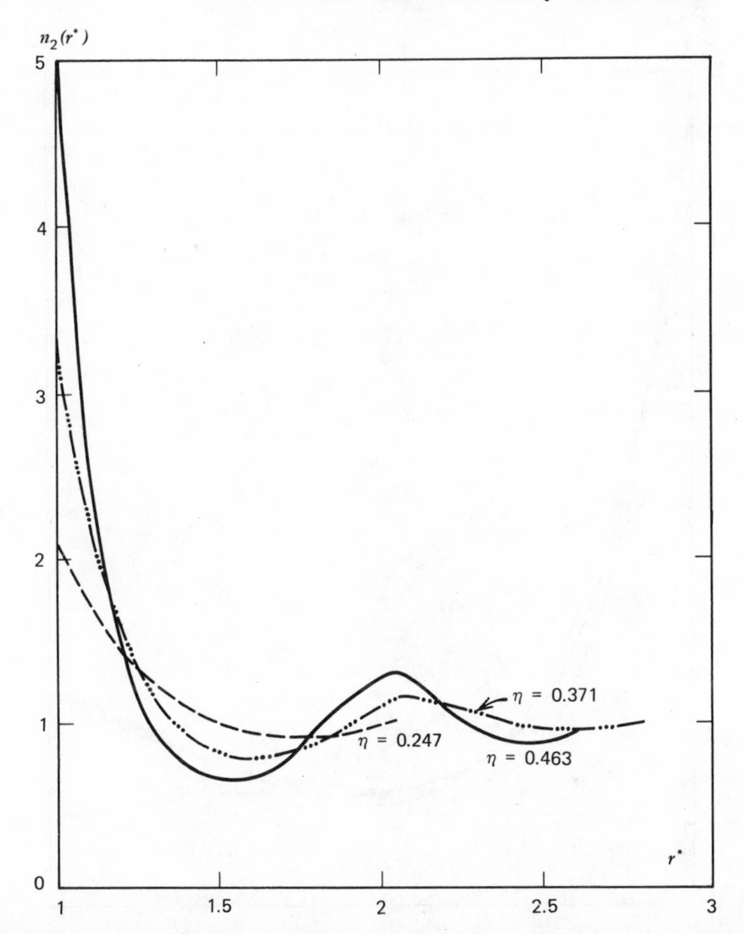

Figure 8.6.2. Pair distribution of hard spheres at various reduced densities, $\eta = (1/6)\pi d_0^3 n$, as a function of the reduced distance $r^* = r/d_0$. Graphs based on data from the MD calculations of Alder and Hecht (1969).

is provided by the Percus–Yevick (P–Y) equation and its analytic Wertheim–Theile (W–T) solution. The agreement is not, of course, quite quantitative: the first maximum is too low and the oscillation of the W–T solution is somewhat out of phase compared to the exact result; but these details can be improved by an adequate fitting of parameters. It remains quite amazing that the simple P–Y equation gives such a good picture even at high density. We also see in this figure the illustration of the "paradox" mentioned in Section 8.3. The HNC equation, although retaining more diagrams than the P–Y equation, gives poorer results.

As a general remark, we may note that the radial distribution for hard

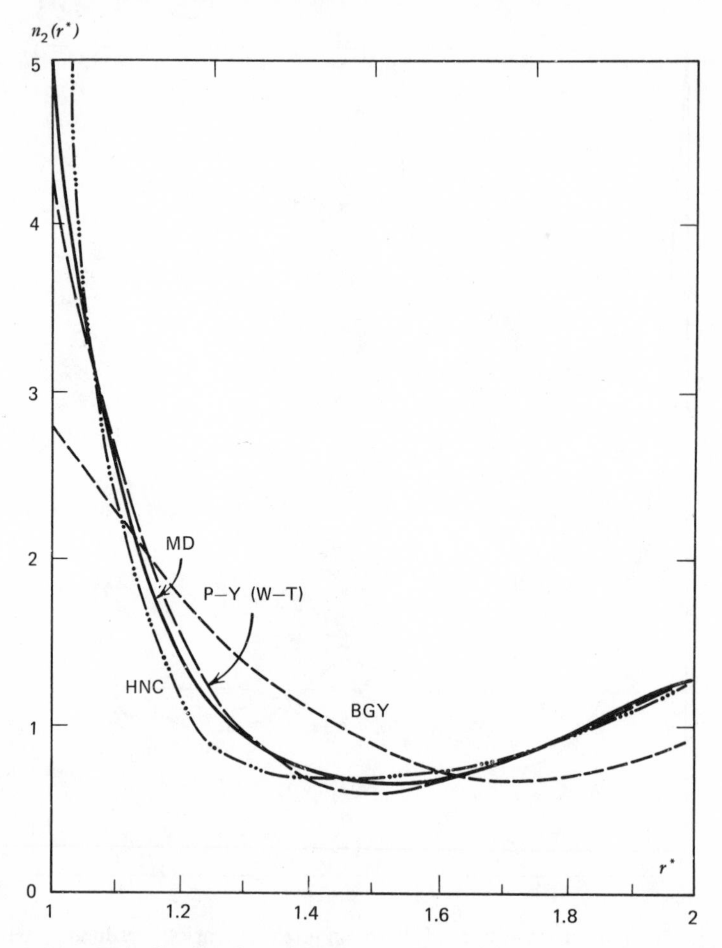

Figure 8.6.3. Pair distribution of hard spheres at density $\eta = 0.463$, as obtained from molecular dynamics and from various approximate theories. (Data from Alder and Hecht, 1969.)

spheres depends only on the density, not on the temperature. This results clearly from the singular nature of the potential. Indeed, the temperature enters the definition of $n_2(r)$ only through the Boltzmann factor $\exp(-\beta V)$, which, however, is either 0 or 1, depending on the distance but regardless of the temperature. It then follows that the thermodynamic quantities are especially simple. We will discuss now the most extensively studied among these, that is, the *equation of state*. In particular, we consider the "compressibility factor" $(\beta P/n)$, Eq. (6.4.14), which is a function of the density alone (see Fig. 8.6.4).

The various theories can be checked in great detail against the extensive results of Alder and co-workers, obtained by the method of molecular dynamics, as well as against many other numerical results obtained both by molecular dynamics and by the Monte Carlo method.

The curve representing the equation of state begins, as it should, at the ideal gas value, $\beta P/n = 1$ for $\eta = 0$. It then starts rising monotonously. At

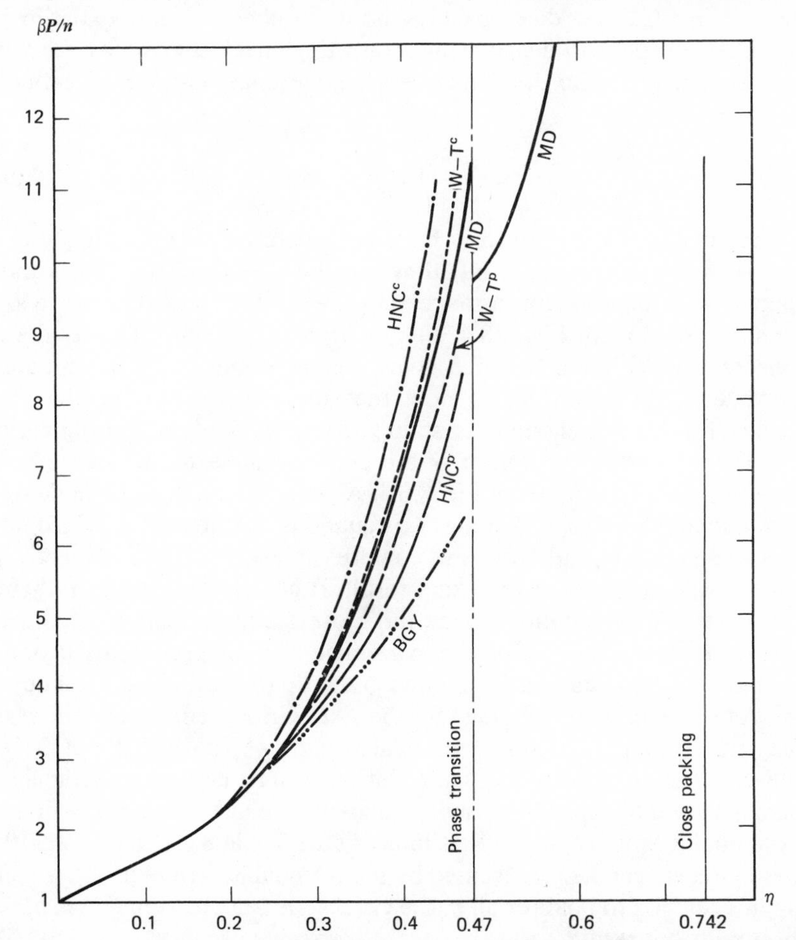

Figure 8.6.4. Equation of state of a system of hard spheres. MD: Molecular dynamics. (Data from Alder and Wainwright, 1969.) W–T^c: Wertheim–Thiele solution of the P–Y equation: compressibility equation of state (8.4.36). W–T^p: Wertheim–Thiele solution of the P–Y equation: pressure equation of state (8.4.34). HNC^c, HNC^p: analogous equations for the hypernetted-chains approximation. (Data from Rice and Gray, 1965.) BGY: Born–Green–Yvon equation. (Data from Rice and Gray, 1965.)

low densities (until $\eta \approx 0.2$) all the theories agree. In particular, the virial expansion limited to three terms yields a good description in this range, which corresponds essentially to a gas phase.* The agreement between the various theories is not an accident: we know from Section 8.2, that both the P-Y and the HNC (as well as the BGY) approximations retain all the diagrams corresponding to the second and third virial coefficient. Beyond $\eta = 0.2$ the discrepancies between theories and experiments begin. At low enough densities an interesting check is given by the virial coefficients. We write the density expansion (6.4.9) in reduced variables as

$$\frac{\beta P}{n} = 1 + \sum_{j=1}^{\infty} b_{j+1} \eta^j \tag{8.6.2}$$

The coefficients b_{j+1} up to b_6 have been determined numerically quite precisely. On the other hand, they can be deduced from the various approximate theories [for instance, they are easily obtained from the W–T equations of state (8.4.34), (8.4.36) by a trivial expansion]. The results are collected in Table 8.6.1, which gives a good summary of the situation. However, it is important to note that the mere values of the virial coefficients are not enough for stating the success of an approximation. The most important criterion is the general shape of the curve at all densities, as it appears from Fig. 8.6.4. A general remark, applicable to all approximate theories, is that the two equations of state, obtained from the pressure equation and from the compressibility equation, respectively, fall on opposite sides of the exact result. It is clear that the P–Y theory, that is, the W–T solution, gives the closest approximation among the "classical" integral equations. In particular, the compressibility equation of state (8.4.36) is qualitatively quite good, up to the highest densities (in spite of the poor value obtained for the sixth virial coefficient!). The result can actually be improved by empirical fittings, which we shall not discuss in detail. Let us only mention an equation of state derived empirically by Carnahan and Stirling. They approximated the exact value of the first six virial coefficients by the closest integer (see Table 8.6.1) and noted that these integers can be represented by the formula $b_n = (n^2 + n - 2)$. Assuming this relation to hold for all n, the virial series can be summed in closed form with the result

$$\frac{\beta P}{n} = \frac{1 + \eta + \eta^2 - \eta^3}{(1 - \eta)^3} \tag{8.6.3}$$

* There is no sharp gas–liquid transition in a hard-sphere fluid, as can be seen from Fig. 8.6.4. Our statement is based on the shape of the pair distribution, which, at $\eta \approx 0.2$, does not yet show a significant second peak characteristic of short-range order.

Table 8.6.1. Virial Coefficients for the Hard-Sphere Fluid

	b_2	b_3	b_4	b_5	b_6
MD	4	10	18.36	28.26	39.53
W–T^c	4	10	19	31	48
W–T^p	4	10	16	22	28
HNC^c	4	10	28.5	37.1	39.1
HNC^p	4	10	13.39	12.6	28.8
BGY^c	4	10	21.9	34.2	
BGY^p	4	10	14.4	12.2	
CS	4	10	18	28	40

This equation is very similar to the W–T^c equation (8.4.36). In spite of its empirical nature, Eq. (8.6.3) represents the hard-sphere compressibility factor with a surprising accuracy throughout the whole density domain, up to $\eta = 0.47$.

At this density, the numerical computations show the occurrence of a quite dramatic change: there appears a sudden drop in the compressibility factor, and a new branch of the function starts toward higher densities. This phenomenon is interpreted as a liquid-to-solid phase transition. The P–Y equation does *not* account for such a phase transition.* Clearly, its analytic solution (8.4.36) is perfectly regular for all $\eta < 1$. The pole at $\eta = 1$ is unphysical. The value $\eta = 0.742$ is a natural limit of the density corresponding to the close-packing arrangement of the spheres: the system cannot be further compressed. The approximation is, therefore, valid only within the fluid phase. The explanation of a phase transition requires a subtler theory.

We now consider the case of a realistic interaction potential between the particles. The most extensively studied potential is the *Lennard–Jones potential*, which we encountered already earlier (6.1.9):

$$V(r) = 4\varepsilon\left[\left(\frac{\sigma}{r}\right)^{12} - \left(\frac{\sigma}{r}\right)^{6}\right] \tag{8.6.4}$$

In this case we have the possibility of a double check of the theories: these may be compared with the computer calculations, as for hard spheres, but now also directly with experimental data. The best real systems for these comparisons are the rare gases, the molecules of which

* It may be mentioned that the Kirkwood integral equation, which makes use of the superposition principle (7.4.6), possesses no solutions beyond a critical density close to 0.47. Kirkwood himself interpreted this fact as a phase transition and Alder and Wainwrights' MD calculations came later as a confirmation of his prediction. However, it is by no means clear whether Kirkwood's equation really explains the phase transition.

are monoatomic and perfectly symmetrical; we must, however, exclude the lighter members of the family, helium and neon, in which quantum effects are important. We are, therefore, essentially left with argon and xenon as good candidates for comparisons with the theory. The thermodynamic properties of these systems have been extensively studied and can be considered as very well-known experimentally. As the Lennard–Jones (L–J) potential contains two parameters, σ and ε, it is natural to take these as the units of energy and of length, and therefore to define reduced temperature, length, and density by

$$\begin{aligned} T^* &= \frac{k_B T}{\varepsilon} \\ r^* &= \frac{r}{\sigma} \\ n^* &= n\sigma^3 \end{aligned} \tag{8.6.5}$$

For argon: $\varepsilon/k_B = 119.8°\text{K}$, $\sigma = 3.405$ Å;

for xenon: $\varepsilon/k_B = 225.3°\text{K}$, $\sigma = 4.04$ Å.

There is an important difference between the L–J and the hard-sphere systems. In the latter, because of the singular nature of the potential, the temperature plays practically no role in the physical quantities; this is because of the relation (8.4.2), which implies that the pair distribution is only a function of the density, as is the macroscopic compressibility factor. In a real gas, however, the temperature plays a decisive role. We know from elementary physics that the compressibility factor as a function of the density (or equivalently, the pressure as a function of volume) behaves differently at different temperatures: this behavior is described by a set of curves called *isotherms* in the $(\beta P/n, n)$ plane (or in the P–V plane). Moreover, the range of temperatures is separated into two qualitatively different regions by the *critical temperature* T_c. For $T < T_c$, there appears a sharp gas–liquid phase transition at a certain density, followed by a range of densities where vapor and liquid coexist, and finally by a range of densities where the fluid is in a truly liquid state. We discuss the difficult problems related to critical phenomena and phase transitions in Chapters 9 and 10.

We first consider Figs. 8.6.5 and 8.6.6, in which the pair correlation function, computed by molecular dynamics, is shown at various densities and temperatures. The general shape of these curves is not very dissimilar to those of Fig. 8.6.2: We see again the complete exclusion at very short distances, a rise (which now is not sudden as for hard spheres) toward a strong first maximum, followed by one or more secondary maxima. As we see from Fig. 8.6.5, an increase in density has the effect of sharpening the

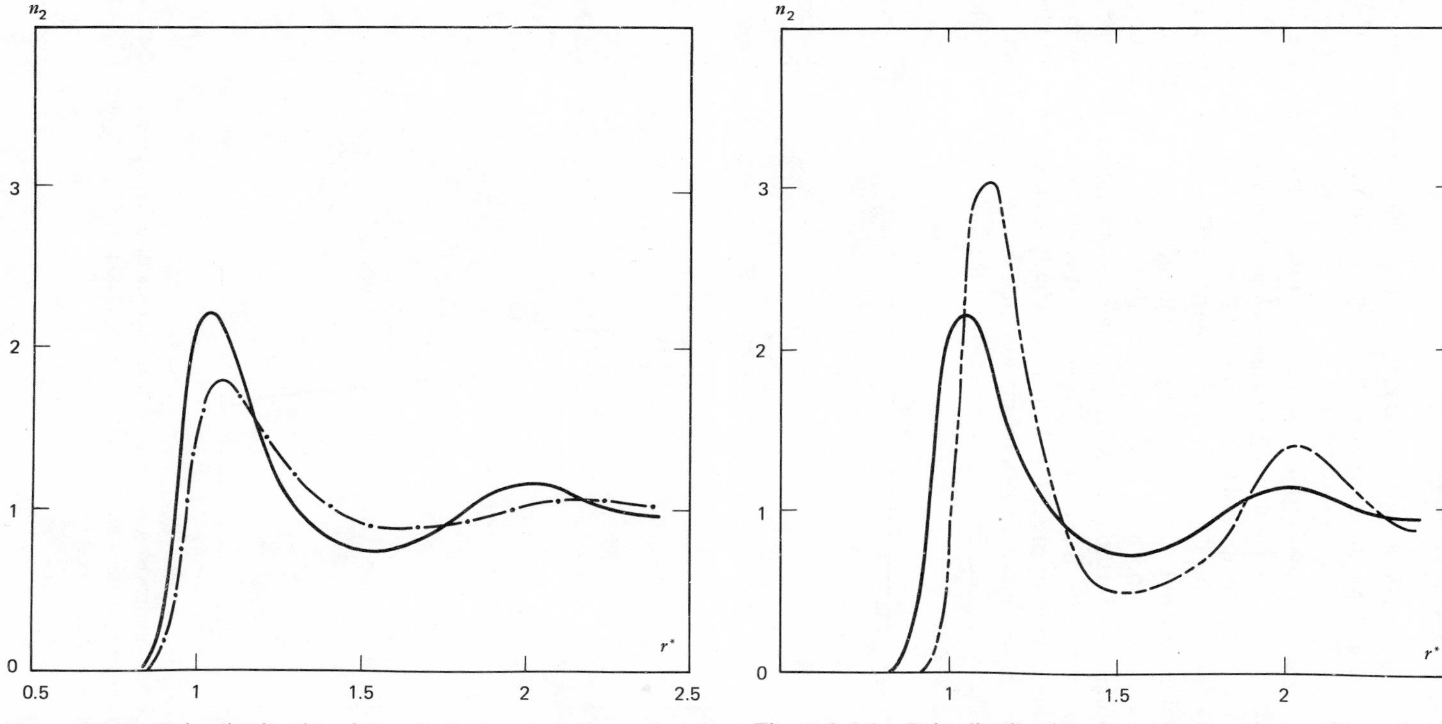

Figure 8.6.5. Pair distribution for a L–J system: effect of the density. ———, $T^* = 2.89$, $n^* = 0.85$; — · —, $T^* = 2.64$, $n^* = 0.55$. (Data based on MD calculations of Verlet, 1968.)

Figure 8.6.6. Pair distribution for a L–J system: effect of the temperature. ———, $T^* = 2.89$, $n^* = 0.85$; — - - —, $T^* = 0.66$, $n^* = 0.85$. (Data based on MD calculations of Verlet, 1968.)

structure, just as for hard spheres: All the peaks become higher; moreover, the slope of the first rise is steeper and the position of the maximum is displaced to the left. In Fig. 8.6.6 we see that a decrease of the temperature also favors ordered structure in the fluid: the first maximum is increased but is now displaced to the right.

The P–Y equation applied to a L–J system yields a good approximation to $n_2(r)$ at low densities, but fails at high densities. This can be seen in Fig. 8.6.7, which corresponds to a true liquid state ($T^* = 0.88$, $n^* = 0.85$). The first maximum is too high and too much to the left. However, as the density increases further, a new, very surprising feature appears: the pair distribution can be fitted by a *hard-sphere* pair distribution with an astounding degree of precision. This is strikingly seen when the Fourier transform of the pair correlation function, that is, the structure factor $a_\mathbf{k}$, Eq. (8.1.5), is studied. In Fig. 8.6.8 the structure factor of a hard-sphere

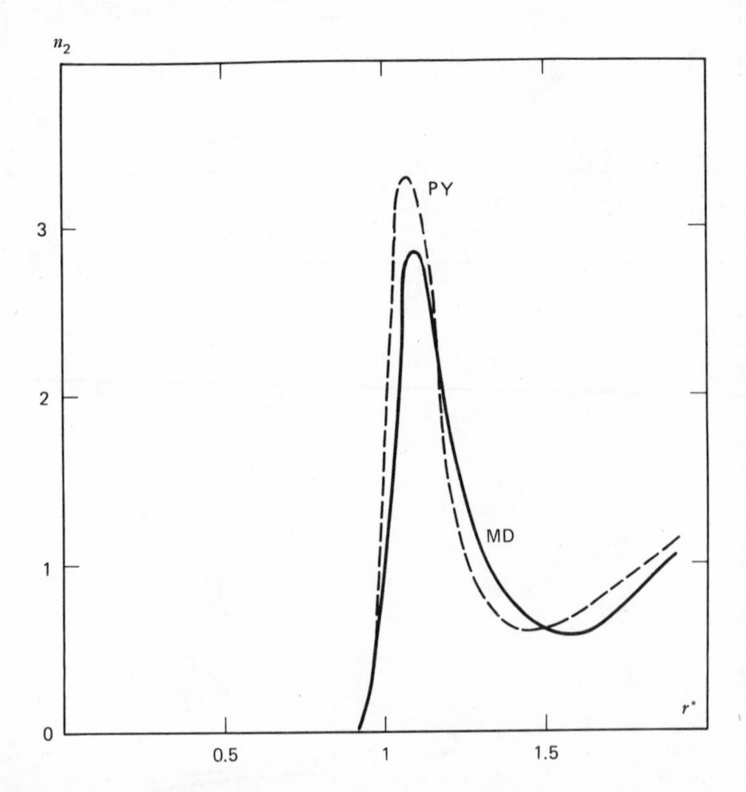

Figure 8.6.7. Comparison of the pair distributions of a L-J system, as obtained from molecular dynamics and from the P–Y equation, at $n^* = 0.85$ and $T^* = 0.88$. (Data from Verlet, 1968.)

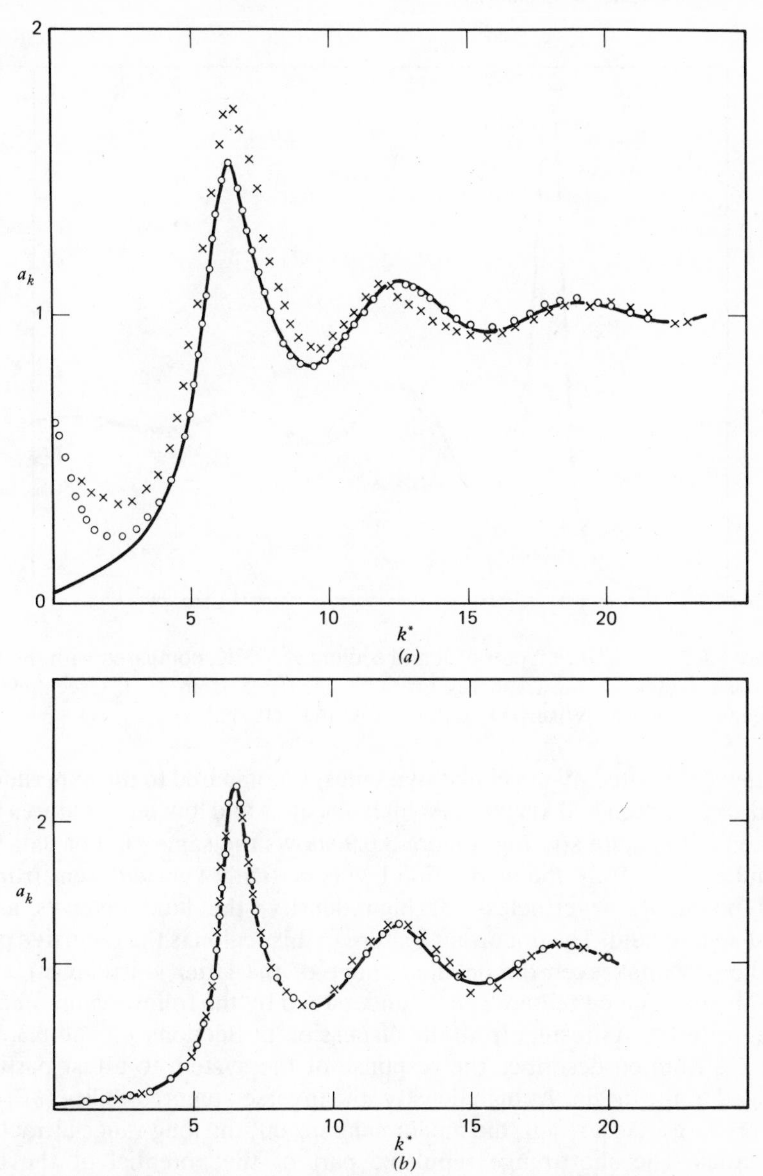

Figure 8.6.8. Structure factor at high densities. (*a*) ———, Structure factor of hard spheres of diameter $d_0 = 1.0$ Å at $n^* = 0.57$. ∘ ∘ ∘ ∘, MD structure factor for L–J system, at $n^* = 0.54$ and $T^* = 1.33$. × × ×, X-ray experiment on argon, at $n^* = 0.77$, $T^* = 1.28$. (*b*) ———, Structure factor of hard spheres of diameter $d_0 = 1.03$ Å at $n^* = 0.82$. ∘ ∘ ∘ ∘, MD structure factor for L-J system, at $n^* = 0.75$, $T^* = 0.83$. × × ×, Neutron experiment on krypton, at $n^* = 0.77$, $T^* = 0.77$. (Data from Verlet, 1968.)

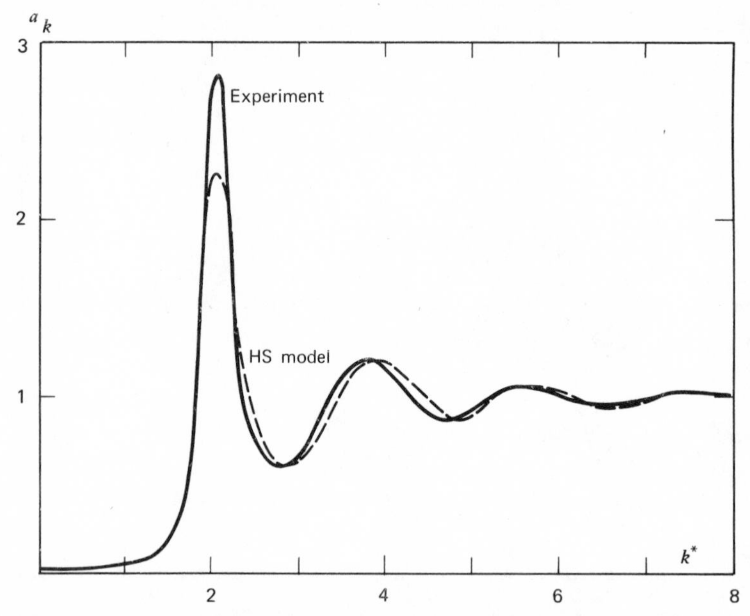

Figure 8.6.9. Structure factor of liquid sodium at 373°K, compared with the WT structure factor for hard spheres for $\eta = 0.45$. [Data from A. J. Greenfield, J. Wellendorf, and N. Wiser, *Phys. Rev.* A **4,** 1607 (1971).]

system (computed by molecular dynamics) is compared to the experimental data for argon and krypton. At high densities and low temperatures the agreement is quite striking. Figure 8.6.9 shows the same kind of data for liquid sodium. Here the interaction law is certainly very different from a L–J potential; nevertheless, at high density, the fluid behaves as a hard-sphere fluid. The important feature in this region is the repulsive part of the potential; even the detailed shape of the latter is irrelevant.

This unexpected feature can be understood by the following argument, due to Verlet. As results from the discussion of Sections 7.5 and 8.3, the pair distribution describes the response of the system to a test particle located at the origin. At high density, the inverse compressibility $(\partial P/\partial n)$ is very large. As a result, the medium damps out the long-range attractive potential. The short-range repulsive part of the potential of the test particle remains the only nontrivial factor determining the pair distribution.

This feature of simplicity immediately leads to an idea that became the basis of the modern theory of liquids. In order to describe the properties of a dense fluid we define a *reference system* as a system of hard spheres with a properly fitted diameter (the latter is temperature and density dependent). This reference system approximates the effect of the repulsive part of the potential. One then starts a perturbation expansion of the thermodynamic quantities, in which the attractive tail of the interactions plays the role of the perturbation. The most elegant—and most successful—form of this treatment appears in the recent works of Weeks, Chandler, and Andersen and of Verlet and Weiss. These authors are able to derive very accurate analytic approximations for the pair distribution and for the thermodynamic quantities of L–J fluids, valid at high and intermediate densities. We have no space for going into the details of these theories, but refer the reader to the literature quoted at the end of the chapter. Suffice it to say that they give remarkably good results, as will be seen below. In the past one tried to start the liquid theory from a reference state very much resembling a crystalline lattice (cell theories, free volume theories). However, the degree of order of such model systems is too high and the results are rather poor. The hard-sphere fluid, on the other hand, is an excellent reference system, because it is a limiting

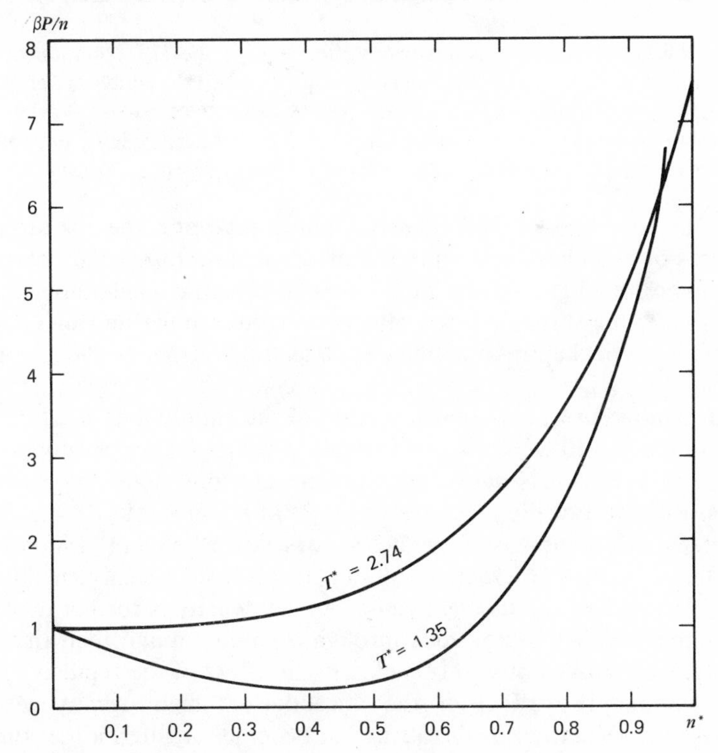

Figure 8.6.10. Equation of state of the L–J fluid. (MD calculations of Verlet and Weiss, 1971.)

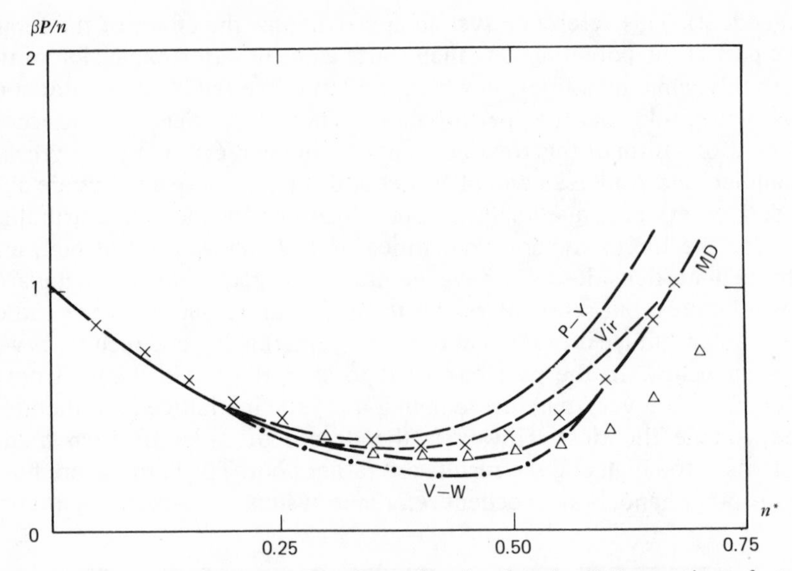

Figure. 8.6.11. Comparison of various theories for the L–J equation of state. Isotherm $T^* = 1.35$. ———, MD calculation; — — —, P-Y equation for L–J potential; - - - -, 5-term virial expansion (see Section 6.3); -·-·-·, Verlet–Weiss perturbation theory; × × ×, experimental data for argon; △ △ △, experimental data for xenon. [Experimental data are from J. Levelt, *Physica* **26,** 361 (1960).]

state for the fluid at high density, while retaining the characteristic features of symmetry and structure of a true fluid rather than of a solid. Moreover, the hard-sphere fluid is by now quite well known, either numerically, or through somewhat poorer but simple and qualitatively correct analytical approximations, such as the Wertheim–Thiele approximation.

Let us now turn to the consideration of the *equation of state* of a L–J fluid. In Fig. 8.6.10 we show two typical isotherms (at temperatures above T_c). Their behavior is quite characteristic of a real fluid. The isotherms necessarily start at the ideal gas value $\beta P/n = 1$ at $n = 0$. As the density increases, the compressibility factor goes down below 1, which means that the attractive interactions dominate at small density, making the pressure smaller than its ideal value. As the density is further increased, the compressibility factor goes through a minimum and then rises very steeply, well above unity: this is clearly the effect of the repulsive forces that overtake the effect of the attraction at high density. At high temperature the effect of the attraction becomes smaller, a fact that can be expected from the behavior of the second virial coefficient, discussed in Section 6.4.

In Fig. 8.6.11 we show the comparison of the results of various approximations, together with the MD results and the experimental results. We see that the best fit is obtained by the Verlet–Weiss theory, which is a perturbation theory of the type described above, starting from the hard-sphere reference fluid. Its characteristic feature is its excellent fit, both in the low-density and in the high-density regions. Only in the intermediate-density region are there small deviations. It is worth noting that $T^* = 1.35$ is near the critical point. The Verlet–Weiss theory is actually successful in the liquid state as far down in temperature as the triple point.

To conclude this chapter, we see that the theory of dense fluids and of liquids is in a pretty good state, as far as numerical accuracy is involved. However, it cannot be said that we have at present a truly fundamental theory of dense fluids. The surprising success of the simple semiphenomenological approximations will probably receive a significant explanation in the future.

BIBLIOGRAPHICAL NOTES

General references for this chapter are

S. Rice and P. Gray, *Statistical Mechanics of Simple Liquids*, Wiley, New York, 1965.

H. L. Frisch and J. L. Lebowitz, *The Equilibrium Theory of Classical Fluids*, Benjamin, New York, 1965.

I. Z. Fisher, *Statistical Theory of Liquids*, Univ. Chicago Press, Chicago, Illinois, 1964.

J. de Boer, *Rep. Progr. Phys.* **12,** 305 (1949).

These problems are also thoroughly covered in the books of Hill and of Münster quoted in the Bibliographical Notes to Chapter 4.

The *cluster expansion for the pair distribution* was introduced in

J. Mayer and E. W. Montroll, *J. Chem. Phys.* **9,** 2 (1941).

J. de Boer (quoted above).

The techniques of graph summations and renormalizations appropriate for high-density problems were developed in

J. M. van Leeuwen, J. Groeneweld, and J. de Boer, *Physica* **25,** 792 (1959) (derivation of the HNC equation).

E. Meeron, *J. Math. Phys.* **1,** 192 (1960) (derivation of the "nodal" expansion).

G. S. Rushbrooke, *Physica* **26,** 259 (1960).

The *P–Y equation* was derived in J. K. Percus and G. J. Yevick, *Phys. Rev.* **110,** 1 (1958). The derivation of the P–Y equation given in the text follows J. L. Lebowitz and J. K. Percus, *J. Math. Phys.* **4,** 116 (1963).

The analytic solution of this equation for hard spheres was found in

E. Thiele, *J. Chem. Phys.* **39,** 474 (1963).

M. S. Wertheim, *Phys. Rev. Lett.* **10,** 321 (1963); *J. Math. Phys.* **5,** 643 (1964).

A good introduction to the methods of *numerical experiments* is given in H. L. Frisch, *Advances in Chemical Physics*, Vol. 6 (I. Prigogine, ed.), Wiley-Interscience, New York, 1964.

The *Monte Carlo method* was introduced by M. N. Rosenbluth and A. W. Rosenbluth, *J. Chem. Phys.* **31,** 369 (1959).

The *molecular dynamics* method was introduced by B. J. Alder and T. Wainwright, *J. Chem. Phys.* **27,** 1209 (1957). In this paper, the equation of state of hard spheres was investigated, and a phase transition was found, as suggested by J. G. Kirkwood and E. Monroe, *J. Chem. Phys.* **9,** 514 (1941).

Other important numerical experiments are

W. W. Wood and F. R. Parker, *J. Chem. Phys.* **27,** 720 (1957) (L–J equation of state; MC method).

A. Rahman, *Phys. Rev.* **136,** A405 (1964) (L–J pair distribution; MD method).

L. Verlet, *Phys. Rev.* **159,** 98 (1967) (L–J equation of state; MD).

L. Verlet, *Phys. Rev.* **165,** 201 (1968) (L–J pair distribution; MD).

B. J. Alder and C. E. Hecht, *J. Chem. Phys.* **50,** 2032 (1969) (pair distribution of hard spheres + perturbation, MD).

The extension of the superposition approximation and its application to the hard-sphere equation of state is due to S. A. Rice and J. Lekner, *J. Chem. Phys.* **42,** 3559 (1965).

The empirical equation of state (8.6.3) for hard spheres was derived by N. F. Carnahan and K. E. Stirling, *J. Chem. Phys.* **51,** 635 (1969).

The structure factor of metals and its relation to the hard-sphere problem was investigated by N. W. Ashcroft and J. Lekner, *Phys. Rev.* **145,** 83 (1966).

The *older theories of liquids* (density expansions, superposition approximation, cell, and cell-cluster theories) are reviewed in J. M. H. Levelt and E. G. D. Cohen, "A Critical Study of some Theories of the Liquid State, Including a Comparison with Experiment," in *Studies in Statistical Mechanics*, Vol. 2 (J. de Boer and G. E. Uhlenbeck, eds.), North Holland, Amsterdam, 1964.

The latest theories of liquids, of the type of *perturbations around the hard-spheres fluid*, are in

J. D. Weeks, D. Chandler, and H. C. Andersen, *J. Chem. Phys.* **54,** 5237 (1971).

H. C. Andersen, J. D. Weeks, and D. Chandler, *Phys. Rev.* A **4,** 1597 (1971).

L. Verlet and J. J. Weiss, *Phys. Rev.* A **5,** 939 (1972).

CHAPTER 9
PHASE TRANSITIONS

9.1. QUALITATIVE DESCRIPTION OF PHASE TRANSITIONS

The theories developed in Chapters 5–8 cover a wide range of states of matter, as can be found under very different conditions of density and temperature. We saw that the description of these homogeneous phases is at present in a fairly satisfactory state.

We know, however, from everyday experience, as well as from our freshman physics course, that the phenomena associated with the passage from one phase to another are quite complex. Let us briefly recall some of these phenomena.

Consider the equation of state of some pure substance: $P = P(T, n)$. In previous chapters we discussed the behavior of this function in finite regions of the (T, n) plane, where some approximations can be made. We now want to look at this function over a much wider range of values of the parameters. We consider isotherms starting with a pretty high temperature. These isotherms start at $n = 0$ as a straight line with a slope proportional to T (ideal gas limit); as the gas is compressed, the curve deviates progressively from the straight line, as a result of the intermolecular interactions (see Fig. 9.1.1). As the temperature is lowered, the deviations become more clear-cut. For a given density the pressure is first lower than the ideal gas value, but as the system is further compressed, the pressure eventually rises very steeply. This behavior is easily understood from our previous study: At moderate densities the attractive part of the interactions dominates, lowering the pressure. At high densities the steep repulsive part of the potential takes over and leads to a behavior similar to that of a hard-sphere system.

As the temperature is further lowered we finally reach a value of T for which the isotherm is so strongly bent by the attractive forces as to get a horizontal inflection point, at which

$$\frac{\partial P(T, n)}{\partial n} = 0, \qquad \frac{\partial^2 P(T, n)}{\partial n^2} = 0 \tag{9.1.1}$$

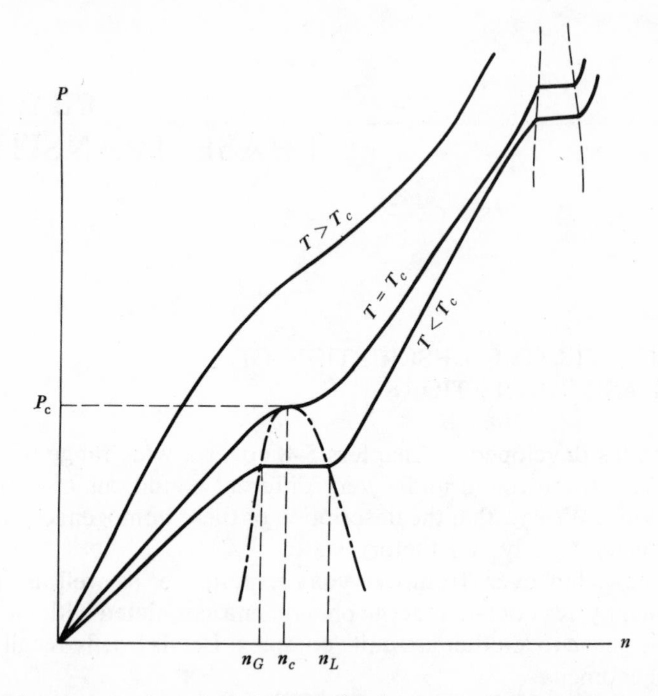

Figure 9.1.1. A set of typical (schematic) isotherms in the pressure–density plane, showing the gas–liquid phase-transition region and the liquid–solid phase-transition region.

The point (P_c, T_c, n_c) defined by these equations is called the *critical point*. Its neighborhood has many exotic properties, which will be discussed in detail further on.

Below the critical point the compression process takes a quite different aspect. When a certain density n_G is reached, a further compression does not lead to a homogeneous increase in density; instead, there appear droplets of liquid having density n_L with $n_L > n_G$. The overall pressure of the system is constant over the whole range of densities $n_G \leqslant n \leqslant n_L$, where the coexistence between gas and liquid is possible. The value of this "vapor pressure" only depends on the temperature. As the system is further compressed, the relative amount of the two coexisting phases changes in favor of the liquid, until the overall density reaches the value n_L. At this point the last gas bubble disappears, and the system is in a homogeneous liquid phase.

Further compression gives rise eventually to crystallization. This new phase transition is very similar to condensation (as far as the shape of the isotherms is concerned), but there is an important difference: as far as we know, there exists no critical point, and hence no possibility of a continuous transition from fluid to solid.

The most striking fact about these isotherms is the presence of sharp edges, reflecting *discontinuities* in various thermodynamic quantities. Also, the presence of the critical point is associated with *divergences* in various thermodynamic quantities, to be discussed below. The main challenge to statistical mechanics is then the understanding of how such peculiar behavior can develop from the interplay of molecules interacting through realistic forces.

A rather analogous behavior can be found if some, apparently very different, phenomena are analyzed in terms of a proper language. A typical and very intensely studied problem is the behavior of *magnetic crystals*. The analogy appears when we associate n with the magnetization, that is, the magnetic moment per particle, M, and P with the magnetic field $\mathcal{H}$. The magnetic equation of state is then the equation relating the magnetization to the magnetic field and to the temperature in equilibrium:

$$M = M(\mathcal{H}, T) \tag{9.1.2}$$

The microscopic picture of these magnetic crystals can be understood in first approximation as a set of atoms located at the vertices of a crystal lattice, each atom having an individual spin, hence an individual magnetic

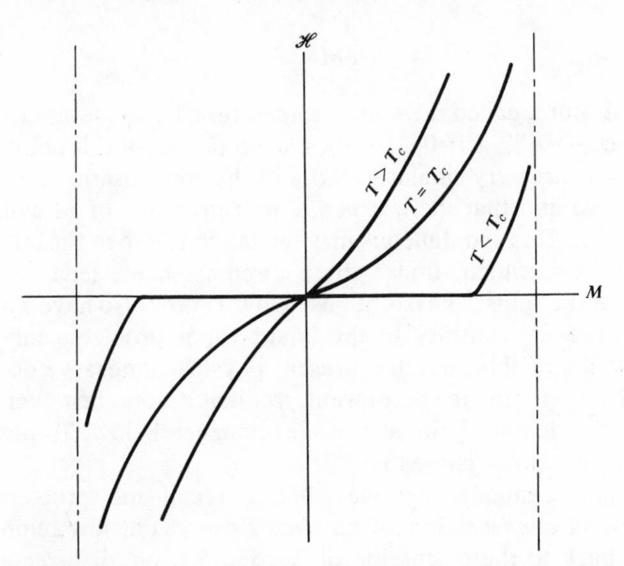

Figure 9.1.2. Isotherms representing the magnetic equation of state of a ferromagnetic crystal.

moment. The atoms interact with each other through forces depending not only on the distance, but also on the value and the mutual orientation of the spins. In the presence of an external magnetic field, the spins tend to orient themselves parallel to the field; but this ordering is counteracted by the thermal agitation. As a result, a typical isotherm, at sufficiently high temperature, looks as follows (see Fig. 9.1.2): At zero field, $M=0$, because the orientations of the spins are completely random. As the magnetic field is increased, the spins get partially oriented, and hence a nonzero average macroscopic moment appears. This moment increases with increasing $\mathcal{H}$ till the maximum possible value is reached, corresponding to complete orientation of all the spins along the magnetic field (saturation). The curve is, of course, symmetric with respect to the origin when it is continued for $\mathcal{H}<0$. At lower temperature, the S-shape of the curve is more pronounced because it is easier to orient the particles in this case, and the saturation value is reached for weaker fields.

This general shape of the isotherms is characteristic of all *paramagnetic* substances, down to $T=0°\text{K}$. For *ferromagnetic* substances, however, there exists a temperature at which the inflection point at $M=0$ becomes horizontal:

$$\frac{\partial \mathcal{H}}{\partial M}=0$$
$$\frac{\partial^2 \mathcal{H}}{\partial M^2}=0 \qquad (9.1.3)$$

This temperature, called the *Curie temperature* T_c, defines a critical point with coordinates (T_c, $\mathcal{H}_c=0$, $M_c=0$). The properties of this point and of its neighborhood are very similar to those of the condensation critical point. Below T_c, we find that there exists a nonzero value of M even at zero magnetic field. This spontaneous magnetization is due to the intermolecular interactions, which, under these circumstances, lead to a partial alignment of the spins. The isotherms below T_c now also have a horizontal stretch. However, contrary to the liquid–vapor problem, only the two extreme points of this stretch represent physical states: we do not have two coexisting phases in the present problem. (Note, however, that the presence of "domains" in a real ferromagnet below T_c bears some analogy to coexisting phases.)

As a third example, we may finally recall the problem of the *Bose–Einstein condensation* of an ideal Bose gas at low temperatures. Referring back to the discussion of Section 5.7, one will recognize the analogy of that problem with the general features described here.

Summarizing the situation we find the following general picture in all these cases: The possible equilibrium states of the system in the thermodynamic plane are of two kinds according to whether $T > T_c$ or $T < T_c$. Below T_c, one may define an *order parameter*, which, at least in the neighborhood of T_c, is a monotonous-decreasing function of T, approaching zero as $T \to T_c$. Above T_c the order parameter is identically zero for all $T > T_c$. The order parameter for the liquid–vapor problem is the difference of densities $n_L - n_G$ of the coexisting phases; for the magnetic problem it is the magnetization at zero magnetic field; in the degenerate boson gas it is the fraction of particles in the ground state $\langle n_0 \rangle / N$. The appearance of a nonvanishing order parameter is a manifestation of a *symmetry breaking* at the microscopic level. Above T_c all the equilibrium states are translationally invariant (i.e., homogeneous) in the first case, and rotationally invariant (i.e., isotropic, for $\mathcal{H} = 0$) in the second case. Below T_c, there exist equilibrium states that do not possess this symmetry, hence the possibility of nonvanishing values for the order parameter.

As the liquid–gas critical point is approached from above, the isothermal compressibility diverges:

$$\chi_T = n^{-1} \left(\frac{\partial n}{\partial P} \right)_T \to \infty \tag{9.1.4}$$

But we know from Chapter 4 that this quantity is related to the fluctuations in density. These fluctuations become very large near the critical point: the system thus prepares the way to the appearance of macroscopic droplets beyond the critical point. These large density fluctuations can be literally seen in the phenomenon of critical opalescence, which was mentioned in Section 8.1.

A similar discussion applies to the magnetic case, where the role of the compressibility is played by the isothermal magnetic susceptibility:

$$\kappa_T = \left(\frac{\partial M}{\partial \mathcal{H}} \right)_T \tag{9.1.5}$$

These are a few of the most prominent qualitative features characterizing the phase transitions. The object of this chapter and of the next one is to understand as clearly as possible the molecular origin of these phenomena.

The three phenomena discussed above are not the only possible examples of phase transitions. They are not even representative of the most general case. The freezing or the sublimation processes, for instance, have quite different features. They have no critical point. Moreover, the change in symmetry is even stronger in these cases (although this change is more gradual than one would expect from the macroscopic appearance: see the discussion of Section 8.6). However, the present state of understanding of these problems is not as advanced as in the theory of critical phenomena. In order to save space, we decided to discuss in this and the following chapters only the liquid–vapor phase transition and the ferromagnetic transition.

9.2. WEISS' MEAN FIELD THEORY OF FERROMAGNETISM

We begin our investigation by introducing one of the simplest phenomenological theories of phase transitions.

We first prepare the way by considering a system of N noninteracting molecules, each having a spin $S\hbar$. The total Hamiltonian can be split into a nonmagnetic term H_{nm}, independent of the spins, and a magnetic term H_m:

$$H = H_{nm} + H_m \tag{9.2.1}$$

If we assume that the Boltzmann statistics is a sufficient approximation, the two terms give rise to additive contributions to the free energy [see Eq. (5.2.12)]. As we are interested in magnetic properties, only the second term is relevant; it can be written as

$$H_m = -\sum_{j=1}^{N} \bar{\mu}\mathbf{S}_j \cdot \mathcal{H} \tag{9.2.2}$$

The spin $\mathbf{S}$ is thus related to the molecular magnetic moment $\boldsymbol{\mu}$ through $\boldsymbol{\mu} = \bar{\mu}\mathbf{S} = g\mu_B\mathbf{S}$, where g is the Landé factor and $\mu_B = e\hbar/2mc$ is the Bohr magneton. From quantum mechanics we know that the product $\mathbf{S}_j \cdot \mathcal{H}$ can only assume the values

$$\mathbf{S}_j \cdot \mathcal{H} = m_j\,|\mathcal{H}|, \qquad m_j = -S, -S+1, \ldots, S-1, S \tag{9.2.3}$$

m_j being the magnetic quantum number.

The method of Section 5.2 is now applied to the calculation of the magnetic factor of the partition function:

$$Z_m = Z_1^N \tag{9.2.4}$$

with

$$\begin{aligned} Z_1 &= \sum_{m=-S}^{S} e^{mx} = e^{-xS}\,\frac{1-e^{(2S+1)x}}{1-e^x} \\ &= \frac{\sinh(S+\frac{1}{2})x}{\sinh(\frac{1}{2}x)} \end{aligned} \tag{9.2.5}$$

where $x = \beta\bar{\mu}\mathcal{H}$. The summation is simple because it represents a finite geometric series. The magnetic contribution to the free enthalpy* is now

$$G(T, \mathcal{H}) = -Nk_B T \ln Z_1(T, \mathcal{H}) \tag{9.2.6}$$

from which the magnetization can be calculated through the following formula:

$$M(T, \mathcal{H}) = -\left(\frac{\partial G}{\partial \mathcal{H}}\right)_T \tag{9.2.7}$$

We omit here the proof of this formula, but note its analogy to the equation for the volume, $\mathcal{V} = -(\partial G/\partial P)_T$. In the present case, using Eqs. (9.2.5)–(9.2.7) we get, after a little algebra,

$$M(T, \mathcal{H}) = M_0 B_S(S\beta\bar{\mu}\mathcal{H}) \tag{9.2.8}$$

where

$$M_0 = M(T=0, \mathcal{H}=0) = NS\bar{\mu} \tag{9.2.9}$$

and $B_S(y)$ is called the *Brillouin function* for spin S:

$$B_S(y) = \frac{2S+1}{2S}\coth\left(\frac{2S+1}{2S}y\right) - \frac{1}{2S}\coth\left(\frac{1}{2S}y\right) \tag{9.2.10}$$

The magnetization curve has a characteristic shape, as shown in Fig. 9.2.1: it is of the same general shape as the isotherms of Fig. 9.1.2 for $T > T_c$ (various temperatures correspond to various scales on this graph).

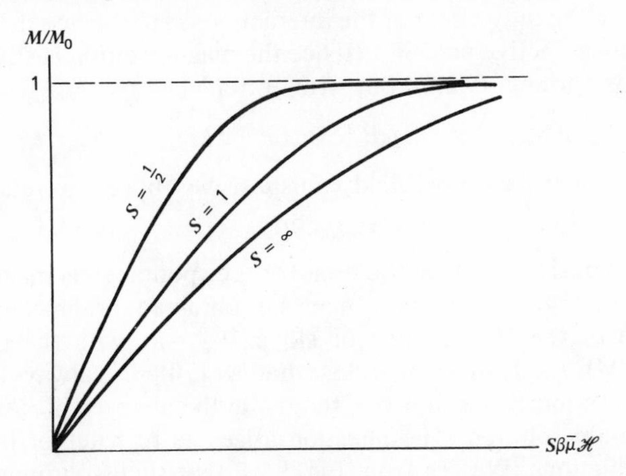

Figure 9.2.1. Magnetic equation of state in reduced variables, for a noninteracting spin system.

* Note that the partition function in the present problem expresses the thermodynamic potential in the variables T and $\mathcal{H}$. As $\mathcal{H}$ corresponds to the pressure (as indicated in Section 9.1), this thermodynamic potential must be identified with the free enthalpy G, rather than with the free energy A. This is, however, a rather secondary question of terminology.

Hence this system behaves like a *paramagnetic system*. This should be expected, since we know that the ferromagnetic behavior is crucially linked to the presence of interactions.

Weiss' idea for a theory of ferromagnetism (which goes back to 1907) consists in taking account as simply as possible of one feature of the interactions, hopefully the more important one. He notes that the set of spins on the lattice is the source of a magnetic field. Therefore the real magnetic field felt by a particular spin is the *sum* of the external field $\mathcal{H}$ and a *molecular field* $\mathcal{H}_m$. If the spins ignore each other, there can be no such field: it stems directly from the interactions. Also, if the spins are oriented at random, their mutual effects cancel. There can only exist a molecular field if some average polarization is already present: it then gives rise to an added field, which in turn increases the polarization, and so on. This cascade mechanism is, of course, limited by the thermal energy, which counteracts the ordering effect of the interactions. We have here a typical case of a so-called *cooperative effect*. In view of this discussion, we may assume in first approximation that the molecular field is proportional to the magnetization, and hence the total effective field is

$$\mathcal{H}_m = \mathcal{H} + \lambda M(T, \mathcal{H}) \tag{9.2.11}$$

where the phenomenological proportionality constant λ is called the molecular field parameter. The simplest theory is now obtained by assuming that the *only* effect of the interactions is to replace the magnetic field $\mathcal{H}$ by the effective field $\mathcal{H}_m$. Hence the magnetization is still given by the ideal gas formula (9.2.8), but with $\mathcal{H}$ replaced by $\mathcal{H}_m$:

$$M = M_0 B_S[S\beta\bar{\mu}(\mathcal{H} + \lambda M)] \tag{9.2.12}$$

In particular, if the external field vanishes, we obtain the relation:

$$M = M_0 B_S(S\beta\bar{\mu}\lambda M) \tag{9.2.13}$$

This is the basic nonlinear equation for the spontaneous magnetization in Weiss' theory. Its solution may be obtained graphically by the intersection of the straight line of slope 1: $y = M$ with the curve $y = M_0 B_S(S\beta\bar{\mu}\lambda M)$. As $B_S(0) = 0$, it is clear that $M(T, 0) = 0$ is always a solution to Eq. (9.2.13); moreover, if $\lambda = 0$, that is, in the absence of interactions, this is the only solution. The question arises as to whether there exist nontrivial solutions. We see from Fig. 9.2.2 that such solutions do exist whenever the initial slope of the function in the rhs of Eq. (9.2.13) is larger than 1. The initial slope of $B_S(y)$ is easily found from Eq. (9.2.10), and using also Eq. (9.2.9) we may write this criterion as

$$C\frac{\lambda}{T} > 1 \tag{9.2.14}$$

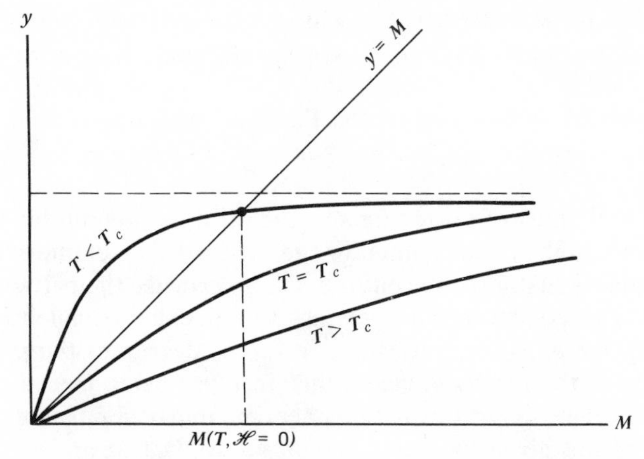

Figure 9.2.2. Graphical determination of the remanent magnetization from Eq. (9.2.13).

with the Curie constant given by

$$C = \frac{N\bar{\mu}^2 S(S+1)}{3k_B} \tag{9.2.15}$$

Hence there exists a nonzero critical temperature:

$$T_c = \lambda C$$

below which a spontaneous magnetization is possible.

It is truly remarkable that with such simple ingredients we were able to get a quite acceptable model of a ferromagnet, presenting many of the real features, such as the existence of a sharp critical point, with the necessary divergences in its neighborhood (these will be discussed later), and so on. Of course, we are unable to calculate the crucial constant λ; also we have no hint as to why some substances are ferromagnetic and others are not. In addition, the more detailed behavior predicted by Weiss' theory (e.g., the type of the divergences near T_c) is not quite correct as compared to real behavior. Nevertheless, one has the feeling that Weiss' theory has touched on an essential ingredient of nature and that any more sophisticated theory should account for the presence of an average molecular field, even if it is corrected by various other effects.

9.3. THE VAN DER WAALS THEORY OF CONDENSATION

We now discuss the problem of liquid–gas phase transitions. Here too there exists a very old theory (1873) that was originally derived in a quite phenomenological way, but that had a surprising success. The heuristic derivation of the van der Waals equation of state can be found in every textbook of freshman physics, so we do not come back to it, but only quote the result:

$$P = \frac{Nk_BT}{\mathcal{V} - Nb} - \frac{\alpha N^2}{\mathcal{V}^2} \tag{9.3.1}$$

There are two kinds of corrections to the ideal gas law in this equation. The first term in the rhs accounts for the fact that the volume available to the molecules is really not the entire volume, because the particles have a hard core. The volume per particle at close packing is called b. On the other hand, the existence of attractive forces decreases the pressure in comparison to the ideal gas value, this decrease being measured by the second term. The values of the two phenomenological positive constants α and b are not given by the theory in its original form.

We will now show that the van der Waals equation of state can be obtained quite simply from a molecular model by making a few strong but clear simplifications. We follow here the presentation of Brout, which is very simple. This method is actually just an example of the perturbation methods developed in the theory of dense fluids and mentioned in Section 8.6.

We consider a fluid described by the usual Hamiltonian $H = H_0 + H'^{\text{tot}}$, and we further assume that the interaction potential consists of a sum of two terms [see also Eq. (6.4.1)]:

$$V_{ij}^{\text{tot}} = V_{ij}^{\text{HS}} + V_{ij} \tag{9.3.2}$$

where V_{ij}^{HS} is a hard-sphere potential (6.4.11), whereas V_{ij} is an attractive potential, whose maximum absolute value is small compared to k_BT.

The logarithm of the configuration integral (6.2.9) is now written as

$$\begin{aligned} \ln Q &= \ln\left\{\mathcal{V}^{-N}\int d^Nq \exp(-\beta H'^{\text{HS}} - \beta H')\right\} \\ &= \ln\left\{Q_{\text{HS}}^{-1}\,\mathcal{V}^{-N}\int d^Nq \exp(-\beta H'^{\text{HS}})\exp(-\beta H')\right\} + \ln Q_{\text{HS}} \\ &= \ln\langle e^{-\beta H'}\rangle_{\text{HS}} + \ln Q_{\text{HS}} \end{aligned} \tag{9.3.3}$$

In the second step we divided and multiplied Q by the hard-sphere configuration integral, so as to exhibit the average of $\exp(-\beta H')$ weighted with the hard-sphere distribution.

Because of the smallness of the attractive tail we now assume that the exponential can be linearized:

$$\begin{aligned} \ln\langle e^{-\beta H'}\rangle_{\text{HS}} &\approx \ln\langle 1 - \beta H'\rangle_{\text{HS}} \approx -\beta\langle H'\rangle_{\text{HS}} \\ &= -\tfrac{1}{2}\beta nN\int d\mathbf{r}\; V(r)n_2^{\text{HS}}(r) \end{aligned} \tag{9.3.4}$$

The last equation is derived in a way similar to Eq. (7.2.3). Add now the translational (ideal-gas) contribution to $-k_B T \ln Q$, and note that

$$A_{tr} - k_B T \ln Q_{HS} = A_{HS}$$

A_{HS} being the exact free energy of a hard-sphere system; we finally obtain the total free energy as

$$A = A_{HS} + \tfrac{1}{2}Nn\int d\mathbf{r}\; V(r)n_2^{HS}(r) \tag{9.3.5}$$

Differentiating both sides with respect to the volume we obtain

$$P = P_{HS} + \tfrac{1}{2}n^2\int d\mathbf{r}\; V(r)n_2^{HS}(r) + \tfrac{1}{2}n^3\int d\mathbf{r}\; V(r)\frac{\partial}{\partial n}\, n_2^{HS}(r) \tag{9.3.6}$$

This equation is more general than (9.3.1). To obtain the van der Waals equation we must assume that the hard-sphere pair distribution is practically independent of the density. In that case, the third term in the rhs is zero and in the second term we may define a density-independent positive constant α:

$$\alpha = -\frac{1}{2}\int d\mathbf{r}\; V(r)n_2^{HS}(r) \tag{9.3.7}$$

(α is positive because the attractive tail $V(r)$ is everywhere negative.) The equation of state then becomes

$$P = P_{HS}(n) - \alpha n^2 \tag{9.3.8}$$

This is essentially the van der Waals equation. The remaining difference with (9.3.1) lies only in a rough approximation of P_{HS} in the latter equation. Contrary to the phenomenological equation (9.3.1), we now have precise expressions for the various parameters of the equation of state. These expressions involve the pressure and the pair distribution of the reference hard-sphere fluid. These are not known analytically, but very good analytical approximations and excellent numerical values for this system are available at present (see Chapter 8).

If we limit ourselves to the phenomenological van der Waals (vdW) equation (9.3.1) and plot isotherms we note the following properties. First, the vdW equation yields an isotherm with a horizontal inflection point. Hence it predicts a critical temperature T_c. The coordinates of the critical point are obtained by combining the two equations (9.1.1) with (9.3.1). A simple calculation yields the solution:

$$\begin{aligned} P_c &= \frac{\alpha}{27b^2} \\ n_c &= \tfrac{1}{3}b \\ k_B T_c &= \frac{8\alpha}{27b} \end{aligned} \tag{9.3.9}$$

The shapes of the isotherms above the critical point are in reasonably good agreement with experiment, at least for not too high densities. An important consequence of the vdW equation results when the phenomenological constants α and b are eliminated through (9.3.9) in terms of P_c, n_c, T_c. We may then define reduced dimensionless variables:

$$\tilde{P}=\frac{P}{P_c}, \quad \tilde{n}=\frac{n}{n_c}, \quad \tilde{T}=\frac{T}{T_c} \tag{9.3.10}$$

The vdW equation (9.3.1) becomes then, in terms of these variables,

$$\tilde{P}=\frac{8\tilde{n}\tilde{T}}{3-\tilde{n}}-3\tilde{n}^2 \tag{9.3.11}$$

This equation implies that, *in properly reduced variables, all the substances* (to which the vdW equation applies) *obey the same equation of state.* This famous *law of corresponding states* appears experimentally to be valid for large classes of substances, *even if the equation is not of the form (9.3.11).*

If we now consider temperatures below T_c we immediately see a conspicuous failure. Instead of the characteristic horizontal plateau corresponding to the coexistence of phases, the vdW isotherms have a smooth wiggle. Such a wiggle cannot represent physically acceptable states, because any state on the ascending part of the wiggle would have negative compressibility (i.e., their pressure would increase as the volume increases) and would be thermodynamically unstable.

This failure of the vdW equation was soon cured by Maxwell. His argument went as follows: Below T_c we must simply modify the vdW equation by replacing the wiggle with a horizontal segment, representing the coexisting phases. We thus need a rule for fixing the position of this plateau. As the latter represents two phases in equilibrium, thermodynamics tells us that they must have equal chemical potential:

$$0=\mu_L-\mu_G=\int_L^G d\mu=\int_L^G \mathcal{V}\,dP=\int_{\mathcal{V}_L}^{\mathcal{V}_G} d\mathcal{V}\,\mathcal{V}\left(\frac{\partial P}{\partial \mathcal{V}}\right)_T \tag{9.3.12}$$

We used the fact that at constant temperature (i.e., along the isotherm), we simply have $d\mu=\mathcal{V}\,dP$. The meaning of this equation is that the two shaded areas in Fig. 9.3.1 must be equal. The complete form of the vdW equation, combined with the Maxwell rule (*the vdW–M equation*) is therefore

$$\begin{aligned} P(n)&=P_{HS}(n)-\alpha n^2, && T>T_c, \ \text{all } n \\ & && T<T_c,\ n>n_L,\ n<n_G \\ P(n)&=P_{sat}, && T<T_c,\ n_G<n<n_L \end{aligned} \tag{9.3.13}$$

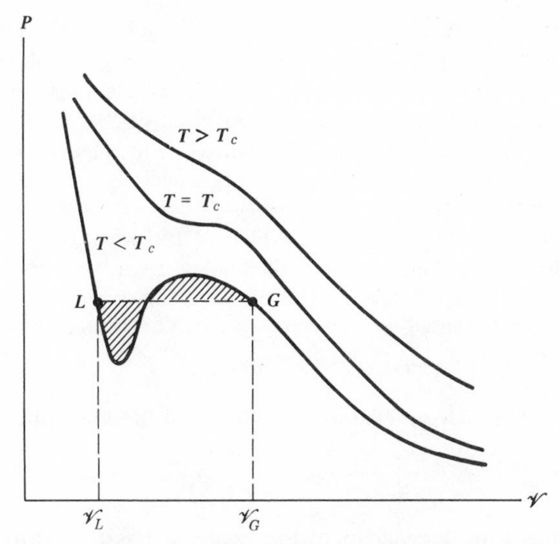

Figure 9.3.1. van der Waals isotherms and the Maxwell construction.

The Maxwell construction does not provide a molecular explanation of the phase transition below T_c: it is merely an *ad hoc* trick that works, provided we accept *a priori* the existence of a coexistence region. Hence, the vdW theory is not a theory of the phase transitions. Even the more refined equation (9.3.6) gives rise to isotherms with wiggles (as shown by Coopersmith and Brout) and must therefore be implemented by a Maxwell construction. There exists, however, one system for which a complete and rigorous theory shows the existence of a true condensation. In this case the vdW–M equation (9.3.13) follows exactly from the partition function. This system will be described in Section 9.4.

*9.4. WEAK, INFINITE-RANGE INTERACTIONS AND THE VAN DER WAALS–MAXWELL EQUATION

We now consider a system characterized by interactions of the same form as (9.3.2), but we make additional restrictions on the form of the attractive part V_{ij}. Roughly speaking, we consider a limit in which its *depth goes to zero*, but at the same time its *range goes to infinity*. In a quite important work (in 1963), Kac, Uhlenbeck, and Hemmer showed that for a one-dimensional system of this type (in which, moreover, the potential V_{ij} was assumed to be exponential) the partition function can be calculated in detail. The result is the van der Waals equation of state *combined* with the Maxwell rule (9.3.13). Hence this model exhibits a true first-order phase transition. This result came somewhat as a surprise, since it was known for a long time that one-dimensional systems cannot exhibit phase transitions whenever the interactions are of finite range. Later, van Kampen showed that the van der Waals–Maxwell behavior could be obtained in more general cases. Finally, Lebowitz and Penrose in 1966 further generalized the class of potentials and showed rigorously how the van der Waals–Maxwell theory of phase transitions results directly from the classical partition function; together with Lieb, they generalized the result to quantum systems. Hence, the "van der Waals problem" can now be added to the very small list of exactly solved problems of statistical mechanics.

In this section we present a simplified sketch of the work of Lebowitz and Penrose. As we do not aim here at full mathematical rigor, we leave out many technical details, for which the reader is directed to the original paper.

We first specify the potential $V_{ij}^{tot} = V^{tot}(r_{ij})$ of Eq. (9.3.2) as follows:

$$V^{tot}(r) = V^{HS}(r) + V(r;\gamma) \tag{9.4.1}$$

where $V^{HS}(r)$ is the hard-core potential as before, and the long-range attractive part is of the form:

$$V(r;\gamma) = \gamma^d \varphi(\gamma r) \tag{9.4.2}$$

where d is the number of dimensions of the space. We assume that

$$\lim_{\gamma\to 0} V(r;\gamma) = 0 \qquad \text{for all } r \tag{9.4.3}$$

But on the other hand:

$$\frac{1}{2}\int d\mathbf{r}\, V(r;\gamma) = \frac{1}{2}\int d\mathbf{x}\, \varphi(x) = -\alpha \tag{9.4.4}$$

The constant α, which because of (9.4.2) is independent of γ and of d, is assumed finite and positive (see Fig. 9.4.1). $V(r;\gamma)$ is called a *Kac potential* after its inventor's name.

The method of Lebowitz and Penrose consists in finding upper and lower bounds to the partition function $Z(\mathcal{V}, N; \gamma)$ and to the free energy* $A(\mathcal{V}, N; \gamma)$, both of which depend on the parameter γ. These bounds are obtained by using an extension of the method described in Section 4.7. Assume that the system is enclosed in a cubic box of volume $\mathcal{V}$, which is subdivided as shown in Fig. 9.4.2. We thus define a network of M congruent cubic cells ω_i, of side $(s+t)$. Since these cubes completely fill the volume, we have

$$\mathcal{V} = M(s+t)^d \tag{9.4.5}$$

We assume that the side of each cell is much smaller than the range of the attractive forces γ^{-1}, but much larger than the hard-core diameter d_0; moreover, we assume that the range γ^{-1} is much smaller than the side of the system's container. Hence we have the following ordering of the characteristic lengths of our problem:

$$d_0 \ll (s+t) \ll \gamma^{-1} \ll \mathcal{V}^{1/d} \tag{9.4.6}$$

* The dependence on the temperature is not written down explicitly.

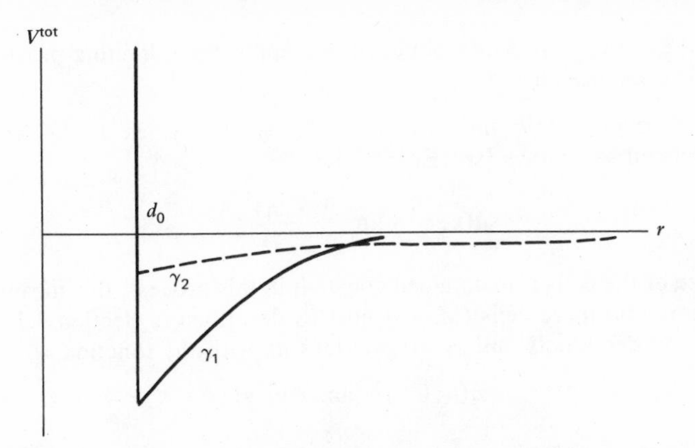

Figure 9.4.1. A typical Kac potential, for two different values of the parameter γ: $\gamma_2 < \gamma_1$.

Inside each cell ω we construct a cell ω'; any two neighboring cells ω' are thus separated by a corridor of width t. We assume that the width t is larger than the hard-core diameter d_0:

$$t > d_0 \tag{9.4.7}$$

Moreover, we require that when we go to the limit $s \to \infty$ (see below), the relative importance of the corridors goes to zero, that is:

$$\lim_{s\to\infty} \frac{t}{s} = 0 \tag{9.4.8}$$

We will see below that this geometrical construction is indeed a clever one.

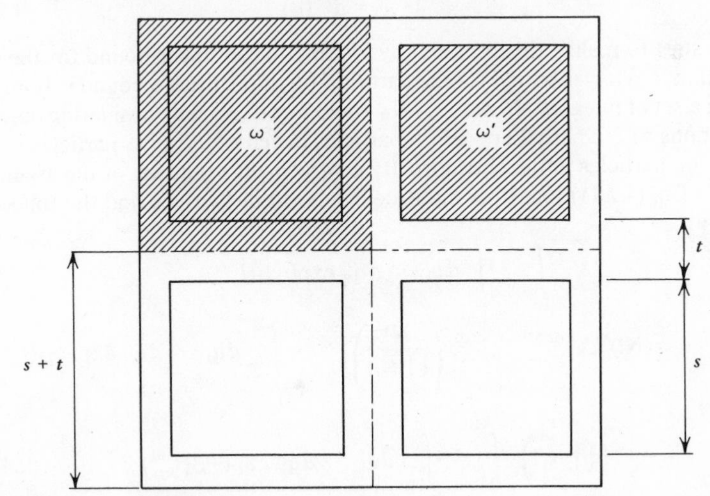

Figure 9.4.2. Construction of the cells for the derivation of the van der Waals equation.

When the bounds on A are obtained, we apply three limiting processes in a well-defined succession:

1. *The thermodynamic limit*, which we know well, gives us the free-energy density defined as follows [see Eq. (4.7.1)]:

$$\tilde{a}(n;\gamma)=\lim_{\mathcal{V}\to\infty}\frac{A(\mathcal{V},n\mathcal{V};\gamma)}{\mathcal{V}} \tag{9.4.9}$$

As the size of the cells is maintained constant in this process, this means that we pile up more and more cells ($M\to\infty$), just as described in Section 4.7.

2. *The van der Waals limit* $\gamma\to 0$ provides us with the function

$$\tilde{a}(n;0+)=\lim_{\gamma\to 0}\tilde{a}(n;\gamma) \tag{9.4.10}$$

3. *The limit* $\omega\to\infty$ finally suppresses the dependence of the result on the cell size.

It is important to note that the order of these limits is not immaterial. This order is fixed by the physical requirement (9.4.6). Actually, had we done the van der Waals limit for a finite system, that is, before the T-limit, then because of (9.4.3) the attractive potential would simply have disappeared from the configuration integral and we would have found the free energy of a *reference system of hard spheres:*

$$\lim_{\gamma\to 0}A(\mathcal{V},N;\gamma)=A^{HS}(\mathcal{V},N)\qquad \mathcal{V},N\text{ finite} \tag{9.4.11}$$

$$\lim_{\mathcal{V}\to\infty}\lim_{\gamma\to 0}\frac{A(\mathcal{V},N;\gamma)}{\mathcal{V}}=\lim_{\mathcal{V}\to\infty}\frac{A^{HS}(\mathcal{V},n\mathcal{V})}{\mathcal{V}}$$
$$=\tilde{a}^{HS}(n) \tag{9.4.12}$$

We now start to realize this program by looking for an upper bound for the free energy, that is, a lower bound for the partition function. Such a bound is found by choosing a set of integers $N_1, N_2, \ldots, N_M$ adding up to N, and considering only the contributions to Z where there are N_1 particles in cell $\omega_1', \ldots, N_M$ particles in ω_M': there are no particles in the corridors. By a very slight extension of the argument leading to Eq. (4.7.11) (considering M cells instead of two) we find the following inequality:

$$Z(\mathcal{V},n\mathcal{V};\gamma)=(N!)^{-1}\Lambda^{-dN}\int_{\mathcal{V}}\cdots\int_{\mathcal{V}}d\mathbf{q}_1\cdots d\mathbf{q}_N\exp(-\beta H'^{\text{tot}})$$

$$\geqslant(N!)^{-1}\Lambda^{-dN}\sum_{N_1}\cdots\sum_{\substack{N_M\\(\Sigma N_j=N)}}\left(\frac{N!}{\prod_j N_j!}\right)\int_{\omega_1'}\cdots\int_{\omega_M'}d\mathbf{q}_1\cdots d\mathbf{q}_N\exp(-\beta H'^{\text{tot}})$$

$$\geqslant\Lambda^{-dN}\left(\prod_j N_j!\right)^{-1}\int_{\omega_1'}\cdots\int_{\omega_M'}d\mathbf{q}_1\cdots d\mathbf{q}_N\exp(-\beta H'^{\text{tot}}) \tag{9.4.13}$$

where the first N_1 d-fold integrations are taken over the region ω_1', the next N_2 over ω_2', and so on. The second inequality is obtained by noting that in the sum over N_j's, all the terms are positive, hence the inequality is strengthened by keeping only one of the terms, corresponding to an (arbitrary) *fixed* choice of N_j's.

We now write

$$H'^{\text{tot}} = H'_1 + H'_2$$

where H'_1 is the total potential energy for pairs of particles within the same cell, and H'_2 is the contribution of particles that are in different cells; we also call H'_{max} an upper bound of H'_2, thus

$$Z \geqslant \prod_j [(N_j!)^{-1}\Lambda^{-dN_j}] \int_{\omega'_1} \cdots \int_{\omega'_M} d\mathbf{q}_1 \cdots d\mathbf{q}_N \exp(-\beta H'_1 - \beta H'_{\text{max}})$$

$$= \left[\prod_j Z(\omega', N_j; \gamma)\right] \exp(-\beta H'_{\text{max}}) \tag{9.4.14}$$

Here $Z(\omega', N_j; \gamma)$ is the partition function for N_j particles in a cube ω'. We further note that H'_2 only contains attractive interactions, because of (9.4.7). Hence, calling $\mathbf{k}_{ij}$ the vector from the center of cell ω_i to the center of ω_j, we have

$$H'_2 \leqslant \frac{1}{2} \sum_{i \neq j}\sum N_i N_j V_{\text{max}}(\mathbf{k}_{ij})$$

with

$$V_{\text{max}}(\mathbf{k}_{ij}) = \max_{\substack{\mathbf{x} \in \omega_i \\ \mathbf{y} \in \omega_j}} V(\mathbf{x} - \mathbf{y}; \gamma) = \max_{\mathbf{r} \in \omega_0} V(\mathbf{k}_{ij} - 2\mathbf{r}; \gamma) \tag{9.4.15}$$

Using all these partial results we find finally

$$A(\mathcal{V}, N; \gamma) \leqslant \sum_j A(\omega', N_j; \gamma) + \frac{1}{2} \sum_{i \neq j}\sum N_i N_j V_{\text{max}}(\mathbf{k}_{ij}) \tag{9.4.16}$$

This bound is valid whatever the choice of $N_1, \ldots, N_M$; in particular it holds in the (most unfavorable) case where all N_j are equal:

$$N_1 = N_2 = \cdots = N_M = \frac{N}{M} = n(s+t)^d \tag{9.4.17}$$

In that case (9.4.16) becomes

$$\mathcal{V}^{-1} A(\mathcal{V}, n\mathcal{V}; \gamma) \leqslant M \mathcal{V}^{-1} A(\omega', n(s+t)^d; \gamma) + \tfrac{1}{2} n^2 (s+t)^{2d} \mathcal{V}^{-1} \sum_{i \neq j}\sum V_{\text{max}}(\mathbf{k}_{ij}) \tag{9.4.18}$$

The following lemma can be understood intuitively or proved rigorously:

$$\lim_{\mathcal{V} \to \infty} [\mathcal{V}^{-1}(s+t)^d] \sum_{i \neq j}\sum V_{\text{max}}(\mathbf{k}_{ij}) = \lim_{\mathcal{V} \to \infty} M^{-1} \sum_{i \neq j}\sum V_{\text{max}}(\mathbf{k}_{ij})$$

$$= \sum_{\mathbf{k}}{}' \, V_{\text{max}}(\mathbf{k})$$

The sum in the rhs is over the complete infinite lattice of vectors $\mathbf{k}_{ij}$ except $\mathbf{k} = 0$. The thermodynamic limit of (9.4.18) now becomes

$$\tilde{a}(n; \gamma) \leqslant (s+t)^{-d} A(\omega', n(s+t)^d; \gamma) + \tfrac{1}{2} n^2 (s+t)^d \sum_{\mathbf{k}}{}' \, V_{\text{max}}(\mathbf{k}) \tag{9.4.19}$$

In order to do the van der Waals limit $\gamma \to 0$, we note that, using (9.4.2) and (9.4.15),

$$(s+t)^d \sum_{\mathbf{k}}' V_{\max}(\mathbf{k}) = 2^{-d} \sum_{\mathbf{n}}' \Delta \max_{\mathbf{x} \in \Delta_n} \varphi(\mathbf{x}) \tag{9.4.20}$$

where the sum goes over all nonzero vectors **n** with integer components, and Δ_n stands for the cube of side $2\gamma(s+t)$ centered at the lattice point **n** and of volume Δ. As $\gamma \to 0$, the sides of these cubes become smaller, but there are more of them. By using the familiar definition of the Riemann integral it can then be shown that

$$\lim_{\gamma \to 0} (s+t)^d \sum_{\mathbf{k}}' V_{\max}(\mathbf{k}) = \int d\mathbf{x}\, \varphi(x) = -2\alpha \tag{9.4.21}$$

We also note that in the first term on the rhs of (9.4.19), the limit $\gamma \to 0$ is taken at a fixed volume ω'; hence, by (9.4.11) the result is the free energy of the reference system, thus:

$$\tilde{a}(n; 0+) \leq (s+t)^{-d} A^{HS}(\omega', n(s+t)^d) - n^2\alpha \tag{9.4.22}$$

The final limiting process, $s \to \infty$, is simple; using (9.4.8) and (9.4.12) we get

$$\tilde{a}(n; 0+) \leq \tilde{a}^{HS}(n) - n^2\alpha \tag{9.4.23}$$

This result can be further strengthened in a nontrivial way. To do so we use the general property (4.7.21) that the free-energy density, for every potential with finite γ, is a *convex function* of n. As the limit of a sequence of convex functions is also convex, $\tilde{a}(n; 0+)$ is convex. On the other hand, $\tilde{a}^{HS}(n)$ and αn^2 are both convex, but their difference is not necessarily so. We now define the *convex envelope:* CE$\{g(n)\}$ of a function $g(n)$, as the maximal function that is convex and does not exceed g. Analytically:

$$\mathrm{CE}\{g(n)\} = \max \phi(n) \qquad \text{for each value of } n$$

with

$$\phi(n) \leq g(n)$$

and

$$\phi(n) \text{ convex} \tag{9.4.24}$$

The concept of a convex envelope is most easily grasped from a graph (Fig. 9.4.3). If $g(n)$ is convex, CE$\{g(n)\} = g(n)$. If not, then we draw a double tangent in the region where it is concave; CE$\{g(n)\}$ consists of segments of $g(n)$ (where it is convex) and of segments of double tangents.

Returning to our argument, we note that $\tilde{a}(n; 0+)$ must satisfy (9.4.23) but also be convex; it then follows that it cannot exceed the convex envelope of the rhs of (9.4.23):

$$\tilde{a}(n; 0+) \leq \mathrm{CE}\{\tilde{a}^{HS}(n) - \alpha n^2\} \tag{9.4.25}$$

This is our final form for the upper bound of the van der Waals free-energy density.

We now consider the lower bound for the free energy, that is, the upper bound for Z. We again distribute the particles among the cells of Fig. 9.4.2, but now we

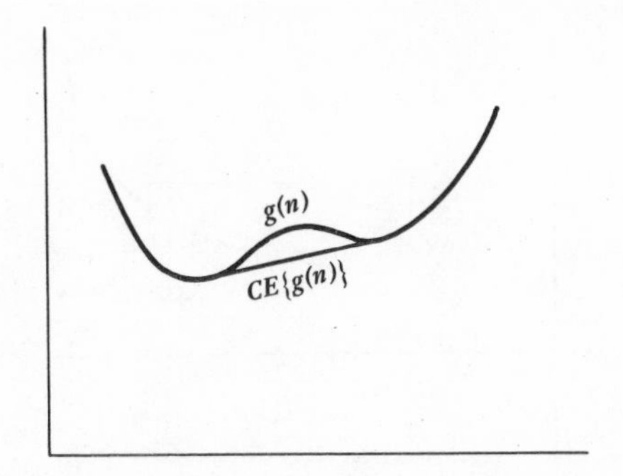

Figure 9.4.3. A function $g(n)$ and its convex envelope CE$\{g(n)\}$.

fill the corridors as well; hence there will be N_1 particles in cell ω_1 (not ω_1'!), N_2 in ω_2, and so on. We may then write

$$Z(\mathcal{V}, N; \gamma) = \sum_{N_1, \ldots, N_M} \cdots \sum Z(N_1, \ldots, N_M)$$

$$\leq \left(\frac{(N+M-1)!}{N!\,(M-1)!}\right) \max_{N_1, \ldots, N_M} Z(N_1, \ldots, N_M) \tag{9.4.26}$$

We now subdivide the total interaction energy in a different way:

$$H'^{\text{tot}} = W_1 + W_2 + H'$$

where W_1 is the contribution to H'^{tot} from hard-core interactions of particles that are in the same cell, W_2 the hard-sphere interactions of particles that are in different cells, and H' is the total attractive interaction. If $W_{2,\min}$ and $H'_{\min}$ are lower bounds on W_2 and H', respectively, a calculation similar to the previous one yields [see also Eq. (4.7.9)]

$$Z(N_1, \ldots, N_M) \leq \left\{\prod_i Z^{\text{HS}}(\omega, N_i)\right\} \exp[-\beta W_{2,\min} - \beta H'_{\min}] \tag{9.4.27}$$

where $Z^{\text{HS}}(\omega, N_i)$ is the partition function of N_i hard spheres in a volume ω. We shall not go into the details of the estimation of $W_{2,\min}$ and $H'_{\min}$, nor shall we do the three limits since no new argument enters in these calculations. We simply quote the very remarkable final result

$$\tilde{a}(n; 0+) \geq \text{CE}\{\tilde{a}^{\text{HS}}(n) - \alpha n^2\} \tag{9.4.28}$$

When this is combined with (9.4.25) we obtain the *equality:*

$$\tilde{a}(n; 0+) = \text{CE}\{\tilde{a}^{\text{HS}}(n) - \alpha n^2\} \tag{9.4.29}$$

It is now very easy to show that this form of the free energy implies the vdW–M equation of state. Indeed, from Table 4.4.1 we find that the pressure can be expressed as follows:

$$P(n) = n^2\left(\frac{\partial a}{\partial n}\right) = n^2 \frac{\partial n^{-1}\tilde{a}(n)}{\partial n} = n\frac{\partial \tilde{a}(n)}{\partial n} - \tilde{a}(n) \tag{9.4.30}$$

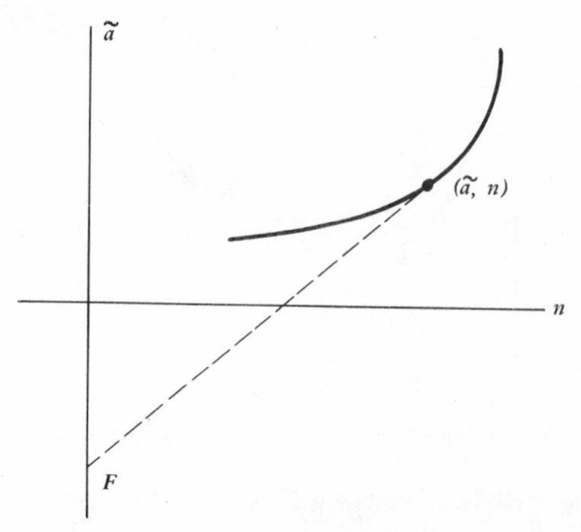

Figure 9.4.4. Geometrical interpretation of Eq. (9.4.29).

This expression, valid for finite γ, can be shown to hold also in the limit $\gamma \to 0$. This equation has a very simple geometric interpretation. Consider a curve represented by a function $\tilde{a}(n)$ with the tangent at the point n; this tangent intercepts the $\tilde{a}$ axis at a point F (see Fig. 9.4.4). The equation of the tangent is

$$\frac{\tilde{a}(n)-F}{n-0}=\frac{\partial \tilde{a}(n)}{\partial n}$$

or

$$-F=n\frac{\partial \tilde{a}(n)}{\partial n}-\tilde{a}(n)$$

Hence the pressure is simply $P=-F$.

We now see that for all temperatures for which $\tilde{a}^{\mathrm{HS}}(n)-\alpha n^2$ is a convex function the pressure is given by

$$\begin{aligned} P_{\mathrm{vdW}}(n) &= \left(n\frac{\partial}{\partial n}-1\right)\tilde{a}^{\mathrm{HS}}(n)-\alpha n^2 \\ &\equiv P^{\mathrm{HS}}(n)-\alpha n^2, \qquad T>T_c \end{aligned} \tag{9.4.31}$$

We interpret this domain as the one above the critical point. When $\tilde{a}^{\mathrm{HS}}(n)-\alpha n^2$ becomes concave, it must be replaced by its convex envelope, hence its graph has a straight segment of double tangent. By the previous geometrical interpretation, the pressure is constant for all densities lying between the two points of contact. The chemical potential $\mu=[\tilde{a}(n)+P(n)]/n=\partial\tilde{a}(n)/\partial n$ is also constant in this domain. Hence, this is clearly the coexistence region characteristic of a phase transition (Fig. 9.4.5). The two points of contact of the double tangent are thus identified with the densities n_L, n_G of the coexisting phases. It is also easy to check, using Eq. (9.4.31) as a definition of the symbol P_{vdW} below T_c:

$$\int_{n_G}^{n_L} dn^{-1}[P_{\mathrm{vdW}}(n)-P_{\mathrm{sat}}]=\left\{\frac{d}{dn}[\tilde{a}^{\mathrm{HS}}(n)-\alpha n^2]\right\}_{n_G}^{n_L}=0$$

which is precisely the Maxwell rule.

Lebowitz and Penrose have also shown that in the coexistence region the pair distribution function is a linear superposition of two hard-sphere distributions evaluated at $n = n_L$ and $n = n_G$, respectively. This is a very direct and elegant proof of the simultaneous presence of two phases in the coexistence region. Hence we have *derived* quite directly the van der Waals equation of state *together* with the Maxwell rule from the partition function. The only feature left implicit is the determination of the critical temperature, which appears as the value at which the function $\tilde{a}^{HS} - \alpha n^2$ starts developing a hump. This clearly depends on the hard-sphere free energy, which is not calculated in the theory but left implicit as a reference state.

We stress again the fact that the vdW–M equation appears as an *exact* property of a system of particles interacting through a Kac potential, in the limit $\gamma \to 0$. No expansion in powers of the density or of any other parameter has been used. It can be argued that a Kac potential is not realistic; the extension of these results to

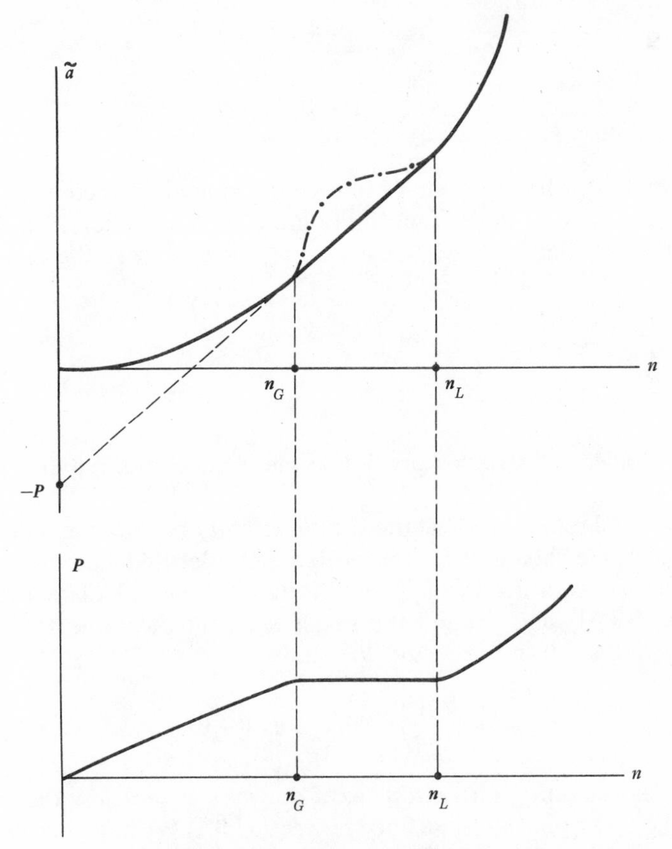

Figure 9.4.5. Construction of the pressure graph from the free-energy density graph.

more physical potentials is still an open problem. However, the detailed rigorous proof of the existence of a first-order phase transition, even for a model system, must be viewed as a great landmark in the development of statistical mechanics.

9.5. MACROSCOPIC PROPERTIES NEAR THE CRITICAL POINT

We now examine in some detail the behavior of various thermodynamic functions in the neighborhood of the critical point, as predicted by the classical theories studied in this section.

We start with the vdW equation. We introduce the following convenient variables:

$$\theta = \frac{T - T_c}{T_c}$$

$$\upsilon = \frac{n - n_c}{n_c} \tag{9.5.1}$$

$$p = \frac{P - P_c}{P_c}$$

We may first ask what is the shape of the *critical isotherm near the critical point*. To see this we start from the reduced vdW equation (9.3.11) and expand all variables as $\tilde{P} = 1 + p$, $\tilde{n} = 1 + \upsilon$, $\tilde{T} = 1 + \theta$ with the result:

$$p \approx \tfrac{3}{2}\upsilon^3 + \cdots + \theta(4 + 2\upsilon + \cdots)$$

On the critical isotherm $\theta = 0$, hence

$$p \sim \upsilon^3 \qquad \text{as } \upsilon \to 0 \tag{9.5.2}$$

This is a precise and detailed prediction that can be checked experimentally.

We have said repeatedly that the *compressibility* becomes infinite at the critical point; we should, however, need a more detailed statement about the precise way in which this happens. To find this we calculate $\tilde{n}(\partial \tilde{P}/\partial \tilde{n})$ from Eq. (9.3.11) and expand the result again in the same fashion as above, setting $\upsilon = 0$ in the result. We find

$$\chi_T \equiv n\left(\frac{\partial P}{\partial n}\right) \sim \theta^{-1} \qquad \text{as } \theta \to 0+ \tag{9.5.3}$$

An important quantity, characteristic of the substance below the critical point is the *order parameter* defined in Section 9.1. We saw that its main property is to vanish above T_c; we may ask how the order parameter $n_L - n_G$ approaches zero as $T \to T_c$ from below. We can calculate this quantity from the vdW–M equation (9.3.13) and find

$$n_L - n_G \sim (-\theta)^{1/2} \qquad \text{for } \theta \to 0- \tag{9.5.4}$$

Consider finally the *specific heat* at constant volume, c_V evaluated at the critical density. It has been known for a long time that at the critical point this quantity has an odd behavior. The vdW theory predicts a finite jump in c_V. More precisely, if we denote by $c_V^0 = \frac{3}{2}k_B$ the specific heat of an ideal gas, we find

$$\begin{aligned} c_V - c_V^0 &= \tfrac{9}{2}k_B(1 - \tfrac{28}{25}\theta + \cdots), \qquad && T < T_c \\ &= 0, && T > T_c \end{aligned} \tag{9.5.5}$$

The set of relations (9.5.2)–(9.5.5) provides a quite detailed description of the neighborhood of the critical point. The behavior described by these equations is characteristic of the vdW–M theory. It can be checked experimentally; but we postpone the comparison to the next chapter.

We now turn to the Weiss theory of ferromagnetism, based on the magnetic equation of state (9.2.12). We will see that, in spite of the apparent dissimilarity between this equation and the vdW–M equation, the behavior predicted by the Weiss theory near the critical point is identical with the latter. To illustrate these calculations, we consider the case of spin $S = \frac{1}{2}$, in which case the Brillouin function reduces to the hyperbolic tangent:

$$\frac{M}{M_0} = \tanh\left(\frac{1}{2}\frac{\bar{\mu}\mathcal{H}}{k_B T} + \frac{M/M_0}{T/T_c}\right) \tag{9.5.6}$$

If this equation is expanded around $M = 0$ and $\mathcal{H} = 0$, we find, after some straightforward algebra, the following equation in terms of the variables $m = M/M_0$, $h = \bar{\mu}\mathcal{H}/2k_BT$ and $\tilde{T} = T/T_c$:

$$h + O(h^3) = m(1 - \tilde{T}^{-1}) + m^3[(3\tilde{T}^3)^{-1} + (1 - \tilde{T}^{-1})\tilde{T}^{-1}] + O(m^5) \tag{9.5.7}$$

From here we immediately find the behavior of the critical isotherm ($\tilde{T} = \tilde{1}$):

$$\tfrac{1}{2}\beta_c\bar{\mu}\mathcal{H} \approx \frac{1}{3}\left(\frac{M^3}{M_0^3}\right)$$

or

$$\mathcal{H} \sim M^3 \qquad \text{as } M \to 0 \tag{9.5.8}$$

an equation analogous to (9.5.2). (Remember that the pressure is the analog of the magnetic field, and the density corresponds to the magnetization.) Similarly, we find the behavior of the isothermal magnetic susceptibility:

$$\kappa_T \sim \theta^{-1} \qquad \text{as } \theta \to 0+ \tag{9.5.9}$$

The order parameter, that is, the zero-field magnetization, behaves below the critical point as

$$M \sim (-\theta)^{1/2} \qquad \text{as } \theta \to 0- \tag{9.5.10}$$

As for the specific heat at constant field, the Weiss theory also predicts a finite jump. Hence, as stated above, we see that all corresponding quantities behave in the same way near the critical point in both so-called "classical" theories. This is not an accident. Indeed, the main physical idea underlying both models is the existence of *long-range forces.* Actually, Kac showed very elegantly that if we consider a simple lattice with one-dimensional spins (Ising model, see Section 10.2) in which all spins interact equally whatever their distance, we obtain precisely the Weiss equation of state. Therefore the vdW and the Weiss theories are, so to say, "isomorphic." The analogy of the two theories also appears very clearly in Landau's theory of phase transitions. This author starts from the free energy and expands it near the critical point: making similar assumptions one gets either vdW or Weiss. We have no space for going into Landau's theory, excellent accounts of which can be found elsewhere (see, however, Section 10.4).

Knowing that the model of infinite-range forces underlying both classical theories is not very realistic, we may expect that the laws (9.5.2)–(9.5.5) and (9.5.8)–(9.5.10) should not be exact. More realistic theories should predict a different behavior near the critical point. Also, the experiments give different types of power laws for real systems, as will be seen in Section 10.1. It is very convenient to check any theory by the predictions it makes for these asymptotic laws, valid near the critical point. We therefore define six *critical exponents*, denoted by the letters α, α', β, γ, γ', and δ, which by now have become standard.

The coefficient α characterizes the behavior of the *specific heat* (both for fluids and for magnetic systems):

$$\begin{aligned} c_V &\sim \theta^{-\alpha} \\ c_H &\sim \theta^{-\alpha}, \qquad \theta \to 0+ \end{aligned} \tag{9.5.11}$$

These powers laws characterize the behavior as $T \to T_c$ from above; one may also introduce a coefficient α', characterizing the behavior of the specific heat when $T \to T_c$ from below:

$$\begin{aligned} c_V &\sim (-\theta)^{-\alpha'} \\ c_H &\sim (-\theta)^{-\alpha'}, \qquad \theta \to 0- \end{aligned} \tag{9.5.12}$$

α and α' are not necessarily equal.

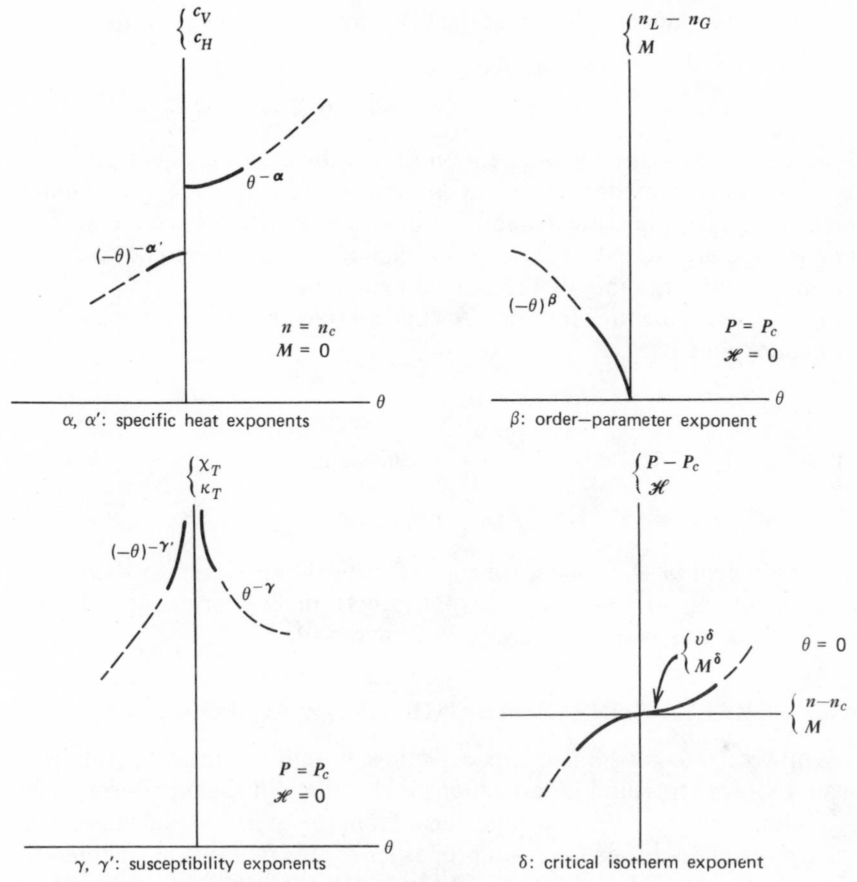

Figure 9.5.1. The macroscopic critical exponents.

The coefficient β characterizes the behavior of the *order parameter:*

$$\begin{aligned} n_L - n_G &\sim (-\theta)^{\beta} \\ M &\sim (-\theta)^{\beta}, \qquad \theta \to 0- \end{aligned} \tag{9.5.13}$$

This law is, of course, meaningful only below the critical point.

The coefficients γ and γ′ characterize the divergence of the *compressibility* or *susceptibility* as the critical point is approached from above or from below, respectively:

$$\begin{aligned} \chi_T &\sim \theta^{-\gamma} \\ \kappa_T &\sim \theta^{-\gamma}, \qquad \theta \to 0+ \\ \chi_T &\sim (-\theta)^{-\gamma'} \\ \kappa_T &\sim (-\theta)^{-\gamma'}, \qquad \theta \to 0- \end{aligned} \tag{9.5.14}$$

Finally, the exponent δ describes the behavior of the *critical isotherm:*

$$\begin{aligned} P - P_c &\sim (n - n_c)^\delta \\ \mathcal{H} &\sim M^\delta, \qquad \theta = 0, n \to n_c, M \to 0 \end{aligned} \tag{9.5.15}$$

We, moreover, make the convention of ascribing an exponent equal to *zero* whenever we have either a *finite discontinuity* or a *logarithmic divergence* near the critical point. It can be shown that this convention is mathematically sound. Figure 9.5.1 should be a mnemonic aid for visualizing the meaning of the critical exponents.

The classical theories thus can be characterized by the following set of critical exponents:

$$\begin{aligned} \alpha = \alpha' &= 0 \\ \beta &= \tfrac{1}{2} \qquad \text{(classical)} \\ \gamma = \gamma' &= 1 \\ \delta &= 3 \end{aligned} \tag{9.5.16}$$

The main goal of any modern theory of critical phenomena is the study and evaluation of the critical exponents from first principles. These modern theories will be reviewed in Chapter 10.

9.6. CORRELATIONS NEAR THE CRITICAL POINT

We now wish to get some microscopic insight into the critical point. The main tool for structural investigations is the study of the pair correlation function, both from the theoretical and from the experimental sides. We have already described the main properties of this function in Chapters 7 and 8. Let us recall the equation (7.2.12), which relates the pair correlation to the compressibility:

$$nk_B T \chi_T(T) - 1 = n \int d\mathbf{r}\, \nu_2(r; T) \tag{9.6.1}$$

Our present notation stresses the fact that the equilibrium pair correlation depends parametrically on the temperature. (It also depends on the density, but this is left implicit.) In terms of the Fourier transform $\tilde{\nu}_k(T)$ defined by Eq. (7.5.22), this equation reads

$$nk_B T \chi_T(T) - 1 = n\tilde{\nu}_0(T) \tag{9.6.2}$$

These two simple equations tell us already the most important features of the critical behavior. Indeed, the salient macroscopic fact at the critical point is the infinite value of the compressibility: $\chi_T(T_c) = \infty$. This implies that at the critical temperature, the integral on the right side of (9.6.1) must diverge. But we know that for realistic potentials with a hard core, $\nu_2(r)$ behaves regularly for short distances; hence we are led to the conclusion that $\nu_2(r; T_c)$ must develop a very long tail, which makes the integral divergent. Thus, a system at the critical point is characterized by *correlations of infinite range*, even if the interactions are of finite range. In other words, in the critical region every molecule feels the action of a large number of other molecules; this influence is not direct (because the interactions are of finite range) but proceeds via a long chain of neighboring molecules that tend to act coherently. Another way of expressing this is through Eq. (9.6.2): the *Fourier transform of the pair correlation at zero wave vector* (i.e., infinite wavelength) *is infinite* at the critical point.

As usual when infinities enter a theory, we try to shift our attention to some related quantity that behaves more smoothly and can, therefore, be handled more conveniently. We already used this idea in the cluster expansion (Section 6.4) when we went over from the potential V_{ij} to the function f_{ij}, which is finite even when the potential has an infinite hard core. In the present problem we direct our attention to the *direct correlation function* $C(r; T)$ defined by Eq. (7.5.21) or its Fourier transform $\tilde{C}_{\mathbf{k}}(T)$ defined by (7.5.24), or equivalently:

$$\tilde{C}_{\mathbf{k}}(T) = \frac{\tilde{\nu}_{\mathbf{k}}(T)}{1 + n\tilde{\nu}_{\mathbf{k}}(T)} \tag{9.6.3}$$

We see immediately that even at the critical temperature, when $\tilde{\nu}_0(T) \to \infty$, the direct correlation function at zero wave vector remains finite:

$$\tilde{C}_0(T_c) = n^{-1} \tag{9.6.4}$$

Actually, this function was first introduced in physics in 1914 by Ornstein and Zernike precisely in connection with their theory of critical phenomena, which we now briefly review.

The basic assumption of Ornstein and Zernike (O–Z) is that $\tilde{C}_{\mathbf{k}}(T)$ *is an even analytic function of $k(=|\mathbf{k}|)$ in the neighborhood of $k = 0$, for all temperatures T, including T_c*. Hence we may write

$$\tilde{C}_{\mathbf{k}}(T) = \tilde{C}_0(T) + \tilde{c}_2(T)k^2 + O(k^4) \tag{9.6.5}$$

Moreover, the O–Z approximation consists in truncating this series after the quadratic term. It is customary to introduce the parameter R^2, having dimensions of the square of a length (R is sometimes called the Debye persistence length):

$$R^2 = -n\tilde{c}_2(T) \sim \int dr\, r^2 C(r; T) \tag{9.6.6}$$

We also introduce the important parameter ξ:

$$\xi^{-2} = R^{-2}[1 - n\tilde{C}_0(T)] \tag{9.6.7}$$

From (9.6.4) we see that $\xi^{-1} \to 0$ as $T \to T_c$.

In order to check this theory against experiments we calculate the *structure factor* a_k, which is directly related to the observable intensity of radiation scattered by the fluid, as shown in Section 8.1. From Eqs. (8.1.5) and (9.6.3) we get

$$a_k(T) = 1 + n\tilde{\nu}_k(T) = [1 - n\tilde{C}_k(T)]^{-1} \tag{9.6.8}$$

and from (9.6.5)–(9.6.7):

$$a_k(T) \approx \frac{R^{-2}}{\xi^{-2} + k^2} \tag{9.6.9}$$

This equation tells us that in the O–Z approximation the scattering intensity has a Lorentzian shape. This can be checked by plotting the inverse measured values of the structure factor against k^2 ("OZD plot"): the result should be a straight line. This prediction is quite well verified for some systems, such as argon (Fig. 9.6.1).

The pair correlation is obtained by inverse Fourier transformation of (9.6.8).* In three-dimensional space a Lorentzian is the Fourier transform of a Debye function, just as in the plasma problem of Section 6.5:

$$\nu_2(r; T) \sim R^{-2} \frac{e^{-r/\xi}}{r} \tag{9.6.10}$$

We now clearly see that ξ measures the range of the correlations. It is customary to call ξ the *correlation length:* it is a function of temperature and of density.

From Eq. (9.6.4) we know that $\xi^{-1} \to 0$ as $T \to T_c$, hence the correlation length tends to infinity, as expected. The pair correlation then tends towards $1/r$, which is not integrable; also $\tilde{\nu}_k(T_c) \to k^{-2}$, which diverges at the origin. Besides these results, the O–Z theory does not tell us how ξ diverges in detail. We may, therefore, introduce phenomenologically a new pair of *critical exponents* ν, ν' defined as follows:

$$\begin{aligned} \xi &\sim \theta^{-\nu}, \qquad && \theta \to 0+ \\ &\sim (-\theta)^{-\nu'}, && \theta \to 0- \end{aligned} \tag{9.6.11}$$

Although we do not know the value of ν, we can relate it, within the O–Z theory, to another exponent. Indeed, combining Eqs. (9.6.7), (9.6.3), and

* Here and further on in this chapter we shall always delete an irrelevant "self-correlation" term proportional to $\delta(r)$ and coming from the term "1" in Eq. (9.6.8).

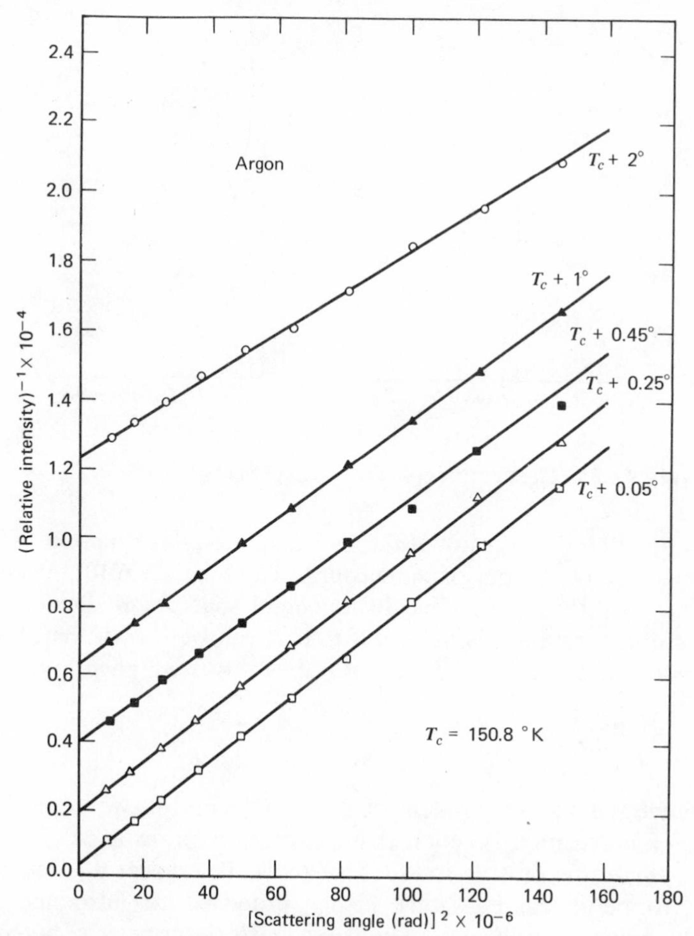

Figure 9.6.1. Experimental OZD graph for argon. [From J. E. Thomas and P. W. Schmidt, *J. Chem. Phys.* **39**, 2506 (1963).]

(9.6.2) we get

$$\left(\frac{R}{\xi}\right)^2 = [1 + n\tilde{v}_0(T)]^2 = (nk_B T\chi_T)^{-1} \tag{9.6.12}$$

But the behavior of the compressibility is measured by the critical exponent γ, Eq. (9.5.14); hence

$$\xi^{-2} \sim \theta^{\gamma}$$

and therefore

$$\begin{aligned} \nu &= \tfrac{1}{2}\gamma \\ \nu' &= \tfrac{1}{2}\gamma' \qquad \text{(O–Z)} \end{aligned} \tag{9.6.13}$$

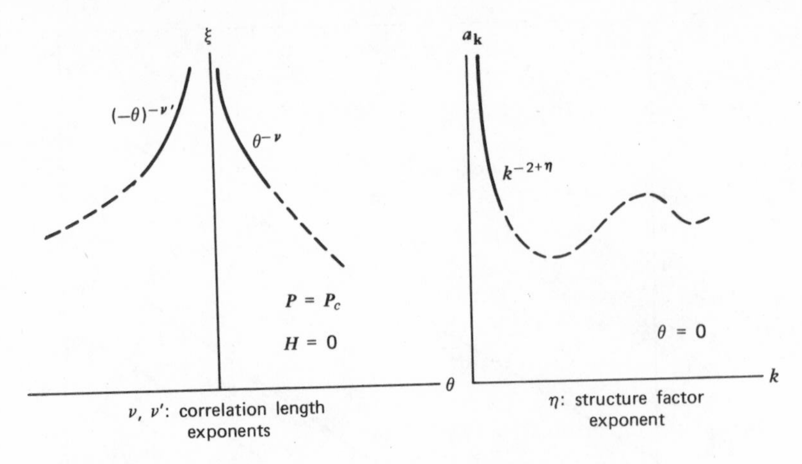

Figure 9.6.2. The microscopic critical exponents.

Equation (9.6.9) is a result of the O–Z analyticity assumption: it is valid in spaces of any number of dimensions. The result (9.6.10), however, is only true in three dimensions. In a general space of d dimensions the Fourier transformation is more difficult. It can be shown that, for large values of r and at the critical point (i.e., $\xi^{-1}=0$) we have

$$\begin{aligned} \nu_2(r;\, T_c) &\sim r^{-(d-2)}, \qquad d \geqslant 3 \\ &\sim \ln r, \qquad d = 2 \end{aligned} \tag{9.6.14}$$

Here we see a serious failure of the O–Z theory in two dimensions: It predicts a correlation function that increases with distance! The theory cannot, therefore, be true for such systems. To correct this failure (and others to be discussed later) Fisher proposed to introduce a new phenomenological critical exponent η, which describes the behavior of the *structure factor* at the critical temperature and for small wave vector:

$$a_k(T_c) \sim k^{-2+\eta}, \qquad k \to 0 \tag{9.6.15}$$

Correspondingly, we get

$$\nu_2(r;\, T_c) \sim r^{-(d-2+\eta)}, \qquad r \to \infty \tag{9.6.16}$$

Clearly, $\eta = 0$ corresponds to the O–Z theory. We should note that if $\eta \neq 0$, the relations (9.6.13) are no longer valid. We have now added to our list of critical exponents the numbers ν, ν', η, which could be called the *microscopic critical exponents* because they describe the behavior of the correlation function (see Fig. 9.6.2).

All the results described in this section can be extended to the magnetic case as well. The role of the pair correlation function is then played by the two-spin correlation function, which has properties quite similar to $\nu_2(r)$. We do not enter this problem here as we have not yet discussed the magnetic systems from a microscopic point of view (see Chapter 10).

BIBLIOGRAPHICAL NOTES

The following *general references* cover the matters discussed in this chapter:

G. E. Uhlenbeck, Brandeis Summer Institute 1962, *Statistical Physics*, Vol. 3, Benjamin, New York, 1963.

R. Brout, *Phase Transitions*, Benjamin, New York, 1965.

H. E. Stanley, *Introduction to Phase Transitions and Critical Phenomena*, Oxford Univ. Press, Oxford, 1971.

A. Münster (quoted in Chapter 4).

The *Weiss model of ferromagnetism* was introduced in P. Weiss, *J. Phys. Radium* **6**, 667 (1907).

The *van der Waals equation of state* was derived in J. D. van der Waals, Ph.D. thesis, Univ. Leiden, 1873.

The *Kac potential* was introduced in M. Kac, *Phys. Fluids* **2**, 8 (1959). The one-dimensional system of particles with Kac interactions was solved in M. Kac, G. E. Uhlenbeck, and P. C. Hemmer, *J. Math. Phys.* **4**, 216 (1963).

The next important developments in this problem were N. G. van Kampen, *Phys. Rev.* **135**, A362 (1964).

J. Lebowitz and O. Penrose, *J. Math. Phys.* **7**, 98 (1966).

E. Lieb, *J. Math. Phys.* **7**, 1016 (1966) (this is the quantum version of the previous paper).

The *critical exponents* are defined and discussed in

M. Fisher, *Rep. Progr. Phys.* **30**, 615 (1967).

H. E. Stanley (see above).

The *Ornstein–Zernike equation* appeared in L. S. Ornstein and F. Zernike, *Proc. Kon. Akad. Wet.* **17**, 793 (1914).

CHAPTER 10

MODERN THEORIES OF CRITICAL PHENOMENA

10.1. METHODS OF INVESTIGATION OF CRITICAL PHENOMENA

In Chapter 9 we examined the problem of phase transitions and saw that, since the very beginning of molecular physics, the great masters (van der Waals, Ornstein, Zernike, Weiss) grasped some essential features of the phenomenon. The very simple theories they formulated are still a reference and a starting point for modern developments. However, it could not be expected that the "classical theories" (as they are now called) could describe all the aspects of the phase transitions and of the critical phenomena. This became very clear with the important work of Kac, Uhlenbeck, Hemmer, Lebowitz, Penrose, and Lieb, discussed in Section 9.4. It then appeared that the classical theories describe systems of particles interacting through attractive forces of infinitely long range, a feature that is not at all in agreement with the real forces.

Since the early 1960s new facts have accumulated at an increasing rate in this field, thus stimulating an ever-growing interest; today, the problem of phase transitions and critical phenomena has become one of the hottest topics of statistical physics, especially as we do not yet possess the final answer.

The new facts alluded to above are essentially of two kinds. From the theoretical side the most important developments came from the study of "models." These are artificial systems, constructed in a purely theoretical way, which should have two properties: (*a*) They should be sufficiently simple so that they can be solved exactly (in the ideal case) or at least in a very good approximation. (*b*) Their simplicity should not imply triviality. Although the results from these models cannot be expected to be compared directly with experimental results, their study is extremely important in indicating which feature of a real system is determining a given observed phenomenon, and also for studying the overall structure and internal consistency of the theory. Indeed, the exact solutions provide us with a "pure" reference for checking more general, but necessarily approximate, theories; by "pure" we mean that these results are not contaminated by interference with possible extraneous phenomena, as sometimes happens with experimental data. The situation is logically much the same as in the role played by the numerical experiments discussed in Chapter 8.

One of the main results of these developments was to show that the critical exponents obtained from the exact solutions do not agree at all with the classical values. We collected, in Table 10.1.1, a set of known numerical values for the critical exponents for various models, as well as a few typical experimental data. An interesting feature appearing here is the dependence of the critical exponents (and hence of the nature of the underlying phenomena) on the number of dimensions d of the space. This dependence is absent in the classical theories. This is a typical instance of an incompleteness of the classical theories that cannot be detected by experiments (for which d is fixed and equal to 3), but only by theoretical models.

The second source of facts that enriched the field in the last few years is the improvement of experimental techniques and the accumulation of

Table 10.1.1. Values of Critical Exponents

Systems	α	α'	β	γ	γ'	δ	ν	ν'	η
Classical	0 (disc.)	0 (disc.)	$\frac{1}{2}$	1	1	3	$\frac{1}{2}$	$\frac{1}{2}$	0
Models (Exact)									
Ising, $d=2$	0 (log)	0 (log)	$\frac{1}{8}$	$\frac{7}{4}$	$\frac{7}{4}$	15	1	1	$\frac{1}{4}$
Spherical, $d=3$	-1 (cusp)		$\frac{1}{2}$	2		5	1		0
Models (Numerical computations)									
Ising, $d=3$	0.125	0.125	0.313	1.250	1.250	5	0.638		0.041
Heisenberg, $d=3$, $s=\frac{1}{2}$	0(?)		0.345	1.375		5	0.702		0.043
Fluids (Experiments)									
Argon, Ar	<0.4	<0.25	0.33	0.6	1.1				
Xenon, Xe		<0.2	0.350	1.26		4.4			
Carbon dioxide, CO_2		<0.1	0.350	1.26	1.0	4.2			
Helium-4			0.359	1.24					
Ferromagnets (Experiments)									
Iron, Fe	≤0.17	≤0.13	0.35	1.33			0.64		0.07
Nickel, Ni			{0.51, 0.33}	{1.35, 1.29}		4.2			
$YFeO_3$			{0.55, 0.35}	1.33		2.8			

good experimental data. The difficulties in this connection are considerable. For the gas–liquid problem the main technique consists of accurate measurements of pressure, density, and temperature (equation of state) as well as of specific heats. It appears that the power-law behavior allowing one to define critical exponents only applies very close to the critical point, say, $\theta < 10^{-2}$. Even the determination of the critical parameters T_c, P_c, n_c with an accuracy consistent with the needs of the experiment is extremely hard. Hence we need a very precise temperature control, typically of one part in 10^4 of T_c. In addition, owing to the large heat capacities (theoretically divergent c_V), the times necessary for the equilibration of the system are very long (of the order of days). The large compressibility also creates serious problems: The effect of gravity on the apparatus becomes enormous; it creates density gradients that must be taken into account in great detail. For magnetic systems the experimental difficulties in measuring magnetization and susceptibility and in performing neutron-scattering experiments are also quite important and demand great care and skill. We cannot go into the details here, but refer the reader to the original literature and to the review papers.

10.2. MODEL SYSTEMS

Most of the models that were extensively studied in the past fifty years in connection with critical phenomena can be expressed as particular instances of a general Hamiltonian introduced by Stanley. The system considered here is a *discontinuous* one, in contrast with the systems studied in the rest of this book. In other words, the degrees of freedom it describes are not allowed to travel throughout the space, but rather are restricted to occupy the sites of a given lattice in d dimensions. Besides its dimensionality d, the lattice is further characterized by its symmetry (e.g., for $d = 3$, simple cubic, face-centered cubic, body-centered cubic, etc.). On each lattice site sits a "molecule" characterized by a single parameter, its "spin." The spin on site i is represented mathematically by a D-dimensional unit vector $\mathbf{S}_i^{(D)}$. Note that D is not necessarily equal to d. (The four cases that can be drawn on flat paper are shown in Fig. 10.2.1.) We may also consider that an external magnetic field is present, interacting with the spins in the usual way.

The Hamiltonian of the system is now assumed to be of the form:

$$H = -\sum_{i<j}\sum J_{ij}\mathbf{S}_i^{(D)} \cdot \mathbf{S}_j^{(D)} - \mathcal{H} \cdot \sum_i \mathbf{S}_i^{(D)} \tag{10.2.1}$$

The summations are carried out over all the lattice sites. J_{ij} is a coupling parameter depending on the distance between the sites i and j and possibly also on the orientation of the vector $\mathbf{r}_i - \mathbf{r}_j$ if the system is

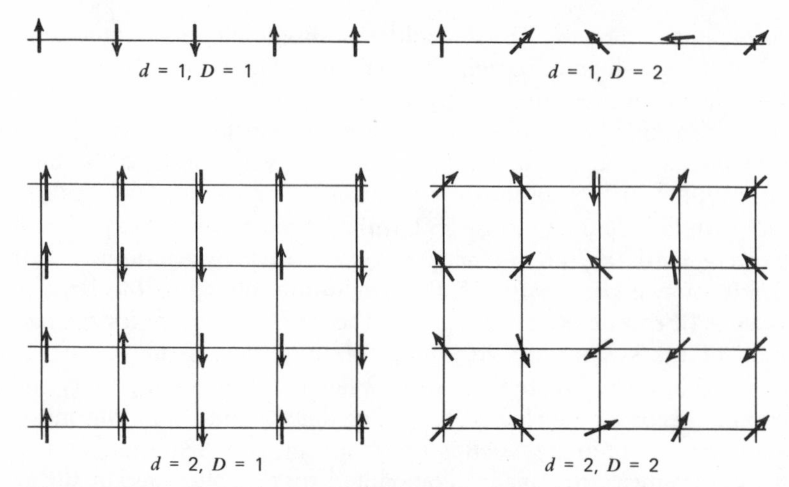

Figure 10.2.1. Four special cases of the model system.

anisotropic. The first term, which is physically the most important one, simulates the exchange interaction of two molecules. The model can even be treated quantum mechanically, in which case the spin is quantized and can only assume $2S+1$ discrete values (S being the spin quantum number). The classical limit is obtained as $S \to \infty$. Recently, however, it has been shown that the critical properties are only very weakly dependent on the quantum number S. Hence a classical theory is quite sufficient as a first approximation to these problems.

Although the Hamiltonian (10.2.1) is already pretty simple, the corresponding thermodynamic properties can only be calculated exactly for special choices of the parameters. The most "realistic" case is the one with $d = D = 3$. It is then called the *Heisenberg model.* This case cannot, however, be solved exactly: only numerical computations are available. In order to obtain exact results, one must first assume that the system is homogeneous and isotropic. One interesting case is the one in which $D \to \infty$. Stanley found that in this limit the problem can be solved completely for $d = 1, 2, 3$, in the presence of an arbitrary magnetic field and coupling either between nearest neighbors only, or with long-range interactions $[J(r) \sim r^{d+x}]$. This is a mathematically interesting model, but it apparently has no physical relevance. The noteworthy fact about this case is its equivalence to the so-called *spherical model,* conceived and solved by Kac in 1947, in which the spins may have arbitrary length, constrained only by the condition $\sum s_j^2 = N$.

Another extreme limiting case in which the problem can be solved exactly is the case of *one-dimensional systems,* $d = 1$. For $\mathcal{H} = 0$ and nearest-neighbor interactions the problem can be solved for all D.

However, these simple systems are too trivial: they display no phase transition (except for infinite-range interactions, see Section 9.4).

By far the most celebrated special case of the Hamiltonian (10.2.1) is, however, the case $D=1$. It is then called the *Ising model* (although it was actually conceived by Lenz in 1920). Restricting ourselves to nearest-neighbor interactions, the Ising-model Hamiltonian can be written as

$$H=-J\sum_{\langle ij\rangle}\sum s_i s_j-\bar{\mu}\mathcal{H}\sum_i s_i \qquad (10.2.2)$$

The variables s_i are now simply numbers assuming either of the two values $s_i=\pm 1$. The summation in the first term is over all couples of nearest-neighbor lattice sites. [In other words, J_{ij} in Eq. (10.2.1) is assumed to vanish, except if i and j are nearest neighbors, in which case $J_{ij}=J$.] The corresponding partition function is then

$$Z=\sum_{\{s_i=\pm 1\}}\exp\left(\beta\bar{\mu}\mathcal{H}\sum_i s_i+\beta J\sum_{\langle ij\rangle}\sum s_i s_j\right) \qquad (10.2.3)$$

The solution of the one-dimensional Ising problem ($d=1$), both in the absence and the presence of a magnetic field, was found by Ising in 1925: It is fairly elementary, but as said above, it leads to no phase transition.

A great landmark in statistical mechanics occurred when Onsager published his paper in 1944: It contained the exact solution of the *two-dimensional* ($d=2$) *Ising problem* for a square lattice with nearest-neighbor interactions, in the absence of a magnetic field. For the first time in history he was able to exhibit an exactly solved model in which a true phase transition appears in the thermodynamic limit $N\to\infty$. Onsager's original paper was very difficult, as he made use of advanced mathematical techniques. Since then, much simpler methods have been found for solving the problem, but the result is still pretty complex and will not be given here. The interested reader is referred to Landau and Lifshitz or to Stanley for an exposition of Vdovichenko's recent simple solution of the two-dimensional Ising problem.

The *three-dimensional Ising problem* cannot be solved exactly, even in the absence of an external field. However, in recent years extremely accurate numerical approximation methods have been developed for the study of this problem. The idea is to obtain the coefficients of Taylor-series expansions, valid either at high or at low temperatures. These coefficients are obtained by graphical methods, involving quite complex combinatorial problems. The use of computers has been of crucial importance in determining the progress of this field. At present, in many cases very long series are known (in some problems 30 to 80 terms are known!). It is no luxury to spend so much labor in calculating such long series. Indeed, it appears that the coefficients in these series are extremely irregular; hence, if the series converge at all, the convergence is very slow. Just to get a feeling of what is involved here, we give here the first terms of the low-temperature expansion (in powers of $u = e^{-4J/k_BT}$) of the zero-field magnetization of the $d = 3$ Ising model, for the face-centered-cubic lattice, obtained by Fisher in 1965:

$$
\begin{aligned}
M(T) = 1 &- 2u^6 - 24u^{11} + 26u^{12} - 48u^{15} - 252u^{16} + 720u^{17} \\
&- 438u^{18} - 192u^{19} - 984u^{20} - 1008u^{21} + 12924u^{22} \\
&- 19536u^{23} + 3062u^{24} - 8280u^{25} + \cdots
\end{aligned}
$$

Therefore, in order to perform an extrapolation that can be trusted, as many coefficients as possible should be calculated. In spite of these difficulties, the numerical techniques available today are astoundingly precise. A more detailed survey of these techniques can be found in Stanley's book.

Before leaving the subject, let us note that the Ising model can also provide a schematic description of a fluid. Consider, indeed, a so-called *lattice gas*. We imagine the physical space divided into a large number N of cells, each labeled by its center; these centers are the sites of a lattice. Each cell can accomodate one and only one molecule (this is an expression of the hard core whose size thus equals the size of the cell). The state of the system is therefore described by specifying the "occupation number" e_i of each cell, where $e_i = 1$ if the cell is occupied, and $e_i = 0$ if it is empty. Assuming that there exists a constant interaction potential equal to $-J$ if two neighboring cells are both occupied, it is easily seen that the grand partition function for this system is

$$
\Xi = \sum_{\{e_i\}} \exp\left(\beta\mu \sum_{i=1}^{N} e_i + \beta J \sum_{\langle ij \rangle}\sum e_i e_j\right) \tag{10.2.4}
$$

This expression is quite analogous to the Ising-model partition function. This analogy can be made even stronger if "spin variables" assuming the values $s_i = \pm 1$ are used through the relation:

$$
e_i = \tfrac{1}{2}(1 + s_i) \tag{10.2.5}
$$

Hence, as remarked by Lee and Yang in 1952, every Ising-model result can be translated into a lattice-gas result.

Besides the models of the class (10.2.1) other types of models have been solved exactly in recent years, in particular by Lieb. These include, among others, two-dimensional lattice models of ferroelectric and antiferroelectric systems. These exhibit extremely interesting and unusual critical properties: We can only refer the reader to the original papers for details.

The main surprise that came from the solution of these models was the serious disagreement between the values of *all* the critical exponents and the corresponding classical values. Some of these values are listed in Table 10.1.1. The behavior of the specific heat is very typical: Instead of a finite jump, as predicted by the classical theory, it presents a logarithmic divergence in the $d=2$ Ising model. In Lieb's ferroelectric model, we have an exact example of a system in which $\alpha'=0$ and $\alpha=\frac{1}{2}$, hence $\alpha \neq \alpha'$.

We stress again the strong dependence on the space dimensionality d. This dependence is by no means simple. Take for instance the Ising model: in going from $d=2$ to $d=3$ the index β increases from 0.125 to 0.313, whereas δ decreases from 15 to 5! We clearly have here a numerological game that presents a challenge to the physicist. Understanding the physics behind these strange numbers is one of the main purposes of the modern theory of critical phenomena.

10.3. EXPONENT INEQUALITIES

A first attempt at finding some underlying order in the critical indices was to look whether thermodynamical arguments could provide any relations among them. The result of these investigations is only partly rewarding: Thermodynamics provides *inequalities*, but no equality allowing one to reduce the number of independent indices. These inequalities are consequences of the thermodynamic stability conditions, the best known of which is the nonnegative character of the specific heat.

Let us derive in detail the simplest of these inequalities, due to Rushbrooke (1963). We start with the following identity for the derivatives of the entropy of a magnetic system:

$$\left(\frac{\partial S}{\partial T}\right)_H = \left(\frac{\partial S}{\partial T}\right)_M + \left(\frac{\partial S}{\partial M}\right)_T \left(\frac{\partial M}{\partial T}\right)_H \tag{10.3.1}$$

On the other hand, we note the following identity from the theory of implicit functions:

$$\left(\frac{\partial M}{\partial \mathscr{H}}\right)_T \left(\frac{\partial \mathscr{H}}{\partial T}\right)_M \left(\frac{\partial T}{\partial M}\right)_H = -1 \tag{10.3.2}$$

Using, moreover, the Maxwell relation:

$$\left(\frac{\partial \mathscr{H}}{\partial T}\right)_M = -\left(\frac{\partial S}{\partial M}\right)_T \tag{10.3.3}$$

we find

$$\left(\frac{\partial M}{\partial \mathscr{H}}\right)_T \left(\frac{\partial S}{\partial M}\right)_T = \left(\frac{\partial M}{\partial T}\right)_H \tag{10.3.4}$$

Using now the definition (9.1.5) together with the definition of the specific heats at constant field or constant magnetization:

$$c_M = T\left(\frac{\partial S}{\partial T}\right)_M, \qquad c_H = T\left(\frac{\partial S}{\partial T}\right)_H \tag{10.3.5}$$

we find the thermodynamic relation:

$$\kappa_T(c_H - c_M) = T\left(\frac{\partial M}{\partial T}\right)_H^2 \tag{10.3.6}$$

We now introduce the stability conditions:

$$c_M \geqslant 0, \qquad \kappa_T \geqslant 0 \tag{10.3.7}$$

This then implies

$$c_H \geqslant T\kappa_T^{-1}\left(\frac{\partial M}{\partial T}\right)_H^2 \tag{10.3.8}$$

Consider now a system at zero magnetic field and at a temperature slightly below T_c. From the definitions (9.5.12)–(9.5.14) of the critical indices we deduce

$$A(-\theta)^{-\alpha'} \geqslant T_c A'(-\theta)^{\gamma'+2(\beta-1)} \tag{10.3.9}$$

We now note that if $f(x) \geqslant g(x)$, then $\ln f(x) \geqslant \ln g(x)$. If $x<1$, then $\ln x<0$, and therefore

$$\frac{\ln f(x)}{\ln x} \leqslant \frac{\ln g(x)}{\ln x}$$

with the inequality sign reversed. But if $f(x) = Ax^{\alpha}$, then

$$\lim_{x\to 0} \frac{\ln f(x)}{\ln x} = \alpha$$

Therefore, (10.3.9) implies

$$-\alpha' \leqslant \gamma' + 2(\beta - 1)$$

or

$$\alpha' + 2\beta + \gamma' \geqslant 2 \tag{10.3.10}$$

This is *Rushbrooke's inequality* relating three critical indices. It is quite general, following directly from the thermodynamic stability condition, which is valid for all physical systems.

Other inequalities can be derived under various conditions; their proof is, however, often more delicate. An inequality as general as Rushbrooke's, as it is solely based on a convexity theorem for the free energy, was derived by Griffith:

$$\alpha' + \beta(1+\delta) \geqslant 2 \tag{10.3.11}$$

Further inequalities can be found only if stronger assumptions are made about the system. To quote an example, if it is assumed that

$$\left(\frac{\partial M}{\partial T}\right)_H \leqslant 0 \qquad \text{for} \quad \mathcal{H} \geqslant 0,$$

$$\left(\frac{\partial^2 M}{\partial \mathcal{H}^2}\right)_T \leqslant 0 \qquad \text{for} \quad \mathcal{H} \geqslant 0,$$

$$\alpha \geqslant \alpha' \tag{10.3.12}$$

then one can show that

$$\gamma' \geqslant \beta(\delta - 1) \tag{10.3.13}$$

The conditions (10.3.12) have been suggested by the Ising model, where they are satisfied; whether they are generally valid for real systems however, is not known.

Another interesting group of inequalities relates the macroscopic to the microscopic critical exponents. Here again we need additional assumptions that are valid in the Ising model, but not necessarily for real systems. These assumptions are: the positivity of the correlation functions and the monotonicity of these functions both with temperature and with magnetic field. Under these conditions, Fisher derived the inequality:

$$(2-\eta)\nu \geqslant \gamma \tag{10.3.14}$$

Another relation interrelates a microscopic and a macroscopic index to the dimensionality of the system:

$$d\frac{\delta - 1}{\delta + 1} \geqslant 2 - \eta \tag{10.3.15}$$

We shall not exhaust the list of known inequalities here. We note only that thermodynamics alone does not allow us to decrease the number of independent critical exponents. However, the two general inequalities (10.3.10) and (10.3.11) provide us with a quite useful consistency test. For instance, if experimental data are available for α', β, and δ for a given system, these data are necessarily obtained from quite different types of measurements. If these data violate Eq. (10.3.11), something must be wrong with some one of the experiments, because this inequality is a general thermodynamic result. We cannot, however, draw such conclusions from the violation of the other inequalities, which may simply imply that the additional conditions are not satisfied for that particular system.

The reader will be tempted, of course, to check these inequalities with the numerical data of Table 10.1.1. The striking fact that then becomes apparent is that, as well for the classical theories as for the two-dimensional Ising model and for the spherical model, they hold as equalities, although the individual critical indices are quite different. Moreover, the experimental data also seem to combine quite closely, within the error limits, to yield equalities. This puzzling feature gave rise to further development and attempts at explanation, which will be considered in the next sections.

10.4. THE SCALING-LAW HYPOTHESIS

As soon as it clearly appeared that the classical values of the critical exponents were not correct for the Ising model or for the real systems, attempts were made at a theoretical study of this problem. The theory put forward by Widom in 1965 proved quite successful. It is a purely phenomenological thermodynamical theory based on a simple assumption for which there is no fundamental basis. Because of its simplicity and of its successful predictions, it triggered further research, which will be considered in the next sections.

To understand the motivation of the scaling-law hypothesis, we consider the free energy of a magnetic substance $A(T, M)$, which is an even function of the magnetization. If this function is analytic, as assumed by *Landau*, it can be expanded as follows:

$$A(T, M) = A_0(T) + A_2(T)M^2 + A_4(T)M^4 + \cdots \tag{10.4.1}$$

For simplicity, we shall only consider positive values for the magnetization in this chapter. If we, moreover, assume analyticity in T, we may further expand the coefficients around $T = T_c$ or, equivalently, around $\theta = 0$:

$$A_j(T) = A_{j0} + A_{j1}\theta + A_{j2}\theta^2 + \cdots \tag{10.4.2}$$

We now write the equation of state by using these assumptions*:

$$\mathscr{H} = \left(\frac{\partial A}{\partial M}\right)_T = 2A_{21}M\theta + \cdots + 4A_{40}M^3 + \cdots \tag{10.4.3}$$

If we set $\mathscr{H} = 0$, it immediately follows that the magnetization near the critical temperature is given by

$$M \sim \theta^{1/2}$$

Hence we find the classical value $\beta = \frac{1}{2}$. The assumption of analyticity (which is at the basis of Landau's theory) thus implies a classical theory.

To get out of this dilemma, we proceed in two steps. We note that the

* It is easily seen that $A_{20} = 0$ because the inverse susceptibility $\kappa_T^{-1} = (\partial^2 A/\partial M^2)_T$ must vanish at the critical point $\theta = 0$.

equation of state (10.4.3) is of the form:

$$\mathcal{H} = b'M\{\theta + c'M^2\}$$

The reason why we get the wrong order-parameter exponent is the occurrence of the exponent 2 in the bracketed term. In order to get the correct (but unspecified) value of β we should replace the former equation of state by

$$\mathcal{H} = bM\{\theta + cM^{1/\beta}\} \tag{10.4.4}$$

where b and c are constants. This form is, however, not yet sufficient; as will be seen below, it does not yield the correct susceptibility exponent. We note, however, that there exist more general functions that are still compatible with $M \sim \theta^\beta$. For instance, we may choose an arbitrary number γ and write

$$\mathcal{H} = bM\{\theta + cM^{1/\beta}\}^\gamma \tag{10.4.5}$$

Even more generally, we may take, instead of the bracket, any homogeneous function of the two variables θ and $M^{1/\beta}$. Thus, *we assume* with Widom that the equation of state, in the neighborhood of the critical point, is of the form:

$$\mathcal{H} = M\,\psi(\theta, M^{1/\beta}) \tag{10.4.6}$$

where *ψ is a homogeneous function of degree γ in the variables θ and $M^{1/\beta}$*. We recall the definition of such a function: If λ is an arbitrary number, the following relation holds:

$$\psi(\lambda\theta, \lambda M^{1/\beta}) = \lambda^\gamma \psi(\theta, M^{1/\beta}) \tag{10.4.7}$$

Equations (10.4.6) and (10.4.7) contain the crux of *Widom's scaling-law hypothesis*. According to this hypothesis, the equation of state near the critical point is determined by two parameters, β and γ, and by the form of the function ψ. However, it will be shown that one can make important predictions about the critical exponents without specifying the form of ψ. We now look for a physical interpretation of the two parameters.

First we check that β can be still identified with the order-parameter exponent, even when the general form (10.4.6) is assumed instead of the particular case (10.4.5). The zero-field magnetization M_0 is defined implicitly as the solution of the equation $\mathcal{H}(\theta, M_0) = 0$, or with (10.4.6):

$$\psi(\theta, M_0^{1/\beta}) = 0 \tag{10.4.8}$$

But Eq. (10.4.7) tells us that

$$\psi(\theta, M^{1/\beta}) = \lambda^{-\gamma}\psi(\lambda\theta, \lambda M^{1/\beta}) \tag{10.4.9}$$

Therefore, (10.4.8) implies

$$\psi(\lambda\theta, \lambda M_0^{1/\beta}) = 0 \tag{10.4.10}$$

Suppose now that the solution of Eq. (10.4.8) is written as

$$M_0^{1/\beta} = \varphi(\theta) \tag{10.4.11}$$

Equation (10.4.10) then yields

$$\lambda M_0^{1/\beta} = \varphi(\lambda\theta)$$

or

$$M_0^{1/\beta} = \lambda^{-1}\varphi(\lambda\theta) \tag{10.4.12}$$

From Eqs. (10.4.11) and (10.4.12) it follows that

$$\varphi(\theta) = \lambda^{-1}\varphi(\lambda\theta) \tag{10.4.13}$$

Hence $\varphi(\theta)$ is a homogeneous function of degree *one* in θ. But as it depends only on one variable, (10.4.13) necessarily implies

$$\varphi(\theta) = a\theta \tag{10.4.14}$$

Hence, from (10.4.11):

$$M_0^{1/\beta} = a\theta$$

or

$$M_0 = a\theta^{\beta} \tag{10.4.15}$$

This confirms the identification of β as the order-parameter exponent.

We now show that γ is the susceptibility exponent (9.5.14). Indeed

$$\left(\frac{\partial \mathscr{H}}{\partial M}\right)_T = \psi(\theta, M^{1/\beta}) + M\frac{\partial \psi(\theta, M^{1/\beta})}{\partial M} \tag{10.4.16}$$

The inverse susceptibility κ_T^{-1} is the value of this derivative at zero field; if $T > T_c$ the condition $\mathscr{H} = 0$ is equivalent to $M = 0$, hence

$$\kappa_T^{-1}(\theta) = \psi(\theta, 0), \qquad \theta > 0 \tag{10.4.17}$$

But the homogeneity condition (10.4.9) implies

$$\kappa_T^{-1} = \lambda^{-\gamma}\psi(\lambda\theta, 0)$$

If we choose in particular the value $\lambda = \theta^{-1}$ for the scale factor, we obtain

$$\kappa_T^{-1} = \theta^{\gamma}\psi(1, 0) = \text{const} \times \theta^{\gamma} \tag{10.4.18}$$

Comparing this with Eq. (9.5.14) we see that γ is, indeed, the critical exponent introduced there.*

* We also see that Eq. (10.4.4) is incompatible with $\gamma \neq 1$.

Let us deduce a few consequences of the functional form (10.4.6). If we calculate the susceptibility *below* T_c, the condition $\mathcal{H}=0$ no longer implies $M=0$, but rather $M=M_0$, the spontaneous magnetization. Hence (10.4.16) becomes

$$\kappa_T^{-1}=\psi(\theta, M_0^{1/\beta})+\left(M\frac{\partial\psi(\theta, M^{1/\beta})}{\partial M}\right)_{M=M_0}, \qquad \theta<0 \tag{10.4.19}$$

But

$$\left(M\frac{\partial\psi(\theta, M^{1/\beta})}{\partial M}\right)_{M=M_0}=M_0\frac{\partial M_0^{1/\beta}}{\partial M_0}\frac{\partial\psi(\theta, M_0^{1/\beta})}{\partial M_0^{1/\beta}}$$

Using this relation and the homogeneity assumption (10.4.9) we find

$$\kappa_T^{-1}=\lambda^{-\gamma}\left\{\psi(\lambda\theta, \lambda M_0^{1/\beta})+\beta^{-1}M_0^{1/\beta}\frac{\partial\psi(\lambda\theta, \lambda M_0^{1/\beta})}{\partial M_0^{1/\beta}}\right\} \tag{10.4.20}$$

But near the critical point, M_0 is related to the temperature through Eq. (10.4.15); hence (10.4.20) becomes

$$\kappa_T^{-1}=\lambda^{-\gamma}\left\{\psi(\lambda\theta, \lambda\theta)+\beta^{-1}a^{1/\beta}\theta\lambda\left(\frac{\partial\psi(\lambda\theta, u)}{\partial u}\right)_{u=\lambda\theta}\right\} \tag{10.4.21}$$

We now choose the particular value $\lambda=(-\theta)^{-1}$:

$$\kappa_T^{-1}=(-\theta)^{\gamma}\left\{\psi(-1, -1)+a^{1/\beta}\beta^{-1}\left(\frac{\partial\psi(-1, u)}{\partial u}\right)_{u=-1}\right\}$$

$$=\text{const}\,(-\theta)^{\gamma}, \qquad \theta<0 \tag{10.4.22}$$

A comparison with the fourth equation (9.5.14) yields a relation between high-temperature and low-temperature susceptibility exponents:

$$\gamma'=\gamma \tag{10.4.23}$$

As another application, let us calculate the exponent δ. We accordingly consider the equation of state (10.4.5) at the critical temperature $\theta=0$ and use again the homogeneity assumption (10.4.9):

$$\mathcal{H}=M\psi(0, M^{1/\beta})=M\lambda^{-\gamma}\psi(0, \lambda M^{1/\beta})$$

Using the particular value $\lambda=M^{-1/\beta}$ we find

$$\mathcal{H}=M^{1+\gamma/\beta}\psi(0, 1)$$

Hence, comparing with (9.5.15) we obtain

$$\delta=1+\frac{\gamma}{\beta}$$

or

$$\gamma=\beta(\delta-1) \tag{10.4.24}$$

We thus obtained Eq. (10.3.13) as an *equality*. Further calculations show that all thermodynamic inequalities of Section 10.3 are transformed into equalities by the homogeneity assumption (10.4.6), (10.4.7). As was mentioned before, both the experimental evidence and the exact-model results are very much in favor of these equalities for many systems. Therefore, it seems that the scaling laws might be the macroscopic manifestation of some deep feature of the critical phenomena. However, one should not overemphasize the generality of the scaling laws. It must be kept in mind that there exist consistent models (e.g., Lieb's ferroelectric models) as well as real systems for which these laws are not satisfied. Hence, the scaling laws define a *class of systems* for which the equation of state is of the form (10.4.6), (10.4.7). It is truly remarkable that within this class we find systems for which the individual values of the exponents are so widely different as for the classical and for the Ising models, for example.

Another interesting aspect of the homogeneity assumption is obtained as follows. Equations (10.4.6) and (10.4.7) can be combined into

$$\mathscr{H} = \lambda^{-\gamma} M\, \psi(\lambda\theta, \lambda M^{1/\beta}) \tag{10.4.25}$$

We now choose $\lambda = |\theta|^{-1}$:

$$\mathscr{H} = |\theta|^{\gamma} M\, \psi\left(\frac{\theta}{|\theta|}, \frac{M^{1/\beta}}{|\theta|}\right)$$

We now introduce the dimensionless variables:

$$\begin{aligned} h &= \mathscr{H}\,|\theta|^{-\beta-\gamma} \equiv \mathscr{H}\,|\theta|^{-\beta\delta} \\ m &= |\theta|^{-\beta} M \end{aligned} \tag{10.4.26}$$

We note that the dimensionless magnetic field h only depends on the scaled magnetization and on the *sign* of the temperature deviation from the critical point. Defining

$$\Phi_{\pm}(m) = m\psi(\pm 1, m) \tag{10.4.27}$$

we find the following reduced equation of state:

$$h = \Phi_{\pm}(m) \tag{10.4.28}$$

where the sign + corresponds to $\theta > 0$ and the sign − to $\theta < 0$. This is a very remarkable result. It tells us that the temperature enters the equation of state only through a scaling factor. After proper reduction, the equation of state is the same for all systems (belonging to the scaling-law class) and for all temperatures: This equation falls into two branches according to whether we are above or below the critical point. This result reminds us of the van der Waals theorem of corresponding states (9.3.11);

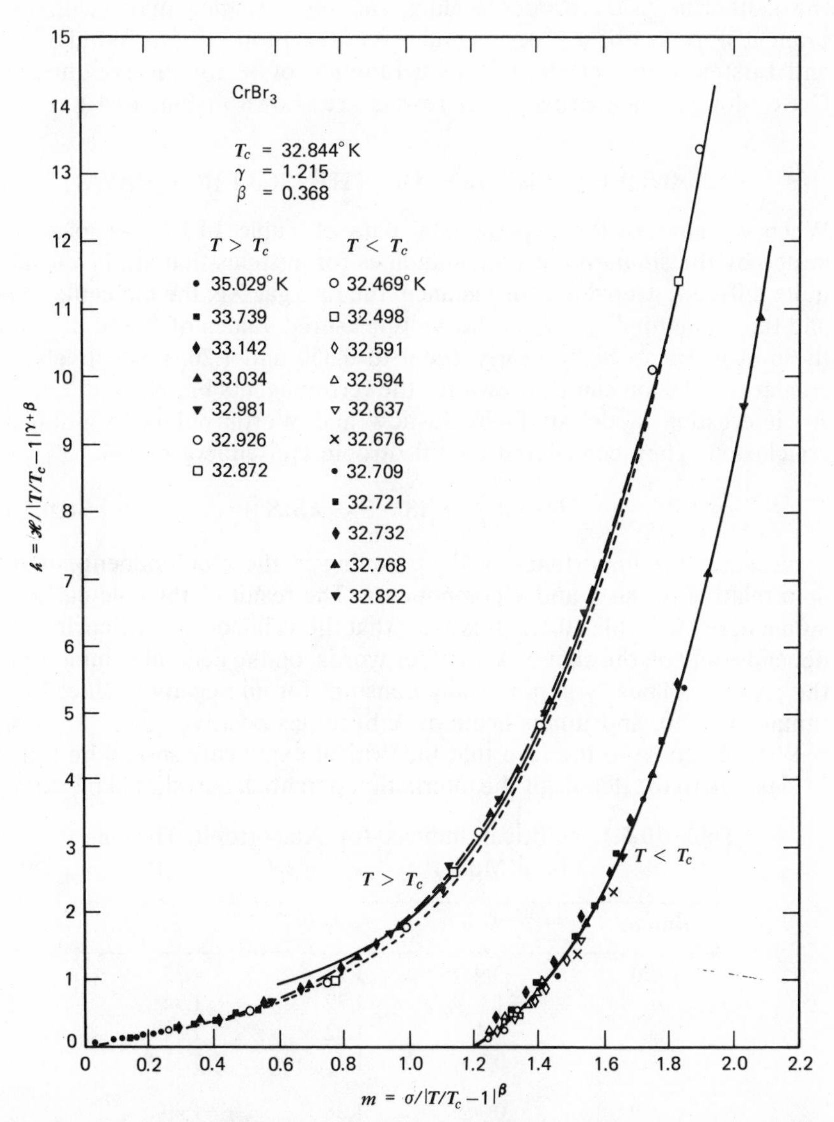

Figure 10.4.1. Scaled magnetic field versus scaled magnetization. Experimental data for the ferromagnet $CrBr_3$. [After J. T. Ho and J. D. Litster, *Phys. Rev. Lett.* **22**, 603 (1969).]

it is, however, stronger because it tells us that *even for a single system* all the isotherms, after proper scaling, fall on a single curve (with two branches). A beautiful experimental test of this property was found by Ho and Litster, who measured M as a function of $\mathcal{H}$ for the ferromagnet $CrBr_3$ along 30 isotherms. Their results are shown in Fig. 10.4.1.

10.5. KADANOFF'S THEORY OF THE SCALING LAWS

When we analyze the experimental data of Table 10.1.1, we might be struck by the similarity of critical indices for systems that are physically quite different. Consider, for instance, the rare gas Xe, the molecule CO_2, and the "quantum" gas He^4: the well-measured values of β and of γ for these systems are very nearly equal to 0.350 and 1.26, respectively. A similar conclusion can be drawn for the ferromagnets Fe, Ni, and $YFeO_3$. An interesting model study by Jasnow and Wortis points to a similar conclusion. They considered an anisotropic Heisenberg model:

$$H = -J \sum_{\langle rr' \rangle}\sum (\mathbf{S}_r \cdot \mathbf{S}_{r'} + \lambda S_r^z S_{r'}^z) \tag{10.5.1}$$

λ measures the importance of the coupling of the z components of the spin relative to the x and y components. The result of the calculation is summarized in Table 10.5.1. It is clear that the value of the critical indices depends only on the *sign* of λ, in other words, on the general symmetry of the problem. Thus, γ is practically constant for all negative values of λ, jumps at $\lambda = 0$, and jumps again as λ becomes positive.

We thus come to the idea that the critical exponents should be rather insensitive to the details of the interaction potential, and should be mainly

Table 10.5.1. Critical Indices for Anisotropic Heisenberg Model[a]

Range	λ	γ	2ν
$\lambda < 0$	−3/2	1.31	1.33
	−1	1.32	1.34
	−0.4	1.32	1.34
	−0.2	1.28(?)	1.31(?)
$\lambda = 0$	0	1.38	1.40
$\lambda > 0$	0.25	1.19	1.20
	0.83	1.24	1.25
	4	1.23	1.25

[a] After D. Jasnow and M. Wortis, *Phys. Rev.* **176**, 739 (1968).

determined by gross features, such as the dimensionality and the overall symmetry of the Hamiltonian. This statement is known as the *universality hypothesis* and was first clearly stated and exploited by Kadanoff in 1967.

Intuitively it is easy to realize that the universality hypothesis must be close to the truth. Indeed, we know from Section 9.6 that the range of correlation near the critical point becomes extremely long compared to the range of the interactions. It seems, therefore, quite natural that a "small" local modification of the interactions would not be felt by the long-range correlations. What is much less obvious is that the scaling laws of the previous section may be "derived" from the universality hypothesis: this was shown by Kadanoff. His arguments are typically those of a theoretical physicist: They are far from rigorous, but are based on a very deep physical intuition, and are intriguingly appealing. We will see in the next sections how they were refined quite recently by Wilson and naturally led to a beautiful (though still nonrigorous) theory.

To fix the ideas, we consider the Ising model defined by the Hamiltonian (10.2.2) and the partition function (10.2.3). Its thermodynamics is characterized by the free enthalpy:

$$G(\theta, h) = -k_B T \ln Z \tag{10.5.2}$$

We choose to express this function (near the critical point) in terms of the reduced variables $\theta = (T - T_c)/T_c$ and $h = \bar{\mu}\mathcal{H}/k_B T$.

It is also immediately found that the magnetization is simply the average value of the spin:

$$M(\theta, h) = \langle s_r \rangle \equiv \sum_{\{s_i = \pm 1\}} \frac{s_r e^{-\beta H}}{Z} \tag{10.5.3}$$

and the spin–spin correlation function is given by

$$\Gamma_2(r; \theta, h) = \langle s_0 s_r \rangle \tag{10.5.4}$$

From a purely energetic point of view, the most favorable configurations are those in which neighboring spins are aligned with each other and in which the spins are pointing along the magnetic field.

Following Kadanoff, we now consider a related problem. We group the initial lattice sites together into identical blocks, each of side La, where a is the initial lattice constant (Fig. 10.5.1). Hence we construct a new lattice of the same symmetry as the initial one, but with lattice constant La.

If we are close to the critical point, the correlation length ξ introduced in Eq. (9.6.10) is much larger than the lattice constant a; we may then find a number L such that

$$1 \ll L \ll \frac{\xi}{a} \tag{10.5.5}$$

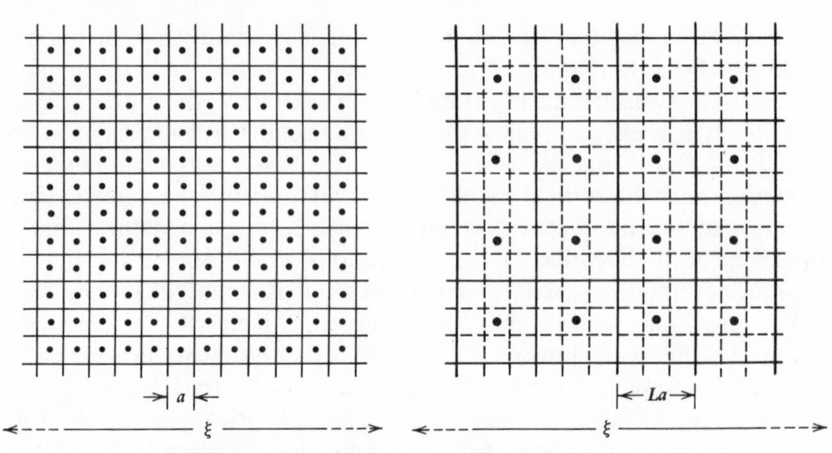

Figure 10.5.1. Scale transformation of the Ising problem.

If this is so, each individual block contains many individual spins, which, moreover, are strongly intercorrelated because they lie well within the correlation length: most spins within a block point predominantly up or down. (We may think of a block as an analog of a magnetic domain.) We may now associate with each cell, labeled α, a local order parameter μ_α, which plays the same role as the individual spin s_r. Among the blocks there are, of course, interactions that produce correlations. The overall symmetry of the block problem is the same as in the original Ising problem. Therefore, *if universality is correct*, the correlations of μ_α's in the block problem should have the same structure as the correlations of spins s_r in the original problem. In other words, the scale parameter L should be considered as an "irrelevant" detail in the Hamiltonian. The only difference between the two problems can be in the *value* of the reduced variables θ and h. Let $\tilde{\theta}$ and $\tilde{h}$ be these values for the block problem: they are related to the original values θ and h through a relation that generally depends on L. As the magnetic field goes to zero, as $T \to T_c$ the effective field $\tilde{h}$ also must vanish; hence in first approximation:

$$\tilde{h} = b(L)h \tag{10.5.6}$$

Moreover, the block problem must become critical at the same reduced temperature as the site problem, hence

$$\tilde{\theta} = c(L)\theta \tag{10.5.7}$$

More specifically, we may assume simple forms for the coefficients $b(L)$ and $c(L)$:

$$\begin{aligned} \tilde{h} &= L^x h \\ \tilde{\theta} &= L^y \theta \end{aligned} \tag{10.5.8}$$

We are now in a position to formulate mathematically what we mean by correlations of identical structure. We require (if universality holds) that the average of the order parameter μ_α in the block problem define a magnetization that depends on $\tilde{\theta}$ and $\tilde{h}$ in the same way as $\langle s_r \rangle$ depends on θ and h in the original problem:

$$\langle \mu_\alpha \rangle = M(\tilde{\theta}, \tilde{h}) \tag{10.5.9}$$

where M is the *same function* as in Eq. (10.5.3). Similarly

$$\langle \mu_\alpha \mu_0 \rangle = \Gamma_2(\tilde{r}; \tilde{\theta}, \tilde{h}) \tag{10.5.10}$$

with Γ_2 the *same function* as in (10.5.4); the scaling of the distance is obvious:

$$\tilde{r} = \frac{r}{L} \tag{10.5.11}$$

Note that we have introduced here very strong (and unproved) assumptions. It is clear that the effective block Hamiltonian generated by the lattice problem cannot be simply of the form (10.2.2) with μ_α replacing s_r. However, Kadanoff argues, the difference between the exact Hamiltonian and its Ising "caricature" should be irrelevant in determining the critical properties, if universality is correct.

In order to close the theory, we still need an extra relation, which we obtain by computing the change in the free enthalpy as the magnetic field is changed by δh. This change must be the same in the two problems (by our general philosophy), hence

$$\delta\left(\frac{G}{k_B T}\right) = -\sum_r \langle s_r \rangle \, \delta h = -\sum_\alpha \langle \mu_\alpha \rangle \, \delta \tilde{h} \tag{10.5.12}$$

As we are interested in long-range, and thus slowly varying phenomena, we may represent the sum over lattice sites as a sum over blocks, times the number of sites per block L^d (d being the dimensionality):

$$\sum_r \to L^d \sum_\alpha$$

Using also Eq. (10.5.8) we obtain

$$\sum_\alpha L^d \langle s_r \rangle \, \delta h = \sum_\alpha \langle \mu_\alpha \rangle L^x \, \delta h$$

This relation can be satisfied in particular if we postulate

$$s_r = L^{x-d} \mu_\alpha \tag{10.5.13}$$

We are now in a position to make definite predictions. Indeed, using

Eqs. (10.5.8), (10.5.13), Eq. (10.5.9) is rewritten as

$$L^{d-x}\langle s_r\rangle = M(L^y\theta, L^x h)$$

Comparing this with Eq. (10.5.3) we obtain

$$M(\theta, h) = L^{x-d}M(L^y\theta, L^x h) \tag{10.5.14}$$

This is nothing other than a *scaling relation.* Indeed, Eq. (10.5.14) expresses the fact that M is a homogeneous function of θ and of h. If we choose $L = |\theta|^{-1/y}$, we find

$$M(\theta, h) = |\theta|^{(d-x)/y} M\left(\frac{\theta}{|\theta|}, h\,|\theta|^{-x/y}\right)$$

Therefore, introducing the reduced variables:

$$\begin{aligned} m &= M\,|\theta|^{(x-d)/y} \\ \hbar &= h\,|\theta|^{-x/y} \end{aligned} \tag{10.5.15}$$

we find that the equation of state has a universal form:

$$m = \varphi_\pm(\hbar) \tag{10.5.16}$$

This is the same equation as (10.4.28), so we have made complete contact with the phenomenological scaling laws. Comparing Eqs. (10.5.15) and (10.4.26) we can express the unknown indices x and y in terms of the critical exponents β and γ:

$$\frac{x-d}{y} = -\beta$$

$$-\frac{x}{y} = -\beta - \gamma$$

or

$$x = d\frac{\beta+\gamma}{2\beta+\gamma}, \qquad y = d\frac{1}{2\beta+\gamma} \tag{10.5.17}$$

Hence all macroscopic critical exponents are expressed in terms of two of them, for example, β and γ. But now we can go one step further. By using an entirely analogous argument, we may analyze the two-spin correlation function and obtain a scaling law in the form:

$$\Gamma_2(r; \theta, h) = L^{2(x-d)}\Gamma_2(L^{-1}r; L^y\theta, L^x h) \tag{10.5.18}$$

Choosing again $L = |\theta|^{-1/y}$ we find

$$\Gamma_2(r; \theta, h) = \theta^{2(d-x)/y} g_\pm(|\theta|^{1/y}\, r; \hbar) \tag{10.5.19}$$

where $g_\pm$ are unknown functions. Alternatively, we may choose $L = r$, in which case we find

$$\Gamma_2(r;\theta,h) = r^{2(x-d)} f(r^y\theta, r^x h) \tag{10.5.20}$$

Returning now to the discussion of Section 9.6, we can identify the microscopic critical exponents. We see from (10.5.19) that the scaling factor for r at $h=0$, which is just the correlation length, behaves as $|\theta|^{1/y}$ both for $\theta>0$ and for $\theta<0$; comparing with (9.6.11) we find

$$\nu = \nu' = y^{-1} \tag{10.5.21}$$

Finally, setting $\theta = h = 0$ in Eq. (10.5.20), we find that the correlation function behaves as $r^{2(x-d)}$ at the critical point; comparison with (9.6.16) yields

$$-d+2-\eta = 2(x-d) \tag{10.5.22}$$

But x and y are expressed in terms of the macroscopic critical exponents: hence we have now derived scaling relations between microscopic and macroscopic exponents:

$$\nu d = \nu' d = 2\beta + \gamma \tag{10.5.23}$$

Also

$$-d+2-\eta = -\frac{2\beta d}{2\beta+\gamma}$$

which, combined with (10.5.23), gives

$$(2-\eta)\nu = \gamma \tag{10.5.24}$$

which is Fisher's inequality (10.3.14) holding as an equality.

Let us now summarize the main aspects of the scaling theory. By assuming (*ad hoc*) a homogeneous form for the equation of state, or by means of the universality hypothesis and Kadanoff's argument, we were able to express the nine critical indices α, α', β, γ, γ', δ, ν, ν', η in terms of only two. There exist, therefore, seven independent relations among them:

$$\begin{aligned}
\alpha &= \alpha' \\
\gamma &= \gamma' \\
\nu &= \nu' \\
2-\alpha &= \gamma + 2\beta \\
2-\alpha &= \beta(\delta+1) \\
2-\alpha &= \nu d \\
(2-\eta)\nu &= \gamma
\end{aligned} \tag{10.5.25}$$

As a final check of these relations we collect in Table 10.5.2 the values of a certain number of combinations of critical exponents. If the scaling relations were satisfied, all these numbers should be equal for a given system. It is seen that the exactly soluble models ($d = 2$ Ising and spherical) satisfy all the relations exactly. The macroscopic relations are also quite well satisfied by all model systems; but the microscopic relations, involving the dimensionality d, are not consistent with the former for the $d = 3$ Ising model nor for the classical theory (in this latter case, in which the interactions are of infinite range, the Kadanoff argument cannot hold). For real systems the agreement among the combinations of exponents is, of course, more fluctuating; the precision of the data is probably not sufficient to provide a very accurate check; however, the orders of magnitude are pretty good.

Table 10.5.2. Test of Scaling-Law Relations among Critical Exponents

System	$2-\alpha$	$2-\alpha'$	$\gamma+2\beta$	$\gamma'+2\beta$	$\beta(\delta+1)$	$d\nu$	$d\nu'$	$d\gamma/(2-\eta)$
Classical	2	2	2	2	2	1.5	1.5	1.5
Models (Exact)								
Ising, $d=2$	2	2	2	2	2	2	2	2
Spherical, $d=3$	3		3		3	3		3
Models (Numerical computations)								
Ising, $d=3$	1.875	1.875	1.876	1.876	1.878	1.914		1.916
Heisenberg, $d=3$, $s=\frac{1}{2}$	2	2	2.065		2.070	2.106		2.108
Experiments								
CO_2		~1.9	1.96		1.82			
Ferromagnets (Average)	1.84	1.84	1.99		1.68	1.95		2.06

10.6. WILSON'S FORMULATION OF THE KADANOFF THEORY

Kadanoff's theory of the scaling laws is certainly very appealing because it highlights the main feature of critical behavior, that is, the infinite correlation length. However, the details of the argument are unsatisfactory. In a recent work, Wilson reformulated Kadanoff's theory in a very original way and showed how the scaling laws and the universality principle could be derived more economically. At the same time his theory sheds a different light on the structure of critical phenomena by stressing new and hitherto unexploited aspects. He could then generalize the theory in a quite nontrivial direction and obtain a fully operational formalism from which critical exponents can be calculated explicitly. We now summarize Wilson's ideas.

We start again from the Ising-model partition function:

$$Z = \sum_{\{s_i\}} \exp\left(K \sum_{\langle ij \rangle}\sum s_i s_j + h \sum_i s_i\right) \tag{10.6.1}$$

where $K = J/k_B T$, $h = \bar{\mu}\mathcal{H}/k_B T$. In a second step, we define a *class of models depending on a scale parameter* L, all of them defined by an Ising-like partition function (10.6.1), but in which the coupling parameters K and h are replaced by L-dependent parameters K_L, h_L. The Kadanoff construction provides an example of such a class; however, we shall no longer assume that L is an integer, nor will L be restricted by the inequalities (10.5.5). Rather, we suppose that L may vary continuously from zero to infinity.

We now *assume*, with Kadanoff, that near the critical point all the models of this class have the same thermodynamic properties. More precisely, we assume that the free enthalpy *per spin* is the same for all systems in the class. As L is a scaling factor of lengths, this implies

$$G(K, h) = L^{-d} G(K_L, h_L) \tag{10.6.2}$$

(In this context it is convenient to use, instead of the temperature T or of θ, the parameter K, which is proportional to the inverse temperature.) Related to this assumption is the fact that the correlation length $\xi(K, h)$ must also be scaled according to

$$\xi(K, h) = L\xi(K_L, h_L) \tag{10.6.3}$$

In order to close the theory we need expressions for K_L and h_L as functions of L: these were guessed by Kadanoff on an intuitive basis [see Eqs. (10.5.8)]. The distinctive feature of Wilson's theory is in replacing this direct guess by the establishment of differential equations for these functions. The argument goes as follows: To go from K_L and h_L to K_{2L} and h_{2L} according to the Kadanoff picture we simply put 2^d blocks together to make a new block. The relative change induced by this process should not depend on the absolute length of the initial blocks. In other words, K_{2L} can only depend on K_L and h_L, but not on L explicitly. [It is easily checked that the Kadanoff choice (10.5.8) does, indeed, satisfy this criterion, as $h_{2L} = 2^x L^x h = 2^x h_L$.] If we repeat the argument for an infinitesimal change of scale $L \to (1+\delta)L$, the corresponding change in K_L is

$$K_{L+L\delta} - K_L = \delta\, L\left(\frac{dK_L}{dL}\right) + \cdots \equiv \delta\, u$$

Our assumption then consists in postulating that u only depends on K_L and on h_L, not on L explicitly. We may even assume that u depends on the field only through h_L^2 because the problem is symmetrical under a change of sign of h_L. Hence we obtain the differential equation:

$$\frac{dK_L}{dL}=\frac{1}{L}u(K_L, h_L^2) \tag{10.6.4}$$

and similarly

$$\frac{dh_L}{dL}=\frac{1}{L}h_L v(K_L, h_L^2) \tag{10.6.5}$$

These two equations may be looked at as defining a group of scale transformations (analogous to the group of translations or the group of motions of classical mechanics). They have been given the name of *renormalization-group equations* (or briefly, RG equations).* These equations, together with Eqs. (10.6.2) and (10.6.3) are the key ingredients of Wilson's theory. To make the theory operational, we will finally assume that *u and v are analytic functions of K_L, h_L^2 even at the critical point.* One of the beautiful features of the theory is to show how the critical singularities can arise quite naturally from a set of differential equations with analytic coefficients.

We start with the initial system (i.e., $L=1$) slightly above the critical point, that is, $K \lesssim K_c$, $h \gtrsim 0$. (Remember that K is inversely proportional to the temperature.) We then increase the scaling factor L. Equation (10.6.3) tells us that $\xi_L \equiv \xi(K_L, h_L)$ must decrease because the product $L\xi_L$ is a constant independent of L. A decreasing correlation length means that we move away from the critical point. This implies that K_L, which was initially smaller than K_c, becomes still smaller.

Consider now an initial condition in which the system is slightly below the critical point, $K \gtrsim K_c$. Equation (10.6.3) tells us again that when we increase L, the correlation length decreases and we move away from the critical point. This can only happen now if K_L increases away from K_c.

Suppose now we start exactly *at* the critical point $K=K_c$, $h=0$. In this case $\xi(K_c, 0)=\infty$, by definition. If we now increase L, the only way to make the right-hand side of (10.6.3) infinite is to have $\xi(K_L, h_L)=\infty$ for all (finite) values of L. This means that the solution of the RG equations for the initial value $K=K_c$, $h=0$ is: $K_L=K_c$, $h_L=0$, independent of L. Note that $h_L=0$ is automatically a stationary solution of (10.6.5); in order that $K_L=K_c$ satisfies (10.6.4), we need to have

$$u(K_c, 0)=0 \tag{10.6.6}$$

To summarize the discussion, if the system has an "initial temperature" K (and, for simplicity, $h=0$), the result of scaling it up is necessarily a motion away from the critical point, whatever the initial value, except if

* The name comes from quantum field theory, where an analogous group occurs. We need not insist on that analogy.

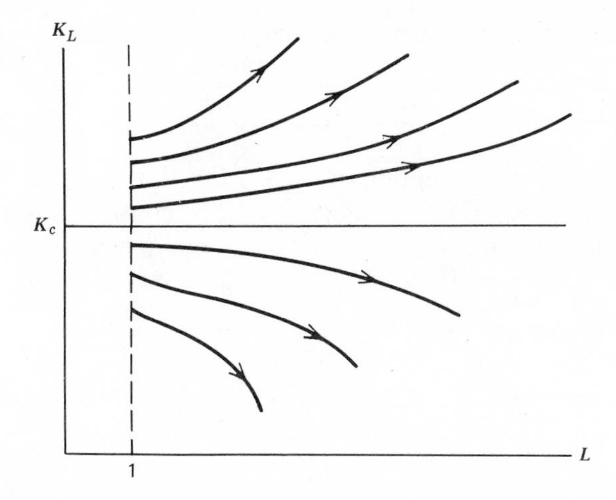

Figure 10.6.1. Solutions of the RG equations (10.6.4), (10.6.5) for $h_L = 0$.

$K = K_c$, in which case the system remains at the critical temperature whatever the scaling factor (see Fig. 10.6.1). The critical point thus appears as an *unstable fixed point of the RG equations*, that is, something analogous to the top of a hill. This picture also shows us how discontinuity can arise from a perfectly innocent, analytic set of differential equations. If we consider two systems starting very close to each other, on the same side of K_c, they will remain close to each other for all L; if, however, they are initially very close neighbors but on different sides of K_c, they move away from each other. K_c is therefore a singularity of the differential equations.

Having determined the critical point, we may solve the RG equations approximately. The first method to be tried is a linearization around the critical point. The equations then assume the form:

$$\frac{dK_L}{dL} = \frac{1}{L}(K_L - K_c)y$$
$$\frac{dh_L}{dL} = \frac{1}{L} h_L x \tag{10.6.7}$$

where

$$y = \frac{\partial u}{\partial K}(K_c, 0)$$
$$x = v(K_c, 0) \tag{10.6.8}$$

These numbers are finite because we assumed analyticity of u and v at the critical point.

These simple equations have the solution:

$$\varepsilon_L = \varepsilon L^y$$
$$h_L = hL^x$$

corresponding to the condition $K_L = K$, $h_L = h$ for $L = 1$. Here $\varepsilon_L = K_c - K_L$, $\varepsilon = K_c - K$. (For simplicity, we only consider the case $\varepsilon > 0$, $h > 0$.) Pick now an arbitrary value of K_L, say $K_L = K_c/2$, and calculate the value of L, say, $\bar{L}$, for which this value is reached:

$$\bar{L} = \left(\frac{K_c}{2\varepsilon}\right)^{1/y} \tag{10.6.9}$$

h_L then has the value

$$h_{\bar{L}} = h\left(\frac{K_c}{2\varepsilon}\right)^{x/y} \tag{10.6.10}$$

The free enthalpy and the correlation length are then given by Eqs. (10.6.2), (10.6.3):

$$\begin{aligned} G(K, h) &= \left(\frac{K_c}{2\varepsilon}\right)^{-d/y} G\left[\frac{K_c}{2}, h\left(\frac{K_c}{2\varepsilon}\right)^{x/y}\right] \\ \xi(K, h) &= \left(\frac{K_c}{2\varepsilon}\right)^{1/y} \xi\left[\frac{K_c}{2}, h\left(\frac{K_c}{2\varepsilon}\right)^{x/y}\right] \end{aligned} \tag{10.6.11}$$

Note that, for $T \approx T_c$, ε is proportional to our former θ. Hence Eqs. (10.6.11) can be rewritten as

$$\begin{aligned} G(\theta, h) &= \theta^{d/y}\psi(h\theta^{-x/y}) \\ \xi(\theta, h) &= \theta^{-1/y}\zeta(h\theta^{-x/y}) \end{aligned} \tag{10.6.12}$$

where ψ and ζ are universal functions whose form is obtained from (10.6.11). The equations (10.6.12) are easily shown to be equivalent to Kadanoff's *scaling-law equations* (10.5.16), (10.5.19), x and y being related to the critical exponents precisely as in Eqs. (10.5.17).

Before concluding, we quickly consider a slightly more general problem, in which the free enthalpy and the correlation length depend also on a third parameter of the Hamiltonian, q, which may be thought of as one of Kadanoff's "irrelevant" variables (such as the ratio of next-nearest-neighbor to nearest-neighbor coupling parameters). Equations (10.6.2), (10.6.3) are now replaced by

$$G(K, h, q) = L^{-d}G(K_L, h_L, q_L) \tag{10.6.13a}$$
$$\xi(K, h, q) = L\xi(K_L, h_L, q_L) \tag{10.6.13b}$$

Extending the previous argument, we may now write a system of three

RG equations:

$$\frac{dK_L}{dL} = \frac{1}{L} u(K_L, q_L, h_L^2)$$
$$\frac{dq_L}{dL} = \frac{1}{L} w(K_L, q_L, h_L^2) \qquad (10.6.14)$$
$$\frac{dh_L}{dL} = \frac{1}{L} h_L v(K_L, q_L, h_L^2)$$

Because we have an additional variable q, there now exists a line of critical points (rather than a single point), defined by the equation $\xi(K_c, q, 0) = \infty$: for each value of q we have a critical temperature $K_c(q)$. Again, if the initial value K, q, h lies on the critical line $[K = K_c(q), h = 0]$, the solution must be on the critical line for all $L: [K_L = K_c(q_L), h_L = 0]$. The equation for q_L then reduces to

$$\frac{dq_L}{dL} = \frac{1}{L} w_c(q_L) \qquad (10.6.15)$$

with $w_c(q_L) = w[K_c(q_L), q_L, 0]$. This equation may predict that, as $L \to \infty$, q_L tends to a limit: $q_L \to q_c$, or that it tends to infinity, or that it oscillates. We assume that the first possibility is realized. Let $K_c(q_c) \equiv K_c$. If initially $K = K_c$, $q = q_c$, $h = 0$, then at all scales, $K_L = K_c$, $q_L = q_c$, $h_L = 0$: We have a *fixed point* for the whole group.

We now discuss again the behavior of the solutions in the neighborhood of the fixed point, in which case the equations reduce to

$$L\frac{dK_L}{dL} = y_{11}(K_L - K_c) + y_{12}(q_L - q_c)$$
$$L\frac{dq_L}{dL} = y_{21}(K_L - K_c) + y_{22}(q_L - q_c) \qquad (10.6.16)$$
$$L\frac{dh_L}{dL} = xh_L$$

The third equation being decoupled from the first two, we may discuss the latter separately. These two equations possess in general two linearly independent solutions:

$$K_L - K_c = L^y, \qquad q_L - q_c = r_y L^y \qquad (10.6.17a)$$
$$K_L - K_c = L^z, \qquad q_L - q_c = r_z L^z \qquad (10.6.17b)$$

Hence, the general solution is a linear combination of these:

$$K_L = K_c - \varepsilon L^y + \eta L^z$$
$$q_L = q_c - \varepsilon r_y L^y + \eta r_z L^z \qquad (10.6.18)$$
$$h_L = hL^x$$

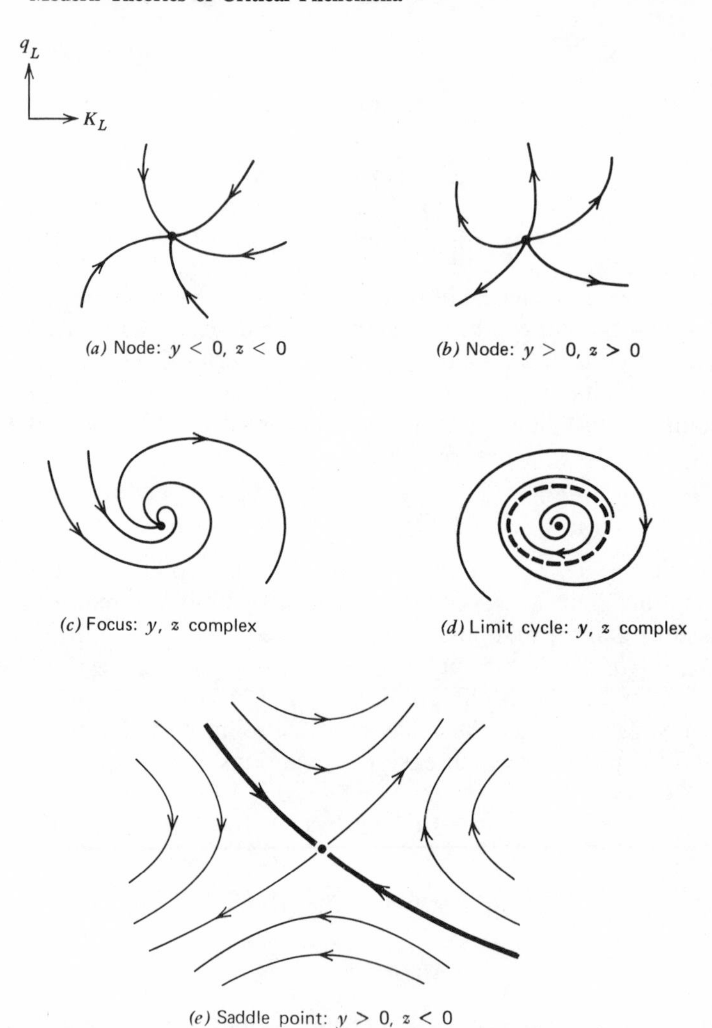

(a) Node: $y < 0$, $z < 0$

(b) Node: $y > 0$, $z > 0$

(c) Focus: y, z complex

(d) Limit cycle: y, z complex

(e) Saddle point: $y > 0$, $z < 0$

Figure 10.6.2. Various types of fixed-point singularities of the RG equations.

where ε and η depend on the initial ($L = 1$) data:

$$\varepsilon = \frac{r_z(K - K_c) - (q - q_c)}{r_y - r_z}$$
$$\eta = \frac{r_y(K - K_c) - (q - q_c)}{r_y - r_z} \tag{10.6.19}$$

We take first $h = h_L = 0$. It is customary in the theory of differential equations to represent the solutions by drawing trajectories in the K–q

plane: these trajectories are oriented in the sense of increasing L. The nature of the singular fixed point depends on the values of the parameters y and z. Some of the possibilities are shown in Fig. 10.6.2.

We now show that only one of these is consistent with our picture. Indeed, we first note as described above, that if initially $K = K_c(q)$ then $K_L \to K_c$, $q_L \to q_c$ as $L \to \infty$. This behavior rules out the cases (b) and (d) of Fig. 10.6.2, because there is no trajectory going into the singular point. On the other hand, Eq. (10.6.13b) requires that if initially $K \neq K_c(q)$, then as L increases we must move away from the critical line. This condition rules out the cases (a) and (c) in which all trajectories eventually go into the critical point. We are thus left with the saddle-point singularity. In this case the fixed point can only be reached if the system starts on the single trajectory drawn in a heavy line: This trajectory is thus identified with the critical line $K = K_c(q)$. Note that the equilibrium point will be reached whatever the initial value of q: the parameter q indeed appears as an "irrelevant variable." To sum up, we have shown that "*normal critical behavior*" *can only be described by a set of RG equations possessing a fixed point of saddle-point type.* This implies that $y > 0$, $z < 0$.

Coming back now to the solution (10.6.18), we proceed as before, starting with values of K, q near the critical line and calculating the value $\bar{L}$ for which $K_{\bar{L}}$ is certainly a large number. The term proportional to η in the first two equations (10.6.18) is quite small because $z < 0$: it can therefore be neglected. We then find, as in Eq. (10.6.9),

$$\bar{L} = \left(\frac{K_c}{2\varepsilon}\right)^{1/y}$$

For this value of L, we have

$$\begin{aligned} q_{\bar{L}} &= q_c - r_y\left(\frac{K_c}{2}\right) \\ h_{\bar{L}} &= h\left(\frac{K_c}{2\varepsilon}\right)^{x/y} \end{aligned} \tag{10.6.20}$$

We note two important features. $q_{\bar{L}}$ is a *constant*, independent of ε, q or h; moreover, $h_{\bar{L}}$ depends on h and ε just as in (10.6.10). Hence, substituting $K_L = K_c/2$ and Eqs. (10.6.20) into Eqs. (10.6.13) we obtain equations of the *same form* as (10.6.12), the only difference being in the form of the functions ψ and ζ. The critical exponents and the relations among them are the same whether there is a parameter q or not in the Hamiltonian. *We have therefore obtained not only the scaling laws but also the universality principle as direct consequences of the renormalization-group equations.*

10.7. THE RENORMALIZATION-GROUP EQUATIONS AND THE PARTITION FUNCTION

The beauty of Wilson's first paper lies in its opening a completely new avenue in the theory of critical phenomena (and possibly in other fields as well). It shows quite clearly which feature follows from which assumption. A very surprising result in this respect is the fact that the scaling laws are only very loosely connected to the Hamiltonian. They appear as being related to intrinsic characteristics of the RG equations. These do not describe the properties of a given Hamiltonian: they only tell us how we relate together two Hamiltonians within an infinite class. The final results of Wilson's first paper were, of course, not different from Kadanoff's. He showed that if we assume:

1. the existence of a class of L-dependent Hamiltonians of Ising type, characterized by two (or three) coupling parameters K_L and h_L (and possibly q_L);
2. the validity of Kadanoff's equations (10.6.2), (10.6.3) [or (10.6.13)] and the definition of the critical point as $\xi(K_c, 0) = \infty$ [or $\xi(K_c(q), q, 0) = \infty$];
3. the existence of a set of RG equations (10.6.4), (10.6.5) [or (10.6.14)] that imprint a structure to the class of Hamiltonians by defining a relation between any two of its members:

then Kadanoff's results follow immediately from the solution of the RG equations, without any reference to the intuitive, but rough construction of blocks of Ising spins.

In his second paper, Wilson attacked the problem of critical phenomena from a more fundamental point of view. The problems he studied there could be formulated as follows:

A. *Can one define an L-scaling process related to the structure of the partition function (rather than to an artificial external construct such as Kadanoff's block picture)?*

B. *Can one find a class of Hamiltonians that, in a certain sense to be defined precisely, is invariant under these transformations?*

C. *If this class exists, one should be able to* **derive** *(rather than postulate) RG equations connecting its members to each other.*

D. *From these equations it should be possible not only to derive scaling relations, but even to compute explicit values for the critical exponents.*

In order to attack this ambitious program Wilson considered again a spin system. However, it turns out that the simple Ising model is not sufficiently flexible (i.e., it does not have enough parameters) in order to satisfy all the conditions: a more general model is necessary in this study. First, it is assumed that the individual spins may take any real value between $-\infty$ and $+\infty$ (instead of only the values -1 and $+1$). Moreover, the discrete lattice is idealized by a continuous distribution of spins throughout the space: This should be an acceptable approximation for long-range phenomena extending over a large number of lattice sites, as is precisely the case for critical phenomena. Hence, instead of a denumerable set of dynamical variables, labeled by the discrete lattice points $\mathbf{r}$: $s_\mathbf{r}$, we now have a continuous set of spin variables, one attached to each point in space, that is, the system is described by a *spin field* $s(\mathbf{x})$. The field $s(\mathbf{x})$, like any function of $\mathbf{x}$, can be expanded in a series of appropriate orthogonal basis functions $\psi_\mathbf{n}(\mathbf{x})$, whose form is irrelevant in this argument. $\mathbf{n}$ is a collection of indices needed for the complete specification of the basis: we choose to call one of these indices l and to write it separately: $\mathbf{n} \equiv (\mathbf{m}, l)$. l will be identified below. The expansion of $s(\mathbf{x})$ now reads

$$s(\mathbf{x}) = \sum_{\mathbf{m}} \sum_{l=0}^{\infty} \psi_{\mathbf{m}l}(\mathbf{x}) s_{\mathbf{m}l} \tag{10.7.1}$$

We now choose the set of numbers $s_{\mathbf{m}l}$ as the dynamical variables describing the system. In discussing questions related to range and length scale, it is always natural to proceed to a Fourier picture. Hence, defining

$$\begin{aligned} \sigma(\mathbf{k}) &= \int d^d\mathbf{x}\, e^{i\mathbf{k}\cdot\mathbf{x}} s(\mathbf{x}) \\ \phi_{\mathbf{m}l}(\mathbf{k}) &= \int d^d\mathbf{x}\, e^{i\mathbf{k}\cdot\mathbf{x}} \psi_{\mathbf{m}l}(\mathbf{x}) \end{aligned} \tag{10.7.2}$$

we transform Eq. (10.7.1) into

$$\sigma(\mathbf{k}) \equiv \sum_{l=0}^{\infty} \sigma_l(\mathbf{k}) = \sum_{l=0}^{\infty} \left\{ \sum_{\mathbf{m}} \phi_{\mathbf{m}l}(\mathbf{k}) s_{\mathbf{m}l} \right\} \tag{10.7.3}$$

where $\sigma_l(\mathbf{k})$ is defined by the bracketed quantity. Conversely:

$$s(\mathbf{x}) = \sum_{l=0}^{\infty} s_l(\mathbf{x}) = \sum_{l=0}^{\infty} (2\pi)^{-d} \int d\mathbf{k}\, e^{-i\mathbf{k}\cdot\mathbf{x}} \sigma_l(\mathbf{k}) \tag{10.7.4}$$

Wilson now introduces the idea of a classification of the various contributions to the field $s(\mathbf{x})$ according to their spatial range. It is well known that a Fourier component of wave vector $\mathbf{k}$ describes a quantity whose range of variation in space is of order $2\pi/k$. We now divide the $\mathbf{k}$ space into cells by constructing an infinite sequence of concentric spheres centered at the origin. The first sphere has a radius $k = 2$, the second $k = 1$, the third $k = \frac{1}{2}, \ldots$, the lth, $k = 2^{-(l-2)}$ (see Fig. 10.7.1).

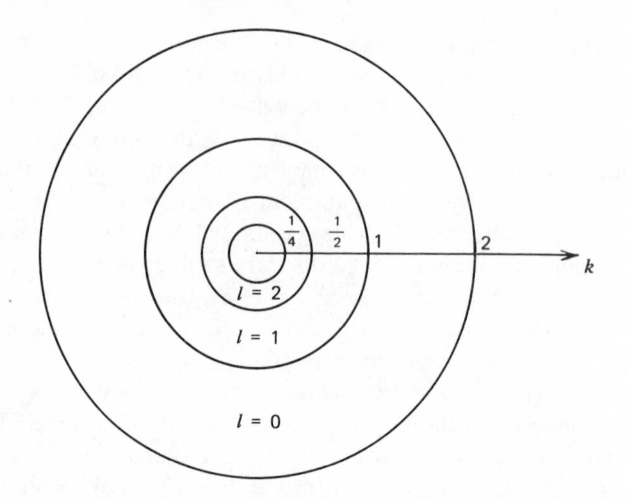

Figure 10.7.1. Subdivision of k space into shells.

We thus define a sequence of spherical shells, which we label by an index l: the lth shell contains all the wave vectors whose length $|\mathbf{k}| \equiv k$ is bounded as follows:

$$l\text{th shell:} \qquad 2^{-l} \leqslant k \leqslant 2 \times 2^{-l} \tag{10.7.5}$$

We now want to associate the index l introduced earlier with the lth shell. To do so, we assume that, to a first approximation the eigenfunction $\phi_{\mathbf{m}l}(\mathbf{k})$ is very nearly a constant, independent of $\mathbf{k}$ and $\mathbf{m}$ (but depending on l) for $\mathbf{k}$ within the lth shell, and vanishes for $\mathbf{k}$ outside of the shell. Hence, in the decomposition (10.7.3), each term $\sigma_l(\mathbf{k})$ corresponds to a definite range in $\mathbf{k}$, defined by Eq. (10.7.5).* Conversely, in Eq. (10.7.4) each term $s_l(\mathbf{x})$ describes contributions that are appreciable only for a range between 2^{l-1} and 2^l in space.

After these preparations, we advance to the main problem. We shall, however, not follow in detail Wilson's complicated calculations but, rather, illustrate his ideas with a trivially simple but instructive example.

Consider an imaginary spin system described in terms of the previously defined spin variables by the following Hamiltonian:

$$\beta H^{(0)} = \sum_{l=0}^{\infty} \sum_{\mathbf{m}} r s_{\mathbf{m}l}^2 \tag{10.7.6}$$

We wish to calculate the corresponding partition function $Z^{(0)}$. Because

* This also implies that $\sigma(\mathbf{k}) = 0$ for $k > 2$, which is reasonable: Large wave numbers correspond to small spatial scales, which are irrelevant in the continuum approximation.

there is no coupling between degrees of freedom, the latter factorizes:

$$Z^{(0)} = \prod_{l=0}^{\infty} \prod_{\mathbf{m}} \int_{-\infty}^{\infty} ds_{\mathbf{m}l} \exp(-rs_{\mathbf{m}l}^2) \tag{10.7.7}$$

We now decide to integrate first over all $s_{\mathbf{m}l}$ with $l = 0$. In the present case this is trivial:

$$Z^{(0)} = \left\{ \prod_{l=1}^{\infty} \prod_{\mathbf{m}} \int_{-\infty}^{\infty} ds_{\mathbf{m}l} \exp(-rs_{\mathbf{m}l}^2) \right\} \prod_{\mathbf{m}'} \int_{-\infty}^{\infty} ds_{\mathbf{m}'0} \exp(-rs_{\mathbf{m}'0}^2) = I_0 Z^{(1)} \tag{10.7.8}$$

where

$$I_0 = \prod_{\mathbf{m}} \left(\frac{\pi}{r}\right)^{1/2}$$

and

$$Z^{(1)} = \prod_{l=1}^{\infty} \prod_{\mathbf{m}} \int_{-\infty}^{\infty} ds_{\mathbf{m}l} \exp(-rs_{\mathbf{m}l}^2) \tag{10.7.9}$$

This quantity looks very much like $Z^{(0)}$, except for the absence of the factors corresponding to $l = 0$. In other words, $Z^{(1)}$ is the partition function for a Hamiltonian that does not contain the short-range contributions from the zeroth shell. However, if we define new field variables through a scale transformation:

$$s_{\mathbf{m}l} = \alpha s'_{\mathbf{m},l-1} \tag{10.7.10}$$

with α left arbitrary at this point, the partition function becomes

$$Z^{(1)} = \prod_{l=0}^{\infty} \prod_{\mathbf{m}} \alpha \int_{-\infty}^{\infty} ds'_{\mathbf{m}l} \exp(-r\alpha^2 s'^2_{\mathbf{m}l}) \tag{10.7.11}$$

Hence, within a constant factor, $Z^{(1)}$ has the same form as $Z^{(0)}$, except that the "coupling constant" r is replaced by $\alpha^2 r$. In other words, the "partial" partition function $Z^{(1)}$ can be expressed as a true partition function:

$$Z^{(1)} = \prod_{l=0}^{\infty} \prod_{\mathbf{m}} \int ds_{\mathbf{m}l} \exp(-\beta H^{(1)}) \tag{10.7.12}$$

corresponding to an effective Hamiltonian $H^{(1)}$; the latter is of the same form as $H^{(0)}$ except for the value of the parameter r:

$$H^{(1)}(r) = H^{(0)}(\alpha^2 r) \tag{10.7.13}$$

This process of rescaling can be continued indefinitely: after $s_{\mathbf{m}0}$, one integrates over $s_{\mathbf{m}1}$, $s_{\mathbf{m}2}$, and so forth. At each step the remaining effective Hamiltonian can be brought back to the form of the previous step:

$$H^{(l+1)}(r_l) = H^{(l)}(r_{l+1}) \tag{10.7.14}$$

with a relation between r_{l+1} and r_l:

$$r_{l+1} = \alpha_l^2 r_l \tag{10.7.15}$$

With this scheme, the evaluation of the partition function is reduced to the evaluation of the first step and to the solution of the recursion relations (10.7.15), which determine the effective parameter r_l as a function of l.

The process of going from $H^{(l)}$ to $H^{(l+1)}$ is very similar to the Kadanoff block construction: it represents essentially a doubling of the length scales, just like Kadanoff's doubling of the block side. The form invariance of the Hamiltonian corresponds to his assumption that the block Hamiltonian is of the same Ising form as the initial one, but with modified parameters. Equation (10.7.15), which relates to one another the parameters of the scaled Hamiltonians, plays the same role as the renormalization group equation (10.6.4), except for the minor point that it appears here as a finite-difference equation (recursion relation) rather than as a differential equation.

It must be stressed, however, that the scaling procedure, that is, the transition from l to $l+1$, has not been forced artificially from the outside but rather appears as being strongly connected to the structure of the partition function.

Although Wilson's procedure is precisely the one described above, it is clear that he starts from a more interesting Hamiltonian, which should model the interactions among spins. The Ising model would be an obvious first guess; but it soon appears that it does not possess the invariance property under scaling (as postulated by Kadanoff). One may therefore try to modify it by adding extra terms (and hence extra coupling parameters). These extra coupling parameters are of the same kind as the parameter q in Section 10.6. Wilson finally added what amounts to an infinite number of extra coupling parameters, in the form of an (unknown) functional $Q[s(\mathbf{x})]$ of the spin field. The role of this extra functional is to cut down the large fluctuations of the spin variables [such fluctuations become possible in a model in which no constraint is set on the length of $s(\mathbf{x})$]. Wilson's Hamiltonian thus reads*

$$\beta H^{(0)} = \frac{1}{2}\int d\mathbf{x}\, [\nabla s(\mathbf{x})] \cdot [\nabla s(\mathbf{x})] + \int d\mathbf{x}\, Q[s(\mathbf{x})] \tag{10.7.16}$$

* To see the connection with the Ising model note that the first term in (10.7.16) reads (in one dimension for simplicity) in the Fourier picture (10.7.2): $(1/2)\int dk\, k^2\sigma(k)\sigma(-k)$. On the other hand, a discrete Ising-like term $\sum_n (-s_n s_{n+1} + s_n^2)$ is Fourier-transformed into $\int dk\,(1-\cos k)\sigma(k)\sigma(-k)$. When only small values of k are relevant, this term reduces to $(1/2)\int dk\, k^2\sigma(k)\sigma(-k)$, thus providing the link with (10.7.16).

Such a Hamiltonian was previously used by Landau and Ginzburg in a different context.

Now, Wilson starts the program as described above; however, the partition function is no longer factorized. Hence, a large part of Wilson's second paper is devoted to an analysis of the orders of magnitude of the coupling terms and to finding approximations that enable one to solve the problem. These approximations are not always very clear; we shall not treat this part of the work in detail. Suffice it to quote the final result. The partial partition function at the lth step can again be written in terms of an effective Hamiltonian that is of the same form as $H^{(0)}$:

$$H^{(l)} = \frac{1}{2} \int d\mathbf{x} \, \{[\nabla s_l(\mathbf{x})] \cdot [\nabla s_l(\mathbf{x})] + Q_l[s_l(\mathbf{x})]\} \tag{10.7.17}$$

The function Q_l is, however, different at each step and is obtained by solving the recursion relation [analogous to (10.7.15)]:

$$Q_{l+1}(y) = -2^d \ln \frac{I_l(2^{1-d/2}y)}{I_l(0)} \tag{10.7.18}$$

where I_l is a functional of Q_l:

$$I_l(z) = \int_{-\infty}^{\infty} dy \, \exp[-y^2 - \tfrac{1}{2}Q_l(y+z) - \tfrac{1}{2}Q_l(-y+z)] \tag{10.7.19}$$

Equations (10.7.18) and (10.7.19) are the *RG equations* of our problem. We stress again the fact that these equations were *derived* (even though approximately) rather than postulated, as in Section 10.6. Once we are in possession of these equations, we may just refer back to Section 10.6 and translate the results obtained there into the language of finite (rather than infinitesimal) scale transformations.

It was argued in the previous section that the critical point can be identified (if it exists) with a fixed point of the RG equations. Hence, the critical point corresponds to an l-independent solution of Eq. (10.7.18): $Q_l(y) = Q_c(y)$.

The following step in Section 10.6 was the study of the linearized RG equations. The corresponding procedure here consists in assuming that $Q_l(z) - Q_c(z)$ is uniformly small; then $Q_{l+1}(z) - Q_c(z)$ should be an approximately linear functional of $Q_l(z) - Q_c(z)$:

$$Q_{l+1}(z) - Q_c(z) = \int_{-\infty}^{\infty} dy \, T(z, y)[Q_l(y) - Q_c(y)] \tag{10.7.20}$$

where the kernel $T(z, y)$ is obtained by expanding the rhs of (10.7.17) to first order in $Q_l - Q_c$: we do not need its explicit form at present. We may look for solutions of this equation in the form:

$$Q_l(z) - Q_c(z) = 2^{lY} q(z) \tag{10.7.21}$$

[This corresponds to Eq. (10.6.17) with $L = 2^l$.] Upon substitution of (10.7.21) into (10.7.20), the constant Y and the function $q(z)$ appear as eigenvalues and eigenfunctions of the kernel T:

$$e^{2Y} q(z) = \int_{-\infty}^{\infty} dy\, T(z, y) q(y) \tag{10.7.22}$$

It was shown in Section 10.6 that the eigenvalues Y were connected to the critical exponents. This continues to be true in the present problem. However, one can now go much further. Indeed, the present method is not based on an initial postulate (10.6.2), (10.6.3) about the form of the free enthalpy and of the correlation length, but was derived in view of the explicit calculation of the partition function. The same method is easily extended to the calculation of the two-spin correlation function. From this calculation, the microscopic critical exponents, η and ν, as well as the susceptibility exponent γ, can be evaluated. The result of this argument (which we do not reproduce here) is, as expected, a relation between critical exponents and the eigenvalues Y:

$$\gamma = \frac{2}{Y_M} \qquad \nu = \frac{1}{Y_M} \qquad \eta = 0 \tag{10.7.23}$$

where Y_M is the largest eigenvalue of (10.7.22). Another important condition must be satisfied, moreover. It may happen that the RG equation (10.7.18) has more than one fixed point. In that case one must make sure that the linearization is performed around a saddle-type fixed point, because this is the only one that can be identified with a critical point (see Section 10.6 and Fig. 10.6.2). The role of this condition will appear clearly in Section 10.8.

The result $\eta = 0$ may seem surprising. Actually, it is a result of the approximations made in deriving the RG equation: more about this point will be found in the next section. We may also note that the values (10.7.23) automatically satisfy the scaling relation (10.5.24).

To conclude we are faced, at this stage, with two problems:

1. In order to locate the critical point, we must find a fixed point of the RG equation (10.7.18) (and make sure it is of saddle-point type).
2. In order to calculate critical indices we must find the largest eigenvalue of the RG equation, linearized around the (saddle-type) fixed point.

10.8. DIMENSIONALITY AS A CONTINUOUS PARAMETER

The solution of the highly nonlinear RG equation is evidently a formidable problem. If we want some indications about its solutions, preferably in analytic form, we must devise some approximation procedure. In his second paper, Wilson initiated such a procedure and also published the results of a numerical treatment of his equation. The most spectacular method, however, appeared in a third paper of the series, by Wilson and Fisher: As a result, the theory becomes so simple that we can describe it here in full detail.

To solve the RG equations (10.7.18), (10.7.19) we must first specify an initial condition. We choose it to be simple but nontrivial:

$$Q_0(y) = r_0 y^2 + u_0 y^4 \tag{10.8.1}$$

where we must take $u_0 \geqslant 0$ in order to ensure convergence. Next, we assume that there is a *small parameter* ε in the problem: we do not yet specify its nature. We suppose that r_0 and u_0 are at least linear in ε:

$$r_0 = O(\varepsilon), \qquad u_0 = O(\varepsilon) \tag{10.8.2}$$

We now try to find a solution of the recursion equation of the form:

$$Q_l(y) = r_l y^2 + u_l y^4 + O(\varepsilon^3) \tag{10.8.3}$$

assuming again

$$r_l = O(\varepsilon), \qquad u_l = O(\varepsilon) \qquad \text{for all } l \tag{10.8.4}$$

If this ansatz works, we have already a very serious reduction of the problem: instead of an arbitrary function $Q_l(y)$ we now only need to calculate two parameters, r_l and u_l, to first order in ε.

We start by substituting (10.8.3) into Eq. (10.7.19), expand the integrand to second order in ε, and integrate explicitly the Gaussian-like expressions:

$$\begin{aligned}
I_l(z) &= \int_{-\infty}^{\infty} dy \, \exp\{-y^2 - \tfrac{1}{2}[r_l(y+z)^2 + u_l(y+z)^4] - \tfrac{1}{2}[r_l(z-y)^2 + u_l(z-y)^4]\} \\
&= \exp(-r_l z^2 - u_l z^4) \int dy \, \exp[-(1+r_l)y^2 - 6u_l y^2 z^2 - u_l y^4] \\
&\approx \exp(-r_l z^2 - u_l z^4) \int dy \, \exp[-(1+r_l)y^2] \\
&\quad \times [1 - u_l(6y^2z^2 + y^4) + \tfrac{1}{2}u_l^2(6y^2z^2 + y^4)^2 + \cdots] \\
&= \exp(-r_l z^2 - u_l z^4)\left(\frac{\pi}{1+r_l}\right)^{1/2} \Big\{1 - u_l[3z^2(1+r_l)^{-1} + \tfrac{3}{4}(1+r_l)^{-2}] \\
&\quad + \tfrac{1}{2}u_l^2\Big[27z^4(1+r_l)^{-2} + \left(\frac{45}{2}\right)z^2(1+r_l)^{-3} + \left(\frac{105}{16}\right)(1+r_l)^{-4}\Big]\Big\}
\end{aligned} \tag{10.8.5}$$

We now substitute this result into the rhs of Eq. (10.7.18):

$$Q_{l+1}(z) = -2^d\{(-r_l 2^{2-d}z^2 - u_l 2^{4-2d}z^4) + \ln[1 - u_l\alpha(2^{1-d/2}z) + u_l^2\beta(2^{1-d/2}z)] - \ln[1 - u_l\alpha(0) + u_l^2\beta(0)]\} \quad (10.8.6)$$

where $\alpha(z)$ and $\beta(z)$ are the coefficients of u_l and u_l^2, respectively, in (10.8.5). We finally expand the logarithms and the factors $(1+r_l)^{-1}$ and find that, to second order in ε, Eq. (10.8.6) is of the form:

$$Q_{l+1}(z) = r_{l+1}z^2 + u_{l+1}z^4$$

with

$$r_{l+1} = 4(r_l + 3u_l - 3r_l u_l - 9u_l^2) \quad (10.8.7)$$

$$u_{l+1} = 16 \times 2^{-d}(u_l - 9u_l^2) \quad (10.8.8)$$

We succeeded in reducing the RG equations from a complicated functional form to a couple of simple algebraic equations. These are, however, not yet definite as long as we do not define the smallness parameter ε. We note that the only disposable parameter left in the equations is d, the number of dimensions. It is already clear from the previous discussions of this chapter that the dimensionality plays a very crucial role in the theory of critical phenomena. The critical exponents, in particular, are quite different in two and in three dimensions. The form of Eq. (10.8.8) suggests that $d=4$ might play a special role. Indeed, when $d<4$, the coefficient of the bracketed function in the right-hand side is larger than 1, and when $d>4$ it is smaller than 1. This may not be a very suggestive argument, but it was clear from the indications of Wilson's second paper that $d=4$ appeared as a turning point; this will be amply confirmed by the subsequent calculations.

At this point, Wilson and Fisher introduce a truly provocative concept. They consider the *dimensionality d as a continuously variable parameter.* Accepting $d=4$ as a turning point they write in general

$$d = 4 - \varepsilon \quad (10.8.9)$$

and take ε as the small expansion parameter introduced above. One now expands r_l and u_l, and eventually the critical exponents as power series in ε (the Wilson–Fisher paper is agressively entitled "Critical Exponents in 3.99 Dimensions"!). These series have no physical meaning for non-integral values of ε; however, one may try to set $\varepsilon=1$ in these expressions: it turns out that the result is amazingly accurate! We now carry out this program.

We have first

$$16 \times 2^{-d} = 16 \times 2^{-4+\varepsilon} = 2^{\varepsilon} = e^{\varepsilon \ln 2} \approx 1 + \varepsilon \ln 2 + \cdots$$

Hence, Eq. (10.8.8) becomes, to second order in ε:

$$u_{l+1} = (1 + \varepsilon \ln 2)u_l - 9u_l^2 \qquad (10.8.10)$$

We now look for *fixed-point solutions:*

$$r_l = r, \qquad u_l = u \qquad \text{for all } l \qquad (10.8.11)$$

As Eq. (10.8.10) is independent of r_l it can be solved separately with the ansatz (10.8.11):

$$u = u + \varepsilon(\ln 2)u - 9u^2$$

which yields two solutions: $u = 0$ and $u = \varepsilon(\ln 2)/9$. Substituting these into (10.8.7) we find that the RG equations have *two fixed points:*

$$P_{\mathrm{I}}: \quad r_{\mathrm{I}} = 0, \quad u_{\mathrm{I}} = 0 \qquad (10.8.12)$$

$$P_{\mathrm{II}}: \quad r_{\mathrm{II}} = -(\tfrac{4}{9})\varepsilon \ln 2, \quad u_{\mathrm{II}} = (\tfrac{1}{9})\varepsilon \ln 2 \qquad (10.8.13)$$

At this stage we do not know yet which of these must be identified with a critical point. Continuing the general procedure, we now linearize Eqs. (10.8.7) and (10.8.10) around each of the singular points.

Calling (r^*, u^*) the coordinates of either P_{I} or P_{II}, we set

$$\begin{aligned} r_l &= r^* + \Delta r_l \\ u_l &= u^* + \Delta u_l \end{aligned} \qquad (10.8.14)$$

and get

$$\begin{aligned} \Delta r_{l+1} &= 4(\Delta r_l + 3\Delta u_l - 3u^*\Delta r_l - 3r^*\Delta u_l - 18u^*\Delta u_l) \\ \Delta u_{l+1} &= (1 + \varepsilon \ln 2)\Delta u_l - 18u^*\Delta u_l \end{aligned} \qquad (10.8.15)$$

We now make the ansatz:

$$\begin{aligned} \Delta r_l &= A\lambda^l \\ \Delta u_l &= B\lambda^l \end{aligned} \qquad (10.8.16)$$

Substituting into (10.8.15) we obtain

$$\begin{aligned} A\lambda^{l+1} &= 4\lambda^l(A + 3B - 3u^*A - 3r^*B - 18u^*B) \\ B\lambda^{l+1} &= \lambda^l(1 + \varepsilon \ln 2 - 18u^*)B \end{aligned} \qquad (10.8.17)$$

After cancellation of a factor λ^l on both sides, we are left with a homogeneous couple of equations for A and B, which possesses nontrivial solutions only if the characteristic determinant vanishes:

$$\begin{vmatrix} 4-12u^*-\lambda & 12-12r^*-72u^* \\ 0 & 1+\varepsilon \ln 2-18u^*-\lambda \end{vmatrix} = 0 \tag{10.8.18}$$

This equation yields two eigenvalues for each choice of u^*:

$$\begin{aligned} \lambda_M^* &= 4-12u^* \\ \lambda_m^* &= 1+\varepsilon \ln 2-18u^* \end{aligned} \tag{10.8.19}$$

or, explicitly, for $u^* = u_{\rm I} = 0$:

$$\begin{aligned} \lambda_M^{\rm I} &= 4 \\ \lambda_m^{\rm I} &= 1+\varepsilon \ln 2 \end{aligned} \tag{10.8.20}$$

and for $u^* = u_{\rm II} = (1/9)\varepsilon \ln 2$:

$$\begin{aligned} \lambda_M^{\rm II} &= 4-\tfrac{4}{3}\varepsilon \ln 2 \\ \lambda_m^{\rm II} &= 1-\varepsilon \ln 2 \end{aligned} \tag{10.8.21}$$

We now have to decide which of the two fixed points is the critical point and which eigenvalue must be used for the calculation of critical exponents. To do so, we study the behavior of the eigenvalues as functions of ε. In principle we should assume $|\varepsilon| \ll 1$, but the conclusions hold even if ε is as big as 2. Consider first $\varepsilon < 0$ (i.e., $d > 4$). Then we have

$$\begin{aligned} &\lambda_M^{\rm I} > 1, \qquad \lambda_M^{\rm II} > 1, \\ &\lambda_m^{\rm I} < 1, \qquad \lambda_m^{\rm II} > 1, \qquad \varepsilon < 0 \end{aligned} \tag{10.8.22}$$

In order to compare this with the previous discussions, we recall that the ansatz (10.8.16) with λ^l corresponds to the ansatz (10.7.21) with 2^{lY} and to (10.6.17) with L^y. Hence, an eigenvalue $\lambda < 1$ corresponds in the language of Section 10.6 to an exponent $y < 0$, and $\lambda > 1$ implies $y > 0$. It then results from (10.8.22) that *for* $\varepsilon < 0$, $P_{\rm I}$ *is a saddle point and* $P_{\rm II}$ *is an unstable node* (Fig. 10.6.2*b*). Hence the critical point must be identified with $P_{\rm I}$ and the critical index must be computed from the largest corresponding eigenvalue $\lambda_M^{\rm I}$, according to Eq. (10.7.23). From the previous remark we have

$$\begin{aligned} \lambda_M^{\rm I} &= 2^{Y_M} \\ Y_M &= \frac{\ln \lambda_M^{\rm I}}{\ln 2}, \qquad \varepsilon < 0 \end{aligned} \tag{10.8.23}$$

As $\lambda_M^I = 4 = 2^2$, we get $Y_M = 2$ and from (10.7.23):

$$\gamma = 1$$
$$\nu = \tfrac{1}{2} \tag{10.8.24}$$
$$\eta = 0, \qquad d > 4$$

This completes our program for the case $\varepsilon < 0$. The very remarkable feature about this result is the fact that we obtained precisely *the "classical" values for the critical exponents* (see Table 10.1.1). Moreover, these values are constant, independent of ε, and therefore valid for all dimensionalities $d > 4$ (provided the theory remains valid). This result confirms earlier conjectures of Brout and of Fisher.

We are, however, condemned to live in a complicated world, so we now turn to the case $\varepsilon > 0$, or $d < 4$. We now see from (10.8.20), (10.8.21) that our former conclusions are reversed:

$$\lambda_M^I > 1, \qquad \lambda_M^{II} > 1$$
$$\lambda_m^I > 1, \qquad \lambda_m^{II} < 1, \qquad \varepsilon > 0 \tag{10.8.25}$$

Thus P_{II} is now a saddle point and P_I is a node. We see that $d = 4$ is indeed a turning point at which the two singularities of the initial nonlinear RG equations cross each other. To calculate the critical exponents we must use now λ_M^{II}, and from Eq. (10.8.23) we get, to first order in ε,

$$Y_M = \frac{\ln \lambda_M^{II}}{\ln 2} = \frac{\ln[4(1-(1/3)\varepsilon \ln 2)]}{\ln 2}$$
$$\approx \frac{2 \ln 2 - (1/3)\varepsilon \ln 2}{\ln 2} = 2 - \tfrac{1}{3}\varepsilon$$

Hence

$$\gamma = \frac{2}{Y_M} = \frac{2}{2[1-(1/6)\varepsilon]} \approx 1 + \tfrac{1}{6}\varepsilon$$

Thus the set of critical exponents becomes

$$\gamma = 1 + \tfrac{1}{6}\varepsilon$$
$$\nu = \tfrac{1}{2} + \tfrac{1}{12}\varepsilon \tag{10.8.26}$$
$$\eta = 0, \qquad d < 4$$

The entire preceding discussion is summarized in Table 10.8.1.

It is, of course, tempting to come down to our real world by setting $\varepsilon = 1$ in Eq. (10.8.26). We then obtain the results

$$\gamma = \tfrac{7}{6} = 1.167$$
$$\nu = \tfrac{7}{12} = 0.583 \tag{10.8.27}$$
$$\eta = 0, \qquad d = 3$$

Table 10.8.1. Discussion of the Singularities of the RG Equations

		Type of P_{I}	Type of P_{II}	Relevant eigenvalue	γ
$\varepsilon<0$	$d>4$	*Saddle*	Node	λ_M^{I}	1
$\varepsilon>0$	$d<4$	Node	*Saddle*	λ_M^{II}	$1+\frac{1}{6}\varepsilon$

These numbers must be compared with the numerical results for the $d=3$ Ising model, taken from Table 10.1.1:

$$\begin{aligned}\gamma &= 1.250\\ \nu &= 0.638\\ \eta &= 0.041, \qquad d=3 \ \text{Ising model}\end{aligned} \tag{10.8.28}$$

Given the many approximations of the method, the agreement is already quite good.

The next thing one should like to do is to push the expansion (10.8.26) to higher orders. This, however, does not work! It turns out that the approximations made in deriving the original RG equation (10.7.18) interfere and make the second-order result incorrect. Accordingly, we should first improve the initial equation, which is a very difficult task.

But one idea implies another! Wilson soon realized that the $d=4-\varepsilon$ trick, which made the RG method so simple, can also be applied to other, more conventional methods. In particular, many authors had studied critical phenomena by doing graphical expansions in powers of a small parameter [e.g., u_0 in Eq. (10.8.1)]; these methods proceed from the philosophy of Chapter 6. The integrals corresponding to the individual graphs are very difficult. However (in a fourth paper of the series), Wilson noted that if these integrals are further expanded in powers of ε, the resulting coefficients can be calculated exactly. Combining this calculation with a renormalization-group argument, he could compute exactly the critical exponents γ and η for a class of models of type (10.2.1) with spins of arbitrary dimensionality D. His results are

$$\begin{aligned}\gamma &= 1+\frac{D+2}{2(D+8)}\varepsilon+\frac{(D+2)(D^2+22D+52)}{4(D+8)^3}\varepsilon^2+O(\varepsilon^3)\\ \eta &= \frac{D+2}{2(D+8)^2}\varepsilon^2+\frac{(D+2)}{2(D+8)^2}\left(\frac{6(3D+14)}{(D+8)^2}-\frac{1}{4}\right)\varepsilon^3+O(\varepsilon^4)\end{aligned} \tag{10.8.29}$$

First, we see that for the Ising model ($D=1$) the exact first-order terms in both γ and η agree with the result (10.8.26). The exponent η starts only in order ε^2: this explains its smallness. Finally, if we try again the (rather

Table 10.8.2. Comparison of Wilson's Results with Model Results

Exponent	D	Model	Eq. (10.8.29)	Best estimate
	1	Ising	1.244	1.250
γ	3	Heisenberg	1.347	1.375
	∞	Spherical	1.750	2 (exact)
	1	Ising	0.037	0.041
η	3	Heisenberg	0.039	0.043
	∞	Spherical	0	0 (exact)

unorthodox) substitution $\varepsilon = 1$ we get the results of Table 10.8.2. The agreement with the known results is amazing!

Wilson's theory will probably not be the last word in the theory of critical phenomena. Many approximations have been introduced into the argument, sometimes in an uncontrollable way. Also, the scaling laws that Wilson derives may not be a general property, although a large class of models and real systems does appear to follow them. It is, however, undeniable that the Wilson theory appears as the conclusion and the culmination of a long effort that we tried to describe in this chapter. The vista it opens into this difficult problem is aesthetically so appealing that it must describe at least part of the truth!

BIBLIOGRAPHICAL NOTES

General references

H. E. Stanley (quoted in Chapter 9).

M. E. Fisher, *Rep. Progr. Phys.* **30,** 615 (1967).

P. Heller, *Rep. Progr. Phys.* **30,** 731 (1967).

L. P. Kadanoff et al., *Rev. Mod. Phys.* **39,** 395 (1967).

The *Ising model* was introduced, and solved in one dimension, in E. Ising, *Z. Phys.* **31,** 253 (1925).

The celebrated solution of the two-dimensional Ising model was obtained by L. Onsager, *Phys. Rev.* **65,** 117 (1944).

A much simpler method for this solution was given by N. V. Vdovichenko, *Zh. Eksp. Teor. Fiz.* **47,** 715 (1944); **48,** 526 (1965) [English transl., *Sov. Phys. JETP* **20,** 477 (1965); **21,** 350 (1965)].

Vdovichenko's method is also reviewed in Landau and Lifshitz, *Statistical Physics* (quoted in Chapter 4) and in Stanley (quoted in Chapter 9).

The *spherical model* was solved by T. H. Berlin and M. Kac, *Phys. Rev.* **86,** 821 (1952).

The Ising model for $D = \infty$ was solved by H. E. Stanley, *Phys. Rev.* **176,** 718 (1968).

Some of Lieb's models are reviewed in E. Lieb, *Lectures in Theoretical Physics,* Vol. 11D, Univ. Colorado Press, Boulder, 1969.

A detailed discussion of *exact models* appears in the recent monograph: C. Domb and M. S. Green (eds.), *Phase Transitions and Critical Phenomena,* Academic Press, New York, 1972.

Landau's theory appeared in: L. D. Landau, *Phys. Z. Sowjun.*, **11,** 26, 545 (1937); see also: *Collected Papers of L. D. Landau* (D. ter Haar, ed.), Gordon and Breach, New York, 1965. The theory is also clearly exposed in Landau and Lifshitz, *Statistical Physics* (quoted in Chapter 4).

The *scaling assumption* was introduced in B. Widom, *J. Chem. Phys.* **43,** 3829, 3898 (1965).

A very clear recent review of all aspects of scaling theory appears in A. Hankey and H. E. Stanley, *Phys. Rev. B* **6,** 3515 (1972).

The *universality assumption* and the block construction appear in L. P. Kadanoff, *Physics* **2,** 263 (1966) and L. P. Kadanoff (quoted above).

Wilson's theory is contained in

K. G. Wilson, *Phys. Rev. B* **4,** 3174, 3184 (1971).

K. G. Wilson and M. E. Fisher, *Phys. Rev. Lett.* **28,** 240 (1972).

K. G. Wilson, *Phys. Rev. Lett.* **28,** 548 (1972).

PART 3

NONEQUILIBRIUM STATISTICAL MECHANICS

CHAPTER 11
INTUITIVE TREATMENT OF NONEQUILIBRIUM PHENOMENA

11.1. INTRODUCTION

Our purpose in the third part of this book is to arrive at an understanding of the process of time evolution in a large system of molecules. We should be able to handle this problem, in principle, by solving the Liouville equation, with appropriate initial and boundary conditions. A detailed analysis of this solution should bring out all the features observed in macroscopic physics. This statement is based on a fundamental philosophy, contained in the following claim: Given a system, described at time $t = 0$ by an arbitrary ensemble distribution, it is possible to explain its subsequent time evolution by the exact laws of classical or quantum mechanics. This statement implies the idea that no qualitative modification of the laws of mechanics is necessary in order to understand why a large collection of molecules behaves, as a whole, in a way that seems to contradict the elementary laws of motion.

We are mainly referring to the celebrated *irreversibility paradox*, which can be illustrated in many elementary ways. Consider, for instance, a box divided into two compartments by a wall; the two halves are filled with gas at two different pressures. If at time $t = 0$ we break the internal wall, the gas will flow from the high-pressure to the low-pressure compartment, until it reaches a state of equal pressure in both half-boxes: at that time an equilibrium state is reached and the macroscopic evolution stops. The inverse flow is never observed.

On the other hand, an individual particle, in the absence of external forces, is not more likely to move to the right rather than to the left. How then can we understand the irreversible, privileged arrow of evolution exhibited by the system as a whole, in terms of the motion of individual molecules subjected to reversible laws?

This is a very difficult problem, which has been the subject of vivid discussions over the last century. In recent years, however, a rather clear understanding is beginning to emerge from a close analysis of the

Liouville equation. The remainder of this book will be devoted to this analysis.

We shall, however, begin, in the present chapter, with a more "traditional" intuitive treatment of a few typical systems. As will be shown, in these elementary treatments, the laws of motion assumed for the molecules deliberately deviate from the laws of exact mechanics: at some point or other, an element of a probabilistic nature is thrown into the argument. It will be our purpose in subsequent chapters to understand and justify such apparent deviations.

In spite of this objection, it is important to study these traditional theories. They provide an excellent orientation as to the types of equations to be obtained from a more general theory. Moreover, the basic equations obtained in this rough way turn out to be perfectly valid (within the range of their applicability) and will be derived, as they stand, from the general theory. This fact by itself should be considered as a tribute to the great physical intuition of the founders of kinetic theory and, above all, to Ludwig Boltzmann.

If the kinetic equations proposed in this chapter are accepted, it is important to visualize clearly the process of evolution that they describe, particular stress being laid on the differences between this image and the image offered by "ordinary mechanics."

A second and equally important question that must be clarified before embarking on a more general, but necessarily more abstract theory, is the following: Given a kinetic equation describing the time evolution of a system, what kind of information can we extract from it? How can we calculate from it the quantities of real physical interest?

These two aspects will be the subject of the following three chapters. After these questions are clarified we will have a good motivation for trying to obtain a better understanding of the more basic aspects of the evolution in time of many-body systems.

11.2. THEORY OF THE BROWNIAN MOTION: THE LANGEVIN EQUATION

We first study a classical problem of nonequilibrium statistical mechanics: The problem of *Brownian motion* relates to the motion of a heavy colloidal particle immersed in a fluid made up of light particles. As will presently be seen, the traditional method for handling this problem departs from exact mechanics in a quite obvious way. At the very first step in the argument one renounces an exact description of the motion, replacing it by a deliberately probabilistic law. This has at least the advantage of clarity.

Let us then consider our Brownian particle, of mass M, immersed in a fluid. If we treated the problem macroscopically, the laws of hydrodynamics would tell us that the particle during its motion undergoes a *friction force* due to the viscosity of the fluid, proportional to its velocity $\mathbf{v}$; hence Newton's equation reads*

$$\dot{v} = -\zeta v \qquad \text{(macroscopic)} \tag{11.2.1}$$

where the friction coefficient ζ will be assumed constant.

We are, however, interested in a more refined description. Therefore we must add to the systematic force the action of all the individual molecules on the Brownian particle, which results in an additional term $A(t)$:

$$\dot{v} = -\zeta v + A(t) \qquad \text{(microscopic)} \tag{11.2.2}$$

The effect of each individual collision is to produce a tiny deflection of the particle away from its macroscopic trajectory. If we do not want to go into the details of the many-body dynamics, the only thing we can say about the collisions is that they are very numerous and very irregular as to their strength and direction. Contrary to our first impression, this statement is by no means negative or discouraging. On the contrary, *provided we are ready to give up a deterministic description of the process*, it supplies us with the necessary basis for an application of the law of large numbers and of the *theory of probabilities*. We cannot specify the force $A(t)$ as a given function of time; but we can make reasonable assumptions about the *average* effect of the collisions over a large number of identical macroscopic situations (i.e., over an ensemble). Similarly, we cannot predict the velocity or the position of the Brownian particle at every time t, but we may be able to predict the average outcome of a large number of experiments performed under identical conditions. Therefore the whole philosophy of solving Eq. (11.2.2) is different from the traditional, deterministic initial-value problem for a differential equation. Equation (11.2.2) is a typical (and celebrated) example of the class of so-called *stochastic* (or *random*) *equations of motion*. It is called, after its discoverer, the *Langevin equation*. Let us see how such an equation is solved.

We must first define quantitatively the random function $A(t)$. We make the following assumptions:

1. The average of $A(t)$ over an ensemble of Brownian particles having the given velocity v_0 at time $t = 0$ vanishes. We write this as follows:

$$\langle A(t)\rangle_{v_0} = 0, \qquad t > 0 \tag{11.2.3}$$

* We restrict ourselves to one-dimensional notations for simplicity. The generalization to three dimensions is not difficult.

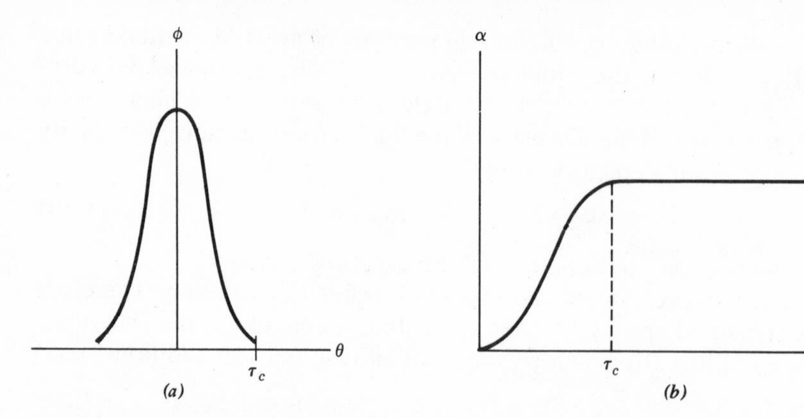

Figure 11.2.1. (*a*) Graph of the function $\phi(\theta)$, Eq. (11.2.4); (*b*) Graph of the function $\alpha(T)$, Eq. (11.2.10).

This condition assures us that the average velocity of the Brownian particle obeys precisely the macroscopic law (11.2.1), that is, that the fluctuations cancel each other on the average.

2. We may express the idea of irregularity by assuming that collisions well separated in time are statistically independent. In other words, the correlation between the values of $A(t)$ at two times t_1, t_2 is different from zero only for time intervals of the order of the duration of a collision τ_c. Explicitly,

$$\langle A(t_1)A(t_2)\rangle_{v_0} = \phi(t_1 - t_2) \tag{11.2.4}$$

where $\phi(t)$ is a function very sharply peaked at $t=0$ and vanishing practically for $|t| > \tau_c$ (see Fig. 11.2.1*a*).

3. We now assume that all higher-order moments of the random variable $A(t)$ can be expressed in terms of the second moments. More precisely:

$$\langle A(t_1)A(t_2)\ldots A(t_{2n+1})\rangle_{v_0} = 0 \tag{11.2.5}$$

$$\langle A(t_1)A(t_2)\ldots A(t_{2n})\rangle_{v_0}$$
$$= \sum \langle A(t_{i_1})A(t_{i_2})\rangle_{v_0}\langle A(t_{i_3})A(t_{i_4})\rangle_{v_0}\ldots\langle A(t_{i_{2n-1}})A(t_{i_{2n}})\rangle_{v_0} \tag{11.2.6}$$

The sum in the last equation is over all the different ways of splitting the $2n$ variables $t_1, \ldots, t_{2n}$ into n pairs. The present assumption can be shown to be equivalent to the assumption that $A(t)$ is distributed according to a *Gaussian* law, which is connected to the law of large numbers (or the central limit theorem) of probability theory.

We are now in a position to solve the Langevin equation very easily. Suppose the Brownian particle starts at $t=0$ with velocity v_0. If we considered (11.2.2) as an ordinary differential equation, its solution would be

$$v = v_0 e^{-\zeta t} + e^{-\zeta t} \int_0^t d\tau \, e^{\zeta \tau} A(\tau) \tag{11.2.7}$$

This equation, as it stands, is empty because $A(t)$ is not known. If, however, we average all its terms over the ensemble defined above, and use Assumption (11.2.3), we obtain

$$\langle v \rangle_{v_0} = v_0 e^{-\zeta t} \tag{11.2.8}$$

This is simply the solution of the macroscopic equation (11.2.1), a result that is by no means surprising. More interesting is the average of the square of the velocity. Squaring the right-hand side of (11.2.7) and averaging the result we obtain

$$\langle v^2 \rangle_{v_0} = v_0^2 e^{-2\zeta t} + e^{-2\zeta t} \int_0^t d\tau_1 \int_0^t d\tau_2 \, e^{\zeta(\tau_1 + \tau_2)} \langle A(\tau_1) A(\tau_2) \rangle_{v_0} \tag{11.2.9}$$

We now make a change of variables of integration:

$$\tau_1 + \tau_2 = \xi, \qquad \tau_2 - \tau_1 = \theta$$

We also use Eq. (11.2.4) and introduce the function $\alpha(T)$:

$$\alpha(T) = \int_{-T}^{T} d\theta \, \phi(\theta) \tag{11.2.10}$$

We then obtain, by elementary calculations,

$$\begin{aligned} J(t) &\equiv \int_0^t d\tau_1 \int_0^t d\tau_2 \, e^{\zeta(\tau_1 + \tau_2)} \langle A(\tau_1) A(\tau_2) \rangle_{v_0} \\ &= \frac{1}{2} \int_0^t d\xi \, e^{\zeta \xi} \alpha(\xi) + \frac{1}{2} \int_t^{2t} d\xi \, e^{\zeta \xi} \alpha(2t - \xi) \end{aligned} \tag{11.2.11}$$

The behavior of the function $\alpha(T)$ gives us a hint for the approximate evaluation of this integral (see Fig. 11.2.1). Our assumption that $\phi(\theta)$ is very sharply peaked around $\theta = 0$ and drops practically to zero for $\theta > \tau_c$ implies that its integral $\alpha(T)$ reaches a constant value $\alpha \equiv \alpha(\infty)$ for $t \sim \tau_c$. Therefore, *if we are interested in times t much longer than the duration of a collision*, we may, without appreciable error, replace $\alpha(\xi)$ and $\alpha(2t - \xi)$ by their constant asymptotic value α in the right-hand side of Eq. (11.2.11). Indeed, the exact integrand differs from the approximate one only in a small domain (of width τ_c) near $\xi = 0$ (for the first integral) and $\xi = 2t$ (for the second integral). Hence, we may write approximately for $t \gg \tau_c$

$$J(t) \approx \tfrac{1}{2} \alpha \int_0^{2t} d\xi \, e^{\zeta \xi} = \alpha \frac{e^{2\zeta t} - 1}{2\zeta} \tag{11.2.12}$$

and, substituting into Eq. (11.2.9):

$$\langle v^2\rangle_{v_0} = v_0^2 e^{-2\zeta t} + \frac{\alpha}{2\zeta}(1 - e^{-2\zeta t}) \qquad (11.2.13)$$

This formula is very interesting in itself. It shows us that for short times, $t \ll (2\zeta)^{-1}$, the velocity fluctuations are mainly determined by the initial value v_0^2. But for longer times the initial value is progressively forgotten, and the average square of the velocity approaches the value $\alpha/2\zeta$, which is solely determined by the mechanism of the collisions and is independent of the initial velocity.

To complete the theory we should evaluate the constant α in terms of a set of molecular parameters. But clearly, our assumptions are only of a qualitative nature and cannot provide us with this information. We can, however, circumvent the difficult problem of evaluating α from first principles by "cheating" in the following way. We have good reason to believe that the end point of the evolution is the state of *thermal equilibrium*. corresponding to the temperature T of the fluid.

In this state, the mean square velocity of the particle is determined by the theorem of equipartition of energy (Section 5.2). Hence, we may *require* that

$$\lim_{t\to\infty} \langle v^2\rangle_{v_0} \equiv \frac{\alpha}{2\zeta} = \frac{k_B T}{M} \qquad (11.2.14)$$

With this requirement we obtain

$$\langle v^2\rangle_{v_0} = \frac{k_B T}{M} + \left[v_0^2 - \frac{k_B T}{M}\right] e^{-2\zeta t} \qquad (11.2.15)$$

This formula completes our calculation.* It describes in very simple and transparent form how the Brownian particles forget their initial velocity and are driven to equilibrium through the effect of the collisions with the fluid particles. All the ingredients of an irreversible evolution are present in this example.

It should be clear, however, that the traditional theory of Brownian motion is only a semiphenomenological theory. The final equilibrium has not been *derived* but, rather, has been *injected* into the theory. The important parameter ζ, which fixes the time scale of the evolution, is assumed to be given. Just like the constant α, it hides all the complicated dynamical processes involved in the collision mechanism.

We should finally stress another aspect, which is important in the

* We do not treat the calculation of higher moments of v, which is quite similar to the calculation studied in detail here. Equations (11.2.5) and (11.2.6) play, of course, an important role in this study.

theories of irreversibility and which will be met again later. In the evaluation of the integral in Eq. (11.2.11) we have used an approximation that is only valid for sufficiently *long times* ($t \gg \tau_c$). The simple equation (11.2.13) must therefore be understood as an *asymptotic* result. We may say that Eq. (11.2.13) does not represent the exact motion (even under the simplifying assumptions accepted from the beginning). We must rather view it as a "coarse-grained" or "time-smoothed" picture, in which the fine details of the motion (over a time scale τ_c) have been wiped out. One is entitled to think that irreversibility has been introduced into the theory precisely through this time-smoothing device. We will see later how this idea must be understood within a general framework.

11.3. RANDOM PROCESSES: THE FOKKER–PLANCK EQUATION

The problem of Brownian motion is a particular example of application of the general theory of *random* or *stochastic processes.* Many other problems can be treated in a semiphenomenological way by applying this theory. Conversely, it will be shown later that the exact analysis of the many-body problem leads to a justification of the validity of these methods in many important cases. It is, therefore, interesting to review several general ideas and methods of the theory of stochastic processes.

Let y denote the variable in which we are interested: it may be, for example, the position or the velocity of a Brownian particle or the current in a noisy electrical circuit. If y were a deterministic quantity, we could construct a function of time $y(t)$, which determines the value of y at every time t, given appropriate initial data at $t = 0$. If, however, y is a random variable, such a function does not exist. At every given time, the variable y can have any value whatsoever within its range of variation. However, to every possible value at time t there is attached a certain probability, which may have any value between zero and one. As we assume y to be a continuous variable, it is easier and shorter to speak about probability densities. To be precise, we shall say that the value y of the random variable has a probability density $W(y; t)$ at time t if there is a probability $W(y; t)\, dy$ of finding the value of the variable in the infinitesimal interval $(y, y + dy)$.

To get an intuitive idea of a random process we may look at Fig. 11.3.1. At successive times, the most probable values of y have been drawn as heavy dots. We may select a most probable trajectory from such a picture. But nothing excludes the possible existence of two or more trajectories of equal probability.

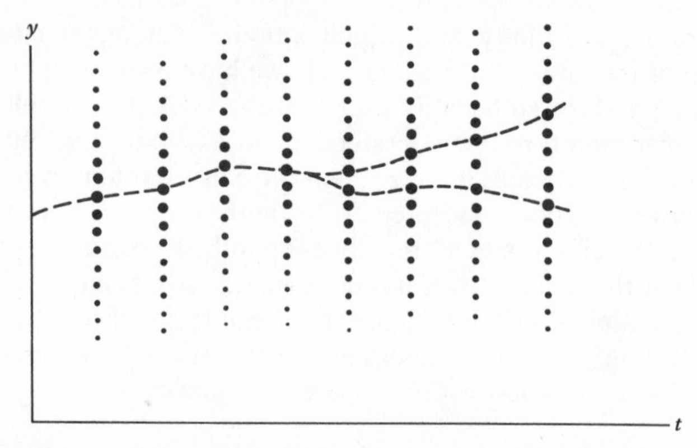

Figure 11.3.1. A representation of a random process. The heavier dots represent the more probable values of y. The broken line is a trajectory of maximum probability.

On the other hand, the mere knowledge of the probability density $W_1(y; t)$ is not sufficient, in general, for the characterization of the process. If we happen to know that the variable has the value y_1 at t_1, this knowledge will influence the probability of finding the value y_2 at time t_2, because the various points are not necessarily independent (i.e., there is a correlation between what happens at t_1 and what happens at t_2). In other words, the *joint probability density* of finding the value y_1 at t_1 and the value y_2 at t_2, $W_2(y_1; t_1 \mid y_2; t_2)$ cannot necessarily be inferred from the knowledge of $W_1(y_1; t_1)$. Hence, for a complete characterization of the random process, we must in principle specify all the joint probabilities, $W_1(y_1; t_1)$, $W_2(y_2; t_2 \mid y_1; t_1)$, $W_3(y_3; t_3 \mid y_2; t_2 \mid y_1; t_1)$, and so on, *ad infinitum*.*

The reader will rightly note at this point the analogy with the problem of reduced distribution functions discussed in Chapter 3. There is, nevertheless, a difference: the probabilities here refer to events occurring at different times. The functions W_n must possess a certain number of obvious properties:

(a) $W_n \geqslant 0$, because they are probability densities.

(b) $W_n(y_n; t_n \mid y_{n-1}; t_{n-1} \mid \cdots \mid y_1; t_1)$ must be symmetric with respect to permutations of the group of variables $(y_1; t_1)$ among each other; this follows from the idea that W_n represents a *joint* probability.

* The limiting situation corresponds to a subdivision of the time axis into smaller and smaller intervals, with corresponding specification of higher and higher joint probabilities. The complete description thus attributes a probability to every *path* in the y–t plane. The probability density becomes a *functional* $W\{y(t)\}$ of the function $y(t)$ (see Section 7.5). The theory of random processes was actually the starting point for the development of the mathematical techniques of functional integration, especially by Wiener. We cannot treat these interesting problems within the framework of this book.

(c) W_n must be compatible with all the lower-order joint probabilities:

$$\int dy_{k+1} \cdots dy_n \, W_n(y_n; t_n | \cdots | y_1; t_1) = W_k(y_k; t_k | \cdots | y_1; t_1) \quad (11.3.1)$$

This relation must hold for all $k < n$, whatever the values of the times $t_{k+1}, \ldots, t_n$.

In order to make further progress, we introduce a *classification* of the random processes. The idea behind this classification is that one may conceive of situations in which the knowledge of one, two, or a finite number of W_n's is sufficient for the characterization of the problem. In other words, a higher-order W_n may, in certain cases, be expressed as a combination of lower-order ones. The idea here is the same as the one underlying the truncation procedures of the hierarchies of reduced distribution functions mentioned in Sections 3.3 and 7.4. We shall not, however, exhaust this classification systematically, but rather consider only the two first classes.

The simplest case (sometimes called the *purely random process*) is one in which the knowledge of $W_1(y; t)$ suffices for the solution of the problem. In particular:

$$W_2(y_2; t_2 | y_1; t_1) = W_1(y_2; t_2) W_1(y_1; t_1) \quad (11.3.2)$$

Hence, correlations in time are here completely nonexistent. This, however, would be a very unrealistic assumption in a *continuous* physical process; indeed, for short enough time intervals there must be a causal relationship between successive events [see the discussion of our earlier assumption (11.2.4)].

The next simplest case is of fundamental importance in statistical physics: it is called the *Markov process*. The whole information is now contained in the first two functions, W_1 and W_2. In order to characterize it precisely it is convenient to introduce the concept of a *transition probability* $w_2(y_2; t_2 | y_1; t_1)$ defined through the relation:

$$W_2(y_2; t_2 | y_1; t_1) = w_2(y_2; t_2 | y_1; t_1) W_1(y_1; t_1) \quad (11.3.3)$$

This relation, defining w_2, tells us that the joint-probability density of finding y_1 at t_1 and y_2 at t_2 equals the probability density of finding y_1 at t_1 times the probability of a transition from y_1 to y_2 in time $t_2 - t_1$.

Relations (a) and (c) above imply the following properties of w_2:

(a′) $w_2(y_2; t_2 | y_1; t_1) \geqslant 0$ (11.3.4)

(b′) $\int dy_2 \, w_2(y_2; t_2 | y_1; t_1) = 1$ (11.3.5)

(c′) $W_1(y_2; t_2) = \int dy_1 \, w_2(y_2; t_2 | y_1; t_1) W_1(y_1; t_1)$ (11.3.6)

We also consider the nth-order transition probability

$$w_n(y_n; t_n \mid y_{n-1}; t_{n-1} | \cdots | y_1; t_1)$$

which is defined as the *conditional* probability density of finding the value y_n at time t_n, given that y had the values $y_{n-1}, y_{n-2}, \ldots, y_1$ at the respective times $t_{n-1}, t_{n-2}, \ldots, t_1$. It is assumed here that the time sequence is ordered: $t_n > t_{n-1} > t_{n-2} > \cdots > t_1$. We now are ready to define a *Markov process* by the condition:

$$w_n(y_n; t_n \mid y_{n-1}; t_{n-1} | \cdots | y_1; t_1) = w_2(y_n; t_n \mid y_{n-1}; t_{n-1}), \qquad n \geqslant 2 \quad (11.3.7)$$

This equation implies that, for a Markov process, the probability of a transition at time t_{n-1} from a value y_{n-1} to a value y_n at time t_n depends (besides on y_n, t_n) only on the value of y at the time t_{n-1} of the transition, and *not at all on the previous history* of the system.

It is now easy to check that for a Markov process, the joint-probability densities W_n for $n \geqslant 3$ are all expressible in terms of W_2 and W_1.

For example:

$$\begin{aligned} W_3(y_3; t_3 | y_2; t_2 | y_1; t_1) &= w_3(y_3; t_3 | y_2; t_2 | y_1; t_1) W_2(y_2; t_2 \mid y_1; t_1) \\ &= w_2(y_3; t_3 \mid y_2; t_2) W_2(y_2; t_2 \mid y_1; t_1) \\ &= \frac{W_2(y_3; t_3 \mid y_2; t_2) W_2(y_2; t_2 \mid y_1; t_1)}{W_1(y_2; t_2)} \end{aligned} \quad (11.3.8)$$

Using now Eqs. (11.3.1) and (11.3.8) we can write

$$\begin{aligned} W_2(y_3; t_3 \mid y_1; t_1) &= \int dy_2\, W_3(y_3; t_3 | y_2; t_2 | y_1; t_1) \\ &= \int dy_2\, w_2(y_3; t_3 \mid y_2; t_2) w_2(y_2; t_2 \mid y_1; t_1) W_1(y_1; t_1) \end{aligned}$$

Combining this with Eq. (11.3.3) we obtain

$$w_2(y_3; t_3 \mid y_1; t_1) = \int dy_2\, w_2(y_3; t_3 \mid y_2; t_2) w_2(y_2; t_2 \mid y_1; t_1) \quad (11.3.9)$$

This important integral equation obeyed by the transition probability is often taken as a definition of a Markov process. It is called the *Chapman–Kolmogorov equation* (or, sometimes, the Smoluchowski equation). The physical interpretation of this equation is clear. The probability of a transition from y_1 at t_1 to y_2 at t_2 can be calculated by taking the product of the probability of transition to some value y_2 at an intermediate time and the probability of a transition from that value to the final one at t_3, and summing over all possible intermediate values.

Let us now specialize the discussion to the physically important situation in which the transition probability $w_2(y_2; t_2 \mid y_1; t_1)$ does not depend on the time t_1 at which the transition occurs, but only on the time interval $t_2 - t_1$:*

$$w_2(y_2; t_2 \mid y_1; t_1) \equiv w(y_2 \mid y_1; t_2 - t_1) \tag{11.3.10}$$

We now show that, under certain assumptions, a partial differential equation can be derived for the transition probability w. Let $R(y)$ be a sufficiently regular, but otherwise arbitrary function of y, and consider the following expression:

$$\begin{aligned}\int dy\, R(y) \frac{\partial w(y \mid x; \tau)}{\partial \tau} &= \lim_{\Delta\tau \to 0} (\Delta\tau)^{-1} \int dy\, R(y) \{w(y \mid x; \tau + \Delta\tau) - w(y \mid x; \tau)\} \\ &= \lim_{\Delta\tau \to 0} (\Delta\tau)^{-1} \int dy\, R(y) \left\{ \int dz\, w(y \mid z; \Delta\tau) w(z \mid x; \tau) - w(y \mid x; \tau) \right\}\end{aligned}$$

where the Chapman–Kolmogorov equation was used. In the first term we interchange the order of integrations and expand $R(y)$ in a Taylor series about z:

$$\begin{aligned}\int dy\, R(y) \frac{\partial w(y \mid x; \tau)}{\partial \tau} = \lim_{\Delta\tau \to 0} (\Delta\tau)^{-1} \Big\{ &\int dz\, R(z) w(z \mid x; \tau) \int dy\, w(y \mid z; \Delta\tau) \\ &+ \int dz \frac{\partial R(z)}{\partial z} w(z \mid x; \tau) \int dy (y - z) w(y \mid z; \Delta\tau) \\ &+ \frac{1}{2} \int dz \frac{\partial^2 R(z)}{\partial z^2} w(z \mid x; \tau) \int dy (y - z)^2 w(y \mid z; \Delta\tau) + \cdots \\ &- \int dy\, R(y) w(y \mid x; \tau) \Big\}\end{aligned} \tag{11.3.11}$$

Because of Eq. (11.3.5), the first and last terms in the right-hand side cancel each other. We now make the following assumptions concerning the behavior of some functions as $\Delta\tau \to 0$:

$$\int dy (y - z) w(y \mid z; \Delta\tau) = A(z)\, \Delta\tau + O[(\Delta\tau)^2] \tag{11.3.12}$$

$$\int dy (y - z)^2 w(y \mid z; \Delta\tau) = B(z)\, \Delta\tau + O[(\Delta\tau)^2] \tag{11.3.13}$$

$$\int dy (y - z)^n w(y \mid z; \Delta\tau) = O[(\Delta\tau)^2], \qquad n \geqslant 3 \tag{11.3.14}$$

* As the two-event transition probability w_2 is the only independent transition probability in a Markov process, the subscript 2 is clearly superfluous and will be dropped henceforth.

These averages of powers of $(y-z)$ are called *transition moments*. The fact that all transition moments of order higher than 2 are proportional at least to $(\Delta\tau)^2$ in the limit $\Delta\tau\to 0$ is suggested by the physical examples to which the theory has been applied (see below). It could easily be generalized, with a more complicated final result.

With these assumptions, the limit in Eq. (11.3.11) can be calculated explicitly, and the result can be integrated by parts:

$$\int dy\, R(y)\, \frac{\partial w(y\,|\,x;\tau)}{\partial\tau}$$

$$= \int dz\, w(z\,|\,x;\tau)\left\{\frac{\partial R(z)}{\partial z}A(z) + \frac{1}{2}\frac{\partial^2 R(z)}{\partial z^2}B(z)\right\}$$

$$= \int dz\, R(z)\left\{-\frac{\partial}{\partial z}[A(z)w(z\,|\,x;\tau)] + \frac{1}{2}\frac{\partial^2}{\partial z^2}[B(z)w(z\,|\,x;\tau)]\right\}$$

After a trivial renaming of the integration variable, this result can be written as

$$\int dy\, R(y)\left\{\frac{\partial w(y\,|\,x;\tau)}{\partial\tau} + \frac{\partial}{\partial y}[A(y)w(y\,|\,x;\tau)] - \frac{1}{2}\frac{\partial^2}{\partial y^2}[B(y)w(y\,|\,x;\tau)]\right\} = 0$$

This equality must hold for every function $R(y)$, which implies

$$\frac{\partial w(y\,|\,x;\tau)}{\partial\tau} = -\frac{\partial}{\partial y}[A(y)w(y\,|\,x;\tau)] + \frac{1}{2}\frac{\partial^2}{\partial y^2}[B(y)w(y\,|\,x;\tau)] \quad (11.3.15)$$

This is the celebrated *Fokker–Planck equation*. It plays a major role in statistical physics; we will have many opportunities of studying it in detail in the forthcoming chapters. Let us note at this stage the analogy in form of this equation with the *diffusion equation* of macroscopic physics:

$$\frac{\partial n(x;t)}{\partial t} = D\frac{\partial^2 n(x;t)}{\partial x^2} \quad (11.3.16)$$

It is well known from elementary mathematical physics that this *parabolic* partial differential equation describes a monotonous, irreversible evolution of any initial density distribution toward the equilibrium state. The additional first-order derivative term in the Fokker–Planck equation describes a systematic slowing-down, called the dynamical friction. Hence, the Fokker–Planck equation describes a superposition of a friction process and of a diffusion process in the space of the variables y.

To obtain an even clearer physical picture of this process, we first note that the one-event distribution function $W_1(y;t)$ also obeys the

Fokker–Planck equation. Indeed, from Eqs. (11.3.1) and (11.3.3) we get

$$W_1(y;t) = \int dx\, w(y \mid x; t - t_0) W_1(x; t_0)$$

where t_0 is an arbitrary constant time ($t > t_0$); hence

$$\frac{\partial W_1(y;t)}{\partial t} = \int dx \frac{\partial w(y \mid x; t - t_0)}{\partial t} W_1(x; t_0)$$

Substituting now the right-hand side of Eq. (11.3.15), we easily find

$$\frac{\partial W_1(y;t)}{\partial t} = -\frac{\partial}{\partial y}[A(y)W_1(y;t)] + \frac{1}{2}\frac{\partial^2}{\partial y^2}[B(y)W_1(y;t)] \quad (11.3.17)$$

We now specialize the Fokker–Planck equation to the case of the Brownian motion problem of Section 11.2. In that case, the variable y is the velocity v of the Brownian particle. We also note that the average of a function of v, given that $v = v_0$ at $t = t_0$, is simply expressed in terms of the transition probability:

$$\langle f(v)\rangle_{v_0} = \int dv\, f(v) w(v \mid v_0; t - t_0) \quad (11.3.18)$$

We can, therefore, immediately use our results obtained from the Langevin equation in order to evaluate the coefficients $A(v)$, $B(v)$ of the Fokker–Planck equation. From Eq. (11.2.8), we obtain

$$\langle v - v_0\rangle_{v_0} = v_0 e^{-\zeta \Delta t} - v_0 = -\zeta v_0 \,\Delta t + O[(\Delta t)^2]$$

$$= -\zeta v \,\Delta t + O[(\Delta t)^2]$$

Hence, from Eq. (11.3.12)

$$A(v) = -\zeta v \quad (11.3.19)$$

Similarly, we obtain from Eq. (11.2.15)

$$B(v) = \frac{2\zeta k_B T}{M} \quad (11.3.20)$$

Therefore, the Fokker–Planck equation for the Brownian motion problem reads

$$\frac{\partial W_1(v;t)}{\partial t} = \zeta \frac{\partial}{\partial v}[v W_1(v;t)] + \zeta \frac{k_B T}{M}\frac{\partial^2}{\partial v^2} W_1(v;t) \quad (11.3.21)$$

The physical mechanism described by this equation can be understood by starting at time zero with a velocity distribution sharply peaked at $v = v_0$. As time passes, the maximum of this distribution is shifted toward smaller velocities, as a result of the systematical friction undergone by the particles. Moreover, the peak broadens progressively as a result of diffusion in velocity space: a finite dispersion of the velocities sets in. The final, time-independent distribution reached by the Brownian particles is nothing other than the Maxwell distribution:

$$W_1(v;\infty) = C \exp\left(-\frac{Mv^2}{2k_BT}\right) \tag{11.3.22}$$

Indeed, this function annuls identically the right-hand side of Eq. (11.3.21).

The picture offered by the Fokker–Planck equation is, of course, in complete agreement with the Langevin equation, together with the statistical assumptions about $A(t)$. The information is, however, presented in a much more compact form in Eq. (11.3.21) and in the similar equation for the transition probability w. If the initial-value problem for this equation is solved (and it can be solved exactly in this case!), the resulting transition probability allows us to calculate immediately *any* average value of any function of v by simple quadratures.

In order to make contact with the more usual concepts of statistical mechanics, discussed in Chapter 3, we note that $W_1(v;t)$ is simply proportional to the reduced one-particle velocity distribution $\varphi(v)$ introduced in Eq. (3.4.2).*

The evolution of the Brownian particles is thus described by a *closed equation for the one-particle distribution*, in contrast to the dynamical BBGKY hierarchy. This feature will be a leitmotiv of the subsequent sections.

11.4. DILUTE GASES: THE BOLTZMANN EQUATION

We now discuss one of the most important equations of nonequilibrium statistical mechanics. It was the first "kinetic equation" in the history of statistical mechanics, derived by Boltzmann in 1872. Since then, it has been the most extensively studied equation of that kind and is of considerable interest from the point of view of both fundamental theory and practical applications.

The reader will be impressed by the difference in method and philosophy between the derivation of the Boltzmann equation given here (which follows essentially Boltzmann's own derivation) and the stochastic approach of Sections 11.2 and 11.3. Here the treatment is "almost" purely mechanical. A probabilistic argument is thrown in only in a very subtle way. In a sense, this derivation stands half-way between a phenomenological stochastic theory and a rigorous dynamical theory.

* The change from the momenta p to the velocities $v = p/m$ is trivial.

The system considered here is a dilute gas of classical point particles of equal mass m, in the absence of any external field. These particles are supposed to interact through a repulsive potential $V(r)$. Let $\bar{f}(\mathbf{q}, \mathbf{v}; t)\, d\mathbf{q}\, d\mathbf{v}$ denote the number of particles located at time t in a small cell of volume $d\mathbf{q}$ around the point $\mathbf{q}$ in space, and having velocities in the range $[\mathbf{v}-(1/2)\, d\mathbf{v},\ \mathbf{v}+(1/2)\, d\mathbf{v}]$. Clearly $\bar{f}(\mathbf{q}, \mathbf{v};\ t)$ is the reduced one-particle distribution function defined in Chapter 3.*

It is customary and convenient in kinetic theory to use the *velocities* $\mathbf{v}_i$ instead of the momenta $\mathbf{p}_i$ as phase-space coordinates. The difference and the passage from one to another is so trivial that it is hardly worth using different symbols in the two cases. The context always shows which variables are used. To pass from one representation to the other, it is just necessary to write a few factors m in the right places. For instance,

$$f(\mathbf{q}, \mathbf{p}; t) = m^3 f(\mathbf{q}, m\mathbf{v}; t) = m^3 \tilde{f}(\mathbf{q}, \mathbf{v}; t) \tag{11.4.1}$$

This relation ensures the validity of the normalization condition in the form:

$$\int d\mathbf{q}\, d\mathbf{p}\, f(\mathbf{q}, \mathbf{p}; t) = \int d\mathbf{q}\, d\mathbf{v}\, \tilde{f}(\mathbf{q}, \mathbf{v}; t) = N \tag{11.4.2}$$

In order to lighten the notations, we shall suppress the wiggle of $\tilde{f}(\mathbf{q}, \mathbf{v}; t)$.

We now want to derive an equation for the rate of change in time of this function. The process of evolution is due to two distinct causes: the free motion of the particles and the intermolecular interactions. If the gas is sufficiently dilute, we may assume (with Boltzmann) that the two mechanisms are clearly separated, and contribute additively to the rate of change of $\bar{f}$.

In the absence of interactions the evolution of the one-particle distribution is given by the corresponding simple BBGKY equation (3.4.9), with $L_j^F = 0$ and $L'_{jn} = 0$:

$$\partial_t \bar{f}(\mathbf{q}, \mathbf{v}; t) = L^0 \bar{f}(\mathbf{q}, \mathbf{v}; \mathrm{t})$$

This is a trivially simple, closed equation. Explicitly, using Eq. (2.4.17):

$$\partial_t \bar{f}(\mathbf{q}, \mathbf{v}; t) + \mathbf{v} \cdot \nabla \bar{f}(\mathbf{q}, \mathbf{v}; t) = 0 \tag{11.4.3}$$

In order to treat the interactions, however, we shall depart from the BBGKY hierarchy. We shall try to obtain a *closed equation* allowing us to calculate $\bar{f}$ instead of having to solve an infinite hierarchy of equations. We therefore proceed to an analysis of the possible effects of the interactions.

* The reason why we write $\bar{f}$ instead of the simpler notation f is rather subtle: it will be discussed in detail later, in Chapters 16 and 17. At the present stage there is no inconvenience to think of $\bar{f}$ simply as the one-particle distribution function.

If the intermolecular forces are of short range, say, r_c, and if the gas is so dilute that the average distance between the particles is much larger than r_c, the following picture prevails. Most of the time the particles move along their free, straight trajectories. When two particles approach each other closely enough, they deflect each other during a short time, after which they continue their free motion along new trajectories and with new velocities. Simultaneous encounters of three or more particles are supposed to be so infrequent as to be negligible. The interaction process is thus idealized as a series of almost discrete events, called *collisions*, each one localized in a small region in space (of volume r_c^3) and occurring in a very short time, say, t_c. The net effect of a collision is an almost abrupt change of the velocities of a pair of particles.

It follows from this qualitative discussion that the collisions affect the rate of change of the distribution function in two ways. First, among the set of particles having velocity $\mathbf{v}$ at time t, some undergo (during a time Δt) collisions with other particles, and their velocity changes to a different value. These particles "disappear" from the set mentioned above: such collisions represent a "loss" to $\bar{f}(\mathbf{q}, \mathbf{v}; t)$. On the other hand, in the same time, there are collisions between particles of different velocities, say $\mathbf{v}'$, $\mathbf{v}_1'$, such that the final velocity of one of the partners is precisely $\mathbf{v}$. These collisions represent a "gain" to $\bar{f}(\mathbf{q}, \mathbf{v}; t)$. The net effect of the collisions is therefore given by the balance $G-L$ of gains and losses. Combining this with Eq. (11.4.3) we obtain

$$\partial_t \bar{f} + \mathbf{v} \cdot \nabla \bar{f} = G - L \tag{11.4.4}$$

We now evaluate the two terms G and L. To do so, we have to review some known results of the theory of two-body motion. (Indeed, through the concept of binary collisions, the general N-body problem has been reduced in this picture to a succession of separate two-body problems.) A two-body problem can in turn be reduced to the motion of a single particle in the field of a fixed force center located at the center of mass of the system.

To solve the problem, we may first use the laws of conservation of momentum and of kinetic energy. If $\mathbf{v}$ and $\mathbf{v}_1$ denote the velocities of the particles in the distant past before the collision, and $\mathbf{v}'$ and $\mathbf{v}_1'$ the velocities of the same particles in the distant future after the collision, we have (for particles of equal mass m)

$$\begin{aligned} \mathbf{v} + \mathbf{v}_1 &= \mathbf{v}' + \mathbf{v}_1' \\ v^2 + v_1^2 &= v'^2 + v_1'^2 \end{aligned} \tag{11.4.5}$$

These four scalar equations are not sufficient for the complete determination of the six numbers defining the unknown vectors $\mathbf{v}'$ and $\mathbf{v}_1'$: two degrees of freedom remain still undetermined. More precisely, one finds from (11.4.5) the following solution:

$$\mathbf{v}' = \tfrac{1}{2}(\mathbf{v} + \mathbf{v}_1 + g\boldsymbol{\varepsilon})$$
$$\mathbf{v}_1' = \tfrac{1}{2}(\mathbf{v} + \mathbf{v}_1 - g\boldsymbol{\varepsilon}) \tag{11.4.6}$$

where g is the relative velocity before the collision:

$$\mathbf{g} = \mathbf{v} - \mathbf{v}_1 \tag{11.4.7}$$

and $\boldsymbol{\varepsilon}$ is a unit vector in an arbitrary direction. The arbitrariness in the orientation of this vector $\boldsymbol{\varepsilon}$ reflects the indetermination mentioned above. It is seen from (11.4.6) that the effect of the collision is a change of direction of the vector $\mathbf{g}$, without change of its magnitude:

$$\mathbf{g}' = \mathbf{v}' - \mathbf{v}_1' = g\boldsymbol{\varepsilon} \tag{11.4.8}$$

In order to determine $\boldsymbol{\varepsilon}$, the geometry of the collision must be more closely defined. If we consider the equivalent one-body problem in the center-of-mass frame, we have the process depicted in Fig. 11.4.1, where the notations are defined.

For a given interaction potential $V(r)$ (depending only on the absolute value of the distance), the angle of deflection χ is entirely determined by

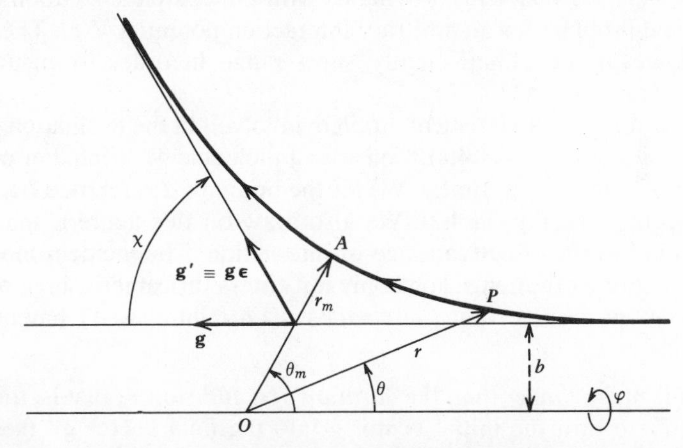

Figure 11.4.1. Geometry of a two-body collision. O is the center of mass. The instantaneous position P of the moving particle is described by polar coordinates r, θ in the plane of the trajectory. r_m, θ_m are the coordinates of the point A of closest approach to the center of mass. In absence of interactions, the distance of closest approach would be b, the impact parameter. b can also be defined as the distance of the center of mass to any of the asymptotes of the trajectory. The angle χ between the initial and the final relative velocity vectors is called the angle of deflection.

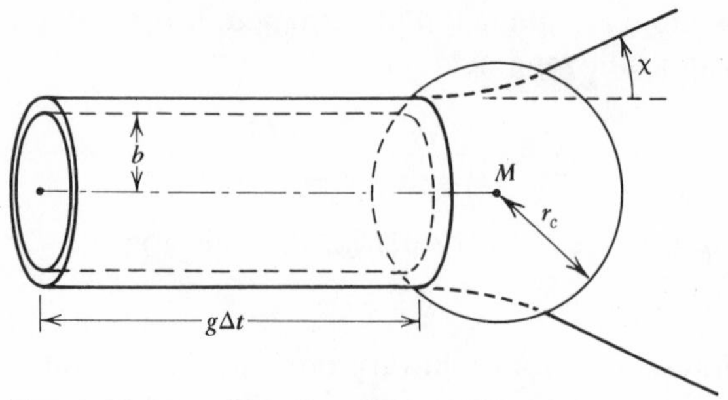

Figure 11.4.2. Counting of the collisions.

the impact parameter b. The angle χ is related to the angle θ_m at the distance of closest approach r_m through

$$\chi = \pi - 2\theta_m \tag{11.4.9}$$

The angle θ_m is in turn obtained from the well-known solution of the two-body problem (see any textbook of mechanics):

$$\theta_m = \int_{r_m}^{\infty} dr \, \frac{b/r^2}{\{1-(b/r)^2-[4V(r)/mg^2]\}^{1/2}} \tag{11.4.10}$$

Equations (11.4.6)–(11.4.10) provide us with the complete solution of the mechanical problem for an arbitrary interaction potential $V(r)$. The latter must, however, be of sufficiently short range in order to ensure the convergence of Eq. (11.4.10).

We now turn to the statistical problem involved in the evaluation of the terms G and L in Eq. (11.4.4). Consider a molecule M located at point $\mathbf{q}$ and having velocity $\mathbf{v}$ at time t. We fix the origin of a reference frame on this molecule (see Fig. 11.4.2). We also draw on this figure a sphere of radius equal to the effective range of interaction. The incident molecule feels the action of the central one only if it enters this spherical region.

Let us now *assume that there exists a time interval* Δt having two properties:

(a) Δt *is much longer than the duration of a collision* τ_c, that is, the time necesssary to turn the initial vector $\mathbf{g}$ into the final vector $g\boldsymbol{\varepsilon}$ (see Fig. 11.4.1).

(b) Δt *is much shorter than the relaxation time* τ_r, that is, the time it takes for the distribution function (at fixed $\mathbf{q}$ and $\mathbf{v}$) to change its value appreciably.

We moreover assume that the system is *not too inhomogeneous in space;* in particular, if $\bar{\mathbf{v}}$ is a typical molecular velocity, we want the distribution in point $\mathbf{q}+\bar{\mathbf{v}}\,\Delta t$ to be nearly equal to the distribution in point $\mathbf{q}$.

These assumptions will be further discussed in Section 11.5. Let us now derive their consequences.

We consider a stream of incident particles being elastically scattered off the central particle. We first note that all the incident particles having initial velocity $\mathbf{g}$ relative to the central one and hitting it within a time Δt, with impact parameters comprised between b and $b+db$, are contained in a cylindrical shell of radii b and $b+db$ and height $g\,\Delta t$. The number of such molecules is

$$\bar{f}(\mathbf{q}, \mathbf{v}_1; t)\, d\mathbf{v}_1\, 2\pi b\, db\, g\, \Delta t \tag{11.4.11}$$

(where, of course, $\mathbf{v}_1 = \mathbf{v} - \mathbf{g}$). Hence, the total number of collisions involving the central particle during Δt is obtained by integrating over all velocities $\mathbf{v}_1$ and all impact parameters b:

$$\int d\mathbf{v}_1\, db\, \bar{f}(\mathbf{q}, \mathbf{v}_1; t) 2\pi g b\, \Delta t \tag{11.4.12}$$

If we further multiply this result by the number density of particles with velocity $\mathbf{v}$, that is, $\bar{f}(\mathbf{q}, \mathbf{v}; t)\, d\mathbf{v}$, we obtain the total number of collisions per unit volume during Δt, involving particles with velocities between $\mathbf{v}$ and $\mathbf{v}+d\mathbf{v}$, at point $\mathbf{q}$;

$$\int d\mathbf{v}_1\, db\, \bar{f}(\mathbf{q}, \mathbf{v}_1; t)\bar{f}(\mathbf{q}, \mathbf{v}; t) 2\pi g b\, d\mathbf{v}\, \Delta t = L\, d\mathbf{v}\, \Delta t \tag{11.4.13}$$

It is clear that this number, divided by $(d\mathbf{v}\, \Delta t)$, represents the loss term of Eq. (11.4.4).

It is now easy to calculate the corresponding gain term. Because of the symmetry of Fig. 11.4.1 around the axis OA, a collision of particles with initial velocities $\mathbf{v}'$, $\mathbf{v}_1'$ and collision parameter b will result in final velocities $\mathbf{v}$, $\mathbf{v}_1$. We can therefore repeat the previous argument and find the gain term in the form:

$$\int d\mathbf{v}_1'\, db\, \bar{f}(\mathbf{q}, \mathbf{v}_1'; t)\bar{f}(\mathbf{q}, \mathbf{v}'; t) 2\pi g b\, d\mathbf{v}'\, \Delta t = G\, d\mathbf{v}\, \Delta t \tag{11.4.14}$$

We now note that the initial and final velocities are related by a canonical transformation; therefore, an application of the Liouville theorem yields the equality:*

$$d\mathbf{v}\, d\mathbf{v}_1 = d\mathbf{v}'\, d\mathbf{v}_1'$$

We next collect the partial results (11.4.13), (11.4.14) and substitute them

* The Liouville theorem takes this simple form only when the two-body problem is isolated, that is, when the collisions are well-separated events in space and time.

into (11.4.4), thus obtaining

$$\partial_t \bar{f}(\mathbf{q}, \mathbf{v}; t) + \mathbf{v} \cdot \nabla \bar{f}(\mathbf{q}, \mathbf{v}; t) = \int d\mathbf{v}_1\, db\, 2\pi g b\, \{\bar{f}(\mathbf{q}, \mathbf{v}'; t)\bar{f}(\mathbf{q}, \mathbf{v}_1'; t) - \bar{f}(\mathbf{q}, \mathbf{v}; t)\bar{f}(\mathbf{q}, \mathbf{v}_1; t)\} \tag{11.4.15}$$

This is the celebrated *Boltzmann equation*, or the *kinetic equation for dilute gases.*

Before discussing its properties, we write it in a somewhat different form. Let ϕ denote the flux of incident particles, that is, the number of particles crossing, per unit time, a unit area element perpendicular to $\mathbf{g}$. It then follows from our arguments that the number of particles scattered per unit time into the solid angle defined by the cones of aperture χ and $\chi + d\chi$ is

$$dN = \phi 2\pi b\, db$$

The ratio dN/ϕ is a number widely used in scattering theory; it is called *the differential cross section $d\sigma$* for the collision process under consideration. Hence

$$d\sigma = 2\pi b\, db \tag{11.4.16}$$

The differential cross section can be expressed in terms of the deflection angle χ. If the potential is such that the relation (11.4.10) provides χ as a monotonously decreasing function of b, we can revert the relation and express $d\sigma$ as

$$d\sigma = b(\chi) \left| \frac{db(\chi)}{d\chi} \right| 2\pi\, d\chi \tag{11.4.17}$$

Noting finally that the solid angle defined above is $d\boldsymbol{\omega} = 2\pi \sin\chi\, d\chi$, we can finally write the differential cross section as

$$d\sigma = \sigma_d\, d\boldsymbol{\omega} \tag{11.4.18}$$

defining the cross section σ_d as

$$\sigma_d = \frac{b(\chi)}{\sin\chi} \left| \frac{db(\chi)}{d\chi} \right| \tag{11.4.19}$$

The Boltzmann equation becomes

$$\partial_t \bar{f}(\mathbf{q}, \mathbf{v}; t) + \mathbf{v} \cdot \nabla \bar{f}(q, v; t) = \int d\mathbf{v}_1\, d\boldsymbol{\omega}\, g\sigma_d\, \{\bar{f}(\mathbf{q}, \mathbf{v}'; t)\bar{f}(\mathbf{q}, \mathbf{v}_1'; t) - \bar{f}(\mathbf{q}, \mathbf{v}; t)\bar{f}(\mathbf{q}, \mathbf{v}_1; t)\} \tag{11.4.20}$$

The advantage of this way of writing is in the fact that Eq. (11.4.20) is valid, as it stands, in quantum mechanics as well. We must simply interpret $\bar{f}(\mathbf{q}, \mathbf{v}; t)$ as the one-particle Wigner function (Chapter 3) and use the correct quantum-mechanical scattering cross section for σ_d. The only limitation is that quantum-statistical effects, that is, effects due to Bose–Einstein or Fermi–Dirac statistics, do not appear in Eq. (11.4.20). It is therefore restricted to nondegenerate quantum gases. We will return later in detail to the more precise discussion of quantum effects (Sections 18.6–18.8).

Before closing this section we note that the Boltzmann equation simplifies if we consider the particular case of a spatially homogeneous gas. The distribution function is then of the form (3.5.2), that is, it is independent of the positions $\mathbf{q}$. Hence the flow term in the Boltzmann equation vanishes identically and we get

$$\partial_t \bar{\varphi}(\mathbf{v}; t) = n \int d\mathbf{v} \int d\boldsymbol{\omega}\, g\sigma_d \{\bar{\varphi}(\mathbf{v}'; t)\bar{\varphi}(\mathbf{v}_1'; t) - \bar{\varphi}(\mathbf{v}; t)\bar{\varphi}(\mathbf{v}_1; t)\} \quad (11.4.21)$$

In a homogeneous system, the collisions are the only driving mechanism of the evolution.

11.5. ASSUMPTIONS MADE IN THE DERIVATION OF THE BOLTZMANN EQUATION

We now come back to the derivation of the Boltzmann equation and discuss in some detail the assumptions used in the argument.

One of the important assumptions, made quite explicitly, is the existence of a time interval Δt such that

$$\tau_c \ll \Delta t \ll \tau_r \quad (11.5.1)$$

Physically speaking, the duration of a collision τ_c is the time during which the trajectory of the particles differs markedly from a straight line (see Fig. 11.4.1). This time must be proportional to the effective range of the interactions r_c. The time τ_r, which is of the order of the time between two collisions, is determined by a quite different parameter of the system: τ_r will be large if the density n is small. In that case, the average distance between particles is large, and collisions are not frequent. It is, therefore, clear that the ratio (τ_c/τ_r) will be small if the following condition is satisfied:

$$\gamma \equiv r_c^3 n \ll 1 \quad (11.5.2)$$

This condition, involving the characteristic dimensionless number γ, is not new to us: We already encountered it in the equilibrium theory of a dilute gas, in Section 6.4. If (11.5.2) is satisfied, Δt can be chosen in agreement with (11.5.1). We now examine the consequences of this assumption.

Physically, the condition (11.5.2) assures us that the collisions are well-defined events in space and time: there is no overlapping between successive collisions involving a given particle. This is essential for our idealization of the interaction process as a succession of simple binary collisions, for our evaluation of the velocities $\mathbf{v}'$ and $\mathbf{v}_1'$ by a solution of the two-body problem, and for the evaluation of the loss–gain balance by a simple counting of collisions.

Equation (11.5.2) has another important, less obvious consequence. When we counted the number of particles hitting the central one, we implicitly assumed that the distribution function does not change much during the time Δt (because of the assumption $\Delta t \ll \tau_r$). Hence we could use in Eq. (11.4.11) the function $\bar{f}(\mathbf{q}, \mathbf{v}_1; t)$ evaluated at the *same time* t at which we calculate the rate of change. If (11.5.1) were not satisfied, we should have used in Eq. (11.4.11) the number of particles at the phase point $(\mathbf{q}, \mathbf{v}_1)$ at an *earlier* time $t-\tau$, where τ represents the time it takes for particle 1 to reach the central particle. As a result, in the final equation (11.4.15), the rate of change of $\bar{f}$ at time t would depend not only on the instantaneous value of $\bar{f}$, but also on its previous history. This memory effect would be characteristic of a *non-Markovian process.* Summarizing the discussion, we see that Condition (11.5.1) ensures the *Markovian character of the Boltzman equation.**

We now come to the second assumption, about the degree of inhomogeneity of the gas. It tells us that the distribution function does not appreciably change *in space* over a range of distances crossed by an average molecule in a time Δt. If this condition were not satisfied, we should have evaluated the function $\bar{f}$ in Eq. (11.4.11) at the point the collision partner is coming from. This would have resulted in a final kinetic equation that is *delocalized* in space. The rate of change of the distribution function at a point $\mathbf{q}$ would be influenced by the spatial neighborhood of that point: this would be analogous to a spatial "non-Markovian" effect. This effect is ruled out, in first approximation, if the local properties of the gas vary appreciably only over distances much longer than the range of the forces.

All these assumptions are pretty innocent. They are statements about the nature (range of the forces) and the state (density, inhomogeneity) of the gas. There is no contradiction with mechanics at this state. Nothing would prevent us from treating a problem in dynamics by making appropriate approximations that take advantage of the simplicity of the particular state of the gas.

* To avoid misunderstandings, let us make it clear that we define here a Markovian process in the wide sense as an evolution process in which the rate of change of a quantity, $\partial a(t)/\partial t$, depends only on its instantaneous value $a(t)$ and not on its previous history. See also Section 11.3.

However, another much less conspicuous assumption has sneaked into the argument, and this one represents a true departure from the laws of mechanics. In going from Eq. (11.4.12) to Eq. (11.4.13) we argued that the total number of collisions is determined by the number of pairs of molecules whose respective velocities are $\mathbf{v}$ and $\mathbf{v}_1$ at time t, and we set this number equal to $\bar{f}(\mathbf{q}, \mathbf{v}; t)\bar{f}(\mathbf{q}, \mathbf{v}_1; t)$. However, we know from the discussion of Section 3.5 that this cannot be true in general. The number of pairs of molecules located simultaneously in two different phase points is fixed by the *two-body reduced distribution function* $f_2(\mathbf{q}, \mathbf{v}; \mathbf{q}, \mathbf{v}_1; t)$. But in general

$$f_2(\mathbf{q}, \mathbf{v}, \mathbf{q}, \mathbf{v}_1; t) \neq f(\mathbf{q}, \mathbf{v}; t) f(\mathbf{q}, \mathbf{v}_1; t) \ .$$

By equating the two sides of this expression we forget the *correlations* existing between the particles. This assumption is in *direct contradiction with mechanics*. Even if we chose a specially prepared system in which correlations could be neglected at time $t = 0$, such correlations will build up later by the effect of motion. We will see this in detail later, but we can understand this effect intuitively. Consider two particles that are far apart initially: they do not feel each other (the forces are of finite, short range) and behave as if they were independent. As they approach each other along straight trajectories, they eventually enter their mutual sphere of interaction. At that time their trajectories are influenced by each other: they no longer behave independently. If the interaction is repulsive, they push each other away. Hence the probability of finding two particles very close together is smaller than the simple product of probabilities of finding a single particle anywhere in the system: *the interactions create correlations.*

The assumption $f_2 = f_1 f_1$ is a very famous one and has been called Boltzmann's *Stosszahlansatz* ("assumption about the number of collisions") or *molecular chaos* assumption (referring to the statistical independence of the molecules). It has caused many polemic discussions ever since Boltzmann's time and, more important, it has stimulated a large amount of work for a century. We believe that the fundamental aspects of this problem have been pretty much settled in recent times. It will be seen in detail in forthcoming chapters how the *Stosszahlansatz* can be justified and interpreted. At this stage, we prefer to see what type of consequences it implies.

A first important consequence can be seen by comparing the Boltzmann equation (11.4.20) to the equation for the rate of change of f_1 provided by the BBGKY hierarchy, Eq. (3.3.9). The *Boltzmann equation is a closed equation for the one-particle distribution function.*

This is in marked contrast with the one-particle component of the generalized Liouville equation: here the rate of change of f_1 depends on the value of the two-particle function f_2, which in turn depends on f_3, and so forth. Thus we see that through Boltzmann's *Stosszahlansatz*, the infinite hierarchy for the distribution functions is cut down to a single equation for f_1.* It is important to note that the Boltzmann equation shares this property of being closed in $\bar{f}_1$ with the Fokker–Planck equation (11.3.17) or (11.3.21), which, however, looks very different at first sight. It will be shown in Sections 11.6 and 11.8 that there actually exists a closer relation between the two equations.

There is, however, a price to be paid for the reduction of the hierarchy to a single equation: *Boltzmann's equation is nonlinear.* Even so, the mathematical simplification is enormous. Powerful approximation techniques have been developed for the solution of the equation and detailed (and favorable) comparisons can be made with experimental results.

The stochastic equations of Sections 11.2 and 11.3 were clearly shown to describe an irreversible evolution. At the present stage, it is not yet clear how the Boltzmann equation is related to that type of evolution. The relation will become clear when we further analyze its properties and the properties of related equations. This will be done in Chapters 12 and 13. Meanwhile, we continue to exhaust the list of "elementary" kinetic equations.

11.6. HOMOGENEOUS WEAKLY COUPLED GASES: THE LANDAU EQUATION

In deriving the Boltzmann equation in Section 11.4, no very stringent assumption was made about the nature of the intermolecular interactions. It was implicitly assumed, however, that their range is sufficiently short in order that the binary collision description be meaningful. The Boltzmann equation is currently used for the description of "ordinary" gases, for which the interaction potential is of the general form shown in Fig. 6.1.1. The collisions are principally due to the strong repulsive part of the potential.

We may now consider a rather idealized type of interaction potential that is uniformly small compared to $k_B T$. This is the case for the *weakly coupled gases* considered, in equilibrium, in Sections 6.2 and 6.3. The Boltzmann equation reduces, for these systems, to a rather different, and much simpler form. One may argue that weakly coupled gases do not really exist in nature; however, the kinetic equation obtained here will

* We may at this point give a provisional justification of our notation $\bar{f}_1$. We reserve the notation f_1 (without a bar) for a distribution function obeying the BBGKY equations, whereas $\bar{f}_1$ denotes a distribution function obeying a closed, nonlinear kinetic equation, which is, supposedly, an approximation to the BBGKY equations. The deeper meaning of $\bar{f}_1$ as compared to f_1 will only be discussed in Chapter 16.

prove to be a very useful first approximation for the study of a certain class of important systems. Moreover, the concept of weakly coupled systems (and the associated expansion procedure) is of considerable theoretical importance and will be a starting point for the treatment of more general systems, in the same spirit as in equilibrium theory.

Let us first rewrite the Boltzmann equation in a slightly more suggestive way. We note that $g\sigma_d$ in Eq. (11.4.20) is simply the probability density of a collision transforming the velocities $\mathbf{v}$, $\mathbf{v}_1$ into $\mathbf{v}'$, $\mathbf{v}_1'$ or the converse. We denote this quantity by the symbol $w(\mathbf{v}, \mathbf{v}_1; \mathbf{v}', \mathbf{v}_1')$. From Eq. (11.4.6) we know that

$$\begin{aligned} \mathbf{v}' &= \mathbf{v} + \mathbf{\Delta} \\ \mathbf{v}_1' &= \mathbf{v}_1 - \mathbf{\Delta} \end{aligned} \tag{11.6.1}$$

with the velocity deviation given by

$$\mathbf{\Delta} = -\tfrac{1}{2}(\mathbf{g} - g\boldsymbol{\varepsilon}) \tag{11.6.2}$$

We now express the probability density of a collision in terms of new variables:

$$\begin{aligned} w(\mathbf{v}, \mathbf{v}_1; \mathbf{v}', \mathbf{v}_1') &\to w[\tfrac{1}{2}(\mathbf{v}+\mathbf{v}'), \tfrac{1}{2}(\mathbf{v}+\mathbf{v}_1'); \mathbf{v}'-\mathbf{v}; \mathbf{v}_1'-\mathbf{v}_1] \\ &= w(\mathbf{v}+\tfrac{1}{2}\mathbf{\Delta}, \mathbf{v}_1-\tfrac{1}{2}\mathbf{\Delta}; \mathbf{\Delta}, -\mathbf{\Delta}) \\ &\equiv w(\mathbf{v}+\tfrac{1}{2}\mathbf{\Delta}, \mathbf{v}_1-\tfrac{1}{2}\mathbf{\Delta}; \mathbf{\Delta}) \end{aligned} \tag{11.6.3}$$

The fact that the probability densities of the direct and inverse collisions are equal implies that w is an even function in the variable $\mathbf{\Delta}$ appearing to the right of the semicolon:

$$w(\mathbf{v}+\tfrac{1}{2}\mathbf{\Delta}, \mathbf{v}_1-\tfrac{1}{2}\mathbf{\Delta}; \mathbf{\Delta}) = w(\mathbf{v}+\tfrac{1}{2}\mathbf{\Delta}, \mathbf{v}_1-\tfrac{1}{2}\mathbf{\Delta}; -\mathbf{\Delta}) \tag{11.6.4}$$

The Boltzmann equation is now rewritten as follows for a homogeneous gas [see Eq. (11.4.21)]:*

$$\begin{aligned} \partial_t \bar{\varphi}(\mathbf{v}) = n\int d\mathbf{v}_1\, d\boldsymbol{\omega}\, & w(\mathbf{v}+\tfrac{1}{2}\mathbf{\Delta}, \mathbf{v}_1-\tfrac{1}{2}\mathbf{\Delta}; \mathbf{\Delta}) \\ & \times\{\bar{\varphi}(\mathbf{v}+\mathbf{\Delta})\bar{\varphi}(\mathbf{v}_1-\mathbf{\Delta}) - \bar{\varphi}(\mathbf{v})\bar{\varphi}(\mathbf{v}_1)\} \end{aligned} \tag{11.6.5}$$

We now assume that *the molecular interactions are very weak compared to the mean kinetic energy of the molecules.* This assumption implies that the *deviation* $\mathbf{\Delta}$ resulting from a collision *is small* (on the average) compared to the initial velocity of the particles:

$$\mathbf{\Delta} \ll \mathbf{v}, \qquad \mathbf{\Delta} \ll \mathbf{v}_1 \tag{11.6.6}$$

As a result, we may approximate the various functions appearing in Eq.

* We shall often omit writing the time argument of the distribution function.

(11.6.5) by their Taylor expansions around $\mathbf{v}$ and $\mathbf{v}_1$:

$$w(\mathbf{v}+\tfrac{1}{2}\Delta, \mathbf{v}_1-\tfrac{1}{2}\Delta; \Delta) \approx w(\mathbf{v}, \mathbf{v}_1; \Delta)+\tfrac{1}{2}\Delta^r\left\{\frac{\partial w(\mathbf{v}, \mathbf{v}_1; \Delta)}{\partial v^r}-\frac{\partial w(\mathbf{v}, \mathbf{v}_1; \Delta)}{\partial v_1^r}\right\}+\cdots$$

$$\bar{\varphi}(\mathbf{v}+\mathbf{\Delta}) \approx \bar{\varphi}(\mathbf{v})+\Delta^r \frac{\partial \bar{\varphi}(\mathbf{v})}{\partial v^r}+\tfrac{1}{2}\Delta^r\Delta^s \frac{\partial^2 \bar{\varphi}(\mathbf{v})}{\partial v^r\, \partial v^s}+\cdots$$

and a similar expression for $\bar{\varphi}(\mathbf{v}_1-\mathbf{\Delta})$.* Substituting these expansions into (11.6.5) and collecting terms through the second order in $\mathbf{\Delta}$ we obtain

$$\begin{aligned}\partial_t\bar{\varphi}(\mathbf{v}) = n\int d\mathbf{v}_1\, d\boldsymbol{\omega}\Bigg\{&\Delta^r w(\mathbf{v}, \mathbf{v}_1; \mathbf{\Delta})\left[\frac{\partial \bar{\varphi}(\mathbf{v})}{\partial v^r}\, \bar{\varphi}(\mathbf{v}_1)-\frac{\partial \bar{\varphi}(\mathbf{v}_1)}{\partial v_1^r}\, \bar{\varphi}(\mathbf{v})\right] \\ &+\tfrac{1}{2}\Delta^r\Delta^s\Bigg[w(\mathbf{v}, \mathbf{v}_1; \mathbf{\Delta})\left(\frac{\partial^2 \bar{\varphi}(\mathbf{v})}{\partial v^r\, \partial v^s}\, \bar{\varphi}(\mathbf{v}_1)+\frac{\partial^2 \bar{\varphi}(\mathbf{v}_1)}{\partial v_1^r\, \partial v_1^s}\, \bar{\varphi}(\mathbf{v})-2\frac{\partial \bar{\varphi}(\mathbf{v})}{\partial v^r}\frac{\partial \bar{\varphi}(\mathbf{v}_1)}{\partial v_1^s}\right) \\ &+\frac{\partial w(\mathbf{v}, \mathbf{v}_1; \mathbf{\Delta})}{\partial v^r}\left(\frac{\partial \bar{\varphi}(\mathbf{v})}{\partial v^s}\, \bar{\varphi}(\mathbf{v}_1)-\frac{\partial \bar{\varphi}(\mathbf{v}_1)}{\partial v_1^s}\, \bar{\varphi}(\mathbf{v})\right) \\ &-\frac{\partial w(\mathbf{v}, \mathbf{v}_1; \mathbf{\Delta})}{\partial v_1^r}\left(\frac{\partial \bar{\varphi}(\mathbf{v})}{\partial v^s}\, \bar{\varphi}(\mathbf{v}_1)-\frac{\partial \bar{\varphi}(\mathbf{v}_1)}{\partial v_1^s}\, \bar{\varphi}(\mathbf{v})\right)\Bigg]\Bigg\}\end{aligned} \tag{11.6.7}$$

Equation (11.6.4) implies that $w(\mathbf{v}, \mathbf{v}_1; \mathbf{\Delta})$ is an even function of $\mathbf{\Delta}$, therefore the first-order terms in Eq. (11.6.7) vanish identically. We now perform an integration by parts with respect to $\mathbf{v}_1$ on the last two terms (involving $\partial w/\partial v_1^r$); we are left with

$$\begin{aligned}\partial_t\bar{\varphi}(\mathbf{v}) = \tfrac{1}{2}n\int d\mathbf{v}_1\, d\boldsymbol{\omega}\, \Delta^r\Delta^s\Bigg\{&w(\mathbf{v}, \mathbf{v}_1; \mathbf{\Delta})\left[\frac{\partial^2 \bar{\varphi}(\mathbf{v})}{\partial v^r\, \partial v^s}\, \bar{\varphi}(\mathbf{v}_1)-\frac{\partial \bar{\varphi}(\mathbf{v})}{\partial v^r}\frac{\partial \bar{\varphi}(\mathbf{v}_1)}{\partial v_1^s}\right] \\ &+\frac{\partial w(\mathbf{v}, \mathbf{v}_1; \mathbf{\Delta})}{\partial v^r}\left[\frac{\partial \bar{\varphi}(\mathbf{v})}{\partial v^s}\, \bar{\varphi}(\mathbf{v}_1)-\bar{\varphi}(\mathbf{v})\frac{\partial \bar{\varphi}(\mathbf{v}_1)}{\partial v_1^s}\right]\Bigg\}\end{aligned}$$

which can be written more compactly as follows:

$$\partial_t\bar{\varphi}(\mathbf{v}) = n\int d\mathbf{v}_1\, \frac{\partial}{\partial v^r}\, G^{rs}\left(\frac{\partial}{\partial v^s}-\frac{\partial}{\partial v_1^s}\right)\bar{\varphi}(\mathbf{v})\bar{\varphi}(\mathbf{v}_1) \tag{11.6.8}$$

(The operators $\partial/\partial v$ are now supposed to act on everything written to their right.) The tensor G^{rs} is defined as

$$G^{rs} = \frac{1}{2}\int d\boldsymbol{\omega}\, w(\mathbf{v}, \mathbf{v}_1; \mathbf{\Delta})\Delta^r\Delta^s \tag{11.6.9}$$

We now show that the tensor G^{rs} can be explicitly expressed in terms of the molecular parameters of the particles.

We note that the velocity increment of the first particle can be

* In these and forthcoming formulas, the superscripts $r, s, \ldots$ denote vector indices (x, y, z); the usual rule of summation over repeated vector indices is implied.

expressed as the time integral of the force due to the second particle, taken along the whole orbit of a collision:

$$\mathbf{\Delta} = -m^{-1}\int_{-\infty}^{\infty} dt\, \frac{\partial V[\mathbf{q}(t)-\mathbf{q}_1(t)]}{\partial \mathbf{q}} \tag{11.6.10}$$

Clearly,

$$\mathbf{q}(t)-\mathbf{q}_1(t) = \mathbf{q}-\mathbf{q}_1 + \int_0^t ds\, \mathbf{g}(s) = \mathbf{a} + \int_0^t ds\, \mathbf{g}(s) \tag{11.6.11}$$

where $\mathbf{a}$ is the distance between the particles at an arbitrary initial time.

We now use the idea of weak coupling in the following way. The time dependence of the relative velocity in (11.6.11) is due to the action of the force; but $\mathbf{\Delta}$ is already proportional to the force; hence, to the first approximation we may consider the relative velocity as a constant and write

$$\mathbf{r}_t \equiv \mathbf{q}(t)-\mathbf{q}_1(t) \approx \mathbf{a}+\mathbf{g}t \tag{11.6.12}$$

This clearly amounts to replacing the slightly bent trajectory of the particle by one of its straight asymptotes in the argument of the force (see Fig. 11.6.1).

In Eq. (2.4.8) we defined the Fourier transform of the potential $V(r)$ by means of the equation:

$$V(r) = \int d\mathbf{k}\, \tilde{V}_k e^{i\mathbf{k}\cdot\mathbf{r}}$$

It is convenient to express the deviation $\mathbf{\Delta}$ in terms of this Fourier transform $\tilde{V}_k$, using the approximation (10.7.12):

$$\begin{aligned}\mathbf{\Delta} &= -m^{-1}\int_{-\infty}^{\infty} dt \int d\mathbf{k}\, i\mathbf{k}\tilde{V}_k e^{i\mathbf{k}\cdot\mathbf{r}_t} \\ &= -m^{-1}\int_{-\infty}^{\infty} dt \int d\mathbf{k}\, i\mathbf{k}\tilde{V}_k e^{i\mathbf{k}\cdot(\mathbf{a}+\mathbf{g}t)}\end{aligned} \tag{11.6.13}$$

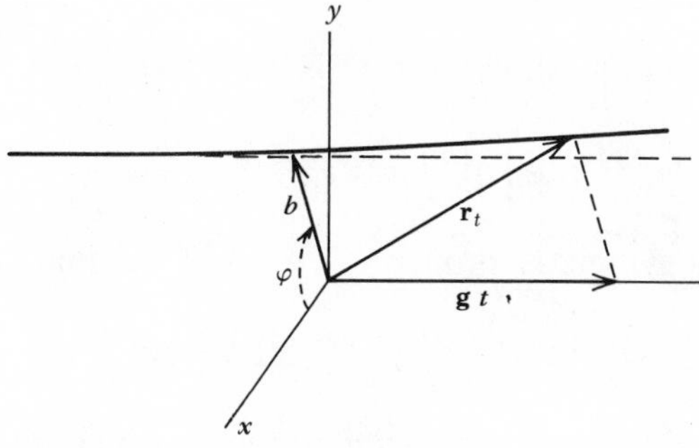

Figure 11.6.1. Geometry of the collision in a weakly coupled gas.

We now return to Eq. (11.6.9). In Fig. 11.6.1 we used a reference frame whose origin is on particle $\mathbf{v}_1$ and whose z axis is parallel to the vector $\mathbf{g}$. The trajectory of particle $\mathbf{v}$ thus meets the x–y plane at the point of closest approach. The latter is (approximately) at a distance b from the origin, and its position vector makes an angle φ with the x axis. We know from Eqs. (11.4.16)–(11.4.18) that, in this geometry:

$$w(\mathbf{v};\mathbf{v}_1;\boldsymbol{\Delta})\,d\boldsymbol{\omega}=g\sigma_d\,d\boldsymbol{\omega}=2\pi gb\,db$$

Hence (11.6.9) can be written as

$$G^{rs}=\tfrac{1}{2}m^{-2}\int_0^\infty db\int_0^{2\pi}d\varphi\,bg\int_{-\infty}^\infty dt_1\int_{-\infty}^\infty dt_2\int d\mathbf{k}_1\int d\mathbf{k}_2\,ik_1^r\,ik_2^s\,\tilde{V}_{\mathbf{k}_1}\tilde{V}_{\mathbf{k}_2}\times\exp(i\mathbf{k}_1\cdot\mathbf{r}_{t_1}+i\mathbf{k}_2\cdot\mathbf{r}_{t_2})\qquad(11.6.14)$$

We now rewrite the exponential as follows:

$$\begin{aligned}\exp(i\mathbf{k}_1\cdot\mathbf{r}_{t_1}+i\mathbf{k}_2\cdot\mathbf{r}_{t_2})&=\exp[i(\mathbf{k}_1+\mathbf{k}_2)\cdot\mathbf{a}]\exp(i\mathbf{k}_1\cdot\mathbf{g}t_1+i\mathbf{k}_2\cdot\mathbf{g}t_2)\\&=\exp[i(\mathbf{k}_1+\mathbf{k}_2)\cdot(\mathbf{a}+\mathbf{g}t_1)]\exp[i\mathbf{k}_2\cdot\mathbf{g}(t_2-t_1)]\\&=\exp[i(\mathbf{k}_1+\mathbf{k}_2)\cdot\mathbf{r}_{t_1}]\exp[i\mathbf{k}_2\cdot\mathbf{g}(t_2-t_1)]\end{aligned}$$

The vector $\mathbf{r}_{t_1}$ has cylindrical coordinates (b,φ,gt_1) in the reference frame of Fig. 11.6.1; therefore

$$\int_0^\infty db\,b\int_0^{2\pi}d\varphi\int_{-\infty}^\infty dt_1\,g\equiv\int d\mathbf{r}_{t_1}$$

Consequently, we may write (dropping the subscript t_1 on $\mathbf{r}_{t_1}$ and putting $t_2-t_1=\tau$)

$$G^{rs}=\tfrac{1}{2}m^{-2}\int d\mathbf{r}\int_{-\infty}^\infty d\tau\int d\mathbf{k}_1\,d\mathbf{k}_2\,ik_1^rik_2^s\tilde{V}_{\mathbf{k}_1}\tilde{V}_{\mathbf{k}_2}e^{i(\mathbf{k}_1+\mathbf{k}_2)\cdot\mathbf{r}}e^{i\mathbf{k}_2\cdot\mathbf{g}\tau}\qquad(11.6.15)$$

The integrations are now very easy. They are performed in the following order: $d\mathbf{r}$, $d\mathbf{k}_1$, $d\tau$, $d\mathbf{k}_2$:

$$\begin{aligned}G^{rs}&=4\pi^3m^{-2}\int_{-\infty}^\infty d\tau\int d\mathbf{k}_1\,d\mathbf{k}_2\,ik_1^rik_2^s\tilde{V}_{\mathbf{k}_1}\tilde{V}_{\mathbf{k}_2}\delta(\mathbf{k}_1+\mathbf{k}_2)e^{i\mathbf{k}_2\cdot\mathbf{g}\tau}\\&=4\pi^3m^{-2}\int_{-\infty}^\infty d\tau\int d\mathbf{k}_2\,k_2^rk_2^s\tilde{V}_{\mathbf{k}_2}^2e^{i\mathbf{k}_2\cdot\mathbf{g}\tau}\end{aligned}$$

Here we used the symmetry property (2.4.10) of $\tilde{V}_{\mathbf{k}}$. Finally:

$$G^{rs}=8\pi^4m^{-2}\int d\mathbf{k}\,k^rk^s\tilde{V}_{\mathbf{k}}^2\,\delta(\mathbf{k}\cdot\mathbf{g})\qquad(11.6.16)$$

We thus obtained a simple expression of the tensor G^{rs} in terms of the interaction potential $\tilde{V}_{\mathbf{k}}$. Combining now Eqs. (11.6.8) and (11.6.16), we obtain

$$\partial_t\bar{\varphi}(\mathbf{v}_1;t)=An\int d\mathbf{v}_2\int d\mathbf{k}\;\tilde{V}_k^2\,\mathbf{k}\cdot\partial_1\,\delta(\mathbf{k}\cdot\mathbf{g})\,\mathbf{k}\cdot\partial_{12}\bar{\varphi}(\mathbf{v}_1;t)\bar{\varphi}(\mathbf{v}_2;t) \quad (11.6.17)$$

We introduced here the following abbreviations (see also (2.4.13)):

$$\partial_i=\frac{\partial}{\partial\mathbf{v}_i}$$

$$\partial_{ij}=\frac{\partial}{\partial\mathbf{v}_i}-\frac{\partial}{\partial\mathbf{v}_j} \quad (11.6.18)$$

$$A=\frac{8\pi^4}{m^2}$$

We now note the following property, to be extensively used in forthcoming calculations. For any function $\Phi(\mathbf{v})$ of the velocity, which vanishes strongly enough at infinity:

$$\int d\mathbf{v}\;\partial\Phi(\mathbf{v})=0 \quad (11.6.19)$$

This is a direct consequence of the Gauss theorem, by which the volume integral can be transformed into a surface integral of Φ, taken over the boundary of the integration volume, that is, at infinity [see also Eq. (3.4.5)].

It then follows that nothing changes if we replace the symbol ∂_1 in the right-hand side of Eq. (11.6.17) by $\partial_{12}=\partial_1-\partial_2$: the subtracted term ∂_2 gives a vanishing contribution to the integral. The equation, however, takes a more symmetric form:

$$\partial_t\bar{\varphi}(\mathbf{v}_1;t)=An\int d\mathbf{v}_2\int d\mathbf{k}\;\tilde{V}_k^2\,\mathbf{k}\cdot\partial_{12}\,\delta(\mathbf{k}\cdot\mathbf{g})\,\mathbf{k}\cdot\partial_{12}\bar{\varphi}(\mathbf{v}_1;t)\bar{\varphi}(\mathbf{v}_2;t) \quad (11.6.20)$$

This equation is called the kinetic equation of a *weakly coupled gas*, or the *Landau equation.* It is frequently used in plasma physics (see Section 11.7).

Although Eq. (11.6.20) is very compact and symmetric, it is sometimes useful to write it in an even more explicit form. We choose a particular reference frame XYZ in which the z axis is directed along the vector $\mathbf{g}$, which is fixed in Eq. (11.6.16) (see Fig. 11.6.2).

Introducing polar coordinates in this frame we may write collectively

$$G\left\{\begin{matrix}x & x\\ y & y\\ z & z\end{matrix}\right\}=8\pi^4m^{-2}\int_0^\infty dk\int_0^\pi d\theta\int_0^{2\pi}d\phi\,k^2\sin\theta\,\tilde{V}_k^2\,\delta(kg\cos\theta)$$

$$\times k^2\begin{Bmatrix}\sin\theta\cos\phi\\ \sin\theta\sin\phi\\ \cos\theta\end{Bmatrix}\begin{Bmatrix}\sin\theta\cos\phi\\ \sin\theta\sin\phi\\ \cos\theta\end{Bmatrix} \quad (11.6.21)$$

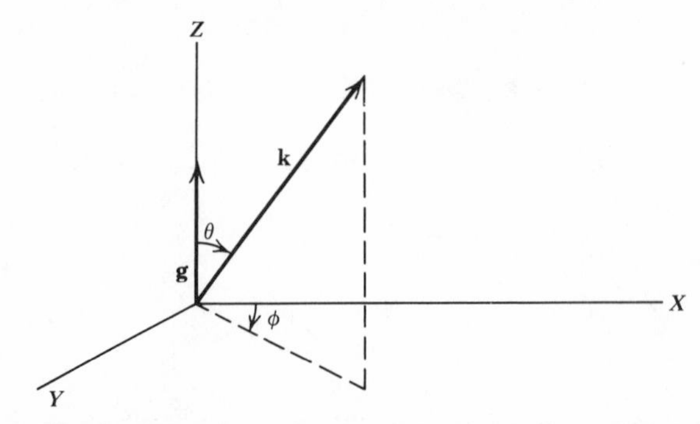

Figure 11.6.2. Reference frame for the evaluation of Eq. (11.6.21).

It is immediately seen, by integration over ϕ, that all nondiagonal components (XY, XZ, YX, YZ, ZX, ZY) vanish. Moreover, the ZZ component also vanishes because of the δ function; finally, $G^{XX} = G^{YY} \equiv G$. Hence the tensor G has the following simple form in this reference frame:

$$\mathsf{G} = \begin{pmatrix} G & 0 & 0 \\ 0 & G & 0 \\ 0 & 0 & 0 \end{pmatrix} \tag{11.6.22}$$

and the single component G can be easily evaluated:

$$G = 8\pi^5 m^{-2} \int_0^\infty dk\, \tilde{V}_k^2 k^4 \int_0^\pi d\theta \sin\theta \sin^2\theta\, \delta(kg\cos\theta) = \frac{B}{gm^2} \tag{11.6.23}$$

where

$$B = 8\pi^5 \int_0^\infty dk\, k^3 \tilde{V}_k^2 \tag{11.6.24}$$

Having evaluated G^{rs} in this particular frame, we must now transform it back to a general frame (because in subsequent operations $\mathbf{g}$ is no longer fixed). This is easily done by noting that the tensor G^{rs} in the present frame is proportional to the three-dimensional unit tensor, minus its ZZ component. The latter is simply the dyadic $\mathbf{1}_Z\mathbf{1}_Z$, where $\mathbf{1}_Z$ is the unit vector along the Z axis:

$$\mathsf{G} = (\mathbf{1} - \mathbf{1}_Z\mathbf{1}_Z)\left(\frac{B}{gm^2}\right) \tag{11.6.25}$$

But $\mathbf{1}_Z$ is simply the unit vector in the direction of the relative velocity:

$$\mathbf{1}_Z = \frac{\mathbf{g}}{g}$$

Hence, in an arbitrary reference frame:

$$G^{rs} = \left(\delta^{rs} - \frac{g^r g^s}{g^2}\right)\left(\frac{B}{gm^2}\right) \tag{11.6.26}$$

Substituting this result into (11.6.20) we get the following explicit form for the *Landau equation:*

$$\partial_t \bar{\varphi}(\mathbf{v}_1; t) = Bnm^{-2} \int d\mathbf{v}_2 \, \partial_{12}^r \frac{g^2 \delta^{rs} - g^r g^s}{g^3} \partial_{12}^s \bar{\varphi}(\mathbf{v}_1; t) \bar{\varphi}(\mathbf{v}_2; t) \quad (11.6.27)$$

In this form we note a very remarkable feature, which was not obvious in the initial form (11.6.20). The mechanism of the time evolution appears to be the same for all systems to which the weak-coupling approximation applies. The evolution operator in the right-hand side of Eq. (11.6.27) has a form independent of the nature of the interactions. The latter only fixes the value of the constant B, which amounts to fixing the time scale of the process. In other words, different systems evolve along the same path, the pace of the evolution being fixed by the interaction potential. This very simple feature is lost when more general systems are considered; it is a limiting behavior valid when the amplitude of the interactions is very small.

The kinetic equation (11.6.21) was derived by Landau in 1936 by following the method exposed in this section. His aim was to obtain an equation valid for a *plasma.* This system is a good approximation to the concept of a weakly coupled gas. Indeed, because of the long range of the Coulomb forces, two particles feel their mutual influence when they are far apart; but these interactions are then very weak. If the plasma is sufficiently dilute, close encounters are rare and the "collisions" may be described in first approximation by the Landau equation.

However, the Landau equation, when applied to a plasma, leads to difficulties, not so much because of the close collisions but because of the "too long" range of the Coulomb potential. We saw in Section 6.5 that this problem arises in equilibrium too. The way out of this difficulty is a typical instance where a systematic theory of kinetic equations is needed, because the simple-minded arguments of this chapter no longer apply. We shall come back to this problem in Sections 20.5 and 20.6. Till then, let us simply mention that in many cases the Landau equation can be used as it stands, provided that one introduces appropriate cutoffs in the limits of the divergent integrals appearing in the theory.

11.7. INHOMOGENEOUS PLASMAS: THE VLASSOV EQUATION

We now consider the extension of the Landau equation to spatially inhomogeneous systems. We shall be particularly interested in the case of a *plasma.* For definiteness we treat the simple model of Section 6.5, that is, a one-component system of electrons of charge $(-e)$ imbedded in a homogeneous neutralizing background. The particles interact through a Coulomb potential, Eq. (6.5.2). [The artificial hard-sphere term V^{HS} of Eq. (6.5.1) is not considered here]. At first sight no special difficulty appears in the extension. We just use the arguments determining the passage from (11.4.20) to (11.4.21) in reversed order. Thus:

(a) $n\bar{\varphi}(\mathbf{v}; t)$ must be replaced by $\bar{f}(\mathbf{q}, \mathbf{v}; t)$.

(b) A flow term must be added to the left-hand side:

$$\mathbf{v} \cdot \nabla \bar{f}(\mathbf{q}, \mathbf{v}; t)$$

(c) In the collision term, the two distribution functions must be evaluated at the same position in space. This is most symmetrically written as follows:

$$n^2\bar{\varphi}(\mathbf{v}_1; t)\bar{\varphi}(\mathbf{v}_2; t) \rightarrow \int d\mathbf{q}_2\, \delta(\mathbf{q}_1 - \mathbf{q}_2)\bar{f}(\mathbf{q}_1, \mathbf{v}_1; t)\bar{f}(\mathbf{q}_2, \mathbf{v}_2; t)$$

With these changes, the Landau equation (11.6.20) becomes

$$\partial_t \bar{f}(\mathbf{q}_1, \mathbf{v}_1; t) + \mathbf{v}_1 \cdot \nabla_1 \bar{f}(\mathbf{q}_1, \mathbf{v}_1; t)$$
$$= A \int d\mathbf{v}_2\, d\mathbf{q}_2 \int d\mathbf{k}\, \delta(\mathbf{q}_1 - \mathbf{q}_2)\tilde{V}_k^2\, \mathbf{k} \cdot \partial_{12}\, \delta(\mathbf{k} \cdot \mathbf{g})\, \mathbf{k} \cdot \partial_{12} \bar{f}(\mathbf{q}_1, \mathbf{v}_1; t)\bar{f}(\mathbf{q}_2, \mathbf{v}_2; t) \tag{11.7.1}$$

Alternatively, Eq. (11.6.27) can be transformed in a similar way.

However, it should be realized that, especially in the case of a plasma, this equation is incomplete, and does not provide an adequate picture of the process of evolution. The reason lies mainly in the long range of the Coulomb interactions. In an inhomogeneous ordinary gas, a particle moves most of the time in a straight line, with uniform velocity until it hits another particle. After a short collision, it resumes its uniform motion in some other direction. This picture results in the superposition of a flow term and of a collision term as in Eq. (11.7.1). In a plasma, however, a particle never moves freely. It constantly feels the influence of a very large number of other particles of the medium. This effect must be separated from the collisions, because these are treated, in the present approximation, as binary encounters that are well separated in space and time. The point is that between two collisions, the particle travels in a medium that is permeated by an electromagnetic field produced by all the other particles.

It is not difficult to formulate this idea in a more quantitative way. We first note that if the plasma is *homogeneous* the average field must be zero; indeed, the overall system is electrically neutral, and the fields produced by individual particles cancel each other on the average. The effective field must therefore be measured by the deviation of the plasma from homogeneity, that is, by the difference: $\bar{f}(\mathbf{q}, \mathbf{v}; t) - n\bar{\varphi}(\mathbf{v}; t)$. The simplest idea consists of introducing an average potential energy $U(\mathbf{q}_1; t)$ defined as

$$U(\mathbf{q}_1; t) = \int d\mathbf{q}_2\, d\mathbf{v}_2\, V(\mathbf{q}_1 - \mathbf{q}_2)[\bar{f}(\mathbf{q}_2, \mathbf{v}_2; t) - n\bar{\varphi}(\mathbf{v}_2; t)] \qquad (11.7.2)$$

According to our discussion, we consider that the potential $U(\mathbf{q}_1; t)$ acts on every particle *as if* it were due to an external field. We therefore add to the equation of evolution a term of the form (2.4.22), with the result:

$$\begin{aligned} &\partial_t \bar{f}(\mathbf{q}_1, \mathbf{v}_1; t) + \mathbf{v}_1 \cdot \nabla_1 \bar{f}(\mathbf{q}_1, \mathbf{v}_1; t) \\ &\qquad - m^{-1}\left\{\nabla_1 \int d\mathbf{q}_2\, d\mathbf{v}_2\, V(\mathbf{q}_1 - \mathbf{q}_2)[\bar{f}(\mathbf{q}_2, \mathbf{v}_2; t) - n\bar{\varphi}(\mathbf{v}_2; t)]\right\} \cdot \partial_1 \bar{f}(\mathbf{q}_1, \mathbf{v}_1; t) \\ &\quad = A \int d\mathbf{v}_2\, d\mathbf{q}_2 \int d\mathbf{k}\, \delta(\mathbf{q}_1 - \mathbf{q}_2)\tilde{V}_k^2 \mathbf{k} \cdot \partial_{12}\, \delta(\mathbf{k} \cdot \mathbf{g})\mathbf{k} \cdot \partial_{12} \bar{f}(\mathbf{q}_1, \mathbf{v}_1; t)\bar{f}(\mathbf{q}_2, \mathbf{v}_2; t) \end{aligned} \qquad (11.7.3)$$

This equation, called the *Vlassov–Landau equation* contains all the effects discussed above. Before analyzing it further, we note that the interaction potential $\dot{V}(r)$ enters to first order in the average field, and to second order in the collision term. If, therefore, the idea of weak coupling is retained, the average field term is quite important, and it is meaningful to discuss an equation in which the collisions are neglected:

$$\begin{aligned} &\partial_t \bar{f}(\mathbf{q}_1, \mathbf{v}_1; t) + \mathbf{v}_1 \cdot \nabla_1 \bar{f}(\mathbf{q}_1, \mathbf{v}_1; t) \\ &\qquad - m^{-1}\left\{\nabla_1 \int d\mathbf{q}_2\, d\mathbf{v}_2\, V(\mathbf{q}_1 - \mathbf{q}_2)[\bar{f}(\mathbf{q}_2, \mathbf{v}_2; t) - n\bar{\varphi}(\mathbf{v}_2; t)]\right\} \cdot \partial_1 \bar{f}(\mathbf{q}_1, \mathbf{v}_1; t) = 0 \end{aligned} \qquad (11.7.4)$$

This is the *Vlassov equation*, an equation of considerable importance in plasma physics.

We note that the Vlassov equation can actually be written in the simpler form:

$$\begin{aligned} &\partial_t \bar{f}(\mathbf{q}_1, \mathbf{v}_1; t) + \mathbf{v}_1 \cdot \nabla_1 \bar{f}(\mathbf{q}_1, \mathbf{v}_1; t) \\ &\qquad - m^{-1}\left\{\nabla_1 \int d\mathbf{q}_2\, d\mathbf{v}_2\, V(\mathbf{q}_1 - \mathbf{q}_2)\bar{f}(\mathbf{q}_2, \mathbf{v}_2; t)\right\} \cdot \partial_1 \bar{f}(\mathbf{q}_1, \mathbf{v}_1; t) = 0 \end{aligned} \qquad (11.7.5)$$

Indeed, the term $n\bar{\varphi}(\mathbf{v}_2; t)$ contributes to the average potential a constant $n\tilde{V}_0$, the gradient of which is identically zero.

We now discuss this equation a little further, particularly in the case when $V(r)$ is the Coulomb potential. The average potential $U(\mathbf{q}; t)$ can then be written in the following suggestive form:

$$e^{-1}U(\mathbf{q};t) = -\int d\mathbf{q}' \frac{1}{|\mathbf{q}-\mathbf{q}'|}\sigma(\mathbf{q}';t) \tag{11.7.6}$$

where

$$\begin{aligned}\sigma(\mathbf{q}';t) &= -e\int d\mathbf{v}'[\bar{f}(\mathbf{q}',\mathbf{v}';t) - n\bar{\varphi}(\mathbf{v}';t)] \\ &= -e[n(\mathbf{q}';t) - n]\end{aligned} \tag{11.7.7}$$

where we used Eq. (3.5.4). The quantity $\sigma(\mathbf{q}';t)$ has a clear physical meaning. As $n(\mathbf{q}';t)$ represents the local density of electrons at point $\mathbf{q}'$ and time t, $-en(\mathbf{q}';t)$ is the corresponding charge density; $(+en)$, on the other hand, is the charge density of the homogeneous positive background that must be assumed present for overall neutrality (see Section 6.5); thus $\sigma(\mathbf{q}';t)$ is the *total charge density* at point $\mathbf{q}'$. It vanishes if the system is homogeneous, that is, $n(\mathbf{q}';t)=n$. Equation (11.7.6) is then precisely the formula giving the electrostatic potential at point $\mathbf{q}$ due to a continuous charge distribution $\sigma(\mathbf{q})$. We note indeed that

$$\begin{aligned}-\nabla^2\int d\mathbf{q}' \frac{1}{|\mathbf{q}-\mathbf{q}'|}\sigma(\mathbf{q}';t) &= -4\pi\int d\mathbf{q}'\, \delta(\mathbf{q}-\mathbf{q}')\sigma(\mathbf{q}';t) \\ &= -4\pi\sigma(\mathbf{q};t)\end{aligned}$$

It then follows that the Vlassov equation (11.7.4) can also be written in the form of *two coupled equations:*

$$\partial_t\bar{f}(\mathbf{q},\mathbf{v};t) + \mathbf{v}\cdot\nabla\bar{f}(\mathbf{q},\mathbf{v};t) - m^{-1}[\nabla U(\mathbf{q};t)]\cdot\partial\bar{f}(\mathbf{q},\mathbf{v};t) = 0 \tag{11.7.8}$$

$$e^{-1}\nabla^2 U(\mathbf{q};t) = -4\pi\sigma(\mathbf{q};t) \tag{11.7.9}$$

Equation (11.7.9) is simply the *Poisson equation* for the electrostatic potential $e^{-1}U(\mathbf{q};t)$. We now clearly see that, although (11.7.8) is formally the Liouville equation of a single particle moving in an external field, the latter is produced by the particles themselves and depends on their instantaneous distribution. $U(\mathbf{q};t)$ is therefore a typical *self-consistent field.* Because of this self-consistency, the Vlassov equation is actually a *nonlinear equation,* as appears clearly in the form (11.7.5). The nonlinearity must be present, since the equation describes interaction processes. In this respect the Vlassov equation differs markedly from the one-particle Liouville equation.

The reader has probably noticed the strong analogy between these ideas and those underlying the equilibrium theory of Debye and Hückel, discussed in Section 6.5. The Vlassov theory actually plays the same role out of equilibrium as the Debye–Hückel theory in equilibrium. It also describes a *dynamical screening* of the interactions. The nature of this screening is more complex than in equilibrium: it is no longer a simple spherically symmetrical polarization cloud but is deformed by the motion of the particles. The Vlassov equation also describes the very important *collective motions*, so characteristic of a plasma. These will be discussed further in Sections 12.7 and 13.6. We cannot go into many details within the framework of the present book. The interested reader should consult the books referred to at the end of this chapter.

11.8. CONNECTION BETWEEN THE LANDAU AND THE FOKKER–PLANCK EQUATIONS

We close this chapter with a brief discussion of the relation existing between the kinetic theory of weakly coupled gases and the theory of Brownian motion.

The fact that both the Landau equation (11.6.27) and the Fokker–Planck equation (11.3.21) are second-order partial-differential equations suggests that some relation should exist among them. There is, however, a very important difference between them: the Landau equation is *nonlinear.* It should be clear by now that *any kinetic equation is necessarily nonlinear.* Indeed, such an equation describes collision processes between two (or more) particles. It must, therefore, depend on the state—and hence on the product of distribution functions—of the two (or more) collision partners.

On the other hand, the Landau equation (11.6.8) can be rewritten in the following form:

$$\partial_t \bar{\varphi}(\mathbf{v}; t) = \partial^r(-A^r + \tfrac{1}{2}\partial^s B^{rs})\bar{\varphi}(\mathbf{v}; t) \tag{11.8.1}$$

with

$$\begin{aligned} A^r &= \int d\mathbf{v}_1 \{G^{rs}\, \partial_1^s \bar{\varphi}(\mathbf{v}_1; t) + \bar{\varphi}(\mathbf{v}_1; t)\, \partial^s G^{rs}\} \\ B^{rs} &= 2\int d\mathbf{v}_1\, G^{rs} \bar{\varphi}(\mathbf{v}_1; t) \end{aligned} \tag{11.8.2}$$

Equation (11.8.1) now appears in the same form as the Fokker–Planck equation (11.3.21) (trivially generalized to three-dimensional notation). We can identify a friction vector A^r and a diffusion tensor B^{rs}. The difference with (11.3.21) is, however, that these coefficients, instead of being constants, depend on the distribution function $\bar{\varphi}(\mathbf{v}; t)$, and therefore, as time goes on, they change in value and in shape in a self-consistent way. This results, of course, from the nonlinearity of the kinetic equation.

We can go one step further toward the Fokker–Planck equation if we study the following idealized problem. Consider a large amount of weakly coupled gas in thermal equilibrium at temperature T. We imagine that, as a result of a fluctuation, there are a few particles (homogeneously distributed) out of equilibrium at time zero. Because of their scarcity, we may assume that these particles never collide among themselves but only with particles of the "thermal bath." Also, we may assume that the global equilibrium of the bath is not affected by the presence of a few nonthermal particles.

How do these distinguished particles return to equilibrium? Clearly, we may write for their distribution function a Landau equation, in which, however, we replace the distribution function of their collision partners by the known, time-independent Maxwellian. This operation has two effects:

1. The Landau equation reduces to a *linear equation.*
2. This linear equation is precisely a Fokker–Planck equation.

Its characteristic transition moments are

$$A_0^r = -\alpha \int d\mathbf{u}\, u^s e^{-u^2} G^{rs}$$

$$B_0^{rs} = \beta \int d\mathbf{u}\, e^{-u^2} G^{rs}$$

where $\mathbf{u}$ is a reduced velocity: $\mathbf{u} = (m/2k_BT)^{1/2}\mathbf{v}$, and α and β are irrelevant constants. These integrals can be evaluated exactly, providing explicit expressions for A^r and B^{rs} in terms of error and Gaussian functions of U.

Without going into details, the physical picture is now clear. The distinguished particles in a weakly coupled gas are driven to equilibrium by a Brownian-motion mechanism. The process is a little bit more complicated than in the simple picture of Section 11.2 and 11.3. The friction coefficient ζ in Eq. (11.4.21) and the diffusion coefficient $2\zeta k_BT/M$ are now replaced by speed-dependent functions; but this is a not too essential feature.

The Landau equation describes the approach to equilibrium without any *ad hoc* procedures of the type needed in Sections 11.3 and 11.4. In the simple treatment of the Langevin equation, we had no details at all about the dynamical mechanism of the interactions, which would have allowed us to evaluate the function $\alpha(T)$ of Eq. (11.2.10); we therefore had to "feed in" the final result through Eq. (11.2.14). On the contrary, by starting from the Boltzmann equation, we used a precise dynamical model for the collision process. Our calculation amounts to an explicit evaluation of the "memory functions" $\phi(\theta)$ and $\alpha(T)$ within the assumed model. It is rewarding that the equilibrium distribution now comes out of the model instead of being fed in. The Boltzmann and the Landau equations, therefore, represent a great step forward toward a microscopic theory of nonequilibrium processes. We must not forget, however, that the Boltzmann equation was not derived in an orthodox way and that the important *Stosszahlansatz* is in apparent contradiction with mechanics. We cannot claim to have a rigorous microscopic theory of irreversibility before this important point is cleared up. The result will come out of a general theory, which, because of its higher degree of abstraction, will be treated in the last part of the book.

BIBLIOGRAPHICAL NOTES

The theory of *Brownian motion* is very well covered by a series of papers contained in N. Wax, *Noise and Stochastic Processes*, Dover, New York. 1954. Among these papers, we highlight two classics, which remain even today among the best introductions to the subject:

G. E. Uhlenbeck and L. S. Ornstein, *Phys. Rev.* **36,** 823 (1930).

S. Chandrasekhar, *Rev. Mod. Phys.* **15,** 85 (1943).

More mathematically minded, but still elementary treatments can be found in D. R. Cox and H. D. Miller, *The Theory of Stochastic Processes*, Methuen, London, 1965.

A culmination of the classical theory of linear stochastic processes is achieved in L. Onsager and S. Machlup, *Phys. Rev.* **91,** 1505, 1512 (1953).

The more modern topic of nonlinear stochastic processes is treated, for instance, in N. G. van Kampen, "Fluctuations in Non-Linear Systems," in *Fluctuation Phenomena in Solids* (R. E. Burgess, ed.), Academic Press, New York, 1965.

An original approach, combining the Liouville equation with stochastic theory, is found in R. Kubo, *J. Math. Phys.* **4,** 174 (1963).

The *Boltzmann equation* first appeared in L. Boltzmann, *Wien, Ber.* **66,** 275 (1872). See also L. Boltzmann, *Vorlesungen über Gastheorie*, Leipzig, 1896 [English transl.: *Lectures on Gas Theory* (S. Brush, transl.), Univ. California Press, Berkeley, 1964].

The first serious and valid criticisms against this equation appeared in

J. Loschmidt, *Wien. Ber.* **73,** 139, (1876); **75,** 67 (1877).

E. Zermelo, *Ann. Phys.* **57,** 485 (1896).

An old, but still very inspiring critical discussion of Boltzmann's ideas, is found in P. and T. Ehrenfest, *Enzyklopaedie d. math. Wiss.*, Vol. IV, Pt. 32, Leipzig, 1911.

A very interesting review of the present state of this problem appears in the Proceedings of the symposium held in Vienna for the 100th anniversary of the Boltzmann equation: *The Boltzmann Equation: Theory and Applications*, E. G. D. Cohen and W. Thirring, eds., Springer, New York, 1973. Other references to the theory of the Boltzmann equation will be found in Chapters 12, 13, 17, and 20.

The *Landau equation* was derived in L. D. Landau, *Phys. Z. Sowj. Union* **10,** 154 (1936).

The *Vlassov equation* was derived in A. Vlassov, *Zh. Eksp. Teor. Fiz.* **8,** 291 (1938).

The latter two equations are extensively discussed in the book of R. Balescu (quoted in Chapter 6).

CHAPTER 12
KINETIC EQUATIONS AND HYDRODYNAMICS

12.1. MOMENT EQUATIONS OF A KINETIC EQUATION

In Chapter 11 we "derived" two equations (Boltzmann and Vlassov–Landau) that are representatives of an important class called the kinetic equations. We define a *kinetic equation* as a *closed, nonlinear equation describing the time evolution and the approach to equilibrium of the one-particle reduced distribution function.**

We now investigate the problem of how we can make the transition from the microscopic to the macroscopic level. In equilibrium theory, this problem was rather simply solved in Chapter 4. Given a microscopic stationary distribution function (such as the canonical ensemble), it was possible to construct, in terms of its characteristic parameters, a quantity that turns out to possess the properties of a thermodynamic potential. Hence the bridge between microscopic theory and macroscopic thermodynamics is obtained in a single shot. No such simple way exists in nonequilibrium theory, because of the variety of phenomena and of the complexity of the evolution processes. We must therefore construct the theory more progressively. In this chapter we start this construction with the derivation of the equations of hydrodynamics, which are typical equations of macroscopic continuum physics. As an orientation, we first describe the idea of the method, which is a very general one, applicable to all kinetic equations.

The characteristic quantities of hydrodynamics are *fields*, that is, functions defined locally in every point $\mathbf{x}$ of physical space and varying in time: $B(\mathbf{x}; t)$ (see the discussion of Section 2.1). Typical examples are the mass density $\rho(\mathbf{x}; t)$, the local velocity $\mathbf{u}(\mathbf{x}; t)$, and so forth. These macroscopic quantities will be defined, according to the general philosophy of Chapters 2 and 3 as averages of microscopic dynamical functions $\mathfrak{b}$ through Eq. (3.1.23) [we define the microscopic function by

*There are *apparent* exceptions to this definition. See the comment on liquids in the Bibliographical Notes to Chapter 20.

the vector of Eq. (3.1.22)]. The time dependence of $B(\mathbf{x}; t)$ is introduced by the distribution function, which is taken as a solution of the kinetic equation.* The dependence on $\mathbf{x}$ must, therefore, come from a parametric dependence of $\mathfrak{b}$ on $\mathbf{x}$. Let us consider the simple (but important) case in which $\mathfrak{b}$ has only a component b_1. The general form of $B(\mathbf{x}, t)$ is then

$$B(\mathbf{x}; t) = \int d\mathbf{q}_1 \, d\mathbf{v}_1 \, b_1(\mathbf{q}_1, \mathbf{v}_1; \mathbf{x}) \bar{f}(\mathbf{q}_1, \mathbf{v}_1; t) \tag{12.1.1}$$

We may further specialize the form of b_1. The important fields of macroscopic physics are *densities* of mass, velocity, energy, and so on. The microscopic dynamical functions describing densities are obtained by the argument of the Sections 3.1 and 3.3. We "test" every point $\mathbf{x}$ of physical space by a delta function $\delta(\mathbf{x}-\mathbf{q}_1)$. If the particle is not in $\mathbf{x}$, it contributes nothing to B; if it is there, then it contributes a dynamical quantity $\beta(\mathbf{v}_1, \mathbf{q}_1)$ [for instance, $\beta(\mathbf{v}_1, \mathbf{q}_1)$ is m or $\mathbf{v}_1$ in the examples quoted above]. Thus b_1 must be of the form:

$$b_1(\mathbf{q}_1, \mathbf{v}_1; \mathbf{x}) = \delta(\mathbf{x}-\mathbf{q}_1)\beta(\mathbf{q}_1, \mathbf{v}_1) \tag{12.1.2}$$

and Eq. (12.1.1) becomes

$$B(\mathbf{x}; t) = \int d\mathbf{q}_1 \, d\mathbf{v}_1 \, \delta(\mathbf{x}-\mathbf{q}_1)\beta(\mathbf{q}_1, \mathbf{v}_1)\bar{f}(\mathbf{q}_1, \mathbf{v}_1; t) \tag{12.1.3}$$

where $\bar{f}(\mathbf{q}_1, \mathbf{v}_1; t)$ is a solution of a kinetic equation.

Dynamical functions of the form (12.1.2) are called *localized quantities.*† In order to abbreviate the notations we will often use the symbol (j) for the set $(\mathbf{q}_j, \mathbf{v}_j)$; thus, Eq. (12.1.3) is written

$$B(\mathbf{x}; t) = \int d1 \, \delta(\mathbf{x}-\mathbf{q}_1)\beta(1)\bar{f}(1; t) \tag{12.1.4}$$

To derive an equation describing the rate of change of $B(\mathbf{x}; t)$, we take the time derivative of Eq. (12.1.3):

$$\partial_t B(\mathbf{x}; t) = \int d1 \, \delta(\mathbf{x}-\mathbf{q}_1)\beta(1) \, \partial_t \bar{f}(1; t) \tag{12.1.5}$$

* The kinetic equation is a closed equation in $\bar{f}_1(x_1; t)$. It will be seen later in great detail that when a function $f_s(x_1, \ldots, x_s; t)$, $s>1$ is needed in the calculation of an average, this function can be expressed (in the framework of kinetic theory) as a functional of $\bar{f}_1$ alone. As $\bar{f}_1$ is the only independent distribution function in the theory, we shall henceforth drop the subscript 1.

† The localization through a function $\delta(\mathbf{x}-\mathbf{q}_1)$ is only correct because the canonical variable $\mathbf{q}_1$ represents the position of a particle. If other dynamical variables are used, or if one treats systems other than material particles (such as oscillators) the localizing δ function has a different argument.

We now substitute for $\partial_t \bar{f}(1)$ the kinetic equation, which we write schematically as

$$\partial_t \bar{f}(1) = -\mathbf{v}_1 \cdot \nabla_1 \bar{f}(1) + \Omega \bar{f}(1) + \mathcal{K} \bar{f}(1) \equiv \mathcal{O} \bar{f}(1) \qquad (12.1.6)$$

where Ω stands for the operator of the Vlassov term and $\mathcal{K}$ stands for the collision operator; $\mathcal{O}$ is an abbreviation for the complete kinetic operator. Hence

$$\partial_t B(\mathbf{x}; t) = \int d1\ \delta(\mathbf{x} - \mathbf{q}_1) \beta(1)\ \mathcal{O} \bar{f}(1; t) \qquad (12.1.7)$$

The idea now is to "transfer" the operator $\mathcal{O}$ from $\bar{f}(1; t)$ to the dynamical function. If $\mathcal{O}$ is a differential operator, this can be achieved by repeated integrations by parts; in other cases this can be done by proper rearrangements. As the function $\bar{f}(1; t)$ is always supposed to vanish strongly at the boundaries (i.e., at infinity) all the surface terms in the partial integrations vanish, and we may write

$$\partial_t B(\mathbf{x}; t) = \int d1\ \bar{f}(1; t)\ [\mathcal{O}^+ \delta(\mathbf{x} - \mathbf{q}_1) \beta(1)] \qquad (12.1.8)$$

where $\mathcal{O}^+$ is the operator "adjoint" to $\mathcal{O}$: it will be constructed below. Define now

$$c_1(\mathbf{q}_1, \mathbf{v}_1; \mathbf{x}) = \mathcal{O}^\dagger \delta(\mathbf{x} - \mathbf{q}_1) \beta(1) \qquad (12.1.9)$$

$c_1(1; \mathbf{x})$ is, formally, a dynamical function.* Thus:

$$\partial_t B(\mathbf{x}; t) = \int d1\ c_1(1; \mathbf{x}) \bar{f}(1; t) \equiv C(\mathbf{x}; t) \qquad (12.1.10)$$

The important result achieved here is that the rate of change of the macroscopic quantity $B(\mathbf{x}; t)$ is expressed entirely in terms of the macroscopic quantity $C(\mathbf{x}; t)$. *We thus achieved the transition from a microscopic equation (12.1.6) to a macroscopic equation (12.1.10).* In other words, the kinetic equation (12.1.6) *induces* the equations of evolution of the averages (12.1.10). These are sometimes called the *moment equations* of the kinetic equation.

We now note that the single kinetic equation induces an infinite number of moment equations, corresponding to all possible dynamical functions $\beta(1)$. These equations form a *hierarchy:* $\partial_t B$ is given by a new function C, $\partial_t C$ will be expressed as a different function D, and so on. It will soon appear that these equations never close on themselves. Thus, we are faced with a problem analogous to the microscopic BBGKY hierarchy,

* Note, however, that $c_1(1; \mathbf{x})$ in general depends on $\bar{f}(\mathbf{q}, \mathbf{v}; t)$ because of the nonlinear character of Ω and of $\mathcal{K}$.

but for quite different reasons. Although the kinetic equations are closed in $\bar{f}(1)$ (contrary to the BBGKY hierarchy), their *moment equations* are not. This feature is well known from macroscopic physics, independently of any kinetic theory. In hydrodynamics, one proceeds to closing the equations by means of phenomenological assumptions and approximations. One of the advantages of kinetic theory is its giving a rational basis to such approximations and finding new ones in situations where the ordinary assumptions fail.

We now reconsider Eq. (12.1.7) and analyze the process of "transfer" of the operator $\mathcal{O}$ in some more detail. Referring to Eq. (12.1.6) we first consider the flow term, which is transformed as follows by partial integration:

$$F \equiv -\int d1\ \delta(\mathbf{x}-\mathbf{q}_1)\beta(1)\ \mathbf{v}_1 \cdot \nabla_1 \bar{f}(1;t)$$

$$= +\int d1\ \bar{f}(1;t)\,\mathbf{v}_1 \cdot \{[\nabla_1 \delta(\mathbf{x}-\mathbf{q}_1)]\beta(1) + \delta(\mathbf{x}-\mathbf{q}_1)\nabla_1 \beta(1)\}$$

We now note the identity:

$$\nabla_1 \delta(\mathbf{x}-\mathbf{q}_1) = -\nabla \delta(\mathbf{x}-\mathbf{q}_1) \tag{12.1.11}$$

where $\nabla \equiv \partial/\partial \mathbf{x}$. Hence

$$F = -\nabla \cdot \int d1\ \delta(\mathbf{x}-\mathbf{q}_1)\,\mathbf{v}_1\,\beta(1)\bar{f}(1;t)$$

$$+\int d1\ \delta(\mathbf{x}-\mathbf{q}_1)[\nabla_1 \beta(1)] \cdot \mathbf{v}_1 \bar{f}(1;t) \tag{12.1.12}$$

We now transform the Vlassov term as follows [see Eq. (11.7.5)]:

$$V \equiv m^{-1}\int d1\ d2\ \delta(\mathbf{x}-\mathbf{q}_1)\beta(1)\,[\nabla_1 V(\mathbf{q}_1-\mathbf{q}_2)]\bar{f}(2;t) \cdot \partial_1 \bar{f}(1;t)$$

$$= -m^{-1}\int d1\ d2\ \delta(\mathbf{x}-\mathbf{q}_1)[\partial_1 \beta(1)] \cdot [\nabla_1 V(\mathbf{q}_1-\mathbf{q}_2)]\bar{f}(2;t)\bar{f}(1;t)$$

$$= -m^{-1}\int d1\ d2\ \delta(\mathbf{x}-\mathbf{q}_1)[\partial_1 \beta(1)] \cdot [\nabla V(\mathbf{x}-\mathbf{q}_2)]\bar{f}(2;t)\bar{f}(1;t)$$

$$= -m^{-1}[\nabla U(\mathbf{x};t)] \cdot \int d1\ \delta(\mathbf{x}-\mathbf{q}_1)[\partial_1 \beta(1)]\bar{f}(1;t) \tag{12.1.13}$$

The term (12.1.12) is of a very general nature: it is the same for all kinetic equations. The Vlassov term (12.1.13) is present whenever the average field plays a role; it is negligible in an ordinary gas of particles with short-range forces. The contribution of the collision term, however, is by far the most complicated one; moreover, it depends crucially on the details of the collision mechanism and will be completely different, say, if the Boltzmann or the Landau equation is used. We shall not evaluate it in detail but only give it a name:

$$\sigma_B^{(3)} = \int d1\, \delta(\mathbf{x}-\mathbf{q}_1)\, \beta(1)\, \mathcal{K}\bar{f}(1;t) \tag{12.1.14}$$

We now introduce also the following definitions:

$$\sigma_B^{(2)} = -m^{-1}[\nabla U(\mathbf{x};t)] \cdot \int d1\, \delta(\mathbf{x}-\mathbf{q}_1)\, [\partial_1 \beta(1)]\, \bar{f}(1;t) \tag{12.1.15}$$

$$\sigma_B^{(1)} = \int d1\, \delta(\mathbf{x}-\mathbf{q}_1)\, [\nabla_1 \beta(1)] \cdot \mathbf{v}_1 \bar{f}(1;t) \tag{12.1.16}$$

$$\sigma_B = \sigma_B^{(1)} + \sigma_B^{(2)} + \sigma_B^{(3)} \tag{12.1.17}$$

σ_B is called the *source of B*. We, moreover, define the *flow of B* as

$$\mathbf{\Phi}_B = \int d1\, \delta(\mathbf{x}-\mathbf{q}_1)\, \mathbf{v}_1 \beta(1)\, \bar{f}(1;t) \tag{12.1.18}$$

Collecting now all the partial results we see that the rate of change of B is quite generally expressed as

$$\partial_t B(\mathbf{x};t) = -\nabla \cdot \mathbf{\Phi}_B(\mathbf{x};t) + \sigma_B(\mathbf{x};t) \tag{12.1.19}$$

This type of equation is very familiar from the physics of continuous media: it is a *local balance equation.* The interpretation of the terms is well known. If $\sigma_B = 0$, Eq. (12.1.19) becomes an equation expressing the *conservation of B*. Indeed, by integrating both terms over the volume in an arbitrary region of space, Gauss's theorem tells us that the rate of change of B in the volume equals the flow of B through the surface of the region. No B can be created or absorbed inside the volume: B can only change because some amount has been brought in or has flown out by some mechanism or other. If the source σ_B is different from zero, it expresses the creation or absorption of B inside the volume, without any crossing of the surface.

We now apply this formalism to the derivation of the most important macroscopic equations.

12.2. BOLTZMANN'S *H* THEOREM AND THE ENTROPY BALANCE: IRREVERSIBILITY AND DISSIPATION

We do not begin our investigation of the macroscopic equations with the simplest case. Rather, we first study the evolution of a quantity that lies at the basis of nonequilibrium thermodynamics. The property to be established here is actually a prerequisite of any macroscopic theory; it consists in deriving the second law of thermodynamics. As is well known, the second law is directly connected to the concept of irreversibility. It claims the existence of a state function, the entropy, that is not conserved. Rather, during the spontaneous evolution of an isolated system, it can only increase in time as a result of the irreversible processes going on in the system. Its increase stops only when the system reaches equilibrium: at that point the entropy attains its maximum value. When formulated locally, the rate of change of the entropy density $\bar{s}(\mathbf{x}; t)$ is expressed by a balance equation of type (12.1.19):

$$\partial_t \bar{s}(\mathbf{x}; t) = -\nabla \cdot \mathbf{\Phi}_s(\mathbf{x}; t) + \sigma_s(\mathbf{x}; t) \tag{12.2.1}$$

The second law of thermodynamics is a statement about the *sign of the source:*

$$\sigma_s(\mathbf{x}; t) \geq 0 \tag{12.2.2}$$

The great success and importance of the kinetic equation derived by Boltzmann in 1872 came from his proof that an entropy can be defined and that it possesses the property (12.2.2) as a result of the kinetic equation. Thus Boltzmann's theory was the first instance in history of an ("almost") mechanical theory of irreversibility. His theorem still bears the famous name of *H theorem:* it stems from Boltzmann's use of the letter H to denote the quantity $[-\bar{s}(\mathbf{x}; t)]$.

We first prove the H theorem for the simpler case of the Landau equation, and consider first a homogeneous system. In that case the kinetic equation is given by Eq. (11.6.20), which we rewrite here:

$$\partial_t \bar{\varphi}(1; t) = An \int d\mathbf{v}_2 \int d\mathbf{k}\, \tilde{V}_k^2\, \mathbf{k} \cdot \partial_{12}\, \delta(\mathbf{k} \cdot \mathbf{g})\, \mathbf{k} \cdot \partial_{12} \bar{\varphi}(1; t) \bar{\varphi}(2; t) \tag{12.2.3}$$

Boltzmann's definition of the entropy density is

$$\bar{s}(t) = -k_B n \int d\mathbf{v}_1\, \bar{\varphi}(\mathbf{v}_1; t) \ln[n\bar{\varphi}(\mathbf{v}_1; t)] + b \tag{12.2.4}$$

where k_B and b are constants, to be determined later. This function depends on time through the distribution function $\bar{\varphi}(\mathbf{v}_1; t)$, which is a solution of the kinetic equation (12.2.3). Let us evaluate its time derivative:

$$\begin{aligned}
\partial_t \bar{s}(t) &= -k_B n \int d\mathbf{v}_1\, [\ln n\bar{\varphi}(1; t) + 1] \partial_t \bar{\varphi}(1; t) \\
&= -Ak_B n^2 \int d\mathbf{v}_1\, d\mathbf{v}_2\, d\mathbf{k}\, [\ln n\bar{\varphi}(1; t) + 1] \tilde{V}_k^2 \\
&\quad \times \mathbf{k} \cdot \partial_{12}\, \delta(\mathbf{k} \cdot \mathbf{g})\, \mathbf{k} \cdot \partial_{12} \bar{\varphi}(1; t) \bar{\varphi}(2; t)
\end{aligned}$$

$$= -Ak_B n^2 \int d\mathbf{v}_1 \, d\mathbf{v}_2 \, d\mathbf{k} \, [\ln n\bar{\varphi}(2;t)+1] \tilde{V}_k^2$$

$$\times \mathbf{k} \cdot \partial_{12} \, \delta(\mathbf{k}\cdot\mathbf{g}) \, \mathbf{k}\cdot\partial_{12}\bar{\varphi}(1;t)\bar{\varphi}(2;t)$$

$$= -\tfrac{1}{2}Ak_B n^2 \int d\mathbf{v}_1 \, d\mathbf{v}_2 \, d\mathbf{k} \, [\ln n\bar{\varphi}(1;t) + \ln n\bar{\varphi}(2;t)+2] \tilde{V}_k^2$$

$$\times \mathbf{k} \cdot \partial_{12} \, \delta(\mathbf{k}\cdot\mathbf{g}) \, \mathbf{k}\cdot\partial_{12}\bar{\varphi}(1;t)\bar{\varphi}(2;t)$$

$$= +\tfrac{1}{2}Ak_B n^2 \int d\mathbf{v}_1 \, d\mathbf{v}_2 \, d\mathbf{k} \, [\bar{\varphi}(1;t)\bar{\varphi}(2;t)]^{-1} \tilde{V}_k^2 \, \delta(\mathbf{k}\cdot\mathbf{g})$$

$$\times \{\mathbf{k}\cdot\partial_{12}\bar{\varphi}(1;t)\bar{\varphi}(2;t)\}^2 \qquad (12.2.5)$$

In going from the second to the third step we made use of the symmetry of the integrand in the kinetic equation ($\mathbf{v}_1$ and $\mathbf{v}_2$ are dummy integration variables). The fourth step represents the half-sum of two equal expressions. In the last step we performed an integration by parts. The remarkable property of the last expression is its semidefinite positive sign. Indeed, the distribution functions $\bar{\varphi}(j;t)$ are nonnegative, and so is the δ function; the other factors under the integral are squares. Therefore:

$$\partial_t \bar{s}(t) \geq 0 \qquad (12.2.6)$$

Hence the function $\bar{s}(t)$ can only increase as time passes.

On the other hand, it can easily be shown that $\bar{s}(t)$ is bounded above. Indeed, $\bar{s}(t)=\infty$ only if the integral in (12.2.4) diverges. We may assume, on a physical basis, that the average kinetic energy exists; this implies that the integral:

$$\int d\mathbf{v} \, \bar{\varphi}(\mathbf{v};t)v^2 < \infty \qquad (12.2.7)$$

converges. We know that as $v\to\infty$, $\bar{\varphi}(\mathbf{v};t)\to 0$, hence $\ln \bar{\varphi}(\mathbf{v};t)\to -\infty$. The question is: how fast does the logarithm go to infinity? If it decreases slower than v^2, the integral converges because of (12.2.7). Divergence might occur only if $-\ln \bar{\varphi}$ tends to infinity faster than v^2; but in that case $\bar{\varphi}$ decays faster than $\exp(-v^2)$, which dominates the increase of $(-\ln \bar{\varphi})$ and causes the integral of $\bar{\varphi} \ln \bar{\varphi}$ to converge.

The conclusion of this discussion is, therefore: Regardless of the initial condition, the distribution function changes in time in such a way that the functional $\bar{s}(t)$ increases monotonously. The evolution reaches an end point, that is, a steady state, when the quantity $\bar{s}(t)$ attains its maximum value (which is finite).

It is easily verified that the following function is a steady-state distribution:

$$\bar{\varphi}^0(\mathbf{v}) = ce^{-a(\mathbf{v}-\mathbf{u})^2} \tag{12.2.8}$$

where c, a, and the vector $\mathbf{u}$ are constants. Indeed

$$\mathbf{k} \cdot \partial_{12}\bar{\varphi}^0(\mathbf{v}_1)\bar{\varphi}^0(\mathbf{v}_2) = -a(\mathbf{k} \cdot \mathbf{g})\bar{\varphi}^0(\mathbf{v}_1)\bar{\varphi}^0(\mathbf{v}_2) \tag{12.2.9}$$

When this is substituted into (12.2.5), we obtain an integrand involving the product $(\mathbf{k} \cdot \mathbf{g})^2\, \delta(\mathbf{k} \cdot \mathbf{g})$, which vanishes identically. It can be shown that (12.2.9) is the only possible stationary distribution. It may be noted now that the constant velocity $\mathbf{u}$ appearing in Eq. (12.2.9) is not really relevant: it can be transformed away by choosing a coordinate system moving with the constant velocity $-\mathbf{u}$ (Galilean transformation). Hence, we can take as the stationary solution of the kinetic equation simply the Maxwellian function:

$$\bar{\varphi}^0(\mathbf{v}) = ce^{-av^2} \tag{12.2.10}$$

This is nothing other than the equilibrium distribution function of the ideal classical gas (7.1.3). The constants have therefore a thermodynamical meaning:

$$\begin{aligned} a &= \frac{m\beta}{2} = \frac{m}{2k_B T} \\ c &= \left(\frac{m}{2\pi k_B T}\right)^{3/2} \end{aligned} \tag{12.2.11}$$

If we now choose the constants in Eq. (12.2.4) as

$$\begin{aligned} k_B &= \text{Boltzmann's constant} \\ b &= nk_B \ln\left(\frac{e}{h^3}\right) \end{aligned} \tag{12.2.12}$$

we see that Eq. (12.2.4) coincides exactly, in equilibrium, with the expression (7.3.12) of the entropy per particle, multiplied by the number density n, that is, with the *entropy density*.

Hence the function $\tilde{s}(t)$ defined by Eq. (12.2.4), the evolution of which is induced by the kinetic equation (12.2.3), has the correct behavior of the entropy out of equilibrium and reduces to the equilibrium entropy deduced from the canonical ensemble when the end point of the evolution is reached. Thus, the collisions drive the system irreversibly to equilibrium.

We now show how the proof of the H theorem is extended to the Boltzmann equation (11.4.21):

$$\partial_t\bar{\varphi}(\mathbf{v}_1; t) = 2\pi n \int d\mathbf{v}_2\, db\, bg\{\bar{\varphi}(\mathbf{v}_1'; t)\bar{\varphi}(\mathbf{v}_2'; t) - \bar{\varphi}(\mathbf{v}_1; t)\bar{\varphi}(\mathbf{v}_2; t)\} \tag{12.2.13}$$

Proceeding as in Eq. (12.2.5) we obtain

$$\partial_t \bar{s}(t) = -2\pi k_B n^2 \int d\mathbf{v}_1\, d\mathbf{v}_2\, db\, bg[\ln n\bar{\varphi}(\mathbf{v}_1; t) + 1]$$
$$\times \{\bar{\varphi}(\mathbf{v}_1'; t)\bar{\varphi}(\mathbf{v}_2'; t) - \bar{\varphi}(\mathbf{v}_1; t)\bar{\varphi}(\mathbf{v}_2; t)\}$$
$$= -2\pi k_B n^2 \int d\mathbf{v}_1\, d\mathbf{v}_2\, db\, bg[\ln n\bar{\varphi}(\mathbf{v}_2; t) + 1]\{\cdots\}$$
$$= -\pi k_B n^2 \int d\mathbf{v}_1\, d\mathbf{v}_2\, db\, bg[\ln n\bar{\varphi}(\mathbf{v}_1; t) + \ln n\bar{\varphi}(\mathbf{v}_2; t) + 2]\{\cdots\}$$
$$= +\pi k_B n^2 \int d\mathbf{v}_1\, d\mathbf{v}_2\, db\, bg[\ln n\bar{\varphi}(\mathbf{v}_1'; t) + \ln n\bar{\varphi}(\mathbf{v}_2'; t) + 2]\{\cdots\}$$
$$= \frac{\pi}{2} k_B n^2 \int d\mathbf{v}_1\, d\mathbf{v}_2\, db\, bg \ln \frac{\bar{\varphi}(\mathbf{v}_1'; t)\bar{\varphi}(\mathbf{v}_2'; t)}{\bar{\varphi}(\mathbf{v}_1; t)\bar{\varphi}(\mathbf{v}_2; t)}$$
$$\times \{\bar{\varphi}(\mathbf{v}_1'; t)\bar{\varphi}(\mathbf{v}_2'; t) - \bar{\varphi}(\mathbf{v}_1; t)\bar{\varphi}(\mathbf{v}_2; t)\} \qquad (12.2.14)$$

In going from one step to another we note that we may interchange the dummy integration variables $\mathbf{v}_1$, $\mathbf{v}_2$: in this process $\mathbf{v}_1' \to \mathbf{v}_2'$, $\mathbf{v}_2' \to \mathbf{v}_1'$, $\mathbf{g} \to -\mathbf{g}$. We then replace the original expression by the half-sum of the two equal quantities. Next we note that we may also substitute under the integral $\mathbf{v}_1 \to \mathbf{v}_1'$, $\mathbf{v}_2 \to \mathbf{v}_2'$. The final symmetrization leads to the expression in the last side of the equality.

We now note the following identity that holds, whatever the values of x and y, provided they are positive*.

$$(x - y) \ln\left(\frac{x}{y}\right) \geq 0 \qquad (12.2.15)$$

It then follows from (12.2.14) that

$$\partial_t \bar{s}(t) \geq 0$$

Thus, Eq. (12.2.6) is valid in this case too. It is not difficult to show that the Maxwell distribution (12.2.8) is the steady-state distribution of the Boltzmann equation. For a distribution to be stationary, Eq. (12.2.14) requires

$$\bar{\varphi}^0(\mathbf{v}_1')\bar{\varphi}^0(\mathbf{v}_2') = \bar{\varphi}^0(\mathbf{v}_1)\bar{\varphi}^0(\mathbf{v}_2)$$

or

$$\ln \bar{\varphi}^0(\mathbf{v}_1') + \ln \bar{\varphi}^0(\mathbf{v}_2') = \ln \bar{\varphi}^0(\mathbf{v}_1) + \ln \bar{\varphi}^0(\mathbf{v}_2) \qquad (12.2.16)$$

* If $x > y$, $(x/y) > 1$ and $\ln(x/y) > 0$; when $x < y$, $(x/y) < 1$ and $\ln(x/y) < 0$.

Substituting Eq. (12.2.8) and using (11.4.6) we obtain

$$-a\,\tfrac{1}{4}(\mathbf{v}_1+\mathbf{v}_2+|\mathbf{v}_1-\mathbf{v}_2|\,\boldsymbol{\varepsilon}-2\mathbf{u})^2-a\,\tfrac{1}{4}(\mathbf{v}_1+\mathbf{v}_2-|\mathbf{v}_1-\mathbf{v}_2|\,\boldsymbol{\varepsilon}-2\mathbf{u})^2$$
$$=-a(\mathbf{v}_1-\mathbf{u})^2-a(\mathbf{v}_2-\mathbf{u})^2$$

an identity that is easily checked.

We now proceed to the richer class of inhomogeneous systems. The basic kinetic equation describes the evolution of a distribution function depending on position, velocity, and time: $\bar{f}(1;t)\equiv\bar{f}(\mathbf{q}_1,\mathbf{v}_1;t)$. The Vlassov–Landau equation (11.7.5) is now

$$\partial_t\bar{f}(1;t)=-\mathbf{v}_1\cdot\nabla_1\bar{f}(1;t)+m^{-1}\left\{\nabla_1\int d2\;V(\mathbf{q}_1-\mathbf{q}_2)\bar{f}(2;t)\right\}\cdot\partial_1\bar{f}(1;t)$$
$$+A\int d2\int d\mathbf{k}\;\delta(\mathbf{q}_1-\mathbf{q}_2)\tilde{V}_k^2\,\mathbf{k}\cdot\partial_{12}\,\delta(\mathbf{k}\cdot\mathbf{g})\,\mathbf{k}\cdot\partial_{12}\bar{f}(1;t)\bar{f}(2;t)\quad(12.2.17)$$

In the Boltzmann case, one drops the Vlassov term and replaces the collision term by the appropriate one. We must now define the entropy density locally, as a field. We choose

$$\bar{s}(\mathbf{x};t)=-k_B\int d\mathbf{v}_1\,d\mathbf{q}_1\;\delta(\mathbf{q}_1-\mathbf{x})\bar{f}(\mathbf{q}_1,\mathbf{v}_1;t)\ln\bar{f}(\mathbf{q}_1,\mathbf{v}_1;t)+b\quad(12.2.18)$$

The time derivative of this function is

$$\partial_t\bar{s}(\mathbf{x};t)=-k_B\int d1\;\delta(\mathbf{q}_1-\mathbf{x})[\ln\bar{f}(1;t)+1]\partial_t\bar{f}(1;t)\quad(12.2.19)$$

We may now use the calculations of Section 12.1, by noting that (12.2.19) is of the form (12.1.5) with

$$\beta(1)=-k_B[\ln\bar{f}(1;t)+1]\quad(12.2.20)$$

Hence the balance equation must be of the form (12.1.19), and we must simply calculate the coefficients. However, some unusual cancellations will occur here, because $\bar{s}(\mathbf{x};t)$ is *not* really the average of a dynamical function: $\beta(1)$ depends on the distribution function. We are already familiar with this feature (see Sections 2.2 and 7.2). We define provisionally a flow $\boldsymbol{\Phi}_s'$ by (12.1.18) and separate the contributions of the two terms in (12.2.20):

$$\boldsymbol{\Phi}_s'(\mathbf{x};t)=\boldsymbol{\Phi}_s(\mathbf{x};t)+\boldsymbol{\varphi}_s(\mathbf{x};t)\quad(12.2.21)$$

$$\boldsymbol{\Phi}_s(\mathbf{x};t)=-k_B\int d1\;\delta(\mathbf{x}-\mathbf{q}_1)\,\mathbf{v}_1[\ln\bar{f}(1;t)]\,\bar{f}(1;t)\quad(12.2.22)$$

$$\boldsymbol{\varphi}_s(\mathbf{x};t)=-k_B\int d1\;\delta(\mathbf{x}-\mathbf{q}_1)\,\mathbf{v}_1\bar{f}(1;t)\quad(12.2.23)$$

Hence

$$-\nabla\cdot\Phi_s' = -\nabla\cdot\Phi_s + k_B\nabla\cdot\int d1\,\delta(\mathbf{x}-\mathbf{q}_1)\,\mathbf{v}_1\bar{f}(1;t) \qquad (12.2.24)$$

We now calculate the source terms. Equation (12.1.16) becomes

$$\begin{aligned}\sigma_s^{(1)} &= -k_B\int d1\,\delta(\mathbf{x}-\mathbf{q}_1)[\nabla_1 \ln \bar{f}(1;t)]\cdot\mathbf{v}_1\bar{f}(1;t)\\ &= -k_B\int d1\,\delta(\mathbf{x}-\mathbf{q}_1)\,\mathbf{v}_1\cdot\nabla_1\bar{f}(1;t)\\ &= +k_B\int d1\,[\nabla_1\,\delta(\mathbf{x}-\mathbf{q}_1)]\cdot\mathbf{v}_1\bar{f}(1;t)\\ &= -k_B\nabla\cdot\int d1\,\delta(\mathbf{x}-\mathbf{q}_1)\,\mathbf{v}_1\bar{f}(1;t)\end{aligned} \qquad (12.2.25)$$

In going from the second to the third step we made a partial integration and then used Eq. (12.1.11). We thus find that $\sigma_s^{(1)}$ exactly cancels the second term in the right-hand side of (12.2.24).

The contribution of the Vlassov term $\sigma_s^{(2)}$ is easily evaluated:

$$\begin{aligned}\sigma_s^{(2)} &= k_B m^{-1}\nabla U\cdot\int d1\,\delta(\mathbf{x}-\mathbf{q}_1)[\partial_1 \ln \bar{f}(1;t)]\bar{f}(1;t)\\ &= k_B m^{-1}\nabla U\cdot\int d1\,\delta(\mathbf{x}-\mathbf{q}_1)\partial_1\bar{f}(1;t)=0\end{aligned} \qquad (12.2.26)$$

The term vanishes because it is a surface integral taken at infinity: *The Vlassov term does not contribute to the entropy source.*

The source term due to the collisions is evaluated as in (12.2.5):

$$\begin{aligned}\sigma_s^{(3)} = \tfrac{1}{2}Ak_B\int d1\,d2\,\delta(\mathbf{x}-\mathbf{q}_1)\,\delta(\mathbf{x}-\mathbf{q}_2)\int d\mathbf{k}\,\tilde{V}_k^2\,[\bar{f}(1;t)\bar{f}(2;t)]^{-1}\\ \times\{\mathbf{k}\cdot\partial_{12}\bar{f}(1;t)\bar{f}(2;t)\}^2\,\delta(\mathbf{k}\cdot\mathbf{g})\end{aligned} \qquad (12.2.27)$$

Collecting these results, we obtain the *entropy balance* in the form:

$$\partial_t\bar{s}(t) = -\nabla\cdot\Phi_s(\mathbf{x};t)+\sigma_s^{(3)}(\mathbf{x};t) \qquad (12.2.28)$$

and we clearly see from (12.2.26) that

$$\sigma_s^{(3)}(\mathbf{x};t)\geqslant 0 \qquad (12.2.29)$$

Hence, we recovered exactly the second law of thermodynamics (12.2.1), (12.2.2). We thus arrived at a complete foundation of irreversible thermodynamics for all the systems for which the Boltzmann and the Landau equations are valid. This is a very remarkable result, especially because the number of known kinetic equations for which an H theorem can be *explicitly* proved is still very limited.

We now note that the most general function annulling the collision term in Eq. (12.2.17) is of the form:

$$\bar{f}^{LE}(\mathbf{q}, \mathbf{v}; t) = \left[\frac{m\beta(\mathbf{q}; t)}{2\pi}\right]^{3/2} n(\mathbf{q}; t) \exp\{-\tfrac{1}{2}m\beta(\mathbf{q}; t)[\mathbf{v} - \mathbf{u}(\mathbf{q}; t)]^2\} \quad (12.2.30)$$

Although this function has the general shape of a Maxwellian, its characteristic parameters $n(\mathbf{q}; t)$, $\mathbf{u}(\mathbf{q}; t)$, $\beta(\mathbf{q}; t)$ are in general functions of the position and of time. As a result, the velocity $\mathbf{u}(\mathbf{q}; t)$ can no longer be transformed away by a Galilean transformation. The distribution (12.2.30) is called a *local equilibrium distribution.* It is important to realize that it is *not* a stationary distribution like the homogeneous Maxwellian (12.2.8). This is due to the fact that the vanishing of the collision term in the kinetic equation (12.2.17) does not imply $\partial_t \bar{f}(1; t) = 0$. The evolution of an inhomogeneous system is due to a superposition of the collisions, on one hand, and of the flow and average field, on the other hand.

These two groups of processes are essentially different. The collision term is the only one responsible for the entropy creation: the collisions are the only source of irreversibility. The free flow and the effect of the average field represent a reversible motion that proceeds without creation of entropy. The effects of this superposition will be further studied in Section 13.2.

It is important at this point to make the following remark. *The presence of the interactions in a system is a necessary but not sufficient condition for the appearance of irreversibility.* Indeed, if the interactions are switched off, the collision term vanishes and the irreversibility disappears. On the other hand, the Vlassov term, although due to the interactions, is a purely reversible process.

In order to avoid useless misunderstandings, it is necessary to state more precisely what is meant by *irreversibility* in this book. An ambiguity may arise because of the existence of a variety of phenomena that might be called "irreversible" in a wide sense. These can be broadly classified in two classes exhibiting rather different features.

As an example of the first class, we consider a cloud of *noninteracting particles,* localized at time zero in a corner of a cubic box with perfectly reflecting walls. It is supposed that the particles of the cloud have different velocities, distributed in all possible directions. It is clear that, after waiting for a sufficiently long time, the cloud spreads out uniformly into the whole box, as a mere result of the free motions of the particles, including the reflections at the walls. In a sense, this is an "irreversible" behavior. Whatever the initial preparation, the system tends toward uniformity. However, the *duration* of the process depends very crucially on the initial preparation. It is easily conceived that if the initial spread of the velocities is very large, the system fills the box very quickly. If, however, all the velocities are concentrated in a narrow cone (i.e., if we consider a *beam* of particles), the homogenization process is very slow. From the microscopic point of view, it is clear that the distribution of velocities is *not* affected by the free flow: it remains constant in time.*

Plasmas described by the *Vlassov equation* exhibit an irreversible behavior that, although much subtler, is of the same general kind. As is well known (see also Sections 12.7 and 13.6), the plasma supports collective local charge oscillations. It turns out that these oscillations are damped, even in the absence of collisions. This *Landau damping* is well described by the Vlassov equation. The characteristic feature of this damping is again its dependence on the initial state. It can be shown that there exist states for which there is no damping at all.†

All the irreversible processes of this class will be called of *phase-mixing type*. They are characterized by the lack of an intrinsic time scale, independent of the initial condition. There is no entropy production associated with these phenomena.

We now consider what happens, for instance, in our first example if the particles can undergo frequent collisions. In this case, because of the irregularity and the large number of these events, the memory of the initial state is rapidly lost, as is shown in Chapter 11. The spreading in space is now of a completely different type (*diffusion*): it is controlled by the intrinsic features of the dynamics of the interactions as well as by global features, such as density and temperature. These intrinsic parameters fix a definite time scale (the relaxation time) that is independent of the initial preparation. The velocity distribution tends toward its Maxwellian equilibrium value, and the process is associated with an increase of entropy. This type of behavior will be further analyzed in the forthcoming chapters.

Irreversible processes in this second class can be called of *dissipative type*. They reflect much deeper properties of the system than the phase-mixing processes.

In this book, whenever we refer briefly to irreversibility, we always mean a behavior of dissipative type.

12.3. THE COLLISIONAL INVARIANTS

In studying the general balance equations in Section 12.1, we found one contribution that is particularly complicated: the contribution of the

* See also the discussion at the end of Section 13.3.

† A detailed discussion of these matters can be found in Balescu's book quoted in Chapter 6.

collisions. We may thus expect that among all possible balance equations there is a class that is particularly simple: it is the one corresponding to dynamical functions $\psi_i(v, q)$ whose average is not affected in time by the collisions. More precisely, referring to Eq. (12.1.14), we require that

$$\sigma_i^{(3)} \equiv \int d1\, \psi_i(1)\, \mathcal{K}\bar{f}(1; t) \equiv 0 \tag{12.3.1}$$

We now show that the functions $\psi_i(\mathbf{v}, \mathbf{q})$, called *collisional invariants*, can be easily found by a systematic research. Considering the Landau collision term in (12.2.17), we find the collisional source:

$$\begin{aligned}
\sigma_i^{(3)} &= A \int d1\, d2 \int d\mathbf{k}\, \psi_i(1)\, \delta(\mathbf{q}_1 - \mathbf{q}_2) \tilde{V}_{\mathbf{k}}^2\, \mathbf{k} \cdot \partial_{12}\, \delta(\mathbf{k} \cdot \mathbf{g})\, \mathbf{k} \cdot \partial_{12} \bar{f}(1; t) \bar{f}(2; t) \\
&= \tfrac{1}{2} A \int d1\, d2 \int d\mathbf{k}\, [\psi_i(1) + \psi_i(2)]\, \delta(\mathbf{q}_1 - \mathbf{q}_2) \tilde{V}_{\mathbf{k}}^2 \\
&\quad \times \mathbf{k} \cdot \partial_{12}\, \delta(\mathbf{k} \cdot \mathbf{g})\, \mathbf{k} \cdot \partial_{12} \bar{f}(1; t) \bar{f}(2; t) \\
&= -\tfrac{1}{2} A \int d1\, d2 \int d\mathbf{k}\, \tilde{V}_{\mathbf{k}}^2 \{\delta(\mathbf{q}_1 - \mathbf{q}_2)\, \delta(\mathbf{k} \cdot \mathbf{g})\, \mathbf{k} \cdot \partial_{12} [\psi_i(1) + \psi_i(2)]\} \\
&\quad \times \mathbf{k} \cdot \partial_{12} \bar{f}(1; t) \bar{f}(2; t)
\end{aligned} \tag{12.3.2}$$

The calculations are similar to (12.2.5). In order that this quantity vanishes whatever the form of the distribution function, the bracketed expression must vanish identically:

$$\delta(\mathbf{q}_1 - \mathbf{q}_2)\, \delta(\mathbf{k} \cdot \mathbf{v}_1 - \mathbf{k} \cdot \mathbf{v}_2)\, [\mathbf{k} \cdot \partial_1 \psi_i(\mathbf{q}_1, \mathbf{v}_1) - \mathbf{k} \cdot \partial_2 \psi_i(\mathbf{q}_2, \mathbf{v}_2)] = 0 \tag{12.3.3}$$

This equation has the following solutions:

(a) $\psi_1(\mathbf{v}, \mathbf{q}) = \psi(\mathbf{q})$, *independent of* $\mathbf{v}$. Then each term in the square brackets vanishes identically.

(b) $\psi_{2,3,4}(\mathbf{q}, \mathbf{v}) = \{v_x, v_y, v_z\} \psi(\mathbf{q})$. These functions are proportional to a component of the velocity. The bracket then reduces to

$$[k^s \psi(\mathbf{q}_1) - k^s \psi(\mathbf{q}_2)], \qquad s = x, y, z$$

which vanishes because of the presence of the delta function $\delta(\mathbf{q}_1 - \mathbf{q}_2)$.

(c) $\psi_5(\mathbf{q}, \mathbf{v}) = v^2 \psi(\mathbf{q})$. This function is proportional to the square of the absolute value of the velocity. The bracket now becomes

$$[\mathbf{k} \cdot \mathbf{v}_1 \psi(\mathbf{q}_1) - \mathbf{k} \cdot \mathbf{v}_2 \psi(\mathbf{q}_2)]$$

which again vanishes because of the two delta functions.

As a result, any linear combination of the five independent collisional invariants: 1, v_x, v_y, v_z, v^2 is also an invariant:

$$\psi(\mathbf{v}, \mathbf{q}) = \gamma(\mathbf{q}) + \boldsymbol{\mu}(\mathbf{q}) \cdot \mathbf{v} + \tfrac{1}{2} \alpha(\mathbf{q}) v^2 \tag{12.3.4}$$

Here $\gamma(\mathbf{q})$, $\boldsymbol{\mu}(\mathbf{q})$, and $\alpha(\mathbf{q})$ are arbitrary functions of $\mathbf{q}$. It will be shown in the Appendix to this chapter that Eq. (12.3.4) is the most general collisional invariant of the Landau equation.

We thus find that, as far as the velocity dependence goes, *there are exactly five independent collisional invariants* for the Landau equation. A similar argument, the development of which we leave as an exercise for the reader, shows that the Boltzmann collision term admits exactly the same five collisional invariants.

The existence of precisely five independent collisional invariants is a very general feature of the kinetic equations, because it is related to the dynamical conservation laws of number of particles, momentum, and energy. There is, however, more in this concept than the mere dynamical conservation law. During a collision, one molecule loses momentum and energy, which are gained by its partner. If the dynamical treatment of the collision process is correct, this conservation law will always be satisfied. But the collisional invariant requires more. Whatever amount of momentum and of energy is lost by a particle *at the point* $\mathbf{x}$ is gained by the second particle *at the same point* $\mathbf{x}$. It is therefore important that in the collision term the two distribution functions are evaluated at the *same point* $\mathbf{q}$. This clearly appears in the crucial role played by the function $\delta(\mathbf{q}_1 - \mathbf{q}_2)$ in the proof. But we know (see Section 11.5) that this is an approximation valid in the "hydrodynamic limit" when the intensive functions vary very little over a length of order of the mean-free path. In a more refined kinetic equation in which the delocalization of the collision process is taken into account, some of the previous functions may well no longer be collisional invariants, although the dynamical conservation laws are not violated. The macroscopic momentum and energy are then no longer conserved *locally*, but only on the average.

We shall, however, not discuss these refinements here and continue our investigation of the kinetic equations in the hydrodynamical limit. The collisional invariants then play a crucial role in the theory as will be seen in the forthcoming discussion. The number "five" will appear everywhere as a kind of magic number.

12.4. THE HYDRODYNAMICAL BALANCE EQUATIONS

The five collisional invariants are related to the mechanical invariants of the system. Hence the corresponding macroscopic balance equations are none other than the five conservation equations of hydrodynamics corresponding to the mass density, the momentum density (a vectorial equation), and the internal energy density. We now derive these equations in detail from kinetic theory.

A. The Mass-Balance Equation

The mass density $\rho(\mathbf{x}; t)$ is defined as

$$\rho(\mathbf{x}; t) = \int d\mathbf{q}_1 \, d\mathbf{v}_1 \, \delta(\mathbf{x} - \mathbf{q}_1) \, m \, \bar{f}(\mathbf{q}_1, \mathbf{v}_1, t) \tag{12.4.1}$$

The corresponding dynamical function $m\delta(\mathbf{x} - \mathbf{q}_1)$ belongs to the first class of collisional invariants, $\psi_1(\mathbf{v}, \mathbf{q})$. We can derive its equation of evolution directly from the results of Section 12.1, taking $\beta(1) = m$. The mass flow $\mathbf{\Phi}_m$ is given by Eq. (12.1.18):

$$\mathbf{\Phi}_m(\mathbf{x}; t) \equiv \rho(\mathbf{x}; t)\mathbf{u}(\mathbf{x}; t) = \int d1 \, \delta(\mathbf{x} - \mathbf{q}_1) \, m\mathbf{v}_1 \, \bar{f}(1; t) \tag{12.4.2}$$

The *local velocity* $\mathbf{u}(\mathbf{x}; t)$ is defined through this equation. Clearly, as $\beta(1)$ is a constant, both $\sigma_m^{(1)}$ and $\sigma_m^{(2)}$ vanish [see Eqs. (12.1.15), (12.1.16)]. Moreover, $\sigma_m^{(3)} = 0$ because m is a collisional invariant. We thus obtain

$$\partial_t \rho(\mathbf{x}; t) = -\nabla \cdot [\rho(\mathbf{x}; t)\mathbf{u}(\mathbf{x}; t)] \tag{12.4.3}$$

This is the well-known hydrodynamic *equation of continuity*.

B. The Momentum-Balance Equation

We see from (12.4.2) that the mass flow is equivalent to the momentum density. The components of this vector belong to the second, third, and fourth collisional invariants. We apply again the results of Section 12.1, with $\beta^r(1) = mv_1^r$, $r = x, y, z$. The flow of momentum is now a second-rank tensor whose components are found from (12.1.18):

$$\Phi^{rs} = \int d1 \, \delta(\mathbf{x} - \mathbf{q}_1) \, mv_1^r v_1^s \, \bar{f}(1; t) \tag{12.4.4}$$

We now note that, by a purely macroscopic argument, the momentum can be transported because a volume element of fluid of density ρ, moving with velocity $\mathbf{u}$, transfers in the x direction (say) an amount of momentum $\rho\mathbf{u}u^x$. Hence, the *convective momentum flow* is $\rho u^r u^s$, but this does not account for the complete mechanism of momentum transfer. Indeed, the individual particles in the gas have velocities v^r, different from the average velocity u^r. Hence each molecule contributes an additional amount $m(\mathbf{v} - \mathbf{u})(v^x - u^x)$ to the transport of momentum in the x direction. This extra amount, when averaged over all molecules, gives rise to the quantity:

$$P^{rs}(\mathbf{x}; t) = \int d1 \, \delta(\mathbf{x} - \mathbf{q}_1) \, m(v_1^r - u^r)(v_1^s - u^s) \, \bar{f}(1, t) \tag{12.4.5}$$

It is easily verified [using the definition (12.4.2) of $\mathbf{u}$] that

$$\Phi^{rs} = P^{rs} + \rho u^r u^s \tag{12.4.6}$$

The tensor P^{rs}, originating from the thermal motions of the particles, is well known in hydrodynamics as the *pressure tensor.* We note that, in the present approximation, the pressure tensor contains no contribution from the intermolecular forces [as in Eq. (7.2.8)]. The *scalar pressure* is defined as one-third of the trace of this tensor:

$$P(\mathbf{x}; t) = \tfrac{1}{3}P^{rr} = \frac{1}{3}\int d1\, \delta(\mathbf{x}-\mathbf{q}_1)\, m[(\mathbf{v}_1-\mathbf{u})\cdot(\mathbf{v}_1-\mathbf{u})]\, \bar{f}(1; t) \tag{12.4.7}$$

It is easily checked that in equilibrium, when $\bar{f}(1; t) = n\varphi^0(1)$ and $\mathbf{u} = 0$, this formula leads to the ideal gas pressure $P^{eq} = nk_BT$.

For an *ordinary gas* we have exhausted all terms of the balance, which reads

$$\partial_t\rho(\mathbf{x}; t)\mathbf{u}(\mathbf{x}; t) = -\nabla\cdot[\rho(\mathbf{x}; t)\mathbf{u}(\mathbf{x}; t)\mathbf{u}(\mathbf{x}; t) + \mathsf{P}(\mathbf{x}; t)] \tag{12.4.8}$$

In the case of a *plasma* we must also take account of the Vlassov field. From Eq. (12.1.15) we now find a nonvanishing source term:

$$\begin{aligned}\sigma^{(2)} &= -m^{-1}\nabla^s U\cdot\int d1\, \delta(\mathbf{x}-\mathbf{q})\, m\, \delta^{sr}\bar{f}(1; t)\\ &= -m^{-1}\rho(\mathbf{x}; t)\nabla^r U(\mathbf{x}; t)\end{aligned} \tag{12.4.9}$$

This is clearly interpreted as the *density of the average force.* As this force is of electrical origin, we may conveniently express it as an *electrical field:*

$$-\nabla^r U(\mathbf{x}; t) = -eE^r(\mathbf{x}; t) \tag{12.4.10}$$

By Eqs. (11.7.7), (11.7.9) this field is determined self-consistently by *Poisson's equation:*

$$\nabla\cdot\mathbf{E} = \left(\frac{4\pi e}{m}\right)[\rho(\mathbf{x}; t) - \rho_0] \tag{12.4.11}$$

where ρ_0 is the average mass density. Thus the momentum-balance equation for a plasma is

$$\begin{aligned}\partial_t\rho(\mathbf{x}; t)\mathbf{u}(\mathbf{x}; t) = &-\nabla\cdot[\rho(\mathbf{x}; t)\mathbf{u}(\mathbf{x}; t)\mathbf{u}(\mathbf{x}; t) + \mathsf{P}(\mathbf{x}; t)]\\ &-\left(\frac{e}{m}\right)\rho(\mathbf{x}; t)\mathbf{E}(\mathbf{x}; t)\end{aligned} \tag{12.4.12}$$

This equation is coupled to Poisson's equation (12.4.11).

C. The Balance of Energy

We first need a correct definition of the total energy density $\rho(\mathbf{x}; t)e'(\mathbf{x}; t)$. This implies that all forms of energy must be included, in such a way that this quantity be exactly conserved. In other words, its balance equation must be of the form:

$$\partial_t \rho e' = -\nabla \cdot \boldsymbol{\Phi}_{e'} \tag{12.4.13}$$

It is readily verified that the following definition is correct for a plasma (in the case of an ordinary gas, the second term is simply dropped):

$$\rho(\mathbf{x};t)e'(\mathbf{x};t) = \tfrac{1}{2}m \int d1\,\delta(\mathbf{x}-\mathbf{q}_1)\,v_1^2\,\bar{f}(1;t)$$
$$+\tfrac{1}{4}\int d1\,d2\,[\delta(\mathbf{x}-\mathbf{q}_1)+\delta(\mathbf{x}-\mathbf{q}_2)]V(\mathbf{q}_1-\mathbf{q}_2)\bar{f}(1;t)\bar{f}(2;t) \tag{12.4.14}$$

As can be checked from Eqs. (12.1.15)–(12.1.18) this quantity obeys Eq. (12.4.13), provided we define the total energy flow as

$$\boldsymbol{\Phi}_{e'}(\mathbf{x};t) = \tfrac{1}{2}m \int d1\,\delta(\mathbf{x}-\mathbf{q}_1)\,\mathbf{v}_1 v_1^2\,\bar{f}(1;t) + m^{-1}\rho\mathbf{u}U \tag{12.4.15}$$

We now note that, from the macroscopic point of view, the total energy of our system is of three different kinds: *potential energy*, which is just the second term in the right side of (12.4.14), *macroscopic kinetic energy*, that is $\frac{1}{2}\rho u^2$, and a remainder due to the *thermal motion*, which must be identified with the thermodynamic *internal energy density*, $\rho(\mathbf{x};t)e(\mathbf{x};t)$. The latter is therefore defined as

$$\rho(\mathbf{x};t)e(\mathbf{x};t) = \tfrac{1}{2}m \int d1\,\delta(\mathbf{x}-\mathbf{q}_1)\,v_1^2\,\bar{f}(1;t) - \tfrac{1}{2}\rho u^2 \tag{12.4.16}$$

A balance of the macroscopic kinetic energy is easily obtained by multiplying both sides of Eq. (12.4.12) by $\mathbf{u}$. A little algebra then leads to the internal energy-balance equation, which takes the form (valid both for ordinary gases and for plasmas):

$$\partial_t \rho(\mathbf{x};t)e(\mathbf{x};t) = -\nabla\cdot[\rho(\mathbf{x};t)e(\mathbf{x};t)\mathbf{u}(\mathbf{x};t) + \mathbf{J}(\mathbf{x};t)] - \mathsf{P}(\mathbf{x};t):\nabla\mathbf{u}(\mathbf{x};t) \tag{12.4.17}$$

The term $\rho e\mathbf{u}$ represents a *convective flow* of the internal energy, whereas $\mathbf{J}$ is identified with the *heat-flow* vector, and is defined by

$$\mathbf{J}(\mathbf{x};t) = \tfrac{1}{2}m \int d1\,\delta(\mathbf{x}-\mathbf{q}_1)\,|\mathbf{v}_1-\mathbf{u}|^2\,(\mathbf{v}_1-\mathbf{u})\,\bar{f}(1;t) \tag{12.4.18}$$

The internal energy is not a conserved quantity. A source term appears in Eq. (12.4.17), expressing the transformation of internal into kinetic energy (and vice versa). It is the work done by the fluid against the pressure forces.

We have now completed the derivation of the hydrodynamical balance equations. Their hierarchical structure, discussed in Section 12.1, appears quite clearly. In deriving the equations for the moments ρ, $\mathbf{u}$, e we have introduced higher moments P and $\mathbf{J}$. In the next section we review the classical, macroscopic arguments allowing one to close the hierarchy approximately.

12.5. THE PHENOMENOLOGICAL TREATMENT OF THE HYDRODYNAMICAL EQUATIONS

Let us collect here the three fundamental conservation laws of hydrodynamics, that is, Eqs. (12.4.3), (12.4.12), and (12.4.17):

$$\partial_t \rho = -\nabla \cdot \rho \mathbf{u} \tag{12.5.1}$$

$$\partial_t \rho \mathbf{u} = -\nabla \cdot (\rho \mathbf{u}\mathbf{u} + \mathsf{P}) - \left(\frac{e}{m}\right) \rho \mathbf{E} \tag{12.5.2}$$

$$\partial_t \rho e = -\nabla \cdot (\rho e \mathbf{u} + \mathbf{J}) - \mathsf{P} : \nabla \mathbf{u} \tag{12.5.3}$$

In the case of an ordinary gas the charge e must be set equal to zero.*

These equations are useless as long as we do not possess expressions for the unknown quantities P and $\mathbf{J}$. In phenomenological hydrodynamics one argues as follows.

In a homogeneous system in equilibrium, the pressure tensor is diagonal:

$$P_0^{rs} = P\,\delta^{rs} \tag{12.5.4}$$

where P is the scalar, hydrostatic pressure. The heat flow vanishes in such a system:

$$\mathbf{J}_0 = 0 \tag{12.5.5}$$

If the system is inhomogeneous,† there generally appears a correction to the pressure tensor and a nonzero heat flow. These corrections must be measured by the deviations from homogeneity; in first approximation they are *linear functions of the gradients of the intensive quantities describing the thermodynamic state of the system*. The gas under consideration is adequately described by the density ρ, the velocity $\mathbf{u}$, and the temperature T. Symmetry considerations allow us to restrict the possible kinds of relations.

Consider first the pressure tensor, which we write as

$$P^{rs} = P\delta^{rs} + \Pi^{rs} \tag{12.5.6}$$

* No confusion should arise between the two (standard) notations e for the electrical charge and for the internal energy.

† This statement will be made more precise in Section 13.2.

The added term Π^{rs}, being a second-rank tensor, can only depend (linearly) on the gradient of velocity:

$$\Pi^{rs} = \alpha^{rsjk} \nabla^j u^k \tag{12.5.7}$$

A second-rank tensor T^{rs} can always be represented as the sum of a symmetric traceless tensor, an antisymmetric tensor, and a diagonal tensor:

$$T^{rs} = \tfrac{1}{2}(T^{rs} + T^{sr} - \tfrac{2}{3}T^{kk}\delta^{rs}) + \tfrac{1}{2}(T^{rs} - T^{sr}) + \tfrac{1}{3}T^{kk}\delta^{rs} \tag{12.5.8}$$

It is well known that the pressure tensor must be symmetric (to ensure conservation of angular momentum). Hence, the only possible form of Eq. (12.5.7) in an isotropic, one-component fluid is

$$\Pi^{rs} = -\eta(\nabla^r u^s + \nabla^s u^r - \tfrac{2}{3}\delta^{rs}\nabla^k u^k) - \zeta\delta^{rs}\nabla^k u^k \tag{12.5.9}$$

η and ζ are phenomenological constants called, respectively, the *shear viscosity* (or simply, *viscosity*) *coefficient*, and the *bulk viscosity coefficient*.

Turning now to the heat flow **J**, which is a vector, a similar argument provides us with the following linear relation:

$$\mathbf{J} = -\lambda \nabla \rho - \kappa \nabla T \tag{12.5.10}$$

The constant κ is the *thermal conductivity;* λ has no standard name.

A further restriction on the relationships (12.5.9), (12.5.10) is provided by the second law of thermodynamics. We derived in Eq. (12.2.27) an entropy-balance equation; but we could not express the quantities Φ_s and σ_s explicitly in terms of hydrodynamical quantities. In order to do so we need an extra assumption. We know that in equilibrium the entropy can be expressed in terms of the energy and of the density through the *Gibbs relation:*

$$T\,ds = de + P\,d(\rho^{-1}) \tag{12.5.11}$$

We now assume that this relation still holds when the system is out of equilibrium. This assumption can only be valid if the deviations from equilibrium are not too large; but in practice, the range of validity of Eq. (12.5.11) turns out to be quite wide. With the present assumptions we are allowed to translate the Gibbs relation into a balance equation:

$$T\partial_t s = \partial_t e - \left(\frac{P}{\rho^2}\right)\partial_t \rho \tag{12.5.12}$$

Substituting in the right-hand side for $\partial_t e$ and $\partial_t \rho$ the appropriate expressions taken from Eqs. (12.5.1), (12.5.3), we find, after some algebra

$$\rho\left(\frac{ds}{dt}\right)\equiv\rho(\partial_t+\mathbf{u}\cdot\nabla)s$$

$$=-\nabla\cdot\left(\frac{\mathbf{J}}{T}\right)-T^{-1}\mathbf{\Pi}:\nabla\mathbf{u}-T^{-2}\mathbf{J}\cdot\nabla T \qquad (12.5.13)$$

The entropy balance is expressed here as a substantial derivative: for this reason there is no convective term in the argument of the divergence. The entropy source is now quite explicit. The second law requires this source to be nonnegative. Substituting the expressions (12.5.9) and (12.5.10), we find

$$\sigma_s=\eta T^{-1}(\nabla^r u^s+\nabla^s u^r-\tfrac{2}{3}\nabla^k u^k\delta^{rs})(\nabla^r u^s+\nabla^s u^r-\tfrac{2}{3}\nabla^l u^l\delta^{rs})$$
$$+3\zeta T^{-1}(\nabla\cdot\mathbf{u})^2+\kappa T^{-2}(\nabla T)^2+\lambda T^{-2}(\nabla T)\cdot(\nabla\rho)\geqslant 0 \qquad (12.5.14)$$

It is obvious that, in order to satisfy the sign requirement, we must have*

$$\begin{aligned}\eta&\geqslant 0\\ \zeta&\geqslant 0\\ \kappa&\geqslant 0\\ \lambda&=0\end{aligned} \qquad (12.5.15)$$

The coefficients η, ζ, κ, λ are called collectively *transport coefficients.* Their calculation from first principles is one of the main objects of nonequilibrium statistical mechanics. The signs imposed by the second law are just as general and as important as the analogous equilibrium stability conditions ($c_v\geqslant 0$, $\chi_T\geqslant 0$).

We now use Eqs. (12.5.6), (12.5.9), (12.5.10), and (12.5.15) in Eqs. (12.5.1)–(12.5.3) and obtain

$$\partial_t\rho=-\nabla\cdot\rho\mathbf{u} \qquad (12.5.16)$$

$$\partial_t\rho u^r=-\nabla^s\rho u^r u^s-\nabla^r P+\eta\nabla^2 u^r+(\zeta+\tfrac{1}{3}\eta)\nabla^r\nabla^s u^s-\left(\frac{e}{m}\right)\rho E^r \qquad (12.5.17)$$

$$\partial_t\rho e=-\nabla\cdot(\rho e\mathbf{u}-\kappa\nabla T)-P\nabla\cdot\mathbf{u}+\eta[\nabla^r u^s+\nabla^s u^r]\cdot\nabla^r u^s+(\zeta-\tfrac{2}{3}\eta)(\nabla\cdot\mathbf{u})^2 \qquad (12.5.18)$$

We still have too many unknown functions. In particular, it should be useful to eliminate the derivatives of the pressure and of the internal energy in terms of other variables. This is done by assuming that *the pressure and the internal energy are functions of the temperature and of*

* Note that in a relativistic theory $\lambda\neq 0$; it can, however, be shown that it is not independent of κ.

the density; the functional relationship is, moreover, assumed to be the same as in equilibrium, that is, it is fixed by the *equations of state:*

$$P = P(\rho, T) \tag{12.5.19}$$

$$e = e(\rho, T) \tag{12.5.20}$$

Furthermore, it is assumed that in an inhomogeneous, nonstationary system, the pressure and the internal energy depend on space and time only through the functions $\rho(\mathbf{x}; t)$ and $T(\mathbf{x}, t)$. The latter is a pretty strong assumption. It can be shown, however, that it is microscopically justified (see Section 13.2) under certain conditions. With these assumptions we write

$$\nabla^r P = \left(\frac{\partial P}{\partial \rho}\right)_T \nabla^r \rho + \left(\frac{\partial P}{\partial T}\right)_\rho \nabla^r T \tag{12.5.21}$$

$$\begin{Bmatrix} \nabla^r \\ \partial_t \end{Bmatrix} e = \left(\frac{\partial e}{\partial \rho}\right)_T \begin{Bmatrix} \nabla^r \\ \partial_t \end{Bmatrix} \rho + \left(\frac{\partial e}{\partial T}\right)_\rho \begin{Bmatrix} \nabla^r \\ \partial_t \end{Bmatrix} T \tag{12.5.22}$$

A simple thermodynamic calculation yields

$$\left(\frac{\partial e}{\partial \rho}\right)_T = -\frac{1}{\rho^2} T \left(\frac{\partial P}{\partial T}\right)_\rho + \frac{P}{\rho^2} \tag{12.5.23}$$

On the other hand, the second thermodynamic coefficient in (12.5.22) is just the specific heat at constant volume (per unit mass):

$$\left(\frac{\partial e}{\partial T}\right)_\rho = c_V \tag{12.5.24}$$

Taking account of these relations, the hydrodynamic equations are brought into the following form:

$$\partial_t \rho = -\boldsymbol{\nabla} \cdot \rho \mathbf{u} \tag{12.5.25}$$

$$\begin{aligned} \partial_t u^r = &-u^s \nabla^s u^r - \frac{1}{\rho}\left(\frac{\partial P}{\partial \rho}\right)_T \nabla^r \rho - \frac{1}{\rho}\left(\frac{\partial P}{\partial T}\right)_\rho \nabla^r T \\ &+ \frac{1}{\rho} \eta \nabla^2 u^r + \frac{1}{\rho}(\zeta + \tfrac{1}{3}\eta) \nabla^r \nabla^s u^s - \frac{e}{m} E^r \end{aligned} \tag{12.5.26}$$

$$\begin{aligned} \partial_t T = &-\frac{T}{\rho c_V}\left(\frac{\partial P}{\partial T}\right)_\rho \boldsymbol{\nabla} \cdot \mathbf{u} - \mathbf{u} \cdot \boldsymbol{\nabla} T + \frac{\kappa}{\rho c_V} \nabla^2 T \\ &+ \frac{\eta}{\rho c_V} [\nabla^r u^s + \nabla^s u^r] \nabla^r u^s + \frac{1}{\rho c_V} (\zeta - \tfrac{2}{3}\eta)(\boldsymbol{\nabla} \cdot \mathbf{u})^2 \end{aligned} \tag{12.5.27}$$

We have now a self-consistent set of five equations for the five independent variables ρ, u^r, and T. This is the most general form of the hydrodynamic equations consistent with our assumptions.

These equations are still complicated because of their nonlinearity. We now make an extra simplifying assumption. The density and temperature are assumed to deviate only slightly from an equilibrium value, which is independent of space and time. The equilibrium value of the velocity is taken to be zero:

$$\begin{aligned} \rho(\mathbf{x}; t) &= \rho_0 + \nu(\mathbf{x}; t) \\ \mathbf{u}(\mathbf{x}; t) &= \qquad \mathbf{u}(\mathbf{x}; t) \\ T(\mathbf{x}; t) &= T_0 + \theta(\mathbf{x}; t) \end{aligned} \tag{12.5.28}$$

Substituting these forms into Eqs. (12.5.25)–(12.5.27) and keeping only terms linear in ν, $\mathbf{u}$, θ, we obtain

$$\partial_t \nu = -\rho \nabla \cdot \mathbf{u} \tag{12.5.29}$$

$$\begin{aligned} \partial_t u^r = &-\frac{1}{\rho}\left(\frac{\partial P}{\partial \rho}\right)_T \nabla^r \nu - \frac{1}{\rho}\left(\frac{\partial P}{\partial T}\right)_\rho \nabla^r \theta + \frac{\eta}{\rho} \nabla^2 u^r \\ &+ \frac{1}{\rho}(\zeta + \tfrac{1}{3}\eta)\nabla^r(\nabla \cdot \mathbf{u}) - \frac{e}{m} E^r \end{aligned} \tag{12.5.30}$$

$$\partial_t \theta = -\frac{T}{\rho c_V}\left(\frac{\partial P}{\partial T}\right)_\rho \nabla \cdot \mathbf{u} + \frac{\kappa}{\rho c_V} \nabla^2 \theta \tag{12.5.31}$$

For simplicity, we now dropped the subscripts 0 on ρ_0 and T_0. In the case of plasmas, the Poisson equation (12.4.11) must be added for the determination of the field E^r:

$$\nabla \cdot \mathbf{E} = -\left(\frac{4\pi e}{m}\right)\nu \tag{12.5.32}$$

These *linearized hydrodynamical equations* are the basic tool for the study of small-amplitude motions. Their structure is rather remarkable. We can clearly recognize two types of terms. In a first class, the coefficients depend only on equilibrium thermodynamic properties: ρ, $(\partial P/\partial \rho)$, $(\partial P/\partial T)$, c_V. In the second class, the terms depend on nonequilibrium properties, that is, on the transport coefficients η, ζ, κ. If the latter are put equal to zero, we get the equations of an ideal or *nondissipative fluid:*

$$\begin{aligned} \partial_t \nu &= -\rho \nabla \cdot \mathbf{u} \\ \partial_t u^r &= -\frac{1}{\rho}\left(\frac{\partial P}{\partial \rho}\right)_T \nabla^r \nu - \frac{1}{\rho}\left(\frac{\partial P}{\partial T}\right)_\rho \nabla^r \theta - \frac{e}{m} E^r \\ \partial_t \theta &= -\frac{T}{\rho c_V}\left(\frac{\partial P}{\partial T}\right)_\rho \nabla \cdot \mathbf{u} \end{aligned} \tag{12.5.33}$$

The additional terms are responsible for the *dissipation* through heat conduction and viscosity. They are also the only ones giving a nonzero entropy production.

12.6. THE HYDRODYNAMICAL NORMAL MODES

We now investigate the types of motion that are possible in a fluid. The simplest, and also the most interesting motions are those with a very long characteristic scale. As we assumed the gradients very small in order to write linear phenomenological laws (12.5.9) and (12.5.10), we are actually bound, for consistency, to study only these motions. We will be led quite naturally to formulate an eigenvalue problem for the linear system of equations (12.5.29)–(12.5.31). If the problem is solved we may express all the long-range motions as a superposition of these elementary ones.

Besides its intrinsic interest, this problem turns out to be of great importance as a basis for a microscopic theory of the transport coefficients. This will appear clearly in Chapter 13.

In this section we study an ordinary fluid, in which the electrical charge $e=0$. We start from the linearized hydrodynamic equations (12.5.29)–(12.5.31), assuming that they describe an infinite fluid. We may ask what are the eigenfunctions and the eigenvalues of this set of linear equations. More specifically, we investigate whether this set admits solutions of the following form:

$$\begin{aligned} \nu(\mathbf{x}, t) &= \exp(i\mathbf{k}\cdot\mathbf{x}+\Lambda t)\nu_{\mathbf{k}} \\ \mathbf{u}(\mathbf{x}, t) &= \exp(i\mathbf{k}\cdot\mathbf{x}+\Lambda t)\mathbf{u}_{\mathbf{k}} \\ \theta(\mathbf{x}, t) &= \exp(i\mathbf{k}\cdot\mathbf{x}+\Lambda t)\theta_{\mathbf{k}} \end{aligned} \tag{12.6.1}$$

Substituting these forms into (12.5.29)–(12.5.31) we find the following set of linear algebraic equations:

$$\begin{aligned} \Lambda\nu_{\mathbf{k}}+i\rho k^s u^s_{\mathbf{k}} &= 0 \\ \Lambda u^r_{\mathbf{k}}+iCk^r\nu_{\mathbf{k}}+iAk^r\theta_{\mathbf{k}}+\eta' k^2 u^r_{\mathbf{k}}+\mu k^r k^s u^s_{\mathbf{k}} &= 0 \\ \Lambda\theta_{\mathbf{k}}+iBk^s u^s_{\mathbf{k}}+\chi k^2\theta_{\mathbf{k}} &= 0 \end{aligned} \tag{12.6.2}$$

We used here the following abbreviations:

$$\begin{aligned} A &= \frac{1}{\rho}\left(\frac{\partial P}{\partial T}\right)_\rho, & \eta' &= \frac{\eta}{\rho} \\ B &= \frac{T}{\rho c_V}\left(\frac{\partial P}{\partial T}\right)_\rho, & \mu &= \frac{1}{\rho}(\zeta+\tfrac{1}{3}\eta) \\ C &= \frac{1}{\rho}\left(\frac{\partial P}{\partial \rho}\right)_T, & \chi &= \frac{\kappa}{\rho c_V} \end{aligned} \tag{12.6.3}$$

The calculations can be done quite generally starting from (12.6.2); however, they are considerably simpler if we choose to orient the **x** axis of our reference frame along the fixed vector **k**:

$$\mathbf{k} = k\mathbf{1}_x$$
$$k_x = k, \qquad k_y = k_z = 0 \tag{12.6.4}$$

It can be checked that this assumption does not alter the eigenvalues; hence nothing is lost in generality.

Coming back to (12.6.2), this homogeneous system of linear equations has a nontrivial solution only if its characteristic determinant vanishes. This condition is written as follows, taking account of (12.6.4):

$$\begin{vmatrix} \Lambda & i\rho k & 0 & 0 & 0 \\ iCk & \Lambda+(\eta'+\mu)k^2 & 0 & 0 & iAk \\ 0 & 0 & \Lambda+\eta' k^2 & 0 & 0 \\ 0 & 0 & 0 & \Lambda+\eta' k^2 & 0 \\ 0 & iBk & 0 & 0 & \Lambda+\chi k^2 \end{vmatrix} = 0 \tag{12.6.5}$$

The expansion of this determinant yields

$$(\Lambda+\eta' k^2)^2\{\Lambda^3+\Lambda^2(\chi+\eta'+\mu)k^2$$
$$+\Lambda[\chi(\eta'+\mu)k^4+(AB+\rho C)k^2]+\chi\rho Ck^4\}=0 \tag{12.6.6}$$

Through this equation, the eigenvalues appear as certain well-defined functions of k. An equation of the type (12.6.6) is called a *dispersion equation.*

Among the five roots of this equation, two are obvious; they will be called (for convenience) Λ_3 and Λ_4:

$$\Lambda_3 = \Lambda_4 = -\eta' k^2 \tag{12.6.7}$$

To determine the others, we must solve an equation of the third degree. In writing the equations (12.5.29)–(12.5.31) we assumed that the gradients are small; this implies, in the Fourier picture, (12.6.2), that the wave vector k is small. Therefore, rather than solving (12.6.6) exactly, we try an approximate solution of the form*:

$$\Lambda_i = a_i k + b_i k^2, \qquad i = 1, 2, 5 \tag{12.6.8}$$

We may thus identify the two forms of the third-degree polynomial:

$$(\Lambda-\Lambda_1)(\Lambda-\Lambda_2)(\Lambda-\Lambda_5)$$
$$= \Lambda^3+\Lambda^2(\eta'+\mu+\chi)k^2+\Lambda[\chi(\eta'+\mu)k^4+(AB+\rho C)k^2]+\chi\rho Ck^4 \tag{12.6.9}$$

* One may also write the exact solutions of the cubic equation in (12.6.6) and expand the resulting (complicated) expressions in powers of k through order k^2.

Substituting (12.6.8) and expanding the left side, we get a polynomial of degree 3 in Λ, and of degree 6 in k. We then identify the coefficients of every separate power of Λ and k in both sides. The terms of degree k^5 and k^6 on the left are not significant in the present approximation and must be disregarded. In this way we obtain the following six independent equations.

$$\begin{aligned} a_1+a_2+a_5&=0\\ b_1+b_2+b_5&=-(\eta'+\mu+\chi)\\ a_1a_2+a_1a_5+a_2a_5&=AB+\rho C\\ a_1b_2+a_2b_1+(a_1+a_2)b_5+a_5(b_1+b_2)&=0\\ a_1a_2b_5+a_1a_5b_2+a_2a_5b_1&=\chi\rho C\\ a_1a_2a_5&=0 \end{aligned} \tag{12.6.10}$$

These equations are readily solved:

$$a_1=-a_2=i(AB+\rho C)^{1/2}$$

$$a_5=0$$

$$b_1=b_2=-\frac{1}{2}\left(\eta'+\mu+\frac{AB}{AB+\rho C}\chi\right) \tag{12.6.11}$$

$$b_5=-\frac{\rho C}{AB+\rho C}\chi$$

We can make these expressions more explicit. From the well-known thermodynamical relation:

$$c_P-c_V=-T\frac{(\partial P/\partial T)_\rho^2}{(\partial P/\partial \rho)_T} \tag{12.6.12}$$

we derive

$$\frac{c_P}{c_V}=1+\frac{T}{\rho^2c_V}\frac{(\partial P/\partial T)_\rho^2}{(\partial P/\partial \rho)_T}=1+\frac{AB}{\rho C} \tag{12.6.13}$$

Hence:

$$\frac{\rho C}{AB+\rho C}=\frac{c_V}{c_P} \tag{12.6.14}$$

$$\frac{AB}{AB+\rho C}=1-\frac{c_V}{c_P} \tag{12.6.15}$$

$$AB+\rho C=\rho C\frac{c_P}{c_V}=\frac{c_P}{c_V}\left(\frac{\partial P}{\partial \rho}\right)_T=U_s^2 \tag{12.6.16}$$

where U_2 is the *speed of sound.*

We have now completed the solution of the eigenvalue problem, and have collected the results in Table 12.6.1.

Table 12.6.1. The Hydrodynamical Normal Modes

Ideal (nondissipative) fluid	Dissipative fluid	
$\Lambda_1 = iU_s k$	$\Lambda_1 = iU_s k - \frac{1}{2\rho}\left[\left(\frac{1}{c_V} - \frac{1}{c_P}\right)\kappa + \frac{4}{3}\eta + \zeta\right]k^2$	Sound mode
$\Lambda_2 = -iU_s k$	$\Lambda_2 = -iU_s k - \frac{1}{2\rho}\left[\left(\frac{1}{c_V} - \frac{1}{c_P}\right)\kappa + \frac{4}{3}\eta + \zeta\right]k^2$	Sound mode
$\Lambda_3 = 0$	$\Lambda_3 = -\frac{\eta}{\rho}k^2$	Shear mode
$\Lambda_4 = 0$	$\Lambda_4 = -\frac{\eta}{\rho}k^2$	Shear mode
$\Lambda_5 = 0$	$\Lambda_5 = -\frac{\kappa}{\rho c_P}k^2$	Thermal mode

With each of these five eigenvalues is associated a specific type of motion, or a "normal mode." We note that Λ_1 and Λ_2 are complex eigenvalues, describing damped oscillations, whereas Λ_3, Λ_4, and Λ_5 correspond to pure decay. The form of the coefficients clearly indicates the physical nature of the processes involved. The modes 1 and 2 describe the propagation of a sound wave, in two opposite directions, with a velocity U_s (12.2.16) and damping due to dissipation by viscosity, bulk viscosity, and heat conduction. The damping coefficient is

$$\Gamma = \frac{1}{2\rho}\left[\left(\frac{1}{c_V} - \frac{1}{c_P}\right)\kappa + \frac{4}{3}\eta + \zeta\right] \tag{12.6.17}$$

The modes 3 and 4 are degenerate. They both relate to the dissipation of an initial perturbation through shear viscosity. Finally, the fifth mode describes the dissipation due to the heat conduction.

We also gave in the first column of Table 12.6.1 the eigenvalues obtained for a nondissipative fluid, described by Eq. (12.5.33). They are, of course, much simpler. The shear and thermal modes have zero eigenvalues. The only possible long-range motion in such a fluid is the undamped propagation of sound waves.

To us, the most important feature in this result is the derivation of an *intrinsic* property of the equations of hydrodynamics, namely, the low-lying part of the spectrum, which is intimately connected with the transport coefficients.

12.7. THE PLASMADYNAMICAL NORMAL MODES

We now briefly discuss the extension of the theory of the previous section to the case of an electron plasma in a continuous positive background. The problem is formulated in exactly the same way, but in the equations we must take account of the nonvanishing charge $(-e)$ (for electrons). Hence instead of Eq. (12.6.2) we must use the following ones, derived from (12.5.29)–(12.5.32):

$$\Lambda \nu_{\mathbf{k}} + i\rho k u_{\mathbf{k}}^{x} = 0$$

$$\Lambda u_{\mathbf{k}}^{x} + iCk\nu_{\mathbf{k}} + iAk\theta_{\mathbf{k}} + (\eta' + \mu)k^2 u_{\mathbf{k}}^{x} + \left(\frac{e}{m}\right)E_{\mathbf{k}} = 0$$

$$\Lambda u_{\mathbf{k}}^{y} + \eta' k^2 u_{\mathbf{k}}^{y} = 0 \tag{12.7.1}$$

$$\Lambda u_{\mathbf{k}}^{z} + \eta' k^2 u_{\mathbf{k}}^{z} = 0$$

$$\Lambda \theta_{\mathbf{k}} + iBku_{\mathbf{k}}^{x} + \chi k^2 \theta_{\mathbf{k}} = 0$$

$$ikE_{\mathbf{k}} = -4\pi e m^{-1} \nu_{\mathbf{k}} \tag{12.7.2}$$

We oriented again the reference frame in such a way that $\mathbf{k} = k\mathbf{1}_x$. The only difference with Eq. (12.6.2) is the presence of the extra term $+(e/m)E_{\mathbf{k}}$ in the second equation. We will see, however, that this term completely alters the nature of the solutions. In a nonrelativistic theory, the electric field is purely longitudinal: in our reference frame $\mathbf{E}_{\mathbf{k}} = E_{\mathbf{k}}\mathbf{1}_x$. (In a relativistic theory the hydrodynamical equations must be coupled to the full set of Maxwell equations, the electric field then also possesses transverse components, and, moreover, it is time dependent. We shall not discuss these problems here.) From the Poisson equation (12.7.2) the electric field can be immediately expressed in terms of the density. Substituting the result into the second equation (12.7.1) the latter becomes

$$\Lambda u_{\mathbf{k}}^{x} + iCk\nu_{\mathbf{k}} + iAk\theta_{\mathbf{k}} + (\eta' + \mu)k^2 u_{\mathbf{k}}^{x} + i\left(\frac{4\pi e^2}{m^2}\right)k^{-1}\nu_{\mathbf{k}} = 0 \tag{12.7.3}$$

The remarkable feature in this equation is the appearance of a coefficient proportional to k^{-1}. It is because of this singular behavior that the normal modes are so deeply altered.

We now write the characteristic determinant of Eqs. (12.7.1) and (12.7.3) and proceed as in Section 12.6. We obtain the following dispersion equation:

$$(\Lambda + \eta' k^2)^2 \{\Lambda^3 + (\chi + \eta' + \mu)k^2\Lambda^2 + [(\eta' + \mu)\chi k^4 + (AB + \rho C)k^2 + \omega_p^2]\Lambda + (\rho C\chi k^4 + \omega_p^2 \chi k^2)\} = 0 \tag{12.7.4}$$

We introduced here the parameter ω_p having dimensions of a frequency:

$$\omega_p^2 = \frac{4\pi e^2 \rho}{m^2} \tag{12.7.5}$$

ω_p is a fundamental quantity of plasma physics: it is called the *plasma frequency*, or the Langmuir frequency of the electrons. Its characteristic feature is the proportionality to the square root of the density.

In order to investigate the eigenvalues it is convenient to proceed in two steps. First we consider a nondissipative plasma, that is, set $\chi = \eta' = \mu = 0$. The dispersion equation then simplifies to

$$\Lambda^3\{\Lambda^2 + [(AB + \rho C)k^2 + \omega_p^2]\} = 0 \tag{12.7.6}$$

Taking into account Eq. (12.6.16) we find the eigenvalues:

$$\begin{aligned} \Lambda_1 &= i(\omega_p^2 + U_s^2 k^2)^{1/2} \\ \Lambda_2 &= -i(\omega_p^2 + U_s^2 k^2)^{1/2} \\ \Lambda_3 &= \Lambda_4 = \Lambda_5 = 0 \end{aligned} \tag{12.7.7}$$

We see that the sound modes are now replaced by a very different kind of eigenvalues, which we call the *plasma modes*. For large k, the term ω_p^2 can be neglected and we are led back to the sound modes $\pm iU_s k$. But for small k (i.e., in the truly "hydrodynamical" regime) the eigenvalue tends to a nonzero constant $\pm i\omega_p$ (see Fig. 12.7.1). If Eq. (12.7.7) is expanded in powers of k, we find

$$\Lambda_{1,2} \simeq \pm i\left(\omega_p + \frac{U_s^2}{2\omega_p} k^2\right) \tag{12.7.8}$$

Compare these modes to the sound mode of an ordinary fluid:

$$\Lambda_{1,2} = \pm iU_s k \tag{12.7.9}$$

we see that the behavior in the limit $k \to 0$ is radically different. In the plasma mode there is no term linear in k, instead there is a constant and a quadratic term. It is worth noting that in Eq. (12.7.8) we cannot recover the sound waves by letting $\omega_p \to 0$. Indeed, ω_p appears in the denominator

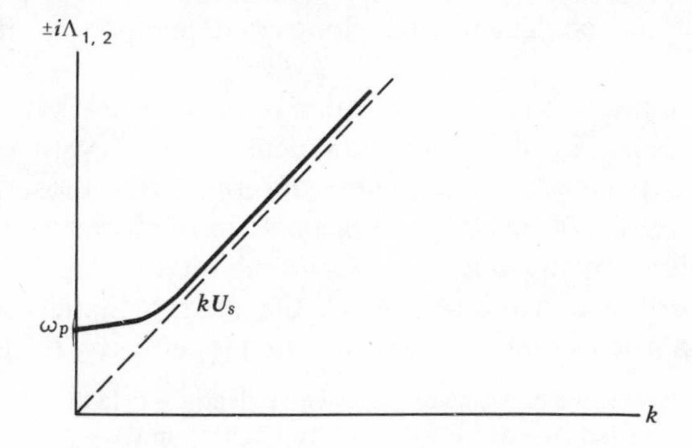

Figure 12.7.1. The plasma dispersion law.

of the second term. The limiting value can be obtained from the exact equation (12.7.7), but again, no useful expansion in powers of ω_p is possible: the resulting series diverges as $k \to 0$:

$$\Lambda_{1,2} \sim \pm ikU_s\left(1 + \frac{1}{2}\frac{\omega_p^2}{k^2U_s^2} + \cdots\right) \tag{12.7.10}$$

This odd behavior is due to the fact that $\Lambda_{1,2}$ are not analytic functions of ω_p. $\omega_p = 0$ is a branch point of these functions.

The physical mechanism of the plasma oscillations is easily visualized in the limit of $k \to 0$. A fluctuation in the electron density ν produces a local deviation of the charge density (because the positive background remains homogeneous at all times): there appears a local space charge. As a result, the electrostatic interaction between electrons and background produces a restoring force that tends to bring the electrons back to equilibrium. But the electrons have kinetic energy, and will actually go past the equilibrium position: the process then starts again in the opposite direction. Hence an oscillation process—which is due to the long-range Coulomb interactions—completely dominates and suppresses the sound waves in the limit $k \to 0$.

We now consider the dissipative plasma. The complete dispersion equation (12.7.4) is solved by a procedure similar to the one of Section 12.6. The results are given in Table 12.7.1. Here we also find some very remarkable features. The shear modes are the same as in an ordinary fluid. This is easily understandable, as these modes correspond to transverse motions. On the contrary, the electric field is longitudinal and only affects the longitudinal motions: there is no coupling between the two types of motion.*

A very strange effect appears in the thermal mode. Although this eigenvalue does not explicitly depend on the charge, its value differs from the corresponding value for a fluid by the finite factor c_P/c_V. Also, the sound-absorption coefficient no longer depends on the thermal conductivity.

We see again that it is not possible to make a continuous transition from a fluid to a plasma. Even for very low densities (i.e., very small ω_p) the properties of a plasma are completely different from those of a normal fluid. The types of motions that are possible in the former are dictated at small wave vector by the long-range Coulomb interaction. They represent a clear macroscopic manifestation of the characteristic collective behavior of the particles that are, so to say, tied together by the interactions.

* In a relativistic theory, the presence of a transverse electrical field deeply alters the shear modes. One actually finds $\Lambda_3 = \Lambda_4 = 0$ for these modes, even in the presence of dissipation [see R. Balescu, L. Brenig, and I. Paiva-Veretennicoff, *Physica* **74,** 447 (1974)].

Table 12.7.1. The Plasmadynamical Normal Modes

Nondissipative plasma	Dissipative plasma	Mode
$\Lambda_1 = i\omega_p\left(1+\frac{U_s^2}{2\omega_p^2}k^2\right)$	$\Lambda_1 = i\omega_p\left(1+\frac{U_s^2}{2\omega_p^2}k^2\right)-\frac{1}{2\rho}(\frac{4}{3}\eta+\zeta)k^2$	Plasma mode
$\Lambda_2 = -i\omega_p\left(1+\frac{U_s^2}{2\omega_p^2}k^2\right)$	$\Lambda_2 = -i\omega_p\left(1+\frac{U_s^2}{2\omega_p^2}k^2\right)-\frac{1}{2\rho}(\frac{4}{3}\eta+\zeta)k^2$	Plasma mode
$\Lambda_3 = 0$	$\Lambda_3 = -\frac{\eta}{\rho}k^2$	Shear mode
$\Lambda_4 = 0$	$\Lambda_4 = -\frac{\eta}{\rho}k^2$	Shear mode
$\Lambda_5 = 0$	$\Lambda_5 = -\frac{\kappa}{\rho c_V}k^2$	Thermal mode

We see here a justification of the well-known claim that a plasma is to be considered as a "fourth state of matter", rather than as a special gas.

APPENDIX: UNIQUENESS THEOREM FOR THE COLLISIONAL INVARIANTS

It was shown in Section 12.3 that the function

$$\psi^0(\mathbf{v}) = \gamma + \boldsymbol{\mu}\cdot\mathbf{v} + \tfrac{1}{2}\alpha v^2 \qquad (12.A.1)$$

is a collisional invariant of the Landau equation. Here γ, $\boldsymbol{\mu}$, α are coefficients, independent of $\mathbf{v}$, but which can be arbitrary functions of $\mathbf{q}$. As the latter dependence is irrelevant in the present argument, it will not be exhibited. We now proceed to strengthen the result by showing that *every collisional invariant of the Landau equation is of the form (12.A.1)*. The following proof is due to Résibois*: For simplicity we only consider here a two-dimensional system, but the proof is easily extended to three dimensions.

Consider the (two-dimensional) vector $\mathbf{X}(\mathbf{v})$ defined as the gradient of a function $\psi(\mathbf{v})$:

$$\mathbf{X}(\mathbf{v}) \equiv \partial\psi(\mathbf{v}) \qquad (12.A.2)$$

We call $\mathbf{X}(\mathbf{v})$ a "good vector" if the equality

$$\mathbf{k}\cdot\mathbf{v}_1 = \mathbf{k}\cdot\mathbf{v}_2 \qquad (12.A.3)$$

implies the equality

$$\mathbf{k}\cdot\mathbf{X}(\mathbf{v}_1) = \mathbf{k}\cdot\mathbf{X}(\mathbf{v}_2) \qquad (12.A.4)$$

* P. Résibois, private communication, 1973.

Clearly, as follows from Eq. (12.3.3), the gradient of any collisional invariant is a good vector. In particular, if we choose $\psi(\mathbf{v}) = \psi^0(\mathbf{v})$, the corresponding vector

$$\mathbf{X}^0(\mathbf{v}) = \boldsymbol{\mu} + \alpha \mathbf{v} \tag{12.A.5}$$

is a good vector. The following lemma will be proven below:

Lemma. *Let $Y(v_x, v_y) \equiv [Y_1(v_x, v_y), Y_2(v_x, v_y)]$ be a good vector having the following properties:*

$$\begin{aligned} Y_1(0, 0) &= 0 \\ Y_2(0, 0) &= 0 \\ Y_1(1, 0) &= 0 \end{aligned} \tag{12.A.6}$$

Then

$$Y(v_x, v_y) = 0 \qquad \textit{for all } v_x, v_y$$

Before proving the lemma, we show that it implies our desired result. Indeed, let $\mathbf{X}(\mathbf{v})$ be an *arbitrary* good vector. We can always construct another good vector:

$$\mathbf{Y}(\mathbf{v}) = \mathbf{X}(\mathbf{v}) - \boldsymbol{\mu} - \alpha \mathbf{v} \tag{12.A.7}$$

which satisfies Eqs. (12.A.6). Indeed, we can always fix the three available constant μ_x, μ_y, α in such a way as to satisfy the three equations (12.A.6). Hence $\mathbf{Y}(\mathbf{v}) \equiv 0$ and

$$\mathbf{X}(\mathbf{v}) = \boldsymbol{\mu} + \alpha \mathbf{v}$$

which proves our theorem.

In order to prove the lemma, we proceed in three steps.

(a) *We show that* $Y_2(1, 0) = 0$. We choose in Eqs. (12.A.3) and (12.A.4) the values $\mathbf{v}_1 = (0, 0)$, $\mathbf{v}_2 = (1, 0)$. Then (12.A.3) becomes

$$k_x \cdot 0 + k_y \cdot 0 = k_x \cdot 1 + k_y \cdot 0$$

which can be satisfied by the choice $k_x = 0$, $k_y \neq 0$. As $\mathbf{Y}(\mathbf{v})$ is a good vector, this equation implies

$$0 \cdot Y_1(0, 0) + k_y Y_2(0, 0) = 0 \cdot Y_1(1, 0) + k_y Y_2(1, 0)$$

But, from Eq. (12.A.6) we have $Y_2(0, 0) = 0$, hence

$$Y_2(1, 0) = 0 \qquad \text{Q.E.D.}$$

(b) *We show that* $\mathbf{Y}(a, b) = 0$ *for all a and for all $b \neq 0$.* We first choose $\mathbf{v}_1 = (a, b)$, $\mathbf{v}_2 = (0, 0)$. Then (12.A.3) implies

$$k_x a + k_y b = k_x \cdot 0 + k_y \cdot 0$$

or

$$k_y = -\left(\frac{a}{b}\right) k_x \tag{12.A.8}$$

Equation (12.A.4) then becomes

$$k_x Y_1(a, b) + k_y Y_2(a, b) = k_x Y_1(0, 0) + k_y Y_2(0, 0)$$

which, using (12.A.8) and (12.A.6) yields

$$k_x\left\{Y_1(a,b)-\left(\frac{a}{b}\right)Y_2(a,b)\right\}=0 \tag{12.A.9}$$

A similar reasoning, involving the choice $\mathbf{v}_1=(a,b)$, $\mathbf{v}_2=(1,0)$ yields the relation:

$$k_y\{Y_1(a,b)-\left(\frac{a-1}{b}\right)Y_2(a,b)\}=0 \tag{12.A.10}$$

Equations (12.A.9) and (12.A.10) are compatible with each other if and only if

$$Y_1(a,b)=0, \qquad Y_2(a,b)=0; \qquad b\neq 0$$

(c) *We show that* $\mathbf{Y}(a,0)=0$ *for all* $a\neq 0$. The proof proceeds as in (b), but we must choose the points (0, 1) and (1, 1) instead of (0, 0) and (1, 0) as reference points. From the result of (b) we know that $\mathbf{Y}(\mathbf{v})$ vanishes in these points.

From these three steps it follows that

$$\mathbf{Y}(\mathbf{v})\equiv 0 \quad \text{everywhere} \qquad \text{Q.E.D.}$$

The same uniqueness theorem can be proven for the Boltzmann equation. The proof is due to Grad* and is reproduced in Sommerfeld's book, quoted in Chapter 4.

BIBLIOGRAPHICAL NOTES

The *moment equations* were first introduced by J. C. Maxwell, *Phil. Trans. Roy. Soc.* (London) **157,** 49 (1867)

The *H theorem* was first proved in L. Boltzmann, *Wien. Ber.* **66,** 275 (1872)

For a discussion of the *equations of hydrodynamics,* see

L. D. Landau and E. N. Lifshitz, *Fluid Mechanics,* Pergamon Press, Oxford, 1963.

S. R. de Groot and P. Mazur, *Nonequilibrium Thermodynamics,* North Holland, Amsterdam, 1962.

The equations of hydrodynamics can be generalized in various ways; see

D. Burnett, *Proc. London Math. Soc.* **39,** 385 (1935); **40,** 382 (1935).

H. Grad, *Comm. Pure Appl. Math.* **2,** 331 (1949).

E. Ikenberry and C. Truesdell, *J. Rat. Mech. Anal.* **5,** 1 (1956).

* H. Grad, *Commun. Pure Appl. Math.* **2,** 311 (1949).

CHAPTER 13

EIGENVALUES OF THE KINETIC EQUATIONS AND THE THEORY OF TRANSPORT COEFFICIENTS

13.1. EIGENVALUES OF THE LINEARIZED KINETIC EQUATION FOR A HOMOGENEOUS GAS

This chapter is devoted to the study of the eigenvalues (or the spectrum) of the kinetic equations. The spectrum of an operator is, of course, an extremely important property, which yields a considerable amount of physical information (as we know, for instance, from quantum mechanics). There is, however, a serious difference between the basic equation of quantum mechanics and the kinetic equations: the former is linear, the latter are not. The concept of eigenvalues is therefore not defined for the kinetic equations in general.

However, in the neighborhood of the equilibrium state, we may approximate the kinetic equation by an equation of evolution that is linear in the deviation from equilibrum. The study of this *linearized kinetic equation* allows us to get a good picture of the evolution process, at least in its final stage of approach to equilibrium. Moreover, as will be presently seen, the linearized kinetic equation provides us with all the information needed for the calculation of crucial quantities such as the transport coefficients.

In the present section we study *spatially homogeneous systems.* This problem, besides its intrinsic interest, will be a starting point for the study of more complex systems. For simplicity, moreover, we assume that the system's evolution is governed by the Landau equation (11.6.27). We assume the distribution function to be of the form:

$$\bar{\varphi}(\mathbf{v}; t) = \varphi^0(v)\{1 + \chi(\mathbf{v}; t)\} \tag{13.1.1}$$

where $\varphi^0(v)$ is the equilibrium distribution:

$$\varphi^0(v) = \left(\frac{m}{2\pi k_B T}\right)^{3/2} \exp\left[-\left(\frac{mv^2}{2k_B T}\right)\right] \tag{13.1.2}$$

and $\chi(\mathbf{v}; t)$ measures the deviation from equilibrium and is supposed small compared to unity:

$$\chi(\mathbf{v}; t) \ll 1 \tag{13.1.3}$$

We substitute Eq. (13.1.1) into the kinetic equation in the form (11.6.27) and keep only terms linear in χ. We then obtain

$$\partial_t[\varphi^0(v_1)\chi(\mathbf{v}_1; t)] = \frac{Bn}{m^2}\int d\mathbf{v}_2\, \partial_{12}^r(g^2\delta^{rs} - g^r g^s)g^{-3} \times \partial_{12}^s \varphi^0(v_1)\varphi^0(v_2)[\chi(\mathbf{v}_1; t) + \chi(\mathbf{v}_2; t)] \tag{13.1.4}$$

It is very easy to check the following commutation relations:

$$\begin{aligned}\partial_{12}^r(g^2\delta^{rs} - g^r g^s)g^{-3}\partial_{12}^s\varphi^0(v_1)\varphi^0(v_2)F(\mathbf{v}_1, \mathbf{v}_2) &= \partial_{12}^r\varphi^0(v_1)\varphi^0(v_2)(g^2\delta^{rs} - g^r g^s)g^{-3}\partial_{12}^s F(\mathbf{v}_1, \mathbf{v}_2)\\ &= \varphi^0(v_1)\varphi^0(v_2)\partial_{12}^r(g^2\delta^{rs} - g^r g^s)g^{-3}\partial_{12}^s F(\mathbf{v}_1, \mathbf{v}_2)\end{aligned} \tag{13.1.5}$$

where $F(\mathbf{v}_1, \mathbf{v}_2)$ is any function of the velocities. Hence we obtain from (13.1.4) the following equation for the deviation:

$$\begin{aligned}\partial_t\chi(\mathbf{v}_1; t) &= Bnm^{-2}\int d\mathbf{v}_2\, \varphi^0(v_2)\partial_{12}^r(g^2\delta^{rs} - g^r g^s)g^{-3}\partial_{12}^s[\chi(\mathbf{v}_1; t) + \chi(\mathbf{v}_2; t)]\\ &\equiv K\chi\end{aligned} \tag{13.1.6}$$

The right side of this equation defines a linear integrodifferential operator K. Equation (13.1.6) is called the *linearized kinetic equation*, and K is called the *linearized collision operator*. We now study some of its properties.

We define a *scalar product* (χ_A, χ_B) of two real functions of $\mathbf{v}$ by the following expression:

$$(\chi_A, \chi_B) = \int d\mathbf{v}\, \varphi^0(v)\chi_A(\mathbf{v})\chi_B(\mathbf{v}) \tag{13.1.7}$$

It is quite easy to show that the operator K is *self-adjoint* with respect to this scalar product:

$$(\chi_A, K\chi_B) = (K\chi_A, \chi_B) \tag{13.1.8}$$

Indeed, a calculation similar to (12.3.2), making use of (13.1.5), yields

$$\begin{aligned}(\chi_A, K\chi_B) &= Bnm^{-2}\int d\mathbf{v}_1\, d\mathbf{v}_2\, \chi_A(\mathbf{v}_1)\varphi^0(v_1)\varphi^0(v_2)\\ &\quad\times \partial_{12}^r(g^2\delta^{rs} - g^r g^s)g^{-3}\partial_{12}^s[\chi_B(\omega_1) + \chi_B(\mathbf{v}_2)]\\ &= \tfrac{1}{2}Bnm^{-2}\int d\mathbf{v}_1\, d\mathbf{v}_2\, [\chi_A(\mathbf{v}_1) + \chi_A(\mathbf{v}_2)]\,\varphi^0(v_1)\varphi^0(v_2)\\ &\quad\times \partial_{12}^r(g^2\delta^{rs} - g^r g^s)g^{-3}\partial_{12}^s[\chi_B(\mathbf{v}_1) + \chi_B(\mathbf{v}_2)]\end{aligned}$$

$$= \tfrac{1}{2} Bnm^{-2} \int d\mathbf{v}_1 \, d\mathbf{v}_2 \{\partial^r_{12} \varphi^0(v_1)\varphi^0(v_2)(g^2\delta^{rs} - g^r g^s) g^{-3}$$
$$\times \partial^s_{12}[\chi_A(\mathbf{v}_1) + \chi_A(\mathbf{v}_2)]\} \, [\chi_B(\mathbf{v}_1) + \chi_B(\mathbf{v}_2)]$$
$$= (K\chi_A, \chi_B) \tag{13.1.9}$$

The fourth expression is obtained from the third by two partial integrations. An important consequence follows from the self-adjoint character: *the operator K has real eigenvalues.*

We can easily derive an even stronger property. Consider the expression:

$$(\chi_A, K\chi_A) = \tfrac{1}{2} Bnm^{-2} \int d\mathbf{v}_1 \, d\mathbf{v}_2 \, [\chi_A(\mathbf{v}_1) + \chi_A(\mathbf{v}_2)] \, \varphi^0(v_1)\varphi^0(v_2)$$
$$\times \partial^r_{12}(g^2\delta^{rs} - g^r g^s) g^{-3} \partial^s_{12} [\chi_A(\mathbf{v}_1) + \chi_A(\mathbf{v}_2)]$$
$$= -\tfrac{1}{2} Bnm^{-2} \int d\mathbf{v}_1 \, d\mathbf{v}_2 \, \varphi^0(v_1)\varphi^0(v_2) \, (g^2\delta^{rs} - g^r g^s) g^{-3}$$
$$\times \{\partial^r_{12}[\chi_A(\mathbf{v}_1) + \chi_A(\mathbf{v}_2)]\} \{\partial^s_{12}[\chi_A(\mathbf{v}_1) + \chi_A(\mathbf{v}_2)]\} \tag{13.1.10}$$

We introduce the shorthand notation $\mathbf{a}$ for the following vector function:

$$\mathbf{a} \equiv \partial_{12}[\chi_A(\mathbf{v}_1) + \chi_A(\mathbf{v}_2)] \tag{13.1.11}$$

and we call θ the angle between the vectors $\mathbf{a}$ and $\mathbf{g}$. Equation (13.1.10) can then be written as follows:

$$(\chi_A, K\chi_A) = -\tfrac{1}{2} Bnm^{-2} \int d\mathbf{v}_1 \, d\mathbf{v}_2 \, \varphi^0(v_1)\varphi^0(v_2) \, g^{-3}[g^2 a^2 - (\mathbf{g} \cdot \mathbf{a})^2]$$

But

$$g^2 a^2 - (\mathbf{g} \cdot \mathbf{a})^2 = g^2 a^2 - g^2 a^2 \cos^2\theta = g^2 a^2 \sin^2\theta$$

Hence

$$(\chi_A, K\chi_A) = -\tfrac{1}{2} Bnm^{-2} \int d\mathbf{v}_1 \, d\mathbf{v}_2 \, \varphi^0(v_1)\varphi^0(v_2) \, g^{-3} g^2 a^2 \sin^2\theta \leq 0 \tag{13.1.12}$$

This definite sign of the scalar product $(\chi_A, K\chi_A)$ implies that *the eigenvalues of K are real and nonpositive.* Indeed, let λ_n^0 be an eigenvalue and $\phi_n(\mathbf{v})$ be the corresponding eigenfunction of K*:

$$K\phi_n = \lambda_n^0 \phi_n \tag{13.1.13}$$

We assume that ϕ_n is normalized as follows:

$$(\phi_n, \phi_n) = \int d\mathbf{v} \, \varphi^0(v) \, [\phi_n(\mathbf{v})]^2 = 1 \tag{13.1.14}$$

* We use here, for simplicity, a notation appropriate for a *discrete* spectrum. It should, however, be realized that the spectrum of a kinetic equation (which is generally unknown) may also be continuous or partly discrete and partly continuous.

Then Eq. (13.1.12) implies

$$(\phi_n, K\phi_n) = \lambda_n^0(\phi_n, \phi_n) = \lambda_n^0 \leq 0 \tag{13.1.15}$$

It is easily seen that $\lambda^0 = 0$ *is indeed an eigenvalue;* it is actually a *fivefold degenerate eigenvalue.* This property is an immediate consequence of the existence of the five collisional invariants and is proved by a very slight modification of the calculations of Section 12.3. The five independent polynomials: 1, v_x, v_y, v_z, v^2 are indeed solutions of (13.1.13) with $\lambda^0 = 0$. They are readily reorganized into five mutually orthogonal and normalized functions, tabulated in Table 13.1.1. Indeed, the mutual orthogonality, defined as

$$(\phi_n, \phi_m) \equiv \int d\mathbf{v}\, \varphi^0(v)\, \phi_n(\mathbf{v})\phi_m(\mathbf{v}) = \delta_{nm} \tag{13.1.16}$$

is a general property of the set of eigenfunctions of the self-adjoint operator K.

The set of eigenfunctions ϕ_n is complete. Hence, any function of $\mathbf{v}$ can be expanded as a series of eigenfunctions (see Section 1.4). In particular, if the initial condition $\chi(\mathbf{v}; 0)$ for the equation (13.1.6) is expanded in the form:

$$\chi(\mathbf{v}; 0) = \sum_{n=1}^{\infty} \gamma_n \phi_n(\mathbf{v}) \tag{13.1.17}$$

then the solution is expressed as

$$\chi(\mathbf{v}; t) = \sum_{n=1}^{\infty} \gamma_n \phi_n(\mathbf{v}) e^{\lambda_n^0 t} \tag{13.1.18}$$

Table 13.1.1. The Eigenfunctions of the Linearized Collision Operator Corresponding to the Eigenvalue $\lambda^0 = 0$

$\phi_1 = 1$
$\phi_2 = \left(\frac{m}{k_B T}\right)^{1/2} v_x$
$\phi_3 = \left(\frac{m}{k_B T}\right)^{1/2} v_y$
$\phi_4 = \left(\frac{m}{k_B T}\right)^{1/2} v_z$
$\phi_5 = \left(\frac{2}{3}\right)^{1/2} \left[-\left(\frac{3}{2}\right) + \left(\frac{m}{2k_B T}\right) v^2\right]$

The property (13.1.15) gives a nice picture of the evolution. An initial disturbance (13.1.17) evolves monotonously towards the final state: the exponentials in (13.1.18) are all decaying to zero, except for the first five, which are constants. The full distribution function is given by

$$\bar{\varphi}(\mathbf{v}; t) = \varphi^0(v)\left[1 + \sum_{n=1}^{\infty} \gamma_n \phi_n(\mathbf{v}) e^{\lambda_n^0 t}\right] \tag{13.1.19}$$

We thus conclude that the end point of the evolution is

$$\lim_{t\to\infty} \bar{\varphi}(\mathbf{v}; t) = \varphi^0(v)[1 + \gamma_1\phi_1(\mathbf{v}) + \cdots + \gamma_5\phi_5(\mathbf{v})] \tag{13.1.20}$$

The five states remaining when $t \to \infty$ are, of course, those corresponding to the eigenvalue $\lambda^0 = 0$. We may ask a question at this point about the unicity of the equilibrium state. We chose in Eq. (13.1.2) a particular Maxwellian as a reference state around which we linearized the solution of the kinetic equation. However, we know from Eq. (12.2.8) that the most general stationary solution depends on five arbitrary parameters: c, $\mathbf{u}$, and a. The particular numerical values of these parameters depend on the initial condition. It may well be, in general, that the values of the parameters of the Maxwellian corresponding to the particular initial condition (13.1.17) are not those of the reference state. In that case it is always possible to choose the values of the five parameters in a function of the form (12.2.8) in such a way that, when the solution is expanded around that function, the five constants $\gamma_1, \ldots, \gamma_5$ be zero.

We shall henceforth always assume that this is done, and that the reference state $\varphi^0(v)$ has been adapted to the initial condition. In other words, the constants $\gamma_1, \ldots, \gamma_5$ in Eq. (13.1.18) can always be absorbed in the reference state $\varphi^0(v)$, and we can write, without loss of generality,

$$\chi(\mathbf{v}; t) = \sum_{n=6}^{\infty} \gamma_n \phi_n(\mathbf{v}) e^{\lambda_n^0 t} \tag{13.1.21}$$

Hence, Eq. (13.1.20) is then replaced by

$$\lim_{t\to\infty} \bar{\varphi}(\mathbf{v}; t) = \varphi^0(v) \tag{13.1.22}$$

The eigenvalues λ_n^0 ($n \geq 6$) measure the rate of the approach to equilibrium. We introduce the *relaxation times* τ_n through

$$\tau_n = -(\lambda_n^0)^{-1} > 0; \qquad n = 6, 7, \ldots \tag{13.1.23}$$

Not much is known about the exact value of the τ_n's. However, we may assume, on physical grounds, that *the eigenvalues do not accumulate near the origin.* In other words, the eigenvalue λ_6^0, closest to zero, is separated from the origin by a finite distance. It then follows that there exists a *finite maximum relaxation time* τ_6 that fixes the overall duration of the relaxation toward equilibrium. This time will be called the "relaxation time" and be denoted by τ_r.

Although τ_r is not known exactly, a simple dimensional argument provides valuable information. Indeed, in the linearized kinetic equation there exists a natural unit for the velocity: $(k_B T/m)^{1/2}$. All the relevant functions, that is, $\varphi^0(v)$ and $\phi_n(\mathbf{v})$, depend on the velocities only through the dimensionless variable:

$$\tilde{\mathbf{v}} = \left(\frac{m}{k_B T}\right)^{1/2} \mathbf{v}$$

We also define a dimensionless distribution function:

$$\tilde{\varphi}_0(\tilde{v}) = (k_B T/m)^{3/2} \varphi^0(v)$$

Consider then Eqs. (13.1.5) and (13.1.12); they can be written as

$$\begin{aligned}(\phi_6, K\phi_6) &= \lambda_6^0 \\ &= -\tfrac{1}{2} Bnm^{-2}\left(\frac{m}{k_B T}\right)^{3/2} \int d\tilde{\mathbf{v}}_1\, d\tilde{\mathbf{v}}_2\, \tilde{\varphi}^0(\tilde{v}_1)\tilde{\varphi}^0(\tilde{v}_2)\, \tilde{g}^{-3}\tilde{g}^2\tilde{a}^2 \sin^2\theta\end{aligned}$$

The integral on the right side is a dimensionless number. Hence

$$-\lambda_6^0 = \tau_r^{-1} = nBm^{-1/2}(k_B T)^{-3/2}\delta \tag{13.1.24}$$

where δ is a finite dimensionless number. Hence we have derived the dependence of the relaxation time on the density and on the temperature. We find that a low density and a high temperature correspond to a long relaxation time. Moreover, τ_r is inversely proportional to the square of the interaction strength through the parameter B, Eq. (11.6.24).

We have now obtained a fairly clear picture of the nature of the eigenvalues of the collision operator K by using very simple arguments. The actual calculation of the eigenvalues, that is, the evaluation of the constant δ in Eq. (13.1.24), leads to a very complex problem, which is not solved analytically at present. It is, however, remarkable that the simple results of the present section are sufficient as a starting point for the theory of the transport coefficients, as will be seen in detail in the forthcoming sections.

13.2. THE "NORMAL" STATE OF AN INHOMOGENEOUS GAS

Our picture of the relaxation of a homogeneous gas toward equilibrium is now becoming clear. The Boltzmann picture of this evolution introduces two characteristic time scales τ_c and τ_r, which, in the Boltzmann or

Landau approximations, are widely separated [see Eq. (11.5.1)]:

$$\tau_c \ll \tau_r \tag{13.2.1}$$

The molecular collisions bring the system to equilibrium in a time of order τ_r.

We now consider a spatially inhomogeneous system. We may always introduce a characteristic length scale L_h, associated with the inhomogeneity, defining it (at a given time) as

$$L_h^{-1} = \max \frac{|\nabla \bar{f}(\mathbf{q}, \mathbf{v}; t)|}{\bar{f}(\mathbf{q}, \mathbf{v}; t)} \tag{13.2.2}$$

L_h represents the typical range over which the distribution function changes appreciably. Together with L_h we can introduce a characteristic time τ_h, by dividing it by the average velocity $\bar{v}$:

$$\tau_h = \frac{L_h}{\bar{v}} \tag{13.2.3}$$

In general, the evolution of an inhomogeneous system is extremely intricate because the inhomogeneity may influence the collision process itself: an example of this influence will be discussed in Section 18.4. The situation is very much simplified if we can assume that

$$\tau_c \ll \tau_h \tag{13.2.4}$$

In this case, the effect of the inhomogeneity and the effect of the collisions appear in two separate terms of the kinetic equation. Moreover, the two factors $\bar{f}$ appearing in the collision term are evaluated at the same point in space. This aspect was already discussed in Section 11.5.

A further very important simplification occurs when the characteristic time τ_h is much longer than the relaxation time τ_r; we then have a sequence of three widely separated time scales:

$$\tau_c \ll \tau_r \ll \tau_h \tag{13.2.5}$$

In parallel, we have an analogous ordering of the length scales:

$$L_c \ll L_r \ll L_h \tag{13.2.6}$$

where L_c is the range of the interactions and L_r is of the order of the average distance between the molecules, or of the mean-free path.

Before going further, it should be noted that the situation described in Eqs. (13.2.5) and (13.2.6) is by no means exceptional. Indeed, in many interesting cases, the inhomogeneity is due to an external device of macroscopic size. For instance, a sample of the fluid is placed between two walls that are at different temperatures, or that are in relative motion.

Such setups are characteristic of all experiments in hydrodynamics, and for this reason we say that systems characterized by Eqs. (13.2.5) and (13.2.6) are in a *hydrodynamic regime.* As an example, such a hydrodynamic experiment performed with hydrogen at normal pressure and temperature corresponds to the following orders of magnitude:

$$\begin{aligned} L_h &\sim 1 \text{ cm}, \qquad & \tau_h &\sim 5\times 10^{-6} \text{ sec} \\ L_r &\sim 10^{-5} \text{ cm}, \qquad & \tau_r &\sim 5\times 10^{-11} \text{ sec} \\ L_c &\sim 10^{-8} \text{ cm}, \qquad & \tau_c &\sim 5\times 10^{-14} \text{ sec} \end{aligned}$$

We now analyze the evolution in such a system qualitatively. The rate of change $\partial_t \bar{f}$ is given by the sum of a flow term $\mathbf{v}\cdot\nabla\bar{f}$ and of a collision term.* In the hydrodynamic regime, the former describes a *slow process,* of order $\bar{f}/\tau_h$, and the latter describes a *fast process,* of order $\bar{f}/\tau_r$. Hence, starting from an arbitrary initial condition, the system first tends to relax quickly towards equilibrium in velocity space, as a result of the collisions. We know, however, from Section 12.2 that if the system is initially inhomogeneous, the end point of the "collisional" evolution is not necessarily the true equilibrium, but rather a state of *local equilibrium* defined by Eq. (12.2.29), which we rewrite here:

$$\bar{f}^{\text{LE}}(\mathbf{q},\mathbf{v};t) = \left(\frac{m}{2\pi k_B T(\mathbf{q};t)}\right)^{3/2} n(\mathbf{q};t) \times \exp\left\{-\left(\frac{m}{2k_B T(\mathbf{q};t)}\right)[\mathbf{v}-\mathbf{u}(\mathbf{q};t)]^2\right\} \tag{13.2.7}$$

This state is completely characterized by the five (scalar) functions of space and time $n(\mathbf{q};t)$, $\mathbf{u}(\mathbf{q};t)$, and $T(\mathbf{q};t)$, that is, the macroscopic quantities associated with the collisional invariants.

An important property that the local equilibrium shares with the global equilibrium is the absence of dissipation in that state. What we mean precisely is that the heat flow vanishes in local equilibrium [see (12.4.18)]:

$$\mathbf{J}^{\text{LE}} = \tfrac{1}{2}m\int d\mathbf{q}\, d\mathbf{v}\, |\mathbf{v}-\mathbf{u}|^2 (\mathbf{v}-\mathbf{u})\, \delta(\mathbf{x}-\mathbf{q})\, \bar{f}^{\text{LE}}(\mathbf{q},\mathbf{v};t) = 0 \tag{13.2.8}$$

as results from the odd character of the integrand in $(\mathbf{v}-\mathbf{u})$. Similarly, from Eq. (12.4.5) we find that the pressure tensor reduces to the diagonal local pressure of an ideal gas:

$$\begin{aligned} P_{rs}^{\text{LE}} &= m\int d\mathbf{q}\, d\mathbf{v}\, (v_r - u_r)(v_s - u_s)\, \delta(\mathbf{x}\times\mathbf{q})\, \bar{f}^{\text{LE}}(\mathbf{q},\mathbf{v};t) \\ &= n(\mathbf{x};t) k_B T(\mathbf{x};t)\, \delta_{rs} \end{aligned} \tag{13.2.9}$$

* We only discuss here ordinary gases. The more complex case of plasmas and the effect of the Vlassov term are investigated in Section 13.6.

Hence the mere *presence of spatial inhomogeneities is not sufficient for the onset of transport phenomena.* Here we have a typical inhomogeneous state in which there is no dissipation. It immediately follows that the thermal conductivity and the viscosity can only be connected to deviations of the distribution function from local equilibrium.

We now note that the system actually never reaches the local equilibrium, or if it accidentally starts in that state it does not remain in it. Indeed, $\bar{f}^{LE}$ is *not* a stationary solution of the complete kinetic equation. The drive toward this state is opposed by the flow term (indeed, $\mathbf{v} \cdot \nabla \bar{f}^{LE} \neq 0$). The effect of this term is to smooth out all spatial inhomogeneities. However, because of the assumed smallness of the gradients, this process acts much more slowly than the collisions. As a result, at all times, the instantaneous, time-dependent distribution function is very close to the local equilibrium:

$$\bar{f}(\mathbf{q}, \mathbf{v}; t) = \bar{f}^{LE}(\mathbf{q}, \mathbf{v}; t)\,[1 + \chi(\mathbf{q}, \mathbf{v}; t)], \qquad \chi(\mathbf{q}, \mathbf{v}; t) \ll 1 \quad (13.2.10)$$

We can go even further. We note that the local equilibrium distribution depends on space and time, but not in an arbitrary way: its dependence is entirely determined by the five quantities $n(\mathbf{q}; t)$, $\mathbf{u}(\mathbf{q}; t)$, $T(\mathbf{q}; t)$. These quantities play a privileged role, because they—and only they—are related to the five collisional invariants. It follows that, in the local equilibrium state, all other average macroscopic quantities must be functions of these five quantities (and possibly of their derivatives). We now *assume* that, even in the instantaneous state (13.2.10), which is very close to local equilibrium, the dependence on $\mathbf{q}$ and t is again through the five invariants alone. In other words, we are led to the idea that, after a time of order τ_r, the collisions bring the system into a state $\bar{f}_{nor}$ of the form:

$$\bar{f}_{nor}(\mathbf{q}, \mathbf{v}; t) = \bar{f}[n(\mathbf{q}; t), \mathbf{u}(\mathbf{q}; t), T(\mathbf{q}; t), \mathbf{v}] \quad (13.2.11)$$

[Equivalently, the function $\chi(\mathbf{q}; t)$ in Eq. (13.2.10) is of the same form.] Any distribution function that depends on $\mathbf{q}$ and t only through the macroscopic parameters n, $\mathbf{u}$, T will be called a *normal distribution function.* The solution of the kinetic equation in the hydrodynamical regime thus amounts to finding an appropriate normal distribution function satisfying the equation.

The concept of "normal" distribution functions is at the heart of the "traditional" methods of solution of the Boltzmann equation (or of other kinetic equations). It was introduced in 1912 by Hilbert: to this great mathematician the Boltzmann equation appeared as a beautiful example of a nonlinear integrodifferential equation, and he treated it from the mathematician's point of view. His method of solution was not very convenient for physical applications. The problem was reconsidered from a similar point of view by Chapman and by Enskog (independently). Their methods (which differed slightly in detail) give identical results and were unified since in the famous Chapman–Enskog method of solution. This method consists essentially in a systematic construction of a "normal" solution as an expansion around the local equilibrium state. The expansion parameter is essentially the strength of the gradients; the expansion is, however, not a trivial Taylor series (which would lead to difficulties) but, rather, a more subtle procedure. As a final result, one gets directly, in the first-order approximation, expressions for the transport coefficients that can be calculated explicitly for various intermolecular potentials and that yield numerical values in excellent agreement with experiment in many important cases.

We shall, however, not give here an exposition of the Chapman–Enskog theory. Excellent and exhaustive monographs exist already on this subject. We, rather, consider in the next chapter an alternative recent theory of the transport coefficients, due to Résibois. For the Boltzmann equation, it yields the same results as the Chapman–Enskog theory. It has, however, extra important advantages that will be discussed in due time.

13.3. EIGENVALUES OF THE LINEARIZED KINETIC EQUATION FOR AN INHOMOGENEOUS GAS

In Section 13.1 we studied some general properties of the spectrum of eigenvalues of the homogeneous linearized kinetic equation. We now extend this study in order to include inhomogeneous systems. The complete one-particle distribution function is defined as

$$\bar{f}(\mathbf{q}, \mathbf{v}; t) = n\varphi^0(v)[1 + \chi(\mathbf{q}, \mathbf{v}; t)] \tag{13.3.1}$$

By a straightforward generalization of Eq. (13.1.6) we obtain the *linearized kinetic equation:*

$$\partial_t \chi(\mathbf{q}_1, \mathbf{v}_1; t) + \mathbf{v}_1 \cdot \nabla_1 \chi(\mathbf{q}_1, \mathbf{v}_1; t)$$
$$= Bnm^{-2} \int d\mathbf{v}_2\, \varphi^0(v_2)\, \partial_{12}^r (g^2 \delta^{rs} - g^r g^s) g^{-3}\, \partial_{12}^r\, [\chi(\mathbf{q}_1, \mathbf{v}_1; t) + \chi(\mathbf{q}_1, \mathbf{v}_2; t)] \tag{13.3.2}$$

We consider here the case of an ordinary gas. The influence of the Vlassov term in a plasma will be studied in Section 13.6. We Fourier-transform the deviation χ as follows:

$$\chi(\mathbf{q}, \mathbf{v}; t) = \int d\mathbf{k}\, e^{i\mathbf{k}\cdot\mathbf{q}} \chi_{\mathbf{k}}(\mathbf{v}; t) \tag{13.3.3}$$

The equation for $\chi_{\mathbf{k}}(\mathbf{v}; t)$ then reads

$$\partial_t \chi_{\mathbf{k}}(\mathbf{v}; t) + i\mathbf{k} \cdot \mathbf{v}\chi_{\mathbf{k}}(\mathbf{v}; t) = K\chi_{\mathbf{k}}(\mathbf{v}; t) \tag{13.3.4}$$

where the operator K is defined by Eq. (13.1.6). Equation (13.3.4) will be our starting point. We shall be particularly interested in the associated eigenvalue problem:

$$(K - i\mathbf{k} \cdot \mathbf{v})\psi_n(\mathbf{v}) = \lambda_n \psi_n(\mathbf{v}) \tag{13.3.5}$$

Clearly, the eigenvalues λ_n and the eigenfunctions $\psi_n(\mathbf{v})$ depend now on $\mathbf{k}$. For $\mathbf{k} = 0$, the problem (13.3.5) reduces to the homogeneous problem (13.1.13). We assume that the homogeneous eigenfunctions and eigenvalues are given, even though their explicit form is unknown. We do know, however, that there exists a fivefold degenerate eigenvalue $\lambda_\alpha^0 = 0$, $\alpha = 1, \ldots, 5$, and we know the corresponding eigenfunctions $\phi_\alpha(\mathbf{v})$. It turns out that these five eigenfunctions will, indeed, play a dominant role as compared to all the others. They will therefore be distinguished by a special notation:

$$\begin{aligned} &\lambda_\alpha^0, \phi_\alpha: \qquad \alpha = 1, 2, 3, 4, 5 \\ &\lambda_\beta^0, \phi_\beta: \qquad \beta = 6, 7, \ldots \end{aligned} \tag{13.3.6}$$

The set of indices α is thus the set $(1, 2, 3, 4, 5)$, whereas β denotes the set $(6, 7, \ldots)$.

Proceeding to the problem (13.3.5), we see that $(-i\mathbf{k} \cdot \mathbf{v})$ appears as a *pertubation* with respect to K. The new operator on the left side of (13.3.5) is still a symmetric operator:

$$(\chi_A, [K - i\mathbf{k} \cdot \mathbf{v}]\chi_B) = ([K - i\mathbf{k} \cdot \mathbf{v}]\chi_A, \chi_B) \tag{13.3.7}$$

but it is no longer Hermitian. Hence *its eigenvalues may be imaginary.*

We now consider $(-i\mathbf{k} \cdot \mathbf{v})$ as a *small pertubation.* Moreover, in order to simplify the calculation, we assume, as in Section 12.6, that we chose the x axis of our reference frame to be oriented along the fixed vector $\mathbf{k}$:

$$\mathbf{k} = \mathbf{1}_x k, \qquad \mathbf{k} \cdot \mathbf{v} = kv_x \tag{13.3.8}$$

The assumption that $\mathbf{k} \cdot \mathbf{v}$ is small is equivalent to assuming that the gradients are small, which is consistent with the phenomenological treatment of the hydrodynamic equations in Section 12.5.

From here on, we determine the eigenvalues λ_n by a quite standard pertubation theory, just as in elementary quantum mechanics. We assume

$$\lambda_n = \lambda_n^0 + \lambda_n^{(1)} + \lambda_n^{(2)} + \cdots \tag{13.3.9}$$

In particular, for $n \in (\alpha)$, $\lambda_\alpha^0 = 0$ and

$$\lambda_\alpha = \lambda_\alpha^{(1)} + \lambda_\alpha^{(2)} + \cdots \tag{13.3.10}$$

As the eigenvalues $\lambda_\alpha^0 = 0$ are degenerate, we must be careful in starting from a proper basis that avoids the appearance of vanishing denominators. As shown in every textbook in quantum mechanics, we must first solve exactly the eigenvalue problem in the subspace spanned by ϕ_α.

We thus consider in a first step the problem:

$$(K - ikv_x)\bar{\phi}_\alpha = \bar{\lambda}_\alpha \bar{\phi}_\alpha \tag{13.3.11}$$

with

$$\bar{\phi}_\alpha = \sum_{\alpha'=1}^{5} c_{\alpha\alpha'} \phi_{\alpha'} \tag{13.3.12}$$

Substituting (13.3.12) into (13.3.11), and taking account of the relation:

$$K\phi_{\alpha'} = 0$$

we obtain the following set of algebraic equations:

$$-i\sum_{\alpha''=1}^{5} W_{\alpha'\alpha''} c_{\alpha\alpha''} = \bar{\lambda}_\alpha c_{\alpha\alpha'}, \quad \alpha = 1, \ldots, 5, \quad \alpha' = 1, \ldots, 5 \tag{13.3.13}$$

where

$$W_{\alpha\alpha'} = \int d\mathbf{v}\, \varphi^0(v)\, \phi_\alpha(\mathbf{v})\, kv_x\, \phi_{\alpha'}(\mathbf{v}) \tag{13.3.14}$$

Using the explicit form of the unperturbed eigenfunctions, (Table 13.1.1), we find that the only nonvanishing matrix elements are

$$\begin{aligned} W_{12} &= W_{21} = k\bar{u} \\ W_{25} &= W_{52} = (\tfrac{2}{3})^{1/2} k\bar{u} \end{aligned} \tag{13.3.15}$$

with

$$\bar{u} = \left(\frac{k_B T}{m}\right)^{1/2} \tag{13.3.16}$$

To ensure the existence of nontrivial solutions for the system (13.3.13), $\bar{\lambda}$ has to obey the dispersion equation:

$$\begin{vmatrix} -\bar{\lambda} & -ik\bar{u} & 0 & 0 & 0 \\ -ik\bar{u} & -\bar{\lambda} & 0 & 0 & -i(\frac{2}{3})^{1/2}k\bar{u} \\ 0 & 0 & -\bar{\lambda} & 0 & 0 \\ 0 & 0 & 0 & -\bar{\lambda} & 0 \\ 0 & -i(\frac{2}{3})^{1/2}k\bar{u} & 0 & 0 & -\bar{\lambda} \end{vmatrix} = 0 \tag{13.3.17}$$

The determinant is readily expanded, with the result:

$$-\bar{\lambda}^3[\bar{\lambda}^2 + \tfrac{5}{3}k^2\bar{u}^2] = 0$$

Table 13.3.1. First Approximation to the Eigenmodes of the Linearized Kinetic Equation

$\bar{\lambda}_1 = +ikU_s$	$\bar{\phi}_1 = 2^{-1/2}[(\frac{3}{5})^{1/2}\phi_1 - \phi_2 + (\frac{2}{5})^{1/2}\phi_5]$
$\bar{\lambda}_2 = -ikU_s$	$\bar{\phi}_2 = 2^{-1/2}[(\frac{3}{5})^{1/2}\phi_1 + \phi_2 + (\frac{2}{5})^{1/2}\phi_5]$
$\bar{\lambda}_3 = 0$	$\bar{\phi}_3 = \phi_3$
$\bar{\lambda}_4 = 0$	$\bar{\phi}_4 = \phi_4$
$\bar{\lambda}_5 = 0$	$\bar{\phi}_5 = (\frac{2}{5})^{1/2}[-\phi_1 + (\frac{3}{2})^{1/2}\phi_5]$

from which we immediately get the eigenvalues:

$$\begin{aligned} \bar{\lambda}_1 = -\bar{\lambda}_2 = ik\left(\frac{5}{3}\frac{k_B T}{m}\right)^{1/2} \\ \bar{\lambda}_3 = \bar{\lambda}_4 = \bar{\lambda}_5 = 0 \end{aligned} \tag{13.3.18}$$

By substitution of these values into (13.3.13) we easily obtain the five corresponding eigenvectors. The results are collected in Table 13.3.1.

It is interesting to note already at this stage that the eigenvalues found here correspond closely to the hydrodynamical values (Table 13.2.1), in which, however, the dissipation is neglected, that is, all the transport coefficients η, ζ, κ are set equal to zero. Another way of looking at the matter is to note that the eigenvalues $\bar{\lambda}_\alpha$ correspond to the hydrodynamical ones, up to the first order in k. Indeed, we know from equilibrium statistical mechanics that, for an ideal classical gas, the following relations are valid:

$$\left(\frac{\partial P}{\partial \rho}\right)_T = \frac{k_B T}{m}$$

$$\frac{c_P}{c_V} = \frac{5}{3}$$

Hence, the sound velocity is:

$$U_s = \left(\frac{5}{3}\frac{k_B T}{m}\right)^{1/2} \tag{13.3.19}$$

Hence the eigenvalues (13.3.18) exactly agree with the nondissipative hydrodynamical modes. The eigenfunctions $\bar{\phi}_\alpha$ constitute a properly normalized set of mutually orthogonal functions:

$$(\bar{\phi}_\alpha, \bar{\phi}_{\alpha'}) = \int d\mathbf{v}\, \varphi^0(v)\, \bar{\phi}_\alpha(\mathbf{v})\bar{\phi}_{\alpha'}(\mathbf{v}) = \delta_{\alpha,\alpha'} \tag{13.3.20}$$

Although the eigenvalue $\bar{\lambda} = 0$ is still triply degenerate, there are no matrix elements of the pertubation connecting the corresponding eigenfunctions:

$$\bar{W}_{34} = \bar{W}_{45} = \bar{W}_{35} = 0 \tag{13.3.21}$$

where we introduce the notation:

$$\bar{W}_{mn} = \langle \bar{\phi}_m | \, kv_x \, | \bar{\phi}_n \rangle$$
$$= \int d\mathbf{v} \; \varphi^0(v) \; \bar{\phi}_m \, kv_x \, \bar{\phi}_n \tag{13.3.22}$$

We may thus use, as a proper basis for starting the pertubation theory, the set of functions ψ_m defined as follows*:

$$\psi_m : \begin{array}{ll} \psi_\alpha = \bar{\phi}_\alpha, & \alpha = 1, 2, \ldots, 5 \\ \psi_\beta = \phi_\beta, & \beta = 6, 7, \ldots \end{array} \tag{13.3.23}$$

Applying now second-order perturbation theory, as found in every textbook on quantum mechanics, we get the result:

$$\lambda_\alpha^{(2)} = \sum_{\beta \geq 6} \langle \psi_\alpha | \, kv_x \, | \psi_\beta \rangle \frac{1}{\lambda_\beta^0} \langle \psi_\beta | \, kv_x \, | \psi_\alpha \rangle \tag{13.3.24}$$

This form is, however, not yet very convenient, because it involves a summation only over eigenvalues *other* than the five known ones. Yet we may lift the restriction on the summation by subtracting the additional terms introduced if we sum over all values. This is realized if we replace the matrix elements in (13.3.24) by the following expression:

$$\langle \psi_\beta | \, kv_x \, | \psi_\alpha \rangle \to \langle \psi_\beta | \, kv_x - i\bar{\lambda}_\alpha \, | \psi_\alpha \rangle$$

and a similar expression with α and β interchanged. Indeed, for $\beta \neq \alpha$, nothing is changed:

$$\langle \psi_\beta | \, kv_x - i\bar{\lambda}_\alpha \, | \psi_\alpha \rangle = \langle \psi_\beta | \, kv_x \, | \psi_\alpha \rangle - i\bar{\lambda}_\alpha \langle \psi_\beta \mid \psi_\alpha \rangle = \langle \psi_\beta | \, kv_x \, | \psi_\alpha \rangle$$

On the other hand, for $\beta = \alpha$

$$\langle \psi_\alpha | \, kv_x - i\bar{\lambda}_\alpha \, | \psi_\alpha \rangle = \langle \psi_\alpha | \, kv_x \, | \psi_\alpha \rangle - i\bar{\lambda}_\alpha \langle \psi_\alpha \mid \psi_\alpha \rangle = i\bar{\lambda}_\alpha - i\bar{\lambda}_\alpha = 0$$

Also, in order to avoid possible divergence difficulties, we subtract a small positive number ε from the denominator λ_β^0 in (13.3.24). We do not go into a detailed technical justification of this procedure, because ε actually plays only a minor role. It will be shown in Section 13.4 that in the final results one simply sets $\varepsilon = 0$. Collecting all the results of this discussion, we now write Eq. (13.3.24) in the form:

$$\lambda_\alpha^{(2)} = \lim_{\varepsilon \to 0} \sum_{\text{all } m} \langle \chi_\alpha | \, kv_x - i\bar{\lambda}_\alpha \, | \psi_m \rangle \frac{1}{\lambda_m^0 - \varepsilon} \langle \psi_m | \, kv_x - i\bar{\lambda}_\alpha \, | \psi_\alpha \rangle \tag{13.3.25}$$

* We assume that, if there are further degeneracies among the ϕ_β ($\beta \geq 6$), they have been lifted in the same way as for the ϕ_α.

Applying a well-known formal relation:

$$\lim_{\varepsilon\to 0} \sum_m |\psi_m\rangle \frac{1}{\lambda_m^0 - \varepsilon} \langle\psi_m| = \lim_{\varepsilon\to 0} \frac{1}{K-\varepsilon} \tag{13.3.26}$$

we finally get

$$\lambda_\alpha^{(2)} = \lim_{\varepsilon\to 0} \langle\chi_\alpha| (kv_x - i\bar{\lambda}_\alpha) \frac{1}{K-\varepsilon} (kv_x - i\bar{\lambda}_\alpha) |\psi_\alpha\rangle \tag{13.3.27}$$

This is our final result, which will be analyzed in the next section. For convenience of reference, we list the complete eigenvalues explicitly below:

$$\lambda_1 = ikU_s + \tfrac{1}{2}k^2 \langle(\tfrac{3}{5})^{1/2}\phi_1 - \phi_2 + (\tfrac{2}{5})^{1/2}\phi_5| (v_x + U_s) \frac{1}{K-\varepsilon} (v_x + U_s) |(\tfrac{3}{5})^{1/2}\phi_1 - \phi_2 + (\tfrac{2}{5})^{1/2}\phi_5\rangle \tag{13.3.28}$$

$$\lambda_2 = -ikU_s + \tfrac{1}{2}k^2 \langle(\tfrac{3}{5})^{1/2}\phi_1 + \phi_2 + (\tfrac{2}{5})^{1/2}\phi_5| (v_x - U_s) \frac{1}{K-\varepsilon} (v_x - U_s) |(\tfrac{3}{5})^{1/2}\phi_1 + \phi_2 + (\tfrac{2}{5})^{1/2}\phi_5\rangle \tag{13.3.29}$$

$$\lambda_3 = k^2 \langle\phi_3| v_x \frac{1}{K-\varepsilon} v_x |\phi_3\rangle \tag{13.3.30}$$

$$\lambda_4 = k^2 \langle\phi_4| v_x \frac{1}{K-\varepsilon} v_x |\phi_4\rangle \tag{13.3.31}$$

$$\lambda_5 = \tfrac{2}{5}k^2 \langle -\phi_1 + (\tfrac{3}{2})^{1/2}\phi_5| v_x \frac{1}{K-\varepsilon} v_x |-\phi_1 + (\tfrac{3}{2})^{1/2}\phi_5\rangle \tag{13.3.32}$$

For brevity, the superscripts (2) have been dropped in these and the forthcoming formulas. The limit $\varepsilon \to 0$ is understood in all these equations.

We now show that there is a one-to-one correspondence between these eigenvalues and the five hydrodynamical normal modes of Table 12.6.1.

It is already clear that the expressions of λ_1 and λ_2 correspond to the eigenvalues Λ_1 and Λ_2 of Table 12.6.1. Indeed, they both contain a linear and a quadratic term in k, and the coefficient of the linear term has already been identified as the sound velocity.

For the three remaining eigenvalues we see that a correspondence exists with $\Lambda_3, \Lambda_4, \Lambda_5$ of Table 12.6.1. In order to make the correspondence precise, we analyze the equations in more detail. Consider first λ_3, which can be evaluated explicitly by using Eqs. (13.3.30) and (13.3.22) and Table 13.1.1:

$$\lambda_3 = k^2 \frac{m}{k_B T} \lim_{\varepsilon\to 0} \int d\mathbf{v}\, \varphi^0(v)\, v_y v_x \frac{1}{K-\varepsilon} v_y v_x \tag{13.3.33}$$

Similarly:

$$\lambda_4 = k^2 \frac{m}{k_B T} \lim_{\varepsilon\to 0} \int d\mathbf{v}\, \varphi^0(v)\, v_z v_x \frac{1}{K-\varepsilon} v_z v_x \tag{13.3.34}$$

A very simple argument shows that these two numbers are equal; hence they correspond to a doubly degenerate eigenvalue, just as Λ_3, Λ_4 of Table 12.6.1. Finally, λ_5 necessarily corresponds to Λ_5.

The one-to-one correspondence between hydrodynamical modes and kinetic modes has a great practical importance as a basis for a theory of transport coefficients, as will be seen in Section 13.4. We may, however, think of it from a more "philosophical" point of view.* A gas is a collection of particles moving in a quite disordered fashion. The results obtained here, however, show that, in the limit of long wavelengths, the only possible modes of motion of the gas are quite ordered motions, such as a sound-wave propagation. These motions involve the coordinated action of enormous numbers of molecules. The existence of this order superimposed on the basic randomness of individual motions is one of the most surprising features of statistical mechanics. The fundamental reason for this situation is in the dominant effect of the collisions. These bring the system very rapidly to a local-equilibrium state (see Section 13.2) that is highly organized; the flow terms can then only bring about slow variations in space and time of this basic state.

Mathematically this fact came in by our obtaining the hydrodynamic modes as *perturbations* of the collisional invariants. In a *collisionless system* the eigenvalue analysis would have been completely different. We would no longer be entitled to take the homogeneous eigenfunctions ϕ_n as a basis for a continuation into a ψ_n. Rather, we should attack directly the eigenvalue problem†

$$-ikv\psi_\nu(v) = \lambda_\nu \psi_\nu(v) \tag{13.3.35}$$

The solution of this problem is very simple:

$$\begin{aligned} &\lambda_\nu = -ik\nu \\ &\psi_\nu(v) = \delta(v-\nu), \qquad -\infty < \nu < +\infty \end{aligned} \tag{13.3.36}$$

The spectrum is quite different. Instead of having at least five discrete eigenvalues as in the hydrodynamical case, we now have a continuous spectrum for every value of k. Moreover, the eigenfunctions are singular Dirac functions.

If we consider an initial function $\chi_k(v; 0)$, represented as

$$\chi_k(v; 0) = \int d\nu \, c_k(\nu) \, \delta(\nu - v) \tag{13.3.37}$$

* See also the discussion at the end of Section 12.2.

† We now drop the subscript x, considering only motions in one dimension.

its value at time t is

$$\chi_k(v;t) = \int d\nu\, c_k(\nu) e^{-ik\nu t}\, \delta(\nu - v)$$
$$= \chi_k(v;0) e^{-ikvt} \tag{13.3.38}$$

The corresponding perturbation in physical space is

$$\chi(q, v; t) = \int dk\, e^{ikq - ikvt} \chi_k(v;0)$$
$$= \chi(q - vt, v; 0) \tag{13.3.39}$$

Each particle carries away the perturbation with its own individual velocity. As a result, an initial coherent fluctuation will rapidly be torn apart. This can be seen by calculating the local density at point q:

$$\chi(q;t) = \int dv\, \chi(q, v; t)$$
$$= \int dk\, e^{ikq} \int dv\, e^{-ikvt} \chi_k(v;0) \tag{13.3.40}$$

By a well-known theorem of Riemann and Lebesgue, the v integral goes to zero as $t \to \infty$. Hence, $\chi(q;t) \to 0$ in a characteristic time depending on the spread of velocities in the initial fluctuations $\chi_k(v;0)$. This disappearance cannot be described as a dissipation in the hydrodynamical sense, because its time scale is not fixed by an intrinsic property of the system independent of the initial condition, such as a transport coefficient. It is a typical example of a "phase-mixing" process, as defined in Section 12.2.

To appreciate the difference with a hydrodynamical behavior we compare this situation with a sound mode, in which

$$\chi_k(v;t) = \chi_k(v;0) e^{-ikU_s t} \tag{13.3.41}$$

The sound velocity is now an intrinsic constant, independent of the velocities of the particles. Hence, instead of Eq. (13.3.40), we now have

$$\chi(q;t) = \int dk\, e^{ikq - ikU_s t} \int dv\, \chi_k(v;0)$$
$$= \chi(q - U_s t; 0) \tag{13.3.42}$$

In this case the initial density disturbance is propagated without distortion at the speed of sound. We have here a truly collective motion. We thus clearly see that only the dominant effect of the collisions can explain the organized hydrodynamic behavior of a gas.

13.4. THE MICROSCOPIC EXPRESSIONS OF THE TRANSPORT COEFFICIENTS

Having established the one-to-one correspondence between kinetic and hydrodynamic modes, we now proceed to the evaluation of the transport coefficients. Identifying the coefficients of k^2 in (13.3.33) and in Λ_3 we get a microscopic expression for the shear viscosity:

$$\eta = -\frac{m^2 n}{k_B T} \int d\mathbf{v}\, \varphi^0(v)\, v_x v_y \frac{1}{K} v_x v_y \tag{13.4.1}$$

In this expression we took the limit $\varepsilon \to 0$ by just dropping the ε in the denominator. Indeed, it is easily checked that the function $v_x v_y$ is orthogonal to all the five eigenvectors of K corresponding to the eigenvalue zero. (A shorter way of expressing this is: "$v_x v_y$ has no component in the null space of K.")

$$(\phi_\alpha, v_x v_y) = \int d\mathbf{v}\, \varphi^0(v)\, \phi_\alpha(\mathbf{v})\, v_x v_y = 0, \qquad \alpha = 1, \ldots, 5 \tag{13.4.2}$$

It then follows that $K v_x v_y \neq 0$ and $K^{-1} v_x v_y \neq \infty$; the expression in the right-hand side of (13.3.33) is, therefore, convergent for $\varepsilon \to 0$.

Going over to λ_5, we obtain

$$\lambda_5 = \tfrac{2}{5} k^2 \lim_{\varepsilon \to 0} \int d\mathbf{v}\, \varphi^0(v) \left[-1 + \frac{1}{2}\left(\frac{mv^2}{k_B T}\right) - \frac{3}{2}\right] v_x \frac{1}{K - \varepsilon} v_x \left[-1 + \frac{1}{2}\left(\frac{mv^2}{k_B T}\right) - \frac{3}{2}\right] \tag{13.4.3}$$

Here it is no longer immediately obvious that the function acted upon by the operator $(K - \varepsilon)^{-1}$ has no component in the null space. Indeed, the term $[-(5/2)v_x]$ is proportional to $\phi_2(\mathbf{v})$. But the function $(m/2k_B T) v_x v^2$ also has a component along $\phi_2(\mathbf{v})$, which, as it happens, exactly cancels the former. We may check by direct integration that

$$\int d\mathbf{v}\, \varphi^0(v)\, \phi_\alpha(\mathbf{v}) \left[\left(\frac{m}{2k_B T}\right) v_x v^2 - \tfrac{5}{2} v_x\right] = 0, \qquad \alpha = 1, \ldots, 5 \tag{13.4.4}$$

We may thus again drop the ε, and obtain

$$\lambda_5 = \frac{2}{5} \frac{k^2}{(k_B T)^2} \int d\mathbf{v}\, \varphi^0(v)\, (\tfrac{1}{2} m v^2 - \tfrac{5}{2} k_B T) v_x \frac{1}{K} (\tfrac{1}{2} m v^2 - \tfrac{5}{2} k_B T) v_x \tag{13.4.5}$$

We know, however, from equilibrium statistical mechanics that the specific heat at constant pressure (per unit mass) for an ideal gas is

$$c_P = \frac{5}{2} \frac{k_B}{m} \tag{13.4.6}$$

It then follows from a comparison with Λ_5 that the microscopic expression for the heat conductivity is

$$\kappa = -\frac{n}{k_B T^2} \int d\mathbf{v}\, \varphi^0(v)\, (\tfrac{1}{2}mv^2 - \tfrac{5}{2}k_B T) v_x \frac{1}{K} v_x (\tfrac{1}{2}mv^2 - \tfrac{5}{2}k_B T) \quad (13.4.7)$$

By a similar, somewhat more involved calculation, one obtains from a comparison of Λ_1, λ_1, η, and κ the result

$$\zeta = 0 \quad (13.4.8)$$

This is a well-known result: the bulk viscosity of a weakly nonideal gas is zero. This transport coefficient appears only in higher orders of approximation.

Equations (13.4.1) and (13.4.7) conclude the formal part of the theory of transport coefficients. They are very compact expressions providing a beautiful link between kinetic theory and macroscopic physics. In order to do explicit calculations, these equations are written in a somewhat different form:

$$\eta = -\frac{m^2 n}{k_B T} \int d\mathbf{v}\, \varphi^0(v)\, v_x v_y\, \chi_{xy} \quad (13.4.9)$$

$$\kappa = -\frac{n}{k_B T^2} \int d\mathbf{v}\, \varphi^0(v)\, (\tfrac{1}{2}mv^2 - \tfrac{5}{2}k_B T) v_x\, \chi_x \quad (13.4.10)$$

Comparing these equations to (13.4.1) and (13.4.7), we immediately see that the functions χ_{xy}, χ_x must satisfy the following inhomogeneous kinetic equations:

$$K\chi_{xy} = v_x v_y \quad (13.4.11)$$

$$K\chi_x = (\tfrac{1}{2}mv^2 - \tfrac{5}{2}k_B T) v_x \quad (13.4.12)$$

The actual calculation of the transport coefficients is thus reduced to the solution of two linear integrodifferential equations. These equations, (13.4.9)–(13.4.12) have been known for a long time. They had first been derived by Chapman and Enskog* around 1916. However, their original theory (which has been the basis of kinetic theory since that time) is much more involved than the present derivation. Moreover, its foundations, although quite reasonable for weakly nonideal systems, are very difficult to extend beyond their range of validity. The present theory of Résibois does not suffer from these weaknesses: it can be extended to cover systems of arbitrary density, and type of interactions.

* In the Chapman–Enskog theory the term $-(5/2)k_B T$ is not present in Eq. (13.4.10). It can be shown pretty easily that this term actually contributes zero to κ. The details of this calculation shall not be given here, but the fact can be checked in a particular case studied in Section 13.5.

The expressions obtained for the transport coefficients, such as the viscosity, can also be interpreted as follows. Consider an evolution operator $\exp(Kt)$ associated with the linearized kinetic equation (13.1.6), such that its solution is expressed as

$$\chi(\mathbf{v}; t) = \exp(Kt)\chi(\mathbf{v}; 0) \tag{13.4.13}$$

(see Section 2.2 for the analogous operator of the Liouville equation). We also define an adjoint operator $K^{\dagger}$ that describes the evolution of the dynamical functions $b(\mathbf{v})$ in a "Heisenberg representation" in this approximation.

As we know that the collision operator is self-adjoint, $K^{\dagger} = K$, and we have

$$b(\mathbf{v}; t) = \exp(Kt) b(\mathbf{v}; 0) \tag{13.4.14}$$

Consider now the particular dynamical function

$$\mathrm{j}_{xy}(\mathbf{v}) = m v_x v_y \tag{13.4.15}$$

We rewrite the viscosity coefficient (13.4.1) as follows:

$$\begin{aligned}\eta &= -m^2 n\beta \int d\mathbf{v}\, \varphi^0(v)\, v_x v_y\, K^{-1}\, v_x v_y \\ &= n\beta \int d\mathbf{v}\, \varphi^0(v) \int_0^{\infty} d\tau\, \mathrm{j}_{xy}(\mathbf{v})\, e^{K\tau}\, \mathrm{j}_{xy}(\mathbf{v})\end{aligned}$$

The upper limit contributes zero because K is a semidefinite negative operator, and $v_x v_y$ is not an eigenvector corresponding to its eigenvalue zero. Hence we may further write

$$\eta = n\beta \int_0^{\infty} d\tau \int d\mathbf{v}\, \varphi^0(v)\, \mathrm{j}_{xy}(\mathbf{v})\, \mathrm{j}_{xy}(\mathbf{v}; \tau) \tag{13.4.16}$$

or, even more briefly

$$\eta = n\beta \int_0^{\infty} d\tau\, \langle \mathrm{j}_{xy}(\mathbf{v})\, \mathrm{j}_{xy}(\mathbf{v}; \tau)\rangle^0 \tag{13.4.17}$$

where $\langle\ \rangle^0$ denotes an average computed with the equilibrium distribution. The particular average appearing here is of an especially important type: it is the average of a product of the values of the same dynamical function at two different times. Such an expression is called an *autocorrelation function.* It is a measure of the degree of memory of the system. We shall devote a whole chapter to the more systematic study of these functions. Suffice it at present to formulate our result as follows: *the viscosity is proportional to the time integral of the equilibrium autocorrelation of the momentum flow.* This result will be seen to be of a great generality: *all* transport coefficients can be expressed as time integrals of autocorrelation functions of appropriate flows: these expressions are called the *Green–Kubo formulas.* They will be studied in detail in Chapter 21.

The expressions obtained here for the transport coefficients are still formal. In order to compute explicit values of these quantities, we must solve Eqs. (13.4.11) and (13.4.12). In the next section we illustrate the method of solution in a simple case.

13.5. EXPLICIT CALCULATION OF A TRANSPORT COEFFICIENT

We now pursue the calculation of the thermal conductivity by showing how Eq. (13.4.12) can be effectively solved. The calculation of the viscosity is completely similar and will not be considered here.

The method of solution is pretty standard. The unknown function χ_x is expanded into a series of properly chosen orthogonal polynomials: As a result, the equation is transformed into an (infinite) set of algebraic equations for the coefficients of this series. Approximate solutions to this set are obtained by truncating it at various levels.

The appropriate orthogonal polynomials to be used here are the *Sonine polynomials* $S_m^n(x)$. They are defined as follows:

$$S_m^n(x) = \sum_{p=0}^{n} (-1)^p \frac{\Gamma(m+n+1)}{\Gamma(m+p+1)(n-p)!\,p!} x^p \tag{13.5.1}$$

where x and m are real numbers, and n is an integer; $\Gamma(x)$ is Euler's gamma-function. In particular:

$$\begin{aligned} S_m^0(x) &= 1 \\ S_m^1(x) &= m+1-x \end{aligned} \tag{13.5.2}$$

The Sonine polynomials have the following orthogonality property:

$$\int_0^\infty dx\, e^{-x} x^m S_m^p(x) S_m^q(x) = \frac{\Gamma(m+p+1)}{p!} \delta_{p,q} \tag{13.5.3}$$

The natural variable to be used in our problem is the dimensionless velocity:

$$\mathbf{u} = (\tfrac{1}{2}\beta m)^{1/2} \mathbf{v} \tag{13.5.4}$$

We assume that the vector χ_x is expanded as follows*:

$$\chi_x = \sum_{r=1}^{\infty} a_r\, S_{3/2}^r(u^2)\, u_x \tag{13.5.5}$$

* Note that $S_{3/2}^0(u^2)u_x \equiv u_x$ is a collisional invariant. It then follows from Eqs. (13.4.4) and (13.4.5) that the term with $r=0$ does not contribute to the heat conductivity. We may therefore set $a_0 = 0$.

where a_r are constants, to be determined. Substituting this expansion into Eq. (13.4.12) (properly transformed to dimensionless variables), multiplying both sides by $e^{-u^2}S^s_{3/2}(u^2)u_x$, and integrating over $\mathbf{u}$ we obtain a system of linear algebraic equations for the coefficients a_r:

$$\sum_{r=1}^{\infty} A_{sr}a_r = \alpha_s \tag{13.5.6}$$

Here A_{sr} are the matrix elements of the operator K in the "Sonine representation":

$$A_{sr} = \int d\mathbf{u}\, e^{-u^2} S^s_{3/2}(u^2)u_x\, K\, S^r_{3/2}(u^2)u_x \tag{13.5.7}$$

The coefficients α_s can be easily evaluated (in spherical coordinates):

$$\begin{aligned}\alpha_s &= \beta^{-1}\int d\mathbf{u}\, e^{-u^2} S^s_{3/2}(u^2)u_x^2(u^2-\tfrac{5}{2})\\ &= \frac{4\pi}{3}\beta^{-1}\int_0^{\infty} du\, e^{-u^2} S^s_{3/2}(u^2)u^4(u^2-\tfrac{5}{2})\\ &= \frac{2\pi}{3}\beta^{-1}\int_0^{\infty} dy\, e^{-y} S^s_{3/2}(y)y^{3/2}(y-\tfrac{5}{2})\end{aligned} \tag{13.5.8}$$

In order to use Eq. (13.5.3) we express y in terms of Sonine polynomials. It is immediately seen from Eq. (13.5.2) that

$$y-\tfrac{5}{2} = -S^1_{3/2}(y) \tag{13.5.9}$$

We now get from Eqs. (13.5.3), (13.5.8), and (13.5.9)

$$\begin{aligned}\alpha_s &= \frac{2\pi}{3}\beta^{-1}[-\Gamma(\tfrac{7}{2})\,\delta_{s,1}]\\ &= -\tfrac{5}{4}\pi^{1/2}\beta^{-1}\delta_{s,1}\end{aligned} \tag{13.5.10}$$

Equation (13.5.6) becomes

$$\sum_{r=1}^{\infty} A_{sr}a_r = \tfrac{5}{4}\pi^{1/2}\beta^{-1}\delta_{s,1} \tag{13.5.11}$$

We know that $S^0_{3/2}(u^2)u_x = u_x$ is a collisional invariant. It then follows from Eq. (13.5.7) that

$$A_{0r} = A_{r0} = 0$$

Therefore, the first equation in (13.5.11), for $s = 0$, reduces to $0 = 0$.

We now agree to retain, as a first approximation, only one term in the series (13.5.5), namely, $a_1 \neq 0$, $a_r = 0$ for $r > 1$. The set (13.5.11) then reduces to a single equation from which we immediately obtain

$$a_1 \approx -\tfrac{5}{4}\pi^{1/2}(\beta A_{11})^{-1} \tag{13.5.12}$$

and, from (13.5.5) and (13.5.2):

$$\chi_x = -\tfrac{5}{4}\pi^{1/2}(\beta A_{11})^{-1}(u^2 - \tfrac{5}{2})u_x$$

Substituting this result into Eq. (13.4.10), we obtain for the thermal conductivity*

$$\begin{aligned}\kappa &= -\frac{n}{k_B T^2}\left(\frac{2k_B T}{m}\right)^{2/2} k_B T \pi^{-3/2} \int d\mathbf{u}\, e^{-u^2}(u^2 - \tfrac{5}{2})u_x \\ &\quad\times \tfrac{5}{4}\pi^{1/2}\left(\frac{k_B T}{A_{11}}\right)(u^2 - \tfrac{5}{2})u_x \\ &= -nk_B\left(\frac{2k_B T}{m}\right)\tfrac{25}{16}\pi^{1/2}A_{11}^{-1}\end{aligned} \tag{13.5.13}$$

We next evaluate the matrix element A_{11} which, for the Landau equation, can be done completely for every interaction potential. From the definition (13.5.7) and from Eq. (13.1.6), we obtain (after the usual symmetrization)

$$\begin{aligned}A_{11} &= \pi^{-3}Bnm^{-2}\left(\frac{m}{2k_B T}\right)^{3/2}\int d\mathbf{u}_1\, d\mathbf{u}_2\, e^{-u_1^2-u_2^2} \\ &\quad\times \tfrac{1}{2}[S_{3/2}^1(u_1^2)u_{1x} + S_{3/2}^1(u_2^2)u_{2x}]\,\partial_{12}^r(g^2\delta^{rs} - g^r g^s)g^{-3}\partial_{12}^s \\ &\quad\times [S_{3/2}^1(u_1^2)u_{1x} + S_{3/2}^1(u_2^2)u_{2x}] \\ &= -\pi^{-3}Bnm^{-2}\left(\frac{m}{2k_B T}\right)^{3/2}\frac{1}{2}\int d\mathbf{u}_1\, d\mathbf{u}_2\, e^{-u_1^2-u_2^2}(g^2\delta^{rs} - g^r g^s)g^{-3} \\ &\quad\times \{\partial_{12}^r[S_{3/2}^1(u_1^2)u_{1x} + \cdots]\}\{\partial_{12}^s[S_{3/2}^1(u_1^2)u_{1x} + \cdots]\} \\ &= -\pi^{-3}Bnm^{-2}\left(\frac{m}{2k_B T}\right)^{3/2}\left(\frac{1}{2}\right)\left(\frac{1}{3}\right)\int d\mathbf{u}_1\, d\mathbf{u}_2\, e^{-u_1^2-u_2^2}(g^2\delta^{rs} - g^r g^s)g^{-3} \\ &\quad\times \{\partial_{12}^r[S_{3/2}^1(u_1^2)u_{1n} + S_{3/2}^1(u_2^2)u_{2n}]\} \\ &\quad\times \{\partial_{12}^s[S_{3/2}^1(u_1^2)u_{1n} + S_{3/2}^1(u_2^2)u_{2n}]\}\end{aligned} \tag{13.5.14}$$

The second step results from an integration by parts; the third step results from isotropy: the value of the $x-x$ component of the tensor A_{11} equals $\frac{1}{3}$ of its trace. We now note

$$\begin{aligned}&\partial_{12}^s[S_{3/2}^1(u_1^2)u_{1n} + S_{3/2}^1(u_2^2)u_{2n}] \\ &\qquad = \partial_{12}^s[\tfrac{5}{2}u_{1n} - u_1^2 u_{1n} + \tfrac{5}{2}u_{2n} - u_2^2 u_{2n}] \\ &\qquad = -2u_{1s}u_{1n} + (\tfrac{5}{2} - u_1^2)\delta_{sn} + 2u_{2s}u_{2n} - (\tfrac{5}{2} - u_2^2)\delta_{sn}\end{aligned} \tag{13.5.15}$$

* It can be checked here that, dropping the term $(-\frac{5}{2})$ in the first bracket under the integral does not change the value of κ, as announced in Section 13.4.

After some algebra, using the fact that $\mathbf{g}=\mathbf{u}_1-\mathbf{u}_2$, we obtain

$$A_{11}=-D\int d\mathbf{u}\int d\mathbf{w}\, e^{-u^2-w^2}g^{-3}\{g^2(10u^4-12u^2w^2+10w^4)$$
$$+(-8u^6+4u^4w^2+4u^2w^4-8w^6)+\cos\theta\,(16u^5w-24u^3w^3+16w^5u)$$
$$+\cos^2\theta\,(4u^2w^4+4u^4w^2)-8u^3w^3\cos^3\theta-8g^2\cos^2\theta\, u^2w^2\} \tag{13.5.16}$$

In order to avoid heavy subscripts we now called $\mathbf{u}_1\equiv\mathbf{u}$, $\mathbf{u}_2\equiv\mathbf{w}$; θ is the angle between $\mathbf{u}$ and $\mathbf{w}$ ($\mathbf{u}\cdot\mathbf{w}=uw\cos\theta$). We also set

$$D=Bnm^{-2}(6\pi^3)^{-1}\left(\frac{m}{2k_BT}\right)^{3/2}$$

We do the $\mathbf{u}$ integration first, introducing polar coordinates in a reference frame, the z axis of which coincides with the vector $\mathbf{w}$. The apparent difficulty is in the complicated θ dependence introduced by the factor g^{-3}. It can, however, be overcome by choosing g instead of $\cos\theta$ as an integration variable:

$$\cos\theta=\frac{w^2+u^2-g^2}{2uw}$$
$$d\cos\theta=-\left(\frac{g}{uw}\right)dg \tag{13.5.17}$$

We now obtain

$$A_{11}=-D\int d\mathbf{w}\, e^{-w^2}J(w) \tag{13.5.18}$$

with

$$J(w)=2\pi\int_0^\infty du\, u^2e^{-u^2}\int_{|w-u|}^{w+u} dg(uwg^2)^{-1}$$
$$\times\{g^2(u^4-2w^2u^2+w^4)+2g^4(u^2+w^2)-g^6\}$$

The g integration is quite elementary:

$$\int_{|w-u|}^{w+u} dg(uwg^2)^{-1}\{\cdots\}=4u(u^2-w^2)^2,\qquad u<w$$
$$=4w(u^2-w^2)^2,\qquad u>w \tag{13.5.19}$$

Hence,

$$J(w)=8\pi\left\{w^{-1}\int_0^w du\, e^{-u^2}u^2(u^2-w^2)^2+\int_w^\infty du\, e^{-u^2}u(u^2-w^2)^2\right\} \tag{13.5.20}$$

Substituting this result into (13.5.18) we obtain, through a series of elementary steps:

$$A_{11} = -D(4\pi)(8\pi)\left\{\int_0^\infty dw\, e^{-w^2} w \int_0^w du\, e^{-u^2} u^2 (u^2 - w^2)^2 + \int_0^\infty dw\, e^{-w^2} w^2 \int_w^\infty du\, e^{-u^2} u (u^2 - w^2)^2\right\}$$

$$= -D(4\pi)(8\pi)2 \int_0^\infty dw\, e^{-w^2} w^2 \int_w^\infty du\, e^{-u^2} (u^5 - 2u^3 w^2 + u w^4)$$

$$= -D(4\pi)(8\pi)2 \int_0^\infty dw\, e^{-2w^2} w^2$$

$$= -8\pi^2 D \left(\frac{\pi}{2}\right)^{1/2}$$

$$= -\frac{1}{3\pi^{1/2}} B \frac{n}{m^{1/2}} \frac{1}{(k_B T)^{3/2}} \tag{13.5.21}$$

And the final expression for the thermal conductivity, from (13.5.13), is

$$\kappa = \tfrac{75}{8}\pi k_B \frac{(k_B T)^{5/2}}{m^{1/2} B} \tag{13.5.22}$$

The transport coefficient is thus positive definite, as it should be in any stable situation in order to avoid conflict with the second law of thermodynamics [see Eq. (12.5.15)]. We also note the important fact that if $B = 0$ the thermal conductivity is infinite. This confirms our earlier statement that transport theory depends crucially on the interactions, which are the physical source of the resistance to the flows and hence of the dissipation. An ideal gas has infinite transport coefficients. We also note that the conductivity is a function of temperature, but is independent of the density. The latter fact is a consequence of the Landau (and also of the Boltzmann) approximation. One would, therefore, be tempted to perform a series expansion in powers of the density, just as we did for the equation of state in equilibrium. Such a program involves the derivation of increasingly accurate kinetic equations, going beyond the Landau or the Boltzmann approximation. In the forthcoming chapters, a general framework will be developed that allows us to deal with these and similar other problems in a quite straightforward manner. This does not mean, of course, that every problem is solved and that we shall find no surprises on the way. The hypothetical "virial expansion" of the transport coefficients will lead us to one of these surprises, as will be seen later on. We are confident, however, in our present ability of constructing at least a guiding path to which we can stick in advancing through the forest.

13.6. EIGENVALUES OF THE LINEARIZED KINETIC EQUATION FOR AN INHOMOGENEOUS PLASMA

We now consider the application of Résibois' method to the problem of an electron plasma. The main difference here is the presence of the *Vlassov term* in the kinetic equation. On the basis of *a priori* intuition one may argue that the Vlassov term, being purely reversible (as we know from Section 12.2) should play no important role in the determination of the transport coefficients, which are a measure of the dissipation. On the other hand, we know from the macroscopic theory of Section 12.7 that the self-consistent Coulomb field deeply alters the spectrum of normal modes. We shall now see that both these statements, although apparently conflicting, are actually true.

We start from the Vlassov–Landau equation (11.7.3), which we linearize with the ansatz (13.3.1):

$$\partial_t \chi(\mathbf{v}_1, \mathbf{q}_1; t) = -\mathbf{v}_1 \cdot \nabla_1 \chi(\mathbf{v}_1, \mathbf{q}_1; t) + m^{-1}[\varphi^0(v_1)]^{-1}[\partial_1 \varphi^0(v_1)] \cdot \nabla_1 \int d\mathbf{v}_2\, d\mathbf{q}_2\, V(\mathbf{q}_1 - \mathbf{q}_2)\, n\varphi^0(v_2)\, \chi(\mathbf{q}_2, \mathbf{v}_2; t) + K\chi \quad (13.6.1)$$

We now Fourier-transform the deviation χ according to (13.3.3) and also the Coulomb potential according to (6.5.3). Using the convolution theorem for Fourier transforms we obtain the basic linearized Vlassov–Landau equation, which is our starting point.

$$\partial_t \chi_{\mathbf{k}}(\mathbf{v}; t) = -i\mathbf{k} \cdot \mathbf{v}\chi_{\mathbf{k}}(\mathbf{v}; t) - m\beta i\mathbf{k} \cdot \mathbf{v}\left(\frac{\omega_p^2}{k^2}\right) \int d\mathbf{v}'\, \varphi^0(v')\, \chi_{\mathbf{k}}(\mathbf{v}'; t) + K\chi_{\mathbf{k}} \quad (13.6.2)$$

where ω_p is the plasma frequency, Eq. (12.7.5). The Vlassov term thus involves a pure integral operator. As compared with Eq. (13.3.4), this equation possesses two new features.

A. The Vlassov operator is no longer a symmetric, that is, self-adjoint operator. We must therefore be careful in treating the eigenvalue problem: We have to calculate separately its eigenvectors and the eigenvectors of the adjoint operator (i.e., kets and bras, respectively): these are no longer the same.

B. There appears the following difficulty, which is responsible for the unique features of the plasma. Both the flow term and the Vlassov term are different from zero only if the system is inhomogeneous. We are thus tempted to treat these two terms as a perturbation to the homogeneous equation, just as in Section 13.3. However, it is clear that the Vlassov term is *singular* as $k \to 0$. It would seem, therefore, that the method of

Résibois cannot be applied to this problem. The surprising fact, however, is that the procedure still works, provided we are careful enough! The reason therefore is that the eigenvalues of the operator (13.6.2) are regular for $k \to 0$.

In order to avoid doing too unorthodox operations, we will start by studying a related problem, just as in equilibrium, in Section 6.5. We replace the true Coulomb potential by the modified potential $\tilde{V}^{(\gamma)}(k)$ defined in Eq. (6.5.19). We then define the solution to the real problem (13.6.2) as the limit of the artificial one as $\gamma \to 0$. The limit is to be taken at an appropriate stage in the calculation.

We now assume again that $\mathbf{k} = k \cdot \mathbf{1}_x$ and formulate the eigenvalue problem for the modified kinetic equation (13.6.2) as follows:

$$[K - ikv_x - i\Omega^{(\gamma)}]\psi_n^{(\gamma)}(\mathbf{v}) = \lambda_n^{(\gamma)}\psi_n^{(\gamma)}(\mathbf{v}) \qquad (13.6.3)$$

where the (modified) Vlassov operator $\Omega^{(\gamma)}$ is defined as follows:

$$\Omega^{(\gamma)}\chi(\mathbf{v}) = m\beta k v_x \frac{\omega_p^2}{k^2 + \gamma^2} \int d\mathbf{v}' \, \varphi^0(v')\chi(\mathbf{v}') \qquad (13.6.4)$$

The new operator tends to zero as $k \to 0$, hence we are allowed to do perturbation theory as in Section 13.3. We define the eigenvalues λ_n of the real Coulomb problem as

$$\lambda_n = \lim_{\gamma \to 0} \lambda_n^{(\gamma)} \qquad (13.6.5)$$

We now proceed exactly as in Section 13.3. We first define a proper basis for perturbation theory by solving exactly the eigenvalue equation in the subspace spanned by the five collisional invariants, in other words, we define

$$\bar{\phi}_\alpha^{(\gamma)} = \sum_{\alpha'=1}^{5} c_{\alpha\alpha'}^{(\gamma)} \phi_{\alpha'} \qquad (13.6.6)$$

and write for the coefficients the following equations:

$$-i\sum_{\alpha''=1}^{5} (W_{\alpha'\alpha''} + \Omega_{\alpha'\alpha''}^{(\gamma)}) c_{\alpha\alpha''}^{(\gamma)} = \bar{\lambda}_\alpha^{(\gamma)} c_{\alpha\alpha'}^{(\gamma)}, \quad \alpha = 1, \ldots, 5, \quad \alpha' = 1, \ldots, 5 \qquad (13.6.7)$$

where $W_{\alpha\alpha'}$ is defined by Eq. (13.3.14), and

$$\begin{aligned} \Omega_{\alpha\alpha'}^{(\gamma)} &= \int d\mathbf{v} \, \varphi^0(v) \, \phi_\alpha(\mathbf{v}) \, \Omega^{(\gamma)} \, \phi_{\alpha'}(\mathbf{v}) \\ &= m\beta k \omega_p^2 k_{(\gamma)}^{-2} \int d\mathbf{v} \, \varphi^0(v) \, \phi_\alpha(\mathbf{v}) v_x \int d\mathbf{v}' \, \varphi^0(v') \, \phi_{\alpha'}(\mathbf{v}') \\ &= \delta_{\alpha,2}\delta_{\alpha',1} \, k\omega_p^2 k_{(\gamma)}^{-2} \, (m\beta)^{1/2} \end{aligned} \qquad (13.6.8)$$

where we introduced the useful abbreviation

$$k_{(\gamma)} = (k^2 + \gamma^2)^{1/2} \tag{13.6.9}$$

The eigenvalues are now determined just as in Section 13.3, by solving the characteristic determinant. One finds the following dispersion equation:

$$\bar{\lambda}^{(\gamma)3}\left\{\bar{\lambda}^{(\gamma)2} + k^2\bar{u}^2\left(\frac{5}{3} + \frac{\omega_p^2}{k_{(\gamma)}^2\bar{u}^2}\right)\right\} = 0 \tag{13.6.10}$$

from which we obtain the eigenvalues:

$$\begin{aligned} \bar{\lambda}_1^{(\gamma)} &= -\bar{\lambda}_2^{(\gamma)} = i\{\omega_p^2 + \tfrac{5}{3}(k^2+\gamma^2)\bar{u}^2\}^{1/2}\frac{k}{(k^2+\gamma^2)^{1/2}} \\ \bar{\lambda}_3^{(\gamma)} &= \bar{\lambda}_4^{(\gamma)} = \bar{\lambda}_5^{(\gamma)} = 0 \end{aligned} \tag{13.6.11}$$

These values will be discussed below. We first complete the solution by determining the eigenfunctions from Eqs. (13.6.6):

$$\begin{aligned} \bar{\phi}_{\substack{1\\2}} &= c_{15}\left[\left(\frac{3}{2}\right)^{1/2}\phi_1 \mp \left(\frac{3}{2}\right)^{1/2}\frac{(\omega_p^2 + \frac{5}{3}k_{(\gamma)}^2\bar{u}^2)^{1/2}}{k_{(\gamma)}\bar{u}}\phi_2 + \phi_5\right] \\ \bar{\phi}_{\substack{3\\4}} &= \phi_{\substack{3\\4}} \\ \bar{\phi}_5 &= c_{55}\left[-\left(\frac{2}{3}\right)^{1/2}\frac{k_{(\gamma)}^2\bar{u}^2}{\omega_p^2 + k_{(\gamma)}^2\bar{u}^2}\phi_1 + \phi_5\right] \end{aligned} \tag{13.6.12}$$

where c_{15} and c_{55} are as yet arbitrary constants. These functions are not normalized. In order to define orthonormalization properties, we must find the eigenfunctions of the *adjoint Vlassov equation:*

$$[K - ikv_x - i\Omega^{\dagger(\gamma)}]\tilde{\psi}_n^{(\gamma)} = \lambda_n^{(\gamma)}\psi_n^{(\gamma)} \tag{13.6.13}$$

with

$$\Omega^{\dagger(\gamma)}\chi(\mathbf{v}) = m\beta\frac{\omega_p^2}{k^2+\gamma^2}\int d\mathbf{v}'\,\varphi^0(v')\,kv_x'\,\chi(\mathbf{v}') \tag{13.6.14}$$

It is indeed easily checked that, with the scalar product (13.1.7) we have

$$(\chi_A, \Omega^{(\gamma)}\chi_B) = (\Omega^{\dagger(\gamma)}\chi_A, \chi_B) \tag{13.6.15}$$

Proceeding as before, we find for the adjoint equation the same eigenvalues (of course) and the following adjoint eigenfunctions:

$$\begin{aligned} \tilde{\phi}_{\substack{1\\2}} &= c_{15}\left[\left(\frac{3}{2}\right)^{1/2}\left(1 + \frac{\omega_p^2}{k_{(\gamma)}^2\bar{u}^2}\right)\phi_1 \mp \left(\frac{3}{2}\right)^{1/2}\frac{\omega_p^2 + \frac{5}{2}k_{(\gamma)}^2\bar{u}^2}{k_{(\gamma)}\bar{u}}\phi_2 + \phi_5\right] \\ \tilde{\phi}_{\substack{3\\4}} &= \phi_{\substack{3\\4}} \\ \tilde{\phi}_5 &= c_{55}[-(\tfrac{2}{3})^{1/2}\phi_1 + \phi_5] \end{aligned} \tag{13.6.16}$$

We may easily check that these functions satisfy the orthogonality property:

$$(\tilde{\phi}_\alpha, \bar{\phi}_{\alpha'}) = 0, \qquad \alpha \neq \alpha' \tag{13.6.17}$$

We now determine the remaining constants c_{15} and c_{55} by imposing the normalization:

$$(\tilde{\phi}_\alpha, \bar{\phi}_\alpha) = 1 \tag{13.6.18}$$

We then obtain

$$\begin{aligned}
\bar{\phi}_{\substack{1\\2}} &= \left\{ \frac{1}{2^{1/2}} \frac{k_{(\gamma)}\bar{u}}{(\omega_p^2 + \frac{5}{3}k_{(\gamma)}^2\bar{u}^2)^{1/2}}\,\phi_1 \mp \frac{1}{2^{1/2}}\,\phi_2 + \frac{1}{3^{1/2}} \frac{k_{(\gamma)}\bar{u}}{(\omega_p^2 + \frac{5}{2}k_{(\gamma)}^2\bar{u}^2)^{1/2}}\,\phi_5 \right\} \\
\bar{\phi}_{\substack{3\\4}} &= \phi_{\substack{3\\4}} \qquad\qquad\qquad\qquad\qquad\qquad\qquad\qquad\qquad\qquad (13.6.19) \\
\bar{\phi}_5 &= \left\{ -\left(\frac{2}{3}\right)^{1/2} \frac{k_{(\gamma)}^2\bar{u}^2}{[(\omega_p^2 + k_{(\gamma)}^2\bar{u}^2)(\omega_p^2 + \frac{5}{3}k_{(\gamma)}^2\bar{u}^2)]^{1/2}}\,\phi_1 + \frac{(\omega_p^2 + k_{(\gamma)}^2\bar{u}^2)^{1/2}}{(\omega_p^2 + \frac{5}{3}k_{(\gamma)}^2\bar{u}^2)^{1/2}}\,\phi_5 \right\}
\end{aligned}$$

and the adjoint eigenfunctions

$$\begin{aligned}
\tilde{\phi}_{\substack{1\\2}} &= \left\{ \frac{1}{2^{1/2}} \frac{\omega_p^2 + k_{(\gamma)}^2\bar{u}^2}{k_{(\gamma)}\bar{u}(\omega_p^2 + \frac{5}{3}k_{(\gamma)}^2\bar{u}^2)^{1/2}}\,\phi_1 \mp \frac{1}{2^{1/2}}\,\phi_2 + \frac{1}{3^{1/2}} \frac{k_{(\gamma)}\bar{u}}{(\omega_p^2 + \frac{5}{3}k_{(\gamma)}^2\bar{u}^2)^{1/2}}\,\phi_5 \right\} \\
\tilde{\phi}_{\substack{3\\4}} &= \phi_{\substack{3\\4}} \qquad\qquad\qquad\qquad\qquad\qquad\qquad\qquad\qquad\qquad (13.6.20) \\
\tilde{\phi}_5 &= \left\{ -\left(\frac{2}{3}\right)^{1/2} \frac{(\omega_p^2 + k_{(\gamma)}^2\bar{u}^2)^{1/2}}{(\omega_p^2 + \frac{5}{3}k_{(\gamma)}^2\bar{u}^2)^{1/2}}\,\phi_1 + \frac{(\omega_p^2 + k_{(\gamma)}^2\bar{u}^2)^{1/2}}{(\omega_p^2 + \frac{5}{3}k_{(\gamma)}^2\bar{u}^2)^{1/2}}\,\phi_5 \right\}
\end{aligned}$$

Before going further, let us analyze the two nonzero eigenvalues in Eq. (13.6.11). First, assume $\gamma \neq 0$. Expanding $\bar{\lambda}_1^{(\gamma)}$ in powers of k, we find

$$\bar{\lambda}_1^{(\gamma)} = i\left\{ \tfrac{5}{3}\bar{u}^2 + \frac{\omega_p^2}{\gamma^2} \right\}^{1/2} k + O(k^3) \tag{13.6.21}$$

This is a typical sound mode, just as for an ordinary gas. The only difference with Eq. (13.3.18) lies in a correction of the sound velocity:

$$U_s \to U_s\left\{ 1 + \frac{\omega_p^2}{U_s^2\gamma^2} \right\}^{1/2}$$

It is easily seen that the following conclusion is very general:

If the Fourier transform of the potential $\tilde{V}_k$ is finite for $k \to 0$ [or, equivalently, if $V(r)$ is integrable], the gas described by the Vlassov-kinetic equation exhibits ordinary hydrodynamic behavior. The only effect of the Vlassov term is a modification of the sound velocity.

For the real plasma, however, the situation is radically different. Setting $\gamma = 0$ in Eq. (13.6.11), and then expanding in powers of k we find

$$\bar{\lambda}_1 = i(\omega_p^2 + \tfrac{5}{3}k^2\bar{u}^2)^{1/2} = i\omega_p\left(1 + \frac{5}{6}\frac{k^2\bar{u}^2}{\omega_p^2}\right) + O(k^4) \tag{13.6.22}$$

This is a typical *plasma mode*, corresponding exactly to the macroscopic mode found in Eq. (12.7.7). It is noteworthy that, in spite of the singular nature of the Vlassov term, we find finite eigenvalues for $k \to 0$. *The result of the Coulomb singularity, however, is to profoundly modify the nature of the eigenvalues* as compared to those of an ordinary fluid. *In particular, the reorganized eigenvalues do not tend to zero as* $k \to 0$.

We now go over to the expression of the eigenvalues to second order in perturbation theory. To do so we keep $\bar{\lambda}_\alpha^{(\gamma)}$, $\bar{\phi}_\alpha$, and $\tilde{\phi}_\alpha$ in their unexpanded form, (13.6.11), (13.6.19), and (13.6.20). The calculation is a trivial extension of the argument of Section 13.3, and we immediately find the following result, corresponding to Eq. (13.3.27):

$$\lambda_\alpha^{(\gamma)} = \lim_{\varepsilon\to 0} \langle \tilde{\phi}_\alpha^{(\gamma)} | (kv_x + \Omega^{(\gamma)} - i\bar{\lambda}_\alpha^{(\gamma)}) \frac{1}{K-\varepsilon} (kv_x + \Omega^{(\gamma)} - i\bar{\lambda}_\alpha^{(\gamma)}) | \bar{\phi}_\alpha^{(\gamma)} \rangle \quad (13.6.23)$$

Let us analyze these expressions, beginning with $\lambda_5^{(\gamma)}$: we show that it corresponds to the thermal mode. We start by letting $\gamma \to 0$. We then rewrite Eq. (13.6.23) as a scalar product, using Eq. (13.6.15):

$$\lambda_5 = \lim_{\varepsilon\to 0} ([kv_x + \Omega^\dagger]\tilde{\phi}_5, (K-\varepsilon)^{-1}[kv_x + \Omega]\bar{\phi}_5) \quad (13.6.24)$$

We now note

$$\Omega\bar{\phi}_5 = \left(\frac{\omega_p^2}{k^2\bar{u}^2}\right) kv_x \int d\mathbf{v}'\, \varphi^0(v')\, \bar{\phi}_5(v') = \left(\frac{\omega_p^2}{k^2\bar{u}^2}\right) \bar{A}\, kv_x \quad (13.6.25)$$

where $\bar{A}$ is the coefficient of ϕ_1 in Eq. (13.6.19) for $\bar{\phi}_5$. Similarly we find

$$\Omega^\dagger \tilde{\phi}_5 = \left(\frac{\omega_p^2}{k^2\bar{u}^2}\right) \int d\mathbf{v}'\, \varphi^0(v')\, kv'_x\, \tilde{\phi}_5(\mathbf{v}') = 0 \quad (13.6.26)$$

Hence

$$\lambda_5 = \lim_{\varepsilon\to 0} k^2 \left(\tilde{\phi}_5, v_x (K-\varepsilon)^{-1} v_x \left[\bar{\phi}_5 + \left(\frac{\omega_p^2}{k^2\bar{u}^2}\right)\bar{A} \right] \right) \quad (13.6.27)$$

We must now study the behavior of this function as $k \to 0$. If it has the right behavior of a thermal mode, the scalar product must tend to a nonzero constant. Substituting for $\tilde{\phi}_5$, $\bar{\phi}_5$, and $\bar{A}$ from Eqs. (13.6.19) and (13.6.20) as well as for the functions ϕ_1, ϕ_5 from Table 13.1.1:

$$\begin{aligned} \lambda_5 = \lim_{\varepsilon\to 0} k^2 \int d\mathbf{v}\, \varphi^0(v) &\left[\left(\frac{2}{3}\right)^{1/2} \frac{(\omega_p^2 + k^2\bar{u}^2)^{1/2}}{[\omega_p^2 + \frac{5}{3}k^2\bar{u}^2]^{1/2}} \right] (-1 - \tfrac{3}{2} + \tfrac{1}{2}m\beta v^2) \\ &\times v_x \frac{1}{K-\varepsilon} v_x \left\{ -\left(\frac{2}{3}\right)^{1/2} \frac{k^2\bar{u}^2}{\{(\omega_p^2 + k^2\bar{u}^2)[\omega_p^2 + \frac{5}{3}k^2\bar{u}^2]\}^{1/2}} \right. \\ &+ \left(\frac{2}{3}\right)^{1/2} \frac{(\omega_p^2 + k^2\bar{u}^2)^{1/2}}{[\omega_p^2 + \frac{5}{3}k^2\bar{u}^2]^{1/2}} (-\tfrac{3}{2} + \tfrac{1}{2}m\beta v^2) \\ &\left. - \frac{\omega_p^2}{k^2\bar{u}^2} \left(\frac{2}{3}\right)^{1/2} \frac{k^2\bar{u}^2}{\{(\omega_p^2 + k^2\bar{u}^2)[\omega_p^2 + \frac{5}{3}k^2\bar{u}^2]\}^{1/2}} \right\} \end{aligned} \quad (13.6.28)$$

At this stage the change of behavior from an ordinary gas to a plasma appears most dramatically. Consider first the former case, for which $\omega_p^2 = 0$. The expression in square brackets becomes

$$[\cdots] = \left(\frac{2}{3}\right)^{1/2} \frac{k\bar{u}}{(\frac{5}{3})^{1/2} k\bar{u}} = \left(\frac{2}{5}\right)^{1/2}, \qquad \omega_p = 0 \tag{13.6.29}$$

In the factor in curly brackets, the third term [which comes directly from the Vlassov operator (13.6.25)] vanishes and we are left with

$$\begin{aligned}\{\cdots\} &= -\left(\frac{2}{3}\right)^{1/2} \frac{k^2\bar{u}^2}{k\bar{u}(\frac{5}{3})^{1/2} k\bar{u}} + \left(\frac{2}{3}\right)^{1/2} \frac{k\bar{u}}{(\frac{5}{3})^{1/2} k\bar{u}} (-\tfrac{3}{2} + \tfrac{1}{2} m\beta v^2) \\ &= (\tfrac{2}{5})^{1/2} (-1 - \tfrac{3}{2} + \tfrac{1}{2} m\beta v^2)\end{aligned} \tag{13.6.30}$$

Substituting these results into (13.6.28) we recover exactly Eq. (13.4.2), that is, the thermal mode for an ordinary gas.

We now consider a plasma, with $\omega_p \neq 0$, and expand the coefficients in powers of k, keeping only the zeroth order. The coefficient in square brackets becomes

$$[\cdots] = \left(\frac{2}{3}\right)^{1/2} \frac{\omega_p}{\omega_p} + O(k^2) \approx \left(\frac{2}{3}\right)^{1/2} \tag{13.6.31}$$

In the coefficient in curly brackets a spectacular compensation appears. The first term, which was finite for an ordinary gas, now becomes of order k^2 and vanishes as $k \to 0$. However, it is replaced by the third, Vlassov term, which tends to a finite value in this limit:

$$\begin{aligned}\{\cdots\} &= O(k^2) + \left(\frac{2}{3}\right)^{1/2} \left[\frac{\omega_p}{\omega_p} + O(k^2)\right] (-\tfrac{3}{2} + \tfrac{1}{2} m\beta v^2) - \left(\frac{2}{3}\right)^{1/2} \left[\frac{\omega_p^2}{\omega_p \omega_p} + O(k^2)\right] \\ &\approx (\tfrac{2}{3})^{1/2} (-1 - \tfrac{3}{2} + \tfrac{1}{2} m\beta v^2)\end{aligned} \tag{13.6.32}$$

Compare now Eqs. (13.6.31), (13.6.32) to (13.6.29), (13.6.30). After the subtle cancellations occurring in the plasma, the net result is the mere replacement of the factor $(\frac{2}{5})^{1/2}$ by $(\frac{2}{3})^{1/2}$. In particular, the function in parentheses in Eqs. (13.6.30) and (13.6.32) is the same. This is very important, since it ensures the very existence of the limit $\varepsilon \to 0$ in (13.6.28): The argument of Eq. (13.4.4) is based on the value $(-\frac{5}{2})$ for the constant in the bracketed function in (13.6.32). We now substitute Eqs. (13.6.31), (13.6.32) into (13.6.28) and find

$$\lambda_5 = \frac{2}{3} \frac{k^2}{(k_B T)^2} \int d\mathbf{v}\, \varphi^0(v)\, (\tfrac{1}{2} m v^2 - \tfrac{5}{2} k_B T) v_x\, K^{-1}\, v_x (\tfrac{1}{2} m v^2 - \tfrac{5}{2} k_B T) \tag{13.6.33}$$

This is practically the same expression as (13.4.5) except for the factor $\frac{2}{3}$ replacing $\frac{2}{5}$. But we know that the factor $\frac{2}{5}$ represents the specific heat at constant pressure (13.4.5). We also know that the specific heat at constant volume (per unit mass) of an ideal gas is

$$c_V = \frac{3}{2}\frac{k_B}{m} \tag{13.6.34}$$

Hence we may write

$$\lambda_5 = -\frac{\kappa}{\rho c_V} k^2 \tag{13.6.35}$$

where the thermal conductivity κ is defined by the same formula (13.4.7) as for an ordinary gas. Equation (13.6.35) and the plasmadynamical thermal mode Λ_5 of Table 12.7.1 are thus *identical.*

We discussed this problem in great detail because it is an extremely illuminating example of the radical difference that exists between a gas and a plasma. It also shows how subtly the Vlassov term comes into play, shuffling the whole structure, and then slipping out again from Eq. (13.6.33)! This peculiar behavior explains the paradox mentioned at the beginning of this section. The Vlassov term in a plasma is of utmost importance in determining the nature of the eigenmodes, and at the same time it has no influence on the transport coefficients.

We now briefly show that the Vlassov term has no influence on the viscosity either. Here the situation is much simpler than for the thermal conductivity. A shear mode is given by Eq. (13.6.23) (for $\gamma \to 0$):

$$\lambda_3 = \lim_{\varepsilon \to 0} \langle \tilde{\phi}_3 | [kv_x + \Omega] (K - \varepsilon)^{-1} [kv_x + \Omega] | \bar{\phi}_3 \rangle$$

$$= \lim_{\varepsilon \to 0} ([kv_x + \Omega^\dagger]\tilde{\phi}_3, (K - \varepsilon)^{-1}[kv_x + \Omega]\bar{\phi}_3)$$

But it is immediately checked that

$$\Omega\bar{\phi}_3 = \Omega^\dagger\tilde{\phi}_3 = 0$$

Moreover, from (13.6.19) and (13.6.20):

$$\tilde{\phi}_3 = \bar{\phi}_3 = \phi_3$$

Hence the viscous mode is

$$\lambda_3 = \lim_{\varepsilon \to 0} \langle \phi_3 | kv_x (K - \varepsilon)^{-1} kv_x | \phi_3 \rangle \tag{13.6.36}$$

that is, the same expression as (13.3.30). It then follows that the viscosity of a plasma is again given by the same formula (13.4.1) as for an ordinary gas. Here, however, the situation is trivial: the Coulomb field is purely longitudinal and therefore uncoupled from the transverse shear modes.

We may note that the eigenvalues obtained here come out of our assumption that the *collisions dominate in the system.* The discussion of the end of Section 13.3 applies here as well. Had we solved the eigenvalue problem for the Vlassov equation in the absence of collisions, we would have obtained a quite different eigenvalue spectrum. We have no space for going into this problem here. Its solution is well known and we refer the reader to the classical papers of Van Kampen and of Case for a thorough discussion of this problem. Let us only say that the eigenvalues also form a continuous spectrum in this case, and that the eigenfunctions are also Schwartz distributions, as in Eq. (13.3.36) (but more complicated). There is, however, an important difference between a collisionless Vlassov plasma and a system of free particles: the former can support collective plasma oscillations. The Coulomb interactions are the cause of this strong coherence in the system.

BIBLIOGRAPHICAL NOTES

An excellent recent survey of the properties of the Boltzmann equation, with stress laid on its spectral properties, is C. Cercignani, *Mathematical Methods in Kinetic Theory*, Plenum Press, New York, 1969.

The spectrum of the homogeneous kinetic equations is discussed in

H. Grad, in *Third Symposium on Rarefied Gases*, Academic Press, New York, 1963.

J. L. Delcroix, *Introduction to the Theory of Ionized Gases*, Wiley-Interscience, New York, 1960.

C. H. Su, *J. Math. Phys.* **8**, 248 (1967).

The first rigorous calculation of the transport coefficients was done by *Maxwell:* J. C. Maxwell, *Phil. Trans. Roy. Soc.* **157**, 49 (1867). In this work he discovered that if the molecules interact through a potential $V(r) = Ar^{-4}$ (*Maxwell potential*), and only in that case, the transport coefficients can be calculated without an explicit knowledge of the distribution function. The "magic" property of this potential was confirmed when Boltzmann wrote down his equation.

Boltzmann himself made heroic attempts toward a solution of his equation and the calculation of transport coefficients. Three long papers were devoted to the viscosity: L. Boltzmann, *Wien. Ber.* **81,** 117 (1880); **84,** 40, 1230 (1881); and two to the diffusion: *Wien Ber.* **86,** 63 (1882); **88,** 835 (1883). But the "elementary" methods he used led to such complications that the results were completely useless. Twenty years later, in 1899 (Wiss. Abhandl. **3,** No. 123, reference quoted in Chapter 11) Boltzmann appeared quite discouraged about the practical use of his equation:

> *Ich erhielt so Formeln für Reibungs- Diffusions- und Wärmeleitungs-koeffizienten von denen ich mir eine Zeitland einbildete, sie seien exakt . . . . Aber vor ihre Publikation entdeckte ich dass in diesen meinen neuen Formeln auch wieder Glieder von der Ordnung der ausschlaggebenden vernachlässigt waren. Auss Ärger liess ich alles unpubliziert liegen. Glücklicherweise nahm*

*ich dann die Schlussformel und den Wert eines der bestimmten Integrale in einem im Jahre 1881 publizierte Abhandlung auf.**

and further

Ich ... gelangte dadurch zu einer Reihenentwicklung, auf Grund deren jedoch wegen ihrer Weitläufigkeit und der fehlenden Gewissheit der Konvergenz die Reibungs- Diffusions- und Wärmeleitungskonstante ***wohl kaum je numerisch berechnet werden wird.***†

He concludes by offering his tribute of admiration to Maxwell for his discovery of the only case where the calculation was possible.

The next advance in the field came through *Lorentz*' work: H. A. Lorentz, *Proc. Amsterdam Acad.* **7,** 438, 585, 684 (1905). He imagined the limiting case of a mixture of two gases, one, very dilute, made of light molecules of mass m and the other of heavy molecules of mass M. If the mutual interactions of the light molecules can be neglected, the transport coefficients can be calculated exactly in the limit $m/M \to 0$.

It is not widely known that *Niels Bohr* wrote his Ph.D. dissertation (Copenhagen, 1911) on a generalization of the Lorentz gas and the calculation of electrical and thermal conductivities.

The decisive step in this field was, however, the introduction of the concept of *normal solutions* and of the expansion methods derived from it. These methods opened the way for a systematic calculation of transport coefficients for arbitrary interaction potentials. Moreover, these methods can be applied or adapted to kinetic equations other than Boltzmann's.

The most important works in this direction were

D. Hilbert, *Götting. Nachr.* 355 (1910); Math. Ann. **72,** 562 (1912).

S. Chapman, *Trans. Roy. Soc.* (London) **A216,** 279 (1916); **A217,** 115 (1917).

D. Enskog, Dissertation, Uppsala, 1917; *Svensk. Vet. Akad., Arkiv. Mat., Ast. Fys.* **16,** 1 (1921).

The classical reference book for this theory is S. Chapman and T. Cowling, *Mathematical Theory of Non-Uniform Gases,* Cambridge Univ. Press, 1939 (2nd ed., 1952).

Other extensive discussions are found in

Hirschfelder, Curtiss, and Byrd (quoted in Chapter 6).

L. Waldmann, *Handbuch Phys.* **12,** 295 (1958).

* "I obtained in this way some formulas for the coefficients of viscosity, diffusion, and thermal conductivity, which I imagined for a certain time to be exact.... But before their publication I discovered that in these new formulas of mine some terms of the same order as the leading ones were neglected. I got so angry that I left everything unpublished. Fortunately, I included in a paper of 1881 the final formula and the value of one of the definite integrals."

† "I thus obtained a series expansion; it was, however, very lengthy and I had no assurance about its convergence. Therefore the viscosity, diffusion, and heat conduction coefficients could hardly be calculated numerically."

A more recent method for the calculation of transport coefficients is the so-called "thirteen-moment method" due to H. Grad, *Commun. Pure Appl. Math.* **2,** 311 (1949).

Volume 5 of the *Studies in Statistical Mechanics* (J. de Boer and G. E. Uhlenbeck, eds.), North Holland, Amsterdam and American Elsevier, New York, 1970, is devoted to two important papers in kinetic theory:

C. S. Wang Chang and G. E. Uhlenbeck, "The Kinetic Theory of Gases."

J. D. Foch and G. W. Ford, "The Dispersion of Sound in Monoatomic Gases."

Résibois' method was published in P. Résibois, *Phys. Lett.* **A30,** 465 (1969); *J. Stat. Phys.* **2,** 21 (1970); *Bull. Cl. Sci. Acad. Roy. Belg.* **56,** 160 (1970); *Physica* **49,** 591 (1970).

See also P. Résibois, in *Irreversibility in the Many-Body Problem* (J. Biel and J. Rae, eds.), Plenum Press, New York, 1972.

There are some interesting historical points connected with the use of *orthogonal polynomials* for the solution of the Boltzmann equation. Maxwell was the first who noted that the collisional rate of change of any spherical harmonic of the velocity can be calculated quite easily. He published these results in a very concise way, in the last months of his life (*Cambr. Phil. Trans.* **1879,** 231). Boltzmann was very interested in his method; he says in his 1899 paper (quoted above) that he did the calculation by Maxwell's method, but his results did not agree with Maxwell's. He then recounts that, when he went to England (in 1895), he tried to get hold of Maxwell's original manuscript, but the latter could no longer be found.

It was only in 1935 that Burnett demonstrated the special role played by the Sonine polynomials for the Chapman–Enskog theory: D. Burnett, *Proc. London Math. Soc.* **39,** 385 (1935); **40,** 382 (1935).

The normal modes of the *Vlassov equation* in the absence of collisions were derived by

N. G. van Kampen, *Physica* **21,** 949 (1955); **23,** 641 (1957).

K. M. Case, *Ann. Phys.* (N.Y.) **7,** 349 (1959).

See also R. Balescu (quoted in Chapter 6).

CHAPTER 14

DYNAMICS OF CORRELATIONS

14.1. TIME EVOLUTION OF THE CLASSICAL CORRELATION PATTERNS

The three previous chapters were devoted to a first study of the kinetic equations of simple gases. We only gave the elementary derivation of these equations, which was admittedly inaccurate from the mechanical point of view. We did this because we feel it important first to become rapidly acquainted with the properties of these equations and with the use that can be made of them.

We are now ready for a general attack on the Liouville equation and for the more rigorous foundation of nonequilibrium statistical mechanics. Before starting, however, we still need a refinement of our mathematical tools.

In order to study the dynamics of large systems, the Liouville equation in the reduced distribution function representation developed in Section 3.4 is a quite natural starting point. However, this representation is not yet sufficiently detailed. From the qualitative discussion of Section 11.5 it already seems that the concept of *correlations*, rather than the distribution functions should play an important role in the theory. We saw, for instance, that the two-body correlation function does not appear in the Boltzmann equation, whereas it does play a role in the exact BBGKY hierarchy. To understand how the Boltzmann equation (and other kinetic equations) can be derived from exact mechanics, we must therefore develop a formalism in which the various correlation patterns are explicitly exhibited.

We briefly recall that in Chapter 3 the reduced distribution functions were represented as a linear superposition of *correlation patterns* [Eq. (3.5.15)]:

$$f_s(x_1, \ldots, x_s) = \sum_{\Gamma_s} \pi_s(x_1, \ldots, x_s; \Gamma_s)$$

where Γ_s is an index characterizing a given partition of the set of s particles. In an alternative system of notations, the partitions are explicitly materialized by bars; for instance [see Eq. (3.5.17)]:

$$f_2(12) = \pi_2(1\,|\,2) + \pi_2(12)$$

$$f_3(123) = \pi_3(1\,|\,2\,|\,3) + \pi_3(1\,|\,23) + \pi_3(2\,|\,13) + \pi_3(3\,|\,12) + \pi_3(123)$$

and so forth.

The cluster representation (3.5.9)–(3.5.11) or (3.5.15) or (3.5.17) is the starting point of our study. The aim is to derive equations of evolution for the individual correlation patterns $\pi_s(\Gamma_s)$. This set of equations must, of course, be equivalent to the BBGKY hierarchy (3.4.14).

This is a rather easy problem, which can be solved by constructing the equations successively, starting from the lowest order and going up in the hierarchy. The one-particle function $f_1(x_1) \equiv \pi_1(1)$ contains a single correlation pattern. Hence, it simply obeys the BBGKY equation (3.4.9); however, the function f_2 appearing in the right-hand side must be expressed in terms of the patterns $\pi_2(1\,|\,2)$ and $\pi_2(12)$ by means of Eq. (3.5.17):

$$(\partial_t - L_1^0 - L_1^F)\pi_1(1) = \int dx_2\, \{L'_{12}\pi_2(1\,|\,2) + L'_{12}\pi_2(12)\} \tag{14.1.1}$$

We then proceed to the two-particle function:

$$f_2(12) = \pi_2(1\,|\,2) + \pi_2(12) \tag{14.1.2}$$

We know that it obeys Eq. (3.4.10), which is again expressed in terms of correlation patterns:

$$\begin{aligned}(\partial_t - L_1^0 - L_1^F - L_2^0 - L_2^F)&[\pi_2(1\,|\,2) + \pi_2(12)]\\ &= L'_{12}[\pi_2(1\,|\,2) + \pi_2(12)] + \int dx_3\,(L'_{13} + L'_{23})\\ &\quad\times[\pi_3(1\,|\,2\,|\,3) + \pi_3(1\,|\,23) + \pi_3(2\,|\,13) + \pi_3(3\,|\,12) + \pi_3(123)]\end{aligned} \tag{14.1.3}$$

But we also know that

$$\pi_2(1\,|\,2) = \pi_1(1)\pi_1(2) \tag{14.1.4}$$

Hence we can write the rate of change of this pattern by using (14.1.1):

$$\begin{aligned}\partial_t\pi_2(1\,|\,2) &= \pi_1(1)\partial_t\pi_1(2) + \pi_1(2)\partial_t\pi_1(1)\\ &= \pi_1(1)\left\{(L_2^0 + L_2^F)\pi_1(2) + \int dx_3\, L'_{23}[\pi_2(2\,|\,3) + \pi_2(23)]\right\}\\ &\quad + \pi_1(2)\left\{(L_1^0 + L_1^F)\pi_1(1) + \int dx_3\, L'_{13}[\pi_2(1\,|\,3) + \pi_2(13)]\right\}\end{aligned} \tag{14.1.5}$$

Keeping in mind Eq. (14.1.4), as well as the properties:

$$\begin{aligned}\pi_3(1\,|\,2\,|\,3) &= \pi_1(1)\pi_1(2)\pi_1(3)\\ \pi_3(1\,|\,23) &= \pi_1(1)\pi_2(23)\end{aligned}\tag{14.1.6}$$

we rewrite the equation for the pattern $\pi_2(1\,|\,2)$ as

$$\begin{aligned}(\partial_t - L_1^0 - L_1^F - L_2^0 - L_2^F)\pi_2(1\,|\,2) = \int dx_3\,\{&L'_{13}\pi_3(1\,|\,2\,|\,3)\\ &+ L'_{13}\pi_3(2\,|\,13) + L'_{23}\pi_3(1\,|\,2\,|\,3) + L'_{23}\pi_3(1\,|\,23)\}\end{aligned}\tag{14.1.7}$$

Combining now Eqs. (14.1.7) and (14.1.3) we obtain the equation for $\pi_2(12)$ by a simple subtraction:

$$\begin{aligned}(\partial_t - L_1^0 - L_1^F - L_2^0 - L_2^F)\pi_2(12) = {}& L'_{12}\pi_2(1\,|\,2) + L'_{12}\pi_2(12)\\ &+ \int dx_3\,\{L'_{13}\pi_3(1\,|\,23) + L'_{23}\pi_3(2\,|\,13)\\ &+ (L'_{13} + L'_{23})\pi_3(3\,|\,12) + (L'_{13} + L'_{23})\pi_3(123)\}\end{aligned}\tag{14.1.8}$$

This procedure can, of course, be continued for successively higher values of the number of particles. It is perfectly well defined, and we may therefore consider that the hierarchy of equations for the correlation patterns is known. In practice, however, this method of construction soon becomes very tedious. We, therefore, introduce a graphical method that is very helpful in enabling one to write automatically the equations for any given pattern, as well as in providing a vivid picture of the processes involved. This method is based on a refinement of the diagrams of Section 3.4.

We represent again the "state" of the system by s superposed labeled lines. However, the mere statement of the number of particles is no longer sufficient: we must introduce a graphical element to express the type of correlation. We agree to *represent correlated particles* [i.e., particles belonging to the same subset in the particular partition of the set (s)] *by a set of connected lines.* In absence of further information about the physical origin of the correlation, the connection is materialized by an arc connecting the lines at the right-hand side of the diagram (see Fig. 14.1.1). These new diagrams will be drawn with continuous lines, in order to distinguish them from those of Section 3.4.

We now consider the representation of the dynamics. We first note that the unperturbed motion affects the various patterns of s particles in the same way: this is clearly seen from Eqs. (14.1.7) and (14.1.8). The reader will easily convince himself that this is a quite general feature.

Moreover, the unperturbed Liouvillian can never connect a given

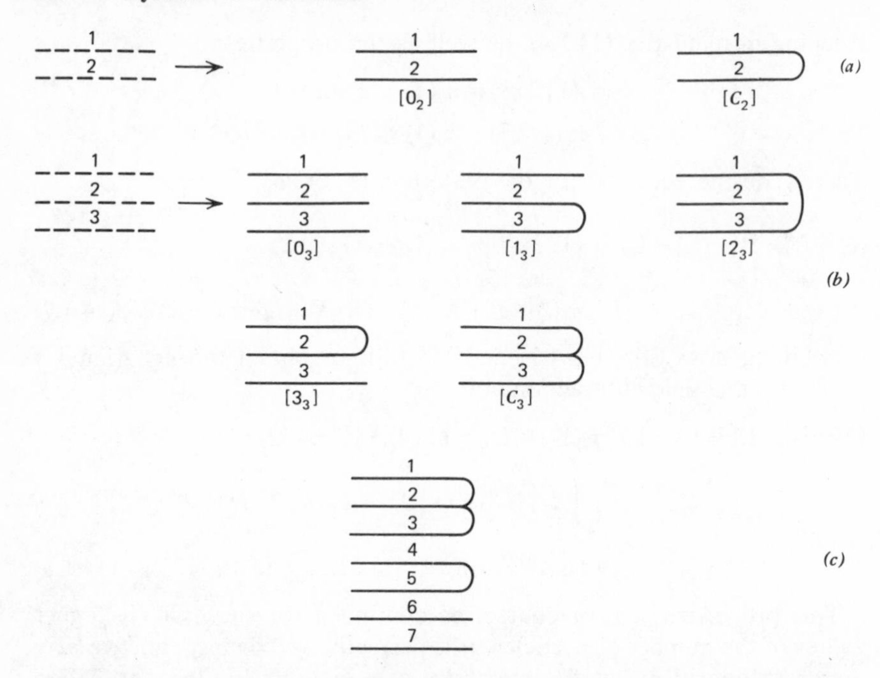

Figure 14.1.1. Graphical representation of the correlation patterns: (*a*) the set of all two-particle correlation patterns; (*b*) the set of all three-particle correlation patterns; (*c*) an example of a complex correlation pattern: $\pi_7(123|45|6|7)$.

pattern to a different one. We may say that the unperturbed Liouvillian is an operator, diagonal in the number of particles *s*, and also in the correlation index Γ_s: It cannot change the pattern of correlation of a group of particles. For this reason, it is unnecessary to introduce any graphical element to represent it.

The interaction Liouvillian, on the other hand, introduces a coupling between patterns of different kinds. This operator will be represented again by the two vertices of Table 3.4.1. But now each diagram of Fig 3.4.1 is replaced by a sum, according to the particular pattern it acts upon.

For instance, the equation for $\pi_1(1) \equiv f_1(1)$ is represented by the diagrams of Fig. 14.1.2, obtained from a combination of Figs. 3.4.1 and 14.1.1. Clearly, the two diagrams correspond to the two terms of the rhs of Eq. (14.1.1) in a quite obvious way. Going on to $s = 2$, we first draw the diagrams of Fig. 3.4.1 (see Fig. 14.1.3*a*), and then express each of the right states in terms of correlation patterns [this corresponds to Eq. (14.1.3)].

We note that the four diagrams in the box in Fig. 14.1.3*b* are different from the others. In the state coming out after the interaction (i.e., the state appearing on the left) the two lines 1 and 2 are not connected. The

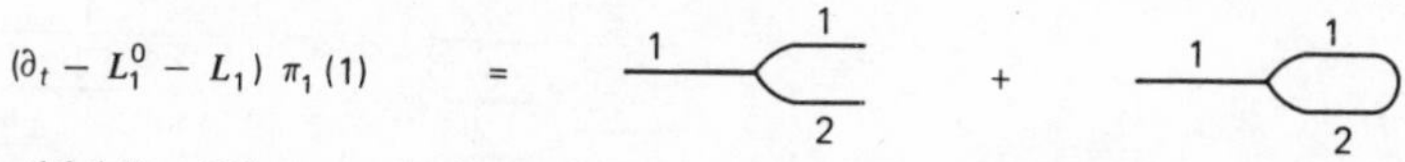

$(\partial_t - L_1^0 - L_1)\ \pi_1(1)$ =

Figure 14.1.2. Diagrammatic representation of the equation of evolution of the correlation pattern $\pi_1(1)$.

diagram consists of two pieces, ending with lines 1 and 2, which can be taken apart. One is therefore tempted to identify these diagrams with the contributions to $\pi_2(1|2)$. If we now compare these diagrams with Eq. (14.1.7), we immediately see that this suggestion is correct. All the other diagrams are contributions to $\pi_2(12)$, as can be checked again with Eq. (14.1.8).

We now formulate the following practical rule. To obtain the equation of evolution for a given pattern $\pi_s(\Gamma_s)$:

1. Draw all the diagrams contributing to $\partial_t f_s$.
2. Replace each diagram by a sum of diagrams corresponding to all possible patterns describing the "initial" state (on the right).
3. Among all the diagrams, pick out those that are made out of connected pieces corresponding precisely to the partition Γ_s.

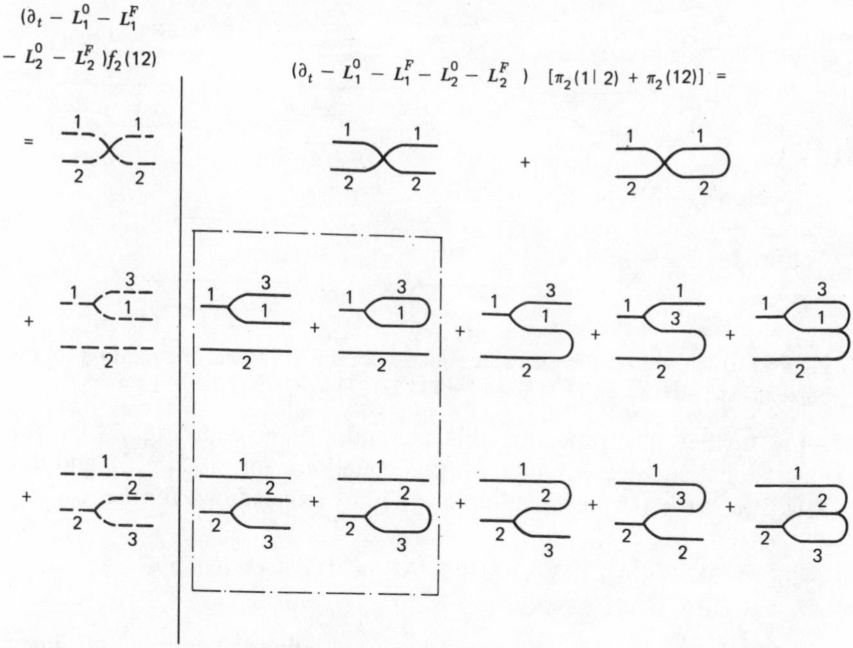

Figure 14.1.3. Graphical construction of the matrix elements in the correlation dynamics.

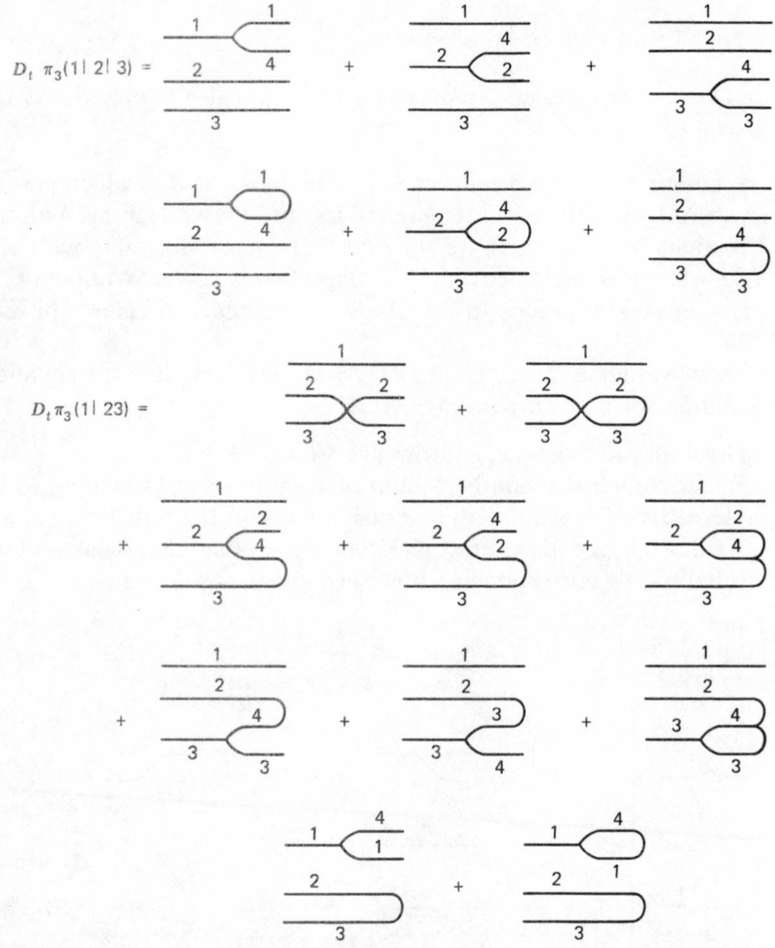

Figure 14.1.4. Diagrammatic representation of the equations of evolution for the correlation patterns $\pi_3(1|2|3)$ and $\pi_3(1|23)$ $(D_t \equiv \partial_t - \sum L_i^0 - \sum L_i^F)$.

As a further illustration of this method, we give in Fig. 14.1.4 the diagrammatic representation of the equations for $\pi_3(1|2|3)$ and for $\pi_3(1|23)$. We leave the equation for $\pi_3(123)$ as an exercise for the reader.

14.2. CLASSICAL DYNAMICS AS A DYNAMICS OF CORRELATIONS

In the previous section we obtained an infinite hierarchy of linear equations describing the time evolution of the correlation patterns. The solution of this hierarchy is, of course, equivalent to the solution of the initial BBGKY hierarchy. In other words, a knowledge of all the correlation patterns completely characterizes the state of the system. The equations were constructed by starting from the cluster representation of the reduced distribution function, (3.5.9)–(3.5.11). As a result, we found that the individual correlation patterns—even those corresponding to the same number of particles—obey different equations of motion.

But once the system of equations obeyed by the correlation patterns is known, we may start thinking in the reverse direction. The system of equations (14.1.1), (14.1.7), (14.1.8), and so on is a more general concept than the cluster representation (3.5.9)–(3.5.11). In other words, the cluster functions $\pi_s(\Gamma_s)$ are a particular solution of our set of equations, but *they are not the only solution.* Any other solution, however, will *not* be factorized in the peculiar "cluster" fashion. It is, therefore, tempting to introduce a new broader definition of the concept of correlations. In this new definition the basic criterion is the *type of evolution in time*, rather than the functional form (i.e., the factorization properties) of the various correlation patterns. These generalized correlation patterns will be denoted either by compact symbols $p_s(\Gamma_s)$, or by the explicit symbols $p_s(i_1 \mid i_2 \mid i_3 i_4 \mid \cdots i_s)$, with the same conventions as in Section 3.4.

The set $\{p_s(\Gamma_s)\}$ is now defined as *any* solution of the hierarchy of equations derived in the previous section, subject to a few constraints to be defined further below.

In order to make this definition more precise, we reformulate the dynamical problem systematically from the beginning.

A dynamical system consisting of identical interacting particles as described in Section 2.4* is represented by a *distribution vector* $\mathfrak{f}$. In the thermodynamic limit, this vector has an infinite number of components. However, instead of considering the components of $\mathfrak{f}$ to be the reduced distribution functions f_s, as in Eq. (3.1.21), we consider $\mathfrak{f}$ to be the collection of all the correlation patterns:

$$\mathfrak{f} = \{p_s(x_1, \ldots, x_s; [\Gamma_s])\} \tag{14.2.1}$$

This point of view simply represents a more refined description of the same system as in Section 3.1. In other words, the mere specification of the number of particles turns out to be insufficient for our purposes: we must also specify the type of correlation Γ_s.†

The distribution vector, as a function of time, obeys the (generalized)

* The modifications necessary for covering other types of systems (many component systems, other types of degrees of freedom) can easily be worked out along the same line.

† We may think of a (somewhat superficial) quantum-mechanical analogy: The state characterized by the "quantum number" s is *degenerate:* to each value of s (≥ 2) there correspond several different values of the second "quantum number" Γ_s.

Liouville equation:

$$\partial_t \mathfrak{f}(t) = \mathcal{L}\mathfrak{f}(t) \tag{14.2.2}$$

This equation, and in particular the operator $\mathcal{L}$, is intrinsically the same as Eq. (3.4.12). However, it will now be resolved into component equations for the correlation patterns, rather than equations for the reduced distributions as in (3.4.14):

$$\partial_t p_s([\Gamma_s]; t) = \sum_{r=0}^{\infty} \sum_{\Gamma'_r} \langle\!\langle (s)[\Gamma_s] | \mathcal{L}^0 + \mathcal{L}^F + \mathcal{L}' | (r)[\Gamma'_r] \rangle p_r([\Gamma'_r]; t) \tag{14.2.3}$$

The matrix elements of $\mathcal{L}^0$ are particularly simple. They are diagonal in s and in Γ_s; moreover, they are independent of Γ_s:

$$\langle\!\langle (s)[\Gamma_s] | \mathcal{L}^0 | (r)[\Gamma'_r] \rangle = \delta_{s,r}\ \delta_{\Gamma_s,\Gamma'_r} \left(\sum_{j=1}^{s} L_j^0 \right) \tag{14.2.4}$$

This property is extremely important. It implies that the *unperturbed Liouvillian $\mathcal{L}^0$ cannot change the state of correlation of a set of particles.* Physically, this property is quite clear: the free-motion part of the Liouvillian affects each particle independently of all the others. It cannot introduce or destroy any link between different particles.

The diagonal character in both indices s and Γ_s is shared by the external field Liouvillian $\mathcal{L}^F$:

$$\langle\!\langle (s)[\Gamma_s] | \mathcal{L}^F | (r)[\Gamma'_r] \rangle = \delta_{s,r}\ \delta_{\Gamma_s,\Gamma'_r} \left(\sum_{j=1}^{s} L_j^F \right) \tag{14.2.5}$$

Hence, the action of an external field does not change the state of correlation of a group of particles.

The matrix elements of the operator $\mathcal{L}'$ in this representation are constructed most conveniently by the method of diagrams explained in Section 14.1.

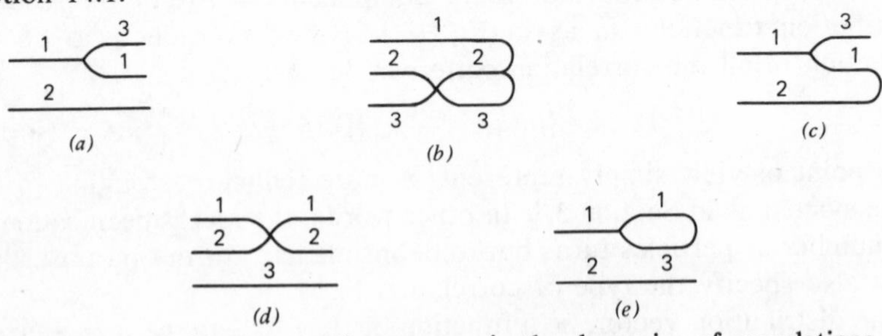

Figure 14.2.1. Various types of processes in the dynamics of correlations. (*a*) *Propagation:* an uncorrelated pattern $p_3(1|2|3)$ goes over into an uncorrelated pattern $p_2(1|2)$. (*b*) *Propagation:* the correlated pattern $p_3(123)$ goes over into itself. (*c*) *Transformation:* the correlated pattern $p_3(3|12)$ goes over into another correlated pattern $p_2(12)$. (*d*) *Creation:* the uncorrelated pattern $p_3(1|2|3)$ goes over into the correlated pattern $p_3(3|12)$. (*e*) *Destruction:* the correlated pattern $p_3(2|13)$ goes over into the uncorrelated pattern $p_2(1|2)$.

As a concrete illustration, we give the complete list of matrix elements contributing to the one-particle and to the two-particle correlation patterns. They can be immediately read off from Fig. 14.1.3 or from Eqs. (14.1.7), (14.1.8). (We now use, of course, the explicit notation for the correlation patterns.)

$$\langle 1|\,\mathcal{L}'\,|1\rangle = 0$$

$$\langle 1|\,\mathcal{L}'\,|1\,|\,2\rangle = \int dx_2\, L'_{12} \tag{14.2.6}$$

$$\langle 1|\,\mathcal{L}'\,|12\rangle = \int dx_2\, L'_{12}$$

$$\langle 1\,|\,2|\,\mathcal{L}'\,|1\,|\,2\rangle = \langle 1\,|\,2|\,\mathcal{L}'\,|12\rangle = 0$$

$$\langle 1\,|\,2|\,\mathcal{L}'\,|1\,|\,2\,|\,3\rangle = \int dx_3\,(L'_{13} + L'_{23})$$

$$\langle 1\,|\,2|\,\mathcal{L}'\,|1\,|\,23\rangle = \int dx_3\, L'_{23} \tag{14.2.7}$$

$$\langle 1\,|\,2|\,\mathcal{L}'\,|2\,|\,13\rangle = \int dx_3\, L'_{13}$$

$$\langle 1\,|\,2|\,\mathcal{L}'\,|3\,|\,12\rangle = \langle 1\,|\,2|\,\mathcal{L}'\,|123\rangle = 0$$

$$\langle 12|\,\mathcal{L}'\,|1\,|\,2\rangle = L'_{12}; \qquad \langle 12|\,\mathcal{L}'\,|12\rangle = L'_{12}$$

$$\langle 12|\,\mathcal{L}'\,|1\,|\,2\,|\,3\rangle = 0$$

$$\langle 12|\,\mathcal{L}'\,|1\,|\,23\rangle = \int dx_3\, L'_{13}; \qquad \langle 12|\,\mathcal{L}'\,|2\,|\,13\rangle = \int dx_3\, L'_{23} \tag{14.2.8}$$

$$\langle 12|\,\mathcal{L}'\,|3\,|\,12\rangle = \int dx_3\,(L'_{13} + L'_{23})$$

$$\langle 12|\,\mathcal{L}'\,|123\rangle = \int dx_3\,(L'_{13} + L'_{23})$$

We now see that the interactions deeply affect the correlation patterns. A variety of processes are possible. Indeed, every operator L'_{ij} creates a correlation between particles i and j. This process corresponds to the physical idea that the behavior (say, the trajectory) of particle i, interacting with j, is clearly dependent on the parameters of particle j; the two particles cannot be independent.

This mechanism, which always creates (or propagates) correlations, is, however, combined with the averaging operation. This operation always "eliminates" one particle; it can thus transform, say, a configuration of three correlated particles into a pattern of two uncorrelated ones.

The complex processes resulting from the interactions and averagings are vividly represented by the diagrams. A few examples of characteristic processes are drawn in Fig. 14.2.1.

We have now constructed a fascinating picture of the process of evolution as a set of transitions from one correlation pattern to another. In the course of time correlation patterns are created, others are destroyed, others are transformed. This justifies the name (due to Prigogine) of *dynamics of correlations* given to this picture of mechanics.

We must still fix the constraints which the correlation patterns must satisfy. These must be chosen in such a way as to ensure the complete equivalence of the correlation patterns description and of the reduced distributions picture. This is obtained by imposing *at the (arbitrary) initial time* $t=0$ the condition:

$$\sum_{\Gamma_s} p_s(x_1, \ldots, x_s; [\Gamma_s]) = f_s(x_1, \ldots, x_s) \qquad \text{all } s \tag{14.2.9}$$

It is clear, from the construction of the equations in Section 14.1, that if (14.2.9) is satisfied at time zero, it will hold at all times.

We also impose another constraint at time zero. This second condition introduces a clear distinction between uncorrelated and truly correlated patterns. We have seen in Section 3.5 that, in the cluster representation, the uncorrelated pattern contains the whole normalization of the distribution function [see Eqs. (3.5.19) and (3.5.20)]. We require this condition to be obeyed, *at time zero*, by the dynamical patterns as well:

$$\text{T-lim } N^{-s} \int dx_1 \cdots dx_s \, p_s(x_1, \ldots, x_s; [0_s]) = 1 \tag{14.2.10}$$

$$\text{T-lim } N^{-s} \int dx_1 \cdots dx_s \, p_s(x_1, \ldots, x_s; [\Gamma_s]) = 0, \qquad \Gamma_s \neq 0_s \tag{14.2.11}$$

Again it is easy to verify that the normalization of the patterns is preserved in time by the equations of motion.

We finally note once more that, if *at the initial time*, we choose for the dynamical patterns the form given by the cluster expansion:

$$p_s(x_1, \ldots, x_s; [\Gamma_s]) = \pi_s(x_1, \ldots, x_s; [\Gamma_s]) \tag{14.2.12}$$

this form will remain true at all times. Indeed, the equations have been constructed precisely in such a way as to ensure this property.

It is important to realize that, if we choose the particular initial conditions (14.2.12), many of the equations in the set (14.2.3) become redundant. Indeed all patterns π_s except the fully correlated ones ($\Gamma_s = C_s$) can be expressed as products of less correlated patterns. If we eliminate the redundant patterns, we obviously obtain a set of *nonlinear* equations for $\pi_s([C_s])$. This feature would be very inconvenient in a general theory. On the contrary, if we adopt the *dynamical* definition of correlations, as was done in the present section, there is no *a priori* relationship between the correlation patterns, except for the (linear) condition (14.2.9). In that case there is no redundancy, and the fundamental Liouville equation correctly appears as a linear equation, as it was originally derived in Chapters 2 and 3.

To conclude the picture, we say a few words about the expression of averages in terms of correlation patterns. Substitution of Eq. (14.2.9) into (3.1.23) yields

$$\langle b\rangle = \sum_{s=0}^{\infty} \sum_{\Gamma_s} (s!)^{-1} \int dx_1 \cdots dx_s\, b_s(x_1, \ldots, x_s) p_s(x_1, \ldots, x_s; [\Gamma_s]) \quad (14.2.13)$$

We may note that it is impossible to decompose in a nontrivial way the dynamical functions into components corresponding to the various correlation patterns. The concept of correlations is related to the state of the system, not to the dynamical functions!

14.3. QUANTUM DYNAMICS OF CORRELATIONS

In Sections 3.6–3.8 we have shown that quantum-mechanical systems can be treated within the same framework as classical ones, provided we use Wigner functions for their description. We now further pursue this idea. We show that a dynamics of correlations can also be constructed in quantum mechanics. It has again the same structure as in Section 14.2, but significant differences will appear in the concrete realization of this structure.

We start directly with the *dynamical definition of the correlation patterns*, as in Section 14.2. The state of the system is represented by a *distribution vector* $\mathfrak{f}$, considered as the collection of all *unsymmetrized correlation patterns* $p_s(\xi_1, \ldots, \xi_s; [\Gamma_s])$*:

$$\mathfrak{f} = \{p_s(\xi_1, \ldots, \xi_s; [\Gamma_s])\} \quad (14.3.1)$$

These patterns are related to the Wigner functions through the relation generalizing Eq. (3.8.14):

$$f_s(\xi_1, \ldots, \xi_s) = \sum_{\Gamma_s} P_s(\Gamma_s) p_s(\xi_1, \ldots, \xi_s; [\Gamma_s]) \quad (14.3.2)$$

where $P_s(\Gamma_s)$ is the symmetrization operator defined by Eqs. (3.8.15) and (3.8.16).†

The occurrence of these operators will introduce the main difference between classical and quantum-statistical systems. We also assume that

* There is no point in using different notations in quantum and classical mechanics: we will even drop the superscript w. The context always makes clear which realization of the general structure must be used.

† Actually, this relation is only postulated to hold at $t = 0$: it is propagated in time.

the normalization conditions (14.2.10), (14.2.11) hold, just as in the classical case. The distribution vector $\mathfrak{f}(t)$ is a solution of the *Liouville equation:*

$$\partial_t \mathfrak{f}(t) = \mathscr{L}\mathfrak{f}(t) \tag{14.3.3}$$

This vector equation is decomposed into component equations for the correlation patterns:

$$\partial_t p_s([\Gamma_s]; t) = \sum_{r=0}^{\infty} \sum_{\Gamma'_r} \langle (s)[\Gamma_s] |\, \mathscr{L}^0 + \mathscr{L}' \,| (r)[\Gamma'_r] \rangle p_r([\Gamma'_r]; t) \tag{14.3.4}$$

For simplicity we assume that there is no external field: $\mathscr{L}^F \equiv 0$.

We now proceed with the construction of these equations. The idea is similar to the classical case. We require that these equations be compatible with the particular solution:

$$p_s(\xi_1, \ldots, \xi_s; [\Gamma_s]) = \pi_s(\xi_1, \ldots, \xi_s; [\Gamma_s]) \tag{14.3.5}$$

where $\pi_s(\xi_1, \ldots, \xi_s; [\Gamma_s])$ are the (unsymmetrized) correlation patterns defined by the cluster expansion in Section 3.8. (In constructing the equations obeyed by these patterns we adopt again the explicit notation explained in Section 3.5.)

The one-particle pattern obeys the first BBGKY equation (3.7.10) whose right-hand side must be expressed in terms of $\pi_2(1\,|\,2)$ and $\pi_2(12)$:

$$(\partial_t - L_1^0)\pi_1(1) = \int d\xi_2\, \delta(\mathbf{k}_2)\{L'_{12}P_2(1\,|\,2)\pi_2(1\,|\,2) + L'_{12}P_2(12)\pi_2(12)\} \tag{14.3.6}$$

It then follows, comparing with (14.3.4):

$$\begin{aligned}
\langle 1 |\, \mathscr{L}^0 \,| 1 \rangle &= L_1^0 \\
\langle 1 |\, \mathscr{L}' \,| 1 \rangle &= 0 \\
\langle 1 |\, \mathscr{L}' \,| 1\,|\,2 \rangle &= \int d\xi_2\, \delta(\mathbf{k}_2)\, L'_{12}P_2(1\,|\,2) \\
\langle 1 |\, \mathscr{L}' \,| 12 \rangle &= \int d\xi_2\, \delta(\mathbf{k}_2)\, L'_{12}P_2(12)
\end{aligned} \tag{14.3.7}$$

We note once again that the unperturbed Liouvillian is exactly the same as in classical mechanics. The interaction part has the same type of nonvanishing elements. They are expressed in terms of the *quantum mechanical* operator L'_{12}, Eq. (3.7.8). The novel feature is the appearance of the symmetrization operators $P_2(1\,|\,2)$ and $P_2(12)$ in the matrix elements of the Liouvillian. These operators are responsible of the correct fermion or boson symmetry of the theory.

Following the model of Section 14.1, we go over to the two-particle

Wigner function. However, we note that this function is now related to the correlation patterns through Eq. (14.3.2), which involves the symmetrizers;

$$f_2(12) = P_2(1\,|\,2)\pi_2(1\,|\,2) + P_2(12)\pi_2(12) \tag{14.3.8}$$

We also know that the symmetrizer $P_2(12)$ (corresponding to the irreducible correlation function) is simply the identity operator.

With these facts in mind we rewrite the BBGKY equation (3.7.9) as

$$\begin{aligned}(\partial_t - L_1^0 - L_2^0)&[P_2(1\,|\,2)\pi_2(1\,|\,2) + \pi_2(12)]\\ &= L'_{12}[P_2(1\,|\,2)\pi_2(1\,|\,2) + P_2(12)\pi_2(12)]\\ &\quad + \int d\xi_3\, \delta(\mathbf{k}_3)\, (L'_{13} + L'_{23})\{P_3(1\,|\,2\,|\,3)\pi_3(1\,|\,2\,|\,3)\\ &\quad + P_3(1\,|\,23)\pi_3(1\,|\,23) + P_3(2\,|\,13)\pi_3(2\,|\,13)\\ &\quad + P_3(3\,|\,12)\pi_3(3\,|\,12) + P_3(123)\pi_3(123)\}\end{aligned} \tag{14.3.9}$$

We now use the factorization property of the (unsymmetrized) pattern:

$$\pi_2(1\,|\,2) = \pi_1(1)\pi_1(2)$$

We therefore obtain

$$\begin{aligned}\partial_t\pi_2(1\,|\,2)\\ &= \pi_1(1)\left\{L_2^0\pi_1(2) + \int d\xi_3\, \delta(\mathbf{k}_3)\, L'_{23}[P_2(2\,|\,3)\pi_2(2\,|\,3) + P_2(23)\pi_2(23)]\right\}\\ &\quad + \pi_1(2)\left\{L_1^0\pi_1(1) + \int d\xi_3\, \delta(\mathbf{k}_3)\, L'_{13}[P_2(1\,|\,3)\pi_2(1\,|\,3) + P_2(13)\pi_2(13)]\right\}\end{aligned}$$

or, using (14.1.6)

$$\begin{aligned}\partial_t\pi_2(1\,|\,2) &= (L_1^0 + L_2^0)\pi_2(1\,|\,2)\\ &\quad + \int d\xi_3\, \delta(\mathbf{k}_3)\, \{L'_{13}P_2(1\,|\,3)\pi_3(1\,|\,2\,|\,3) + L'_{23}P_2(2\,|\,3)\pi_3(1\,|\,2\,|\,3)\\ &\quad + L'_{13}P_2(13)\pi_3(2\,|\,13) + L'_{23}P_2(23)\pi_3(1\,|\,23)\}\end{aligned} \tag{14.3.10}$$

Before subtracting this equation from (14.3.9) we must first symmetrize it, by multiplying all terms on the left by the operator $P_2(1\,|\,2)$. We now note a very important property relating the unperturbed Liouvillian and the symmetrizer:

$$P_2(1\,|\,2)(L_1^0 + L_2^0) = (L_1^0 + L_2^0)P_2(1\,|\,2) \tag{14.3.11}$$

This property is easily checked by using the explicit forms (3.7.7) and

(3.8.10) (we set here $\hbar = m = 1$ for simplicity):

$$P_2(1\,|\,2)(L_1^0+L_2^0)\pi_2(1\,|\,2)$$

$$= -i\int d\mathbf{k}_1'\, d\mathbf{k}_2'\, \{\delta(-\mathbf{k}_1'+\mathbf{k}_1)\,\delta(-\mathbf{k}_2'+\mathbf{k}_2)+\theta\delta(-\mathbf{k}_1'+(\mathbf{p}_2-\mathbf{p}_1)+\tfrac{1}{2}(\mathbf{k}_2+\mathbf{k}_1)]$$
$$\times\,\delta[-\mathbf{k}_2'+(\mathbf{p}_1-\mathbf{p}_2)+\tfrac{1}{2}(\mathbf{k}_1+\mathbf{k}_2)]\}\exp[\tfrac{1}{2}(\mathbf{k}_1'-\mathbf{k}_1)\cdot\partial_1+\tfrac{1}{2}(\mathbf{k}_2'-\mathbf{k}_2)\cdot\partial_2]$$
$$\times(\mathbf{k}_1'\cdot\mathbf{p}_1+\mathbf{k}_2'\cdot\mathbf{p}_2)\pi_1(\mathbf{k}_1',\mathbf{p}_1)\pi_1(\mathbf{k}_2',\mathbf{p}_2)$$
$$= -i\int d\mathbf{k}_1'\, d\mathbf{k}_2'\,\{(\mathbf{k}_1\cdot\mathbf{p}_1+\mathbf{k}_2\cdot\mathbf{p}_2)\,\delta(-\mathbf{k}_1'+\mathbf{k}_1)\,\delta(-\mathbf{k}_2'+\mathbf{k}_2)$$
$$+[\mathbf{k}_1'\cdot(\mathbf{p}_1+\tfrac{1}{2}(\mathbf{k}_1'-\mathbf{k}_1))+\mathbf{k}_2'\cdot(\mathbf{p}_2+\tfrac{1}{2}(\mathbf{k}_2'-\mathbf{k}_2))]$$
$$\times\,\theta\delta(-\mathbf{k}_1'+\cdots)\,\delta(-\mathbf{k}_2'+\cdots)\}\exp[\cdots]\pi_1\pi_1$$
$$= -i\int d\mathbf{k}_1'\, d\mathbf{k}_2'\,\{(\mathbf{k}_1\cdot\mathbf{p}_1+\mathbf{k}_2\cdot\mathbf{p}_2)\,\delta(-\mathbf{k}_1'+\mathbf{k}_1)\,\delta(-\mathbf{k}_2'+\mathbf{k}_2)$$
$$+[\tfrac{1}{2}(\mathbf{p}_2-\mathbf{p}_1+\tfrac{1}{2}\mathbf{k}_1+\tfrac{1}{2}\mathbf{k}_2)\cdot(\mathbf{p}_2+\mathbf{p}_1+\tfrac{1}{2}\mathbf{k}_2-\tfrac{1}{2}\mathbf{k}_1)$$
$$+\tfrac{1}{2}(\mathbf{p}_1-\mathbf{p}_2+\tfrac{1}{2}\mathbf{k}_1+\tfrac{1}{2}\mathbf{k}_2)\cdot(\mathbf{p}_1+\mathbf{p}_2+\tfrac{1}{2}\mathbf{k}_1-\tfrac{1}{2}\mathbf{k}_2)]$$
$$\times\,\theta\delta(-\mathbf{k}_1'+\cdots)\,\delta(-\mathbf{k}_2'+\cdots)\exp[\cdots]\pi_1\pi_1$$
$$= -i(\mathbf{k}_1\cdot\mathbf{p}_1+\mathbf{k}_2\cdot\mathbf{p}_2)\int d\mathbf{k}_1'\, d\mathbf{k}_2'\,\{\delta(-\mathbf{k}_1'+\mathbf{k}_1)\,\delta(-\mathbf{k}_2'+\mathbf{k}_2)$$
$$+\theta\delta(-\mathbf{k}_1'+\cdots)\,\delta(-\mathbf{k}_2'+\cdots)\}\exp[\cdots]\pi_1\pi_1$$
$$= (L_1^0+L_2^0)P_2(1\,|\,2)\pi_2(1\,|\,2)$$

It is easily seen from the general form of the symmetrizer, (3.8.16) that Eq. (14.3.11) is but a particular case of a quite general property:

$$P_s(\Gamma_s)\Big(\sum_{j=1}^{s} L_j^0\Big)=\Big(\sum_{j=1}^{s} L_j^0\Big)P_s(\Gamma_s)\qquad \text{all } \Gamma_s \tag{14.3.12}$$

The symmetrizer for all the correlation patterns commutes with the unperturbed Liouvillian. Because of this property, the equation for the *symmetrized* correlation pattern $\pi_2^S(1\,|\,2)$ can now be written as follows*:

$$(\partial_t-L_1^0-L_2^0)P_2(1\,|\,2)\pi_2(1\,|\,2)$$
$$=\int d\xi_3\,\delta(\mathbf{k}_3)P_2(1\,|\,2)\{L_{13}'P_2(1\,|\,3)\pi_3(1\,|\,2\,|\,3)+L_{23}'P_2(2\,|\,3)\pi_3(1\,|\,2\,|\,3)$$
$$+L_{13}'P_2(13)\pi_3(2\,|\,13)+L_{23}'P_2(23)\pi_3(1\,|\,23)\} \tag{14.3.13}$$

* At this stage it is clear why the unsymmetrized patterns are a better choice for the representation of $\mathfrak{f}$ than the symmetrized patterns. The latter do not obey a closed system of equations. Indeed, in the rhs of (14.3.13) appear, for instance, the combinations $P_2(1\,|\,3)\pi_3(1\,|\,2\,|\,3)$ and $P_2(2\,|\,3)\pi_3(1\,|\,2\,|\,3)$: None of these is the completely symmetrized pattern $P_3(1\,|\,2\,|\,3)\pi_3(1\,|\,2\,|\,3)$. We will see that precisely this difference gives rise to important effects.

We now obtain the equation for the correlation pattern $\pi_2(12)$ by subtracting term by term Eq. (14.3.13) from (14.3.9):

$$
\begin{aligned}
&(\partial_t - L_1^0 - L_2^0)\pi_2(12) \\
&\quad = L'_{12}P_2(1\,|\,2)\pi_2(1\,|\,2) + L'_{12}P_2(12)\pi_2(12) \\
&\qquad + \int d\xi_3\; \delta(\mathbf{k}_3)\{[L'_{13}P_3(1\,|\,2\,|\,3) - P_2(1\,|\,2)L'_{13}P_2(1\,|\,3) \\
&\qquad + L'_{23}P_3(1\,|\,2\,|\,3) - P_2(1\,|\,2)L'_{23}P_2(2\,|\,3)]\pi_3(1\,|\,2\,|\,3) \\
&\qquad + [L'_{13}P_3(1\,|\,23) + L'_{23}P_3(1\,|\,23) - P_2(1\,|\,2)L'_{23}P_2(23)]\pi_3(1\,|\,23) \\
&\qquad + [L'_{23}P_3(2\,|\,13) + L'_{13}P_3(2\,|\,13) - P_2(1\,|\,2)L'_{13}P_2(13)]\pi_3(2\,|\,13) \\
&\qquad + (L'_{13} + L'_{23})P_3(3\,|\,12)\pi_3(3\,|\,12) + (L'_{13} + L'_{23})P_3(123)\pi_3(123)\}
\end{aligned}
\tag{14.3.14}
$$

This step concludes our study of the two-particle correlation patterns. We now *define* the dynamical correlation patterns as being *any* solution of Eqs. (14.3.10), (14.3.14); comparing these equations with the general form (14.3.4) we obtain the list of matrix elements. We first note the quite important property:

$$
\langle (s)[\Gamma_s]|\, \mathscr{L}^0 \,|(r)[\Gamma'_r]\rangle = \delta_{s,r}\, \delta_{\Gamma_s,\Gamma'_r}\left(\sum_{j=1}^{s} L_j^0\right) \tag{14.3.15}
$$

This relation is quite general. It is *the same as the classical relation* (14.2.4). Once again the free-motion quantum Liouvillian is diagonal both in the number of particles s and in the correlation index Γ_s.

Proceeding to the interactions, we obtain from (14.3.10) the following matrix elements contributing to $\pi_2(1\,|\,2)$:

$$
\begin{aligned}
&\langle 1\,|\,2|\, \mathscr{L}' \,|1\,|\,2\rangle = \langle 1\,|\,2|\, \mathscr{L}' \,|12\rangle = 0 \\
&\langle 1\,|\,2|\, \mathscr{L}' \,|1\,|\,2\,|\,3\rangle = \int d\xi_3\; \delta(\mathbf{k}_3)[L'_{13}P_2(1\,|\,3) + L'_{23}P_2(2\,|\,3)] \\
&\langle 1\,|\,2|\, \mathscr{L}' \,|1\,|\,23\rangle = \int d\xi_3\; \delta(\mathbf{k}_3)L'_{23}P_2(2\,|\,3) \\
&\langle 1\,|\,2|\, \mathscr{L}' \,|2\,|\,13\rangle = \int d\xi_3\; \delta(\mathbf{k}_3)L'_{13}P_2(1\,|\,3) \\
&\langle 1\,|\,2|\, \mathscr{L}' \,|3\,|\,12\rangle = \langle 1\,|\,2|\, \mathscr{L}' \,|123\rangle = 0
\end{aligned}
\tag{14.3.16}
$$

These matrix elements are very similar to the classical ones, Eq. (14.2.7): the only difference is the occurrence of symmetrizers, in conjunction with every interaction process, just as in Eqs. (14.3.7). Consider now the matrix elements contributing to the pattern $\pi_2(12)$:

$$
\begin{aligned}
\langle 12|\,\mathscr{L}'\,|1\,|\,2\rangle &= L'_{12}P_2(1\,|\,2)\\
\langle 12|\,\mathscr{L}'\,|12\rangle &= L'_{12}P_2(12)\\
\langle 12|\,\mathscr{L}'\,|1\,|\,2\,|\,3\rangle &= \int d\xi_3\; \delta(\mathbf{k}_3)\{[L'_{13}P_3(1\,|\,2\,|\,3) - P_2(1\,|\,2)L'_{13}P_2(1\,|\,3)]\\
&\qquad + [L'_{23}P_3(1\,|\,2\,|\,3) - P_2(1\,|\,2)L'_{23}P_2(2\,|\,3)]\}\\
\langle 12|\,\mathscr{L}'\,|1\,|\,23\rangle &= \int d\xi_3\; \delta(\mathbf{k}_3)\{L'_{13}P_3(2\,|\,13)\\
&\qquad + [L'_{23}P_3(1\,|\,23) - P_2(1\,|\,2)L'_{23}P_2(23)]\}\\
\langle 12|\,\mathscr{L}'\,|2\,|\,13\rangle &= \int d\xi_3\; \delta(\mathbf{k}_3)\{L'_{23}P_3(1\,|\,23)\\
&\qquad + [L'_{13}P_3(2\,|\,13) - P_2(1\,|\,2)L'_{13}P_2(13)]\}\\
\langle 12|\,\mathscr{L}'\,|3\,|\,12\rangle &= \int d\xi_3\; \delta(\mathbf{k}_3)(L'_{13} + L'_{23})P_3(3\,|\,12)\\
\langle 12|\,\mathscr{L}'\,|123\rangle &= \int d\xi_3\; \delta(\mathbf{k}_3)(L'_{13} + L'_{23})P_3(123)
\end{aligned}
\tag{14.3.17}
$$

Here we see for the first time the appearance of dramatic new effects; *there are four contributions that have no classical counterpart.* They are the terms enclosed in square brackets.

Most typically, the classical matrix element $\langle 12|\,\mathscr{L}'\,|1\,|\,2\,|\,3\rangle$ is zero [see Eq. (14.2.8)] whereas for quantum systems it does not vanish. The origin of these terms is very clear; they occur because of the different symmetrizers appearing in Eqs. (14.3.9) and (14.3.13). If the symmetrizers were simply the identity operator, the two terms in the square brackets would cancel, and one would recover the classical result. The presence of these terms is thus a quite clear signature of the Bose–Einstein or of the Fermi–Dirac statistics.

The presence of these terms (there are more and more of them as we go to larger numbers of particles) introduces a considerable (and unavoidable) complication in the theory of quantum-statistical time evolution. It is, however, possible to disentangle these effects by introducing again a diagrammatic technique.

In classical mechanics, the dynamics of correlations is adequately represented by the graphical system explained in Section 14.1. The basic idea there is that the only source of correlations in the system is the presence of interactions. The "vertex" is a good graphical representation of this idea, because it introduces connections between two lines and thus represents the transformations of the correlation patterns into one another by means of the interactions.

In quantum mechanics, besides the interactions *there exists an additional mechanism for the creation of correlations: the constraints imposed by the Pauli principle.* This is precisely the origin of the extra terms. We, therefore, need another graphical element to represent this mechanism. We note that all the "extra" terms correspond, one to one, to the classical disconnected diagrams.

For instance, the first term in the matrix element $\langle 12|\mathcal{L}'|1|2|3\rangle$:

$$\int d\xi_3\, \delta(\mathbf{k}_3)[L'_{13}P_3(1|2|3) - P_2(1|2)L'_{13}P_2(1|3)]$$

is clearly related to the interaction L'_{13}; however, the interaction process between particles 1 and 3 is influenced by the presence of particle 2: as a result, a "quantum-statistical" correlation between particles 1 and 2 appears in the final state. We may understand this process physically as follows: In our formalism, the particles are *labeled* by indices $1, 2, 3, \ldots$. This is, of course, no error: the symmetrizers see to it that the Pauli principle is satisfied. However, in some cases, this picture of labeled particles is too extreme. In particular in the process under discussion, we have an interaction between two (labeled) particles in presence of a "loose" one: this would be an inconsistent picture, because the particles 1, 2, and 3 are physically indistinguishable. Hence, the mere presence of particle 2 must influence the interaction process. The result is the creation of a correlation between the three particles.

Such a process is adequately represented by a diagram in which the lines 1 and 3 are connected by a vertex, and an extra connection (a wavy line) links this vertex to the line 2. With these ideas in mind, the diagrams representing the two equations (14.3.10) and (14.3.14) are drawn in Fig. 14.3.1.

As we go on to correlation patterns with more particles the contributions become, of course, more and more complicated. There appear new "statistical" contributions. They are always generated by disconnected diagrams for the reasons just explained. An example is given in Fig. 14.3.2. We now have the possibility of one, two, three statistical links. It should be stressed however that there can only be one link between the vertex and a connected part of a diagram (not a link to each line of that part). We now give the rules for the construction of the matrix elements. These rules can be inferred from the analysis of higher order patterns and by inductive reasoning: we do not give formal proofs.

In order to analyze the contributions to the correlation patterns of s particles:

1. Draw all the possible "classical" diagrams.

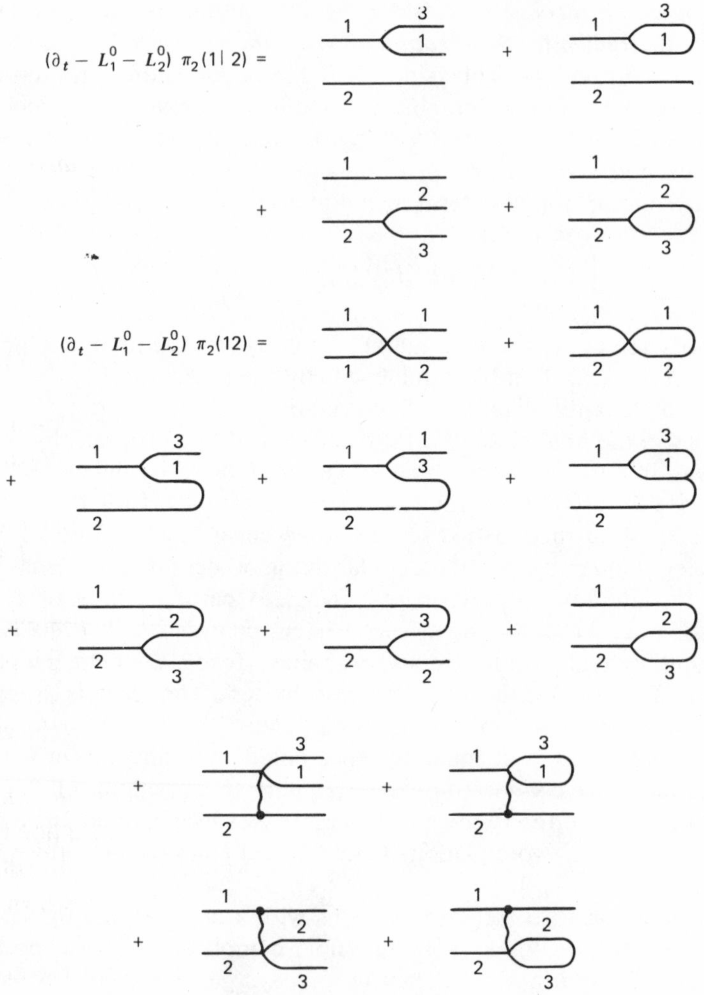

$(\partial_t - L_1^0 - L_2^0)\,\pi_2(1|2) =$

$(\partial_t - L_1^0 - L_2^0)\,\pi_2(12) =$

Figure 14.3.1. Diagrammatic representation of the equation of evolution of the quantum-statistical two-body correlation patterns.

2. Starting with each disconnected diagram, draw all the quantum-statistical diagrams it generates by connecting the vertex to the various connected parts, in all possible ways (see Fig. 14.3.2).

3. Classify the diagrams as contributions to various correlation patterns according to their connectivity.

In order to write down the contribution corresponding to an arbitrary diagram, we fix the following system of notations. A general diagram is

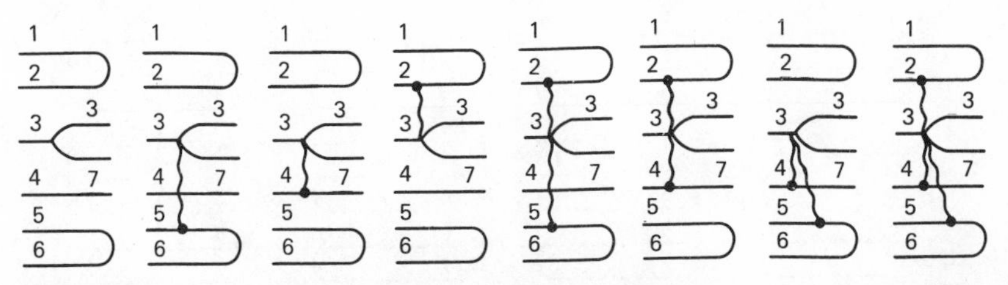

Figure 14.3.2. The set of all quantum-statistical diagrams generated by a disconnected diagram. These diagrams contribute, respectively, to the equations of evolution of: $\pi_6(3|4|12|56)$, $\pi_6(4|12|356)$, $\pi_6(12|34|56)$, $\pi_6(4|56|123)$, $\pi_6(4|12356)$, $\pi_6(56|1234)$, $\pi_6(12|3456)$, $\pi_6(123456)$.

made up of $(K+1)$ connected components, which in the global figure are either disconnected or connected by wavy lines. One of these parts contains the vertex joining particles i and j: it will be called S; the other components are numbered I, II, III, . . . , K (see Fig. 14.3.3).

Disregarding the quantum links, each of the components I, II, . . . , K taken isolately, represents a certain pattern that we denote by (I), (II), ... In the component S, the pattern is in general different to the right of the vertex (S'), and to the left (S''). The notation (II | S') denotes the pattern formed by the union—without mutual correlation—of the particles in the components II and S'. In the example of Fig. 14.3.3, (II | S') denotes the pattern (8 | 45 | 67). The notation (IS'') denotes the pattern formed by the union of the particles in the components I and S'', considered as mutually correlated [in our example, (IS'') = (12345)]. With these notations we may write the operators corresponding to the diagrams as follows:

A. Diagrams with No Wavy Line

$$L'_{ij}P(S') \tag{14.3.18}$$

B. Diagrams with One Wavy Line Between the Components S and I

$$L'_{ij}P(\mathrm{I}\,|\,\mathrm{II}\,|\cdots|\,K\,|\,S') - P(\mathrm{I}\,|\,S'')L'_{ij}P(S') \tag{14.3.19}$$

C. Diagrams with Two Wavy Lines Between Components S, I, and II

$$\begin{aligned}
&L'_{ij}P(\mathrm{I}\,|\,\mathrm{II}\,|\cdots|\,K\,|\,S') - P(\mathrm{I}\,|\,\mathrm{II}\,|\,S'')L'_{ij}P(S') \\
&\qquad - P(\mathrm{I}\,|\,\mathrm{II}S'')[L'_{ij}P(\mathrm{II}\,|\,S') - P(\mathrm{II}\,|\,S'')L'_{ij}P(S')] \\
&\qquad - P(\mathrm{II}\,|\,\mathrm{I}S'')[L'_{ij}P(\mathrm{I}\,|\,S') - P(\mathrm{I}\,|\,S'')L'_{ij}P(S')]
\end{aligned}$$

In these expressions, if the pattern to the right of the vertex has one more particle than to the left (i.e., if the vertex is of type B), an integration over the extra particle is understood implicitly.

The reader can immediately check that these rules are indeed in agreement with Eqs. (14.3.16), (14.3.17). To illustrate them in a more

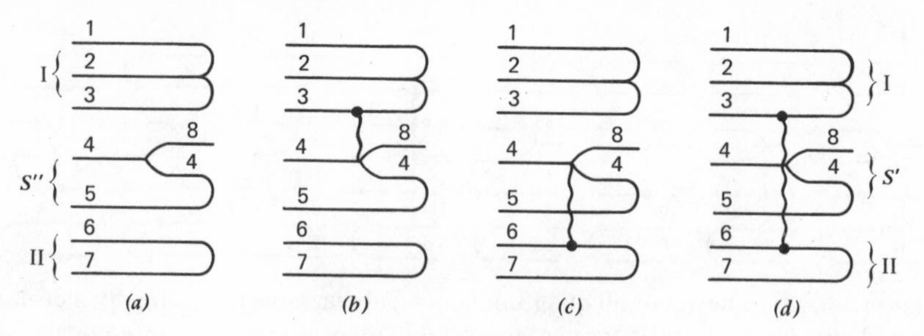

Figure 14.3.3. Analysis of a complex diagram. (I) = (123), (II) = (67), (S′) = (8 | 45), (S″) = (45).

complicated case we write down the contributions of the diagrams of Fig. 14.3.3:

(a) $\int d8\, L'_{48}P(8 \mid 45)$

(b) $\int d8\, \{L'_{48}P(8 \mid 45 \mid 67 \mid 123) - P(45 \mid 123)L'_{48}P(8 \mid 45)\}$

(c) $\int d8\, \{L'_{48}P(8 \mid 45 \mid 67 \mid 123) - P(45 \mid 67)L'_{48}P(8 \mid 45)\}$

(d)
$$\begin{aligned}\int d8\, \{&L'_{48}P(8 \mid 45 \mid 67 \mid 123) - P(45 \mid 67 \mid 123)L'_{48}P(8 \mid 45)\\ &- P(123 \mid 4567)[L'_{48}P(8 \mid 45 \mid 67) - P(45 \mid 67)L'_{48}P(8 \mid 45)]\\ &- P(67 \mid 12345)[L'_{48}P(8 \mid 45 \mid 123) - P(45 \mid 123)L'_{48}P(8 \mid 45)]\}\end{aligned}$$

Diagrams with three or more wavy lines give rise to increasingly complicated contributions. The careful reader has certainly understood the general formation law of these contributions. We shall not write them down, because they will not be used in this book. We insist, however, that, with a little patience, any diagram can be translated systematically into a well-defined mathematical expression.

As a result of this analysis, we are entitled to consider the Liouvillian $\mathcal{L}$ as a known operator: its matrix elements can be constructed by a systematic method. The *explicit* form of these matrix elements in quantum mechanics is considerably more complicated than in classical mechanics. This unavoidable complexity is due to the Pauli principle. However, we cannot insist enough on the *identity of structure* of the two cases. Equation (14.3.4) holds in both cases, as it stands: Only the translation of the symbols is different. *This fundamental fact allows us to develop the general theory of time evolution in Chapters 15–17 without even asking ourselves if we are doing classical or quantum mechanics: The theory is exactly the same in both cases.* Only in Chapters 18 and 19 shall we consider the specific translation of the theory in the two languages.

BIBLIOGRAPHICAL NOTES

The *BBGKY hierarchy* was transformed into a nonlinear system of equations for the irreducible correlation functions by several authors. Let us mention, among others:

M. S. Green, *J. Chem. Phys.* **25,** 836 (1956).

E. G. D. Cohen, *Physica* **28,** 1025 (1956).

The *correlation patterns,* as defined here, were introduced in R. Balescu, *Physica* **54,** 1 (1971) (classical systems); *Physica* **62,** 485 (1972) (quantum systems).

The concept was also applied to relativistic systems and to systems of oscillators (such as the electromagnetic field): see R. Balescu and I. Paiva-Veretennicoff, *Bull. Cl. Sci. Acad. Roy. Belg.* **62,** 457 (1971).

The idea of *dynamics of correlations* is due to Prigogine, who introduced it in a somewhat different representation (Fourier components of the N-body distribution function): I. Prigogine, *Non-Equilibrium Statistical Mechanics,* Wiley-Interscience, New York, 1963.

CHAPTER 15

DYNAMICS AND SUBDYNAMICS OF IDEAL SYSTEMS

15.1. SOLUTION OF THE UNPERTURBED LIOUVILLE EQUATION

The simplicity of a system of noninteracting particles is such that the problem of its evolution in time can be solved exactly. Such a system is not very interesting in itself. Indeed, it manifests none of the characteristic features of macroscopic systems; in particular, it shows no drive toward a thermal equilibrium of the type described in the first part of this book. Because of its simplicity, the ideal gas will, however, be taken as a reference system, just as in equilibrium theory.

The Liouville equation, in its generalized matrix form, has been derived and discussed in detail in Chapter 3. For an ideal system it is simply written as

$$\partial_t \mathfrak{f}(t) = \mathscr{L}^0 \mathfrak{f}(t) \qquad (15.1.1)$$

$\mathfrak{f}(t)$ stands for the *distribution vector*, which can be represented either as a set of reduced distribution functions $f_s(x_1, \ldots, x_s; t)$ (see Sections 3.3 and 3.6) or as a set of dynamical correlation patterns, $p_s(x_1, \ldots, x_s; [\Gamma_s]; t)$ (see Sections 14.2 and 14.3).

Let the initial condition be denoted by $\mathfrak{f}(0)$: it is not further specified (at present). The formal solution of the initial-value problem for Eq. (15.1.1) can be written in the form:

$$\mathfrak{f}(t) = \mathscr{U}^0(t)\mathfrak{f}(0) \qquad (15.1.2)$$

This equation *defines* the operator $\mathscr{U}^0(t)$, called the unperturbed *Greenian* or *propagator*. It could be written formally as

$$\mathscr{U}^0(t) = \exp(t\mathscr{L}^0) \qquad (15.1.3)$$

However, this form is not yet very useful: we must give a meaning to the exponential. Another characterization consists in noting that the operator $\mathscr{U}^0(t)$ must satisfy the Liouville equation:

$$\partial_t \mathscr{U}^0(t) = \mathscr{L}^0 \mathscr{U}^0(t) \qquad (15.1.4)$$

with the initial condition:

$$\mathcal{U}^0(0) = I \tag{15.1.5}$$

where I is the identity operator. Our purpose now is to construct useful representations of the propagator $\mathcal{U}^0(t)$.

The simplicity of the unperturbed case lies mainly in the fact that the Liouvillian $\mathcal{L}^0$ is *diagonal* in all the representations of Chapters 3 and 14. Hence the component equations are decoupled from each other. Consider first the representation

$$\mathfrak{f}(t) = \{f_s(x_1, \ldots, x_s; t)\} \tag{15.1.6}$$

The components of Eq. (15.1.1) are then

$$\partial_t f_s(x_1, \ldots, x_s; t) = \left(\sum_{j=1}^{s} L_j^0\right) f_s(x_1, \ldots, x_s; t) \tag{15.1.7}$$

We stress again here the important fact that the s-particle Liouvillian is a *sum* of one-particle Liouvillians. This property is characteristic of ideal systems.

We recall the form of the one-particle Liouvillian, defined by Eq. (2.4.17) as

$$L_j^0 = -\mathbf{v}_j \cdot \nabla_j \tag{15.1.8}$$

where

$$\mathbf{v}_j = \frac{\mathbf{p}_j}{m} \tag{15.1.9}$$

is the velocity of particle j. We also recall that this form is valid both for classical and for quantum systems [see Eq. (3.7.7)].

The formal solution of equation (15.1.7) is now written as

$$f_s(x_1, \ldots, x_s; t) = \exp\left\{t \sum_{j=1}^{s} L_j^0\right\} f_s(x_1, \ldots, x_s; 0) \tag{15.1.10}$$

The exponential operator appearing in this equation has a very simple meaning. We first note that the operators L_j^0, each referring to a single particle, *commute* with each other. But we know that the exponential of the sum of mutually commuting operators equals the *product* of the exponentials of the individual terms. Hence

$$\exp\left\{t \sum_{j=1}^{s} L_j^0\right\} = \prod_{j=1}^{s} \exp(tL_j^0) \tag{15.1.11}$$

We now recall the important identity [see Eq. (3.6.7)]:

$$\exp(a\nabla)g(q) = g(q+a)$$

Hence, because of the form (15.1.11) of the operator L_j^0, each individual factor is a simple *finite displacement operator:*

$$\exp(tL_j^0)f_s(\mathbf{q}_1, \ldots, \mathbf{q}_j, \mathbf{p}_j, \ldots, \mathbf{p}_s) = \exp(-t\mathbf{v}_j \cdot \nabla_j)f_s(\cdots \mathbf{q}_j, \mathbf{p}_j, \cdots)$$
$$= f_s(\mathbf{q}_1, \ldots, \mathbf{q}_j - \mathbf{v}_j t, \mathbf{p}_j, \ldots, \mathbf{p}_s) \quad (15.1.12)$$

The two equations (15.1.11), (15.1.12) completely define the action of the operator $\exp(t \sum L_j^0)$ on any function of q and p.

Let us introduce the following simpler notation:

$$\exp\left\{t \sum_{j=1}^{s} L_j^0\right\} = U_{1,\ldots,s}^0(t) \quad (15.1.13)$$

Equation (15.1.11) can then be written as

$$U_{1,\ldots,s}^0(t) = \prod_{j=1}^{s} U_j^0(t) \quad (15.1.14)$$

We may now sum up this discussion by giving a complete definition of the global unperturbed propagator $\mathcal{U}^0(t)$. Writing Eq. (15.1.2) explicitly in the representation (15.1.6) we obtain

$$f_s(t) = \sum_{r=0}^{\infty} \langle\!\langle (s)|\, \mathcal{U}^0(t)\, |(r)\rangle\!\rangle f_r(0) \quad (15.1.15)$$

Equation (15.1.10) shows that the operator $\mathcal{U}^0(t)$ is diagonal:

$$\langle\!\langle (s)|\, \mathcal{U}^0(t)\, |(r)\rangle\!\rangle = \delta_{rs} \prod_{j=1}^{s} U_j^0(t), \qquad s > 0 \quad (15.1.16)$$

which, for completeness, must be supplemented with

$$\langle\!\langle (0)|\, \mathcal{U}^0(t)\, |(r)\rangle\!\rangle = \delta_{r0} \quad (15.1.17)$$

Hence, the s-particle diagonal element of the operator is simply the *product* of s one-particle propagators.

Finally, we write the explicit form of the solution of the unperturbed Liouville equation (15.1.15):

$$f_s(\mathbf{q}_1, \mathbf{p}_1, \ldots, \mathbf{q}_s, \mathbf{p}_s; t) = U_{1,\ldots,s}^0(t) f_s(\mathbf{q}_1, \mathbf{p}_1, \ldots, \mathbf{q}_s, \mathbf{p}_s; 0)$$
$$= f_s(\mathbf{q}_1 - \mathbf{v}_1 t, \mathbf{p}_1, \ldots, \mathbf{q}_s - \mathbf{v}_s t, \mathbf{p}_s; 0) \quad (15.1.18)$$

This equation implies that the distribution functions are constant along the straight trajectories of the noninteracting particles: this is precisely the content of the classical Liouville theorem.

We now proceed to the correlation patterns representation of the distribution vector (see Sections 14.2 and 14.3):

$$\mathfrak{f} = \{p_s(x_1, \ldots, x_s; [\Gamma_s]; t)\} \quad (15.1.19)$$

Clearly, nothing new appears here. Indeed, we know from the results of Sections 14.2 and 14.3 that the unperturbed Liouvillian is diagonal in the partition indices Γ_s. We therefore immediately obtain the matrix elements of the operator $\mathcal{U}^0(t)$ in this representation:

$$\langle (s)[\Gamma_s] | \mathcal{U}^0(t) | (r)[\Gamma'_r] \rangle = \delta_{sr}\, \delta_{\Gamma_s \Gamma'_r} \prod_{j=1}^{s} U_j^0(t) \tag{15.1.20}$$

The solution of the initial-value problem for the individual correlation patterns is derived in the same way as Eq. (15.1.18):

$$p_s(\mathbf{q}_1, \mathbf{p}_1, \ldots, \mathbf{q}_s, \mathbf{p}_s; [\Gamma_s]; t) = p_s(\mathbf{q}_1 - \mathbf{v}_1 t, \mathbf{p}_1, \ldots, \mathbf{q}_s - \mathbf{v}_s t, \mathbf{p}_s; [\Gamma_s]; 0) \tag{15.1.21}$$

Before closing this section, let us recall how one solves an equation slightly more general than (15.1.1), namely, the *inhomogeneous Liouville equation:*

$$\partial_t \mathfrak{f}(t) = \mathcal{L}^0 \mathfrak{f}(t) + \mathfrak{g}(t) \tag{15.1.22}$$

where $\mathfrak{g}(t)$ is a given, time-dependent vector, that is, a "source term." It is well known from the theory of linear differential equations that if the propagator $\mathcal{U}^0(t)$ is known, the solution of the initial-value problem to Eq. (15.1.22) is

$$\mathfrak{f}(t) = \mathcal{U}^0(t)\mathfrak{f}(0) + \int_0^t d\tau\, \mathcal{U}^0(\tau)\mathfrak{g}(t-\tau) \tag{15.1.23}$$

or, equivalently (after a trivial change of integration variables):

$$\mathfrak{f}(t) = \mathcal{U}^0(t)\mathfrak{f}(0) + \int_0^t d\tau\, \mathcal{U}^0(t-\tau)\mathfrak{g}(\tau) \tag{15.1.24}$$

This solution is easily checked by time differentiation of both sides and use of Eqs. (15.1.4) and (15.1.5):

$$\begin{aligned}\partial_t \mathfrak{f}(t) &= \mathcal{L}^0 \mathcal{U}^0(t)\mathfrak{f}(0) + \int_0^t d\tau\, \mathcal{L}^0 \mathcal{U}^0(t-\tau)\mathfrak{g}(\tau) + \mathcal{U}^0(0)\mathfrak{g}(t) \\ &= \mathcal{L}^0 \left\{ \mathcal{U}^0(t)\mathfrak{f}(0) + \int_0^t d\tau\, \mathcal{U}^0(t-\tau)\mathfrak{g}(\tau) \right\} + \mathfrak{g}(t)\end{aligned}$$

which agrees with (15.1.22).

Equations (15.1.23) and (15.1.24) are very important and will frequently be used, in particular in perturbation theory.

15.2. THE RESOLVENT OPERATOR

An alternative, often very useful way of expressing the solution of Eq. (15.1.1) is in terms of the *resolvent operator*, which will be presently

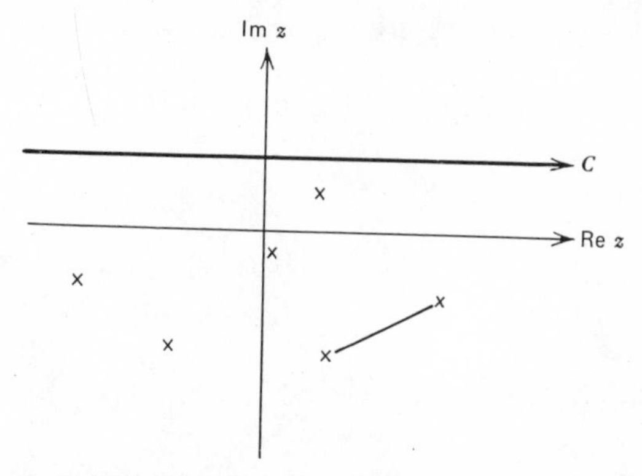

Figure 15.2.1. Contour for the Laplace inversion formula. The crosses represent singularities of $f(z)$.

defined. This concept is introduced through the *Laplace transformation** of the time-dependent functions. The Laplace transform $\tilde{f}(z)$ of a (sufficiently regular) function of time $f(t)$ is defined by

$$\tilde{f}(z) = \int_0^\infty dt\, e^{izt} f(t) \equiv \text{Lap}\{f(t)\} \tag{15.2.1}$$

where z is a *complex* variable. The well-known inversion formula is

$$f(t) = (2\pi)^{-1} \int_C dz\, e^{-izt} \tilde{f}(z) \tag{15.2.2}$$

where C is a contour in the complex z plane, parallel to the real axis, and lying above all the singularities of $\tilde{f}(z)$ (Fig. 15.2.1).

We come back later to the study of the properties of $\tilde{f}(z)$. At present we just note that by means of the Laplace transformation the image of the time differentiation is a simple algebraic operation:

$$\begin{aligned}\text{Lap}\{\partial_t f(t)\} &= \int_0^\infty dt\, e^{izt} \partial_t f(t) \\ &= [e^{izt} f(t)]_0^\infty - \int_0^\infty dt\, e^{izt} iz f(t) \\ &= -f(0) - iz\tilde{f}(z)\end{aligned} \tag{15.2.3}$$

(It is assumed that the boundary term at $t = \infty$ vanishes, which may imply that z has a sufficiently large positive imaginary part.)

* Also called *one-sided Fourier transformation*. The notation adopted here is closer to the Fourier-transform notation than to the standard Laplace-transform notation. The link between these two notations is trivial: replace in our formulas the variable iz by $-s$; replace our terms "upper half-plane," "above · · ·" by "right half-plane," "to the right of · · ·".

Consider now Eq. (15.1.7); its Laplace image is

$$-iz\tilde{f}_s(x_1,\ldots,x_s;z)=f_s(x_1,\ldots,x_s;0)+\sum_{j=1}^{s}L_j^0\tilde{f}_s(x_1,\ldots,x_s;z) \quad (15.2.4)$$

Hence, we can write, formally

$$\tilde{f}_s(x_1,\ldots,x_s;z)=\left[-\sum_{j=1}^{s}L_j^0-iz\right]^{-1}f_s(x_1,\ldots,x_s;0) \quad (15.2.5)$$

and by (15.2.1)

$$f_s(x_1,\ldots,x_s;t)=(2\pi)^{-1}\int_C dz\, e^{-izt}\left[-\sum_{j=1}^{s}L_j^0-iz\right]^{-1}f_s(x_1,\ldots,x_s;0) \quad (15.2.6)$$

We therefore obtain a new form of the solution of the initial-value problem, equivalent to (15.1.10). Let us introduce the *resolvent operator* defined formally as

$$\mathcal{R}^0(z)=(-\mathcal{L}^0-iz)^{-1} \quad (15.2.7)$$

Its matrix elements are

$$\langle(s)|\,\mathcal{R}^0(z)\,|(r)\rangle=\delta_{sr}\left[-\sum_{j=1}^{s}L_j^0-iz\right]^{-1} \quad (15.2.8)$$

and the solution (15.2.6) can be written as

$$\mathfrak{f}(t)=(2\pi)^{-1}\int_C dz\, e^{-izt}\mathcal{R}^0(z)\mathfrak{f}(0) \quad (15.2.9)$$

Comparing this equation to (15.1.2) we see that the resolvent is the Laplace transform of the propagator:

$$\mathcal{U}^0(t)=(2\pi)^{-1}\int_C dz\, e^{-izt}\mathcal{R}^0(z) \quad (15.2.10)$$

To summarize our results, we have seen that the unperturbed problem can be completely and explicitly solved. Two equivalent methods can be used to derive the solution: the method of the propagator $\mathcal{U}^0(t)$ and the method of the resolvent $\mathcal{R}^0(z)$, each one having its own advantages. Both operators are diagonal in all representations used in this section.

15.3. VACUUM AND CORRELATIONS: THE CONCEPT OF SUBDYNAMICS

The idea of correlation patterns provides us with a deep insight into the structure of the evolution process. We discuss this question now at the level of the ideal gas. These ideas pave the way to the far-reaching results to be discussed later, in Chapter 16.

We first note that, among all possible correlation patterns of s particles, one pattern is privileged. This is the pattern:

$$p_s(x_1, \ldots, x_s; [0_s]) \equiv p_s(1 \mid 2 \mid \cdots \mid s)$$

This pattern corresponds to the maximum partition of the set s or, physically, to completely uncorrelated particles. Its importance can be understood from its normalization property: it is normalized to *one*, whereas all other patterns are normalized to zero [see Eqs. (14.2.10), (14.2.11)]. As $p_s([0_s])$ carries the whole normalization of f_s, *it can never vanish*. On the contrary, correlations $p_s([\Gamma_s])$ with $\Gamma_s \neq 0_s$ may or may not be present in a given statistical state of the system.

We now collect all the correlation patterns $p_s([0_s])$, for $s = 0, 1, 2, \ldots$, and arrange them into an ordered set, of the same form as the distribution vector $\mathfrak{f}$. This set is clearly a subset of $\mathfrak{f}$. It will be called the *vacuum component* of the distribution vector (or simply the *vacuum*)* and will be denoted by $V\mathfrak{f}$. The remaining correlation patterns form a complementary subset, called the *correlation component* of the distribution vector (or simply the *correlations*) and denoted by $C\mathfrak{f}$. Technically, it is more convenient however for $V\mathfrak{f}$ and $C\mathfrak{f}$ to have the same dimension as $\mathfrak{f}$ itself: this is easily obtained by writing zeros for the missing components.

Hence we may express the decomposition as follows:

$$\begin{bmatrix} p_0 \\ p_1(1) \\ p_2(1 \mid 2) \\ p_2(12) \\ p_3(1 \mid 2 \mid 3) \\ p_3(1 \mid 23) \\ p_3(2 \mid 13) \\ p_3(3 \mid 12) \\ p_3(123) \\ \cdot \\ \cdot \\ \cdot \end{bmatrix} = \begin{bmatrix} p_0 \\ p_1(1) \\ p_2(1 \mid 2) \\ 0 \\ p_3(1 \mid 2 \mid 3) \\ 0 \\ 0 \\ 0 \\ 0 \\ \cdot \\ \cdot \\ \cdot \end{bmatrix} + \begin{bmatrix} 0 \\ 0 \\ 0 \\ p_2(12) \\ 0 \\ p_3(1 \mid 23) \\ p_3(2 \mid 13) \\ p_3(3 \mid 12) \\ p_3(123) \\ \cdot \\ \cdot \\ \cdot \end{bmatrix}$$

This vector equation is expressed compactly as

$$\mathfrak{f} = V\mathfrak{f} + C\mathfrak{f} \tag{15.3.1}$$

* This terminology is inspired from quantum field theory. In the latter it denotes a state without particles (see Section 1.5). Here it denotes a state without correlations.

$V\mathfrak{f}$ and $C\mathfrak{f}$ are vectors whose components are denoted by $Vp_s([\Gamma_s])$ and $Cp_s([\Gamma_s])$:

$$\begin{aligned} V\mathfrak{f} &= \{Vp_s(x_1, \ldots, x_s; [\Gamma_s])\} \\ C\mathfrak{f} &= \{Cp_s(x_1, \ldots, x_s; [\Gamma_s])\} \end{aligned} \tag{15.3.2}$$

The components are defined as follows:

$$Vp_s(x_1, \ldots, x_s; [\Gamma_s]) = \delta_{\Gamma_s, 0_s} p_s(x_1, \ldots, x_s; [0_s]) \tag{15.3.3}$$

$$Cp_s(x_1, \ldots, x_s; [\Gamma_s]) = (1 - \delta_{\Gamma_s, 0_s}) p_s(x_1, \ldots, x_s; [\Gamma_s]) \tag{15.3.4}$$

With this definition we clearly have

$$Vp_s(x_1, \ldots, x_s; [\Gamma_s]) + Cp_s(x_1, \ldots, x_s; [\Gamma_s]) = p_s(x_1, \ldots, x_s; [\Gamma_s]) \tag{15.3.5}$$

We may also define a similar separation of the reduced distribution functions. We derive from Eq. (14.2.9) a consistent definition of this decomposition as follows:

$$Vf_s(x_1, \ldots, x_s) = \sum_{\Gamma_s} Vp_s(x_1, \ldots, x_s; [\Gamma_s]) = p_s(x_1, \ldots, x_s; [0_s]) \tag{15.3.6}$$

$$Cf_s(x_1, \ldots, x_s) = \sum_{\Gamma_s} Cp_s(x_1, \ldots, x_s; [\Gamma_s]) = \sum_{\substack{\Gamma_s \\ (\neq 0_s)}}' p_s(x_1, \ldots, x_s; [\Gamma_s]) \tag{15.3.7}$$

Again, we have

$$Vf_s(x_1, \ldots, x_s) + Cf_s(x_1, \ldots, x_s) = f_s(x_1, \ldots, x_s) \tag{15.3.8}$$

We note that the separation (15.3.1) can be performed by acting on the distribution vector with two *operators* V and C. The effect of the operator V is to select in $\mathfrak{f}$ all the uncorrelated components and annull the others. The effect of the operator C is complementary. Let us study some properties of these operators.

We first note that these operators are *linear:* given two distribution vectors $\mathfrak{f}^{(1)}$ and $\mathfrak{f}^{(2)}$ we have

$$V(a\mathfrak{f}^{(1)} + b\mathfrak{f}^{(2)}) = aV\mathfrak{f}^{(1)} + bV\mathfrak{f}^{(2)} \tag{15.3.9}$$

where $a\mathfrak{f}^{(1)} + b\mathfrak{f}^{(2)}$ is the set whose components are*

$$a\mathfrak{f}^{(1)} + b\mathfrak{f}^{(2)} = \{ap_s^{(1)}(x_1, \ldots, x_s; [\Gamma_s]) + bp_s^{(2)}(x_1, \ldots, x_s; [\Gamma_s])\}$$

The linearity follows from the fact that the dynamical correlation patterns $p_s(x_1, \ldots, x_s; [0_s])$ are defined as the solutions of a set of linear differential equations (see Section 14.2). Any linear combination of particular solutions of these equations is still a solution.

* We may note that such a linear combination has a physical meaning only if $a + b = 1$. Otherwise the normalization properties of $a\mathfrak{f}^{(1)} + b\mathfrak{f}^{(2)}$ would be violated.

Equation (15.3.1) implies that

$$V + C = I \tag{15.3.10}$$

This relation simply means that the two vectors are complementary. We next note the properties:

$$\begin{aligned} V^2 &= V \\ C^2 &= C \\ VC &= 0 \\ CV &= 0 \end{aligned} \tag{15.3.11}$$

These equations simply express the fact that the two subsets do not overlap. If the components $\Gamma_s = 0_s$ are selected by a first application of V, a subsequent application of V will not change the result: A subsequent application of C, however, yields zero because there are no correlated components in the vector Vf.

The properties (15.3.9)–(15.3.11) entitle us to call V and C (somewhat loosely) *projection operators* (or *projectors*) on the vacuum and on the correlations, respectively. We can write an explicit representation of the matrix elements of these operators:

$$\langle (s)[\Gamma_s] | \, V \, | (r)[\Gamma'_r] \rangle = \delta_{sr} \, \delta_{\Gamma_s,0_s} \, \delta_{\Gamma'_r,0_r} \tag{15.3.12}$$

$$\langle (s)[\Gamma_s] | \, C \, | (r)[\Gamma'_r] \rangle = \delta_{sr} \, \delta_{\Gamma_s,\Gamma'_r} (1 - \delta_{\Gamma_s,0_s}) \tag{15.3.13}$$

It is easily verified that these definitions are precisely equivalent to the operational definitions (15.3.2)–(15.3.4). Indeed, if the expression $V\mathfrak{f}$ is written as a matrix product, it becomes

$$\begin{aligned} (V\mathfrak{f})(x_1, \ldots, x_s; [\Gamma_s]) &= \sum_{r=0}^{\infty} \sum_{\Gamma'_r} \langle (s)[\Gamma_s] | \, V \, | (r)[\Gamma'_r] \rangle p_r(x_1, \ldots, x_r; [\Gamma'_r]) \\ &= \sum_{r=0}^{\infty} \sum_{\Gamma'_r} \delta_{s,r} \, \delta_{\Gamma_s,0_s} \, \delta_{\Gamma'_s,0_s} p_r(x_1, \ldots, x_r; [\Gamma'_r]) \\ &= \delta_{\Gamma_s,0_s} p_s(x_1, \ldots, x_s; [0_s]) = V p_s(x_1, \ldots, x_s; [\Gamma_s]) \end{aligned} \tag{15.3.14}$$

which is consistent with Eqs. (15.3.2) and (15.3.3). The properties (15.3.9)–(15.3.11) can be formally verified in a similar way.

One of the most important properties of the correlation patterns is their *invariance under the unperturbed motion.* Let us clearly explain what is meant by this sentence. An individual correlation pattern generally changes its shape under the unperturbed motion, as can be seen from Eq. (15.1.21). But it does not change its character (specified by the partition index Γ_s). Starting from a given $p_s([\Gamma_s])$, the unperturbed motion cannot generate a $p_s([\Gamma'_s])$ with $\Gamma'_s \neq \Gamma_s$. Inversely, the evolution of $p_s([\Gamma_s])$ is unaffected by the other correlation patterns. The knowledge of the initial value of the pattern of index Γ_s: $p_s([\Gamma_s]; t=0)$ is sufficient for the determination of its value at any other time. All these properties are consequences of the diagonal character of $\mathcal{L}^0$.

The invariance property of individual correlation patterns *a fortiori* applies to the sets of correlation patterns $V\mathfrak{f}$ and $C\mathfrak{f}$. It is, indeed, easily seen that the equations of evolution of the vectors $V\mathfrak{f}$ and $C\mathfrak{f}$ are

$$\partial_t V\mathfrak{f} = \mathcal{L}^0 V\mathfrak{f} \tag{15.3.15}$$

$$\partial_t C\mathfrak{f} = \mathcal{L}^0 C\mathfrak{f} \tag{15.3.16}$$

The proof is very simple. Using Eqs. (14.2.4) and (14.3.14) the right-hand side of Eq. (15.3.15) is written, in matrix form, as follows:

$$\begin{aligned}
&\sum_{r=0}^{\infty}\sum_{\Gamma'_r} \langle (s)[\Gamma_s]|\, \mathcal{L}^0\, |(r)[\Gamma'_r]\rangle V p_r(x_1, \ldots, x_r; [\Gamma'_r]) \\
&\qquad = \sum_{r=0}^{\infty}\sum_{\Gamma'_r} \delta_{sr}\, \delta_{\Gamma_s,\Gamma'_r} \left(\sum_{j=1}^{s} L_j^0\right) \delta_{\Gamma'_r,0_r} p_r(x_1, \ldots, x_r; [\Gamma'_r]) \\
&\qquad = \left(\sum_{j=1}^{s} L_j^0\right) \delta_{\Gamma_s,0_s} p_s(x_1, \ldots, x_s; [0_s]) = \delta_{\Gamma_s,0_s} \partial_t p_s(x_1, \ldots, x_s; [0_s]) \\
&\qquad = \partial_t V p_s(x_1, \ldots, x_s; [\Gamma_s])
\end{aligned}$$

This result precisely equals the rate of change of the s-particle component of $V\mathfrak{f}$.

Equations (15.3.15) and (15.3.16) clearly bring out the complete mutual independence of the components Vf and Cf. Hence the separation (15.3.1), obtained by means of the projectors V and C, defines a *decomposition of the set of distribution vectors* $\mathfrak{f}$ *into two subsets that are invariant under the unperturbed motion.* Any element of the subset $\{V\mathfrak{f}\}$ *remains* in $\{V\mathfrak{f}\}$ under the effect of the motion.

This property can also be expressed in a different, quite suggestive way. Consider any arbitrary matrix operator $\mathcal{P}$ acting on the distribution vectors $\mathfrak{f}$. This operator can be uniquely decomposed in the following fashion, by making use of (15.3.10):

$$\mathcal{P} = I\mathcal{P}I = (V+C)\mathcal{P}(V+C)$$

hence

$$\mathcal{P} = V\mathcal{P}V + V\mathcal{P}C + C\mathcal{P}V + C\mathcal{P}C \tag{15.3.17}$$

We express this by saying that every operator has a vacuum-to-vacuum part, a correlation-to-vacuum part, a vacuum-to-correlation part and a correlation-to-correlation part.

In particular, the unperturbed Liouville equation (15.1.1) can be written as

$$\partial_t \mathfrak{f}(t) = (V\mathscr{L}^0 V + V\mathscr{L}^0 C + C\mathscr{L}^0 V + C\mathscr{L}^0 C)\mathfrak{f}(t)$$

Applying now the operators V or C to both sides and using Eqs. (15.3.11) we obtain, respectively,

$$\begin{aligned} \partial_t V\mathfrak{f}(t) &= V\mathscr{L}^0 V\mathfrak{f}(t) + V\mathscr{L}^0 C\mathfrak{f}(t) \\ \partial_t C\mathfrak{f}(t) &= C\mathscr{L}^0 V\mathfrak{f}(t) + C\mathscr{L}^0 C\mathfrak{f}(t) \end{aligned} \tag{15.3.18}$$

These equations must be identical to Eqs. (15.3.15), (15.3.16). It therefore follows that

$$\begin{aligned} V\mathscr{L}^0 C &= 0 \\ C\mathscr{L}^0 V &= 0 \\ V\mathscr{L}^0 V &= \mathscr{L}^0 V \\ C\mathscr{L}^0 C &= \mathscr{L}^0 C \end{aligned} \tag{15.3.19}$$

These equations are equivalent to the following set:

$$\begin{aligned} \mathscr{L}^0 V &= V\mathscr{L}^0 \\ \mathscr{L}^0 C &= C\mathscr{L}^0 \end{aligned} \tag{15.3.20}$$

as can be easily seen from Eq. (15.3.11). Hence, *the operators $\mathscr{L}^0$ and V (and therefore also $\mathscr{L}^0$ and C) commute with each other.* Equation (15.3.20) provides the most general expression of *the decoupling of the subsets $\{V\mathfrak{f}\}$ and $\{C\mathfrak{f}\}$.*

Let us, finally, mention the following trivial consequences of Eq. (15.3.20):

$$\begin{aligned} \mathscr{U}^0(t)V &= V\mathscr{U}^0(t) \\ \mathscr{U}^0(t)C &= C\mathscr{U}^0(t) \qquad \text{all } t \end{aligned} \tag{15.3.21}$$

and similarly

$$\begin{aligned} \mathscr{R}^0(z)V &= V\mathscr{R}^0(z) \\ \mathscr{R}^0(z)C &= C\mathscr{R}^0(z) \qquad \text{all } z \end{aligned} \tag{15.3.22}$$

These equations hold for arbitrary values of the parameters t and z.

The main significance of these results is the following. The problem of the complete time evolution of $\mathfrak{f}(t)$ has been split into two independent problems, that is, the study of $V\mathfrak{f}(t)$ and the study of $C\mathfrak{f}(t)$. If it so happens that in a given problem we are only interested in the value of $V\mathfrak{f}(t)$, we can completely forget about $C\mathfrak{f}(t)$ in its evaluation. The component $V\mathfrak{f}(t)$ obeys its own "subdynamics" [and so does $C\mathfrak{f}(t)$].

The reduction of the dynamical problem is pretty trivial in the case of the unperturbed motion, because the problem can anyway be solved exactly. The very unexpected, and highly nontrivial, result is that a similar reduction leading to independent subdynamics exists even in a system of interacting particles. This result will be progressively derived in the coming chapters. It constitutes the backbone of the entire nonequilibrium theory.

BIBLIOGRAPHICAL NOTES

The dynamics of nonideal systems is a very standard subject. Any one of the general references quoted in Chapter 17 treats it.

The *resolvent* was extensively used by Prigogine (quoted in Chapter 14), Balescu (quoted in Chapter 6) and P. Résibois, in *Many Particle Physics* (E. Meeron, ed.) Gordon and Breach, New York, 1967.

The use of *projection operators* in statistical mechanics was initiated by R. Zwanzig, in *Lectures in Theoretical Physics*, Vol. 3 (Summer Institute of Theor. Phys., Univ. Colorado, 1960), Wiley-Interscience, New York, 1961.

The vacuum and correlation projectors, as used here, were introduced in R. Balescu, *Physica* **38,** 98 (1968).

Similar definitions of these operators were given by

M. Baus, *Bull. Cl. Sci. Acad. Roy. Belg.* **53,** 1332 (1967).

G. Severne, *Transp. Theory Stat. Phys.* **1,** 145 (1971).

CHAPTER 16

DYNAMICS AND SUBDYNAMICS OF INTERACTING SYSTEMS

16.1. FORMAL SOLUTION OF THE LIOUVILLE EQUATION

We now proceed to the study of more realistic systems including the interactions in the picture. The basic Liouville equation has been discussed in detail in Chapters 3 and 14:

$$\partial_t \mathfrak{f}(t) = \mathscr{L}\mathfrak{f}(t) \tag{16.1.1}$$

where the Liouvillian $\mathscr{L}$ consists of two terms:

$$\mathscr{L} = \mathscr{L}^0 + \mathscr{L}' \tag{16.1.2}$$

For simplicity we only consider here systems in absence of external fields, $\mathscr{L}^F \equiv 0$.

The solution of the linear equation (16.1.1) can again be expressed in terms of a greenian or *propagator* $\mathscr{U}(t)$, defined by the relation:

$$\mathfrak{f}(t) = \mathscr{U}(t)\mathfrak{f}(0) \tag{16.1.3}$$

Alternatively, the Greenian can be defined as the solution of the equation:

$$\partial_t \mathscr{U}(t) = \mathscr{L}\mathscr{U}(t) \tag{16.1.4}$$

with the initial condition:

$$\mathscr{U}(0) = I \tag{16.1.5}$$

We may write the solution of this equation as in Eq. (15.1.3):

$$\mathscr{U}(t) = \exp(t\mathscr{L}) = \exp[t(\mathscr{L}^0 + \mathscr{L}')] \tag{16.1.6}$$

However, it is no longer possible to give an explicit meaning to the operator $\mathscr{U}(t)$. (This reflects the fact that the equations of motion for interacting systems cannot be solved explicitly.)

$\mathscr{U}(t)$ is not even equal to the product $(\exp t\mathscr{L}^0)(\exp t\mathscr{L}')$ because the operators $\mathscr{L}^0$ and $\mathscr{L}'$ do not commute with each other. One may define

$\mathcal{U}(t)$ by its power-series expansion:

$$\mathcal{U}(t)=\sum_{r=0}^{\infty}(r!)^{-1}t^r\mathcal{L}^r \tag{16.1.7}$$

This expansion, however, is not useful in practice, except for very small (physically uninteresting) values of t. Indeed, if the series is truncated after a finite number of terms, it is clear that the neglected terms become unboundedly large as time passes. The expansion (16.1.7) will, however, be used in certain formal proofs.

To obtain useful approximate forms of the propagator, we may start again from its definition (16.1.4):

$$\partial_t\mathcal{U}(t)=\mathcal{L}^0\mathcal{U}(t)+\mathcal{L}'\mathcal{U}(t) \tag{16.1.8}$$

and treat it in a slightly different way. Equation (16.1.8) is of the same form as the inhomogeneous Liouville equation (15.1.22), provided we consider formally the interaction term as a source term. Taking account of the initial condition (16.1.5), Eq. (15.1.24) provides us immediately with the "solution" of (16.1.8):

$$\mathcal{U}(t)=\mathcal{U}^0(t)+\int_0^t d\tau\,\mathcal{U}^0(t-\tau)\mathcal{L}'\mathcal{U}(\tau) \tag{16.1.9}$$

As the second term in the rhs of (16.1.8) is not a true source term, but involves the unknown $\mathcal{U}(t)$, it is clear that (16.1.9) is not a complete solution. Rather, it is an *integral equation* equivalent to the *set* of equations (16.1.8) and (16.1.5). This integral form has, however, the advantage of being very easily soluble by the method of successive iterations. Indeed, the zeroth-order approximation is obtained by neglecting the second term in the rhs:

$$^{(0)}\mathcal{U}(t)=\mathcal{U}^0(t)$$

The next approximation is obtained by substituting this form into the integrand of Eq. (16.1.9):

$$^{(1)}\mathcal{U}(t)=\mathcal{U}^0(t)+\int_0^t d\tau\,\mathcal{U}^0(t-\tau)\mathcal{L}'\mathcal{U}^0(\tau)$$

The next approximation is similarly obtained:

$$^{(2)}\mathcal{U}(t)=\mathcal{U}^0(t)+\int_0^t d\tau\,\mathcal{U}^0(t-\tau)\mathcal{L}'\mathcal{U}^0(\tau)$$
$$+\int_0^t d\tau\int_0^\tau d\tau'\,\mathcal{U}^0(t-\tau)\mathcal{L}'\mathcal{U}^0(\tau-\tau')\mathcal{L}'\mathcal{U}^0(\tau')$$

and so on. The structure of the terms is now apparent. They involve multiple convolutions of operators $[\mathcal{U}^0(t)\mathcal{L}']$.

We recall that a *convolution* of two time-dependent functions is usually denoted by the symbol $A*B$ [or $(A*B)_t$ if it is necessary to specify the time variable] and is defined as

$$A*B \equiv (A*B)_t = \int_0^t d\tau\, A(t-\tau)B(\tau) = \int_0^t d\tau\, A(\tau)B(t-\tau) \qquad (16.1.10)$$

The convolution has all the algebraic properties of an associative product operation:

$$\begin{aligned} A*(B*C) &= (A*B)*C = A*B*C \\ A*(B+C) &= A*B + A*C \end{aligned} \qquad (16.1.11)$$

With this abbreviated notation, the second-order approximation can be written as

$$^{(2)}\mathcal{U} = \mathcal{U}^0 + \mathcal{U}^0*(\mathcal{L}'\mathcal{U}^0) + \mathcal{U}^0*(\mathcal{L}'\mathcal{U}^0)*(\mathcal{L}'\mathcal{U}^0)$$

This form gives us the hint for writing at once the general term of the expansion:

$$\mathcal{U} = \sum_{n=0}^{\infty} \mathcal{U}^0(*\,\mathcal{L}'\mathcal{U}^0)^n \qquad (16.1.12)$$

This equation gives the formal solution of the Liouville equation in terms of an infinite series, each term of which is a known and calculable expression, involving only the operators $\mathcal{U}^0(t)$, $\mathcal{L}'$. This expansion, moreover, coincides with the formal *perturbation expansion in terms of the interaction strength.* Of course, in order to have a precise meaning, this should be a convergent series. The convergence can, in general, not be proven, and we shall leave it open, treating Eq. (16.1.12) as a kind of "raw material," to be further transformed (see also Section 6.1).

The derivation of Eqs. (16.1.9) or (16.1.12) has been very easy owing to our use of a very compact notation. In explicit calculations, we have to evaluate matrix elements of the various operators appearing in these equations. This is very simply done by using the explicit expressions for the matrix elements of $\mathcal{U}^0(t)$ (Section 15.1) and of $\mathcal{L}'$ (Sections 3.4, 3.7, and 14.2) in any desired representation. These elements are combined by the rules of matrix multiplication. As an illustration we evaluate all matrix elements of the second-order contribution to $\mathcal{U}(t)$ ending with a one-particle state at left. We get, in the representation (15.1.6):

$$\begin{aligned} &\langle(1)|\, \mathcal{U}^0*\mathcal{L}'\mathcal{U}^0*\mathcal{L}'\mathcal{U}^0\, |(2)\rangle \\ &\quad = \int_0^t d\tau \int_0^\tau d\tau' \langle(1)|\, \mathcal{U}^0(t-\tau)\, |(1)\rangle\langle(1)|\, \mathcal{L}'\, |(2)\rangle \\ &\qquad \times \langle(2)|\, \mathcal{U}^0(\tau-\tau')\, |(2)\rangle\langle(2)|\, \mathcal{L}'\, |(2)\rangle\langle(2)|\, \mathcal{U}^0(\tau')\, |(2)\rangle \end{aligned}$$

and

$$\langle(1)|\,\mathcal{U}^0 * \mathcal{L}'\mathcal{U}^0 * \mathcal{L}'\mathcal{U}^0\,|(3)\rangle$$

$$= \int_0^t d\tau \int_0^\tau d\tau' \langle(1)|\,\mathcal{U}^0(t-\tau)\,|(1)\rangle\langle(1)|\,\mathcal{L}'\,|(2)\rangle$$

$$\times \langle(2)|\,\mathcal{U}^0(\tau-\tau')\,|(2)\rangle\langle(2)|\,\mathcal{L}'\,|(3)\rangle\langle(3)|\,\mathcal{U}^0(\tau')\,|(3)\rangle$$

It is easily seen that these are the only matrix elements of this operator (to second order) having a single particle at left. Indeed, we know from Section 3.4 that, in going from left to right, the operator $\mathcal{L}'$ can either leave the number of particles constant or increase it by one. Hence, from a 1-particle state one can reach either a 2-particle or a 3-particle state by two successive actions of $\mathcal{L}'$.

The reader is urged to work out similar examples in order to acquaint himself with the routine of the calculations.* It should be apparent that the matrix $\mathcal{U}(t)$ has a triangular shape; in other words:

$$\langle(s)|\,\mathcal{U}(t)\,|(s')\rangle = 0 \qquad \text{for } s' < s \tag{16.1.13}$$

$$\mathcal{U}(t) = \begin{pmatrix} 1 & 0 & 0 & 0 & 0 & 0 \\ 0 & & & & & \\ 0 & & & & & \\ 0 & & & & & \\ 0 & \mathbf{0} & & & & \\ 0 & & & & & \end{pmatrix}$$

The resolvent operator can also be defined for the complete Liouville operator by an extension of Eq. (15.2.9):

$$\mathfrak{f}(t) = (2\pi)^{-1} \int_C dz\, e^{-izt} \mathcal{R}(z) \mathfrak{f}(0) \tag{16.1.14}$$

The relation between propagator and resolvent is clearly

$$\mathcal{U}(t) = (2\pi)^{-1} \int_C dz\, e^{-izt} \mathcal{R}(z) \tag{16.1.15}$$

The resolvent can be defined again formally as

$$\mathcal{R}(z) = [-\mathcal{L}^0 - \mathcal{L}' - iz]^{-1} \tag{16.1.16}$$

In order to give a meaning to this equation we have to use once more a relation between the complete and the unperturbed resolvent. To derive it, we note the following identity, valid even for noncommuting operators

* A systematic treatment of this "translation" process appears in Chapter 19.

A, B:

$$\frac{1}{A+B}=\frac{1}{A}\left(I-B\frac{1}{A+B}\right) \tag{16.1.17}$$

Applying this relation to (16.1.16) and keeping in mind Eq. (15.2.7), we obtain

$$\mathcal{R}(z)=\mathcal{R}^0(z)+\mathcal{R}^0(z)\mathcal{L}'\mathcal{R}(z) \tag{16.1.18}$$

This equation is the counterpart of Eq. (16.1.9) and could have been obtained from the latter by using the relation (16.1.15). It reflects the important, well-known property that the *Laplace image of a convolution is an ordinary product*. Indeed, if

$$\tilde{a}(z)=\text{Lap}\{a(t)\}$$
$$\tilde{b}(z)=\text{Lap}\{b(t)\}$$

then

$$\text{Lap}\{(a*b)_t\}=\tilde{a}(z)\cdot\tilde{b}(z) \tag{16.1.19}$$

or explicitly

$$\int_0^\infty dt\, e^{izt}\int_0^t d\tau\, a(t-\tau)b(\tau)=\tilde{a}(z)\tilde{b}(z) \tag{16.1.20}$$

and, inversely

$$(2\pi)^{-1}\int_C dz\, e^{-izt}\tilde{a}(z)\tilde{b}(z)=\int_0^t d\tau\, a(t-\tau)b(\tau) \tag{16.1.21}$$

The formal perturbation solution of the operator equation (16.1.18) is

$$\mathcal{R}(z)=\sum_{n=0}^{\infty}[\mathcal{R}^0(z)\mathcal{L}']^n\mathcal{R}^0(z) \tag{16.1.22}$$

This relation is simpler than (16.1.12), because only ordinary products are involved (rather than convolutions). The various terms can be evaluated explicitly by again resolving these expressions into products of matrix elements.

16.2. PROGRAM FOR A KINETIC THEORY OF IRREVERSIBILITY

If we look back at the equations of evolution considered in Chapter 11 (Boltzmann, Fokker–Planck, Landau), we discover a certain number of common features:

(a′) They are *closed Markovian** equations for the one-particle distribution function.

* We recall that the term "Markovian" is used here in a very broad sense; it means that the rate of change of a function at time t depends only on the value of this function and of its position and momentum derivatives at the *same* time t.

(b) They lead irreversibly to a well-defined thermal equilibrium state.

The property (a′) can be reformulated in a more general way in terms of the correlation patterns:

(a) They are *closed Markovian equations for the vacuum component of the distribution vector.*

Indeed, if the particular form (14.2.12) is chosen for the vacuum patterns, a closed equation for the vacuum component reduces to a (nonlinear) closed equation for the one-particle distribution function.

Now, we know from the equations of the dynamics of correlations that Property (a) cannot be true for the distribution vector describing a system of interacting particles. We know from Sections 14.2 and 14.3, that the Liouville operator $\mathscr{L}$ has matrix elements connecting the vacuum components $p_s([0_s])$ to the correlated components $p_r([\Gamma_r])$. This can be seen in a compact form when the Liouville equation (16.1.1), (16.1.2) is projected on the vacuum by means of the projector defined in Section 15.3:

$$\partial_t V\mathfrak{f} = V(\mathscr{L}^0 + \mathscr{L}')\mathfrak{f}$$

or, using (15.3.10) and (15.3.20):

$$\partial_t V\mathfrak{f} = \mathscr{L}^0 V\mathfrak{f} + V\mathscr{L}' V\mathfrak{f} + V\mathscr{L}' C\mathfrak{f} \tag{16.2.1}$$

Hence, the equation for the vacuum component $V\mathfrak{f}$ involves the correlations $C\mathfrak{f}$, and vice versa. We have localized here, in a nutshell, the crux of the difficulty: As long as this apparent contradiction is not understood, the problem of irreversibility remains open. It is clear that this question is directly related to Boltzmann's *Stosszahlansatz* discussed in Section 11.5. We shall not discuss here the various approaches to this problem in historical order; some of the more important earlier theories will be discussed later. Rather, we shall directly introduce the ideas used in this book: they have the advantage of generality and of unification.

As the *Stosszahlansatz* cannot be true in exact dynamics, the next simplest guess we can make is the following: We may assume that the distribution vector can be decomposed into two terms:

$$\mathfrak{f}(t) = \bar{\mathfrak{f}}(t) + \hat{\mathfrak{f}}(t) \tag{16.2.2}$$

The splitting would be such that the term $\bar{\mathfrak{f}}(t)$ would have the characteristics described in Chapters 11–13. In particular, its vacuum part would obey a closed kinetic equation, describing irreversible approach to equilibrium. Moreover, hopefully, the remainder $\hat{\mathfrak{f}}(t)$ should be unimportant, at least for the problems studied in kinetic theory.

Many of the older approaches could be formulated in this manner. They consist in approximating the exact Liouville equation by means of various devices (coarse graining in phase space, time smoothing, asymptotic approximation, truncation of the hierarchy, etc.). The result of these approximations is a kinetic equation. Hence, by these methods one replaces the exact $\mathfrak{f}(t)$ by an approximate distribution vector obeying a kinetic equation and playing the role of $\bar{\mathfrak{f}}(t)$. In none of these theories was there much thought given to the complementary part $\hat{\mathfrak{f}}(t)$.

It must be realized, however, that, without further specification, the separation (16.2.2) into a kinetic and a nonkinetic part is trivial and does not prove anything. It is always possible to write a number A in terms of any other number B as $A = B + (A - B)$.

In order to make Eq. (16.2.2) the basis of a true theory, we must require it to reflect an *intrinsic and self-consistent structure* that is forced upon us (rather than the desired result being forced *by* us into the theory).

As a first property, we may ask that the separation (16.2.2) have *a geometrical meaning*. Considering the set of all possible distribution vectors $\mathfrak{f}$, it may be possible to find two time-independent operators, say $\boldsymbol{\Pi}$ and $\hat{\boldsymbol{\Pi}}$, which separate the set into two complementary subsets, one containing all kinetic parts, the other containing all nonkinetic parts. These operators, applied to an arbitrary element of the set $\mathfrak{f}(t)$, would automatically perform the separation (16.2.2):

$$\boldsymbol{\Pi}\mathfrak{f}(t) = \bar{\mathfrak{f}}(t) \tag{16.2.3}$$

$$\hat{\boldsymbol{\Pi}}\mathfrak{f}(t) = \hat{\mathfrak{f}}(t) \tag{16.2.4}$$

$$\mathfrak{f}(t) = \boldsymbol{\Pi}\mathfrak{f}(t) + \hat{\boldsymbol{\Pi}}\mathfrak{f}(t) \tag{16.2.5}$$

The two terms would then appear as true components of the vector $\mathfrak{f}(t)$.

For self-consistency, the operators must have the properties of "projection operators":

$$\begin{aligned} \boldsymbol{\Pi}^2 &= \boldsymbol{\Pi} \\ \hat{\boldsymbol{\Pi}}^2 &= \hat{\boldsymbol{\Pi}} \\ \boldsymbol{\Pi}\hat{\boldsymbol{\Pi}} &= 0 \\ \hat{\boldsymbol{\Pi}}\boldsymbol{\Pi} &= 0 \end{aligned} \tag{16.2.6}$$

together with the completeness relation:

$$\boldsymbol{\Pi} + \hat{\boldsymbol{\Pi}} = \mathsf{I} \tag{16.2.7}$$

These properties ensure that the two subsets $\{\bar{\mathfrak{f}}\}$ and $\{\hat{\mathfrak{f}}\}$ are completementary and not overlapping.

But the most important property, which would really convince us of the deep nature of the theory, should be an *invariance property*. Here is what

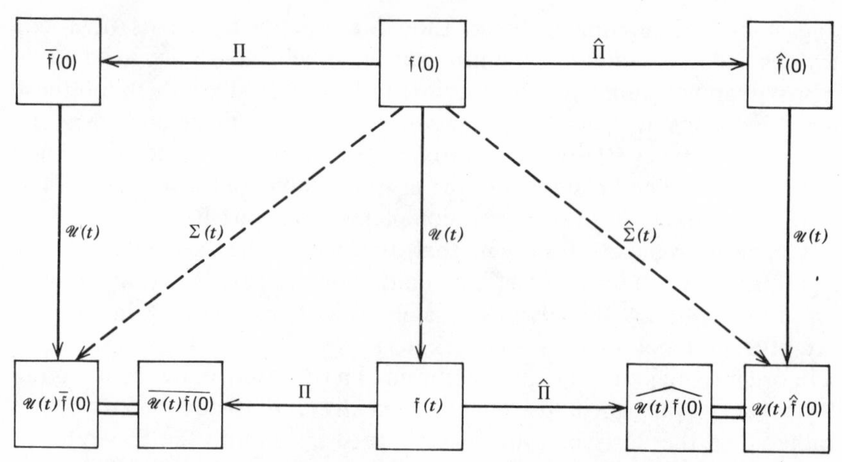

Figure 16.2.1. Time-translation invariance of the basic decomposition of the distribution vector. The operators $\Sigma(t)$ and $\hat{\Sigma}(t)$ will be defined in Section 16.4.

we mean (see Fig. 16.2.1). Consider a system described at time zero by a distribution vector $\mathfrak{f}(0)$. We split it into two components according to (16.2.2) and (16.2.5):

$$\mathfrak{f}(0) = \bar{\mathfrak{f}}(0) + \hat{\mathfrak{f}}(0)$$
$$\bar{\mathfrak{f}}(0) = \mathbf{\Pi}\mathfrak{f}(0)$$
$$\hat{\mathfrak{f}}(0) = \hat{\mathbf{\Pi}}\mathfrak{f}(0)$$

We now let the system evolve. At time t its distribution vector is obtained by the action of the propagagator $\mathcal{U}(t)$ on $\mathfrak{f}(0)$ [see Eq. (16.1.3)]:

$$\mathfrak{f}(t) = \mathcal{U}(t)\mathfrak{f}(0)$$

The vector $\mathfrak{f}(t)$ is then decomposed according to (16.2.2) with the result:

$$\begin{aligned} \mathfrak{f}(t) &= \bar{\mathfrak{f}}(t) + \hat{\mathfrak{f}}(t) \\ \bar{\mathfrak{f}}(t) &= \mathbf{\Pi}\mathfrak{f}(t) = \mathbf{\Pi}\mathcal{U}(t)\mathfrak{f}(0) \\ \hat{\mathfrak{f}}(t) &= \hat{\mathbf{\Pi}}\mathfrak{f}(t) = \hat{\mathbf{\Pi}}\mathcal{U}(t)\mathfrak{f}(0) \end{aligned} \tag{16.2.8}$$

On the other hand, we may wonder what happens to the separate components $\bar{\mathfrak{f}}(0)$ and $\hat{\mathfrak{f}}(0)$ when they are taken as an initial condition and when their evolution is followed in time. Clearly:

$$\begin{aligned} \bar{\mathfrak{f}}(0) &\to \mathcal{U}(t)\bar{\mathfrak{f}}(0) = \mathcal{U}(t)\mathbf{\Pi}\mathfrak{f}(0) \\ \hat{\mathfrak{f}}(0) &\to \mathcal{U}(t)\hat{\mathfrak{f}}(0) = \mathcal{U}(t)\hat{\mathbf{\Pi}}\mathfrak{f}(0) \end{aligned} \tag{16.2.9}$$

At this point it should be clear that the theory can only be self-consistent if, at time t, the $\mathbf{\Pi}$ component of $\mathfrak{f}(t)$ coincides with the result of the evolution of the initial $\mathbf{\Pi}$ component; in other words, if the right-hand sides of Eqs. (16.2.8) and (16.2.9) are identical. If this were not the case, the separation would depend on the time at which it is performed; there would be a privileged instant of time. But such an instant could not be singled out by any special physical property: hence such a theory would be physically untenable.

Comparing Eqs. (16.2.8) and (16.2.9) we see that the condition of *invariance of the separation (16.2.2) under time translation* is expressed in the following simple form:

$$\begin{aligned} &\mathbf{\Pi}\mathcal{U}(t) = \mathcal{U}(t)\mathbf{\Pi} \\ &\hat{\mathbf{\Pi}}\mathcal{U}(t) = \mathcal{U}(t)\hat{\mathbf{\Pi}} \qquad \text{all } t \end{aligned} \tag{16.2.10}$$

The operators $\mathbf{\Pi}$ and $\hat{\mathbf{\Pi}}$ must commute with the propagator $\mathcal{U}(t)$ for all values of t.

As a direct consequence of this commutation relation, the components $\bar{\mathfrak{f}}(t)$ and $\hat{\mathfrak{f}}(t)$ obey *separate* equations of evolution: there is no mixing between the subset $\{\bar{\mathfrak{f}}\}$ and $\{\hat{\mathfrak{f}}\}$. Indeed, using Eqs. (16.1.4) and (16.2.10):

$$\partial_t\bar{\mathfrak{f}}(t) = \partial_t\mathbf{\Pi}\mathcal{U}(t)\mathfrak{f}(0) = \partial_t\mathcal{U}(t)\mathbf{\Pi}\mathfrak{f}(0) = \mathcal{L}\mathcal{U}(t)\mathbf{\Pi}\mathfrak{f}(0)$$

or

$$\partial_t\bar{\mathfrak{f}}(t) = \mathcal{L}\bar{\mathfrak{f}}(t) \tag{16.2.11}$$

and similarly

$$\partial_t\hat{\mathfrak{f}}(t) = \mathcal{L}\hat{\mathfrak{f}}(t) \tag{16.2.12}$$

Hence the components $\bar{\mathfrak{f}}$ and $\hat{\mathfrak{f}}$ evolve in time, ignoring each other.* We shall say that each component obeys its own *subdynamics*.

We now note that, in the particular case of a *noninteracting system*, we have already found a pair of operators having precisely all the properties listed above: they are the operators V and C. Indeed, it was shown in Section 15.3 that the vacuum and correlation parts of $\mathfrak{f}$ are disconnected (15.3.15), that V and C are projection operators (15.3.11), and that they commute with the unperturbed propagator (15.3.20). Hence, it is reasonable to expect that, if an operator $\mathbf{\Pi}$ having all the required properties can be constructed, it should reduce to the operator V when the interactions are switched off:

$$\begin{aligned} &\mathbf{\Pi} \xrightarrow[\text{no interactions}]{} V \\ &\hat{\mathbf{\Pi}} \xrightarrow{\qquad\quad} C \end{aligned} \tag{16.2.13}$$

* One may ask at this stage the following question. The component $\bar{\mathfrak{f}}(t)$ obeys the Liouville equation, just like the total $\mathfrak{f}(t)$. How can it then describe irreversibility? The rather subtle answer to this question has to await further developments. The question just indicates that the theory is far from trivial.

Summarizing this discussion, we may start on the following program. Given a system of interacting degrees of freedom, to construct an operator Π having the following properties†:

A. Π commutes with $\mathcal{U}(t)$.
B. Π reduces to V in the limit of no interactions.
C. Π is idempotent (i.e., Π is a projector).
D. The vacuum part of $\Pi\mathfrak{f}(t)$ obeys a closed equation of evolution.
E. The stationary solution of this evolution equation coincides with the equilibrium distribution vector.
F. The complementary component $\hat{\Pi}\mathfrak{f}(t)$ is irrelevant in certain well-defined problems.

This program will be developed in the forthcoming sections. But we may already announce the great surprise coming out of this investigation. We will see that actually, *Condition* A (*combined with* B) *is so strong that it completely determines a unique operator* Π. Hence, we are given no freedom: Either this unique operator does possess the remaining properties C–F, or it does not. This unexpected feature now completely changes our strategy. We can no longer make a choice among a number of items "on the market" and take the one that pleases us most; on the contrary, the requirement of invariance offers us a single possibility. It is now up to us to *prove* that this possibility has anything to do with kinetic theory.

We cannot insist enough on the fact that the requirements A and B have nothing to do with irreversibility. Hence, if we do succeed (and we shall!) in proving that the unique operator Π defined by Conditions A and B possesses all the properties C–F, we shall be in presence of a truly objective theory of irreversibility. The result can be reformulated as follows:

The only invariant decomposition of the distribution vector with respect to time translation, is one in which the component $\mathfrak{f}(t)$ *has a time evolution of kinetic type.*

Let us conclude this section with the following remark. It would be completely unreasonable if we could construct an operator Π for *all* possible systems of interacting particles. A system of three interacting particles does not approach equilibrium; a system of particles interacting through unscreened long-range forces (e.g., gravitational forces) most probably does not reach equilibrium either. There must be some restriction to the validity of our scheme. Sure enough, we shall find that a nontrivial operator Π exists only if certain *additional conditions* are met

† It is clearly sufficient to construct Π: the operator $\hat{\Pi}$ is then given by $I-\Pi$.

by the system. These conditions precisely involve the thermodynamic limit, on one hand, and the nature of the interactions, on the other.

We are now ready to start carrying out our program.

16.3. FURTHER TRANSFORMATION OF THE RESOLVENT OPERATOR: THE IRREDUCIBLE EVOLUTION OPERATOR $\tilde{\mathscr{E}}(z)$—THE MASTER EQUATION

If we are to carry out the program of Section 16.2, it is clear that the vacuum and correlation projectors V and C must play an important role. This appears obviously in Condition B of Section 16.2, as well as from the fact that V and C precisely play the role of Π and $\hat{\Pi}$ for noninteracting systems.

But in the representations (16.1.12) and (16.1.22) these projectors do not appear at all. Hence, in order to start our program we first need a representation of the propagator $\mathscr{U}(t)$ and of the resolvent $\mathscr{R}(z)$ where the projectors V and C are explicitly exhibited.

It is not difficult to introduce these operators into the theory. Indeed, combining Eqs. (16.1.18) with (15.3.10) we have*

$$\mathscr{R}(z) = \mathscr{R}^0(z) + \mathscr{R}^0(z)\mathscr{L}'(V+C)\mathscr{R}(z) \tag{16.3.1}$$

or, separating the terms:

$$\mathscr{R} = \mathscr{R}^0 + \mathscr{R}^0\mathscr{L}'V\mathscr{R} + \mathscr{R}^0\mathscr{L}'C\mathscr{R} \tag{16.3.2}$$

(We omit writing the argument z in intermediate calculations.). The resolvent is clearly determined by both its vacuum and correlation parts. Let us *eliminate the correlation term* in the rhs of this equation. To do so, we project both sides on the correlation subspace:

$$C\mathscr{R} = C\mathscr{R}^0 + C\mathscr{R}^0\mathscr{L}'V\mathscr{R} + C\mathscr{R}^0\mathscr{L}'C\mathscr{R} \tag{16.3.3}$$

This is a linear inhomogeneous equation for the component $C\mathscr{R}$. It is of the form:

$$C\mathscr{R} = a + bC\mathscr{R}$$

The equation can be formally solved by successive iterations, just as Eq. (16.1.18) [see Eq. (16.1.22)]:

$$C\mathscr{R} = \sum_{m=0}^{\infty} (C\mathscr{R}^0\mathscr{L}')^m [C\mathscr{R}^0 + C\mathscr{R}^0\mathscr{L}'V\mathscr{R}]$$

* One should never forget, in the coming calculations, that $\mathscr{L}'V \neq V\mathscr{L}'$, $\mathscr{L}'C \neq C\mathscr{L}'$.

Substituting this result into the rhs of Eq. (16.3.2) we obtain

$$\mathcal{R} = \mathcal{R}^0 + \mathcal{R}^0\mathcal{L}' \sum_{m=0}^{\infty} (C\mathcal{R}^0\mathcal{L}')^m C\mathcal{R}^0 + \mathcal{R}^0\mathcal{L}' \sum_{m=0}^{\infty} (C\mathcal{R}^0\mathcal{L}')^m C\mathcal{R}^0\mathcal{L}' V\mathcal{R} + \mathcal{R}^0\mathcal{L}' V\mathcal{R}$$

which can be rearranged as follows:

$$\mathcal{R} = \mathcal{R}^0 + \mathcal{R}^0 \sum_{m=0}^{\infty} \mathcal{L}'(C\mathcal{R}^0\mathcal{L}')^m C\mathcal{R}^0 + \mathcal{R}^0 \sum_{m=0}^{\infty} \mathcal{L}'(C\mathcal{R}^0\mathcal{L}')^m V\mathcal{R} \quad (16.3.4)$$

We now introduce the following operator, which appears naturally in this equation:

$$\tilde{\mathscr{E}}(z) = \sum_{m=0}^{\infty} \mathcal{L}'[C\mathcal{R}^0(z)\mathcal{L}']^m \quad (16.3.5)$$

This operator, which plays a fundamental role in the theory, will be called the *irreducible evolution operator.* It has the same general structure as the resolvent [see Eq. (16.1.22)], with the important difference that the projector C prevents transitions through the vacuum in intermediate states. The operator $\tilde{\mathscr{E}}(z)$ can also be defined by the following equation:

$$\tilde{\mathscr{E}}(z) = \mathcal{L}' + \mathcal{L}' C\mathcal{R}^0(z)\tilde{\mathscr{E}}(z) \quad (16.3.6)$$

An equivalent equation is also

$$\tilde{\mathscr{E}}(z) = \mathcal{L}' + \tilde{\mathscr{E}}(z)C\mathcal{R}^0\mathcal{L}' \quad (16.3.7)$$

The formal iterative solution of both these equations is precisely (16.3.5).

The resolvent equation (16.3.4) can now be written as

$$\mathcal{R}(z) = \mathcal{R}^0(z) + \mathcal{R}^0(z)\tilde{\mathscr{E}}(z)C\mathcal{R}^0(z) + \mathcal{R}^0(z)\tilde{\mathscr{E}}(z)V\mathcal{R}(z) \quad (16.3.8)$$

Only the vacuum part of $\mathcal{R}(z)$ occurs now in the right-hand side. This implies that the correlation part of $\mathcal{R}$ is explicitly expressed in terms of the vacuum by the correlation component of this equation:

$$C\mathcal{R}(z) = C\mathcal{R}^0(z) + C\mathcal{R}^0(z)\tilde{\mathscr{E}}(z)C\mathcal{R}^0(z) + C\mathcal{R}^0(z)\tilde{\mathscr{E}}(z)V\mathcal{R}(z) \quad (16.3.9)$$

On the other hand, the vacuum part appears as the solution of the following *closed* equation (obtained from 16.3.8):

$$V\mathcal{R}(z) = V\mathcal{R}^0(z) + V\mathcal{R}^0(z)\tilde{\mathscr{E}}(z)C\mathcal{R}^0(z) + V\mathcal{R}^0(z)\tilde{\mathscr{E}}(z)V\mathcal{R}(z) \quad (16.3.10)$$

We now transform this equation further, to bring it in a particularly useful form. Multiplying both sides by $(-iz)$ we obtain*

$$(-iz)V\mathcal{R} = -iz\mathcal{R}^0 V - iz\mathcal{R}^0 V\tilde{\mathscr{E}}C\mathcal{R}^0 - iz\mathcal{R}^0 V\tilde{\mathscr{E}}V\mathcal{R} \quad (16.3.11)$$

* Equation (15.3.22) is repeatedly used in these calculations.

As a consequence of the definition (15.2.7) of the unperturbed resolvent, we have the following very useful identity:

$$-iz\mathcal{R}^0(z) = 1 + \mathcal{L}^0\mathcal{R}^0(z) \tag{16.3.12}$$

Hence

$$-izV\mathcal{R} = (1 + \mathcal{L}^0\mathcal{R}^0)(V + V\tilde{\mathscr{E}}C\mathcal{R}^0) + V\tilde{\mathscr{E}}V\mathcal{R} + \mathcal{L}^0\mathcal{R}^0V\tilde{\mathscr{E}}V\mathcal{R} \tag{16.3.13}$$

But the factor $\mathcal{R}^0V\tilde{\mathscr{E}}V\mathcal{R}$ in the last term on the rhs can be taken from Eq. (16.3.10):

$$-izV\mathcal{R} = (1 + \mathcal{L}^0\mathcal{R}^0)(V + V\tilde{\mathscr{E}}C\mathcal{R}^0) + V\tilde{\mathscr{E}}V\mathcal{R} + \mathcal{L}^0V\mathcal{R} - \mathcal{L}^0\mathcal{R}^0(V + V\tilde{\mathscr{E}}C\mathcal{R}^0)$$

Hence, we are left with the following equation for $V\mathcal{R}(z)$:

$$V\mathcal{R}(z) = (-iz)^{-1}[V + V\tilde{\mathscr{E}}(z)C\mathcal{R}^0(z)] + (-iz)^{-1}[V\tilde{\mathscr{E}}(z)V + \mathcal{L}^0V]V\mathcal{R}(z) \tag{16.3.14}$$

The iterative solution of this equation is

$$V\mathcal{R} = \sum_{n=0}^{\infty}(-iz)^{-n-1}[V\tilde{\mathscr{E}}V + \mathcal{L}^0V]^n(V + V\tilde{\mathscr{E}}\mathcal{R}^0C) \tag{16.3.15}$$

If we now combine Eq. (16.3.15) with (16.3.9), we can write the complete resolvent in the form:

$$\mathcal{R}(z) = \sum_{n=0}^{\infty}(-iz)^{-n-1}(V + C\mathcal{R}^0\tilde{\mathscr{E}}V)(V\tilde{\mathscr{E}}V + \mathcal{L}^0V)^n(V + V\tilde{\mathscr{E}}\mathcal{R}^0C) + C\mathcal{R}^0 + C\mathcal{R}^0\tilde{\mathscr{E}}\mathcal{R}^0C \tag{16.3.16}$$

The main result to be retained from these, somewhat tedious, calculations is the following. Suppose we decompose the resolvent in the form:

$$\mathcal{R} = V\mathcal{R}V + V\mathcal{R}C + C\mathcal{R}V + C\mathcal{R}C \tag{16.3.17}$$

It follows from Eq. (16.3.16) that *the four components are not independent: They can all be expressed in terms of the single* $V\mathcal{R}(z)V$:

$$V\mathcal{R}(z)C = [V\mathcal{R}(z)V][V\tilde{\mathscr{E}}(z)\mathcal{R}^0(z)C] \tag{16.3.18}$$

$$C\mathcal{R}(z)V = [C\mathcal{R}^0(z)\tilde{\mathscr{E}}(z)V][V\mathcal{R}(z)V] \tag{16.3.19}$$

$$C\mathcal{R}(z)C = [C\mathcal{R}^0(z)\tilde{\mathscr{E}}(z)V][V\mathcal{R}(z)V][V\tilde{\mathscr{E}}(z)\mathcal{R}^0(z)C] + C\mathcal{R}^0(z) + C\mathcal{R}^0(z)\tilde{\mathscr{E}}(z)\mathcal{R}^0(z)C \tag{16.320}$$

As for the V–V component, it is represented as

$$V\mathcal{R}(z)V = \sum_{n=0}^{\infty}(-iz)^{-n-1}[V\mathcal{L}^0 + V\tilde{\mathscr{E}}(z)V]^n \tag{16.3.21}$$

This very peculiar structure of the resolvent will play a considerable role in the forthcoming developments.

Before going further in our program, we make a digression at this point. Equation (16.3.14) can be rewritten as follows:

$$-izV\mathcal{R}(z) - V = \mathcal{L}^0 V\mathcal{R}(z) + V\tilde{\mathcal{E}}(z)V\mathcal{R}(z) + V\tilde{\mathcal{E}}(z)\mathcal{R}^0(z)C \quad (16.3.22)$$

We now write $\mathfrak{f}(0)$ to the right of every operator in this equation and perform an inverse Laplace transform. Using Eqs. (15.2.3) and (16.1.21) we find

$$\partial_t V\mathfrak{f}(t) = \mathcal{L}^0 V\mathfrak{f}(t) + \int_0^t d\tau\, V\mathcal{E}(\tau)V\mathfrak{f}(t-\tau)$$
$$+ \int_0^t d\tau\, V\mathcal{E}(\tau)\mathcal{U}^0(t-\tau)C\mathfrak{f}(0) \quad (16.3.23)$$

This is an important equation, first derived by Prigogine and Résibois in 1961. It is called the *master equation.* As appears clearly from its derivation, no approximation whatever is involved here: it is an exact equation. We note that it is an integrodifferential equation in time for the function $V\mathfrak{f}(t)$, formally *closed* in $V\mathfrak{f}$, with a source term. One would, therefore, be tempted to think that our search for a closed equation for the vacuum is accomplished. This is, however, an illusion. Indeed, the source term depends on the value of the complete correlation component, evaluated at time zero. Thus, in order to solve the initial value problem for this equation we need to specify not only $V\mathfrak{f}(0)$ but also $C\mathfrak{f}(0)$. We thus did not achieve a complete separation of vacuum and correlations. Moreover, Eq. (16.3.23) is definitely non-Markovian.

Nevertheless, this equation played a very important role in the recent developments of statistical mechanics. It can be shown that in some important systems, the source term (or *destruction fragment*) tends rapidly to zero. We then are left with an approximate equation of evolution, valid for the description of phenomena that vary slowly on the scale determined by the destruction fragment.* This equation is†

$$\partial_t V\mathfrak{f}(t) \approx \mathcal{L}^0 V\mathfrak{f}(t) + \int_0^\infty d\tau\, V\mathcal{E}(\tau)V\mathfrak{f}(t-\tau) \quad (16.3.24)$$

This is now a truly closed equation for $V\mathfrak{f}$. We will actually derive later (Section 17.2) precisely the same equation. However, its interpretation will be different. Instead of having the status of an approximate equation for $V\mathfrak{f}(t)$, it will appear as an exact, universal equation for *part* of the distribution function. In other words, the validity of (16.3.24) as a

* This separation of time scales was already discussed in the case of the Brownian-motion problem and of the Boltzmann equation in Chapter 11.

† It can be shown that in this approximation the limit of integration t can be pushed to infinity.

"subdynamical" equation is not bound to the stringent condition of the existence of two well-separated time scales. We now proceed with our program.

16.4. CONSTRUCTION OF THE OPERATOR $\boldsymbol{\Pi}$

We now attack our main problem. We want to construct an operator $\boldsymbol{\Pi}$ acting on the set of distribution vectors $\{\mathfrak{f}\}$. The vector resulting from the action of the operator $\boldsymbol{\Pi}$ on the distribution vector $\mathfrak{f}$ will be denoted by $\bar{\mathfrak{f}}$:

$$\bar{\mathfrak{f}} = \boldsymbol{\Pi}\mathfrak{f} \tag{16.4.1}$$

At this stage we require $\boldsymbol{\Pi}$ to satisfy the two conditions A and B of Section 16.2:

$$\boldsymbol{\Pi}\mathcal{U}(t) = \mathcal{U}(t)\boldsymbol{\Pi} \qquad \text{all } t \tag{16.4.2}$$

$$\boldsymbol{\Pi} \xrightarrow[\text{no interactions}]{} V \tag{16.4.3}$$

It is more convenient, actually, to work with the resolvent $\mathcal{R}(z)$ rather than the propagator $\mathcal{U}(t)$. In view of the definition (16.1.15), Eq. (16.4.2) implies

$$\boldsymbol{\Pi}\mathcal{R}(z) = \mathcal{R}(z)\boldsymbol{\Pi} \tag{16.4.4}$$

As follows from Eq. (16.1.15), this relation must hold for all values of z on the contour C; but this contour can be freely deformed, provided it avoids the singularities of $\mathcal{R}(z)$. Hence Eq. (16.4.4) must hold for all values of z within the domain of regularity of the function $\mathcal{R}(z)$.

Applying now the decomposition (15.3.17) to both sides, Eq. (16.4.4) is made equivalent to four component equations. Let us first analyze the $V-V$ component:

$$V\boldsymbol{\Pi}\mathcal{R}(z)V = V\mathcal{R}(z)\boldsymbol{\Pi}V \tag{16.4.5}$$

which can be further decomposed by inserting $I = V + C$ between the factors $\boldsymbol{\Pi}$ and $\mathcal{R}(z)$:

$$V\boldsymbol{\Pi}V\mathcal{R}(z)V + V\boldsymbol{\Pi}C\mathcal{R}(z)V = V\mathcal{R}(z)V\boldsymbol{\Pi}V + V\mathcal{R}(z)C\boldsymbol{\Pi}V \tag{16.4.6}$$

Using Eqs. (16.3.19) and (16.3.18), we rewrite this as

$$V\boldsymbol{\Pi}V\mathcal{R}(z)V - V\mathcal{R}(z)V\boldsymbol{\Pi}V + V\boldsymbol{\Pi}C\mathcal{R}^0(z)\mathcal{E}(z)V\mathcal{R}(z)V - V\mathcal{R}(z)V\tilde{\mathcal{E}}(z)\mathcal{R}^0(z)C\boldsymbol{\Pi}V = 0 \tag{16.4.7}$$

[Equations (15.3.22) and (15.3.11) were used in this transformation.]

Finally, we substitute Eq. (16.3.21) for $V\mathcal{R}(z)V$:

$$V\Pi V\left\{\sum_{n=0}^{\infty}(-iz)^{-n-1}\psi^n(z)\right\}-\left\{\sum_{n=0}^{\infty}(-iz)^{-n-1}\psi^n(z)\right\}V\Pi V$$
$$+V\Pi C\mathcal{R}^0(z)\tilde{\mathscr{E}}(z)V\left\{\sum_{n=0}^{\infty}(-iz)^{-n-1}\psi^n(z)\right\}$$
$$-\left\{\sum_{n=0}^{\infty}(-iz)^{-n-1}\psi^n(z)\right\}V\tilde{\mathscr{E}}(z)\mathcal{R}^0(z)C\Pi V=0 \quad (16.4.8)$$

where we introduced the provisional abbreviation:

$$V\mathscr{L}^0+V\tilde{\mathscr{E}}(z)V\equiv\psi(z) \quad (16.4.9)$$

A completely similar analysis of the V–C component of Eq. (16.4.4):

$$V\Pi\mathcal{R}(z)C=V\mathcal{R}(z)\Pi C \quad (16.4.10)$$

leads to

$$V\Pi V\left\{\sum_{n=0}^{\infty}(-iz)^{-n-1}\psi^n(z)V\tilde{\mathscr{E}}(z)\mathcal{R}^0(z)C\right\}-\left\{\sum_{n=0}^{\infty}(-iz)^{-n-1}\psi^n(z)\right\}V\Pi C$$
$$+V\Pi C\left\{\sum_{n=0}^{\infty}(-iz)^{-n-1}C\mathcal{R}^0(z)\tilde{\mathscr{E}}(z)V\psi^n(z)V\tilde{\mathscr{E}}(z)\mathcal{R}^0(z)C\right\}$$
$$+V\Pi C\{C\mathcal{R}^0(z)+C\mathcal{R}^0(z)\tilde{\mathscr{E}}(z)\mathcal{R}^0(z)C\}$$
$$-\left\{\sum_{n=0}^{\infty}(-iz)^{-n-1}\psi^n(z)V\tilde{\mathscr{E}}(z)\mathcal{R}^0(z)C\right\}C\Pi C=0 \quad (16.4.11)$$

Equations (16.4.8) and (16.4.11) provide us with two relations for the determination of the components of Π. We could obtain two more relations by using the C–V and the C–C components of Eq. (16.4.4), but we shall soon see that they are redundant.

The disadvantage of Eqs. (16.4.8), (16.4.11) is in their providing a relation between components of $\mathcal{R}(z)$ and components of Π that, by definition, are independent of z. It would be helpful for the solution to derive a relation involving only operators independent of z. The particular structure of the resolvent is a useful guide for this operation. It is immediately clear that the value $z=0$ plays a privileged role: All terms in our equations (except one) have an explicitly exhibited singularity (i.e., a multiple pole) in $z=0$. Of course, we do not know *a priori* the nature of the other z-dependent operators appearing in the equations, and in particular the location of their singularities. We are, therefore, forced to make additional assumptions at this point. These assumptions may seem arbitrary at first sight; let us calm the reader's apprehensions. What appears here as a rigid postulate is actually the crystallization of a long experience of many people with these operators. We shall show later that there exist nontrivial systems that do satisfy them.

Coming back to our problem, the simplest case is realized when, in every term of Eqs. (16.4.8), (16.4.11) the only singularity in $z=0$ is the multiple pole $(-iz)^{-n-1}$. This is so if the following conditions are satisfied*:

$V\tilde{\mathscr{E}}(z)V$

$V\tilde{\mathscr{E}}(z)\mathscr{R}^0(z)C$ — are regular functions of z in the neighborhood of $z=0$ and do not vanish in $z=0$

$C\mathscr{R}^0(z)\tilde{\mathscr{E}}(z)V$

$$C\mathscr{R}^0(z)+C\mathscr{R}^0(z)\tilde{\mathscr{E}}(z)\mathscr{R}^0(z)C \tag{16.4.12}$$

These auxiliary conditions define a class of dynamical systems; the latter are the only ones to be considered in this book.

We may now use these conditions as follows. We integrate all the terms of Eqs. (16.4.8), (16.4.11) over z along a small circle centered at the origin.

We then use the Cauchy residue theorem in the form†:

$$(2\pi)^{-1}\oint dz(-iz)^{-n-1}f(z)=\lim_{z\to 0}\{(n!)^{-1}\,\partial^n f(z)\} \tag{16.4.13}$$

where we use the abbreviation:

$$\partial \equiv i\frac{\partial}{\partial z} \tag{16.4.14}$$

* These conditions will be further discussed; see, for example, Section 18.3.

† We write

$$\lim_{z\to 0}\{\partial^n f(z)\}$$

rather than $\{\partial^n f(z)\}_{z=0}$ for the following technical reason. The function $f(z)$ may have singularities other than poles, particularly branch cuts in the upper half-plane or on the real axis. If this happens, the value in $z=0$ may be ambiguous and must be specified more precisely. The ambiguity disappears by realizing that the resolvent, being a Laplace transform, is regular in a half-plane lying above the contour C of Eq. (16.1.15). As we now let z approach the real axis, we must be careful, if we cross any cut, to choose the particular branch which is the analytic continuation of the function $f(z)$ originally defined in its domain of regularity. To sum up, $\lim_{z\to 0}$ must be understood as: "limit for $z\to 0$ *from above*." We do not want to insist too much on these technical points, although they may become important in some problems.

With these operations, Eqs. (16.4.8), (16.4.11) become*

$$
\begin{aligned}
V\Pi V\Big\{\lim_{z\to 0}\sum_{n=0}^{\infty}(n!)^{-1}\partial^n\psi^n(z)\Big\}
&-\Big\{\lim_{z\to 0}\sum_{n=0}^{\infty}(n!)^{-1}\partial^n\psi^n(z)\Big\}V\Pi V\\
&+V\Pi C\Big\{\lim_{z\to 0}\sum_{n=0}^{\infty}(n!)^{-1}\partial^n C\mathscr{R}^0(z)\tilde{\mathscr{E}}(z)V\psi^n(z)\Big\}\\
&-\Big\{\lim_{z\to 0}\sum_{n=0}^{\infty}(n!)^{-1}\partial^n\psi^n(z)V\tilde{\mathscr{E}}(z)\mathscr{R}^0(z)C\Big\}C\Pi V=0 \qquad (16.4.15)
\end{aligned}
$$

$$
\begin{aligned}
V\Pi V\Big\{\lim_{z\to 0}\sum_{n=0}^{\infty}(n!)^{-1}\partial^n\psi^n(z)V\tilde{\mathscr{E}}(z)\mathscr{R}^0(z)C\Big\}
&-\Big\{\lim_{z\to 0}\sum_{n=0}^{\infty}(n!)^{-1}\partial^n\psi^n(z)\Big\}V\Pi C\\
&+V\Pi C\Big\{\lim_{z\to 0}\sum_{n=0}^{\infty}(n!)^{-1}\partial^n C\mathscr{R}^0(z)\tilde{\mathscr{E}}(z)V\psi^n(z)V\tilde{\mathscr{E}}(z)\mathscr{R}^0(z)C\Big\}\\
&-\Big\{\lim_{z\to 0}\sum_{n=0}^{\infty}(n!)^{-1}\partial^n\psi^n(z)V\tilde{\mathscr{E}}(z)\mathscr{R}^0(z)C\Big\}C\Pi C=0 \qquad (16.4.16)
\end{aligned}
$$

At this stage, the form is so symmetrical that the solutions are immediately obvious. First, we have a completely trivial solution:

$$V\Pi V=V\Pi C=C\Pi V=C\Pi C=0$$

which implies

$$\Pi=0 \qquad (16.4.17)$$

This solution does not satisfy Condition (16.4.3) and must therefore be rejected. A second solution can be seen to be

$$
\begin{aligned}
V\Pi V&=V\\
C\Pi C&=C\\
C\Pi V&=V\Pi C=0
\end{aligned}
$$

From these components, the operator Π is reconstructed by (15.3.17):

$$\Pi=V+C=I \qquad (16.4.18)$$

This solution also violates (16.4.3) and must be rejected. We may note that both (16.4.17) and (16.4.18) are trivial in the sense that, on substitution into Eq. (16.2.5) they provide no separation at all: in the former case we get $\mathfrak{f}(t)=\hat{\mathfrak{f}}(t)$, and in the latter $\mathfrak{f}(t)=\bar{\mathfrak{f}}(t)$.

* We recall that, here and in the forthcoming formulas, the operator ∂^n acts on everything written to its right.

Finally, we obtain a nontrivial solution to our equations from a first glance:

$$V\Pi V = \lim_{z\to 0} \sum_{n=0}^{\infty} (n!)^{-1} \partial^n [V\mathscr{L}^0 + V\tilde{\mathscr{E}}(z)V]^n \quad (16.4.19)$$

$$V\Pi C = \lim_{z\to 0} \sum_{n=0}^{\infty} (n!)^{-1} \partial^n [V\mathscr{L}^0 + V\tilde{\mathscr{E}}(z)V]^n V\tilde{\mathscr{E}}(z)\mathscr{R}^0(z)C \quad (16.4.20)$$

$$C\Pi V = \lim_{z\to 0} \sum_{n=0}^{\infty} (n!)^{-1} \partial^n C\mathscr{R}^0(z)\tilde{\mathscr{E}}(z)V[V\mathscr{L}^0 + V\tilde{\mathscr{E}}(z)V]^n \quad (16.4.21)$$

$$C\Pi C = \lim_{z\to 0} \sum_{n=0}^{\infty} (n!)^{-1} \partial^n C\mathscr{R}^0(z)\tilde{\mathscr{E}}(z)V[V\mathscr{L}^0 + V\tilde{\mathscr{E}}(z)V]^n V\tilde{\mathscr{E}}(z)\mathscr{R}^0(z)C \quad (16.4.22)$$

On substitution into Eqs. (16.4.15), (16.4.16), all the terms in the left-hand side cancel each other pairwise. The reader may now verify that the two remaining components, $C-V$ and $C-C$, of Eq. (16.4.4) add nothing new: the solution (16.4.19)–(16.4.22) also satisfies those equations.* The complete operator Π can now be reconstructed as follows:

$$\Pi = \lim_{z\to 0} \sum_{n=0}^{\infty} (n!)^{-1} \partial^n [V + C\mathscr{R}^0(z)\tilde{\mathscr{E}}(z)V] \times [V\mathscr{L}^0 + V\tilde{\mathscr{E}}(z)V]^n [V + V\tilde{\mathscr{E}}(z)\mathscr{R}^0(z)C] \quad (16.4.23)$$

It remains to be shown that this operator satisfies the boundary condition (16.4.3). In order to do the limit formally, we replace the interaction Liouvillian $\mathscr{L}'$ by $\lambda\mathscr{L}'$, where λ is a dimensionless scaling parameter, and then let λ go to zero. From Eq. (16.3.5) we see first that

$$\lim_{\lambda\to 0} \tilde{\mathscr{E}}(z) = 0 \quad (16.4.24)$$

Hence:

$$\lim_{\lambda\to 0}\Pi = \lim_{z\to 0} \sum_{n=0}^{\infty} (n!)^{-1} \partial^n V(\mathscr{L}^0)^n V$$
$$= \lim_{z\to 0}(0!)^{-1} \partial^0 V(\mathscr{L}^0)^0 V = V \quad (16.4.25)$$

The second step follows from the fact that $\mathscr{L}^0$ is independent of z, hence only the term $n = 0$ in the sum can give a nonvanishing contribution.

We now possess the solution to Eqs. (16.4.15), (16.4.16). These equations are, however, weaker than the original equation (16.4.2). We must now *prove* that the operator Π obtained in Eq. (16.4.23) commutes with $\mathscr{U}(t)$ for arbitrary values of t. The detailed proof is given in Appendix 1 to

* A comparison between Eqs. (16.4.19)–(16.4.22) and (16.3.18)–(16.3.21) reveals a striking similarity. Clearly, the structure of the resolvent has been impressed on the operator Π. This analogy will become even deeper at a later stage.

this chapter. The idea is the following. One first shows that the product $\mathbf{\Pi}\mathcal{U}(t)$ can be written as an operator $\mathbf{\Sigma}(t)$:

$$\mathbf{\Pi}\mathcal{U}(t) = \mathbf{\Sigma}(t) \tag{16.4.26}$$

The form of this operator is very similar to (16.4.23):

$$\mathbf{\Sigma}(t) = \lim_{z\to 0} \sum_{n=0}^{\infty} (n!)^{-1}(t+\partial)^n [V + C\mathcal{R}^0(z)\tilde{\mathscr{E}}(z)V] \times [V\mathcal{L}^0 + V\tilde{\mathscr{E}}(z)V]^n [V + V\tilde{\mathscr{E}}(z)\mathcal{R}^0(z)C] \tag{16.4.27}$$

In a second step we show that we also have

$$\mathcal{U}(t)\mathbf{\Pi} = \mathbf{\Sigma}(t) \tag{16.4.28}$$

Equations (16.4.26) and (16.4.28) then constitute a proof of Eq. (16.4.2).*

We note that Eq. (16.4.2) also implies the commutation of the operator $\mathbf{\Pi}$ with the Liouvillian:

$$\mathbf{\Pi}\mathcal{L} = \mathcal{L}\mathbf{\Pi} \tag{16.4.29}$$

This relation immediately follows from an application of Eq. (16.4.2) to the infinitesimal time translation:

$$\mathcal{U}(t) \sim 1 + t\mathcal{L}$$

We have now completed the points A and B of the program of Section 16.2. We have been able to construct a unique operator $\mathbf{\Pi}$, and hence its complement:

$$\hat{\mathbf{\Pi}} = \mathsf{I} - \mathbf{\Pi} \tag{16.4.30}$$

These operators provide us with a separation of the distribution vector, $\mathfrak{f}(t)$, *invariant under time translations:*

$$\mathfrak{f}(t) = \bar{\mathfrak{f}}(t) + \hat{\mathfrak{f}}(t)$$
$$\bar{\mathfrak{f}}(t) = \mathbf{\Pi}\mathfrak{f}(t) \tag{16.4.31}$$
$$\hat{\mathfrak{f}}(t) = \hat{\mathbf{\Pi}}\mathfrak{f}(t)$$

We now show that the *operator* $\mathbf{\Pi}$ *is an idempotent operator:*

$$\mathbf{\Pi}^2 = \mathbf{\Pi} \tag{16.4.32}$$

* The proof of Appendix 1 helps understanding the following question that one may ask: "Why should $z = 0$ play a special role in the argument of this section?" In spite of its obvious role in Eqs. (16.4.8), (16.4.11), one might be tempted to calculate their residue in some other point $z_0 \neq 0$. This would provide a different operator, involving necessarily $\tilde{\mathscr{E}}(z_0)$, $\mathcal{R}^0(z_0)$ and other functions evaluated in $z = z_0$; but such an operator would certainly *not* commute with $\mathcal{U}(t)$ for all values of t. This property involves the value $z = 0$ in an essential way, as can be seen in Appendix 1.

This property was announced in Section 16.2 (Proposition C). It is necessary for the perfection of the geometrical interpretation of the separation (16.4.31). The sets $\{\bar{\mathfrak{f}}\}$ and $\{\hat{\mathfrak{f}}\}$ can now be regarded as "orthogonal subspaces" of the set of vectors $\{\mathfrak{f}\}$. The operator $\boldsymbol{\Pi}$ having the property (16.4.32) appears as a *projection operator*, or briefly, a "projector" on the subspace $\{\bar{\mathfrak{f}}\}$. Moreover, from Eq. (16.4.30) follows immediately the complete set of relations:

$$\begin{gathered} \boldsymbol{\Pi} + \hat{\boldsymbol{\Pi}} = \mathsf{I} \\ \boldsymbol{\Pi}^2 = \boldsymbol{\Pi} \\ \hat{\boldsymbol{\Pi}}^2 = \hat{\boldsymbol{\Pi}} \\ \boldsymbol{\Pi}\hat{\boldsymbol{\Pi}} = \hat{\boldsymbol{\Pi}}\boldsymbol{\Pi} = 0 \end{gathered} \tag{16.4.33}$$

The form of Eqs. (16.4.23) clearly shows that the property (16.4.32) is far from trivial. It results from very subtle cancellations, for which every detail of the form (16.4.23) plays a role. The proof of Relation (16.4.32) is not too difficult, but is very lengthy. We give it in Appendix 2.

Before closing this section, we note that the operator $\boldsymbol{\Sigma}(t)$ defined by Eqs. (16.4.26)–(16.4.28) provides us with an alternative definition of the component $\bar{\mathfrak{f}}(t)$:

$$\bar{\mathfrak{f}}(t) = \boldsymbol{\Pi}\mathfrak{f}(t) = \boldsymbol{\Pi}\mathcal{U}(t)\mathfrak{f}(0) = \boldsymbol{\Sigma}(t)\mathfrak{f}(0)$$

If we also introduce the operator:

$$\hat{\boldsymbol{\Sigma}}(t) = \mathcal{U}(t) - \boldsymbol{\Sigma}(t) = \hat{\boldsymbol{\Pi}}\mathcal{U}(t) = \mathcal{U}(t)\hat{\boldsymbol{\Pi}} \tag{16.4.34}$$

we have

$$\hat{\mathfrak{f}}(t) = \mathfrak{f}(t) - \bar{\mathfrak{f}}(t) = [\mathcal{U}(t) - \boldsymbol{\Sigma}(t)]\mathfrak{f}(0) = \hat{\boldsymbol{\Sigma}}(t)\mathfrak{f}(0)$$

Hence, we obtain the following important formulas:

$$\begin{gathered} \bar{\mathfrak{f}}(t) = \boldsymbol{\Sigma}(t)\mathfrak{f}(0) \\ \hat{\mathfrak{f}}(t) = \hat{\boldsymbol{\Sigma}}(t)\mathfrak{f}(0) \end{gathered} \tag{16.4.35}$$

Whereas the definition (16.4.31) is of a *geometrical* type (because it provides an operator that splits every element $\mathfrak{f}(t)$ into two components), the definition (16.4.35) is of a *dynamical* type. Indeed it provides two operators that, applied to the initial condition $\mathfrak{f}(0)$, give directly the two separate components at time t (see Fig. 16.2.1). They play the same role as the propagator $\mathcal{U}(t)$; actually, they provide an invariant decomposition of this operator:

$$\mathcal{U}(t) = \boldsymbol{\Sigma}(t) + \hat{\boldsymbol{\Sigma}}(t) \tag{16.4.36}$$

In the limit of no interactions, these operators reduce to known operators:

$$\lim_{\lambda\to 0}\boldsymbol{\Sigma}(t)=\mathcal{U}^0(t)V$$
$$\lim_{\lambda\to 0}\hat{\boldsymbol{\Sigma}}(t)=\mathcal{U}^0(t)C \tag{16.4.37}$$

We finally note the obvious relations:

$$\boldsymbol{\Sigma}(0)=\boldsymbol{\Pi}$$
$$\hat{\boldsymbol{\Sigma}}(0)=\hat{\boldsymbol{\Pi}} \tag{16.4.38}$$

We now sum up our results. Combining Eqs. (16.4.26), (16.4.28), (16.4.31), and (16.4.35) we obtain

$$\boldsymbol{\Pi}\mathfrak{f}(t)=\mathcal{U}(t)\boldsymbol{\Pi}\mathfrak{f}(0)$$
$$\hat{\boldsymbol{\Pi}}\mathfrak{f}(t)=\mathcal{U}(t)\hat{\boldsymbol{\Pi}}\mathfrak{f}(0) \tag{16.4.39}$$

These equations express the fact that *the function* $\boldsymbol{\Pi}\mathfrak{f}(t)$ *is the result of the exact time evolution of* $\boldsymbol{\Pi}\mathfrak{f}(0)$. [A similar statement holds for $\hat{\boldsymbol{\Pi}}\mathfrak{f}(t)$.] As time proceeds, there is no mixing between the two components: they evolve independently of each other. This property of independent time evolution of the two subspaces is expressed by saying that *each component obeys the laws of its own subdynamics.* This subdynamics is a natural continuation of the unperturbed subdynamics studied in Section 15.3. Its very existence, in spite of the interactions, is an extremely remarkable feature of the time-evolution process of a many-body system. Indeed, intuition would only suggest that the interactions mix up the whole structure of the system. A deep analysis was necessary in order to uncover the present invariance property.

APPENDIX 1: PROOF OF THE RELATION $\boldsymbol{\Sigma}(t)=\boldsymbol{\Pi}\mathcal{U}(t)$

We note first that it is sufficient to prove*

$$V\boldsymbol{\Pi}\mathcal{U}(t)=V\boldsymbol{\Sigma}(t) \tag{16.A.1}$$

Using Eqs. (16.1.7) and (16.4.23) the left side is

$$V\boldsymbol{\Pi}\mathcal{U}(t)=\lim_{z\to 0}\sum_{r=0}^{\infty}\sum_{p=0}^{\infty}\frac{t^r}{r!\,p!}\,\partial^p(V\mathcal{L}^0+V\tilde{\mathscr{E}}V)^p(V+V\tilde{\mathscr{E}}\mathscr{R}^0C)\mathcal{L}^r$$

On the other side, from (16.4.27) and the binomial expansion of $(t+\partial)^n$:

$$V\boldsymbol{\Sigma}(t)=\lim_{z\to 0}\sum_{r=0}^{\infty}\sum_{p=0}^{\infty}\frac{t^r}{r!\,p!}\,\partial^p(V\mathcal{L}^0+V\tilde{\mathscr{E}}V)^{p+r}(V+V\tilde{\mathscr{E}}\mathscr{R}^0C)\mathcal{L}^r$$

* Indeed, nothing is changed in the proof if a factor $(V+C\mathscr{R}^0\tilde{\mathscr{E}}V)$ is inserted to the right of the factor ∂^p in the expression below.

It is therefore sufficient to prove the following relation:

$$\lim_{z\to 0}\sum_{p=0}^{\infty}(p!)^{-1}\partial^p(V\mathcal{L}^0+V\tilde{\mathscr{E}}V)^p(V+V\tilde{\mathscr{E}}\mathcal{R}^0C)\mathcal{L}^r$$
$$=\lim_{z\to 0}\sum_{p=0}^{\infty}(p!)^{-1}\partial^p(V\mathcal{L}^0+V\tilde{\mathscr{E}}V)^{p+r}(V+V\tilde{\mathscr{E}}\mathcal{R}^0C), \qquad r=0,1,2,\ldots \tag{16.A.2}$$

The proof is by induction. The theorem is trivially true for $r=0$. Assume now that (16.A.2) is true for given r and consider

$$\lim_{z\to 0}\sum_{p=0}^{\infty}(p!)^{-1}\partial^p(V\mathcal{L}^0+V\tilde{\mathscr{E}}V)^p(V+V\tilde{\mathscr{E}}\mathcal{R}^0C)\mathcal{L}^{r+1}$$
$$=\lim_{z\to 0}\sum_{p=0}^{\infty}(p!)^{-1}\partial^p(V\mathcal{L}^0+V\tilde{\mathscr{E}}V)^{p+r}(V+V\tilde{\mathscr{E}}\mathcal{R}^0C)\mathcal{L} \tag{16.A.3}$$

Expanding $\mathcal{L}$ in the form (15.3.17) and using (16.1.2) we can write

$$(V+V\tilde{\mathscr{E}}\mathcal{R}^0C)\mathcal{L}=V(\mathcal{L}^0+\mathcal{L}'+\tilde{\mathscr{E}}\mathcal{R}^0C\mathcal{L}')V$$
$$+V(\mathcal{L}'+\tilde{\mathscr{E}}\mathcal{R}^0C\mathcal{L}'+\tilde{\mathscr{E}}\mathcal{R}^0C\mathcal{L}^0)C \tag{16.A.4}$$

Making use of Eqs. (16.3.7) and (16.3.12) we can simplify (16.A.4) to

$$(V+V\tilde{\mathscr{E}}\mathcal{R}^0C)\mathcal{L}=(V\mathcal{L}^0+V\tilde{\mathscr{E}}V)-izV\tilde{\mathscr{E}}\mathcal{R}^0C \tag{16.A.5}$$

Substituting this result into the right-hand side of Eq. (16.A.3) we get

$$\lim_{z\to 0}\sum_{p=0}^{\infty}(p!)^{-1}\partial^p(V\mathcal{L}^0+V\tilde{\mathscr{E}}V)^p(V+V\tilde{\mathscr{E}}\mathcal{R}^0C)\mathcal{L}^{r+1}$$
$$=\lim_{z\to 0}\sum_{p=0}^{\infty}(p!)^{-1}\partial^p(V\mathcal{L}^0+V\tilde{\mathscr{E}}V)^{p+r}(V\mathcal{L}^0+V\tilde{\mathscr{E}}V)$$
$$+\lim_{z\to 0}\sum_{q=0}^{\infty}(q!)^{-1}\,\partial^q\{(-iz)(V\mathcal{L}^0+V\tilde{\mathscr{E}}V)^{q+r}V\tilde{\mathscr{E}}\mathcal{R}^0C\} \tag{16.A.6}$$

Using the well-known Leibnitz differentiation rule:

$$\partial^n[a(z)b(z)]=\sum_{r=0}^{n}\frac{n!}{r!\,(n-r)!}[\partial^r a(z)][\partial^{n-r}b(z)] \tag{16.A.7}$$

The second term can be written as

$$\lim_{z\to 0}\sum_{q=0}^{\infty}\sum_{n=0}^{q}\frac{1}{n!\,(q-n)!}\{\partial^n(-iz)\}\{\partial^{q-n}(V\mathcal{L}^0+V\tilde{\mathscr{E}}V)^{q+r}V\tilde{\mathscr{E}}\mathcal{R}^0C\}$$
$$=\lim_{z\to 0}\sum_{q=1}^{\infty}\frac{1}{(q-1)!}\,\partial^{q-1}(V\mathcal{L}^0+V\tilde{\mathscr{E}}V)^{q+r}V\tilde{\mathscr{E}}\mathcal{R}^0C$$

The last step is obtained by noting that the first bracketed factor differs from zero only for $n=1$ and $n=0$. Moreover, it follows from the regularity assumptions (16.4.12) that the contribution of the term $n=0$ vanishes in the limit $z\to 0$: we are, therefore, left with the contribution of $n=1$. Changing the summation index q to

$p = q - 1$ and substituting the result into the right side of (16.A.6) we get

$$\lim_{z\to 0}\sum_{p=0}^{\infty}(p!)^{-1}\partial^p(V\mathscr{L}^0 + V\tilde{\mathscr{E}}V)^p(V + V\tilde{\mathscr{E}}\mathscr{R}^0C)\mathscr{L}^{r+1}$$
$$= \lim_{z\to 0}\sum_{p=0}^{\infty}(p!)^{-1}\partial^p(V\mathscr{L}^0 + V\tilde{\mathscr{E}}V)^{p+r+1}(V + V\tilde{\mathscr{E}}\mathscr{R}^0C) \quad (16.A.8)$$

The induction hypothesis (16.A.2) is thus extended to the value $r+1$, and hence the theorem (16.A.1) is proved.

The proof of the relation $\mathscr{U}(t)\mathbf{\Pi} = \mathbf{\Sigma}(t)$ is quite analogous to the present one. It is again sufficient to prove that

$$\mathscr{U}(t)\mathbf{\Pi}V = \mathbf{\Sigma}(t)V$$

from which it follows that we must verify the identity

$$\lim_{z\to 0}\sum_{p=0}^{\infty}(p!)^{-1}\partial^p\mathscr{L}^r(V + C\mathscr{R}^0\tilde{\mathscr{E}}V)(V\mathscr{L}^0 + V\tilde{\mathscr{E}}V)^p$$
$$= \lim_{z\to 0}\sum_{p=0}^{\infty}(p!)^{-1}\partial^p(V + C\mathscr{R}^0\tilde{\mathscr{E}}V)(V\mathscr{L}^0 + V\tilde{\mathscr{E}}V)^{p+r}$$

The argument is the same as above, except that Eq. (16.A.5) is replaced by

$$\mathscr{L}(V + C\mathscr{R}^0\tilde{\mathscr{E}}V) = (V\mathscr{L}^0 + V\tilde{\mathscr{E}}V) - izC\mathscr{R}^0\tilde{\mathscr{E}}V \quad (16.A.9)$$

which results from Eqs. (16.3.6) and (16.3.12).

APPENDIX 2: PROOF OF THE RELATION $\mathbf{\Pi}^2 = \mathbf{\Pi}$

This rather involved proof is done in several steps.*

A. It Is Sufficient to Prove That

$$V\mathbf{\Pi}^2V = V\mathbf{\Pi}V \quad (16.A.10)$$

The justification for this claim is in the fact, to be proved later, that the various components of $\mathbf{\Pi}$ are not independent: they can all be expressed in terms of $V\mathbf{\Pi}V$ alone. The suspicious reader will interpret the present Appendix as the mere proof of Eq. (16.A.10). He will find the three lines necessary for the completion of the proof of $\mathbf{\Pi}^2 = \mathbf{\Pi}$ in a footnote to Section 17.3.

B. A Lemma on the Derivatives of $\tilde{\mathscr{E}}(z)$

It immediately follows from the definition (15.2.7) of the unperturbed resolvent that

$$\partial\mathscr{R}^0(z) = -\mathscr{R}^0(z)\mathscr{R}^0(z) \quad (16.A.11)$$

[Remember the notation: $\partial \equiv i(\partial/\partial z)$.] Using this result, we can easily calculate

* The proof given here closely follows the paper of C. George, *Bull. Cl. Sci. Acad. Roy. Belg.* **56**, 386 (1970).

the derivative of the irreducible evolution operator $\tilde{\mathscr{E}}(z)$, defined by Eq. (16.3.5):

$$\partial\tilde{\mathscr{E}}(z) = -\sum_{m=1}^{\infty}\sum_{s=0}^{m-1}\mathscr{L}'(C\mathscr{R}^0\mathscr{L}')^s(C\mathscr{R}^0\mathscr{R}^0\mathscr{L}')(C\mathscr{R}^0\mathscr{L}')^{m-1-s}$$

$$= -\sum_{p=0}^{\infty}\sum_{s=0}^{\infty}[\mathscr{L}'(C\mathscr{R}^0\mathscr{L}')^s\mathscr{R}^0C][C\mathscr{R}^0\mathscr{L}'(C\mathscr{R}^0\mathscr{L}')^p]$$

Hence

$$\partial\tilde{\mathscr{E}}(z) = -(\tilde{\mathscr{E}}\mathscr{R}^0C)(C\mathscr{R}^0\tilde{\mathscr{E}}) \tag{16.A.12}$$

This result can be generalized and yields the following relation:

$$(\partial^m\tilde{\mathscr{E}}\mathscr{R}^0C)(\partial^nC\mathscr{R}^0\tilde{\mathscr{E}}) = -\frac{m!\,n!}{(m+n+1)!}(\partial^{m+n+1}\tilde{\mathscr{E}}) \tag{16.A.13}$$

The proof of this lemma is rather tedious and will not be given here. The reader may easily verify it on special examples, using repeatedly the relations (16.A.11) and (16.A.12).

C. Transformation of the Expression of $V\Pi^2V$

We start from the explicit expression for the components of Π, Eqs. (16.4.19), (16.4.21):

$$V\Pi^2V = V\Pi(V+C)\Pi V$$

$$= \lim_{z\to 0}\sum_{s=0}^{\infty}\sum_{r=0}^{\infty}(s!\,r!)^{-1}\{\partial^s[\psi^s(V+V\tilde{\mathscr{E}}\mathscr{R}^0C)]\}\{\partial^r[(V+C\mathscr{R}^0\tilde{\mathscr{E}}V)\psi^r]\}$$

$$= \lim_{z\to 0}\sum_{s=0}^{\infty}\sum_{r=0}^{\infty}\sum_{m=0}^{s}\sum_{n=0}^{r}\frac{1}{(s-m)!\,(r-n)!\,m!\,n!}(\partial^{s-m}\psi^s)$$

$$\times\{\partial^m(V+V\tilde{\mathscr{E}}\mathscr{R}^0C)\}\{\partial^n(V+C\mathscr{R}^0\tilde{\mathscr{E}}V)\}(\partial^{r-n}\psi^r) \tag{16.A.14}$$

where we used the abbreviation (16.4.9) and Leibnitz's rule (16.A.7).

We now apply the lemma (16.A.13) and obtain

$$V\Pi^2V = \lim_{z\to 0}\sum_{s=0}^{\infty}\sum_{r=0}^{\infty}\sum_{m=0}^{s}\sum_{n=0}^{r}\frac{1}{m!\,n!\,(s-m)!\,(r-n)!}(\partial^{s-m}\psi^s)$$

$$\times\left(\delta_{m,0}\delta_{n,0}-\frac{m!\,n!}{(m+n+1)!}\partial^{m+n+1}\psi\right)(\partial^{r-n}\psi^r) \tag{16.A.15}$$

We have thus succeeded in expressing $V\Pi^2V$ entirely in terms of the derivatives of the operator ψ. In order to prove our theorem, this expression must be shown to reduce to Eq. (16.4.19):

$$V\Pi V = \lim_{z\to 0}\sum_{s=0}^{\infty}(s!)^{-1}\,\partial^s\psi^s \tag{16.A.16}$$

D. A Second Lemma

Consider two complex variables, z and z_1. Given a function $\psi(z)$, regular in $z=0$, we introduce the following abbreviations: $\psi = \psi(z)$, $\psi_1 = \psi(z_1)$, $\partial = i(\partial/\partial z)$,

$\partial_1 = i(\partial/\partial z_1)$. We now show that

$$\lim_{z_1\to z}\frac{1}{\mu!\,\nu!}\,\partial^{\mu}\partial_1^{\nu} i\,\frac{\psi_1-\psi}{z_1-z}=\frac{1}{(\mu+\nu+1)!}\,\partial^{\mu+\nu+1}\psi \tag{16.A.17}$$

We first expand ψ_1 in a Taylor series around $z_1 = z$:

$$i\,\frac{\psi_1-\psi}{z_1-z}=\sum_{n=0}^{\infty}\frac{1}{(n+1)!}\,i^{-n}(z_1-z)^n(\partial^{n+1}\psi) \tag{16.A.18}$$

We then use the simple relations

$$\begin{aligned}\partial_1^{\nu}[i^{-n}(z_1-z)^n]&=\frac{n!}{(n-\nu)!}\,i^{\nu-n}(z_1-z)^{n-\nu}, &\quad n\geqslant\nu\\ &=0, &\quad n<\nu\end{aligned} \tag{16.A.19}$$

$$\begin{aligned}\partial^{\nu}[i^{-n}(z_1-z)^n]&=(-)^{\nu}\frac{n!}{(n-\nu)!}\,i^{\nu-n}(z_1-z)^{n-\nu}, &\quad n\geqslant\nu\\ &=0, &\quad n<\nu\end{aligned} \tag{16.A.20}$$

These formulas, together with the Leibnitz rule, allow us to calculate the derivatives $\partial^{\mu}\partial_1^{\nu}$ explicitly:

$$\frac{1}{\mu!\,\nu!}\,\partial^{\mu}\partial_1^{\nu} i\,\frac{\psi_1-\psi}{z_1-z}=\sum_{n=0}^{\infty}\sum_{p=0}^{\mu}\frac{(-)^p i^{p-n+\nu}}{(n+1)\nu!\,p!\,(\mu-p)!\,(n-\nu-p)!}(z_1-z)^{n-\nu-p}(\partial^{\mu-p+n+1}\psi)$$

Because of the factor $[(n-\nu-p)!]^{-1}$, only values of n, ν, p such that $n-\nu-p\geqslant 0$ can contribute to the sum. It follows that the exponent of (z_1-z) in the right-hand side is nonnegative. As we now take the limit $z_1\to z$, all terms in the sum will vanish, except those for which $n=\nu+p$. We thus obtain

$$\lim_{z_1\to z}\frac{1}{\mu!\,\nu!}\,\partial^{\mu}\partial_1^{\nu} i\,\frac{\psi_1-\psi}{z_1-z}=\frac{1}{\nu!}\left\{\sum_{p=0}^{\mu}\frac{(-)^p}{(\nu+p+1)p!\,(\mu-p)!}\right\}\partial^{\mu+\nu+1}\psi$$

Hence the lemma (16.A.17) will be proved if we can show that

$$\sum_{p=0}^{\mu}\frac{(-)^p}{(\nu+p+1)p!\,(\mu-p!)}=\frac{\nu!}{(\nu+\mu+1)!} \tag{16.A.21}$$

To evaluate the sum in the left-hand side, we consider the following integral:

$$A_{\nu,\mu}=\int_0^1 dx\,x^{\nu}(1-x)^{\mu}$$

where ν, μ are nonnegative integers. Clearly:

$$A_{\nu,0}=\int_0^1 dx\,x^{\nu}=\frac{1}{\nu+1}$$

The integral $A_{\nu,\mu}$ can be evaluated by using Newton's binomial formula:

$$\begin{aligned}A_{\nu,\mu}&=\int_0^1 dx\,x^{\nu}\sum_{p=0}^{\mu}\frac{\mu!}{p!\,(\mu-p)!}(-)^p x^p\\ &=\sum_{p=0}^{\mu}(-)^p\frac{\mu!}{p!\,(\mu-p)!}\frac{1}{\nu+p+1}\end{aligned} \tag{16.A.22}$$

We can also evaluate this integral by partial integration:

$$A_{\nu,\mu} = \int_0^1 dx \left(\frac{1}{\nu+1} \frac{d}{dx} x^{\nu+1} \right) (1-x)^{\mu}$$

$$= \int_0^1 dx \frac{\mu}{\nu+1} x^{\nu+1} (1-x)^{\mu-1}$$

$$= \frac{\mu}{\nu+1} A_{\nu+1,\mu-1}$$

By iterating this recurrence relation we get

$$A_{\nu,\mu} = \frac{\mu}{\nu+1} A_{\nu+1,\mu-1} = \left(\frac{\mu}{\nu+1} \right) \left(\frac{\mu-1}{\nu+2} \right) A_{\nu+2,\mu-2}$$

$$= \cdots = \frac{\mu!}{(\nu+1)(\nu+2)\cdots(\nu+\mu)} A_{\nu+\mu,0}$$

$$= \frac{\mu!\,\nu!}{(\mu+\nu)!} \frac{1}{\nu+\mu+1} \tag{16.A.23}$$

Comparing the right-hand sides of Eqs. (16.A.22) and (16.A.23) we immediately obtain the expected result (16.A.21).

E. Final Stage of the Proof

We now return to Eq. (16.A.15). Noting that $\partial^{\mu}\partial_1^{\nu} 1 = \delta_{\mu,0}\delta_{\nu,0}$ and using the lemma (16.A.17) we can rewrite the rhs of (16.A.15) as

$$V\sqcap^2 V = \lim_{z\to 0} \lim_{z_1 \to z} \sum_{s=0}^{\infty} \sum_{r=0}^{\infty} \sum_{m=0}^{s} \sum_{n=0}^{r} \frac{1}{m!\,(s-m)!\,n!\,(r-m)!}$$

$$\times (\partial^{s-m}\psi^s) \left[\partial^m \partial_1^n \left(1 - i \frac{\psi_1 - \psi}{z_1 - z} \right) \right] (\partial_1^{r-n} \psi_1^r)$$

$$= \lim_{z\to 0} \lim_{z_1 \to z} \sum_{s=0}^{\infty} \sum_{r=0}^{\infty} \frac{1}{s!\,r!} \partial^s \partial_1^r \left[\psi^s \left(1 - i \frac{\psi_1 - \psi}{z_1 - z} \right) \psi_1^r \right]$$

$$= \lim_{z\to 0} \lim_{z_1 \to z} \sum_{s=0}^{\infty} \sum_{r=0}^{\infty} \frac{1}{s!\,r!} \left\{ \partial^s \partial_1^r (\psi^s \psi^r) - \partial^s \partial_1^r \left(\psi^s i \frac{\psi_1 - \psi}{z_1 - z} \psi^r \right) \right.$$

$$\left. + \partial^s \partial_1^r \left[\psi^s \left(1 - i \frac{\psi_1 - \psi}{z_1 - z} \right) (\psi_1^r - \psi^r) \right] \right\} \tag{16.A.24}$$

The first term in the resulting expression provides the expected result:

$$\lim_{z\to 0} \lim_{z_1 \to z} \sum_{s=0}^{\infty} \sum_{r=0}^{\infty} \frac{1}{s!\,r!} \partial^s \partial_1^r \psi^{s+r} = \lim_{z\to 0} \lim_{z_1 \to z} \sum_{s=0}^{\infty} \sum_{r=0}^{\infty} \frac{1}{s!\,r!} \delta_{r,0} \partial^s \psi^{s+r}$$

$$= \lim_{z\to 0} \sum_{s=0}^{\infty} \frac{1}{s!} \partial^s \psi^s = V\sqcap V$$

It then remains to be shown that the additional terms in the rhs of (16.A.24) sum up to zero. A rather tedious calculation of the same type as in Section C above (involving an expansion of ψ_1^r around $z_1 = z$, repeated use of Leibnitz's rule and various rearrangements of the series) leads to the following result:

$$\lim_{z_1 \to z} \sum_{s=0}^{\infty} \sum_{r=0}^{\infty} \frac{1}{s!\, r!} \partial^s \partial_1^r \left\{ \psi^s \left[-i \frac{\psi_1 - \psi}{z_1 - z} \psi^r + \left(1 - i \frac{\psi_1 - \psi}{z_1 - z}\right)(\psi_1^r - \psi^r) \right] \right\}$$

$$= \lim_{z_1 \to z} \sum_{n=0}^{\infty} \sum_{r=0}^{\infty} \sum_{j=1}^{\infty} \sum_{k=0}^{\infty} \frac{(-)^n}{r!\, n!\, k!\, (j-n-r-1)!\, j} \left(\frac{z_1 - z}{i}\right)^{j-n-r} \partial^k (\psi^{k+n} \partial^j \psi^r)$$

Because of the presence of the factor $[(j-n-r-1)!]^{-1}$, only values of j, n, r such that $j-n-r-1 \geqslant 0$ can contribute to the sum. It then follows that the exponent of $(z_1 - z)$ in the right hand side is *definite positive* (zero excluded):

$$j - n - r \geqslant 1$$

Hence, the limit of the right side as $z_1 \to z$ is zero. This achieves the proof of the theorem.

BIBLIOGRAPHIC NOTES

The *projection operator* ⊓ was introduced systematically, in connection with the idea of subdynamics by I. Prigogine, C, George, and F. Henin, *Physica* **45,** 418 (1969); *Proc. Nat. Acad. Sci. USA* **65,** 789 (1970).

The existence of a projection operator of this kind was suggested, in a different context, by K. Haubold in 1967 (unpublished).

The presentation given here appears in

R. Balescu and J. Wallenborn, *Physica* **54,** 504 (1971).

R. Balescu, in *Irreversibility in the Many-Body Problem* (J. Biel and J. Rae, eds.), Plenum, New York, 1972.

The proofs of the formal properties of the operator ⊓ were given in

R. Balescu, P. Clavin, P. Mandel, and J. Turner, *Bull. Cl. Sci. Acad. Roy. Belg.* **55,** 1055 (1969).

C. George, *Bull. Cl. Sci. Acad. Roy. Belg.* **56,** 386, (1970).

R. Balescu and J. Wallenborn (quoted above).

The concept of subdynamics was further developed recently, in particular with the introduction of the *physical particle representation.* This concept was introduced in C. George, *Bull. Cl. Sci. Acad. Roy. Belg.* **53,** 623 (1967).

The consequences of the "physical particle description" have been very extensively developed by I. Prigogine and his co-workers in Brussels in recent years. We have no space here for a detailed exposition of these ideas: they will appear in a monograph prepared by Prigogine *et al.* We shall only quote here a few of the more important papers along this line (other references will be found in Chapter 17).

General expositions of the *causal-dynamics* concept are found in

I. Prigogine, C. George, and F. Henin (reference quoted above).

I. Prigogine, C. George, and F. Henin, P. Mandel, and J. W. Turner, *Proc. Nat. Acad. Sci. USA* **66,** 709 (1970).

C. George, I. Prigogine, and L. Rosenfeld, *Nature* **240,** 25 (1972); *Det Kong. Danske Vid. Selsk, Mat-fys. Medd.* **38,** No. 12 (1972).

I. Prigogine, C. George, F. Henin, and L. Rosenfeld, *Chemica Scripta* **4,** 5 (1973).

Interesting applications of the physical particle formalism were given in equilibrium theory:

P. Allen and G. Nicolis, *Physica* **50,** 206 (1970): nonideal gases.

P. Clavin and J. Wallenborn, *C. R. Acad. Sci.* (Paris) **270,** 717 (1970): classical plasmas.

P. Mandel, *Bull. Cl. Sci. Acad. Roy. Belg.* **56,** 1140 (1970): quantum plasmas.

The $\boldsymbol{\Pi}$ operator and the other related concepts were constructed for a certain number of *exactly soluble models,* without using perturbation theory:

A. Grecos and I. Prigogine, *Physica* **59,** 77 (1972).

M. De Haan and F. Henin, *Physica* **67,** 197 (1973).

G. C. Stey, *Physica* **68,** 243 (1973).

The $\boldsymbol{\Pi}$ operator was shown to be a *relativistic invariant:* R. Balescu and L. Brenig, *Physica* **54,** 504 (1971).

CHAPTER 17

SUBDYNAMICS AND KINETIC THEORY

17.1. THE KINETIC COMPONENT OF THE CORRELATIONS

In Chapter 16 we discovered a very remarkable structure of the time-evolution process. We defined projection operators $\mathbf{\Pi}$ and $\hat{\mathbf{\Pi}}$, which perform a separation of the distribution vector $\mathfrak{f}(t)$ into two components:

$$\begin{aligned}\mathfrak{f}(t) &= \bar{\mathfrak{f}}(t) + \hat{\mathfrak{f}}(t)\\ \bar{\mathfrak{f}}(t) &= \mathbf{\Pi}\mathfrak{f}(t) = \mathbf{\Sigma}(t)\mathfrak{f}(0)\\ \hat{\mathfrak{f}}(t) &= \hat{\mathbf{\Pi}}\mathfrak{f}(t) = \hat{\mathbf{\Sigma}}(t)\mathfrak{f}(0)\end{aligned} \tag{17.1.1}$$

The two components $\bar{\mathfrak{f}}(t)$ and $\hat{\mathfrak{f}}(t)$ evolve independently of each other, each one obeying its own subdynamics.

Our purpose in this chapter is to show that the evolution of one of the components, $\bar{\mathfrak{f}}(t)$, is closely related to the kinetic type of evolution studied in Chapter 11. For this reason, it will be called *the kinetic component of the distribution vector.*

Before starting the study of its properties, let us introduce a few new symbols.

The inverse Laplace transform of the irreducible evolution operator $\tilde{\mathscr{E}}(z)$ will be called $\mathscr{E}(t)$:

$$\mathscr{E}(t) = (2\pi)^{-1} \int_C dz\, e^{-izt} \tilde{\mathscr{E}}(z) \tag{17.1.2}$$

It should be noted that the zeroth-order term in $\tilde{\mathscr{E}}(z)$ is independent of z [see Eq. (16.3.5)]; its Laplace transform is therefore $\mathscr{L}'\, \delta(t)$. It is sometimes convenient to separate this special term from $\mathscr{E}(t)$ and write

$$\mathscr{E}(t) = \mathscr{L}'\, \delta(t) + \mathscr{G}(t) \tag{17.1.3}$$

where the operator $\mathscr{G}(t)$ is (usually) nonsingular as a function of time. It is clear from Eqs. (16.3.5)–(16.3.7) that the Laplace transform $\tilde{\mathscr{G}}(z)$ of this operator has the following properties:

$$\begin{aligned}\tilde{\mathcal{G}}(z) &\equiv \tilde{\mathcal{E}}(z) - \mathcal{L}' \\ &= \mathcal{L}' C\mathcal{R}^0(z)\tilde{\mathcal{E}}(z) \\ &= \tilde{\mathcal{E}}(z)\mathcal{R}^0(z)C\mathcal{L}' \end{aligned} \tag{17.1.4}$$

We shall denote by special symbols the following frequently occurring operators:

$$C\mathscr{C}(t)V = (2\pi)^{-1}\int_C dz\, e^{-izt}C\mathcal{R}^0(z)\tilde{\mathcal{E}}(z)V \tag{17.1.5}$$

$$V\mathcal{D}(t)C = (2\pi)^{-1}\int_C dz\, e^{-izt}V\tilde{\mathcal{E}}(z)\mathcal{R}^0(z)C \tag{17.1.6}$$

$C\mathscr{C}(t)V$ and $V\mathcal{D}(t)C$ are called the *creation operator* and the *destruction operator*, respectively, because their action may be understood as "creating correlations" out of the vacuum, or "destroying correlations." Alternatively, they can be defined by

$$C\mathscr{C}(t)V = \int_0^t d\tau\, C\mathcal{U}^0(t-\tau)\mathcal{E}(\tau)V \tag{17.1.7}$$

$$V\mathcal{D}(t)C = \int_0^t d\tau\, V\mathcal{E}(\tau)\mathcal{U}^0(t-\tau)C \tag{17.1.8}$$

The equivalence of these equations with Eqs. (17.1.5), (17.1.6) follows from the convolution theorem for Laplace transforms (16.1.21).

The following important relation is also easily derived from Eqs. (17.1.4) and (17.1.5):

$$V\mathcal{G}(t)V = V\mathcal{L}'C\mathscr{C}(t)V \tag{17.1.9}$$

We also have the "symmetrical" relation:

$$V\mathcal{G}(t)V = V\mathcal{D}(t)C\mathcal{L}'V \tag{17.1.10}$$

We now come back to the study of the kinetic component $\bar{\mathfrak{f}}(t)$. This component is an element of the set of distribution functions: hence it can always be decomposed into a vacuum and a correlation part according to Eq. (15.3.1):

$$\begin{aligned}\bar{\mathfrak{f}}(t) &= V\bar{\mathfrak{f}}(t) + C\bar{\mathfrak{f}}(t) \\ &= V\Pi\mathfrak{f}(t) + C\Pi\mathfrak{f}(t)\end{aligned} \tag{17.1.11}$$

We must now clearly understand the following crucial point. If an *arbitrary* distribution vector is decomposed into its vacuum and correlation components, the latter are, of course, completely independent of each other. If we choose at random a collection of vectors $\mathfrak{f}$, their

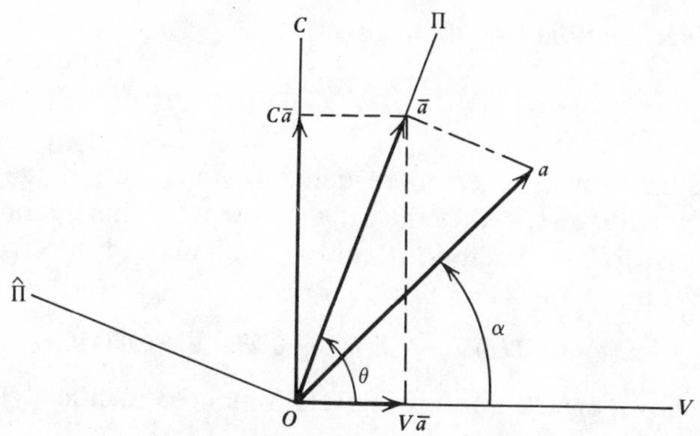

Figure 17.1.1. A geometrical analogy to the relation between correlations and vacuum components of the kinetic part of the distribution vector.

respective components are totally unrelated to each other. But the vector $\bar{\mathfrak{f}}(t)$ in Eq. (17.1.11) is *not* arbitrary: rather, it is the *component in a well-defined subspace*, of an arbitrary vector. One may, therefore, expect that its vacuum and correlation components be interrelated in a well-defined way.

Let us illustrate this statement by means of a simple geometrical analogy in ordinary two-dimensional Cartesian space (see Fig. 17.1.1). Consider a pair of orthogonal reference axes, labeled OV and OC. Consider also another pair of orthogonal axes, with the same origin, labeled $O\Pi$ and $O\hat{\Pi}$; the angle between the axes $O\Pi$ and OV is called θ.

Let $\mathbf{a}$ be a vector originating in O and making an angle α with OV; It has components $\bar{a}$ and $\hat{a}$ along the axes $O\Pi$ and $O\hat{\Pi}$, respectively. [This corresponds to the decomposition (17.1.1).] Its component $\bar{a}$ may be further resolved into components $V\bar{a}$ and $C\bar{a}$ along OV and OC. Clearly:

$$C\bar{a} = \tan\theta\ (V\bar{a})$$

This relation shows that the ratio $C\bar{a}/V\bar{a}$ is independent of the orientation α of the original vector $\mathbf{a}$; it only depends on the fixed relative orientation of the reference axes $O\Pi$ and OV.

Thus, as we said above, the correlation component of the Π component of $\mathbf{a}$ bears a fixed relationship to the vacuum component of the Π component of $\mathbf{a}$, whatever the vector $\mathbf{a}$. On the contrary, the C and V components of the *original* vector $\mathbf{a}$ are interrelated by

$$Ca = \tan\alpha\ (Va)$$

Their ratio is different for every different orientation α: this is what we mean by stating that they are mutually independent.

Coming back now to our original problem, we prove that *the correlation part of the kinetic distribution vector* $\bar{\mathfrak{f}}(t)$ *is related to its vacuum part by the following functional relationship:*

$$C\bar{\mathfrak{f}}(t) = \int_0^\infty d\tau\, C\mathscr{C}(\tau) V\bar{\mathfrak{f}}(t-\tau) \tag{17.1.12}$$

We can actually make even more detailed statements if we study the *kinetic propagator* $\mathbf{\Sigma}(t)$ rather than the vector $\bar{\mathfrak{f}}(t)$. This operator acts on the set of distribution vectors; it can be decomposed into four components according to Eq. (15.3.17):

$$\mathbf{\Sigma}(t) = V\mathbf{\Sigma}(t)V + V\mathbf{\Sigma}(t)C + C\mathbf{\Sigma}(t)V + C\mathbf{\Sigma}(t)C \tag{17.1.13}$$

It will be shown that the following relations hold identically:

$$C\mathbf{\Sigma}(t)V = \int_0^\infty d\tau\, C\mathscr{C}(\tau)V\mathbf{\Sigma}(t-\tau)V \tag{17.1.14}$$

$$C\mathbf{\Sigma}(t)C = \int_0^\infty d\tau\, C\mathscr{C}(\tau)V\mathbf{\Sigma}(t-\tau)C \tag{17.1.15}$$

These two relations can be coupled together in the compact form:

$$C\mathbf{\Sigma}(t) = \int_0^\infty d\tau\, C\mathscr{C}(\tau)V\mathbf{\Sigma}(t-\tau) \tag{17.1.16}$$

This relation clearly implies (17.1.12): indeed

$$\begin{aligned} C\bar{\mathfrak{f}}(t) = C\mathbf{\Sigma}(t)\mathfrak{f}(0) &= \int_0^\infty d\tau\, C\mathscr{C}(\tau)V\mathbf{\Sigma}(t-\tau)\mathfrak{f}(0) \\ &= \int_0^\infty d\tau\, C\mathscr{C}(\tau)V\bar{\mathfrak{f}}(t-\tau) \end{aligned}$$

Two more relations can be derived: They exhibit a characteristic "mirror" symmetry with respect to (17.1.14)–(17.1.16), a feature we shall often meet in this theory:

$$V\mathbf{\Sigma}(t)C = \int_0^\infty d\tau\, V\mathbf{\Sigma}(t-\tau)V\mathscr{D}(\tau)C \tag{17.1.17}$$

$$\begin{aligned} C\mathbf{\Sigma}(t)C &= \int_0^\infty d\tau\, C\mathbf{\Sigma}(t-\tau)V\mathscr{D}(\tau)C \\ &= \int_0^\infty d\tau \int_0^\infty d\tau'\, C\mathscr{C}(\tau')V\mathbf{\Sigma}(t-\tau-\tau')V\mathscr{D}(\tau)C \end{aligned} \tag{17.1.18}$$

These relations can again be put together in the form:

$$\mathbf{\Sigma}(t)C = \int_0^\infty d\tau\, \mathbf{\Sigma}(t-\tau)V\mathscr{D}(\tau)C \tag{17.1.19}$$

Equations (17.1.14), (17.1.17), and (17.1.18) show that all four components of the operator $\mathbf{\Sigma}(t)$ can be expressed in terms of the single component $V\mathbf{\Sigma}(t)V$ by means of universal relationships. This result is thus stronger than (17.1.12).

We now prove our statements. It is clearly sufficient to consider Eqs. (17.1.16) and (17.1.19); we only give the details for the former.

Using Newton's binomial formula and Leibnitz's differentiation formula (16.A.7), we obtain from Eq. (16.4.27):

$$C\mathbf{\Sigma}(t) = \lim_{z\to 0} \sum_{n=0}^{\infty} (n!)^{-1}(t+\partial)^n C\mathcal{R}^0\tilde{\mathscr{E}}V(V\mathscr{L}^0 + V\tilde{\mathscr{E}}V)^n(V + V\tilde{\mathscr{E}}\mathcal{R}^0 C)$$

$$= \lim_{z\to 0} \sum_{n=0}^{\infty}\sum_{p=0}^{n}\sum_{q=0}^{n-p} \left(\frac{t^p}{p!\,q!\,(n-p-q)!}\right)(\partial^q C\mathcal{R}^0\tilde{\mathscr{E}}V) \times\{\partial^{n-p-q}(V\mathscr{L}^0 + V\tilde{\mathscr{E}}V)^n(V + V\tilde{\mathscr{E}}\mathcal{R}^0 C)\} \quad (17.1.20)$$

On the other hand

$$\int_0^\infty d\tau\, C\mathscr{C}(\tau)V\mathbf{\Sigma}(t-\tau)$$

$$= \int_0^\infty d\tau\, C\mathscr{C}(\tau)V \lim_{z\to 0}\sum_{n=0}^{\infty}(n!)^{-1}(t-\tau+\partial)^n(V\mathscr{L}^0 + V\tilde{\mathscr{E}}V)^n(V + V\tilde{\mathscr{E}}\mathcal{R}^0 C)$$

$$= \lim_{z\to 0}\int_0^\infty d\tau\, C\mathscr{C}(\tau)V \sum_{n=0}^{\infty}\sum_{p=0}^{n}\sum_{q=0}^{n-p}\left(\frac{t^p}{p!\,q!\,(n-p-q)!}\right)(-\tau)^q \times\{\partial^{n-p-q}(V\mathscr{L}^0 + V\tilde{\mathscr{E}}V)^n(V + V\tilde{\mathscr{E}}\mathcal{R}^0 C)\} \quad (17.1.21)$$

We now use the following relation connecting a pair of Laplace transforms $\tilde{A}(z)$ and $A(t)$:

$$\left(i\frac{\partial}{\partial z}\right)^n \tilde{A}(z) = \int_0^\infty d\tau\, e^{iz\tau}(-\tau)^n A(\tau) \quad (17.1.22)$$

from which follows, in particular,

$$\{\partial^n \tilde{A}(z)\}_{z=0} = \int_0^\infty d\tau\, (-\tau)^n A(\tau) \quad (17.1.23)$$

We therefore obtain

$$\int_0^\infty d\tau\, C\mathscr{C}(\tau)V\,(-\tau)^q = \{\partial^q C\tilde{\mathscr{C}}(z)V\}_{z=0}$$

$$= \{\partial^q C\mathcal{R}^0(z)\tilde{\mathscr{E}}(z)V\}_{z=0}$$

Substituting this result into the right-hand side of Eq. (17.1.21) we obtain the right-hand side of Eq. (17.1.20).

17.2. THE GENERAL KINETIC EQUATION

It was shown in the previous section that a relation exists between the correlation and vacuum components of the kinetic distribution vector $\bar{\mathfrak{f}}(t)$;

the determination of this vector is therefore reduced to the computation of the single component $V\bar{\mathfrak{f}}(t)$. The latter is, in principle, completely defined by Eqs. (17.1.1) and (16.4.27). This definition, in terms of an infinite power series in t, is, however, not convenient for practical purposes because the convergence of this series is very slow, especially for large values of t. We now prove that the relevant component can be characterized in a different way, namely, as the solution of an integro-differential equation. This equation can then be solved by any of the standard (or less standard) methods particularly adapted to the problem at hand.

The derivation of this equation is very simple and illuminating. From Eq. (16.4.39) follows the fact, already stressed in Eq. (16.2.11) that $\bar{\mathfrak{f}}(t)$ is a solution of the Liouville equation, hence

$$\partial_t \bar{\mathfrak{f}}(t) = (\mathscr{L}^0 + \mathscr{L}')\bar{\mathfrak{f}}(t) \tag{17.2.1}$$

Its vacuum component therefore obeys

$$\partial_t V\bar{\mathfrak{f}}(t) = \mathscr{L}^0 V\bar{\mathfrak{f}}(t) + V\mathscr{L}' V\bar{\mathfrak{f}}(t) + V\mathscr{L}' C\bar{\mathfrak{f}}(t) \tag{17.2.2}$$

There is, however, a serious difference between Eq. (17.2.2) and the apparently similar Liouville equation (16.2.1). In the latter, the components $V\mathfrak{f}(t)$ and $C\mathfrak{f}(t)$ are mutually independent; Eq. (16.2.1) must be completed with an equation for $\partial_t C\mathfrak{f}(t)$ that is coupled to the former.

On the contrary, the correlation component of $\bar{\mathfrak{f}}(t)$ is an explicit, fixed functional of the vacuum component $V\bar{\mathfrak{f}}(t)$. Equation (17.1.12) must be added to Eq. (17.2.2) as a *constraint* to be satisfied at all times.

The correlation component can now be explicitly eliminated from Eq. (17.2.2) by means of the constraint:

$$\partial_t V\bar{\mathfrak{f}}(t) = \mathscr{L}^0 V\bar{\mathfrak{f}}(t) + V\mathscr{L}' V\bar{\mathfrak{f}}(t) + \int_0^\infty d\tau\, V\mathscr{L}' C\mathscr{C}(\tau) V\bar{\mathfrak{f}}(t-\tau)$$

which, by Eq. (17.1.9), can be transformed into

$$\partial_t V\bar{\mathfrak{f}}(t) = V(\mathscr{L}^0 + \mathscr{L}') V\bar{\mathfrak{f}}(t) + \int_0^\infty d\tau\, V\mathscr{G}(\tau) V\bar{\mathfrak{f}}(t-\tau) \tag{17.2.3}$$

Hence the constraint (17.1.12) transforms the coupled Liouville *equations into a single closed equation for the vacuum component of the kinetic part of the distribution vector.* Equation (17.2.3) is, however, not yet a kinetic equation in the strict sense defined in Section 16.2. Although it is closed, it is not a Markovian equation. We shall see, however, that a further transformation leads us to a true kinetic equation.

Indeed, we now show that in many important cases, *Eq. (17.2.3) possesses a quite remarkable exponential solution:*

$$V\bar{f}(t) = \{\exp(tV\Gamma V)\}V\bar{f}(0) \tag{17.2.4}$$

where $V\Gamma V$ is a time-independent operator, having only a V–V component in the representation (15.3.17).

It then follows that the vacuum part of the kinetic distribution vector obeys the equation:

$$\partial_t V\bar{f}(t) = V\Gamma V\bar{f}(t) \tag{17.2.5}$$

This equation must, of course, be equivalent to Eq. (17.2.3). Indeed, *if* the operator $V\Gamma V$ exists, Eq. (17.2.4) is the unique solution of the linear equation (17.2.3).

We now show how the operator $V\Gamma V$ can be explicitly constructed. Substituting (17.2.4) into (17.2.3) we obtain

$$V\Gamma V\bar{f}(t) = V(\mathcal{L}^0 + \mathcal{L}')V\bar{f}(t) + \int_0^\infty d\tau\, V\mathcal{G}(\tau)V[\exp(t-\tau)V\Gamma V]V\bar{f}(0)$$

from which we deduce the following equation for the operator $V\Gamma V$:

$$V\Gamma V = V(\mathcal{L}^0 + \mathcal{L}')V + \int_0^\infty d\tau\, V\mathcal{G}(\tau)V \exp(-\tau V\Gamma V) \tag{17.2.6}$$

This is a rather complex nonlinear integral equation. The existence of a solution is not granted: there may be systems for which $V\Gamma V$ does not exist. In this book we limit ourselves to systems for which we *assume* the existence of a solution to Eq. (17.2.6). We even assume that this solution can be obtained by a meaningful iteration procedure.

There are several possible iteration schemes for solving this equation, each giving different, though equivalent representations of the operator $V\Gamma V$. The method chosen here gives the fastest convergence (especially for inhomogeneous systems) and provides an explicit, relatively simple form for the general term.

The zeroth-order approximation is taken as

$$V\Gamma^{(0)}V = V(\mathcal{L}^0 + \mathcal{L}')V \equiv V\mathcal{L}V \tag{17.2.7}$$

The physical meaning of this equation will become clearer in the next chapters. The operator $V\mathcal{L}V$ turns out to be the *Vlassov operator* of Eq. (11.7.5), as will be shown in Section 18.3. The corresponding exponentiated operator, denoted by $V\tilde{\mathcal{U}}(t)V$, will be called the *Vlassov propagator:*

$$V\tilde{\mathcal{U}}(t)V = V\exp\{tV(\mathcal{L}^0 + \mathcal{L}')V\} \tag{17.2.8}$$

It must be stressed that this operator depends on the interactions. Substituting this expression into the integral of Eq. (17.2.6) we obtain the first approximation:

$$V\mathbf{\Gamma}^{(1)}V = V(\mathcal{L}^0 + \mathcal{L}')V + \int_0^\infty d\tau_2\, V\mathcal{G}(\tau_2)V\tilde{\mathcal{U}}(-\tau_2)V \tag{17.2.9}$$

To go further, we calculate the exponential of this operator, using a well-known identity:

$$e^{t(A+B)} = e^{tA} + \int_0^t d\theta\, e^{(t-\theta)A}Be^{\theta(A+B)} \tag{17.2.10}$$

[This identity can be checked by time differentiation of both sides; it is a straightforward generalization of Eq. (16.1.19).] Calling A and B the first and second term of the right side of (17.2.9), respectively, we get, to first order (i.e., one single $\mathcal{G}$ factor)

$$\{\exp(tV\mathbf{\Gamma}V)\}^{(1)} = V\tilde{\mathcal{U}}(t)V + \int_0^t d\theta\, V\tilde{\mathcal{U}}(t-\theta)V\int_0^\infty d\tau_2\, V\mathcal{G}(\tau_2)V\tilde{\mathcal{U}}(-\tau_2)V\tilde{\mathcal{U}}(\theta)V \tag{17.2.11}$$

With this result the second iteration of (17.2.6) takes the form:

$$V\mathbf{\Gamma}^{(2)}V = V(\mathcal{L}^0 + \mathcal{L}')V + \int_0^\infty d\tau_2\, V\mathcal{G}(\tau_2)V\tilde{\mathcal{U}}(-\tau_2)V + \int_0^\infty d\tau_2 \int_0^\infty d\tau_4 \int_0^{-\tau_2} d\tau_3\, V\mathcal{G}(\tau_2)V\tilde{\mathcal{U}}(-\tau_2-\tau_3)V\mathcal{G}(\tau_4)V\tilde{\mathcal{U}}(\tau_3-\tau_4)V \tag{17.2.12}$$

This iteration process can be continued systematically in the same way. One will soon be convinced that the successive terms have a very regular structure. The complete formal solution can therefore be written as

$$V\mathbf{\Gamma}V = V(\mathcal{L}^0 + \mathcal{L}')V + \sum_{n=1}^\infty V\mathbf{\Gamma}_n V \tag{17.2.13}$$

with

$$\begin{aligned} V\mathbf{\Gamma}_n V = {} & \int_0^\infty d\tau_2 \int_0^\infty d\tau_4 \cdots \int_0^\infty d\tau_{2n} \int_0^{\tau_1-\tau_2} d\tau_3 \int_0^{\tau_3-\tau_4} d\tau_5 \cdots \int_0^{\tau_{2n-3}-\tau_{2n-2}} d\tau_{2n-1} \\ & \times V\mathcal{G}(\tau_2)V\tilde{\mathcal{U}}(\tau_1-\tau_2-\tau_3)V\mathcal{G}(\tau_4)V\tilde{\mathcal{U}}(\tau_3-\tau_4-\tau_5)V\mathcal{G}(\tau_6)V \cdots \\ & V\tilde{\mathcal{U}}(\tau_{2n-3}-\tau_{2n-2}-\tau_{2n-1})V\mathcal{G}(\tau_{2n})V\tilde{\mathcal{U}}(\tau_{2n-1}-\tau_{2n}-\tau_{2n+1})V \end{aligned} \tag{17.2.14}$$

where one has to set $\tau_1 = \tau_{2n+1} = 0$. We preferred to write these two spurious variables in (17.2.14) in order to enhance the basic regularity in the limits of integration of the odd τ's and in the order of succession of the arguments of the $\tilde{\mathcal{U}}$ factors.

We thus obtained an explicit expression for the kinetic operator $V\mathbf{\Gamma}V$

of Eq. (17.2.5):

$$\partial_t V\bar{\mathfrak{f}}(t) = V\mathbf{\Gamma} V\bar{\mathfrak{f}}(t)$$

This equation realizes the point D of our program of Section 16.2: *it is a closed Markovian equation for the vacuum component of* $\bar{\mathfrak{f}}(t)$. All the memory effects appearing in Eq. (17.2.3) are now included in the new operator $V\mathbf{\Gamma}V$. The price to be paid is, of course, in the added complexity of the latter, as compared to Eq. (17.2.3).

Equation (17.2.5) will be called the *general kinetic equation.* This name will be further justified when we show, in Chapters 18 and 20, that the known kinetic equations (such as those studied in Chapter 11) are particular cases of Eq. (17.2.5).

We now see that the kinetic equation entered the theory quite smoothly, without any "cheating" or any contradiction with mechanics. At no point did we use any probabilistic argument: the kinetic equation appears as an exact equation of evolution for the vacuum component of an invariant part of the distribution vector. The only restrictions for its validity are the assumption (16.4.12) and the mathematical existence of the operator $V\mathbf{\Gamma}V$. Both of these are dynamical (not statistical) assumptions.

We close this section by writing a few additional relations that will be used below. The proofs of these relations go along the same lines as above: they are omitted here. It is first easily seen that Eq. (17.2.4) implies the following form for the kinetic propagator:

$$V\mathbf{\Sigma}(t) = \{\exp(tV\mathbf{\Gamma}V)\}V\mathbf{\Sigma}(0) \tag{17.2.15}$$

It can further be shown that the following "mirror-symmetrical" relation holds:

$$\mathbf{\Sigma}(t)V = \mathbf{\Sigma}(0)V\{\exp(tV\mathbf{\Delta}V)\} \tag{17.2.16}$$

The operator $V\mathbf{\Delta}V$ can be represented in a form quite similar to (17.2.13), (17.2.14):

$$V\mathbf{\Delta}V = V(\mathscr{L}^0 + \mathscr{L}')V + \sum_{n=1}^{\infty} V\mathbf{\Delta}_n V \tag{17.2.17}$$

with

$$\begin{aligned} V\mathbf{\Delta}_n V = \int_0^\infty d\tau_2 \int_0^\infty d\tau_4 \cdots \int_0^\infty d\tau_{2n} \int_0^{\tau_1-\tau_2} d\tau_3 \int_0^{\tau_3-\tau_4} d\tau_5 \cdots \int_0^{\tau_{2n-3}-\tau_{2n-2}} d\tau_{2n-1} \\ \times V\tilde{\mathscr{U}}(-\tau_{2n+1}-\tau_{2n}+\tau_{2n-1})V\mathscr{G}(\tau_{2n})V\tilde{\mathscr{U}}(-\tau_{2n-1}-\tau_{2n-2}+\tau_{2n-3})V \\ \times V\mathscr{G}(\tau_{2n-2})V \cdots V\mathscr{G}(\tau_6)V\tilde{\mathscr{U}}(-\tau_5-\tau_4+\tau_3)V\mathscr{G}(\tau_4)V \\ \times V\tilde{\mathscr{U}}(-\tau_3-\tau_2+\tau_1)V\mathscr{G}(\tau_2)V \end{aligned} \tag{17.2.18}$$

where one must again set $\tau_1 = \tau_{2n+1} = 0$. One immediately notes the nice symmetry of "mirror" type existing between the operators $V\mathbf{\Gamma}V$ and $V\mathbf{\Delta}V$.

17.3. TIME-INDEPENDENT FUNCTIONAL RELATIONS

We have derived in Section 17.1 a number of relations among the components of the kinetic propagator $\mathbf{\Sigma}(t)$ and hence between correlation and vacuum components of the kinetic distribution vector $\bar{\mathfrak{f}}(t)$. All these relations involved an integration over the past history of the system. Hence, although they are nice structural features of the theory, they are not easy to use in practice because they presuppose the knowledge of the solution of the kinetic equation. We will now show that these relations can be transformed into an equivalent form, involving, however, no time integration, but only functions evaluated all at the same time. The crux of this remarkable property is the exponential form of the propagator $V\mathbf{\Sigma}(t)V$, Eqs. (17.2.15), (17.2.16).

Consider, indeed, Eq. (17.1.16), and substitute Eq. (17.2.15); we obtain

$$C\mathbf{\Sigma}(t)=\int_0^\infty d\tau\, C\mathscr{C}(\tau)V\{\exp(t-\tau)V\mathbf{\Gamma}V\}V\mathbf{\Sigma}(0)$$

and hence

$$C\mathbf{\Sigma}(t)=C\mathbf{C}V\mathbf{\Sigma}(t) \tag{17.3.1}$$

with

$$C\mathbf{C}V=\int_0^\infty d\tau\, C\mathscr{C}(\tau)V\exp(-\tau V\mathbf{\Gamma}V) \tag{17.3.2}$$

Equation (17.3.1) relates the correlation components $C\mathbf{\Sigma}(t)$ to the vacuum component $V\mathbf{\Sigma}(t)$ evaluated at *the same time*, through the action of the *time-independent* operator $C\mathbf{C}V$.

By using the expansion of the exponential in the same way as in Section 17.2, we can get an explicit series expansion for $C\mathbf{C}V$, which is quite useful in practical calculations:

$$C\mathbf{C}V=\sum_{n=1}^{\infty} C\mathbf{C}_n V \tag{17.3.3}$$

with

$$\begin{aligned}C\mathbf{C}_nV=\delta_{n,1}&\int_0^\infty d\tau_2\, C\mathscr{C}(\tau_2)V\tilde{\mathscr{U}}(-\tau_2)V+(1-\delta_{n,1})\\&\times\int_0^\infty d\tau_2\int_0^\infty d\tau_4\cdots\int_0^\infty d\tau_{2n}\int_0^{\tau_1-\tau_2} d\tau_3\int_0^{\tau_3-\tau_4} d\tau_5\cdots\int_0^{\tau_{2n-3}-\tau_{2n-2}} d\tau_{2n-1}\\&\times C\mathscr{C}(\tau_2)V\tilde{\mathscr{U}}(\tau_1-\tau_2-\tau_3)V\mathscr{G}(\tau_4)V\tilde{\mathscr{U}}(\tau_3-\tau_4-\tau_5)V\mathscr{G}(\tau_6)\\&\times V\cdots V\tilde{\mathscr{U}}(\tau_{2n-3}-\tau_{2n-2}-\tau_{2n-1})V\mathscr{G}(\tau_{2n})V\tilde{\mathscr{U}}(\tau_{2n-1}-\tau_{2n}-\tau_{2n+1})V\\&\qquad[\tau_1=\tau_{2n+1}=0]\end{aligned} \tag{17.3.4}$$

This expansion is closely similar to Eqs. (17.2.13) and (17.2.14). The only difference is in the replacement of the first factor, $V\mathscr{G}(\tau_2)V$, in the integrand by the factor $C\mathscr{C}(\tau_2)V$.

Combining now Eq. (17.2.16) with (17.1.19) we obtain

$$\mathbf{\Sigma}(t)C = \int_0^\infty d\tau\, \mathbf{\Sigma}(0)\{\exp[(t-\tau)V\mathbf{\Delta}V]\}V\mathscr{D}(\tau)C$$

and hence

$$\mathbf{\Sigma}(t)C = \mathbf{\Sigma}(t)V\mathbf{D}C \tag{17.3.5}$$

with

$$V\mathbf{D}C = \int_0^\infty d\tau\, \exp(-\tau V\mathbf{\Delta}V)V\mathscr{D}(\tau)C \tag{17.3.6}$$

Again, the operator $V\mathbf{\Delta}C$ can be expanded in the same way as the previous operators; we do not write down the complete formula, which is very similar to (17.3.4):

$$V\mathbf{D}C = \int_0^\infty d\tau_2\, V\tilde{\mathscr{U}}(-\tau_2)V\mathscr{D}(\tau_2)C + \cdots \tag{17.3.7}$$

We can also obtain a different and very compact form of the kinetic evolution operator $V\mathbf{\Gamma}V$. Starting from Eq. (17.2.2) and expressing now the constraint in the new form (17.3.1), we obtain the kinetic equation:

$$\partial_t V\bar{\mathfrak{f}}(t) = V\mathbf{\Gamma}V\bar{\mathfrak{f}}(t)$$

with

$$V\mathbf{\Gamma}V = V(\mathscr{L}^0 + \mathscr{L}')V + V\mathscr{L}'C\mathbf{C}V \tag{17.3.8}$$

One should not forget, however, that $C\mathbf{C}V$ is itself expressed in terms of $V\mathbf{\Gamma}V$; hence (17.3.8) is really an implicit nonlinear relation involving $V\mathbf{\Gamma}V$. By using Eqs. (17.1.9) and (17.3.2) one easily sees that

$$\begin{aligned} V\mathscr{L}'C\mathbf{C}V &= \int_0^\infty d\tau\, V\mathscr{L}'C\mathscr{C}(\tau)V\exp(-\tau V\mathbf{\Gamma}V) \\ &= \int_0^\infty d\tau\, V\mathscr{G}(\tau)V\exp(-\tau V\mathbf{\Gamma}V) \end{aligned}$$

which establishes the equivalence of Eqs. (17.3.8) and (17.2.3).

Let us now collect, for easy reference, the relationships obtained until now. The components of the kinetic propagator can be expressed as follows:

$$V\mathbf{\Sigma}(t)C = V\mathbf{\Sigma}(t)V\mathbf{D}C \tag{17.3.9}$$

$$C\mathbf{\Sigma}(t)V = C\mathbf{C}V\mathbf{\Sigma}(t)V \tag{17.3.10}$$

$$C\mathbf{\Sigma}(t)C = C\mathbf{C}V\mathbf{\Sigma}(t)V\mathbf{D}C \tag{17.3.11}$$

$$\begin{aligned} V\mathbf{\Sigma}(t)V &= \exp(tV\mathbf{\Gamma}V)V\mathbf{\Sigma}(0)V \\ &= V\mathbf{\Sigma}(0)V\exp(tV\mathbf{\Delta}V) \end{aligned} \tag{17.3.12}$$

From Eq. (16.4.38) it is clear that the projection operator $\mathbf{\Pi}$ has exactly the same structure:

$$V\mathbf{\Pi}C = V\mathbf{\Pi}V\mathbf{D}C \tag{17.3.13}$$

$$C\mathbf{\Pi}V = C\mathbf{C}V\mathbf{\Pi}V \tag{17.3.14}$$

$$C\mathbf{\Pi}C = C\mathbf{C}V\mathbf{\Pi}V\mathbf{D}C \tag{17.3.15}$$

This structure is clearly a remnant and an imprint of the structure (16.3.18)–(16.3.20) of the Liouville resolvent.*

Proceeding now to the kinetic distribution vector, these relations imply that the correlation component $C\bar{\mathfrak{f}}(t)$ is related to the vacuum component $V\bar{\mathfrak{f}}(t)$ through a time-independent functional:

$$C\bar{\mathfrak{f}}(t) = C\mathbf{C}V\bar{\mathfrak{f}}(t) \tag{17.3.16}$$

whereas the vacuum part obeys the differential equation

$$\partial_t V\bar{\mathfrak{f}}(t) = V\mathbf{\Gamma}V\bar{\mathfrak{f}}(t) \tag{17.3.17}$$

These relations completely characterize the kinetic part of the propagator and the kinetic part of the distribution vector. They may be taken as a *definition* of these concepts.

17.4. EVOLUTION OF THE NONKINETIC COMPONENT OF THE DISTRIBUTION VECTOR

We now turn our attention to the nonkinetic part of the propagator $\mathcal{U}(t)$ and show that it has properties analogous, though complementary, to those of $\mathbf{\Sigma}(t)$.

Using Eqs. (16.4.33) and (17.3.13) we obtain

$$0 = V\mathbf{\Pi}\hat{\mathbf{\Pi}} = (V\mathbf{\Pi}V + V\mathbf{\Pi}C)\hat{\mathbf{\Pi}} = V\mathbf{\Pi}V(V + V\mathbf{D}C)\hat{\mathbf{\Pi}}$$

and therefore (as $V\mathbf{\Pi}V \neq 0$)

$$V\hat{\mathbf{\Pi}} = -V\mathbf{D}C\hat{\mathbf{\Pi}} \tag{17.4.1}$$

Similarly

$$\hat{\mathbf{\Pi}}V = -\hat{\mathbf{\Pi}}C\mathbf{C}V \tag{17.4.2}$$

Using these equations together with (16.4.34) we obtain three relations

* Equations (17.3.13)–(17.3.15) show that the proof of the property $V\mathbf{\Pi}^2V = V\mathbf{\Pi}V$ (Appendix to Chapter 16) implies the validity of the relation $\mathbf{\Pi}^2 = \mathbf{\Pi}$. Indeed, we have, for example $V\mathbf{\Pi}^2C = V\mathbf{\Pi}\mathbf{\Pi}V\mathbf{D}C = V\mathbf{\Pi}V\mathbf{D}C = V\mathbf{\Pi}C$ and similar relations for the $(C-V)$ and $(C-C)$ components.

analogous to (17.3.9)–(17.3.11):

$$V\hat{\mathbf{\Sigma}}(t)C = -V\mathbf{D}C\hat{\mathbf{\Sigma}}(t)C \tag{17.4.3}$$

$$C\hat{\mathbf{\Sigma}}(t)V = -C\hat{\mathbf{\Sigma}}(t)C\mathbf{C}V \tag{17.4.4}$$

$$V\hat{\mathbf{\Sigma}}(t)V = V\mathbf{D}C\hat{\mathbf{\Sigma}}(t)C\mathbf{C}V \tag{17.4.5}$$

Hence, we can again eliminate three of the components of $\hat{\mathbf{\Sigma}}(t)$ in terms of the fourth. However, unlike $\mathbf{\Sigma}(t)$, the independent component is now the correlation–correlation part, $C\hat{\mathbf{\Sigma}}(t)C$. It also follows that the *vacuum part of $\hat{\mathfrak{f}}(t)$ is a functional of the correlation,* a situation quite opposite to $\bar{\mathfrak{f}}(t)$:

$$V\hat{\mathfrak{f}}(t) = -V\mathbf{D}C\hat{\mathfrak{f}}(t) \tag{17.4.6}$$

To complete the picture, we now show that the independent component can be represented as

$$C\hat{\mathbf{\Sigma}}(t)C = \exp(tC\hat{\mathbf{\Delta}}C)C\hat{\mathbf{\Pi}}C = C\hat{\mathbf{\Pi}}C\,\exp(tC\hat{\mathbf{\Gamma}}C) \tag{17.4.7}$$

in complete analogy with (17.3.12).

To derive an expression for $C\hat{\mathbf{\Delta}}C$, we start from

$$\begin{aligned}
\partial_t C\hat{\mathbf{\Sigma}}(t)C &= \partial_t C\mathcal{U}(t)\hat{\mathbf{\Pi}}C = C\mathcal{L}\hat{\mathbf{\Sigma}}(t)C \\
&= C\mathcal{L}V\hat{\mathbf{\Sigma}}(t)C + C\mathcal{L}C\hat{\mathbf{\Sigma}}(t)C \\
&= -C\mathcal{L}V\mathbf{D}C\hat{\mathbf{\Sigma}}(t)C + C\mathcal{L}C\mathbf{\Sigma}(t)C
\end{aligned}$$

This equation can therefore be rewritten as

$$\partial_t C\hat{\mathbf{\Sigma}}(t)C = C\hat{\mathbf{\Delta}}C\hat{\mathbf{\Sigma}}(t)C \tag{17.4.8}$$

and provides the definition of $C\hat{\mathbf{\Delta}}C$ as

$$C\hat{\mathbf{\Delta}}C = C(\mathcal{L}^0 + \mathcal{L}')C - C\mathcal{L}'V\mathbf{D}C \tag{17.4.9}$$

This equation looks quite similar to (17.3.8); however, it is in some sense simpler. Indeed, this relation explicitly expresses $C\hat{\mathbf{\Delta}}C$ in terms of the operator $V\mathbf{\Delta}V$, which has been determined earlier. Another equivalent form, analogous to Eq. (17.2.6) is

$$C\hat{\mathbf{\Delta}}C = C(\mathcal{L}^0 + \mathcal{L}')C - \int_0^\infty d\tau\, C\mathcal{L}'V\exp(-\tau V\mathbf{\Delta}V)V\mathcal{D}(\tau)C \tag{17.4.10}$$

Using the methods of Section 17.2, it is a very easy matter to expand the exponential in this equation and obtain an explicit expression for the operator $C\hat{\mathbf{\Delta}}C$ (of which we only give the first two terms):

$$C\hat{\mathbf{\Delta}}C = C(\mathcal{L}^0 + \mathcal{L}')C - \int_0^\infty d\tau_2\, C\mathcal{L}'V\tilde{\mathcal{U}}(-\tau_2)V\mathcal{D}(\tau_2)C - \cdots \tag{17.4.11}$$

The proof of the second part of Eq. (17.4.7) will not be given here.

17.5. THE INITIAL-VALUE PROBLEM IN SUBDYNAMICS

We have now a fairly clear picture of the evolution in the two separate subspaces. Summarizing our findings, we started with a distribution vector $\mathfrak{f}(t)$ obeying the Liouville equation:

$$\partial_t \mathfrak{f}(t) = \mathscr{L}\mathfrak{f}(t) \tag{17.5.1}$$

We then decomposed this vector into two parts in such a way that each part still obeys its own subdynamics:

$$\mathfrak{f}(t) = \bar{\mathfrak{f}}(t) + \hat{\mathfrak{f}}(t) \tag{17.5.2}$$

$$\begin{aligned} \partial_t \bar{\mathfrak{f}}(t) &= \mathscr{L}\bar{\mathfrak{f}}(t) \\ \partial_t \hat{\mathfrak{f}}(t) &= \mathscr{L}\hat{\mathfrak{f}}(t) \end{aligned} \tag{17.5.3}$$

We saw that there is only one way of realizing such a decomposition and that it implies the existence of constraints, to be added to the Liouville equations (17.5.3). As a result, we can further reduce the problem to the study of $V\bar{\mathfrak{f}}(t)$ and of $C\hat{\mathfrak{f}}(t)$, obeying, respectively, the following uncoupled equations:

$$\begin{aligned} \partial_t V\bar{\mathfrak{f}}(t) &= V\boldsymbol{\Gamma} V\bar{\mathfrak{f}}(t) \\ \partial_t C\hat{\mathfrak{f}}(t) &= C\hat{\boldsymbol{\Delta}} C\hat{\mathfrak{f}}(t) \end{aligned} \tag{17.5.4}$$

These are two differential equations (in time) that, under mild conditions, possess a unique solution if the initial values of $V\bar{\mathfrak{f}}(t)$ and of $C\hat{\mathfrak{f}}(t)$ are given.

We are now faced with the following problem. We study a system described at time $t=0$ by a given distribution vector:

$$\mathfrak{f}(0) \equiv \mathfrak{f}_0 \tag{17.5.5}$$

If we could solve the Liouville equation (17.5.1) with this initial condition, we would obtain a unique solution $\mathfrak{f}(t)$. We replaced this problem, however, by the solution of the two uncoupled equations (17.5.4). Clearly, in order to obtain [through (17.5.2)] an equivalent result, we cannot use arbitrary initial conditions for the equations (17.5.4): rather, the initial condition to be used is completely determined by $\mathfrak{f}_0$. The relation between $\mathfrak{f}_0$ and the equivalent initial conditions for (17.5.4) is very easily obtained by projection of $\mathfrak{f}_0$ on the relevant subspaces. Thus, the kinetic equation:

$$\partial_t V\bar{\mathfrak{f}}(t) = V\boldsymbol{\Gamma} V\bar{\mathfrak{f}}(t)$$

must be solved with the initial condition:

$$V\bar{\mathfrak{f}}(0) = V\boldsymbol{\Pi}\mathfrak{f}_0 \tag{17.5.6}$$

Similarly, the nonkinetic equation:

$$\partial_t C\hat{\mathfrak{f}}(t) = C\hat{\boldsymbol{\Delta}}C\hat{\mathfrak{f}}(t)$$

must be solved with the initial condition:

$$C\hat{\mathfrak{f}}(0) = C\hat{\boldsymbol{\Pi}}\mathfrak{f}_0 \tag{17.5.7}$$

Let us stress the peculiar nature of these redefined initial conditions. Writing the right-hand side of (17.5.6) in more detail we find

$$\begin{aligned} V\bar{\mathfrak{f}}(0) &= V\boldsymbol{\Pi}V\mathfrak{f}_0 + V\boldsymbol{\Pi}C\mathfrak{f}_0 \\ &= V\boldsymbol{\Pi}V(V\mathfrak{f}_0 + V\mathbf{D}C\mathfrak{f}_0) \end{aligned} \tag{17.5.8}$$

Hence the initial condition to be used for the kinetic equation is *not* simply the vacuum component of $\mathfrak{f}_0$. Rather, it is a well-defined combination of both vacuum and correlation components of the given initial distribution vector.

Similarly,

$$\begin{aligned} C\mathfrak{f}(0) &= C\hat{\boldsymbol{\Pi}}C\mathfrak{f}_0 + C\hat{\boldsymbol{\Pi}}V\mathfrak{f}_0 \\ &= C\hat{\boldsymbol{\Pi}}C(C\mathfrak{f}_0 - C\mathbf{C}V\mathfrak{f}_0) \end{aligned} \tag{17.5.9}$$

We now see from Eq. (17.5.9) that a very interesting situation arises if

$$C\mathfrak{f}_0 = C\mathbf{C}V\mathfrak{f}_0 \tag{17.5.10}$$

In that case $C\hat{\mathfrak{f}}(0) = 0$ and $C\hat{\mathfrak{f}}(t) = 0$ at all times. Hence, *there exists a class of initial conditions for which the kinetic equation is exactly equivalent to the Liouville equation.* Although this result may seem surprising, it is actually a mere test of the self-consistency of our formalism. Indeed, (17.5.10) precisely tells us that the initial condition is a $\boldsymbol{\Pi}$ function [see Eq. (17.3.16)]: hence, by the law of subdynamics, it remains in the $\boldsymbol{\Pi}$ subspace for all later times. Therefore its *exact* evolution is determined by the kinetic equation.

In general, however, a physical system chosen at random at time zero will have components in both subspaces $\boldsymbol{\Pi}$ and $\hat{\boldsymbol{\Pi}}$. These two components evolve independently of each other and give rise to a vector $\bar{\mathfrak{f}}(t)$ and a vector $\hat{\mathfrak{f}}(t)$ at all times. Our hope can then be expressed as follows. We should like that, at least in certain important problems of statistical physics, we could forget about the nonkinetic component $\hat{\mathfrak{f}}(t)$. In other words, we should like to show that, for these problems,

(a) either $\hat{\mathfrak{f}}(t) \equiv 0$,

(b) or $\hat{\mathfrak{f}}(t) \neq 0$, but it gives a vanishing contribution to the calculation of the particular physical quantities in view.

Although these hopes seem very ambitious, they will turn out to come true. There is one simple, but very important case that will be treated in Section 17.7: a system in thermal equilibrium. We shall see that in this case our first possibility is realized: All equilibrium distribution vectors lie in the $\mathbf{\Pi}$ subspace. Hence, the vast field of equilibrium thermodynamics is formulated exactly in terms of the kinetic distribution vector alone. Other important, nonequilibrium problems will be treated in Section 17.8.

17.6. SPATIALLY HOMOGENEOUS STATES

We know already from our investigations of Chapters 11–13 that spatially homogeneous systems have a particularly simple behavior. For such systems the free-flow term and the Vlassov term vanish identically. We may therefore characterize a homogeneous system quite generally by the following relations:

$$V\mathcal{L}^0 = 0$$
$$V\mathcal{L}'V = 0 \qquad \text{(Homogeneous systems)} \qquad (17.6.1)$$

It then follows that the Vlassov propagator is simply a constant:

$$V\tilde{\mathcal{U}}(t)V = V \qquad \text{(Homogeneous systems)} \qquad (17.6.2)$$

It is of interest to see how the previously derived results specialize in this case. All the expressions become much simpler. In particular, the kinetic equation reduces to its collision term:

$$\partial_t V\bar{\mathfrak{f}}(t) = V\mathbf{\Gamma}V\bar{\mathfrak{f}}(t) \qquad (17.6.3)$$

with

$$V\mathbf{\Gamma}V = \sum_{n=1}^{\infty} V\mathbf{\Gamma}_n V \qquad (17.6.4)$$

Equation (17.2.14) defining $V\mathbf{\Gamma}_n V$ now reduces to

$$V\mathbf{\Gamma}_n V = \int_0^\infty d\tau_2 \int_0^\infty d\tau_4 \cdots \int_0^\infty d\tau_n \int_0^{-\tau_2} d\tau_3 \int_0^{\tau_3-\tau_4} d\tau_5 \cdots \int_0^{\tau_{2n-3}-\tau_{2n-2}} d\tau_{2n-1}$$
$$\times V\mathcal{G}(\tau_2)V\mathcal{G}(\tau_4)V \cdots V\mathcal{G}(\tau_{2n})V \qquad (17.6.5)$$

As the $\mathcal{G}$ factors in the integrand only depend on the even τ variables, it follows that the integrations over the odd variables can be done explicitly and yield simple polynomials in the even variables. The first four terms can then be written as

$$
\begin{aligned}
V\boldsymbol{\Gamma}V &= \int_0^\infty d\tau\, V\mathscr{G}(\tau_2)V + \int_0^\infty d\tau_2\,(-\tau_2)V\mathscr{G}(\tau_2)V\int_0^\infty d\tau_4\, V\mathscr{G}(\tau_4)V \\
&+ \int_0^\infty d\tau_2 \int_0^\infty d\tau_4\,(\tfrac{1}{2}\tau_2^2 + \tau_2\tau_4)V\mathscr{G}(\tau_2)V\mathscr{G}(\tau_4)V\int_0^\infty d\tau_6\, V\mathscr{G}(\tau_6)V \\
&+ \int_0^\infty d\tau_2 \int_0^\infty d\tau_4 \int_0^\infty d\tau_6\,[-\tfrac{1}{6}\tau_2^3 - \tfrac{1}{2}\tau_2^2(\tau_4+\tau_6) - \tau_2(\tfrac{1}{2}\tau_4^2 + \tau_4\tau_6)] \\
&\times V\mathscr{G}(\tau_2)V\mathscr{G}(\tau_4)V\mathscr{G}(\tau_6)V\int_0^\infty d\tau_8\, V\mathscr{G}(\tau_8)V + \cdots
\end{aligned}
\tag{17.6.6}
$$

These terms can be more simply expressed by introducing the Laplace transform of the operator $V\mathscr{G}(\tau)V$, which will be denoted by $\Psi(z)$ (to make contact with a system of notation widely used in the literature):

$$
V\mathscr{G}(\tau)V = (2\pi)^{-1}\int_C dz\, e^{-izt}\Psi(z) \tag{17.6.7}
$$

We also introduce the abbreviations:

$$
i\frac{\partial\Psi(z)}{\partial z} = \Psi'(z), \qquad \left(i\frac{\partial}{\partial z}\right)^2\Psi(z) = \Psi''(z) \tag{17.6.8}
$$

and so on. Then the identity (17.1.23) yields the relation:

$$
\int_0^\infty d\tau\,(-\tau)^n V\mathscr{G}(\tau)V = \Psi^{n\prime}(0) \tag{17.6.9}
$$

Hence the series (17.6.6) can be rewritten as

$$
\begin{aligned}
V\boldsymbol{\Gamma}V &= \Psi(0) + \Psi'(0)\Psi(0) + \tfrac{1}{2}\Psi''(0)\Psi(0)\Psi(0) \\
&+ \Psi'(0)\Psi'(0)\Psi(0) + \tfrac{1}{6}\Psi'''(0)\Psi(0)\Psi(0)\Psi(0) \\
&+ \tfrac{1}{2}\Psi''(0)\Psi'(0)\Psi(0)\Psi(0) + \tfrac{1}{2}\Psi''(0)\Psi(0)\Psi'(0)\Psi(0) \\
&+ \tfrac{1}{2}\Psi'(0)\Psi''(0)\Psi(0)\Psi(0) + \Psi'(0)\Psi'(0)\Psi'(0)\Psi(0) \\
&+ \cdots
\end{aligned}
\tag{17.6.10}
$$

A remarkable feature of this series is the fact that all terms terminate at the right with the operator $\Psi(0)$. Therefore, the operator $V\boldsymbol{\Gamma}V$ can be represented, in the case of homogeneous systems, as a *product of two operators:*

$$
V\boldsymbol{\Gamma}V = \Omega\Psi(0) \tag{17.6.11}
$$

where Ω is the following combination of the operator $\Psi(0)$ and its derivatives, evaluated in $z=0$:

$$\Omega = 1+\Psi'(0)+\tfrac{1}{2}\Psi''(0)\Psi(0)+\Psi'(0)\Psi'(0)$$
$$+\tfrac{1}{6}\Psi'''(0)\Psi(0)\Psi(0)+\tfrac{1}{2}\Psi''(0)\Psi'(0)\Psi(0)$$
$$+\tfrac{1}{2}\Psi''(0)\Psi(0)\Psi'(0)+\tfrac{1}{2}\Psi'(0)\Psi''(0)\Psi(0)$$
$$+\Psi'(0)\Psi'(0)\Psi'(0)+\cdots \qquad (17.6.12)$$

This is a quite complicated series; neither the numerical coefficient nor the structure of the general term can be guessed from the first few terms written out explicitly. This example shows how much more transparent the equations become when expressed in a time formalism. Indeed, one can immediately deduce from Eq. (17.6.5) the simple and compact expansion:

$$\Omega = \sum_{n=0}^{\infty} \Omega_n \qquad (17.6.13)$$

with

$$\Omega_n = \int_0^{\infty} d\tau_2 \int_0^{\infty} d\tau_4 \cdots \int_0^{\infty} d\tau_{2n-2} \int_0^{-\tau_2} d\tau_3 \int_0^{\tau_3-\tau_4} d\tau_5 \cdots \int_0^{\tau_{2n-3}-\tau_{2n-2}} d\tau_{2n-1}$$
$$\times V\mathcal{G}(\tau_2)V\mathcal{G}(\tau_4)V\cdots V\mathcal{G}(\tau_{2n-2})V \qquad (17.6.14)$$

The operators Ω and $\Psi(0)$ have a simple physical interpretation. Consider Eq. (17.2.3) (for homogeneous systems) and suppose we neglected the retardation in the distribution vector under the integral. We would then obtain

$$\partial_t V\bar{\mathfrak{f}}(t) \approx \int_0^{\infty} d\tau\, V\mathcal{G}(\tau)V\bar{\mathfrak{f}}(t)$$

that is:

$$\partial_t V\bar{\mathfrak{f}}(t) \approx \Psi(0)V\bar{\mathfrak{f}}(t) \qquad (17.6.15)$$

The neglect of the retardation means that the *collision described by* $\Psi(0)$ *is considered as an instantaneous process.* We know, however, that (17.6.15) is an incorrect equation. Hence we see that the operator Ω in Eq. (17.6.11) describes the correction to Eq. (17.6.15) *accounting for the finite duration of the collision process.*

It is very easy to derive an expression for the creation operator $C\mathbf{C}C$ in a form parallel to Eq. (17.6.10). Indeed, we know from Eq. (17.3.4) that the only difference between $V\boldsymbol{\Gamma}_n V$ and $C\mathbf{C}_n V$ is the replacement of the first factor $V\mathcal{G}(\tau_2)V$ in Eq. (17.2.14) by a factor $C\mathscr{C}(\tau_2)V$. Hence, if we call $\tilde{\mathscr{C}}(z)$ the Laplace transform of the latter operator, we immediately deduce from (17.6.10) the following series:

$$C\mathbf{C}V = \tilde{\mathscr{C}}(0)+\tilde{\mathscr{C}}'(0)\Psi(0)+\tfrac{1}{2}\tilde{\mathscr{C}}''(0)\Psi(0)\Psi(0)+\tilde{\mathscr{C}}'(0)\Psi'(0)\Psi(0)+\cdots \qquad (17.6.16)$$

This expression is also often found in the literature. Clearly, the time-dependent formula derived from (17.3.4) is again much simpler and more compact.

17.7. EQUILIBRIUM STATES

We now show that the equilibrium distribution vector has very remarkable properties in connection with the subdynamics concept.

An equilibrium distribution vector $\mathfrak{f}^0$ will be very broadly defined as a stationary solution of the Liouville equation (see also Section 4.1):

$$\mathscr{L}\mathfrak{f}^0 = 0 \tag{17.7.1}$$

This equation can also be written as

$$\mathscr{L}^0\mathfrak{f}^0 + \mathscr{L}' V\mathfrak{f}^0 + \mathscr{L}' C\mathfrak{f}^0 = 0 \tag{17.7.2}$$

We now evaluate the vector $\tilde{\mathscr{E}}(z)V\mathfrak{f}^0$: using Eq. (16.3.7), then (17.7.2), and finally (16.3.12) we obtain

$$\begin{aligned}\tilde{\mathscr{E}}(z)V\mathfrak{f}^0 &= \mathscr{L}' V\mathfrak{f}^0 + \tilde{\mathscr{E}}C\mathscr{R}^0\mathscr{L}' V\mathfrak{f}^0 \\ &= -\mathscr{L}^0\mathfrak{f}^0 - \mathscr{L}' C\mathfrak{f}^0 - \tilde{\mathscr{E}}C\mathscr{R}^0\mathscr{L}^0\mathfrak{f}^0 - \tilde{\mathscr{E}}C\mathscr{R}^0\mathscr{L}' C\mathfrak{f}^0 \\ &= -\mathscr{L}^0\mathfrak{f}^0 - \mathscr{L}' C\mathfrak{f}^0 + \tilde{\mathscr{E}}C\mathfrak{f}^0 + iz\tilde{\mathscr{E}}\mathscr{R}^0 C\mathfrak{f}^0 - \tilde{\mathscr{E}}C\mathfrak{f}^0 - \mathscr{L}' C\mathfrak{f}^0\end{aligned}$$

Hence

$$\tilde{\mathscr{E}}(z)V\mathfrak{f}^0 = -\mathscr{L}^0\mathfrak{f}^0 + iz\tilde{\mathscr{E}}(z)\mathscr{R}^0(z)C\mathfrak{f}^0 \tag{17.7.3}$$

From this equation we obtain a first important result by projecting both sides on the vacuum and taking the limit $z \to 0$:

$$\begin{aligned}\lim_{z\to 0} V\tilde{\mathscr{E}}(z)V\mathfrak{f}^0 &= -V\mathscr{L}^0\mathfrak{f}^0 + \lim_{z\to 0}(iz)V\tilde{\mathscr{E}}(z)\mathscr{R}^0(z)C\mathfrak{f}^0 \\ &= -V\mathscr{L}^0\mathfrak{f}^0\end{aligned}$$

where the regularity assumption (16.4.12) has been used. We thus obtain

$$\left\{V\mathscr{L}^0 + \lim_{z\to 0} V\tilde{\mathscr{E}}(z)V\right\}V\mathfrak{f}^0 = 0 \tag{17.7.4}$$

Translated into time-dependent operators, this equation becomes

$$V(\mathscr{L}^0 + \mathscr{L}')V\mathfrak{f}^0 + \int_0^\infty d\tau\, V\mathscr{G}(\tau)V\mathfrak{f}^0 = 0 \tag{17.7.5}$$

A second important lemma is obtained from (17.7.3) by multiplying both sides by $C\mathscr{R}^0(z)$:

$$\begin{aligned}C\mathscr{R}^0\tilde{\mathscr{E}}V\mathfrak{f}^0 &= -C\mathscr{R}^0\mathscr{L}^0\mathfrak{f}^0 + izC\mathscr{R}^0\tilde{\mathscr{E}}\mathscr{R}^0 C\mathfrak{f}^0 \\ &= C\mathfrak{f}^0 + izC\mathscr{R}^0\mathfrak{f}^0 + izC\mathscr{R}^0\tilde{\mathscr{E}}\mathscr{R}^0 C\mathfrak{f}^0\end{aligned}$$

where we used (16.3.12). From (16.4.12), it follows that

$$\lim_{z\to 0} C\mathcal{R}^0(z)\tilde{\mathcal{E}}(z)V\mathfrak{f}^0 = C\mathfrak{f}^0 \tag{17.7.6}$$

This equation is equivalent [see (17.1.5)] to

$$C\mathfrak{f}^0 = \int_0^\infty d\tau\, C\mathscr{C}(\tau)V\mathfrak{f}^0 \tag{17.7.7}$$

Recalling that $\mathfrak{f}^0$ is a time-independent vector, Eqs. (17.7.5), (17.7.7) show that the correlation and vacuum parts of $\mathfrak{f}^0$ are related to each other as the corresponding components of a kinetic vector, [see (17.1.12), (17.2.3)]. We may derive a couple of other results that strengthen this statement.

We note first the property:

$$V\boldsymbol{\Gamma}V\mathfrak{f}^0 = 0 \tag{17.7.8}$$

This property is very easily proved by using Eq. (17.2.6):

$$V\boldsymbol{\Gamma}V\mathfrak{f}^0 = V(\mathscr{L}^0 + \mathscr{L}')V\mathfrak{f}^0 + \int_0^\infty d\tau\, V\mathscr{G}(\tau)V \exp(-\tau V\boldsymbol{\Gamma}V)\mathfrak{f}^0$$

By substituting (17.7.8) as an ansatz, this equation reduces to (17.7.5), which has been proved above.

From Eqs. (17.7.8) and (17.3.2) follows then

$$C\mathbf{C}V\mathfrak{f}^0 = \int_0^\infty d\tau\, C\mathscr{C}(\tau)V \exp(-\tau V\boldsymbol{\Gamma}V)\mathfrak{f}^0 = \int_0^\infty d\tau\, C\mathscr{C}(\tau)V\mathfrak{f}^0$$

and hence, from (17.7.7)

$$C\mathfrak{f}^0 = C\mathbf{C}V\mathfrak{f}^0 \tag{17.7.9}$$

Hence the (time-independent) distribution vector $\mathfrak{f}^0$ obeys both equations (17.3.16) and (17.3.17) defining intrinsically a $\bar{\mathfrak{f}}$ vector. We therefore arrive at the fundamental conclusion:

$$\boldsymbol{\Pi}\mathfrak{f}^0 = \mathfrak{f}^0 \tag{17.7.10}$$

$$\hat{\boldsymbol{\Pi}}\mathfrak{f}^0 = 0 \tag{17.7.11}$$

The set of stationary distribution vectors lies entirely in the $\boldsymbol{\Pi}$ *subspace.*

We may now summarize all the results of our discussion as follows. Starting from an arbitrary initial state, the $\boldsymbol{\Pi}$ and $\hat{\boldsymbol{\Pi}}$ parts of the distribution function evolve quite independently. The $\boldsymbol{\Pi}$ component obeys a generalized kinetic equation (17.3.17) that drives that component to equilibrium; on the other hand, the $\hat{\boldsymbol{\Pi}}$ component has an evolution that could be compared to a phase-mixing process. In the end stage, the $\hat{\boldsymbol{\Pi}}$ component vanishes.

17.8. NONEQUILIBRIUM STATIONARY STATES

We now envisage the situation in which the system (of the type discussed till now) is immersed in a constant external field. As a concrete example, think of a plasma in an external electric field, constant in space and time for $t > 0$. This field acts as a *constraint*, which prevents the system from reaching equilibrium. We know from experiment what will happen under these circumstances. The system relaxes towards a steady state, which no longer changes in time (provided the constraint is itself stationary). This steady state is, so to say, the best the system can reach in order to cope with the constraint at a minimum expense. The state is, of course, *not* an equilibrium state; this can be noticed immediately by the existence of a constant current. This current is the system's response to the external stimulus: it is proportional, in first approximation, to the external field. The coefficient of proportionality is a "mechanical" *transport coefficient* (e.g., the electrical conductivity). We now formulate these ideas mathematically.

The evolution due to the external field is described by an additional Liouvillian $\mathscr{L}^F$; hence the total Liouvillian is now [see Eq. (2.4.15)]

$$\mathscr{L} = \mathscr{L}^0 + \mathscr{L}' + \mathscr{L}^F \tag{17.8.1}$$

We consider first a *time-independent external field*, but do not make too specific assumptions about $\mathscr{L}^F$ at this stage. We note the physically natural idea that an *external* field acts on each particle independently from the other particles. In other words, an external field can produce no correlations. We already noted this property in Section 14.2. Equation (14.2.5) implies the general property:

$$V\mathscr{L}^F = \mathscr{L}^F V \tag{17.8.2}$$

As the external Liouvillian shares this property with $\mathscr{L}^0$, it appears quite natural to group these two operators together. All the evolution operators of the general theory can then be constructed immediately in order to describe the system in the constant external field. In particular, the unperturbed resolvent will now be replaced by

$$\mathscr{R}^F(z) = (-iz - \mathscr{L}^0 - \mathscr{L}^F)^{-1} \tag{17.8.3}$$

The irreducible evolution operator is now defined by

$$\tilde{\mathscr{E}}^F(z) = \mathscr{L}' + \mathscr{L}'\mathscr{R}^F(z)C\tilde{\mathscr{E}}^F(z) \tag{17.8.4}$$

Most important, we can define a projector $\mathbf{\Pi}^F$:

$$\mathfrak{f}(t) = \mathbf{\Pi}^F \mathfrak{f}(t) + \hat{\mathbf{\Pi}}^F \mathfrak{f}(t) \tag{17.8.5}$$

such that

$$\mathbf{\Pi}^F \mathscr{L} = \mathscr{L} \mathbf{\Pi}^F \tag{17.8.6}$$

and which reduces to V in the absence of interactions. Indeed, Eq. (17.8.6) must reduce, in the absence of interactions, to

$$V(\mathcal{L}^0+\mathcal{L}^F)=(\mathcal{L}^0+\mathcal{L}^F)V \tag{17.8.7}$$

which is satisfied because of (17.8.2). The operator $\mathbf{\Pi}^F$ is immediately obtained from Eq. (16.4.23) by the substitution $\mathcal{L}^0\to\mathcal{L}^0+\mathcal{L}^F$ in all operators.

The operator $\mathbf{\Pi}^F$ is thus a true continuation of the operator $\mathbf{\Pi}$, in the same sense as $\mathbf{\Pi}$ is a continuation of V. When the field is switched off, the projector $\mathbf{\Pi}^F$ reduces to $\mathbf{\Pi}$. Hence we note the following sequence of limits:

$$\mathbf{\Pi}^F \xrightarrow[\mathcal{L}^F\to 0]{} \mathbf{\Pi} \xrightarrow[\mathcal{L}'\to 0]{} V \tag{17.8.8}$$

Because of Eq. (17.8.6), the $\mathbf{\Pi}^F$ subset obeys its own subdynamics, independent of $\hat{\mathbf{\Pi}}^F$. This subdynamics is, of course, not the same as the $\mathbf{\Pi}$ subdynamics. In particular, a component $\mathbf{\Pi}^F\mathfrak{f}$ has components both in $\mathbf{\Pi}$ and $\hat{\mathbf{\Pi}}$, just as $\mathbf{\Pi}\mathfrak{f}$ has components in both V and C.

It is easily found that the vacuum component $V\mathbf{\Pi}^F\mathfrak{f}$ of the "field-kinetic" part of the distribution vector obeys a closed *kinetic equation:*

$$\partial_t[V\mathbf{\Pi}^F\mathfrak{f}(t)]=V\mathbf{\Gamma}^F V[V\mathbf{\Pi}^F\mathfrak{f}(t)] \tag{17.8.9}$$

where

$$V\mathbf{\Gamma}^F V=V(\mathcal{L}^0+\mathcal{L}'+\mathcal{L}^F)V+\sum_{n=1}^{\infty}V\mathbf{\Gamma}_n^F V \tag{17.8.10}$$

and $\mathbf{\Gamma}_n^F$ is defined by Eq. (17.2.14), where $\mathcal{L}^0$ is always replaced by $\mathcal{L}^0+\mathcal{L}^F$. The whole structure discussed in Chapters 16 and 17 is carried over smoothly to this case.

It is now a simple matter to investigate the properties of the steady state that (if it exists) must satisfy the equation:

$$\mathcal{L}\mathfrak{f}^{st}\equiv(\mathcal{L}^0+\mathcal{L}^F+\mathcal{L}')\mathfrak{f}^{st}=0 \tag{17.8.11}$$

We now redo the calculations of Section 17.7 with only trivial changes, and obtain the following results:

$$\left\{V\mathcal{L}^0+V\mathcal{L}^F+\lim_{z\to 0}V\tilde{\mathcal{E}}^F(z)V\right\}V\mathfrak{f}^{st}=0 \tag{17.8.12}$$

$$\lim_{z\to 0}C\mathcal{R}^F(z)\tilde{\mathcal{E}}^F(z)V\mathfrak{f}^{st}=C\mathfrak{f}^{st} \tag{17.8.13}$$

Just as in Section 17.7 we conclude from these equations that the *stationary nonequilibrium distributions lie entirely in the* $\mathbf{\Pi}^F$ *subspace:*

$$\mathbf{\Pi}^F\mathfrak{f}^{st}=\mathfrak{f}^{st} \tag{17.8.14}$$

$$\hat{\mathbf{\Pi}}^F\mathfrak{f}^{st}=0 \tag{17.8.15}$$

We note that all information about the "mechanical" transport coefficients is in the stationary distribution vector $\mathfrak{f}^{st}$. It describes the response of the system to the external stimulus. We thus come to the conclusion that the calculation of the *mechanical transport coefficients only involves the study of the subdynamics in the subset* Π^F. This is a very remarkable feature that confirms Point F of our program of Section 16.2. It shows that the theory of transport coefficients, one of the most important topics of statistical mechanics, only involves a study of the kinetic equation (17.8.9).

The relations studied above become quite remarkable in a *linear* response theory, that is, in the approximation relevant to the theory of linear transport coefficients.

Let us linearize all the relevant operators of the theory. The resolvent $\mathcal{R}^F(z)$ obeys the equation:

$$\mathcal{R}^F = \mathcal{R}^0 + \mathcal{R}^0 \mathcal{L}^F \mathcal{R}^F \tag{17.8.16}$$

Hence, to first order

$$\mathcal{R}^{[1]}(z) = \mathcal{R}^0(z) + \mathcal{R}^0(z)\mathcal{L}^F \mathcal{R}^0(z) \tag{17.8.17}$$

The linearization of the operator $\tilde{\mathscr{E}}^F(z)$ is a little more involved. However, the reader will easily convince himself by using Eqs. (17.8.4) and (17.8.17) that

$$\tilde{\mathscr{E}}^{[1]}(z) = \tilde{\mathscr{E}}(z) + \tilde{\mathscr{E}}(z)\mathcal{R}^0(z)C\mathcal{L}^F C\mathcal{R}^0(z)\tilde{\mathscr{E}}(z) \tag{17.8.18}$$

We finally write

$$\mathfrak{f}^{st} = \mathfrak{f}^0 + \mathfrak{f}^{[1]} \tag{17.8.19}$$

We now substitute Eqs. (17.8.17)–(17.8.19) into (17.8.12), linearize the resulting equation, and use (17.7.4). We then obtain

$$\left\{\mathcal{L}^0 \mathsf{V} + \lim_{z\to 0} \mathsf{V}\tilde{\mathscr{E}}(z)\mathsf{V}\right\}\mathsf{V}\mathfrak{f}^{[1]} = -\mathcal{L}^F \mathsf{V}\mathfrak{f}^0 - \lim_{z\to 0} \mathsf{V}\tilde{\mathscr{E}}\mathcal{R}^0 C\mathcal{L}^F C\mathcal{R}^0\tilde{\mathscr{E}}\mathsf{V}\mathfrak{f}^0 \tag{17.8.20}$$

or equivalently

$$\left\{\mathsf{V}(\mathcal{L}^0 + \mathcal{L}')\mathsf{V} + \int_0^\infty d\tau\, \mathsf{V}\mathcal{G}(\tau)\mathsf{V}\right\}\mathsf{V}\mathfrak{f}^{[1]} = \mathsf{V}\mathcal{F}^F\mathsf{V}\mathfrak{f}^0 \tag{17.8.21}$$

with

$$\mathsf{V}\mathcal{F}^F\mathsf{V} = -\mathcal{L}^F\mathsf{V} - \lim_{z\to 0} \mathsf{V}\tilde{\mathscr{E}}\mathcal{R}^0 C\mathcal{L}^F C\mathcal{R}^0\tilde{\mathscr{E}}\mathsf{V} \tag{17.8.22}$$

Hence the linear deviation from equilibrium is obtained by an essentially simple procedure. $\mathsf{V}\mathfrak{f}^{[1]}$ obeys the same equation as the equilibrium distribution (17.7.5), but with an additional source term $\mathsf{V}\mathcal{F}^F\mathsf{V}\mathfrak{f}^0$. We also note the most important fact that this equation does *not* involve the complete kinetic operator $V\mathsf{\Gamma}V$, but only its first approximation $V\mathsf{\Gamma}_1V$ [Eq. (17.2.16)]. Hence, although the steady-state distribution vector does not belong to the internal $\mathsf{\Pi}$ space, its vacuum component is obtained by solving an equation that is even simpler than the full (internal) kinetic equation. The difference between the full operator $V\mathsf{\Gamma}V$ and its first approximation can be understood as corrections due to the final duration of the collisions (see Section 17.6). The remarkable result found here is that these corrections can give no contribution to the linear steady state and hence to the transport coefficients. This property implies a quite significant simplification of the practical calculation of the transport coefficients.

A nontrivial extension of the results described here was obtained by Balescu and Misguich. They considered a system submitted to a *time-dependent external field*, to which corresponds a time-dependent Liouvillian: $\mathscr{L}^F(t)$. It was shown that, under very mild integrability conditions, it is possible to define a time-dependent projection operator $\mathsf{P}(t)$ that is a continuation of the projector $\mathsf{\Pi}$, in a sense similar to (17.8.8):

$$\mathsf{P}(t) \xrightarrow[\mathscr{L}^F(t)\to 0]{} \mathsf{\Pi} \xrightarrow[\mathscr{L}'\to 0]{} V \tag{17.8.23}$$

In spite of the *arbitrary* time dependence of the external field, it is possible to extend most of the properties of the $\mathsf{\Pi}$ subspace to this case too. In particular, there exists a subdynamics in the following sense: if the distribution vector belongs to the subspace $\mathsf{P}(t_0)$ at time t_0, it remains in the *instantaneous* subspace $\mathsf{P}(t)$ at all later times. Equation (17.8.6) is no longer true: it is replaced by the following one, which takes account of the time dependence of the operators:

$$\partial_t\, \mathsf{P}(t) = \mathsf{P}(t)\mathscr{L}(t) - \mathscr{L}(t)\mathsf{P}(t) \tag{17.8.24}$$

A consequence of this relation is the following pseudocommutation relation:

$$\mathscr{U}^F(t, t_1)\mathsf{P}(t_1) = \mathsf{P}(t)\mathscr{U}^F(t, t_1) \tag{17.8.25}$$

where $\mathscr{U}^F(t, t_1)$ is the propagator associated with the time-dependent Liouvillian. Equation (17.8.25) ensures the existence of a subdynamics, in the sense defined above. This subdynamics is ruled by a *generalized kinetic equation* for the component $\rho(t) = V\mathsf{\Pi}\mathsf{P}(t)\mathfrak{f}(t)$:

$$\partial_t\rho(t) = V\mathsf{\Gamma}^F(t)V\rho(t) \tag{17.8.26}$$

which is a nontrivial extension of Eq. (17.8.9).

We cannot go into the details of this theory here, but we cannot refrain from quoting the *main theorem* proved in that work. Consider the following physical situation, which is a good representation of the experimental setup for the measurement of a transport coefficient. The system is left to itself for a long time in absence of the field. It "ages" during that time: all the transient phenomena die away and the system reaches a state described by a distribution vector lying in the subspace $\mathsf{\Pi}$; possibly it even reaches equilibrium. At some time t_0, the field is switched on: the system reacts and adjusts itself to the new environment. The theorem is now formulated as follows: If

$$\mathscr{L}^{F}(t)=0 \qquad \text{for} \quad t<t_0 \tag{17.8.27}$$

and if

$$\mathfrak{f}(t_0)=\mathsf{\Pi}\mathfrak{f}(t_0) \tag{17.8.28}$$

then for all $t>t_0$:

$$\mathfrak{f}(t)=\mathsf{P}(t)\mathfrak{f}(t) \tag{17.8.29}$$

Thus, with the "usual initial condition" described above, the evolution of the system is described *exactly* by the subdynamics in the subspace $\mathsf{P}(t)$. This result is very important, because it shows that the calculation of the mechanical transport coefficients, both in the linear and in the nonlinear regime, is governed by the generalized kinetic equation (17.8.26), *without any approximation.* This property is perhaps the most beautiful illustration of Point F of our program in Section 16.2.

BIBLIOGRAPHICAL NOTES

The problem of the foundations of nonequilibrium statistical mechanics has been a very active field of investigation in the past thirty years. Among the many papers devoted to this subject we shall only quote a few, which are significant of the various tendencies.

The oldest idea in this direction was based on the assumption that the distribution function reaches in time a "plateau value"; in this regime one could define time-smoothed distribution functions that obey kinetic equations: J. G. Kirkwood, *J. Chem. Phys.* **14,** 180 (1946).

The next important development was due to Bogoliubov. It was based on the assumption that after a short time, of the order of the duration of a collision, the correlations become functionals of the one-particle reduced distribution functions, which in turn obey kinetic equations. In a second stage, for times of the order of a hydrodynamical time, the one-particle functions become functionals of macroscopic parameters obeying hydrodynamical equations. These ideas are exposed in N. N. Bogoliubov, *Problemy Dinamicheskoy Teorii v Statisticheskoy Fizike,* Moscow (1946) [English translation: "Problems of a Dynamical Theory in Statistical Physics," in *Studies in Statistical Mechanics,* Vol. 1 (J. de Boer and G. E. Uhlenbeck, eds.), North Holland, Amsterdam, 1962].

Many applications of Bogoliubov's theory were developed in the USSR. Important developments along this line of thought were

S. T. Choh and G. E. Uhlenbeck, *The Kinetic Theory of Dense Gases*, U. Michigan Report, 1958.

E. G. D. Cohen, *Physica* **28,** 1025, 1045, 1060 (1956); *J. Math. Phys.* **4,** 143 (1963).

In 1953, Klein and Prigogine developed a complete microscopic theory of a simple irreversible process: the randomization of phases in an assembly of harmonic oscillators:

G. Klein and I. Prigogine, *Physica* **19,** 1053 (1953).

I. Prigogine and R. Bingen, *Physica* **21,** 299 (1955).

This process is however of the "phase-mixing" type (see Section 12.2) and does not represent a true approach to equilibrium.

A quite distinct approach was taken by Van Hove. He analyzed the quantum weakly coupled gas in the asymptotic limit $t \to \infty$, $\lambda \to 0$, $\lambda^2 t$ = finite. He invoked a property of the interactions, called the *diagonal singularity condition,* which he considered as a condition of dissipativity. By virtue of this condition, he was able to derive an irreversible equation of evolution for the density matrix, valid in the asymptotic limit mentioned above: L. van Hove, *Physica* **21,** 517 (1955). His method was generalized in L. van Hove, *Physica* **23,** 441 (1957).

Shortly after van Hove's paper appeared the paper by Brout and Prigogine which inaugurated a long series of works done by the so-called "Brussels school." The basic idea here was the introduction of a Fourier decomposition of the distribution function and the use of action-angle variables (in classical mechanics) in a systematic way. This representation showed the importance of a separate analysis of the various types of correlations (i.e., the dynamics of correlations). It was shown again that an irreversible "master equation" for the N-particle momentum distribution (which plays the role of the vacuum in the Fourier representation) could be derived asymptotically, in the limit $\lambda \to 0$, $t \to \infty$, $\lambda^2 t$ finite: R. Brout and I. Prigogine, *Physica* **22,** 621 (1956).

The theory was extended systematically to higher orders by means of a diagram representation and a perturbation theory of the same type as in quantum field theory: I. Prigogine and R. Balescu, *Physica* **25,** 281, 302 (1959).

These papers dealt mainly with homogeneous systems. The extension to inhomogeneous systems along this line was done in

I. Prigogine and R. Balescu, *Physica* **26,** 145 (1960).

G. Severne, *Physica* **31,** 877 (1965).

A very important paper along this line is I. Prigogine and P. Résibois, *Physica* **27,** 629 (1961). In this paper the general classical "master equation" (16.3.23) was derived by an analysis of the diagrams. This equation is non-Markovian and contains a term depending on the initial conditions. In the limit of long times it was shown to reduce to the general Markovian kinetic equation discussed in this chapter. The concepts of collision, destruction, and creation operators were introduced here. These results were generalized to quantum mechanics by P. Résibois, *Physica* **27,** 541 (1961).

A presentation of the ideas of the Brussels school in this earlier form can be found in the following three monographs:

I. Prigogine, *Non-Equilibrium Statistical Mechanics*, Wiley-Interscience, New York, 1963.

R. Balescu, *Statistical Mechanics of Charged Particles*, Wiley-Interscience, New York, 1963.

P. Résibois, in *Many Particle Physics* (E. Meeron, ed.), Gordon and Breach, New York, 1967.

A similar derivation of the master equation, but using projection operators was given by R. Zwanzig, in *Lectures in Theoretical Physics*, Vol. 3 (Summer Inst. Theor. Phys., Univ. Colorado, 1960), Wiley-Interscience, New York, 1961.

More recent versions of the theory, also using projection operators, of the kind appearing in this book, were given in

R. Balescu, *Physica* **38,** 98 (1968).

M. Baus, *Bull. Cl. Sci. Acad. Roy. Belg.* **53,** 1291, 1333 (1967).

A very interesting approach to the theory of transport coefficients, based on projection operator methods (of a rather different type) is H. Mori, *Progr. Theor. Phys.* **33,** 423 (1965).

In all these papers, the kinetic equation appeared as an approximate asymptotic equation, valid for sufficiently long times. The important change of viewpoint, according to which one could write a kinetic equation valid for *all* times, independently of any asymptotic consideration, was brought about by the discovery and exploitation of the $\mathbf{\Pi}$-projection operator:

I. Prigogine, C. George, and F. Henin, *Physica* **45,** 418 (1969).

R. Balescu and J. Wallenborn, *Physica* **54,** 504 (1971).

All the references to the "physical particle representation" and the "causal dynamics" quoted in Chapter 16 are relevant to the theory of the kinetic equations. One of the great goals pursued by Prigogine and his co-workers in recent years is a generalized H theorem and a general definition of the entropy out of equilibrium. Much progress has been achieved along these lines. A comprehensive discussion of Prigogine's ideas can be found in I. Prigogine, *Acta Phys. Austr.* Suppl. **X,** 401 (1973).

A rather different type of approach to nonequilibrium statistical mechanics is due to Iu. L. Klimontovich, *Statisticheskaya Teoriya Neravnovesnykh Protsessov v Plazme*, Izd. Mosk. Univ., 1964 (English translation, *Statistical Theory of Non-Equilibrium Processes in Plasmas*. MIT Press, Cambridge, Mass., 1967). The idea here is to write an exact nonlinear equation for the microscopic density function $\sum \delta(q_i - q)\, \delta(p_i - p)$ (in the Heisenberg representation) and to construct various approximations by using truncation procedures. This method was very extensively used in plasma physics.

The problem of nonequilibrium steady states was studied in R. Balescu, *Physica* **27,** 693 (1961).

In this paper appeared the result (17.8.20) according to which even in the most general case, the linear response to a steady external field is given only by the "Boltzmann-like" part of the kinetic equation $V\tilde{\mathscr{E}}(0)V$. This result was also studied by R. J. Swenson, *Physica* **29,** 1174 (1963).

This problem was studied from the point of view of the Π projector in R. Balescu, L. Brenig, and J. Wallenborn, *Physica* **52,** 29 (1971).

The extension of the subdynamics concept to systems in a *time-dependent* external field, with applications to the theory of kinetic equations, the theory of transport coefficients, and the theory of turbulence in plasmas appears in R. Balescu and J. Misguich, *J. Plasma Phys.* **11,** 357, 377 (1974).

CHAPTER 18

THE EVOLUTION IN TIME OF WEAKLY COUPLED GASES

18.1. THE CLASSICAL KINETIC EQUATION

Our purpose now is to show how the general, but abstract results of Chapters 16 and 17 can be applied to specific physical systems and how the equations of evolution for the components of the distribution vector $\mathfrak{f}$ can be transcribed into concrete kinetic equations for the reduced distribution functions. There is no new physical principle involved in these operations: all the physics is already present in the general theory. We must simply develop, on one hand, a set of systematic approximation procedures and, on the other hand, a method of coming down from the abstract to the concrete level.

In the present chapter we illustrate these procedures in the simplest possible case, that is, the weakly coupled gas. This problem has already been studied in Chapter 11 by the classical semi-intuitive method. It is by no means clear at first sight that the Landau equation obtained there has anything to do with the general kinetic equation derived in Chapter 17. It will be easily seen, however, that if the latter is made explicit, by using the correlation-patterns representation of Chapter 14, we find that its weak-coupling limit is precisely the Landau equation.

The weakly coupled classical gas is defined by the usual Liouvillian of Eq. (3.3.13), with $\mathscr{L}^F=0$:

$$\mathscr{L}=\mathscr{L}^0+\lambda\mathscr{L}' \tag{18.1.1}$$

where λ is a formal parameter used to keep track of the orders of magnitude. In the final equations, it can be set equal to 1. The matrix elements of this Liouvillian in the correlation-patterns representation have been studied in detail in Section 14.2. They are all expressed in terms of two simple operators: the one-particle free Liouvillian L_j^0 and the two-particle interaction Liouvillian L'_{jn}. These operators were defined in Eqs. (2.4.17) and (2.4.20), but we rewrite them here:

$$L_j^0=-\mathbf{v}_j\cdot\nabla_j \tag{18.1.2}$$

$$L'_{jn}=(\nabla_j V_{jn})\cdot\partial_{jn} \tag{18.1.3}$$

We now assume λ to be a small parameter and express all quantities as formal power series in λ. Consider first the operator $V\mathbf{\Gamma}V$, defined by Eqs. (17.2.13), (17.2.14). This operator contains a zeroth-order term $V\mathcal{L}^0$, and a first-order term $\lambda V\mathcal{L}'V$, which are rather trivial. The first nontrivial contributions come from $V\mathbf{\Gamma}_n V$, with $n \geqslant 1$. It should be clear that the expansion (17.2.14) is *not* an expansion in powers of λ. Every factor $V\mathcal{G}(\tau)V$ and $V\tilde{\mathcal{U}}(\tau)V$ is still a complicated function of λ, defined by Eqs. (17.1.4), (16.3.5), and (17.2.8). From the two former equations it is clear that $V\mathcal{G}(\tau)V$ begins with a term in λ^2, whereas Eq. (17.2.8) shows that $V\tilde{\mathcal{U}}(\tau)V$ begins with a term in λ^0. Therefore we may write

$$V\mathcal{G}(\tau)V = \sum_{p=2}^{\infty} \lambda^p V\mathcal{G}^{[p]}(\tau)V \tag{18.1.4}$$

$$V\tilde{\mathcal{U}}(\tau)V = \sum_{p=0}^{\infty} \lambda^p V\tilde{\mathcal{U}}^{[p]}(\tau)V \tag{18.1.5}$$

We note that $\tilde{\mathcal{U}}^{[0]}(\tau)$ is simply the unperturbed propagator $\mathcal{U}^0(\tau)$. It is then clear from Eq. (17.2.14) that the term $V\mathbf{\Gamma}_n V$ starts with a power λ^{2n}:

$$V\mathbf{\Gamma}_n V = \sum_{p=2n}^{\infty} \lambda^p V\mathbf{\Gamma}_n^{[p]} V \tag{18.1.6}$$

If we decide to keep, for the weakly coupled gas, only the lowest-order nontrivial term, that is, the term of order λ^2, we get the following approximation for the kinetic operator:

$$V\mathbf{\Gamma}V \approx V(\mathcal{L}^0 + \mathcal{L}')V + V\mathbf{\Gamma}_1^{[2]}V \tag{18.1.7}$$

with

$$V\mathbf{\Gamma}_1^{[2]}V = \int_0^{\infty} d\tau_2\, V\mathcal{G}(\tau_2)V\mathcal{U}^0(-\tau_2)V \tag{18.1.8}$$

Equation (17.1.4) yields the expression for the leading term:

$$V\mathcal{G}^{[2]}(\tau)V = V\mathcal{L}'C\mathcal{U}^0(\tau)\mathcal{L}'V \tag{18.1.9}$$

We therefore obtain, as a first approximation to Eq. (17.3.17), the following equation:

$$\partial_t V\bar{\mathfrak{f}}(t) = V\mathcal{L}^0\bar{\mathfrak{f}}(t) + \lambda V\mathcal{L}'V\bar{\mathfrak{f}}(t) + \lambda^2 \int_0^{\infty} d\tau\, V\mathcal{L}'C\mathcal{U}^0(\tau)\mathcal{L}'\mathcal{U}^0(-\tau)V\bar{\mathfrak{f}}(t) \tag{18.1.10}$$

This generalized kinetic equation defines our model of a weakly coupled gas.

We now write explicitly the equation for the simplest component of $V\bar{\mathfrak{f}}(t)$, that is, $\bar{\varphi}_1(x_1; t)$, the one-particle distribution function. The relevant matrix elements of the operators in the right-hand side can be taken from

Section 14.2. Equation (14.2.4) implies

$$\langle 1|\, V\mathscr{L}^0 V\, |1, \ldots, s\rangle = \delta_{s,1} \langle 1|\, \mathscr{L}^0\, |1\rangle = \delta_{s,1}\, L_1^0 \tag{18.1.11}$$

The vacuum-to-vacuum element of $\mathscr{L}'$ is given in Eq. (14.2.6):

$$\begin{aligned}\langle 1|\, V\mathscr{L}' V\, |1, \ldots, s\rangle &= \delta_{s,2} \langle 1|\mathscr{L}'\, |1 \,|\, 2\rangle \\ &= \delta_{s,2} \int dx_2\, L'_{12}\end{aligned} \tag{18.1.12}$$

For the calculation of $V\mathscr{G}^{[2]}V$, we must start from the left with one particle and make a transition to a *correlated* state that, by Eq. (14.2.6) is necessarily a two-particle state. This transition must be followed by a transition to a vacuum state. From Eq. (14.2.8) we see that the latter can only be a two-particle state. Hence, using also Eqs. (15.1.14), (15.1.20):

$$\begin{aligned}&\langle 1|\, V\mathscr{L}' C\mathscr{U}^0(\tau)\mathscr{L}'\mathscr{U}^0(-\tau)V\, |1, \ldots, s\rangle \\ &\qquad = \delta_{s,2} \langle 1|\, \mathscr{L}'\, |12\rangle\, U_{12}^0(\tau) \langle 12|\, \mathscr{L}'\, |1 \,|\, 2\rangle\, U_{12}^0(-\tau) \\ &\qquad = \delta_{s,2} \int dx_2\, L'_{12} U_{12}^0(\tau) L'_{12} U_{12}^0(-\tau)\end{aligned} \tag{18.1.13}$$

Collecting all these partial results we obtain the explicit form:

$$\begin{aligned}\partial_t \bar{p}_1(x_1; t) &= L_1^0 \bar{p}_1(x_1; t) + \lambda \int dx_2\, L'_{12} \bar{p}_2(x_1 \,|\, x_2; t) \\ &\quad + \lambda^2 \int_0^\infty d\tau \int dx_2\, L'_{12} U_{12}^0(\tau) L'_{12} U_{12}^0(-\tau) \bar{p}_2(x_1 \,|\, x_2; t)\end{aligned} \tag{18.1.14}$$

The reader will easily derive an analogous equation for $\partial_t \bar{p}_2(x_1 \,|\, x_2; t)$ and perhaps for $\partial_t \bar{p}_3(x_1 \,|\, x_2 \,|\, x_3; t)$. By that time he will be convinced that the general equation for the s-particle component of the vacuum is

$$\begin{aligned}&\partial_t \bar{p}_s(x_1 \,|\, x_2 \,|\, \cdots \,|\, x_s; t) = \sum_{j=1}^{s} L_j^0\, \bar{p}_s(x_1 \,|\, x_2 \,|\, \cdots \,|\, x_s; t) \\ &\quad + \lambda \sum_{j=1}^{s} \int dx_{s+1}\, L'_{j\,s+1} \bar{p}_{s+1}(x_1 \,|\, x_2 \,|\, \cdots \,|\, x_{s+1}; t) \\ &\quad + \lambda^2 \sum_{j=1}^{s} \int_0^\infty d\tau \int dx_{s+1}\, L'_{j\,s+1} U^0_{1\cdots s+1}(\tau) L'_{j\,s+1} U^0_{1\cdots s+1}(-\tau) \bar{p}_{s+1}(x_1 \,|\, x_2 \,|\, \cdots \,|\, x_{s+1}; t)\end{aligned} \tag{18.1.15}$$

At this stage, the formal calculations are finished. We are left with an infinite hierarchy of linear equations for the vacuum-correlation patterns. We now note that this hierarchy has a very simple particular solution, which, moreover, is of special physical interest:

$$\bar{p}_s(x_1 \,|\, x_2 \,|\, \cdots \,|\, x_s; t) = \prod_{j=1}^{s} \bar{f}(x_j; t) \tag{18.1.16}$$

This is just a special case of our general theorem on the link between dynamic and static correlation patterns (14.2.12). It can, however, be derived directly by substituting Eq. (18.1.16) into (18.1.15). The right-hand side is a sum of operators relative to the particles j, ($j = 1, \dots, s$). Each term in this sum is seen to be precisely equal to

$$[\partial_t \bar{p}_1(x_j\,; t)] \prod_{\substack{n=1 \\ (n \neq j)}}^{s} \bar{p}_1(x_n\,; t) \tag{18.1.17}$$

where $\partial_t \bar{p}_1(x_j\,; t)$ is given by the right-hand side of Eq. (18.1.14) [with the ansatz (18.1.16)]. Hence, the additive structure of the evolution operator (18.1.15) ensures the propagation in time of the factorization property (18.1.16).

The function $\bar{f}(x_j\,; t)$ is simply the kinetic part of the one-particle reduced distribution function. It obeys an equation derived from (18.1.14) combined with the constraint (18.1.16):

$$\begin{aligned} \partial_t \bar{f}(x_1; t) = L_1^0 \bar{f}(x_1; t) + \lambda \int dx_2\, L'_{12} \bar{f}(x_1; t) \bar{f}(x_2; t) \\ + \lambda^2 \int_0^\infty d\tau \int dx_2\, L'_{12} U_{12}^0(\tau) L'_{12} U_{12}^0(-\tau) \bar{f}(x_1; t) \bar{f}(x_2; t) \end{aligned} \tag{18.1.18}$$

Hence we obtained the following remarkable result:

Provided the relations (18.1.16) are supposed to hold at time zero, the infinite system of linear equations (18.1.15) is reduced to a single nonlinear equation (18.1.18).

The latter is called the (proper) *kinetic equation* (to order λ^2).

18.2. THE CLASSICAL KINETIC CORRELATIONS

We now evaluate Eq. (17.3.16) explicitly for the kinetic correlations in a (classical) weakly coupled gas. The only new operator appearing here is $C\mathscr{C}(\tau)V$, which, by Eqs. (17.1.7) and (16.3.5), has a λ-expansion beginning with a term of order λ:

$$C\mathscr{C}(\tau)V = \sum_{p=1}^{\infty} \lambda^p C\mathscr{C}^{[p]}(\tau)V \tag{18.2.1}$$

with

$$C\mathscr{C}^{[1]}(\tau)V = C\mathscr{U}^0(\tau)\mathscr{L}'V \tag{18.2.2}$$

Let us try to make it clear that the approximation to order λ is sufficient for the correlations, contrary to the kinetic equation where we pushed the approximation to order λ^2. The latter equation describes the *rate of* *change* of the function $\bar{f}$. The function $\bar{f}$ itself must be assumed of order 1 (i.e., λ^0) because, as we know, it must carry the complete normalization of the distribution function, however small the value of λ.* Equation (17.3.17), on the other hand, provides directly the correlation patterns (not their rate of change!). It tells us that the leading term in these patterns is at most of order λ, that is, is very small compared to $\bar{f}$. Any correction of size λ^2 can therefore be consistently neglected.

The correlation component of the kinetic distribution is therefore given, to order λ, by

$$C\bar{\mathfrak{f}}(t) = C\mathbf{C}_1^{[1]}V\bar{\mathfrak{f}}(t) \tag{18.2.3}$$

which is made explicit by use of (17.3.4) and (18.2.1):

$$C\bar{\mathfrak{f}}(t) = \lambda \int_0^\infty d\tau\, C\mathcal{U}^0(\tau)\mathcal{L}' V\mathcal{U}^0(-\tau)V\bar{\mathfrak{f}}(t) \tag{18.2.4}$$

We now resolve the vector $C\bar{\mathfrak{f}}(t)$ into its component correlation patterns. We note that, starting from the vacuum (at right), a single interaction operator [and hence $C\mathscr{C}^{[1]}(\tau)V$] can only introduce a correlation between two particles. Any kinetic pattern in which more than two particles are correlated is, therefore, of higher order in λ. Moreover the matrix element creating the correlations between particles i, j is diagonal in the number of particles and simply equals L'_{ij} as follows from Section 14.2. Hence, from Eq. (18.2.4) we obtain

$$\bar{p}_s(x_1x_2 \,|\, x_3 \,|\cdots|\, x_s; t) = \lambda \int_0^\infty d\tau\, U^0_{1\cdots s}(\tau)$$
$$\times \langle 12 \,|\, 3 \,|\cdots|\, s \,|\, \mathcal{L}' \,|1\,|\,2\,|\,3\,|\cdots|\, s\rangle U^0_{1\cdots s}(-\tau)\bar{p}_s(x_1 \,|\, x_2 \,|\cdots|\, x_s; t) \tag{18.2.5}$$

or, explicitly

$$\bar{p}_s(x_1x_2 \,|\, x_3 \,|\cdots|\, x_s; t)$$
$$= \lambda \int_0^\infty d\tau\, U^0_{1\cdots s}(\tau)L'_{12}U^0_{1\cdots s}(-\tau)\bar{p}_s(x_1 \,|\, x_2 \,|\cdots|\, x_s; t) \tag{18.2.6}$$

We now take into account the product property (15.1.14). We also note that all factors $U^0_j(\tau)$ for $j \neq 1, 2$ can be freely commuted to the right of the operator L'_{12}. They are then canceled by the corresponding operator

* A simple analogy is the following: Consider the equation $\partial_t f(t) = -\lambda^2 a[f(t) - b]$, with a, b constants independent of λ. If the initial condition is $f(0) = f_0 + b = O(\lambda^0)$, the solution is $f(t) = b + f_0 \exp(-\lambda^2 at)$. Although the *rate of decay* is of order λ^2, the function itself is of order λ^0 at all times.

$U_j^0(-\tau)$ because $U_j^0(\tau)U_j^0(-\tau)=1$. We can therefore simplify Eq. (18.2.6) to

$$\bar{\rho}_s(x_1x_2 \mid x_3 \mid \cdots \mid x_s; t) = \lambda \int_0^\infty d\tau\, U_{12}^0(\tau) L'_{12} U_{12}^0(-\tau) \times \bar{\rho}_s(x_1 \mid x_2 \mid \cdots \mid x_s; t), \qquad s = 2, 3, \ldots \quad (18.2.7)$$

Hence, to order λ, the correlations are determined from the vacuum component *at the same time* by the action of a *time-independent* operator. The vacuum component appearing in the right-hand side is a solution of Eq. (18.1.15).

Assuming now a factorized initial condition for the kinetic vacuum, we know that the vacuum components will be of the form (18.1.16) at all times. It then immediately follows that*

$$\bar{\rho}_s(x_1x_2 \mid x_3 \mid \cdots \mid x_s; t) = \bar{g}_2(x_1x_2; t) \prod_{j=3}^{s} \bar{f}(x_j; t) \quad (18.2.8)$$

where $\bar{g}_2(x_1x_2; t)$ is determined by the nonlinear relation:

$$\bar{g}_2(x_1x_2; t) = \lambda \int_0^\infty d\tau\, U_{12}^0(\tau) L'_{12} U_{12}^0(-\tau) \bar{f}(x_1; t) \bar{f}(x_2; t) \quad (18.2.9)$$

Hence, to order λ, all the components of the asymptotic correlation vector $C\bar{f}(t)$ can be expressed in terms of a single two-particle correlation function $\bar{g}_2(x_1x_2; t)$. The latter, in turn, is completely determined by the one-particle distribution function $\bar{f}(x_1; t)$.

18.3. THE SPATIALLY HOMOGENEOUS CLASSICAL WEAKLY COUPLED GAS

The results obtained thus far can be made even more explicit. Let us first consider the particularly simple case of a spatially homogeneous state. We know from Eq. (3.5.2) that the one-particle distribution function can only depend on the momentum, hence

$$\bar{f}(x_1; t) = n\bar{\varphi}(\mathbf{p}_1; t) \quad (18.3.1)$$

with

$$\int d\mathbf{p}\, \bar{\varphi}(\mathbf{p}; t) = 1 \quad (18.4.2)$$

We first note that there is no time evolution due to unperturbed motion in this case:

$$L_1^0 \bar{\varphi}(\mathbf{p}_1; t) = -\mathbf{v}_1 \cdot \nabla_1 \bar{\varphi}(\mathbf{p}_1; t) = 0 \quad (18.3.3)$$

* Clearly, for $s = 2$, this equation reduces to $\bar{\rho}_2(x_1x_2; t) = \bar{g}_2(x_1x_2; t)$.

The first-order term in Eq. (18.1.18) gives the following contribution*:

$$n^2\int dx_2\, L'_{12}\bar{\varphi}(\mathbf{p}_1;t)\bar{\varphi}(\mathbf{p}_2;t)$$

$$= n^2\int d\mathbf{p}_2\, d\mathbf{q}_2\, [\nabla_1 V(\mathbf{q}_1-\mathbf{q}_2)]\cdot\partial_{12}\bar{\varphi}(\mathbf{p}_1;t)\bar{\varphi}(\mathbf{p}_2;t)$$

$$= n^2\left\{\nabla_1\int d\mathbf{q}_2\, V(\mathbf{q}_1-\mathbf{q}_2)\right\}\cdot\partial_{12}\bar{\varphi}(\mathbf{p}_1;t) \tag{18.3.4}$$

As the potential only depends on the relative distance, the integration variable can be changed to $\mathbf{q}_2-\mathbf{q}_1$, and the result of the integration is a constant, independent of $\mathbf{q}_1$. The bracket therefore vanishes identically.

Proceeding to the last term of Eq. (18.1.18) and using the explicit forms for the operators L'_{12} and $U^0_{12}(\tau)$, Eqs. (18.1.3) and (15.1.13), together with the assumed form (18.3.1), we find

$$n^2\int_0^\infty d\tau\int dx_2\, L'_{12}U^0_{12}(\tau)L'_{12}U^0_{12}(-\tau)\bar{\varphi}(\mathbf{p}_1;t)\bar{\varphi}(\mathbf{p}_2;t)$$

$$= n^2\int_0^\infty d\tau\int d\mathbf{p}_2\, d\mathbf{q}_2\, [\nabla_1 V(\mathbf{r}_{12})]\cdot\partial_{12}\exp[-\tau(\mathbf{v}_1\cdot\nabla_1+\mathbf{v}_2\cdot\nabla_2)]$$

$$\times[\nabla_1 V(\mathbf{r}_{12})]\cdot\partial_{12}\exp[\tau(\mathbf{v}_1\cdot\nabla_1+\mathbf{v}_2\cdot\nabla_2)]\bar{\varphi}(\mathbf{p}_1;t)\bar{\varphi}(\mathbf{p}_2;t)$$

$$= n^2\int_0^\infty d\tau\int d\mathbf{p}_2\int d\mathbf{q}_2\, \partial_{12}\{[\nabla_1 V(\mathbf{r}_{12})][\nabla_1 V(\mathbf{r}_{12}-\mathbf{g}\tau)]\}$$

$$\cdot\,\partial_{12}\bar{\varphi}(\mathbf{p}_1;t)\bar{\varphi}(\mathbf{p}_2;t) \tag{18.3.5}$$

where we introduced the standard notation $\mathbf{g}$ for the relative velocity of particles 1 and 2 [see Eq. (11.4.7)]. Since the analysis of orders of magnitude is now completed, we set $\lambda=1$ in Eq. (18.1.18), which reduces to

$$\partial_t\bar{\varphi}(\mathbf{p}_1;t) = n\int d\mathbf{p}_2\, \partial_{12}\cdot\mathbf{G}(\mathbf{g})\cdot\partial_{12}\bar{\varphi}(\mathbf{p}_1;t)\bar{\varphi}(\mathbf{p}_2;t) \tag{18.3.6}$$

where the second-rank tensor $\mathbf{G}(\mathbf{g})$ is defined as

$$\mathbf{G}(\mathbf{g}) = \int_0^\infty d\tau\int d\mathbf{r}\, [\nabla_1 V(\mathbf{r})][\nabla_1 V(\mathbf{r}-\mathbf{g}\tau)] \tag{18.3.7}$$

At this point we already see the remarkable analogy of this equation with the *Landau equation* (11.6.8) derived intuitively in Chapter 11. It only

* Use is made of Eq. (18.3.2) and of the usual boundary condition

$$\lim_{p\to\infty}\bar{\varphi}(\mathbf{p};t)=0$$

remains to be shown that the tensor G appearing in Eq. (18.3.7) is the same as the tensor defined by Eqs. (11.6.9) or (11.6.16).

To evaluate the tensor G it is again convenient to use the Fourier transform of the potential energy $\tilde{V}_k$, as defined by Eq. (2.4.8):

$$V(r) = \int d\mathbf{k}\, \tilde{V}_k \exp(i\mathbf{k}\cdot\mathbf{r})$$

Substituting this representation into (18.3.7) we obtain

$$\begin{aligned}\mathsf{G}(g) &= \int_0^\infty d\tau \int d\mathbf{r} \int d\mathbf{k} \int d\mathbf{k}' \exp[i(\mathbf{k}+\mathbf{k}')\cdot\mathbf{r} - i\mathbf{k}'\cdot\mathbf{g}\tau](i\mathbf{k}\tilde{V}_k)(i\mathbf{k}'\tilde{V}_{k'})\\ &= 8\pi^3 \int_0^\infty d\tau \int d\mathbf{k} \int d\mathbf{k}'\, \delta(\mathbf{k}+\mathbf{k}')\exp(-i\mathbf{k}'\cdot\mathbf{g}\tau)(i\mathbf{k}\tilde{V}_k)(i\mathbf{k}'\tilde{V}_{k'})\\ &= 8\pi^3 \int_0^\infty d\tau \int d\mathbf{k} \exp(i\mathbf{k}\cdot\mathbf{g}\tau)(i\mathbf{k}\tilde{V}_k)(-i\mathbf{k}\tilde{V}_k) \qquad (18.3.8)\end{aligned}$$

We now recall the well-known representation of the distributions ("singular functions") $\delta_\pm(x)$:*

$$\int_0^\infty d\tau \exp(\pm ix\tau) = \pi\, \delta_\pm(x) \equiv \pi\, \delta(x) \pm i\mathcal{P}\left(\frac{1}{x}\right) \qquad (18.3.9)$$

where $\mathcal{P}$ denotes the principal part:

$$\mathcal{P}\int_{-\infty}^{\infty} dx \frac{1}{x} f(x) = \lim_{\varepsilon\to 0}\left\{\int_{-\infty}^{-\varepsilon} dx \frac{1}{x} f(x) + \int_{+\varepsilon}^{\infty} dx \frac{1}{x} f(x)\right\}$$

We note the important fact that $\delta(x)$ is an even function of x, whereas $\mathcal{P}(1/x)$ is an odd function. As the factor $(\mathbf{kk}\tilde{V}_k^2)$ is even in the vector $\mathbf{k}$, we can finally write the tensor $\mathsf{G}(\mathbf{g})$ in the following form:

$$\begin{aligned}\mathsf{G}(\mathbf{g}) &= 8\pi^4 \int d\mathbf{k}\, \delta_+(\mathbf{k}\cdot\mathbf{g})\, \mathbf{kk}\tilde{V}_k\\ &= 8\pi^4 \int d\mathbf{k}\, \delta(\mathbf{k}\cdot\mathbf{g})\, \mathbf{kk}\tilde{V}_k \qquad (18.3.10)\end{aligned}$$

This result is identical with Eq. (11.6.16). We have thus shown that the *Landau equation*, which we rewrite in its final form (11.6.27):†

$$\partial_t\bar{\varphi}(\mathbf{p}_1;t) = Bn\int d\mathbf{p}_2\, \partial_{12}^r \frac{g^2\delta^{rs} - g^r g^s}{g^3}\, \partial_{12}^s \bar{\varphi}(\mathbf{p}_1;t)\bar{\varphi}(\mathbf{p}_2;t) \qquad (18.3.11)$$

* See, for instance, R. Balescu, *Statistical Mechanics of Charged Particles*, Wiley-Interscience, New York, 1963; Appendix 2.

† Note the minor difference with (11.6.27): we work here in the momentum representation; ∂_{12}^r denotes derivatives with respect to momentum, rather than velocity. Hence the absence of the factor m^{-2} of Eq. (11.6.27).

can be derived as the weak-coupling approximation of the general kinetic equation (17.3.17). *We thus bridged the gap between exact dynamics and the semi-intuitive derivations of Chapter 11.* This result is very significant for the general theory. It shows that there is no need at all to depart from exact dynamics in order to derive a kinetic equation. The only fact to be kept in mind is that a kinetic equation does not describe the evolution of the complete distribution function $\varphi(\mathbf{p}; t)$ but only of a part of it, namely, $\bar{\varphi}(\mathbf{p}; t)$. We may recall that the kinetic part $\bar{\varphi}(\mathbf{p}; t)$ has been obtained as the result of the only possible decomposition of the distribution that is invariant under the motion. At no point have we used any probabilistic argument in the present derivation. *The Stosszahlansatz, in particular, does not appear here as an ad hoc stochastic assumption, but as an exact property of the kinetic component $\bar{\varphi}(\mathbf{p}; t)$.*

The only point left open is the validity of the basic assumption of separation, that is, Eq. (16.4.12). In order to check this point we must make our model more specific by choosing a particular form for the interaction potential.

A nice theoretical model for a weakly coupled gas is provided by a Gaussian potential:

$$V(r) = V_0 e^{-\alpha^2 r^2} \tag{18.3.12}$$

Although it is not very realistic (it has no hard core at small distances) it agrees with all the assumptions made in the definition of a weakly coupled gas; in particular, it is bounded for all values of r. Its Fourier transform is

$$\tilde{V}_{\mathbf{k}} = (2\pi^{1/2}\alpha)^{-3} V_0 \exp\left(\frac{-k^2}{4\alpha^2}\right) \tag{18.3.13}$$

The constant B is now very easily calculated from Eq. (11.6.24):

$$B = \frac{\pi^2 V_0^2}{4\alpha^2} \tag{18.3.14}$$

The fact that we obtained a finite constant B has an important meaning for the general theory. Tracing back the steps of the general argument, it implies that, for this model, *the kinetic operator (18.3.6) and hence (17.2.13) exists,* in the mathematical sense.

In particular, its second-order term $V\mathbf{\Gamma}_1^{[2]}V$, defined by Eq. (18.1.8), is finite. As, however, $V\mathcal{U}^0(\tau)V = 1$ in the homogeneous systems we are considering here, we have actually proved the existence of the operator:

$$\int_0^\infty d\tau\, V\mathcal{G}(\tau)V$$

We also note that, because of Eqs. (18.3.4) and (17.1.3), we have

$$V\mathcal{G}(\tau)V = V\mathscr{E}(\tau)V$$

in a homogeneous system. But, from the definition of a Laplace transform, Eq. (15.2.1), we have

$$\int_0^\infty d\tau\, V\mathscr{E}(\tau)V = \lim_{z\to 0}\int_0^\infty d\tau\, e^{izt}\, V\mathscr{E}(\tau)V = \lim_{z\to 0} V\tilde{\mathscr{E}}(z)V \quad (18.3.15)$$

Hence, the harmless-looking result of Eq. (18.3.14) is a proof that, for the present model, the operator $V\tilde{\mathscr{E}}(z)V$ approaches a finite, nonvanishing value in the limit $z\to 0$: the first condition (16.4.12) is therefore explicitly verified in this case. This result is very important in showing that our general theory is not void. There is at least one nontrivial model for which the existence of the concepts introduced in the theory can be proved in detail.

We conclude this section with a few remarks about the kinetic two-particle correlation function. Equation (18.2.9), which defines this function, can be evaluated explicitly in the homogeneous case.

The calculations are quite analogous to the previous ones of this section; hence we give immediately the result:

$$\bar{g}_2(x_1x_2; t) = n^2\left\{\nabla_1 \int_0^\infty d\tau\, V(\mathbf{r}-\mathbf{g}\tau)\right\}\cdot \partial_{12}\bar{\varphi}(\mathbf{p}_1; t)\bar{\varphi}(\mathbf{p}_2; t) \quad (18.3.16)$$

We clearly see that the time dependence of these correlations is entirely determined by the one-particle kinetic distribution function. Again, in the case of the Gaussian potential (18.3.12), the integral can be easily calculated:

$$\int_0^\infty d\tau\, V(\mathbf{r}-\mathbf{g}\tau)$$

$$= \left(\frac{\pi^{1/2}V_0}{2\alpha g}\right)\exp\left[-\alpha^2 r^2 + \alpha^2\left(\frac{\mathbf{r}\cdot\mathbf{g}}{g}\right)^2\right]\left\{1-\Phi\left(\frac{-\alpha\mathbf{r}\cdot\mathbf{g}}{g}\right)\right\} \quad (18.3.17)$$

where $\Phi(x)$ is the error function:

$$\Phi(x) = \left(\frac{2}{\pi^{1/2}}\right)\int_0^x d\xi\, \exp(-\xi^2) \quad (18.3.18)$$

The finiteness of this function again provides a justification for the assumption that the operator $C\tilde{\mathscr{C}}(z)V$ is finite and analytic at $z=0$ [Eq. (16.4.12)].

18.4. THE INHOMOGENEOUS CLASSICAL WEAKLY COUPLED GAS

We now study the more general case of spatially inhomogeneous systems along the same lines as in Section 18.3. The distribution function $\bar{f}(\mathbf{q}_1, \mathbf{p}_1; t)$ now depends on both position and momentum.

Going back to Eq. (18.1.18) (and setting $\lambda = 1$), we note in the right-hand side the presence of two typically inhomogeneous terms:

$$L_1^0\bar{f}(x_1; t) + \int dx_2\, L'_{12}\bar{f}(x_1; t)\bar{f}(x_2; t)$$

These terms have no counterpart in homogeneous systems, where the vacuum-to-vacuum matrix elements of $\mathscr{L}^0$ and $\mathscr{L}'$ vanish identically [see Eqs. (18.3.3), (18.3.4)]. The first term is the familiar *flow term*, which we know quite well from Chapters 11–13:

$$L_1^0\bar{f}(x_1; t) = -\mathbf{v}_1 \cdot \nabla_1 \bar{f}(x_1; t) \tag{18.4.1}$$

The second term has the explicit form:

$$\begin{aligned}\int dx_2\, L'_{12}\bar{f}(x_1; t)\bar{f}(x_2; t) &= \int dx_2\, (\nabla_1 V_{12}) \cdot \partial_{12}\bar{f}(x_1; t)\bar{f}(x_2; t)\\ &= \nabla_1 \cdot \int d\mathbf{q}_2\, d\mathbf{p}_2\, V(\mathbf{q}_1 - \mathbf{q}_2)(\partial_1 - \partial_2)\bar{f}(x_1; t)\bar{f}(x_2; t)\\ &= [\nabla_1 U(\mathbf{q}_1; t)] \cdot \partial_1 \bar{f}(x_1; t)\end{aligned} \tag{18.4.2}$$

where

$$U(\mathbf{q}_1; t) = \int d\mathbf{q}_2\, d\mathbf{p}_2\, V(\mathbf{q}_1 - \mathbf{q}_2)\, \bar{f}(\mathbf{q}_2, \mathbf{p}_2; t) \tag{18.4.3}$$

We immediately recognize the *Vlassov term*, which was discussed in detail in Section 11.7.

In order to evaluate the third term in Eq. (18.1.18) we must be careful. From Eq. (18.3.5) we have

$$\begin{aligned}&\int_0^\infty d\tau \int d\mathbf{p}_2\, d\mathbf{q}_2\, [\nabla_1 V(\mathbf{r}_{12})] \cdot \partial_{12} \exp[-\tau(\mathbf{v}_1 \cdot \nabla_1 + \mathbf{v}_2 \cdot \nabla_2)]\, [\nabla_1 V(\mathbf{r}_{12})] \cdot \partial_{12}\\ &\qquad\qquad \times \exp[\tau(\mathbf{v}_1 \cdot \nabla_1 + \mathbf{v}_2 \cdot \nabla_2)]\, \bar{f}(\mathbf{q}_1, \mathbf{p}_1; t)\bar{f}(\mathbf{q}_2, \mathbf{p}_2; t)\\ &\quad = \int_0^\infty d\tau \int d\mathbf{p}_2\, d\mathbf{q}_2\, [\nabla_1 V(\mathbf{r}_{12})] \cdot \partial_{12}[\nabla_1 V(\mathbf{r}_{12} - \mathbf{g}\tau)]\\ &\qquad\qquad \cdot \{\partial_{12} + \tau\nabla_{12}\}\, U_{12}^0(\tau) U_{12}^0(-\tau)\, \bar{f}(\mathbf{q}_1, \mathbf{p}_1; t)\bar{f}(\mathbf{q}_2, \mathbf{p}_2; t)\end{aligned}$$

The term $\tau\nabla_{12}$ results from the fact that in writing the central propagator $U_{12}^0(\tau)$ to the right of the interaction matrix element, we must take into account that it does not commute with ∂_{12}. This extra term did not appear in the homogeneous case because the distribution functions on which it acts are independent of $\mathbf{q}_1$, $\mathbf{q}_2$.

Collecting now all our results, we obtain

$$\partial_t \bar{f}(\mathbf{q}_1, \mathbf{p}_1; t) = -\mathbf{v}_1 \cdot \nabla_1 \bar{f}(\mathbf{q}_1, \mathbf{p}_1; t) + [\nabla_1 U(\mathbf{q}_1; t)] \cdot \partial_1 \bar{f}(\mathbf{q}_1, \mathbf{p}_1; t)$$
$$+ \int_0^\infty d\tau \int d\mathbf{q}_2\, d\mathbf{p}_2\, \partial_{12} \cdot \{\nabla_1 V(\mathbf{r}_{12})\, \nabla_1 V(\mathbf{r}_{12} - \mathbf{g}\tau)\}$$
$$\cdot (\partial_{12} + \tau \nabla_{12})\, \bar{f}(\mathbf{q}_1, \mathbf{p}_1; t)\, \bar{f}(\mathbf{q}_2, \mathbf{p}_2; t) \tag{18.4.4}$$

This is *the general form of the kinetic equation for weakly coupled gases, valid for arbitrary inhomogeneity.* We note immediately that the third term cannot be written in the form of Eq. (18.3.6) because the integration over $\mathbf{q}_2$ now involves *both* the product of forces and the distribution function $\bar{f}$. This equation is, therefore, more general than the Landau–Vlassov equation (11.7.3). It would be difficult to obtain it from a purely phenomenological treatment.

We must keep in mind, however, that the force has a finite range (in "normal" systems); it falls to zero beyond a characteristic distance L_c, which is very short on the macroscopic scale. On the other hand, the spatial variation of the distribution function is characterized by a hydrodynamical length L_h defined in Eq. (13.2.2):

$$L_h^{-1} = \max \frac{|\nabla \bar{f}(\mathbf{q}, \mathbf{p}; t)|}{\bar{f}(\mathbf{q}, \mathbf{p}; t)}$$

It follows from our discussion of Section 13.2 that in all *hydrodynamical problems* the distribution function $\bar{f}$ is practically constant over the range of the forces, because $L_h \gg L_c$. We may take advantage of this fact in order to simplify the kinetic equation.

We expand the distribution function at point $\mathbf{q}_2$ around the point $\mathbf{q}_1$:

$$\bar{f}(\mathbf{q}_2, \mathbf{p}_2; t) = \bar{f}(\mathbf{q}_1, \mathbf{p}_2; t) - \mathbf{r}_{12} \cdot \nabla_1 \bar{f}(\mathbf{q}_1, \mathbf{p}_2; t) + \cdots \tag{18.4.5}$$

We substitute this expression into Eq. (18.4.3) for the average field:

$$U(\mathbf{q}_1; t) = \int d\mathbf{q}_2\, d\mathbf{p}_2\, V(\mathbf{q}_1 - \mathbf{q}_2) \bar{f}(\mathbf{q}_2, \mathbf{p}_2; t)$$
$$\approx \int d\mathbf{p}_2 \left\{ \int d\mathbf{r}_{12}\, V(\mathbf{r}_{12}) \bar{f}(\mathbf{q}_1, \mathbf{p}_2; t) - \int d\mathbf{r}_{12}\, V(\mathbf{r}_{12}) \mathbf{r}_{12} \cdot \nabla_1 \bar{f}(\mathbf{q}_1, \mathbf{p}_2; t) \right\} \tag{18.4.6}$$

We immediately see that the second term in the right-hand side is of order L_c/L_h compared to the first, and hence can be neglected in the hydrodynamical approximation. Let us introduce the notation $8\pi_3 \tilde{V}_0$ for the unweighted average potential:

$$8\pi^3\tilde{V}_0 = \int d\mathbf{r}\; V(\mathbf{r}) \tag{18.4.7}$$

[This symbol already appeared in the equilibrium theory, see, for example, Eq. (6.2.9).]

Let us also use the notation $\bar{n}(\mathbf{q}; t)$ for the local number density corresponding to the kinetic distribution $\bar{f}(\mathbf{q}, \mathbf{p}; t)$:

$$\bar{n}(\mathbf{q}; t) = \int d\mathbf{p}\; \bar{f}(\mathbf{q}, \mathbf{p}; t) \tag{18.4.8}$$

Then, in the hydrodynamic approximation, the average field is simply given by

$$U(\mathbf{q}; t) = 8\pi^3\tilde{V}_0\bar{n}(\mathbf{q}; t) \tag{18.4.9}$$

We now study the third term in the right-hand side of Eq. (18.4.4). Substituting the expansion (18.4.6), we obtain

$$\begin{aligned}
&\int_0^\infty d\tau \int d\mathbf{p}_2\, d\mathbf{q}_2\, \partial_{12}\cdot\{[\nabla_1 V(\mathbf{r})][\nabla_1 V(\mathbf{r}-\mathbf{g}\tau)]\}(\partial_{12}+\tau\nabla_{12})\bar{f}(\mathbf{q}_1, \mathbf{p}_1; t)\bar{f}(\mathbf{q}_2, \mathbf{p}_2; t)\\
&\approx \int_0^\infty d\tau \int d\mathbf{p}_2 \int d\mathbf{r}\; \partial_{12}\cdot\{[\nabla_1 V(\mathbf{r})][\nabla_1 V(\mathbf{r}-\mathbf{g}\tau)]\}\cdot\partial_{12}\bar{f}(\mathbf{q}_1, \mathbf{p}_1; t)\bar{f}(\mathbf{q}_1, \mathbf{p}_2; t)\\
&\quad - \int_0^\infty d\tau \int d\mathbf{p}_2 \int d\mathbf{r}\; \partial_{12}\cdot\{[\nabla_1 V(\mathbf{r})][\nabla_1 V(\mathbf{r}-\mathbf{g}\tau)]\}\cdot\partial_{12}\bar{f}(\mathbf{q}_1, \mathbf{p}_1; t)\\
&\qquad\qquad\qquad\qquad \times[\mathbf{r}\cdot\nabla_1\bar{f}(\mathbf{q}_1, \mathbf{p}_2; t)]\\
&\quad + \int_0^\infty d\tau \int d\mathbf{p}_2 \int d\mathbf{r}\; \partial_{12}\cdot\{[\nabla_1 V(\mathbf{r})][\nabla_1 V(\mathbf{r}-\mathbf{g}\tau)]\}\tau\nabla_{12}\bar{f}(\mathbf{q}_1, \mathbf{p}_1; t)\bar{f}(\mathbf{q}_2, \mathbf{p}_2; t)
\end{aligned} \tag{18.4.10}$$

It is again clear that the second term and the third term are of order L_c/L_h compared to the first and therefore can be neglected in the hydrodynamical approximation. On the other hand, the leading term involves precisely the same operator $\mathbf{G}(\mathbf{g})$ as in the homogeneous problem, Eq. (18.3.7). This is, of course, not surprising, because in the hydrodynamic limit the system is practically homogeneous over a distance of the order of the interaction range, L_c.

From here on we can, therefore, perform all the transformations of Section 18.3, and we obtain the *final kinetic equation in the hydrodynamic approximation:*

$$\begin{aligned}
\partial_t\bar{f}(\mathbf{q}_1, \mathbf{p}_1; t) &= -\mathbf{v}_1\cdot\nabla_1\bar{f}(\mathbf{q}_1, \mathbf{p}_1; t) + 8\pi^3\tilde{V}_0[\nabla_1\bar{n}(\mathbf{q}_1; t)]\cdot\partial_1\bar{f}(\mathbf{q}_1, \mathbf{p}_1; t)\\
&\quad + B\int d\mathbf{p}_2\, \partial_{12}^r \frac{g^2\delta^{rs} - g^r g^s}{g^3}\, \partial_{12}^s\bar{f}(\mathbf{q}_1, \mathbf{p}_1; t)\bar{f}(\mathbf{q}_1, \mathbf{p}_2; t)
\end{aligned} \tag{18.4.11}$$

The characteristic feature of the hydrodynamic approximation is the occurrence of products of one-particle distribution functions evaluated at the *same* position $\mathbf{q}_1$.

Equation (18.4.11) is of direct application to a true weakly coupled gas, such as the Gaussian gas considered in Section 18.3. In practice, this equation is very often applied to the study of plasmas. This, however, requires some care. First, we note that the hydrodynamic linearization (18.4.5) cannot be applied to the Vlassov term because $\tilde{V}_0$ is infinite for the Coulomb potential. The Vlassov term must, therefore, be kept in its complete original form (18.4.2). In other words, Eq. (11.7.3) rather than (18.4.11) must be used for a plasma. We investigated in Section 13.6 some of the very peculiar effects of the Vlassov term on the behavior of a plasma.

In the collision term the hydrodynamic limit makes no difficulty, but there appears another serious problem even for a homogeneous plasma. The coefficient B defined in Eq. (11.6.24) is infinite for a Coulomb potential. This is a result of the long range of the forces: It is analogous to the difficulty encountered in Section 6.5 in the study of the equilibrium properties of plasmas.

In practice, this problem is treated pragmatically by introducing reasonable, but somewhat arbitrary cutoffs on the limits of integration in Eq. (11.6.24). A more fundamental approach, however, is necessary for a real understanding of the physical problem. The general theory developed here provides us with an appropriate tool for handling this problem. It will be studied in detail in Sections 20.5 and 20.6.

18.5. THE NONKINETIC COMPONENT OF THE DISTRIBUTION VECTOR

We now consider the equations obeyed by the nonkinetic part of the distribution vector. The analysis is very similar to Sections 18.1 and 18.2, hence we may give the results directly. The correlation part obeys, through order λ^2, a closed equation obtained by expanding Eqs. (17.4.11) and (17.4.12):

$$\begin{aligned}\partial_t C\hat{\mathfrak{f}}(t) &= C\hat{\mathbf{\Delta}}C\hat{\mathfrak{f}}(t)\\ &\approx C(\mathcal{L}^0+\mathcal{L}')C\hat{\mathfrak{f}}(t)\\ &\quad -\lambda^2\int_0^\infty d\tau\, C\mathcal{L}'V\mathcal{U}^0(-\tau)V\mathcal{L}'C\mathcal{U}^0(\tau)C\hat{\mathfrak{f}}(t)\end{aligned} \tag{18.5.1}$$

To the leading order, the vacuum part is obtained from Eq. (17.4.6):

$$\begin{aligned}V\hat{\mathfrak{f}}(t) &= -V\mathbf{D}C\hat{\mathfrak{f}}(t)\\ &\approx -\lambda\int_0^\infty d\tau\, V\mathcal{U}^0(-\tau)V\mathcal{L}'C\mathcal{U}^0(\tau)C\hat{\mathfrak{f}}(t)\end{aligned} \tag{18.5.2}$$

From these general equations we can derive particular ones, describing the evolution of the correlation patterns $\hat{\rho}_2(x_1x_2; t)$ and $\hat{\rho}_1(x_1; t)$. These are obtained by evaluating the appropriate matrix elements, just as in Sections 18.1 and 18.2:

$$\partial_t\hat{\rho}_2(x_1x_2; t) = (L_1^0 + L_2^0)\hat{\rho}_2(x_1x_2; t) + \lambda L'_{12}\hat{\rho}_2(x_1x_2; t)$$
$$-\lambda^2\int_0^\infty d\tau\, L'_{12}U_{12}^0(-\tau)\int dx_3\,\{L'_{13}U_{123}^0(\tau)\hat{\rho}_3(x_1x_3\,|\,x_2; t)$$
$$+ L'_{23}U_{123}^0(\tau)\hat{\rho}_3(x_2x_3\,|\,x_1; t)\} \tag{18.5.3}$$

Correspondingly, Eq. (18.5.2) yields

$$\hat{\rho}_1(x_1; t) = -\lambda\int_0^\infty d\tau\int dx_2\, U_1^0(-\tau)L'_{12}U_{12}^0(\tau)\hat{\rho}_2(x_1x_2; t) \tag{18.5.4}$$

These equations clearly show that the problem of determining the nonkinetic part of the correlation patterns is essentially more difficult than the corresponding problem for the kinetic part. We see that the correlation patterns of various types appear to be mixed [e.g., $\partial_t\hat{\rho}_2(x_1x_2)$ depends on $\hat{\rho}_3(x_1x_2\,|\,x_3)$]: we do not escape an infinite hierarchy. The factorization properties of the correlation patterns that allowed us to reduce the hierarchies for the kinetic part to single nonlinear equations cannot hold here. A simple example will convince the reader.* The correlation pattern $\rho_2(x_1\,|\,x_2)$ is decomposed as follows:

$$\rho_2(x_1\,|\,x_2) = \bar{\rho}_2(x_1\,|\,x_2) + \hat{\rho}_2(x_1\,|\,x_2)$$

It is physically natural to assume that both $\rho_2(x_1\,|\,x_2)$ and $\bar{\rho}_2(x_1\,|\,x_2)$ are factorized; hence

$$f_1(x_1)f_1(x_2) = \bar{f}(x_1)\bar{f}_1(x_2) + \hat{\rho}_2(x_1\,|\,x_2)$$

Clearly, $\hat{\rho}_2(x_1\,|\,x_2)$ cannot be a product. Hence all the nonkinetic correlation patterns are interconnected, a striking difference with their kinetic counterparts.

However, if we are interested in the fastest processes only, the

* We have here a clear case showing that the dynamical definition of the correlation patterns, developed in Chapter 14, is more general and more useful than the static definition of Section 3.4, based on the usual cluster expansion. The latter cannot be consistently applied to the nonkinetic component. It is also for this reason that we use here the general notations $\hat{\rho}_s$ rather than the more common notation $\hat{f}_s$, $\hat{g}_s$, which would suggest the cluster structure.

situation becomes much simpler. The leading term in Eq. (18.5.3) is simply

$$\partial_t \hat{\rho}_2(x_1x_2;t) = (L_1^0 + L_2^0)\hat{\rho}_2(x_1x_2;t) \tag{18.5.5}$$

In this approximation, the correlation patterns evolve independently of each other. We shall see that this simple mechanism gives by itself a fairly clear picture of the evolution.

Let us first determine the initial condition to be used consistently for this equation (according to the discussion of Section 17.5). We use the following simple argument. The one-particle function and the two-particle correlation can be split as follows [by writing the constraints (18.2.9) and (18.5.4)]:

$$\begin{aligned} f_1(x_1;0) &= \bar{f}_1(x_1;0) - \lambda \int dx_2 \int_0^\infty d\tau\, U_1^0(-\tau) L'_{12} U_{12}^0(\tau) \hat{\rho}_2(x_1x_2;0) \\ g_2(x_1x_2;0) &= \lambda \int_0^\infty d\tau\, U_{12}^0(\tau) L'_{12} U_{12}^0(-\tau) \bar{f}(x_1;0)\bar{f}_1(x_2;0) + \hat{\rho}_2(x_1x_2;0) \end{aligned} \tag{18.5.6}$$

If we now make the reasonable assumption that, at time zero

$$\begin{aligned} f_1(x_1;0) &= O(\lambda^0) \\ g_2(x_1x_2;0) &= O(\lambda) \end{aligned} \tag{18.5.7}$$

and if we only keep the leading term we obtain

$$\begin{aligned} \bar{f}(x_1;0) &= f_1(x_1;0) + O(\lambda^2) \\ \hat{\rho}_2(x_1x_2;0) &= g_2(x_1x_2;0) - \lambda \int_0^\infty d\tau\, U_{12}^0(\tau) L'_{12} U_{12}^0(-\tau) f_1(x_1;0) f_1(x_2;0) + O(\lambda^2) \end{aligned} \tag{18.5.8}$$

This equation provides us explicitly with the initial condition to be used for the solution of Eq. (18.5.5) [and also for (18.1.18)], expressed in terms of the initial value of the *complete* distribution functions $f_1(x_1;0)$, $g_2(x_1x_2;0)$.

The solution of (18.5.5) is immediate: Making use of (15.1.18) we get*

$$\begin{aligned} \hat{\rho}_2(x_1x_2;t) = U_{12}^0(t) \Big\{ & g_2(x_1x_2;0) \\ & - \int_0^\infty d\tau\, U_{12}^0(\tau) L'_{12} U_{12}^0(-\tau) f_1(x_1;0) f_1(x_2;0) \Big\} \end{aligned}$$

After an obvious change of integration variables in the second term, and

* We may again set $\lambda = 1$ at this stage.

use of the group property $U_{12}^0(t)U_{12}^0(\tau)=U_{12}^0(t+\tau)$, we obtain

$$\hat{\rho}_2(x_1x_2;t)=U_{12}^0(t)g_2(x_1x_2;0) - \int_t^\infty d\theta\, U_{12}^0(\theta)L'_{12}U_{12}^0(t-\theta)f_1(x_1;0)f_1(x_2;0) \quad (18.5.9)$$

Combining this result with (18.5.4) we obtain

$$\hat{\rho}_1(x_1;t)=-\int_t^\infty d\tau \int dx_2\, U_1^0(t-\tau)L'_{12}U_{12}^0(\tau)g_2(x_1x_2;0) + \int_t^\infty d\tau \int_t^\tau d\theta \int dx_2\, U_1^0(\theta-\tau)L'_{12}U_{12}^0(\tau)L'_{12}U_{12}^0(t-\theta)f_1(x_1;0)f_1(x_2;0) \quad (18.5.10)$$

Actually, $\hat{\rho}_1(x_1;t)$ is of order λ^2 and can be neglected in the present approximation, which implies for the complete one-particle distribution

$$f_1(x_1;t)=\bar{f}_1(x_1;t)+O(\lambda^2) \quad (18.5.11)$$

We wrote down the result, however, because it displays rather vividly a quite important and general property of $\hat{\rho}_1(x_1;t)$. It is clearly seen from the integration limits in the right-hand side of (18.5.10) that

$$\lim_{t\to\infty} \hat{\rho}_1(x_1;t)=0 \quad (18.5.12)$$

This shows that *the nonkinetic part of the one-particle distribution function [and more generally, the component* $\mathcal{V}\mathfrak{f}(t)$ *of the distribution vector] is a transient function, which decays to zero after a long enough time.*

This is in agreement with our general theory: The end point of the evolution (i.e., the state of thermal equilibrium) corresponds to $\hat{\mathfrak{f}}(t)=0$, as shown in Section 17.7.

Going back to the two-body correlation function (18.5.9), we see that the second term in the rhs has the same manifestly decaying form as $\hat{\rho}_1(x_1;t)$. However, the properties of the first term are somewhat subtler and are best discussed and understood on the basis of a model.

We specify the initial condition and the interaction potential as follows, by using Gaussian functions, which always give a good insight:

$$\begin{aligned} g_2(x_1x_2;0)&=R_0\exp(-\gamma^2r^2)\phi(\mathbf{p}_1,\mathbf{p}_2)\\ f(x_1;0)&=n\varphi(\mathbf{p}_1;0) \\ V(r)&=V_0\exp(-\alpha^2r^2)\end{aligned} \quad (18.5.13)$$

where α, γ, R_0, V_0 are constant parameters, $\mathbf{r}=\mathbf{q}_1-\mathbf{q}_2$ and $\varphi(\mathbf{p}_1;0)$, $\phi(\mathbf{p}_1,\mathbf{p}_2)$ are properly normalized functions of the momenta. In this case, a pretty straightforward calculation yields the following result:

$$\hat{\rho}_2(x_1x_2;t) = R_0\phi(\mathbf{p}_1,\mathbf{p}_2)\exp[-\gamma^2|\mathbf{r}-\mathbf{g}t|^2] - \frac{n^2V_0}{2\alpha g}\pi^{1/2}\frac{\partial}{\partial\mathbf{r}}$$
$$\times\left\{\exp\left[-\alpha^2\frac{r^2g^2-(\mathbf{r}\cdot\mathbf{g})^2}{g^2}\right]\left[1-\Phi\left(\alpha gt-\alpha\frac{\mathbf{r}\cdot\mathbf{g}}{g}\right)\right]\right\}\cdot\partial_{12}\varphi(\mathbf{p}_1;0)\varphi(\mathbf{p}_2;0) \tag{18.5.14}$$

where $\Phi(x)$ is the error function:

$$\Phi(x) = \left(\frac{2}{\pi^{1/2}}\right)\int_0^x d\xi\,\exp(-\xi^2)$$
$$\approx 1-\frac{\exp(-x^2)}{\pi^{1/2}x}\left(1-\frac{1}{2x^2}+\cdots\right) \tag{18.5.15}$$

If the rather complicated second term of Eq. (18.5.14) is approximated by the first term of the asymptotic expansion of the error function, it becomes

$$-\frac{n^2V_0}{2\alpha^2 g}\left(\frac{\partial}{\partial\mathbf{r}}\frac{\exp[-\alpha^2|\mathbf{r}-\mathbf{g}t|^2]}{gt-(\mathbf{r}\cdot\mathbf{g})/g}\right)\cdot\partial_{12}\varphi(\mathbf{p}_1;0)\varphi(\mathbf{p}_2;0) \tag{18.5.16}$$

Its behavior in time is rather similar to the first term of $\hat{\rho}_2$: the exponential decay is strengthened by an additional factor $[\alpha gt-\alpha(\mathbf{r}\cdot\mathbf{g})/g]^{-1}$. The main difference is in the time scale of the decay, which is fixed by the range of the interactions α^{-1}, rather than by the range of the initial correlations γ^{-1}.

However in most realistic situations these parameters are of comparable orders of magnitude. As we are here interested in qualitative features only, we shall not further discuss this term, but rather concentrate on the first term:

$$\hat{\rho}_2^{(0)}(\mathbf{r},\mathbf{g};t) = R_0\phi\exp[-\gamma^2|\mathbf{r}-\mathbf{g}t|^2] \tag{18.5.17}$$

We first note that, for every fixed distance $\mathbf{r}$ and fixed velocity $\mathbf{g}$, this function tends to zero as $t\to\infty$. It is, however, interesting to follow in some more detail the way in which it does so. We note that, for fixed $\mathbf{g}$, the behavior is not the same in all directions in space: the correlations are anisotropic ($\mathbf{g}$ is a privileged direction). We plotted $\hat{\rho}_2^{(0)}/R_0\phi$ as a function of the dimensionless parameter $\tilde{t}=\gamma gt$ for various values of $\tilde{r}=\gamma r$. In Fig. 18.5.1 we see the behavior in the direction $\mathbf{r}\perp\mathbf{g}$. The correlation function decays regularly with time, at all distances. The decay occurs on a time scale $\tau_c\sim(\gamma\bar{v})^{-1}$, where $\bar{v}$ is a typical average velocity and τ_c represents the time it takes for a typical particle to move through the range of correlation with another particle. As $\gamma\approx\alpha$, this (usually) also represents the time it takes for a particle to go through the range of interaction, that

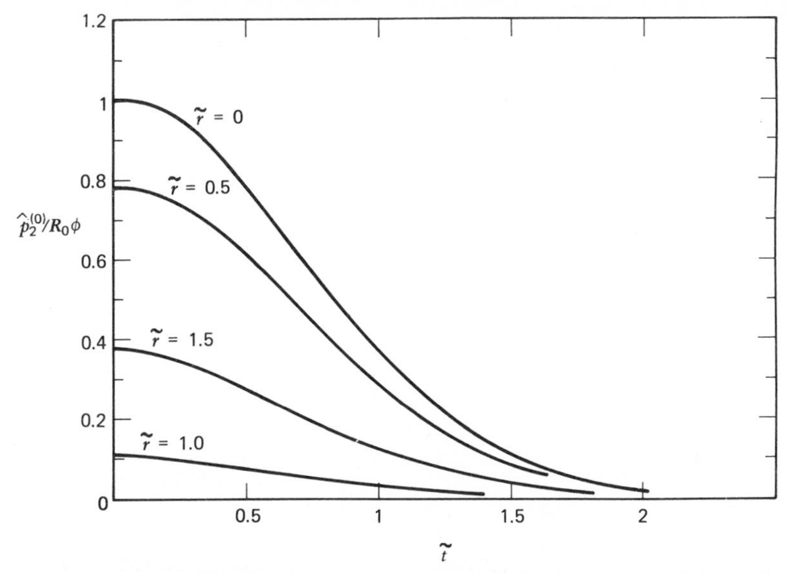

Figure 18.5.1. Time evolution of the nonkinetic part of the correlation function, at various distances, in a direction perpendicular to the relative velocity.

is, *the duration of a collision.* τ_c defines a short time scale in our problem.

If we now look in the direction of the relative velocity, $\mathbf{r} \parallel \mathbf{g}$ (Fig. 18.5.2), the behavior is strikingly different. At $r = 0$ we find the same regular decay as before. But at larger distances, we see first a buildup of the correlation, followed by a decay. The phenomenon is quite similar to the passage of a wave.

This represents the *convection of the correlation* by two free particles, which we can understand by the following mechanism (Fig. 18.5.3). Two particles that became correlated at some time in the past (say, through a collision) move away from each other along straight trajectories: their mutual distance increases linearly in time. In the absence of other events (as implied by the free propagator) the initial correlation pattern is transferred toward increasing distances.*

It should be clear, however, that in many important physical problems this convection effect plays no role. Consider, indeed, the contribution of $\hat{p}_2(\mathbf{r}; t)$ to an average of some function $B(\mathbf{r})$. (We do not write the velocity dependence, which is not very relevant.)

$$\langle B(t)\rangle \simeq \int d\mathbf{r}\, B(\mathbf{r})\hat{p}_2^{(0)}(\mathbf{r}; t) \tag{18.5.18}$$

* In higher orders in λ one expects also a damping of this pattern as one moves toward larger distances, because of collisions with other particles in the medium. This damping, which proceeds on a slower time scale, is produced by the terms of order $\lambda, \lambda^2, \ldots$ in Eq. (18.5.3).

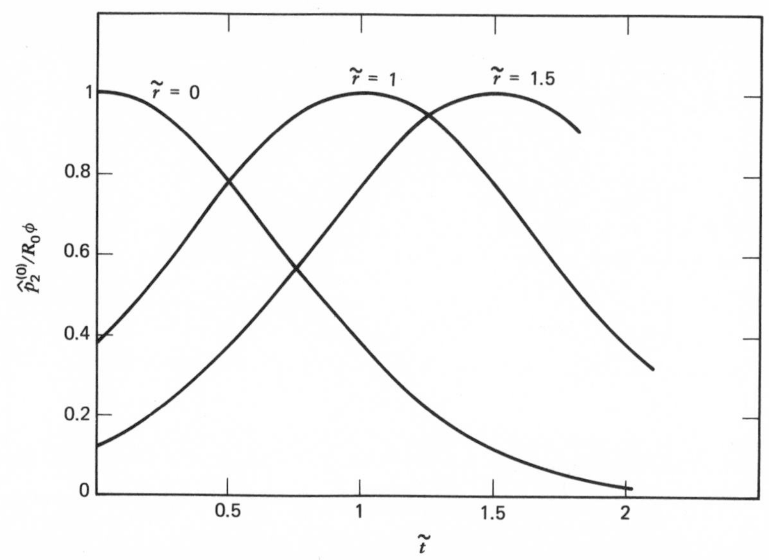

Figure 18.5.2. Time evolution of the nonkinetic part of the correlation function, at various distances, in a direction parallel to the relative velocity.

Most dynamical functions occurring in thermodynamics or in transport theory are functions of the interaction potential, and hence have a finite range, of order L_c. The domain of integration, therefore, is effectively cut off at this distance. It then follows from our discussion that the average is significantly different from zero only when the bulk of the correlations is within the effective domain of integration (Fig. 18.5.4). As t becomes large compared to τ_c, the wave has moved out of this domain; there is no overlap of B and $\hat{p}_2$, and the integral practically vanishes. Hence $\langle B(t)\rangle$ is a monotonously decaying function of time, the characteristic scale being again of the order τ_c.

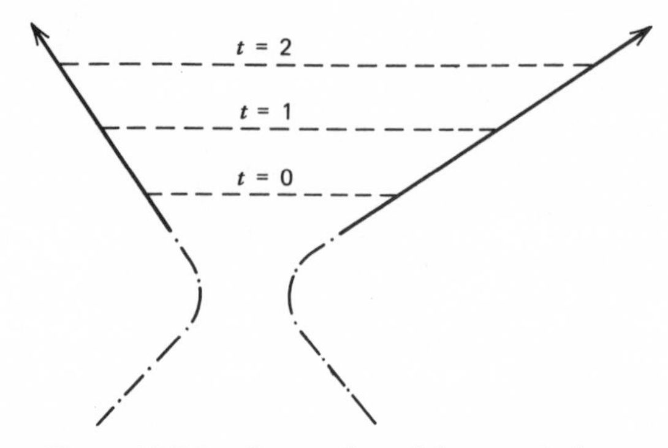

Figure 18.5.3. Convection of the correlations.

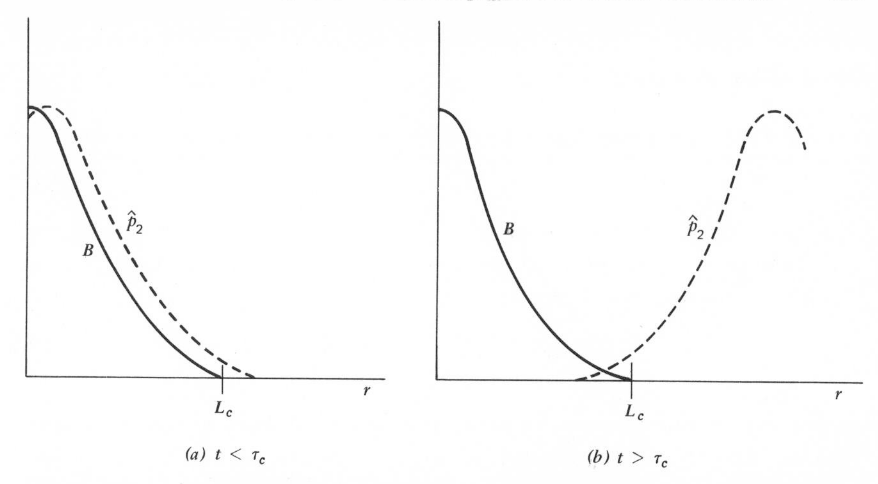

Figure 18.5.4. The factors $B(r)$ *and* $\hat{p}_2^{(0)}(r; t)$ in Eq. (18.5.18), at two typical times.

We have now a pretty clear picture of the evolution of the nonkinetic part of the distribution vector. The main feature is a very fast decay of the correlations due to the unperturbed convection. For all practical purposes the nonkinetic correlations die out after a time of the order of the duration of a collision, τ_c. Although the unperturbed convection cannot describe any damping, it has the effect of driving the correlated region out of the range of interest. The final decay of these correlations is a matter of collisions, and is a slow and complicated process by which the effect of the two-body correlations is transmitted to higher-order correlation patterns and finally dissipated away. We know, indeed, that when the system reaches equilibrium the nonkinetic component of its distribution vector vanishes rigorously.

18.6. THE QUANTUM KINETIC EQUATION AND KINETIC CORRELATIONS

We now study the same physical system as in the previous sections except that we assume its motion to be governed by the laws of quantum mechanics, that is, by the Hamiltonian (2.4.26)–(2.4.28). We shall therefore consider the specific effects of quantum statistics in the simplest possible case.

The derivation of the quantum kinetic equation is quite straightforward. In fact, the treatment of the beginning of Section 18.1 is valid just as it stands, because it is expressed in terms of abstract symbols, without specific reference to the type of mechanics obeyed by the system. Hence, we may write Eq. (18.1.10) for the vacuum component of the kinetic distribution vector:

$$\partial_t V\bar{\mathfrak{f}}(t) = V\mathcal{L}^0\bar{\mathfrak{f}}(t) + \lambda V\mathcal{L}' V\bar{\mathfrak{f}}(t) + \lambda^2 \int_0^\infty d\tau\, V\mathcal{L}' C\mathcal{U}^0(\tau)\mathcal{L}'\mathcal{U}^0(-\tau)V\bar{\mathfrak{f}}(t) \tag{18.6.1}$$

We now translate this symbolism into an explicit quantum kinetic equation for the one-particle kinetic Wigner function. Proceeding as in Section 18.1, we first obtain from Eq. (14.3.7)

$$\langle 1|\, V\mathcal{L}^0 V\, |1, 2, \ldots, s\rangle = \delta_{s,1} L_1^0 \tag{18.6.2}$$

This operator is the same as in the classical case.

The vacuum-to-vacuum element of $\mathcal{L}'$ again involves a reduction from two to one particles [see Eq. (14.3.7)]:

$$\begin{aligned}\langle 1|\, V\mathcal{L}' V\, |1, 2, \ldots, s\rangle &= \delta_{s,2} \langle 1|\, \mathcal{L}'\, |1\,|\,2\rangle \\ &= \delta_{s,2} \int d\xi_2\, \delta(\mathbf{k}_2)\, L'_{12} P_2(1\,|\,2)\end{aligned} \tag{18.6.3}$$

In order to analyze the last term in Eq. (18.6.1) we first note that (starting from the left) the only possible transition from a one-particle (vacuum) state to a correlated state is a transition to the two-particle state (12) [see Eq. (14.3.7)]. We then must go over to a vacuum state. Equation (14.3.17) shows that this is possible *either* by a transition to a two-particle state $(1\,|\,2)$ *or* by a transition to the three-particle state $(1\,|\,2\,|\,3)$. This latter possibility is a characteristic quantum effect:

$$\begin{aligned}&\langle 1|\, V\mathcal{L}' C\mathcal{U}^0(\tau)\mathcal{L}' V\mathcal{U}^0(-\tau)\, |1 \cdots s\rangle \\ &\quad= \delta_{s,2} \langle 1|\, \mathcal{L}'\, |12\rangle\, U_{12}^0(\tau)\, \langle 12|\, \mathcal{L}'\, |1\,|\,2\rangle\, U_{12}^0(-\tau) \\ &\qquad+ \delta_{s,3} \langle 1|\, \mathcal{L}'\, |12\rangle\, U_{12}^0(\tau)\, \langle 12|\, \mathcal{L}'\, |1\,|\,2\,|\,3\rangle\, U_{123}^0(-\tau) \\ &\quad= \delta_{s,2} \int d\xi_2\, \delta(\mathbf{k}_2)\, L'_{12} P_2(12)\, U_{12}^0(\tau)\, L'_{12} P_2(1\,|\,2)\, U_{12}^0(-\tau) \\ &\qquad+ \delta_{s,3} \int d\xi_2\, d\xi_3\, \delta(\mathbf{k}_2)\, \delta(\mathbf{k}_3)\, L'_{12} P_2(12)\, U_{12}^0(\tau) \\ &\qquad\times \{[L'_{13} P_3(1\,|\,2\,|\,3) - P_2(1\,|\,2) L'_{13} P_2(1\,|\,3)] \\ &\qquad+ [L'_{23} P_3(1\,|\,2\,|\,3) - P_2(1\,|\,2) L'_{23} P_2(2\,|\,3)]\} U_{123}^0(-\tau)\end{aligned} \tag{18.6.4}$$

These partial results are substituted into Eq. (18.6.1). By the same argument as used in Section 18.1, it is shown that a consistent form for the vacuum components of the kinetic distribution vectors is

$$\bar{p}_s(\xi_1\,|\,\xi_2\,|\cdots|\,\xi_s; t) = \prod_{j=1}^{s} f(\xi_j; t) \tag{18.6.5}$$

Hence we finally obtain the quantum kinetic equation in the form ($\lambda = 1$):

$$\begin{aligned}
\partial_t \bar{f}(\xi_1; t) = L_1^0 \bar{f}(\xi_1; t) + \int d\xi_2\, \delta(\mathbf{k}_2) L'_{12} P_2(1\,|\,2) \bar{f}(\xi_1; t) \bar{f}(\xi_2; t) \\
+ \int d\xi_2\, \delta(\mathbf{k}_2) \int_0^\infty d\tau\, L'_{12} U_{12}^0(\tau) L'_{12} U_{12}^0(-\tau) P_2(1\,|\,2) \bar{f}(\xi_1; t) \bar{f}(\xi_2; t) \\
+ \int d\xi_2\, d\xi_3\, \delta(\mathbf{k}_2)\, \delta(\mathbf{k}_3) \int_0^\infty d\tau\, L'_{12} U_{12}^0(\tau) \\
\times \{[L'_{13} P_3(1\,|\,2\,|\,3) - P_2(1\,|\,2) L'_{13} P_2(1\,|\,3)] \\
+ [L'_{23} P_3(1\,|\,2\,|\,3) - P_2(1\,|\,2) L'_{23} P_2(2\,|\,3)] \\
\times U_{123}^0(-\tau) \bar{f}(\xi_1; t) \bar{f}(\xi_2; t) \bar{f}(\xi_3; t)
\end{aligned} \qquad (18.6.6)$$

The most striking difference with the classical equation is the stronger nonlinearity of the quantum equation: *Its collision term involves products of two and of three Wigner functions.* The physical meaning of this feature will be analyzed in the next section.

Consider now the *kinetic correlations.* Here again we find a serious difference with classical mechanics. Whereas the kinetic correlations in classical mechanics are of maximum order λ, that is, are due entirely to interactions, we have in quantum mechanics correlation patterns *of all kinds* that are *independent of* λ. These are the correlations resulting from the symmetrization of the correlation patterns $\bar{p}_s$; they are, therefore, specific quantum-statistical effects. Although they are formally included in the vacuum, they are physically correlation functions and must not be forgotten when averages are calculated. As an important example, we consider the two-particle, purely quantum-statistical correlation function that can be read from Eq. (3.8.7):

$$\begin{aligned}
\bar{g}_2^{[0]}(\xi_1, \xi_2) = \theta \bar{f}[\tfrac{1}{2}(\mathbf{k}_1 + \mathbf{k}_2) + \hbar^{-1}(\mathbf{p}_2 - \mathbf{p}_1), \tfrac{1}{2}(\mathbf{p}_1 + \mathbf{p}_2) + \tfrac{1}{4}\hbar(\mathbf{k}_2 - \mathbf{k}_1)] \\
\times \bar{f}[\tfrac{1}{2}(\mathbf{k}_1 + \mathbf{k}_2) + \hbar(\mathbf{p}_1 - \mathbf{p}_2), \tfrac{1}{2}(\mathbf{p}_1 + \mathbf{p}_2) + \tfrac{1}{4}\hbar(\mathbf{k}_1 - \mathbf{k}_2)]
\end{aligned} \qquad (18.6.7)$$

The higher correlation functions of this kind can be found systematically by the following procedure. One writes out the expression of the symmetrized vacuum components $P_s(1\,|\,2\,|\cdots|\,s)\bar{f}(1)\bar{f}(2)\cdots\bar{f}(s)$ by using Eq. (3.8.16) for the symmetrizer. The resulting sum of terms has the structure of the cluster representation (3.5.9)–(3.5.11); therefore, the various correlation patterns—and, in particular, the irreducible correlation functions—are easily identified. For instance, the three-particle symmetrized vacuum component is written in the following form, using

Eqs. (3.8.16) and (18.6.7) (we set $\hbar = 1$ for simplicity):

$$\begin{aligned}
&P_3(1\,|\,2\,|\,3)\bar{f}(\xi_1)\bar{f}(\xi_2)\bar{f}(\xi_3)\\
&\quad= \bar{f}(\xi_1)\bar{f}(\xi_2)\bar{f}(\xi_3) + \bar{f}(\xi_1)\bar{g}_2^{[0]}(\xi_2\xi_3) + \bar{f}(\xi_2)\bar{g}_2^{[0]}(\xi_1\xi_3) + \bar{f}(\xi_3)\bar{g}_2^{[0]}(\xi_1\xi_2)\\
&\qquad + \bar{f}[\tfrac{1}{2}(\mathbf{k}_1+\mathbf{k}_2)+(\mathbf{p}_2-\mathbf{p}_1), \tfrac{1}{2}(\mathbf{p}_1+\mathbf{p}_2)+\tfrac{1}{4}(\mathbf{k}_2-\mathbf{k}_1)]\\
&\qquad \times \bar{f}[\tfrac{1}{2}(\mathbf{k}_2+\mathbf{k}_3)+(\mathbf{p}_3-\mathbf{p}_2), \tfrac{1}{2}(\mathbf{p}_2+\mathbf{p}_3)+\tfrac{1}{4}(\mathbf{k}_3-\mathbf{k}_2)]\\
&\qquad \times \bar{f}[\tfrac{1}{2}(\mathbf{k}_3+\mathbf{k}_1)+(\mathbf{p}_1-\mathbf{p}_3), \tfrac{1}{2}(\mathbf{p}_3+\mathbf{p}_1)+\tfrac{1}{4}(\mathbf{k}_1-\mathbf{k}_3)]\\
&\qquad + \bar{f}[\tfrac{1}{2}(\mathbf{k}_1+\mathbf{k}_2)+(\mathbf{p}_1-\mathbf{p}_2), \tfrac{1}{2}(\mathbf{p}_1+\mathbf{p}_2)+\tfrac{1}{4}(\mathbf{k}_1-\mathbf{k}_2)]\\
&\qquad \times \bar{f}[\tfrac{1}{2}(\mathbf{k}_2+\mathbf{k}_3)+(\mathbf{p}_2-\mathbf{p}_3), \tfrac{1}{2}(\mathbf{p}_2+\mathbf{p}_3)+\tfrac{1}{4}(\mathbf{k}_2-\mathbf{k}_3)]\\
&\qquad \times \bar{f}[\tfrac{1}{2}(\mathbf{k}_3+\mathbf{k}_1)+(\mathbf{p}_3-\mathbf{p}_1), \tfrac{1}{2}(\mathbf{p}_3+\mathbf{p}_1)+\tfrac{1}{4}(\mathbf{k}_3-\mathbf{k}_1)]
\end{aligned} \tag{18.6.8}$$

Clearly, the sum of the last two terms is the quantum-statistical three-body correlation function $g_3^{[0]}(\xi_1\xi_2\xi_3)$.

Proceeding now to the contribution of first order in the interactions, we find again, as in Eq. (18.2.4)

$$C\bar{\mathfrak{f}}(t) = \lambda \int_0^\infty d\tau\, C\mathcal{U}^0(\tau)\mathcal{L}'\, V\mathcal{U}^0(-\tau)V\bar{\mathfrak{f}}(t) \tag{18.6.9}$$

This formal expression can be made explicit by going through the same arguments as in Section 18.2, duly modified as above. We thus find the contribution of the creation fragment to the two-particle correlation function, which we denote by $\bar{g}_2^{[1]}(\xi_1, \xi_2)$:

$$\begin{aligned}
\bar{g}_2^{[1]}(\xi_1\xi_2) &= \lambda \int_0^\infty d\tau\, U_{12}^0(\tau)L'_{12}U_{12}^0(-\tau)\bar{f}(\xi_1; t)\bar{f}(\xi_2; t)\\
&\quad + \lambda \int_0^\infty d\tau \int d\xi_3\, \delta(\mathbf{k}_3)U_{12}^0(\tau)\{[L'_{13}P_3(1\,|\,2\,|\,3) - P_2(1\,|\,2)L'_{13}P_2(1\,|\,3)]\\
&\quad + [L'_{23}P_3(1\,|\,2\,|\,3) - P_2(1\,|\,2)L'_{23}P_2(2\,|\,3)]\}\, U_{123}^0(-\tau)\\
&\quad \times \bar{f}(\xi_1; t)\bar{f}(\xi_2; t)\bar{f}(\xi_3; t)
\end{aligned} \tag{18.6.10}$$

We find again the same kind of quantum effects as in the kinetic equation (18.6.6). They will be analyzed in the next section.

In the classical case, the only kinetic correlation patterns $\bar{p}_s([\Gamma_s])$ of order λ are the patterns of the form $\bar{p}_s(12\,|\,3\,|\cdots|\,s)$ in which only two particles are correlated: all other patterns of $C\bar{\mathfrak{f}}$ are at least of order λ^2. This is no longer true in the quantum case, for the same reason as above. When the patterns are symmetrized, there appear quantum-statistical contributions of first order in λ to *all* the correlation functions. As an example, we consider again the three-particle correlation function. The symmetrization of the correlation pattern $\bar{p}_3(1\,|\,23)$ yields

$$\begin{aligned} P_3(1\,|\,23)\bar{f}(\xi_1)\bar{g}_2^{[1]}(\xi_2\xi_3) &= \bar{f}(\xi_1)\bar{g}_2^{[1]}(\xi_2\xi_3) + \bar{f}[\tfrac{1}{2}(\mathbf{k}_1+\mathbf{k}_2)+(\mathbf{p}_2-\mathbf{p}_1), \tfrac{1}{2}(\mathbf{p}_1+\mathbf{p}_2)+\tfrac{1}{4}(\mathbf{k}_2-\mathbf{k}_1)] \\ &\quad \times \bar{g}_2^{[1]}[\tfrac{1}{2}(\mathbf{k}_1+\mathbf{k}_2)+(\mathbf{p}_1-\mathbf{p}_2), \tfrac{1}{2}(\mathbf{p}_1+\mathbf{p}_2)+\tfrac{1}{4}(\mathbf{k}_1-\mathbf{k}_2); \mathbf{k}_3, \mathbf{p}_3] \\ &\quad + \bar{f}[\tfrac{1}{2}(\mathbf{k}_1+\mathbf{k}_3)+(\mathbf{p}_3-\mathbf{p}_1), \tfrac{1}{2}(\mathbf{p}_1+\mathbf{p}_3)+\tfrac{1}{4}(\mathbf{k}_3-\mathbf{k}_1)] \\ &\quad \times \bar{g}_2^{[1]}[\tfrac{1}{2}(\mathbf{k}_1+\mathbf{k}_3)+(\mathbf{p}_1-\mathbf{p}_3), \tfrac{1}{2}(\mathbf{p}_1+\mathbf{p}_3)+\tfrac{1}{4}(\mathbf{k}_1-\mathbf{k}_3); \mathbf{k}_2, \mathbf{p}_2] \end{aligned} \quad (18.6.11)$$

Obviously, the two last terms are contributions to the three-body correlation function $\bar{g}_3^{[1]}(123)$.

The nature of the kinetic correlations is, therefore, much more complex in quantum-statistical systems than in classical ones. The reason is in the double origin of the correlations: interactions and exclusion principle. For this reason, there are nonzero correlation patterns of all kinds, to all orders in the interaction strength. However, the structure of these patterns is well defined. They are functionals of the unsymmetrized patterns $\bar{f}_1$ and of $\bar{g}_2^{[1]}$, linear in the latter. But $\bar{g}_2^{[1]}$ is, in turn, a functional of $\bar{f}_1$. Hence, although the explicit form is complicated, we find again the important result that *all the kinetic correlation patterns are functionals of the kinetic one-particle Wigner function.* This is true even when quantum-statistical effects are exactly accounted for.

We also realize one more reason for taking the *unsymmetrized patterns* as a basis for our description of quantum systems. Indeed, there are as many nonvanishing unsymmetrized patterns, to any order in λ, as there are in classical mechanics. Hence the construction and classification of these patterns is much simpler.

On the other hand, when needed for explicit computation of averages, the complete, symmetrized patterns are easily obtained from the unsymmetrized ones.

18.7. THE SPATIALLY HOMOGENEOUS QUANTUM WEAKLY COUPLED GAS

We now consider again the homogeneous system as a particularly simple case in which the equations can be written quite explicitly. In this case, we have, from Eq. (3.7.3)

$$\bar{f}(\xi; t) = 8\pi^3 n\, \delta(\mathbf{k})\bar{\varphi}(\mathbf{p}; t) \quad (18.7.1)$$

Just as in the classical case, there is no free-flow term:

$$L_1^0\bar{f}(\xi_1; t) = 8\pi^3 n\, (\mathbf{k}_1 \cdot \mathbf{v}_1)\, \delta(\mathbf{k}_1)\bar{\varphi}(\mathbf{p}_1; t) \quad (18.7.2)$$

Similarly, the "Vlassov" term vanishes for homogeneous systems:

$$\int d\xi_2\, \delta(\mathbf{k}_2) L'_{12} P_2(1\,|\,2) \bar{f}(\xi_1) \bar{f}(\xi_2)$$

$$= \int d\mathbf{p}_2\, d\mathbf{k}_2 \int d\mathbf{l} \int d\mathbf{k}'_1\, d\mathbf{k}'_2\, \delta(\mathbf{k}_2) \tilde{V}_l$$

$$\times [\exp(-\tfrac{1}{2}\mathbf{l} \cdot \boldsymbol{\partial}_{12}) - \exp(\tfrac{1}{2}\mathbf{l} \cdot \boldsymbol{\partial}_{12})] \exp(-\mathbf{l} \cdot \boldsymbol{\delta}_{12})$$

$$\times [\delta(-\mathbf{k}'_1 + \mathbf{k}_1)\, \delta(-\mathbf{k}'_2 + \mathbf{k}_2) + \theta\delta(-\mathbf{k}'_1 + \mathbf{p}_2 - \mathbf{p}_1 + \tfrac{1}{2}\mathbf{k}_1 + \tfrac{1}{2}\mathbf{k}_2)$$

$$\times \delta(-\mathbf{k}'_2 + \mathbf{p}_1 - \mathbf{p}_2 + \tfrac{1}{2}\mathbf{k}_1 + \tfrac{1}{2}\mathbf{k}_2)]\, \delta(\mathbf{k}'_1)\, \delta(\mathbf{k}'_2) \bar{\varphi}(\mathbf{p}_1) \bar{\varphi}(\mathbf{p}_2)$$

$$= \delta(\mathbf{k}_1) \int d\mathbf{p}_2\, d\mathbf{l}\, \tilde{V}_l [\exp(-\tfrac{1}{2}\mathbf{l} \cdot \boldsymbol{\partial}_{12}) - \exp(\tfrac{1}{2}\mathbf{l} \cdot \boldsymbol{\partial}_{12})]$$

$$\times [\delta(\mathbf{l}) + \delta(\mathbf{p}_1 - \mathbf{p}_2)] \bar{\varphi}(\mathbf{p}_1) \bar{\varphi}(\mathbf{p}_2) = 0 \qquad (18.7.3)$$

We are left therefore with the collision term, just as in classical mechanics. To evaluate the latter, we first calculate the rather complicated purely quantum-statistical terms, that is, the matrix element $\langle 12|\, \mathscr{L}'\, |1\,|\,2\,|\,3\rangle$. It is most conveniently represented in the form of a matrix element between states with different wave vectors, that is, in the same representation as the symmetrizers; see Eq. (3.7.9).

Using Eq. (3.7.16) we find

$$\langle \mathbf{k}_1\mathbf{k}_2\mathbf{k}_3|\, (L'_{13} + L'_{23}) P_3(1\,|\,2\,|\,3) - P_2(1\,|\,2)[L'_{13} P_2(1\,|\,3) + L'_{23} P_2(2\,|\,3)] \,|\, \mathbf{k}'_1\mathbf{k}'_2\mathbf{k}'_3\rangle$$

$$= \int d\mathbf{l}\, \bar{V}_l \{[\exp(-\tfrac{1}{2}\mathbf{l} \cdot \boldsymbol{\partial}_{13}) - \exp(\tfrac{1}{2}\mathbf{l} \cdot \boldsymbol{\partial}_{13})]$$

$$\times \theta\delta(-\mathbf{k}'_1 + \mathbf{k}_1 - \mathbf{l})\, \delta(-\mathbf{k}'_2 + \mathbf{p}_3 - \mathbf{p}_2 + \tfrac{1}{2}\mathbf{l})\, \delta(-\mathbf{k}'_3 + \mathbf{p}_2 - \mathbf{p}_3 + \tfrac{1}{2}\mathbf{l})$$

$$+ \exp(-\tfrac{1}{2}\mathbf{l} \cdot \boldsymbol{\partial}_{13})\, \delta(-\mathbf{k}'_1 + \mathbf{p}_3 - \mathbf{p}_1 + \tfrac{1}{2}\mathbf{k}_1)$$

$$\times \delta(-\mathbf{k}'_2 + \mathbf{p}_1 - \mathbf{p}_2 + \tfrac{1}{2}\mathbf{k}_1 - \tfrac{1}{2}\mathbf{l})\, \delta(-\mathbf{k}'_3 + \mathbf{p}_2 - \mathbf{p}_3 + \tfrac{1}{2}\mathbf{l})$$

$$- \exp(\tfrac{1}{2}\mathbf{l} \cdot \boldsymbol{\partial}_{13})\, \delta(-\mathbf{k}'_1 + \mathbf{p}_2 - \mathbf{p}_1 + \tfrac{1}{2}\mathbf{k}_1 - \tfrac{1}{2}\mathbf{l})$$

$$\times \delta(-\mathbf{k}'_2 + \mathbf{p}_3 - \mathbf{p}_2 + \tfrac{1}{2}\mathbf{l})\, \delta(-\mathbf{k}'_3 + \mathbf{p}_1 - \mathbf{p}_3 + \tfrac{1}{2}\mathbf{k}_1)\}$$

$$\times \exp[\tfrac{1}{2}(\mathbf{k}'_1 - \mathbf{k}_1 + \mathbf{l}) \cdot \boldsymbol{\partial}_1 + (\mathbf{k}'_2 - \mathbf{k}_2) \cdot \boldsymbol{\partial}_2 + (\mathbf{k}'_3 - \mathbf{k}_3 - \mathbf{l}) \cdot \boldsymbol{\partial}_3] + [1 \leftrightarrow 2] \qquad (18.7.4)$$

The abreviation $(1 \leftrightarrow 2)$ denotes the same terms with the indices 1 and 2 permuted. The derivation of this result is rather lengthy, but straightforward. Some cancellations occurring in this derivation are not very obvious at first sight. One should note identities of the following kind:

$$\exp(-\tfrac{1}{2}\mathbf{l} \cdot \boldsymbol{\partial}_{13})\, \delta(-\mathbf{k}'_1 + \mathbf{p}_2 - \mathbf{p}_1 - \tfrac{3}{2}\mathbf{l} + \tfrac{1}{2}\mathbf{k}_1 + \tfrac{1}{2}\mathbf{k}_2)\, \delta(-\mathbf{k}'_2 + \mathbf{p}_1 - \mathbf{p}_2 + \tfrac{1}{2}\mathbf{l} + \tfrac{1}{2}\mathbf{k}_1 + \tfrac{1}{2}\mathbf{k}_2)$$

$$\equiv \exp(-\tfrac{1}{2}\mathbf{l} \cdot \boldsymbol{\partial}_{23})\, \delta(-\mathbf{k}'_1 + \mathbf{p}_2 - \mathbf{p}_1 - \tfrac{1}{2}\mathbf{l} + \tfrac{1}{2}\mathbf{k}_1 + \tfrac{1}{2}\mathbf{k}_2)\, \delta(-\mathbf{k}'_2 + \mathbf{p}_1 - \mathbf{p}_2 - \tfrac{1}{2}\mathbf{l} + \tfrac{1}{2}\mathbf{k}_1 + \tfrac{1}{2}\mathbf{k}_2)$$

We are now ready for the final calculation. Substituting the matrix elements taken from Eqs. (14.3.7), (14.3.17), and (18.7.4), and the symmetrizers taken from Eq. (3.8.16), we find

$$
\begin{aligned}
\partial_t \bar{f}(\mathbf{k}_1\mathbf{p}_1) = i^2 \int d\mathbf{p}_2 \int d\mathbf{l}\, d\mathbf{l}' \int d\mathbf{k}_1'\, d\mathbf{k}_2' \int_0^\infty d\tau \\
&\times \tilde{V}_{l'}[\exp(-\tfrac{1}{2}\mathbf{l}' \cdot \partial_{12}) - \exp(\tfrac{1}{2}\mathbf{l}' \cdot \partial_{12})] \\
&\times \exp\{-i[(\mathbf{k}_1 - \mathbf{l}') \cdot \mathbf{v}_1 + i\mathbf{l}' \cdot \mathbf{v}_2]\tau\} \tilde{V}_l \\
&\times \Big\{ [\exp(-\tfrac{1}{2}\mathbf{l} \cdot \partial_{12}) - \exp(\tfrac{1}{2}\mathbf{l} \cdot \partial_{12})][\delta(-\mathbf{k}_1' + \mathbf{k}_1 - \mathbf{l} - \mathbf{l}')\, \delta(-\mathbf{k}_2' + \mathbf{l} + \mathbf{l}') \\
&+ \theta\delta(-\mathbf{k}_1' + \mathbf{p}_2 - \mathbf{p}_1 + \tfrac{1}{2}\mathbf{k}_1)\, \delta(-\mathbf{k}_2' + \mathbf{p}_1 - \mathbf{p}_2 + \tfrac{1}{2}\mathbf{k}_1] \\
&\times \exp[\tfrac{1}{2}(\mathbf{k}_1' - \mathbf{k}_1 + \mathbf{l} + \mathbf{l}') \cdot \partial_1]\, \exp[\tfrac{1}{2}(\mathbf{k}_2' - \mathbf{l} - \mathbf{l}') \cdot \partial_2] \\
&\times \exp[i(\mathbf{k}_1' \cdot \mathbf{v}_1 + \mathbf{k}_2' \cdot \mathbf{v}_2)\tau] \bar{f}(\mathbf{k}_1', \mathbf{p}_1) \bar{f}(\mathbf{k}_2', \mathbf{p}_2) \\
&+ \int d\mathbf{p}_3 \int d\mathbf{k}_3' \{[\exp(-\tfrac{1}{2}\mathbf{l} \cdot \partial_{13}) - \exp(\tfrac{1}{2}\mathbf{l} \cdot \partial_{13})] \\
&\times \theta\delta(-\mathbf{k}_1' + \mathbf{k}_1 - \mathbf{l}' - \mathbf{l})\, \delta(-\mathbf{k}_2' + \mathbf{p}_3 - \mathbf{p}_2 + \tfrac{1}{2}\mathbf{l} + \tfrac{1}{2}\mathbf{l}') \\
&\times \delta(-\mathbf{k}_3' + \mathbf{p}_2 - \mathbf{p}_3 + \tfrac{1}{2}\mathbf{l} + \tfrac{1}{2}\mathbf{l}') \\
&+ \exp(-\tfrac{1}{2}\mathbf{l} \cdot \partial_{13})\, \delta(-\mathbf{k}_1' + \mathbf{p}_3 - \mathbf{p}_1 + \tfrac{1}{2}\mathbf{k}_1 - \tfrac{1}{2}\mathbf{l}') \\
&\times \delta(-\mathbf{k}_2' + \mathbf{p}_1 - \mathbf{p}_2 + \tfrac{1}{2}\mathbf{k}_1 - \tfrac{1}{2}\mathbf{l})\, \delta(-\mathbf{k}_3' + \mathbf{p}_2 - \mathbf{p}_3 + \tfrac{1}{2}\mathbf{l} + \tfrac{1}{2}\mathbf{l}') \\
&- \exp(\tfrac{1}{2}\mathbf{l} \cdot \partial_{13})\, \delta(-\mathbf{k}_1' + \mathbf{p}_2 - \mathbf{p}_1 + \tfrac{1}{2}\mathbf{k}_1 - \tfrac{1}{2}\mathbf{l}) \\
&\times \delta(-\mathbf{k}_2' + \mathbf{p}_3 - \mathbf{p}_2 + \tfrac{1}{2}\mathbf{l}' + \tfrac{1}{2}\mathbf{l})\, \delta(-\mathbf{k}_3' + \mathbf{p}_1 - \mathbf{p}_3 + \tfrac{1}{2}\mathbf{k}_1 - \tfrac{1}{2}\mathbf{l}')\} \\
&\times \exp[\tfrac{1}{2}(\mathbf{k}_1' - \mathbf{k}_1 + \mathbf{l} + \mathbf{l}') \cdot \partial_1 + \tfrac{1}{2}(\mathbf{k}_2' - \mathbf{l}') \cdot \partial_2 + \tfrac{1}{2}(\mathbf{k}_3' - \mathbf{l}) \cdot \partial_3] \\
&\times \exp[i(\mathbf{k}_1' \cdot \mathbf{v}_1 + \mathbf{k}_2' \cdot \mathbf{v}_2 + \mathbf{k}_3' \cdot \mathbf{v}_3)\tau] \bar{f}(\mathbf{k}_1', \mathbf{p}_1) \bar{f}(\mathbf{k}_2', \mathbf{p}_2) \bar{f}(\mathbf{k}_3', \mathbf{p}_3) \\
&+ \int d\mathbf{p}_3 \int d\mathbf{k}_3' \{[\exp(-\tfrac{1}{2}\mathbf{l} \cdot \partial_{23}) - \exp(\tfrac{1}{2}\mathbf{l} \cdot \partial_{23})] \\
&\times \theta\delta(-\mathbf{k}_1' + \mathbf{p}_3 - \mathbf{p}_1 + \tfrac{1}{2}\mathbf{k}_1 + \tfrac{1}{2}\mathbf{l} - \tfrac{1}{2}\mathbf{l}')\, \delta(-\mathbf{k}_2' + \mathbf{l}' - \mathbf{l}) \\
&\times \delta(-\mathbf{k}_3' + \mathbf{p}_1 - \mathbf{p}_3 + \tfrac{1}{2}\mathbf{k}_1 + \tfrac{1}{2}\mathbf{l} - \tfrac{1}{2}\mathbf{l}') \\
&+ \exp(-\tfrac{1}{2}\mathbf{l} \cdot \partial_{23})\, \delta(-\mathbf{k}_1' + \mathbf{p}_2 - \mathbf{p}_1 + \tfrac{1}{2}\mathbf{k}_1 - \tfrac{1}{2}\mathbf{l}) \\
&\times \delta(-\mathbf{k}_2' + \mathbf{p}_3 - \mathbf{p}_2 + \tfrac{1}{2}\mathbf{l}')\, \delta(-\mathbf{k}_3' + \mathbf{p}_1 - \mathbf{p}_3 + \tfrac{1}{2}\mathbf{k}_1 - \tfrac{1}{2}\mathbf{l}' + \tfrac{1}{2}\mathbf{l}) \\
&- \exp(\tfrac{1}{2}\mathbf{l} \cdot \partial_{23})\, \delta(-\mathbf{k}_1' + \mathbf{p}_3 - \mathbf{p}_1 + \tfrac{1}{2}\mathbf{k}_1 - \tfrac{1}{2}\mathbf{l}' + \tfrac{1}{2}\mathbf{l}) \\
&\times \delta(-\mathbf{k}_2' + \mathbf{p}_1 - \mathbf{p}_2 + \tfrac{1}{2}\mathbf{k}_1 - \tfrac{1}{2}\mathbf{l})\, \delta(-\mathbf{k}_3' + \mathbf{p}_2 - \mathbf{p}_3 + \tfrac{1}{2}\mathbf{l}')\} \\
&\times \exp[\tfrac{1}{2}(\mathbf{k}_1' - \mathbf{k}_1 + \mathbf{l}') \cdot \partial_1 + \tfrac{1}{2}(\mathbf{k}_2' - \mathbf{l}' + \mathbf{l}) \cdot \partial_2 + \tfrac{1}{2}(\mathbf{k}_3' - \mathbf{l}) \cdot \partial_3] \\
&\times \exp[i(\mathbf{k}_1' \cdot \mathbf{v}_1 + \mathbf{k}_2' \cdot \mathbf{v}_2 + \mathbf{k}_3' \cdot \mathbf{v}_3)\tau] \\
&\times \bar{f}(\mathbf{k}_1', \mathbf{p}_1) \bar{f}(\mathbf{k}_2', \mathbf{p}_2) \bar{f}(\mathbf{k}_3', \mathbf{p}_3) \Big\} \qquad (18.7.5)
\end{aligned}
$$

We now express the homogeneity of the system by substituting in both sides Eq. (18.7.1). By a somewhat tedious, but quite straightforward calculation, we perform the finite displacement operations, and then eliminate the δ functions by integrating successively over $\mathbf{k}_1'$, $\mathbf{k}_2'$, $\mathbf{k}_3'$, $\mathbf{l}$, $\mathbf{p}_3$. Using also a few obvious symmetry properties, we finally obtain

$$\begin{aligned}\partial_t\bar{\varphi}(\mathbf{p}_1) &= (i\hbar)^{-2}8\pi^3 n\int d\mathbf{p}_2\, d\mathbf{l}\int_{-\infty}^{\infty} d\tau \\ &\times \exp\left\{i\left[\mathbf{l}\cdot(\mathbf{v}_1-\mathbf{v}_2)+\frac{\hbar l^2}{m}\right]\tau\right\}\tilde{V}_l[\tilde{V}_l+\theta\tilde{V}_{\mathbf{l}+(\mathbf{p}_1-\mathbf{p}_2)/\hbar}] \\ &\times\{\bar{\varphi}(\mathbf{p}_1)\bar{\varphi}(\mathbf{p}_2)[1+8\pi^3\hbar^3 n\theta\bar{\varphi}(\mathbf{p}_1+\hbar\mathbf{l})+8\pi^3\hbar^3 n\theta\bar{\varphi}(\mathbf{p}_2-\hbar\mathbf{l})] \\ &-\bar{\varphi}(\mathbf{p}_1+\hbar\mathbf{l})\bar{\varphi}(\mathbf{p}_2-\hbar\mathbf{l})[1+8\pi^3\hbar^3 n\theta\bar{\varphi}(\mathbf{p}_1)+8\pi^3\hbar^3 n\theta\bar{\varphi}(\mathbf{p}_2)]\}\end{aligned} \quad (18.7.6)$$

(We have now restored the $\hbar$ factors in their correct position.) The time integration simply gives a δ function. The equation takes a more suggestive form when we add and subtract the term* $h^6 n^2\bar{\varphi}(\mathbf{p}_1)\bar{\varphi}(\mathbf{p}_2)\bar{\varphi}(\mathbf{p}_1+\hbar\mathbf{l})\bar{\varphi}(\mathbf{p}_2-\hbar\mathbf{l})$:

$$\begin{aligned}\partial_t\bar{\varphi}(\mathbf{p}_1) &= 16\pi^4 n\hbar^{-2}\int d\mathbf{p}_2\int d\mathbf{l} \\ &\times\delta\left[\mathbf{l}\cdot(\mathbf{v}_1-\mathbf{v}_2)+\frac{\hbar l^2}{m}\right][\tilde{V}_l+\theta\tilde{V}_{\mathbf{l}+(\mathbf{p}_1-\mathbf{p}_2)/\hbar}]^2 \\ &\times\{-\bar{\varphi}(\mathbf{p}_1)\bar{\varphi}(\mathbf{p}_2)[1+h^3 n\theta\bar{\varphi}(\mathbf{p}_1+\hbar\mathbf{l})][1+h^3 n\theta\bar{\varphi}(\mathbf{p}_2-\hbar\mathbf{l})] \\ &+\bar{\varphi}(\mathbf{p}_1+\hbar\mathbf{l})\bar{\varphi}(\mathbf{p}_2-\hbar\mathbf{l})[1+h^3 n\theta\bar{\varphi}(\mathbf{p}_1)][1+h^3 n\theta\bar{\varphi}(\mathbf{p}_2)]\}\end{aligned} \quad (18.7.7)$$

18.8. PROPERTIES OF THE QUANTUM KINETIC EQUATION

The quantum kinetic equation for a weakly coupled gas is very similar to the classical Boltzmann equation. We can write it in the following form:

$$\begin{aligned}\partial_t\bar{\varphi}(\mathbf{p}_1) &= n\int d\mathbf{p}_2\, d\mathbf{l}\, W(\mathbf{p}_1+\tfrac{1}{2}\hbar\mathbf{l}, \mathbf{p}_2-\tfrac{1}{2}\hbar\mathbf{l}; \mathbf{l}) \\ &\times\{\bar{\varphi}(\mathbf{p}_1+\hbar\mathbf{l})\bar{\varphi}(\mathbf{p}_2-\hbar\mathbf{l})\bar{\psi}(\mathbf{p}_1)\bar{\psi}(\mathbf{p}_2)-\bar{\varphi}(\mathbf{p}_1)\bar{\varphi}(\mathbf{p}_2)\bar{\psi}(\mathbf{p}_1+\hbar\mathbf{l})\bar{\psi}(\mathbf{p}_2-\hbar\mathbf{l})\}\end{aligned} \quad (18.8.1)$$

where we used the notation, already introduced in Eq. (7.3.18):

$$\bar{\psi}(\mathbf{p}) = 1+\theta h^3 n\bar{\varphi}(\mathbf{p})$$

The definition of W is easily obtained by comparing Eqs. (18.8.1) and (18.7.7):

$$\begin{aligned}&W(\mathbf{p}_1+\tfrac{1}{2}\hbar\mathbf{l}, \mathbf{p}_2-\tfrac{1}{2}\hbar\mathbf{l}; \mathbf{l}) \\ &\quad\equiv 16\pi^4 n\hbar^{-2}[\tilde{V}_l+\theta\tilde{V}_{\mathbf{l}+(\mathbf{p}_1-\mathbf{p}_2)/\hbar}]^2\,\delta\left[\mathbf{l}\cdot(\mathbf{p}_1-\mathbf{p}_2)+\frac{\hbar l^2}{m}\right]\end{aligned} \quad (18.8.2)$$

* Recall that $8\pi^3\hbar^3 = h^3$.

The kinetic equation is now very similar to Eq. (11.6.5).

The first noteworthy remark is the following. In a classical theory, the weak-coupling limit of the kinetic equation is a differential equation in $\mathbf{p}$. This feature comes from the fact that when the interactions are weak, the deflections of the two particles in a collision are very small. Landau's derivation of his equation from the Boltzmann equation was precisely based on this idea, as was shown in Section 11.6. *In a quantum system, the equivalence between the weak-coupling limit and the small-deflection limit is no longer true.* In quantum mechanics, even a weak interaction potential can produce a very strong momentum transfer: this is a consequence of Heisenberg's uncertainty principle. The quantum analog of the full Boltzmann equation is exactly of the same form as (18.8.1): it is known as the *Uehling–Uhlenbeck equation.* The only difference with (18.8.1) is that the function W is now related to the exact cross section for the elastic collisions of the given intermolecular potential. The cross section (18.8.2) corresponds to the first nonzero approximation to the exact cross section, that is, the first Born approximation.

In both the Uehling–Uhlenbeck equation and in its weak-coupling limit, the cross section consists of two terms, corresponding to "direct" and "exchange" collisions. The latter is the term proportional to θ in Eq. (18.8.2). Its presence is an effect of quantum statistics and has no classical analog. As the molecules are indistinguishable, in the calculation of the collision cross section the situation where the outcoming particles are interchanged must be added to the direct process (see Fig. 18.8.1).

But the most spectacular effect of quantum statistics appears in the bracketed function of Eq. (18.8.1), through the presence of the factors $\bar{\psi}$. Let us analyze their role first in Fermi statistics.

Equation (18.8.1) is a gain–loss equation, just as the classical Boltzmann equation. The collision term expresses the rate of change of $\bar{\varphi}(\mathbf{p}_1)$ as a result of two processes: a loss due to collisions of particles of momentum $\mathbf{p}_1$ and $\mathbf{p}_2$ resulting in final momenta $\mathbf{p}_1 + \hbar\mathbf{l}$, $\mathbf{p}_2 - \hbar\mathbf{l}$; and a gain due to the restituting collision, in which the final states are $\mathbf{p}_1$ and $\mathbf{p}_2$. Note that the argument of the δ function in the cross section (18.8.2) of this

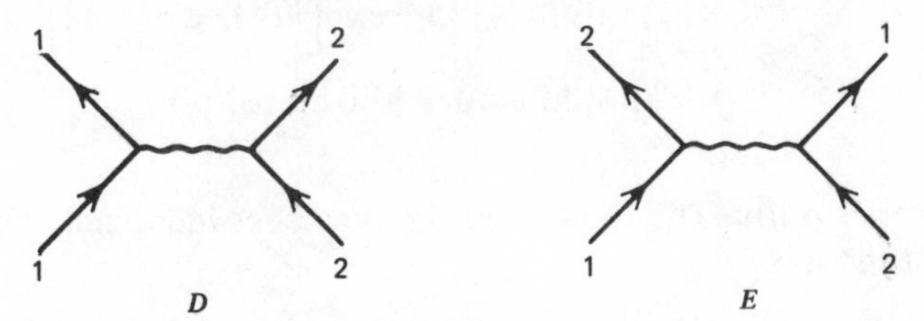

Figure 18.8.1. Feynman diagrams for the direct and for the exchange collisions.

process is just the difference in energies between the initial and the final states. Therefore the collision is possible only if the energy is conserved. The number of collisions of the first class is proportional to the probability of occurence of the momenta $\mathbf{p}_1$ and $\mathbf{p}_2$, that is, $\bar{\varphi}(\mathbf{p}_1)\bar{\varphi}(\mathbf{p}_2)$ (just as in the classical case) but *also* to the probability that the final states $\mathbf{p}_1+\hbar\mathbf{l}$, $\mathbf{p}_2-\hbar\mathbf{l}$ are *empty*, that is, $[1-h^3n\bar{\varphi}(\mathbf{p}_1+\hbar\mathbf{l})][1-h^3n\bar{\varphi}(\mathbf{p}_2-\hbar\mathbf{l})]$. The same argument applies to the restituting collision. This effect is clearly an expression of the Pauli exclusion principle, which dominates the Fermi–Dirac statistics. Thus, the probability of a collision is, in general, smaller in Fermi systems than in the corresponding classical system. A similar discussion leads to the conclusion that the collision probability is correspondingly enhanced in Bose systems.

As a result of this effect, the kinetic equation for a quantum-statistical system is more highly nonlinear than the corresponding classical equation. It involves terms that are cubic in the distribution function.*

These terms, which appear to be absolutely necessary for a consistent quantum-statistical theory, have been introduced in our formalism by the characteristic quantum-statistical matrix elements of $\mathscr{L}'$, as can be seen in Eqs. (18.6.4), (18.6.6). This simple illustration shows the important role played by these new terms. It also shows strikingly how the two quite different equations (18.7.7) and (18.3.11) resulted from the *same* abstract equation (18.6.1), by applying the appropriate rules of translation. This is a unique advantage of the correlation pattern formalism.

We now briefly show that Eq. (18.7.7) possesses the necessary properties of a kinetic equation. We first show that the Fermi or the Bose equilibrium distributions $\varphi_\theta^0(\mathbf{p})$, Eqs. (7.3.15), (7.3.16):

$$\varphi_\theta^0(\mathbf{p})=(h^3n)^{-1}\left\{\exp\left[\beta\left(\frac{p^2}{2m}-\mu\right)\right]-\theta\right\}^{-1} \tag{18.8.3}$$

are stationary solutions of Eq. (18.7.7). We substitute this functional form in the right side of the equation and use the identity (7.3.20):

$$\begin{aligned}\partial_t\varphi_\theta^0(\mathbf{p}_1) &= 16\pi^4 n\hbar^{-2}\int d\mathbf{p}_2\int d\mathbf{l}\;\delta\left[\mathbf{l}\cdot\mathbf{g}+\frac{\hbar l^2}{m}\right]\\ &\quad\times[\tilde{V}_{\mathbf{l}}+\theta\tilde{V}_{\mathbf{l}+m\mathbf{g}/\hbar}]^2\left\{1-\exp\left[\hbar\beta\left(\mathbf{l}\cdot\mathbf{g}+\frac{\hbar l^2}{m}\right)\right]\right\}\\ &\quad\times\varphi_\theta^0(\mathbf{p}_1+\hbar\mathbf{l})\varphi_\theta^0(\mathbf{p}_2-\hbar\mathbf{l})\psi_\theta^0(\mathbf{p}_1)\psi_\theta^0(\mathbf{p}_2)\\ &=0\end{aligned} \tag{18.8.4}$$

where $\mathbf{g}=(\mathbf{p}_1-\mathbf{p}_2)/m$. The right-hand side vanishes identically because of the delta function.

* The quartic terms in (18.7.7) are spurious because they cancel each other.

We now show that an H theorem exists for the quantum kinetic equation. We define the entropy by an extension of the equilibrium formula (7.3.27):

$$s(t) = -k_B n \int d\mathbf{p}_1 \, \{\bar{\varphi}(\mathbf{p}_1) \ln[h^3 n \bar{\varphi}(\mathbf{p}_1)] + (h^3 n)^{-1} \bar{\psi}(\mathbf{p}_1) \ln \bar{\psi}(\mathbf{p}_1)\} \quad (18.8.5)$$

The rate of change of this function is

$$\partial_t s(t) = -k_B n \int d\mathbf{p}_1 \, \{(\ln h^3 n\bar{\varphi} + 1)\, \partial_t \bar{\varphi} + (h^3 n)^{-1} (\ln \bar{\psi} + 1)\, \partial_t \bar{\psi}\}$$

$$= -k_B n \int d\mathbf{p}_1 \, \{\ln h^3 n \bar{\varphi}(\mathbf{p}_1) - \ln \bar{\psi}(\mathbf{p}_1)\}\, \partial_t \bar{\varphi}(\mathbf{p}_1)$$

The calculation is completely similar to Eq. (12.2.14) and will therefore not be repeated here. The result is

$$\begin{aligned} \partial_t s(t) = & -16\pi^4 n^2 \hbar^{-2} k_B \int d\mathbf{p}_1 \, d\mathbf{p}_2 \int d\mathbf{l} \, [\tilde{V}_l + \theta \tilde{V}_{l+m\mathbf{g}/\hbar}]^2 \\ & \times \delta\left(\mathbf{l}\cdot\mathbf{g} + \frac{\hbar l^2}{m}\right) \ln \frac{\bar{\varphi}(\mathbf{p}_1)\bar{\varphi}(\mathbf{p}_2)\bar{\psi}(\mathbf{p}_1 + \hbar\mathbf{l})\bar{\psi}(\mathbf{p}_1 - \hbar\mathbf{l})}{\bar{\varphi}(\mathbf{p}_1 + \hbar\mathbf{l})\bar{\varphi}(\mathbf{p}_2 - \hbar\mathbf{l})\bar{\psi}(\mathbf{p}_1)\bar{\psi}(\mathbf{p}_2)} \\ & \times \{-\bar{\varphi}(\mathbf{p}_1)\bar{\varphi}(\mathbf{p}_2)\bar{\psi}(\mathbf{p}_1 + \hbar\mathbf{l})\bar{\psi}(\mathbf{p}_2 - \hbar\mathbf{l}) \\ & + \bar{\varphi}(\mathbf{p}_1 + \hbar\mathbf{l})\bar{\varphi}(\mathbf{p}_2 - \hbar\mathbf{l})\bar{\psi}(\mathbf{p}_1)\bar{\psi}(\mathbf{p}_2)\} \end{aligned} \quad (18.8.6)$$

The right-hand side is obviously semidefinite positive, because of the identity (12.2.15).

Equation (18.7.7) is thus a proper kinetic equation having the same general features as the classical equations discussed in Chapters 12–14.

BIBLIOGRAPHICAL NOTES

The weakly coupled gas has been treated by many authors, because of its simplicity. Most of the references from the Brussels school quoted in Chapter 17 treat this problem in more or less detail. An extensive discussion is given in the author's book on the *Statistical Mechanics of Charged Particles* quoted there.

The Uehling–Uhlenbeck equation was first derived by E. A. Uehling and G. E. Uhlenbeck, *Phys. Rev.* **43,** 552 (1933).

CHAPTER 19

TRANSLATION OF THE GENERAL FORMALISM INTO EXPLICIT EQUATIONS

19.1. REDUCTION OF THE KINETIC EQUATION

In Chapter 17 we developed a general formalism for the time evolution of a many-body system. We achieved a considerable degree of compactness in this formalism by using abstract notations as $\mathfrak{f}$, V, $\mathscr{L}$, $\mathscr{G}$, and the like. In real problems, we must be able to translate these abstract symbols into kinetic equations, or expressions for the pair-correlation function, and so on. The idea of this translation process has been illustrated in Chapter 18 in the simplest possible case of a weakly coupled gas. For simplicity, we shall discuss here only the kinetic component $\bar{\mathfrak{f}}$ of the distribution function, but a similar discussion can be extended to the nonkinetic part $\hat{\mathfrak{f}}$.

Consider first the *kinetic equation*, that is, the equation of evolution for the vacuum component $V\bar{\mathfrak{f}}$. By the definition of the vacuum, the s-particle component of this vector is the correlation pattern $\bar{p}_s(1\,|\,2|\cdots|s)$. At this point we make the assumption

$$\bar{p}_s(1\,|\,2|\cdots|s)=\prod_{j=1}^{s}\bar{f}(j) \qquad (19.1.1)$$

where $\bar{f}(j)$ is the kinetic component of the one-particle reduced distribution function. The justification of this important assumption is twofold. First, *one can show that the factorization of the vacuum component is compatible with the kinetic equation.* In other words, if we consider the infinite set of equations for the patterns $\bar{p}_s(1\,|\,2|\cdots|s)$, there exists a solution of the form (19.1.1). We have shown this in Section 18.1 for the weak-coupling approximation; the proof can be extended to the general case. Hence, if Eq. (19.1.1) holds at $t=0$ it remains true at all later times.

Furthermore, one can prove a much stronger property. We know from Chapter 14 that the *exact* equations for the (complete) correlation patterns are, by their very construction, compatible with a solution factorized in the cluster fashion. This result was expressed by Eq.

(14.2.12), which we rewrite here:

$$p_s(x_1, \ldots, x_s; [\Gamma_s]) = \pi_s(x_1, \ldots, x_s; [\Gamma_s]) \qquad (19.1.2)$$

We now consider the kinetic part of the distribution vector, $\bar{\mathfrak{f}} = \Pi \mathfrak{f}$, and resolve it into component patterns:

$$\bar{p}_s(x_1, \ldots, x_s; [\Gamma_s])$$
$$= \sum_{r=0}^{\infty} \sum_{\Gamma_r} \langle x_1, \ldots, x_s; [\Gamma_s] | \Pi | x_1, \ldots, x_r; [\Gamma_r] \rangle \, p_r(x_1, \ldots, x_r; [\Gamma_r]) \quad (19.1.3)$$

We now state that, if the complete correlation patterns $p_s([\Gamma_s])$ are of the cluster type (19.1.2), the kinetic components $\bar{p}_s([\Gamma_s])$ are also factorized in the cluster fashion as products of the kinetic parts of the irreducible correlation functions. This property is expressed as follows:

$$\sum_{r=0}^{\infty} \sum_{\Gamma_r} \langle x_1, \ldots, x_s; [\Gamma_s] | \Pi | x_1, \ldots, x_r; [\Gamma_r] \rangle \, \pi_r(x_1, \ldots, x_r; [\Gamma_r])$$
$$= \bar{\pi}_s(x_1, \ldots, x_s; [\Gamma_s]) \quad (19.1.4)$$

The "kinetic cluster patterns" are defined in a pretty obvious way, as is clear from the following examples:

$$\bar{\pi}_s(x_1 \,|\, x_2 | \cdots | x_s) = \prod_{j=1}^{s} \bar{f}(x_j)$$

$$\bar{\pi}_s(x_1 x_2 \,|\, x_3 | \cdots | x_s) = \bar{g}_2(x_1 x_2) \prod_{j=3}^{s} \bar{f}(x_j)$$

$$\bar{\pi}_s(x_1 | \, x_2 x_3 \, | x_4 x_5 x_6 x_7) = \bar{f}(x_1) \bar{g}_2(x_2 x_3) \bar{g}_4(x_4 x_5 x_6 x_7)$$

and so on.

The theorem (19.1.4) is by no means trivial: we must refer the reader to the original paper for its proof. We stress its importance. It shows clearly that *Boltzmann's Stosszahlansatz is an exact property of the dynamics of evolution in the* Π *subspace.* Indeed, if Eq. (19.1.1) is true, then all but the first (one-particle) components of the equation for $V\bar{\mathfrak{f}}$ are redundant: they merely reproduce the equation for $\bar{f}$. Hence, in order to determine $V\bar{\mathfrak{f}}$ we are no longer compelled to think in terms of an infinite set of s-body distribution functions. It is sufficient to determine the one-body kinetic distribution function $\bar{f}$. But this function obeys a *closed* equation, which is obtained from Eq. (17.3.17) by projection on the one-particle state and use of Eq. (19.1.1):

$$\partial_t \bar{f}(x_1) = \sum_{s=1}^{\infty} \langle x_1 | \Gamma | x_1 \,|\, x_2 \,|\cdots|\, x_s \rangle \bar{f}(x_1) \bar{f}(x_2) \cdots \bar{f}(x_s) \quad (19.1.5)$$

We have achieved a formidable reduction of the description of the many-body problem. The price to be paid is in the nonlinearity of the kinetic equation.

If Eq. (19.1.5) is solved, its solution determines all the *correlation functions.* Here too, the theorem (19.1.4) introduces a serious simplification. We do not have to determine separately all the possible correlation patterns with s particles. It suffices to calculate the irreducible correlation function $\bar{g}_s(x_1, \ldots, x_s)$: all other s-particle patterns are obtained as products of r-particle patterns, with $r < s$. The s-particle kinetic correlation function is obtained by the action of the appropriate creation fragment on a product of one-particle functions:

$$\bar{g}_s(x_1, \ldots, x_s) = \sum_{n=s}^{\infty} \langle x_1, \ldots, x_s | \mathbf{C} | x_1 | x_2 | \cdots | x_n \rangle \bar{f}(x_1) \bar{f}(x_2) \cdots \bar{f}(x_n) \quad (19.1.6)$$

In order to measure the importance of the reduction achieved here, let us briefly discuss the nonkinetic part of the reduced distributions. Consider, for example, the vacuum pattern $\hat{p}_2(x_1 | x_2)$:

$$\begin{aligned} \hat{p}_2(x_1 | x_2) &= p_2(x_1 | x_2) - \bar{p}_2(x_1 | x_2) \\ &= f_1(x_1) f_1(x_2) - \bar{f}(x_1) \bar{f}(x_2) \end{aligned}$$

$\hat{p}_2(x_1 | x_2)$ is a difference of two products: it therefore cannot be of the form of a product $\hat{f}(x_1)\hat{f}(x_2)$. Hence, on one hand, $\hat{f}(x_1)$ does not obey a closed equation; on the other hand, even if we had determined $\hat{f}(x_1)$, its knowledge does not help us in calculating the other vacuum components: each $\hat{p}_s(x_1 | x_2 | \cdots | x_s)$ has to be determined separately. The fundamental equation in $\hat{\mathbf{\Pi}}$ space is an equation for the correlations. However, when this equation is resolved into components it yields an infinite hierarchy, which cannot be reduced by a theorem of the type (19.1.4). Thus, in spite of the formal similarity of the two equations (17.5.4), their explicit content is quite different. The structure of the time-evolution process in the $\hat{\mathbf{\Pi}}$ space is by orders of magnitude more complex than in the $\mathbf{\Pi}$ space. These properties were illustrated in Section 18.5 for the weakly coupled gas.

19.2. THE DIAGRAM REPRESENTATION OF THE IRREDUCIBLE EVOLUTION OPERATOR

We have now reduced the dynamics in $\mathbf{\Pi}$ space to Eqs. (19.1.5), (19.1.6). In order to write out these equations explicitly, we must find rules for calculating the matrix elements $\langle x_1 | \mathbf{\Gamma} | x_1 | x_2 | \cdots | x_s \rangle$ and $\langle x_1 \cdots x_s | \mathbf{C} | x_1 | x_2 | \cdots | x_n \rangle$. This calculation is based on the general formulas (17.2.13), (17.2.14), and (17.3.3), (17.3.4). The building stones of these formulas are the operators $V\tilde{\mathcal{U}}(\tau)V$, $V\mathcal{G}(\tau)V$, and $C\mathcal{C}(\tau)V$: the operators $\mathbf{\Gamma}$ and $\mathbf{C}$ are well-defined integrals of products of these operators. Among them, the Vlassov propagator $V\tilde{\mathcal{U}}(\tau)V$ is relatively simple: it will be discussed later. The operators $V\mathcal{G}(\tau)V$ and $C\mathcal{C}(\tau)V$ contain the truly nontrivial part; they are both related through Eqs.

(17.1.3) and (17.1.7) to the irreducible evolution operator $\mathscr{E}(\tau)$, which, by Eq. (16.3.5), is given as

$$\mathscr{E}(\tau)=\sum_{m=0}^{\infty}(2\pi)^{-1}\int dz\, e^{-iz\tau}\mathscr{L}'[C\mathscr{R}^0(z)\mathscr{L}']^m \tag{19.2.1}$$

The operator $\mathscr{E}(\tau)$ can also be defined as the solution of the integral equation:

$$\mathscr{E}(\tau)=\mathscr{L}'\delta(\tau)+\int_0^\tau d\tau'\ \mathscr{L}'C\mathscr{U}^0(\tau')\mathscr{E}(\tau-\tau') \tag{19.2.2}$$

From Eqs. (17.2.14) and (17.3.4) we see that the evaluation of the matrix element of **Γ** and of **C** requires the determination of the following matrix elements of $\mathscr{E}$:

$$\langle x_1 \mid x_2 \mid \cdots \mid x_s \mid \mathscr{E}(\tau)\, | x_1 \mid x_2 \mid \cdots \mid x_{s+r}\rangle, \qquad s=1,2,\ldots, \quad r=1,2,\ldots \tag{19.2.3}$$

$$\langle x_1 x_2 \cdots x_s \mid \mathscr{E}(\tau)\, | x_1 \mid x_2 \mid \cdots \mid x_{s+r}\rangle, \qquad s=2,3,\ldots, \quad r=0,1,2,\ldots \tag{19.2.4}$$

(The ranges of values of s and r given here will be presently discussed.)

Any matrix element of $\mathscr{E}(\tau)$ is constructed, according to Eq. (19.2.1) as a matrix product of operators $\mathscr{L}'$, $\mathscr{U}^0$ (or $\mathscr{R}^0$), C in a definite order. Among these, the operators C and $\mathscr{U}^0$ are simple, because they are diagonal. The factors $\mathscr{L}'$, on the other hand, are nondiagonal and describe transitions from one correlation pattern to another, according to definite selection rules that were discussed in Sections 14.2 and 14.3. In those sections the possible individual transitions were studied. Now we are confronted with a global problem, which can be stated as follows. In order to construct an mth-order approximation to the matrix elements (19.2.3), (19.2.4) according to (19.2.1), we have *to go from an* $(s+r)$*-particle vacuum state* (*at right*) *to an uncorrelated—or to a fully correlated—*s*-particle state* (*at left*) *in* m *steps, admitting only correlated intermediate steps.* In general (when the transition is possible at all), there are many possible paths from the initial to the final state. We are thus faced with a topological problem. Just as in equilibrium (Chapter 6), but for different reasons, the type of connectivity of the path will play the main role in the argument. The simplest way to disentangle this complex problem is again to use a diagram technique, which starts from the diagrams introduced in Sections 14.2 and 14.3.

The particular path (i.e., the succession of intermediate states) in a given contribution to the matrix elements of $\mathscr{E}(\tau)$ depends only on the successive factors $\mathscr{L}'$, not on $\mathscr{U}^0$ or C, which are diagonal. It is therefore sufficient to represent graphically the former. Every mth-order contribution to a matrix element of type (19.2.3), (19.2.4) is represented by a graph

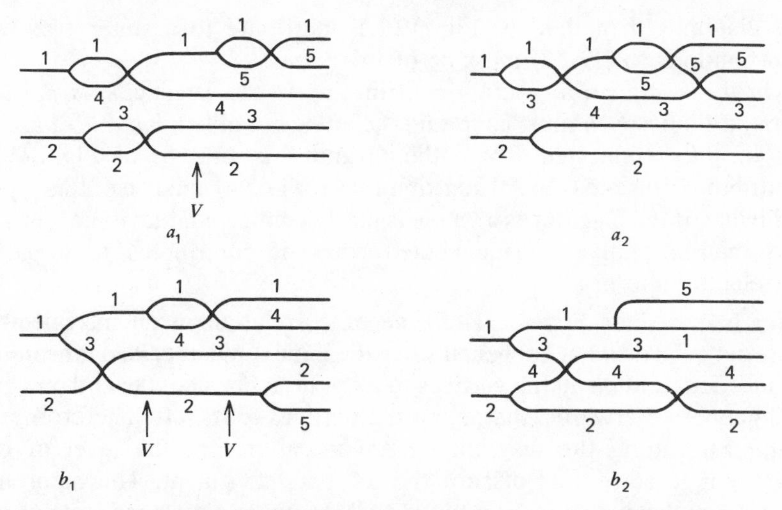

Figure 19.2.1. Diagram representation of matrix elements of operators

a_1, a_2: contributions to $\langle 1\,|\,2|\,\mathcal{L}'\mathcal{U}^0 \cdots \mathcal{U}^0\mathcal{L}'\,|1\,|\,2\,|\,3\,|\,4\,|\,5\rangle$

b_1, b_2: contributions to $\langle 12|\,\mathcal{L}'\mathcal{U}^0 \cdots \mathcal{U}^0\mathcal{L}'\,|1\,|\,2\,|\,3\,|\,4\,|\,5\rangle$

Diagrams a_1 and b_1 do not represent contributions to $\mathscr{E}(\tau)$, because they contain intermediate vacuum states.

with s lines at left and $s+r$ lines at right, connected by a succession of m vertices. In the *classical* case, which we discuss first, the vertices are of the two types, A or B, introduced in Table 3.3.1. A few examples of such diagrams are given in Fig. 19.2.1.

The picture, however, is not yet complete as long as we have no rules for recognizing vacuum and correlated states of different kinds in a diagram. These rules are direct extensions of the rules given for one-vertex diagrams in Section 14.1. The problem looks different according to whether the state occurs at right, at left, or in an intermediate position.

(a) "Right-Hand States" ("Kets"). The particular correlation pattern corresponding to the right-hand state is materialized in the diagram by means of connections between the lines representing the particles, as shown in Fig. 14.1.1. Hence, the diagrams contributing to the matrix elements (19.2.3), (19.2.4) must all end at right with $s+r$ unconnected lines.

(b) "Left-Hand States." By a straightforward generalization of the argument illustrated in Fig. 14.1.3, it is concluded that a diagram contributes to the s-*particle vacuum state* if it has *s lines coming out to the left and if, as a whole, it consists of exactly s mutually disconnected components.* We call such a diagram a *fully disconnected* diagram. Thus, the diagrams a_1 and a_2 of Fig. 19.2.1 contribute to a state $\langle 1 | 2 |$. All contributions to (19.2.3) must be of this type.

On the contrary, a diagram contributing to the *irreducible s-particle correlation function* must have *s lines coming out at left, but must be, as a whole, fully connected.* Thus, the diagrams b_1 and b_2 of Fig. 19.2.1 contribute to a state $\langle 12 |$. All contributions to (19.2.4) must be of this type.

Finally, if the diagram as a whole is partly connected, that is, consists of less than s mutually disconnected parts, it contributes to a partly correlated pattern.

(c) Intermediate States. The type of correlation of an intermediate state depends on what happened since the initial time, that is, on the initial connections and on all the vertices to its right; it does not depend on what will happen afterwards, that is, on the vertices to its left. Therefore, we begin by cutting the diagram by a vertical line at the level of the intermediate state, and discard the left-hand fragment. The nature of the intermediate state is determined as if it were the left-hand state of the remaining fragment, that is by Rule (b) (see Fig. 19.2.2). In particular, an intermediate state with $s+n$ particles is a vacuum state if the remaining fragment consists of exactly $s+n$ mutually disconnected components; it is a correlated state if it consists of less than $s+n$ components. With a little practice, the reader will soon recognize an intermediate vacuum

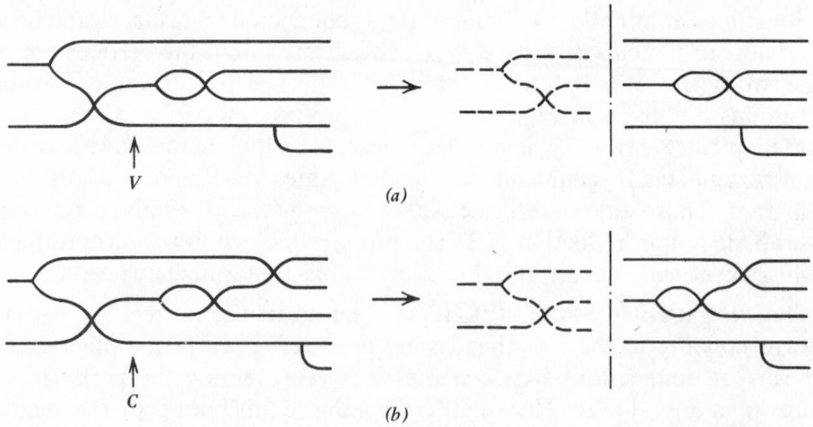

Figure 19.2.2. Identification of intermediate vacuum states in a diagram. (*a*) The state shown by an arrow is a vacuum state, because after discarding the part to its left, there remains a fully disconnected diagram with three lines at left. (*b*) The state shown by an arrow is a correlated state, because after the same operation, there remains a diagram with three lines at left, consisting of two (<3) disconnected fragments.

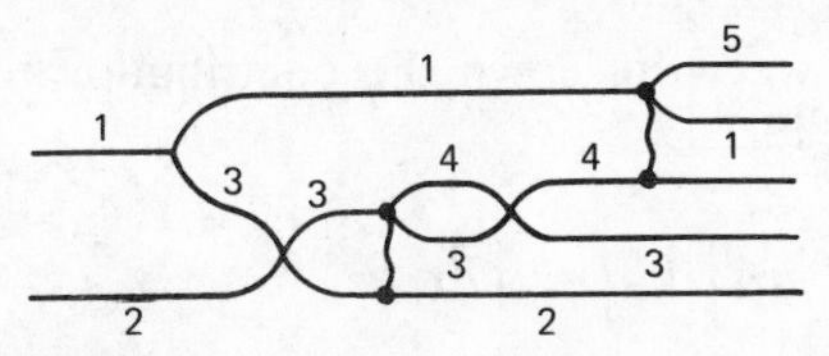

Figure 19.2.3. A contribution to $\langle 12|\mathscr{E}(\tau)|1|2|3|4|5\rangle$ in the quantum case. Note that the classical diagram obtained by suppressing the quantum-statistical links would not be admissible, because the first intermediate state from the right would be a vacuum state.

state at a first glance. This is important, because the constraint expressed by the projectors C in Eq. (19.2.1) means that the contributions to $\mathscr{E}(\tau)$ can contain no intermediate vacuum states.

As a result of this discussion. we may now state the following general rule:

Every mth-order contribution to the matrix element (19.2.3) [(19.2.4)] *is represented by a fully disconnected* [*fully connected*] *m-vertex diagram with s lines at left and* $s+r$ *lines at right, containing no intermediate vacuum state.*

The whole previous discussion is valid as it stands for the *quantum case* as well. The only difference is the possibility of inserting quantum-statistical vertices in addition to the vertices A and B in the diagrams, according to the rules of Section 14.3. We therefore have a larger variety of diagrams contributing to $\mathscr{E}(\tau)$ (see Fig. 19.2.3).

We now have a practical method for writing down explicitly any mth-order contribution to (19.2.3), (19.2.4). We first draw all possible diagrams satisfying the definition given above, by combining vertices A and B (and possibly, quantum links). Each diagram is then translated into a mathematical formula as follows.

1. To every vertex A joining lines labeled i and j corresponds a factor L'_{ij}; to every vertex B corresponds an operator $\int dx_j\, L'_{ij}$. (The corresponding operators in the quantum problem are taken from Section 14.3.)

2. To every intermediate state containing lines labeled $1, 2, \ldots, s+n$ corresponds a factor $U^0_{1,\ldots,s+n}$ or $R^0_{1,\ldots,s+n}$.

3. The two types of factors alternate in the order represented in the diagram.

4. The expression obtained in this way is integrated over intermediate times according to the rule of the convolution product (in the U^0 representation) or it is Laplace-transformed (in the resolvent representation).

As an example, we write down the contribution represented by the diagram a_2 of Fig. 19.2.1:

$$\int_0^{\tau} d\tau_1 \int_0^{\tau_1} d\tau_2 \int_0^{\tau_3} d\tau_4 \int_0^{\tau_4} d\tau_5 \int dx_3 \int dx_4 \int dx_5$$
$$\times L'_{13} U^0_{123}(\tau-\tau_1) L'_{24} U^0_{1234}(\tau_1-\tau_2) L'_{13} U^0_{1234}(\tau_2-\tau_3)$$
$$\times L'_{15} U^0_{12345}(\tau_3-\tau_4) L'_{15} U^0_{12345}(\tau_4-\tau_5) L'_{35} \quad (19.2.5)$$

Before closing this section, we add a few remarks that are direct consequences of our discussion. We note that, in all classical contributions to (19.2.3), (19.2.4), *the right-most vertex must be of type A*. Indeed, if the vertex were of type B, the first intermediate state would be a vacuum state (see, e.g., diagram b_1 of Fig. 19.2.1). On the other hand, every contribution to (19.2.3) *must end at left with a vertex B*, otherwise there would be a correlation between two particles in the final state. Contributions to (19.2.4) may end either with a vertex A or a vertex B. It then follows that an mth-order classical contribution to (19.2.3) has at least one vertex of type B and at most $(m-1)$ such vertices. Similarly, an mth order classical contribution to (19.2.4) has at least zero vertices of type B and at most $(m-1)$ such vertices. In the quantum case, the first vertex at right can be of type B, provided it is quantum statistically linked to the rest of the diagram. Hence the maximum number of B vertices is now m.

We now note that every B vertex introduces a new particle to the set existing at left. Hence *the index r* in Eqs. (19.2.3), (19.2.4) that is the difference between the numbers of particles in the initial and the final state, *equals simply the number of B vertices in the diagram*. But each new particle introduces an additional factor $\bar{f}$ in the right-hand side of the final equations [see Eqs. (19.1.5), (19.1.6)]. This factor, in turn, is a measure of the local density. Hence we may classify the orders of magnitude of the diagrams not only with respect to the coupling strength λ but also with respect to the density n. Summing up our discussion, we see that *a diagram with m vertices, r of which are of type B, is of order $\lambda^m n^r$. The range of values of r is*

$$1 \leqslant r \leqslant m-1 \qquad \text{for contributions to (19.2.3)}$$
$$0 \leqslant r \leqslant m-1 \qquad \text{for contributions to (19.2.4)}$$

This rule is valid for classical or quantum diagrams without quantum-statistical links. In case such links are present, the density dependence is more complicated, as can be seen from Eq. (18.7.7): it will not be discussed here.

19.3. THE DIAGRAM REPRESENTATION OF THE KINETIC EQUATION AND OF THE CORRELATION FUNCTIONS

Having elaborated a diagram technique for the representation of the irreducible evolution operator $\mathscr{E}(\tau)$, it is now a simple matter to extend the diagram representation to the general operators $V\mathbf{\Gamma}V$ and $C\mathbf{C}V$. We only need one more "building stone": the Vlassov propagator $V\tilde{\mathscr{U}}(\tau)V$. A perturbation expansion of this operator is easily obtained by the techniques of Section 16.1. Indeed, it obeys the Vlassov equation:

$$\partial_t V\tilde{\mathscr{U}}(t)V = V\mathscr{L}^0 V\tilde{\mathscr{U}}(t)V + V\mathscr{L}' V\tilde{\mathscr{U}}(t)V \tag{19.3.1}$$

or, alternatively, the integral equation similar to (16.1.12):

$$V\tilde{\mathscr{U}}(t)V = V\mathscr{U}^0(t) + \int_0^t d\tau\, V\mathscr{U}^0(t-\tau)V\mathscr{L}' V\tilde{\mathscr{U}}(\tau)V \tag{19.3.2}$$

This equation can be iterated as usual, with the result:

$$V\tilde{\mathscr{U}}(t)V = \sum_{n=0}^{\infty} V\mathscr{U}^0(*\mathscr{L}' V\mathscr{U}^0)^n \tag{19.3.3}$$

The individual terms in the series are n-fold convolution products. We may also write this formula in Laplace form:

$$V\tilde{\mathscr{U}}(t)V = (2\pi)^{-1}\int_C dz\, e^{-izt} \sum_{n=0}^{\infty} V\mathscr{R}^0(z)[\mathscr{L}' V\mathscr{R}^0(z)]^n \tag{19.3.4}$$

This formula is, in a sense, complementary to (19.2.1). Every term starts and ends with a factor $\mathscr{R}^0$ instead of $\mathscr{L}'$ as in the latter formula. More important, *only vacuum states appear now as intermediate states.*

To construct the matrix elements of $\mathbf{\Gamma}$ and $\mathbf{C}$ we need the matrix elements:

$$\langle x_1 | x_2 | \cdots | x_s | \tilde{\mathscr{U}}(\tau) | x_1 | x_2 | \cdots | x_{s+r}\rangle \tag{19.3.5}$$

These matrix elements can be represented by the same type of diagrams as in the previous section. The new rule defining these diagrams is the following:

> *Every mth-order contribution to the matrix element (19.3.5) is represented by a fully disconnected m-vertex diagram with s lines at left and s + r lines at right, containing only intermediate vacuum states.*

We note that a vertex of type A necessarily introduces a correlation between the "incoming" particles. Hence, *the diagrams contributing to (19.3.5) must be constructed only with B vertices.* In the quantum case, only diagrams without statistical links can be admitted. It then follows

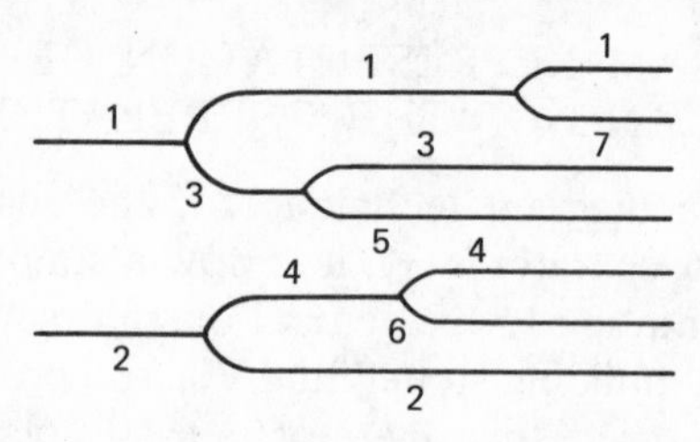

Figure 19.3.1. A contribution to $\langle 1 \,|\, 2 |\, \tilde{\mathscr{U}}(\tau)\, |1\,|\,2\,|\cdots|\,7\rangle$.

that the parameter r is always equal to m:

$$r = m \qquad \text{for contributions to (19.3.5)}$$

Therefore, every m-vertex diagram contributes a term of order $\lambda^m n^m$. A typical diagram is shown in Fig. 19.3.1.

It is now an easy matter to bring together the diagrams for $V\mathscr{G}(t)V$ [which derive directly from $V\mathscr{E}(t)V$], for $C\mathscr{C}(t)V$ [which derive from $C\mathscr{E}(t)V$] and $V\tilde{\mathscr{U}}(t)V$ in order to construct the matrix elements:

$$\langle x_1 |\, \mathbf{\Gamma}\, | x_1 \,|\, x_2 \,|\cdots|\, x_{r+1}\rangle \tag{19.3.6}$$

and

$$\langle x_1 \cdots x_s |\, \mathbf{C}\, | x_1 \,|\, x_2 \,|\cdots|\, x_{r+s}\rangle \tag{19.3.7}$$

These are the operators needed in the final equations (19.1.5) and (19.1.6), determining the dynamics in the $\mathbf{\Pi}$ subspace. As follows from Eqs. (17.2.13), (17.2.14), the operator $\mathbf{\Gamma}$ consists of a succession of factors $\mathscr{G}$ and $\tilde{\mathscr{U}}$. Therefore the diagrams consist of a succession of diagrams for $V\mathscr{E}V$ and for $V\tilde{\mathscr{U}}V$. These two components differ from each other only through the constraints on the intermediate states. Therefore the most general diagram representing the matrix element (19.3.6) is made up of a succession of A and B vertices (and possibly of statistical links), ending at left with one line; successive regions can be identified, involving only correlated or only vacuum states. A diagram with "n correlated regions" contributes to $\mathbf{\Gamma}_n$, Eq. (17.2.14) (see Fig. 19.3.2*a*). A diagram contributing to (19.3.7) has the same general structure, but [as follows from Eq. (17.3.4)] ends (at left) with a creation diagram, that is, with a fully connected diagram with s outcoming lines (Fig. 19.3.2*b*).

The translation of these diagrams into explicit formulas is just as straightforward as before. These formulas will look exactly like Eq. (19.2.5); the only difference is in the prescriptions for the time integrations, which must be taken from Eqs. (17.2.14) or (17.3.4).

The diagrams of Fig. 19.3.2 look extremely complex. Clearly, the corresponding kinetic equations, in very general cases, *are* quite complicated, and have actually never been studied. It is important however to be ready to write down any kinetic equation in any situation: The present diagram formalism allows us to do this just as easily as the Mayer-graph formalism in equilibrium. In the latter case as well, the graphs involving many particles are quite complicated; nevertheless, whenever one such complex contribution is needed in a problem, it can be written down automatically.

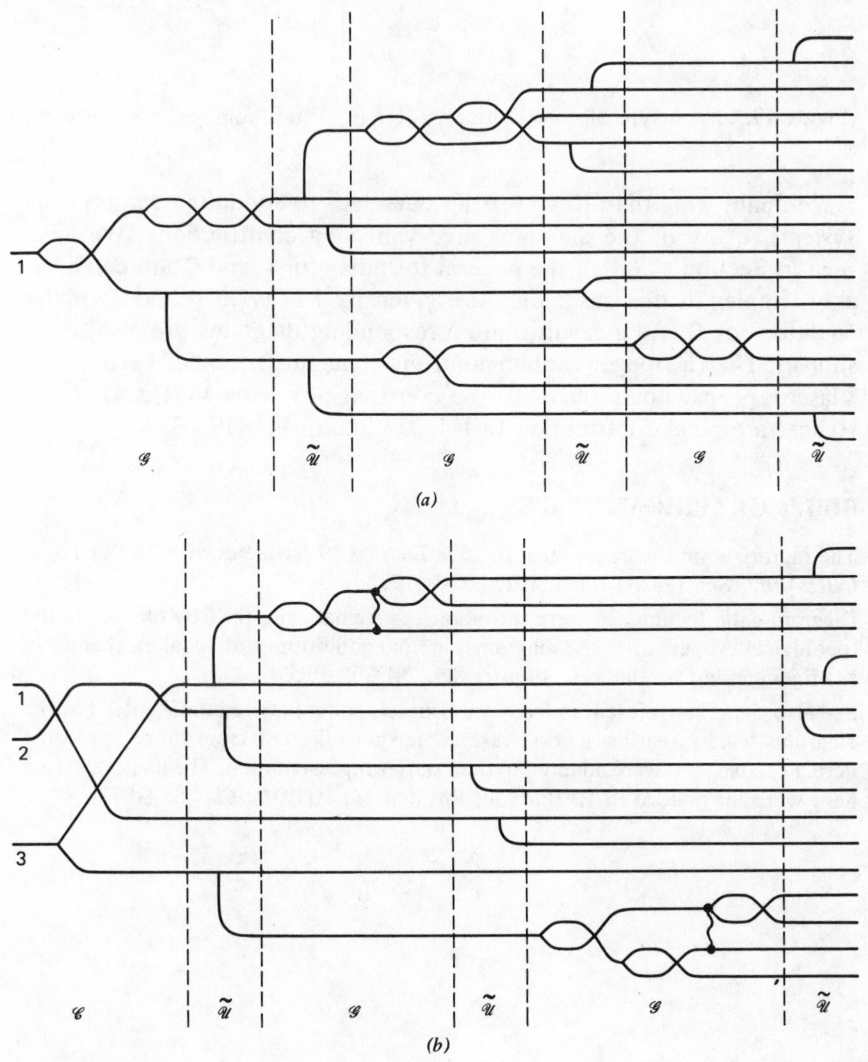

Figure 19.3.2. (*a*) A typical contribution to $\langle 1|\, \mathbf{\Gamma}\, |\cdots\rangle$. (*b*) A typical contribution to $\langle 123|\, \mathbf{C}\, |\cdots\rangle$.

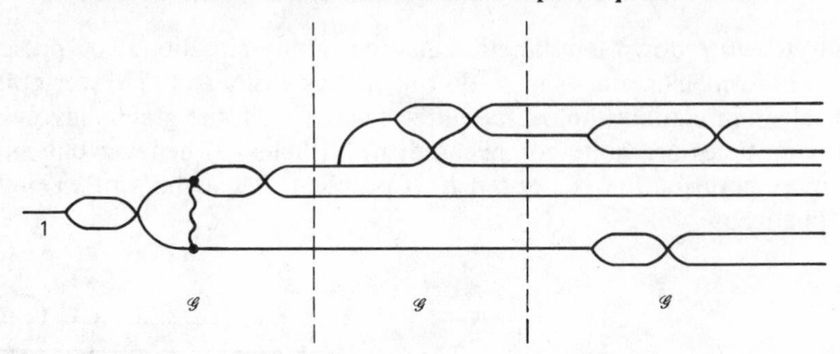

Figure 19.3.3. A typical contribution to $\langle 1|\mathbf{\Gamma}|\cdots\rangle$ in a homogeneous system.

We finally note that, if we restrict ourselves to *spatially homogeneous systems,* many of the diagrams give vanishing contributions. We have seen in Section 17.6 that the general formulas for $\mathbf{\Gamma}$ and $\mathbf{C}$ are considerably simpler in this case: they are given by Eq. (17.6.5) and a similar equation for $\mathbf{C}$. As a result, the corresponding diagrams are also much simpler. They no longer contain nontrivial "vacuum regions," because the Vlassov propagator reduces to the constant projector V [Eq. (17.6.2)]. Hence, a typical contribution to $\mathbf{\Gamma}$ looks like in Fig. 19.3.3.

BIBLIOGRAPHICAL NOTES

The factorization theorems mentioned in Section 19.1 were derived in P. Clavin, *C.R. Acad. Sci.* (Paris) **A274,** 1022, 1085 (1972).

Diagrammatic techniques were introduced systematically (i.e., by exploiting the topological properties of the diagrams) in nonequilibrium statistical mechanics in I. Prigogine and R. Balescu, *Physica* **25,** 281 302 (1959).

See also the other references from the Brussels school quoted in Chapter 17. The diagrams used in earlier works were somewhat different from those appearing here, because they were adapted to the Fourier representation. The diagrams used here were introduced in R. Balescu, *Physica* **56,** 1 (1971); **62,** 485 (1972).

CHAPTER 20

SPECIAL KINETIC EQUATIONS

20.1. REARRANGEMENTS OF THE PERTURBATION EXPANSION: DIVERGENCES

Now that we have developed a fairly general formalism and also have means of translating it into explicit equations, we illustrate the use of the machinery by discussing several applications.

The nature of the formalism developed in this book is a formal perturbation theory carried out to all orders. The parameter of the expansion is, naturally enough in a Hamiltonian theory, the strength of the interactions.

The lowest-order problem, hence the simplest, is the weakly coupled gas that was studied in great detail in Chapter 18. In the spirit of elementary perturbation theory one would go on and calculate corrections, of order $\lambda^3, \lambda^4, \ldots$, to this first result. In some cases, this is indeed an interesting problem, especially for testing general statements or for studying the structure of complex mathematical objects.

However, it should not be forgotten that the weak-coupling approximation is not a very physical concept. We should note particularly that λ is not an externally controllable parameter like the density, the temperature, and the like. The more realistic models involve systems with short-range hard cores, or long-range Coulomb forces, and so on. None of these systems conforms to the weak-coupling idea. Nevertheless, in limiting situations, one can find other small parameters in the problem that make the systems tractable again.

Actually, all these ideas were already introduced and discussed in equilibrium theory in Chapter 6. We also showed there how these problems are handled within this framework. The process of "changing the smallness parameter" involves a rearrangement of the λ expansion and a summation of an infinite partial series (see Section 6.1). An exactly identical procedure will be followed in nonequilibrium problems, as shown in the present chapter.

The nonvalidity of the expansion procedure in a given problem usually manifests itself in a very dramatic way. In "ordinary" perturbation theory, as we know from elementary quantum mechanics, when the parameter increases, it becomes necessary to take into account second-order corrections, then third-order ones, and so on: these corrections come in quite smoothly. In many-body problems we are often confronted with a much more brutal event. When the standard expansion is applied to the "wrong" problem (more precisely, when the wrong expansion is applied to the given problem) the contributions, already in the lowest orders, turn out to be infinite. The occurrence of such a divergence is the surest sign of the need of a resummation. Actually, a procedure that works very often is simply to pick out all the most divergent terms in every order in the expansion and sum the resulting subseries. The result of this method (which essentially consists of a permutation of summation and integration) is usually a convergent expression.

The mathematically minded reader may frown at such unorthodox procedures. Actually, several years ago, a group of courageous people (among which we may quote Ruelle, Lanford, Ginibre, Emch, Lebowitz, Sinai, . . .) started a very ambitious program of treating infinite systems in a flawless mathematical fashion: This is the so-called *C^*-algebra approach* to statistical mechanics. In this approach one does not accept any result that is not a perfectly rigorous mathematical theorem. Although this approach had some beautiful successes in equilibrium theory, the difficulties of the enterprise are such that, out of equilibrium, only the most (physically) trivial problems could be handled (the perfect gas or a system of hard rods in one dimension).

The difficulty lies already in the very first stages of the theory. In an infinite system there always exists an infinite number of initial conditions leading to a catastrophic collapse of the whole system (e.g., all particles with velocities aimed at a single point). To prove the mere existence of a time-evolution operator $\mathcal{U}(t)$ one needs to show that these pathological conditions are, on some sense, of measure zero. This we do not yet know how to do in general.

If we, as physicists, want to go ahead and get some insight into the physically nontrivial problems, we are practically forced to compromise with mathematical rigor. The art of theoretical physics (this may be taken as a definition) is in optimizing the ratio $R(t)$ between nonrigorous and rigorous aspects of the theory in any given problem at a given time in history; this rigor factor $R(t)$ is assumed to be a decaying function of time, as a result of the refinements of the theory. A theory reaches equilibrium and stationarity when $R(t)=0$. But perfect beauty is equivalent to death. A perfect theory can only be contemplated and admired: no further work can be done on it. Fortunately, for all physical theories, $R(t)\neq 0$.

The problems selected for this chapter are the nonequilibrium counterparts of the problems studied in Chapters 6 and 9 in equilibrium, the dilute and moderately dense gas, the plasma, and the van der Waals fluid.* It so happens that most problems that were solved in equilibrium theory were solved, in time, in nonequilibrium theory too. This is, of course, no accident. There are very few soluble problems in physics, and these have some simple features that are exploitable in both cases. However, it will strike everyone that there is an extra order of magnitude in the complication of the nonequilibrium problem. Just to quote an example: the analytical expression of every virial coefficient can be written down by simple inspection of the Mayer graphs. Nothing of this kind is true for transport coefficients: An explicit analytical expression is only known for the first density correction beyond Boltzmann! As far as numerical results are concerned, the situation is even worse. In equilibrium, six virial coefficients are known for hard spheres; out of equilibrium, the "second virial coefficient" is known only for hard disks in two dimensions.

For this reason of increased complexity, we chose to describe only two examples in full detail: the dilute gas and the plasma. For all the other cases, we outlined only the physical ideas and the mathematical methods; technical details were omitted.

20.2. DILUTE GAS OF STRONGLY INTERACTING PARTICLES

As a first example of the systematic use of diagrams in the derivation of explicit kinetic equations, we consider a gas of particles interacting through arbitrarily strong forces. This is, of course, a much more realistic problem than the weakly coupled gas treated in Chapter 18, if only because of the universal presence of a hard, but short-ranged repulsive core in the interaction potential of any kind of molecules. We assume, however, that we still have a small parameter in our problem in order to make it simply tractable. Referring to our discussion of Section 6.4, it is clear that the parameter $\gamma = d_0^3 n$ (where d_0 is the effective hard-core diameter, and n the number density) is a convenient parameter for the case of a dilute gas. We thus are going to treat the following problem: *To derive the kinetic equation of a classical gas of point particles, interacting with arbitrary strength, in the limit* $\gamma \ll 1$.

Let us analyze the general structure of the diagrams for the collision operator in this case. The leading contributions in our problem will be those involving the least number of factors n. But *all* diagrams contributing to the collision operator have at least one vertex of type B, as appears clearly from our discussion of Section 19.2; hence they all involve at least one factor n. We are therefore led to select all the diagrams that involve

* The Brownian-motion problem is trivial in equilibrium.

no other factor n, hence no other vertex B, but that may involve an arbitrary number of vertices of type A, corresponding to an arbitrary number of factors λ, that is, to arbitrarily strong interactions.

We immediately see from Fig. 19.3.2 that, in the general expression for the collision operator, the only diagrams satisfying our criterion are those involving a single irreducible block, that is, a single factor $V\mathcal{G}V$. Indeed, every additional block introduces an additional factor n through its leftmost vertex. We are now left with the collision term

$$\int_0^\infty d\tau \, V\mathcal{G}(\tau)V\tilde{\mathcal{U}}(-\tau)V$$

To our order of approximation, $V\tilde{\mathcal{U}}V$ must be further approximated by $V\mathcal{U}^0V$, because every additional factor $V\mathcal{L}'V$ again introduces a factor n. Finally, within the block representing the operator $V\mathcal{G}(\tau)V$, we can have only A vertices to the right of the leftmost B vertex. This requirement implies that there can only be two lines coming out of the diagram to the right. At this stage, we have pinned down all the contributions to $V\Gamma V$, proportional to $\lambda^p n$ $(p = 1, 2, \ldots)$: they are shown in Fig. 20.2.1, and can be shortly called the *chain diagrams.* These diagrams are the kinetic counterpart of the equilibrium graphs of order n (Table 6.4.1) contributing to the second virial coefficient.

The first diagram in this figure is simply the Vlassov term that we studied already; the second diagram is the Landau collision term, considered in detail in Chapter 18. We are now faced with the problem of summing all the diagrams in this class, a problem quite analogous to the summation of the corresponding equilibrium diagrams of order $\lambda^p n$. However, the technical problem is here much less trivial than for the cluster expansion, because the individual terms of the series are *operators* rather than simple functions.

From our discussion and from Eq. (17.2.11) it follows that the kinetic equation in the dilute gas approximation is written as

$$\begin{aligned}\partial_t \bar{\rho}_1(1;t) = &\langle 1|\,\mathcal{L}^0\,|1\rangle\bar{\rho}_1(1;t) + \langle 1|\,\mathcal{L}'\,|1\,|\,2\rangle\,\bar{\rho}_2(1\,|\,2;t)\\ &+\int_0^\infty d\tau\,\langle 1|\,\mathcal{G}(\tau)\,|1\,|2\rangle\langle 1\,|\,2|\,\mathcal{U}^0(-\tau)\,|1\,|\,2\rangle\,\bar{\rho}_2(1\,|\,2;t) \qquad (20.2.1)\end{aligned}$$

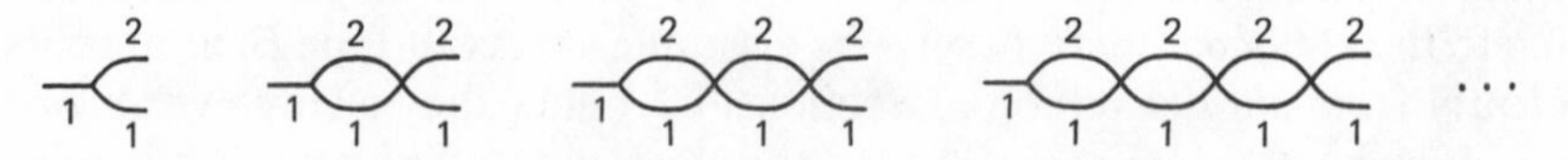

Figure 20.2.1. The chain diagrams for the collision operator of a dilute gas.

Using Eqs. (17.1.4) and (16.3.5), we further write

$$\begin{aligned}\langle 1|\,\mathscr{G}(\tau)\,|1\,|\,2\rangle &= \langle 1|\,\mathscr{L}'\,|12\rangle\,\langle 12|\,\mathscr{U}^0(\tau)\,|12\rangle\,\langle 12|\,\mathscr{L}'\,|1\,|\,2\rangle\\ &\quad+\int_0^\tau d\tau_1 \langle 1|\,\mathscr{L}'\,|12\rangle\,\langle 12|\,\mathscr{U}^0(\tau-\tau_1)\,|12\rangle\,\langle 12|\,\mathscr{L}'\,|12\rangle\\ &\quad\times\langle 12|\,\mathscr{U}^0(\tau_1)\,|12\rangle\,\langle 12|\,\mathscr{L}'\,|1\,|\,2\rangle+\cdots\end{aligned}$$

Finally, using the matrix elements of Eqs. (14.2.6) and (14.2.8), we write Eq. (20.2.1) explicitly as follows:

$$\begin{aligned}&\partial_t\bar{p}_1(1;t)-L_1^0\bar{p}_1(1;t)\\ &\quad=\int dx_2\,L'_{12}\bar{p}_2(1\,|\,2;t)+\int dx_2\int_0^\infty d\tau\Big\{L'_{12}U_{12}^0(\tau)L'_{12}U_{12}^0(-\tau)\\ &\quad+\int_0^\tau d\tau_1\,L'_{12}U_{12}^0(\tau-\tau_1)L'_{12}U_{12}^0(\tau_1)L'_{12}U_{12}^0(-\tau)\\ &\quad+\int_0^\tau d\tau_1\int_0^{\tau_1} d\tau_2\,L'_{12}U_{12}^0(\tau-\tau_1)L'_{12}U_{12}^0(\tau_1-\tau_2)L'_{12}U_{12}^0(\tau_2)L'_{12}U_{12}^0(-\tau)+\cdots\Big\}\\ &\quad\times\bar{p}_2(1\,|\,2;t)\end{aligned}\tag{20.2.2}$$

We now rewrite this equation by introducing an operator $Y_{12}(\tau)$ as follows:

$$\begin{aligned}&\partial_t\bar{p}_1(1;t)-L_1^0\bar{p}_1(1;t)\\ &\quad=\int dx_2\,L'_{12}\bar{p}_2(1\,|\,2;t)+\int dx_2\int_0^\infty d\tau\,L'_{12}Y_{12}(\tau)L'_{12}U_{12}^0(-\tau)\bar{p}_2(1\,|\,2;t)\end{aligned}\tag{20.2.3}$$

The operator $Y_{12}(\tau)$ is defined as the solution of the integral equation:

$$Y_{12}(\tau)=U_{12}^0(\tau)+\int_0^\tau d\tau_1\,U_{12}^0(\tau-\tau_1)L'_{12}Y_{12}(\tau_1)\tag{20.2.4}$$

The iterative solution of this equation clearly reproduces the series of Eq. (20.2.3). The operator $Y_{12}(\tau)$ has a clear and simple physical meaning. It is easily seen that it obeys the differential equation:

$$\partial_\tau Y_{12}(\tau)=(L_1^0+L_2^0+L'_{12})Y_{12}(\tau)\tag{20.2.5}$$

with the initial condition

$$Y_{12}(0)=1\tag{20.2.6}$$

Indeed, the "solution" to this initial-value problem is provided by Eqs. (15.1.22), (15.1.24), projected on the two-particle state, with $g(t)=L'_{12}Y_{12}(t)$: it is precisely Eq. (20.2.4). But Eq. (20.2.5) is the Liouville equation for an isolated two-body system; hence $Y_{12}(t)$ is simply the *exact two-body propagator.* Therefore, in the limit of small density, the collision term of the kinetic equation is entirely determined by the two-body dynamics, a conclusion that agrees with our qualitative discussion of Section 11.4. It simply confirms the intuitive idea that in a dilute gas, three- or more-body collisions are very infrequent events.

A comparison of Eq. (20.2.3) with the Landau equation (18.1.14) is also very suggestive. The only difference is in the appearance of $Y_{12}(t)$ instead of $U^0_{12}(t)$ in the collision term. In the weak-coupling approximation, the collision process is also a binary one, but the propagation "during the collision" is approximated by the free-motion propagator (see Section 11.6) whereas in the present case it is described by the exact two-body propagator, which, in general, is a complicated object.

From this discussion it is pretty obvious that one should expect a relation between kinetic theory in this approximation and the ordinary scattering theory, which is also based on two-body dynamics. This relationship will be developed in the next section.

20.3. TWO-BODY SCATTERING THEORY AND THE BOLTZMANN EQUATION

In order to simplify the forthcoming discussion, we shall limit ourselves to the discussion of a *spatially homogeneous gas*. It then follows from Section 18.3 that

$$\begin{aligned} \bar{p}_1(1;t) &= n\bar{\varphi}(\mathbf{p}_1;t) \\ \bar{p}_2(1\,|\,2;t) &= n^2\bar{\varphi}(\mathbf{p}_1;t)\bar{\varphi}(\mathbf{p}_2;t) \end{aligned} \tag{20.3.1}$$

Moreover,

$$\begin{aligned} L^0_1\bar{p}_1(1;t) &\equiv 0 \\ \int dx_2\, L'_{12}\bar{p}_2(1\,|\,2;t) &\equiv 0 \\ U^0_{12}(\tau)\bar{p}_2(1\,|\,2;t) &= \bar{p}_2(1\,|\,2;t) \end{aligned} \tag{20.3.2}$$

Furthermore, the only dependence on $\mathbf{q}_2$ in the collision term is introduced by the potential V_{12} in the form of the combination $\mathbf{r}=\mathbf{q}_2-\mathbf{q}_1$: we may therefore take $\mathbf{r}$ instead of $\mathbf{q}_2$ as an integration variable in Eq. (20.2.3). Also, instead of the momenta, it is usually convenient to use velocities: $\mathbf{v}_j=\mathbf{p}_j/m$. We are now left with the kinetic equation:

$$\partial_t\bar{p}_1(1;t) = \int d\mathbf{v}_2\, J(\mathbf{v}_1,\mathbf{v}_2) \tag{20.3.3}$$

$$J(\mathbf{v}_1,\mathbf{v}_2) = \int d\mathbf{r}\int_0^\infty d\tau\, L'Y(\tau)L'\bar{p}_2(1\,|\,2;t) \tag{20.3.4}$$

The subscripts 12 on the operators L' and Y are clearly unnecessary, as we only have a single pair of particles in this problem; these indices have therefore been dropped. Instead of the velocities $\mathbf{v}_1$ and $\mathbf{v}_2$ we may also use the center of mass velocity $\mathbf{w}$ and the relative velocity $\mathbf{g}$. In terms of these variables, the interaction Liouvillian becomes

$$L' = [\nabla V(r)] \cdot \partial \tag{20.3.5}$$

with

$$\nabla = \frac{\partial}{\partial \mathbf{r}}, \qquad \partial = \left(\frac{2}{m}\right)\left(\frac{\partial}{\partial \mathbf{g}}\right) \tag{20.3.6}$$

Finally, we note that the unperturbed propagator, when acting on any function h of $\mathbf{r} = \mathbf{q}_1 - \mathbf{q}_2$, reduces to the form:

$$\begin{aligned} U^0(\tau)h(\mathbf{r}) &= \exp[-\mathbf{v}_1 \cdot \nabla_1 - \mathbf{v}_2 \cdot \nabla_2]\tau h(\mathbf{r}) \\ &= \exp(-\tau \mathbf{g} \cdot \nabla)h(\mathbf{r}) \end{aligned} \tag{20.3.7}$$

Hence, all the operators of Eq. (20.3.4) have been reduced to the variables $\mathbf{r}$, $\mathbf{g}$ describing the relative motion. This corresponds to the well-known reduction of the two-body problem to the motion of a single effective particle in a central potential (see Section 11.4).

We now note that, in the *homogeneous* problem, the τ integral in Eq. (20.3.4) only affects the propagator $Y(\tau)$. Let us integrate all terms in Eq. (20.2.4) over τ from 0 to ∞, and then take τ_1 and $\tau - \tau_1 \equiv \theta$ as new integration variables. A simple calculation yields

$$\begin{aligned} \int_0^\infty d\tau\, Y(\tau) &= \int_0^\infty d\tau\, U^0(\tau) + \int_0^\infty d\tau \int_0^\tau d\tau_1\, U^0(\tau - \tau_1)L'Y(\tau_1) \\ &= \int_0^\infty d\tau\, U^0(\tau) + \int_0^\infty d\tau_1 \int_0^\infty d\theta\, U^0(\theta)L'Y(\tau_1) \end{aligned} \tag{20.3.8}$$

We now introduce the time-independent operators:

$$G = \int_0^\infty d\tau\, U^0(\tau) \tag{20.3.9}$$

and

$$Z = \int_0^\infty d\tau\, Y(\tau) \tag{20.3.10}$$

Equation (20.3.8) then becomes

$$Z = G + GL'Z \tag{20.3.11}$$

We thus achieved the transformation of Equation (20.2.4) into a *time-independent* operator equation. We now examine more closely the meaning of these new operators.

We first note that the unperturbed propagator can be written in the form of an integral operator:

$$\begin{aligned} U^0(\tau)h(\mathbf{r},\mathbf{g}) &= \exp(-\tau\mathbf{g}\cdot\nabla)h(\mathbf{r},\mathbf{g}) \\ &= h(\mathbf{r}-\mathbf{g}\tau,\mathbf{g}) \\ &= \int d\mathbf{g}'\int d\mathbf{r}'\ \delta(\mathbf{r}-\mathbf{g}\tau-\mathbf{r}')\delta(\mathbf{g}-\mathbf{g}')h(\mathbf{r}',\mathbf{g}') \end{aligned} \qquad (20.3.12)$$

It then follows that the operator G can also be expressed as an integral kernel:

$$Gh(\mathbf{r},\mathbf{g}) = \int d\mathbf{g}'\ d\mathbf{r}'\ G(\mathbf{r},\mathbf{g};\mathbf{r}',\mathbf{g}')h(\mathbf{r}',\mathbf{g}') \qquad (20.3.13)$$

with

$$G(\mathbf{r},\mathbf{g};\mathbf{r}',\mathbf{g}') = \delta(\mathbf{g}-\mathbf{g}')G^0(\mathbf{r}-\mathbf{r}') \qquad (20.3.14)$$

and

$$G^0(\mathbf{r}-\mathbf{r}') = \int_0^\infty d\tau\ \delta(\mathbf{r}-\mathbf{r}'-\mathbf{g}\tau) \qquad (20.3.15)$$

The operator Z can, of course, also be expressed in the representation (20.3.13); hence, using (20.3.5), Eq. (20.3.11) can be written explicitly as an integral equation:

$$\begin{aligned} Z(\mathbf{r},\mathbf{g};\mathbf{r}',\mathbf{g}') &= G^0(\mathbf{r}-\mathbf{r}')\delta(\mathbf{g}-\mathbf{g}') \\ &\quad + \int d\mathbf{r}''\int d\mathbf{g}''\ G^0(\mathbf{r}-\mathbf{r}'')\delta(\mathbf{g}-\mathbf{g}'')[\nabla''V(\mathbf{r}'')]\cdot\partial''Z(\mathbf{r}'',\mathbf{g}'';\mathbf{r}',\mathbf{g}') \end{aligned} \qquad (20.3.16)$$

Let us now discuss the meaning of the kernel $G^0(\mathbf{r}-\mathbf{r}')$.* If we apply the unperturbed Liouville operator to this function and make use of the Fourier representation of the δ and δ_- functions [Eq. (18.3.9)] we find

$$\begin{aligned} (L_1^0+L_2^0)G^0(\mathbf{r}-\mathbf{r}') &= \mathbf{g}\cdot\nabla\int_0^\infty d\tau\ \delta(\mathbf{r}-\mathbf{g}\tau-\mathbf{r}') \\ &= (8\pi^3)^{-1}\int_0^\infty d\tau\int d\mathbf{k}\ i\mathbf{k}\cdot\mathbf{g}\exp[i\mathbf{k}\cdot(\mathbf{r}-\mathbf{r}'-\mathbf{g}\tau)] \\ &= (8\pi^3)^{-1}\int d\mathbf{k}\ i\mathbf{k}\cdot\mathbf{g}\ \pi\delta_-(\mathbf{k}\cdot\mathbf{g})\exp[i\mathbf{k}\cdot(\mathbf{r}-\mathbf{r}')] \\ &= (8\pi^3)^{-1}\int d\mathbf{k}\exp[i\mathbf{k}\cdot(\mathbf{r}-\mathbf{r}')] = \delta(\mathbf{r}-\mathbf{r}') \end{aligned}$$

* The definition (20.3.15) actually makes sense only if the function $f(\mathbf{r})$ on which the operator acts falls off fast enough at infinity in order that the τ integral exists. Otherwise one should add a convergence factor, that is, change the integrand into $\delta(\mathbf{r}-\mathbf{r}'-\mathbf{g}\tau)\exp(-\varepsilon\tau)$ in Eq. (20.3.15), and take the limit $\varepsilon\to 0$ at the end of the calculations. We shall not insist here on this point.

where we made use of the following identity (valid in the sense of distributions):

$$x\pi\delta_-(x) = \pi x\delta(x) + x\mathcal{P}\left(\frac{1}{x}\right) = 1 \tag{20.3.17}$$

Hence,

$$\mathbf{g}\cdot\nabla G^0(\mathbf{r}-\mathbf{r}') = \delta(\mathbf{r}-\mathbf{r}') \tag{20.3.18}$$

The function $G^0(\mathbf{r}-\mathbf{r}')$ thus obeys the *stationary unperturbed Liouville equation* everywhere except at the source point $\mathbf{r}=\mathbf{r}'$. Equation (20.3.18) is very analogous to the equation of electrostatics defining in the whole space the potential of a point source at $\mathbf{r}=\mathbf{r}'$. Hence, the function $G(\mathbf{r}-\mathbf{r}')$ expresses the influence of a point particle at $\mathbf{r}=\mathbf{r}'$ and $\mathbf{g}=\mathbf{g}'$ on the stationary distribution in the whole phase space: It is called the *Green's function* of the stationary Liouvillian L^0. However, we should stress the difference with the Greenian or propagator $\mathcal{U}^0(t)$ considered earlier: there is no propagation in time. We study here the stationary distribution that is reached after all the time-dependent processes have played their role.

Combining now Eqs. (20.3.18) and (20.3.16) it is easily shown that the function $Z(\mathbf{r}, \mathbf{g}; \mathbf{r}', \mathbf{g}')$ is the Green's function for the complete two-body Liouvillian:

$$\{\mathbf{g}\cdot\nabla - [\nabla V(\mathbf{r})]\cdot\partial\}Z(\mathbf{r}, \mathbf{g}; \mathbf{r}', \mathbf{g}') = \delta(\mathbf{r}-\mathbf{r}')\delta(\mathbf{g}-\mathbf{g}') \tag{20.3.19}$$

We now introduce the simpler function $f(\mathbf{r}, \mathbf{g})$ defined as follows:

$$f(\mathbf{r}, \mathbf{g}) = \bar{p}_2(1\,|\,2; t) + \int d\mathbf{r}'\, d\mathbf{g}'\, Z(\mathbf{r}, \mathbf{g}; \mathbf{r}', \mathbf{g}')[\nabla' V(r')]\cdot\partial'\bar{p}_2(1'\,|\,2'; t) \tag{20.3.20}$$

From Eq. (20.3.19) it follows that $f(\mathbf{r}, \mathbf{g})$ is a *stationary solution of the two-body Liouville equation*, such that when the interaction is switched off, it reduces to $\bar{p}_2(1\,|\,2; t)$:

$$\{\mathbf{g}\cdot\nabla - [\nabla V(r)]\cdot\partial\}f(\mathbf{r}, \mathbf{g}) = 0 \tag{20.3.21}$$

$$\lim_{V\to 0} f(\mathbf{r}, \mathbf{g}) = \bar{p}_2(1\,|\,2; t) \tag{20.3.22}$$

The set of these two equations is also equivalent to the single integral equation:

$$f(\mathbf{r}, \mathbf{g}) = \bar{p}_2(1\,|\,2; t) + \int d\mathbf{r}'\, G^0(\mathbf{r}-\mathbf{r}')[\nabla' V(r')]\cdot\partial' f(\mathbf{r}', \mathbf{g}') \tag{20.3.23}$$

We have reached the point where the link with scattering theory is clear. Consider an incident beam of particles, homogeneous in space, and described by the distribution function $\bar{p}_2(1\,|\,2; t)$. As the beam passes the potential center, its particles are scattered, and, after a sufficient time, a new stationary distribution sets in. This distribution, which is a superposition of the incoming "wave" and of the scattered "wave," is described by the distribution $f(\mathbf{r}, \mathbf{g})$.

It is now easily shown, by combining Eqs. (20.3.4), (20.3.10), and (20.3.20) that the collision term in the kinetic equation can be written as follows:

$$J = \int d\mathbf{r}\, [\nabla V(r)] \cdot \partial \{f(\mathbf{r}, \mathbf{g}) - \bar{p}_2(1 \mid 2; t)\}$$

Noting also that the distance-independent function $\bar{p}_2$ does not contribute to the integral [see Eq. (18.3.4)], and using (20.3.21) we also get

$$J = \int d\mathbf{r}\, \mathbf{g} \cdot \nabla f(\mathbf{r}, \mathbf{g}) \tag{20.3.24}$$

The function $f(\mathbf{r}, \mathbf{g})$ can be easily evaluated if we note that the unperturbed Green's function $G^0(\mathbf{r} - \mathbf{r}')$ has a particularly simple form in a reference frame in which the z axis is taken along the direction of the velocity $\mathbf{g}$:

$$G^0(\mathbf{r} - \mathbf{r}') = g^{-1} \delta(x - x') \delta(y - y') \Theta(z - z') \tag{20.3.25}$$

where $\Theta(x)$ is Heaviside's step function:

$$\begin{aligned} \Theta(x) &= 0, \qquad x < 0 \\ &= 1, \qquad x > 0 \end{aligned}$$

Recalling that $d\Theta/dx = \delta(x)$, it is easily checked that $G^0(\mathbf{r} - \mathbf{r}')$ indeed obeys Eq. (20.3.18). Substituting this form into Eq. (20.3.23) and using (20.3.21) we find

$$\begin{aligned} f(\mathbf{r}, \mathbf{g}) &= \bar{p}_2(1 \mid 2; t) + \int_{-\infty}^{z} dz'\, g^{-1} [\nabla' V(xyz')] \cdot \partial f(xyz', \mathbf{g}) \\ &= \bar{p}_2(1 \mid 2; t) + \int_{-\infty}^{z} dz' \left(\frac{\partial}{\partial z'}\right) f(xyz', \mathbf{g}) \end{aligned} \tag{20.3.26}$$

Calling the integral in the right-hand side $\Phi(z)$ and substituting into (20.3.24), we find

$$\begin{aligned} J = \int d\mathbf{r}\, \mathbf{g} \cdot \nabla f &= \int_{-\infty}^{\infty} dz\, g \left(\frac{\partial}{\partial z}\right) \Phi(z) \\ &= g\{\Phi(\infty) - \Phi(-\infty)\} = g \int_{-\infty}^{\infty} dz' \left(\frac{\partial}{\partial z'}\right) f(xyz', \mathbf{g}) \\ &= g\{f(x, y, +\infty, \mathbf{g}) - f(x, y, -\infty, \mathbf{g})\} \end{aligned} \tag{20.3.27}$$

The function $f(x, y, -\infty, \mathbf{g})$ is easily seen, from (20.3.26), to be simply the incoming distribution of particles of velocity $\mathbf{v}_1$, $\mathbf{v}_2$:

$$f(x, y, -\infty, \mathbf{g}) = \bar{\rho}_2(\mathbf{v}_1 \mid \mathbf{v}_2; t) \tag{20.3.28}$$

As for $f(x, y, +\infty, \mathbf{g})$, it represents the outcoming distribution of particles with relative velocity $\mathbf{g}$. But these particles result from a collision of incoming particles having velocities $\mathbf{v}_1'$. $\mathbf{v}_2'$ such that the corresponding velocities after the collision are precisely $\mathbf{v}_1$, $\mathbf{v}_2$. By the Liouville theorem, the outgoing distribution just equals the incoming distribution, evaluated at the velocities $\mathbf{v}_1'$, $\mathbf{v}_2'$:

$$f(x, y, +\infty, \mathbf{g}) = \bar{\rho}_2(\mathbf{v}_1' \mid \mathbf{v}_2'; t) \tag{20.3.29}$$

We have thus recovered all the concepts introduced heuristically in Section 11.4. If we finally note that the surface element $dx\, dy$ perpendicular to the relative velocity can be written in polar coordinates:

$$dx\, dy = b\, db\, d\varphi$$

and that b is simply the collision parameter defined in Section 11.4, we see that the kinetic equation obtained by combining the results (20.3.1), (20.3.3), and (20.3.27)–(20.3.29) is

$$\partial_t \bar{\varphi}(\mathbf{v}_1; t) = n \int d\mathbf{v}_2 \int d\varphi\, db\, bg\{\bar{\varphi}(\mathbf{v}_1'; t)\bar{\varphi}(\mathbf{v}_2'; t) - \bar{\varphi}(\mathbf{v}_1; t)\bar{\varphi}(\mathbf{v}_2; t)\} \tag{20.3.30}$$

This result, which is identical to Eq. (11.4.21), concludes the derivation of the *Boltzmann equation* for a dilute homogeneous gas.

We now clearly see that the Boltzmann equation describes a particular case of evolution in the $\mathbf{\Pi}$ space. It does *not* represent the evolution of the complete distribution function but only of its kinetic part. Hence, all the difficulties discussed in Section 11.5 disappear. In particular, we have illustrated once more that the *Stosszahlansatz* is *exactly* maintained by the evolution process in the $\mathbf{\Pi}$ space, if it is realized at the initial time (see also the discussion of Section 19.1).

Another characteristic feature, which appears quite clearly from Eq. (20.3.27), is the fact that the collision term depends solely on the incoming and on the outcoming distributions, and not at all on the details of the processes occurring during the collision. It is in this sense that the collision process is to be considered as an instantaneous process. This feature is a result of the low-density approximation.

20.4. KINETIC EQUATIONS FOR MODERATELY DENSE GASES

In order to grasp fully the motivation and the significance of this section, it is interesting to describe the historical context of statistical physics around 1950.* At that time, the only known irreversible kinetic equation was Boltzmann's (and its direct variants such as the Landau equation, the kinetic equation for phonons in solids, etc.). Its theoretical status, its range of validity, and its limitations were pretty well understood; but there was no systematic way of going beyond the Boltzmann approximation. The only attempt of "breaking the barrier" was Enskog's work of 1922, in which he derived an extension of Boltzmann's equation, valid for a dense gas of hard spheres.

The main feature in Enskog's equation is in taking account of the delocalization of the collision process.† Instead of evaluating the distribution functions (in an inhomogeneous system) at the same point, he replaced the curly bracket in Eq. (11.4.20) by

$$\{\bar{f}(\mathbf{q}, \mathbf{v}'; t)\bar{f}(\mathbf{q}+d_0\boldsymbol{\varepsilon}, \mathbf{v}_1'; t) - \bar{f}(\mathbf{q}, \mathbf{v}; t)\bar{f}(\mathbf{q}-d_0\boldsymbol{\varepsilon}, \mathbf{v}_1; t)\} \qquad (20.4.1)$$

where d_0 is the hard-sphere diameter and $\boldsymbol{\varepsilon}$ a unit vector along the line of centers. We already met a similar problem in our analysis of inhomogeneous systems, Eq. (18.4.4). We now see that there is a subtle relation between inhomogeneity and order in density. To understand it, we note that an expansion of the displaced distribution functions around $\mathbf{q}$ yields a term proportional to the spatial gradient, as in Eq. (18.4.5). When the kinetic equation is linearized, this extra term can be grouped together with the free-flow term (which is also proportional to the gradient ∇f). When we now proceed to the hydrodynamical equations, as in Section 12.4, we find that the pressure tensor (for instance) no longer is simply $\langle(\mathbf{v}-\mathbf{u})\times(\mathbf{v}-\mathbf{u})\rangle$, but contains an additional term involving d_0, and therefore the interactions. This term is the nonequilibrium equivalent of the second term in the rhs of Eq. (7.2.8), which describes the effect of the interactions and becomes important at high densities. It is customary to say that this term describes *collisional transfer of momentum*. The Enskog equation thus appears as describing the kinetic behavior of a nonideal gas. Although important, Enskog's work was limited to hard-sphere interactions and was not based on a systematic procedure: hence the barrier was pushed a little further, but was still there!

In 1946, Bogoliubov published his classical book: *Problems of Dynamical Theory in Statistical Physics*. When the news reached the West (a process that, at that time, was less trivial than nowadays) it aroused the enthusiasm of an important group of people, in particular of Uhlenbeck. Here was a way of doing systematically an expansion of the kinetic

* This description is very schematic and, clearly, leaves out many authors, to whom we apologize.

† Enskog also modified the cross section σ by a shielding factor χ that was meant to take roughly into account the effect of triple collisions. This aspect is disregarded here.

equation in powers of the density: the Boltzmann equation would be simply the first term of this expansion. Moreover, following the model of the equilibrium virial expansion of the pressure in powers of the density, we had now a potential method for obtaining virial expansions for the (nonequilibrium) transport coefficients.

Bogoliubov's theory has essentially two aspects.

A. It is based on the fundamental assumption that, in the "kinetic regime" the higher-order correlations are functionals of the one-particle distribution. Hence, a closed kinetic equation for f_1 completely solves the problem. We have seen in Chapter 17 how this assumption fits into the general framework: It corresponds precisely to the subdynamics in the space Π.

B. The explicit determination of the kinetic equation involves a subtle expansion procedure in powers of a small parameter, for example, the density. (This aspect corresponds to the realization procedure described in Chapter 19.)

Bogoliubov only gave an outline of this second aspect and derived the Boltzmann equation. Much remained to be done by ways of obtaining explicitly the higher-density corrections to that equation. A first step was taken quite successfully by Choh and Uhlenbeck in 1958. They derived a kinetic equation valid one order beyond the Boltzmann equation. This equation involves the effect of three-body collisions, as well as the contribution of modified two-body collisions. In the case of hard spheres, the latter contribution corresponds precisely to the Enskog equation.

Let us sketch very briefly the idea of the derivation of the kinetic equations to various orders in the density. A variety of methods were used, but they are all variants of the so-called *binary collision expansion* introduced by Siegert and Teramoto and by Yang and Lee in equilibrium theory and extended by Zwanzig and by Weinstock to nonequilibrium problems. We introduce the symbol $\mathcal{L}'_\alpha$ to denote the contribution to $\mathcal{L}'$ due to the interaction of the *pair* of particles denoted by the Greek symbol α: $\alpha \equiv (a, b)$. Its matrix elements (see Section 3.3) are

$$
\begin{aligned}
\langle a, b, 1, 2, \ldots, s|\, \mathcal{L}'_\alpha \,|a, b, 1, 2, \ldots, s'\rangle &= \delta_{s',s} L'_{ab} \\
\langle a, 1, 2, \ldots, s|\, \mathcal{L}'_\alpha \,|a, b, 1, 2, \ldots, s'\rangle &= \delta_{s',s} \int dx_b\, L'_{ab} \qquad (20.4.2) \\
\langle 1, 2, \ldots, s|\, \mathcal{L}'_\alpha \,|1, 2, \ldots, s'\rangle &= 0, \qquad 1, 2, \ldots, s \neq (a, b)
\end{aligned}
$$

Returning to the expansion (16.1.25) of the resolvent, we may group the terms into subsums according to the number of particles involved in the various interaction processes. More precisely, we consider first all the interactions involving only the pair of particles α (and sum over α); next all interactions involving a pair α followed by interactions involving another pair β, and so forth:

$$\mathcal{R} = \mathcal{R}^0(z) + \sum_{\alpha} \mathcal{R}^0(z)\mathcal{T}_\alpha(z)\mathcal{R}^0(z) + \sum_{\alpha}\sum_{\substack{\beta \\ (\alpha\neq\beta)}} \mathcal{R}^0(z)\mathcal{T}_\alpha(z)\mathcal{R}^0(z)\mathcal{T}_\beta\mathcal{R}^0(z)$$

$$+ \sum_{\alpha}\sum_{\beta}\sum_{\substack{\gamma \\ (\alpha\neq\beta)(\beta\neq\gamma)}} \mathcal{R}^0(z)\mathcal{T}_\alpha(z)\mathcal{R}^0(z)\mathcal{T}_\beta(z)\mathcal{R}^0(z)\mathcal{T}_\gamma(z)\mathcal{R}^0(z) + \cdots \quad (20.4.3)$$

where we denoted by $\mathcal{T}_\alpha(z)$ the operator

$$\mathcal{T}_\alpha(z) = \mathcal{L}'_\alpha + \sum_{n=1}^{\infty} \mathcal{L}'_\alpha[\mathcal{R}^0(z)\mathcal{L}'_\alpha]^n \quad (20.4.4)$$

It is clear that *adjacent* pairs α, β must be different, but nonadjacent ones can be identical. Equation (20.4.3) is the binary collision expansion, from which all relevant contributions can be easily extracted. For instance, in the case of the Choh–Uhlenbeck equation, we must extract all terms involving three (and only three) particles. There are contributions of this kind in all terms of (20.4.3), corresponding to pairs of interactions (12)(13)(23) in all possible orderings. A rather involved analysis leads to a three-body collision term of the following form (for homogeneous systems):

$$[\partial_t\bar{\varphi}(1)]^{(3)} = \int d\mathbf{v}_2\, d\mathbf{v}_3\, d\mathbf{q}_2\, d\mathbf{q}_3\, K_{123}\bar{\varphi}(1;t)\bar{\varphi}(2;t)\bar{\varphi}(3;t) \quad (20.4.5)$$

with

$$K_{123} = \lim_{\tau\to\infty} L'_{12}\{S_{123}(\tau) - S_{12}(\tau)S_{13}(\tau) - S_{12}(\tau)S_{23}(\tau) + S_{12}(\tau)\} \quad (20.4.6)$$

where

$$S_{12\ldots s}(\tau) = U_{12\ldots s}(-\tau)\prod_{j=1}^{s} U_j^0(\tau) \quad (20.4.7)$$

$U_{12\ldots s}(t)$ being the propagator of the isolated s-body problem. We have no space here for going into a detailed analysis of this equation. Let us just make the following comment in order to understand its overall structure. In using the binary collision expansion (20.4.3) we made no difference between intermediate vacuum and correlation states. We know however from Chapters 16 and 17 that it is essential to exclude intermediate vacuum states in the construction of the collision operator. Otherwise, we would put together true three-body collisions and successive isolated binary collisions. If we, therefore, start from the binary collision expansion (which is simpler to handle), we must correct the final result by subtracting the latter contributions: this is the origin of the subtracted terms $S_{12}S_{13}$, $S_{12}S_{23}$, S_{12} in Eq. (20.4.6). Actually, if we express the collision operator in terms of the irreducible operators defined in Section

17.6, it takes the form:

$$K_{123} = \Psi_{123}(0) + \Psi'_{12}(0)[\Psi_{13}(0) + \Psi_{23}(0)]$$
$$\equiv [\Omega\Psi(0)]_{123} \qquad (20.4.8)$$

The notations are defined in Eqs. (17.8.9), (17.6.11), and (17.6.12). The equivalence of Eq. (20.4.8) with the Choh–Uhlenbeck form (20.4.6) was shown by Résibois.

Choh and Uhlenbeck derived formal expressions for the transport coefficients from Eq. (20.4.5), in the form (e.g., for the heat conductivity):

$$\kappa = \kappa^{(0)} + n\kappa^{(1)} \qquad (20.4.9)$$

The correction $\kappa^{(1)}$ to the Boltzmann value contains two clearly separated contributions: One is due to genuine triple collisions, the other is due to "delocalized" binary collisions and was shown to agree precisely (for hard spheres) with the result obtained from Enskog's equation (20.4.1). It was not before 1966, however, that complete numerical calculations, down to numbers, were performed by Sengers for hard spheres: we shall come back to his results.

After this successful determination of the first "nonequilibrium virial coefficient," a number of people started to work on the next corrections, aiming at a general virial expansion as in equilibrium. But, already at the next step, a serious disappointment cooled the enthusiasm. Cohen and Dorfman showed in 1965 that the four-body collision term contains a divergent contribution.* Actually, for a two-dimensional system, the divergence occurs already for the three-body collision term. The origin of the difficulty lies in the fact that the description of a four-body collision contains some processes in which a particle 1 interacts with a particle 2, then with a particle 3, then again with 2, and that the time between "reinteractions" can become very long, thus causing the divergence of the time integrals involved in the collision operators. In higher orders the divergence becomes, of course, worse.

Thus, *a virial expansion of the transport coefficients does not exist.* One can find in the literature very strong claims, saying that this fact implies the breakdown of Bogoliubov's assumption, hence of kinetic theory. These statements are misleading. Only the aspect B of Bogoliubov's method has been shown to fail; but this is a rather minor one. The main assumption A is not disproved. It was simply showed that the ansatz about the functional form, $\kappa = \sum n^s \kappa^{(s)}$ is not correct, that is, the transport coefficients are not analytic functions of the density (why should they be analytic in the first place?).

* This divergence was noted earlier by Weinstock, in 1963, but only in passing and without any detailed analysis. Independently, it was also found by Goldman and Frieman in 1965.

We have already met repeatedly with divergences in series expansions. What one attempts then is a rearrangement of the series and a partial summation of a subseries. This program is precisely what Kawasaki and Oppenheim did in the present case in 1965. They selected the most divergent terms in each order of the expansion and summed the resulting partial series. This idea is very similar to (and was inspired by) the theory of plasmas. We have seen how it worked in equilibrium (Section 6.5) and shall see again how it works out of equilibrium (Section 20.5). As the present problem is technically more complicated than the plasma problem we do not treat it in detail. The most divergent integrals are many-body contributions involving an arbitrary number of particles in a special configuration. They describe the fact that particle 1 (in the example above), on its long path between two recollisions, actually encounters many other particles of the medium. Hence, its motion is not free but, rather, damped by the interaction with the medium. This damping, in turn, makes the integrals convergent. The result of the summation is an expression of the heat conductivity (for instance) of the form:

$$\kappa = \kappa^{(0)} + \kappa^{(1)} n + \kappa^{(a)} n^2 \ln n + \kappa^{(2)} n^2 + \cdots . \qquad d = 3 \quad (20.4.10)$$

Similarly, in two dimensions,

$$\kappa = \kappa^{(0)} + \kappa^{(a)} n \ln n + \kappa^{(1)} n + \cdots , \qquad d = 2 \quad (20.4.11)$$

Thus, as expected, we obtain finite, but nonanalytic expressions for the transport coefficients. In 1966, Sengers calculated explicitly the coefficients $\kappa^{(a)}$ and $\kappa^{(1)}$ for a two-dimensional gas of hard disks. An interesting fact comes out of his results. The regular term $\kappa^{(1)}$ comes from "excluded-volume" effects of Enskog type: It turns out that it accounts for 97% of the correction to $\kappa^{(0)}$, leaving only 3% for the logarithmic term. It is therefore not impossible that, in higher orders and in three dimensions, the excluded-volume effects alone would provide a good approximation to the transport coefficients: this approximation would be analytic! But this is, of course, mere speculation. Some attempts have been made at a comparison with experimental data. However, at present, neither the theoretical estimates of the coefficients in Eq. (20.4.10) for a Lennard–Jones potential, nor the measurements are sufficiently accurate for a reliable test and, in particular, for ascertaining the presence of the logarithmic term.

20.5. CLASSICAL PLASMAS OUT OF EQUILIBRIUM

The next type of system we shall study here is a classical plasma, as defined and described in Section 6.5. More precisely, for simplicity, we idealize the plasma by using again the picture of an electron gas moving in a neutralizing continuous background. We, moreover, consider only the Coulomb interaction between the electrons: The short-range repulsive interactions are not discussed here because we believe that we understand the basic physics of these interactions from the discussion of Sections 20.2 and 20.3. Just as in the equilibrium problem, we concentrate here on the role of the long-range interactions.*

The basic difficulty manifests itself here in the same way as in equilibrium. Because of their long range, the Coulomb interactions are weak; one may try to use the Landau equation to describe the kinetics of a plasma. However, one runs into difficulties because the collision term diverges. Indeed if we evaluate the characteristic constant B, Eq. (11.6.24) for a Coulomb potential, whose Fourier transform is given by Eq. (6.5.3), we find

$$B = \int_0^\infty dk\, k^3 \tilde{V}_k^2 \sim \int_0^\infty dk\, k^3 k^{-4} = \ln \infty - \ln 0 \tag{20.5.1}$$

The integral diverges logarithmically at both ends: We find here a situation similar to the second virial coefficient (see Section 6.5). The divergence at $k \to \infty$ corresponds to short distances and is due to our neglect of the hard core. The more serious problem is posed by the divergence at $k \to 0$, due to the long range of the Coulomb interactions.

Heuristically, one may argue that the divergence being logarithmic, is not too bad; hence if one integrates between a minimum $k_{\min}$ and a maximum $k_{\max}$, the result will not depend too strongly on the precise choice of the value of these cutoffs. Here we, rather, adopt a more fundamental point of view. We shall show that a systematic analysis of the diagrams allows one to derive a kinetic equation that is convergent at large distances. The resulting equation is quite similar under certain circumstances to a Landau equation with a natural cutoff; under other circumstances, however, the behavior of plasmas may be quite different.

The guiding idea here is again the Debye theory, which provides us with a natural screening of the interactions; however, out of equilibrium the screening mechanism is much more complex. In order to construct a theory consistent with the equilibrium theory, we refer the reader to the discussion of Section 6.5. We came there to the conclusion that the dominant contributions to the problem depend on the charge e and on the density through the combination:

$$e^2(e^2 n)^p, \qquad p = 0, 1, 2, \ldots \tag{20.5.2}$$

* The precise combination of a short-range repulsive core with the long-range interactions in a single convergent kinetic equation is actually not a trivial problem, because of the variety of possible mixed effects. We have no space here for the treatment of this problem (see, however, Section 20.7).

We adopt the same criterion for the construction of a kinetic equation.

The dominant contributions to the collision operator $V\Gamma V$ will be taken as the sum of all terms of Eqs. (17.2.13), (17.2.14) depending on the charge and on the density as in (20.5.2), for all values of $p = 0, 1, 2, \ldots$.

The diagram technique of Chapter 19 provides us with a valuable tool for the quick identification of the relevant diagrams. We first note that the role of the parameter λ is played here by the square of the charge e^2. In order to meet the requirement (20.5.2), we must therefore construct diagrams of the type shown in Fig. 19.3.2*a*, by using a single vertex of type A (which is of order e^2), and an arbitrary number of vertices of type B, which introduce factors $e^2 n$. But we saw in Section 19.2 that every factor $V\mathcal{G}(t)V$ necessarily ends at right with a vertex of type A. Hence, in Eqs. (17.2.13), (17.2.14) all terms $V\Gamma_m V$ with $m > 1$ contain at least a factor e^{2m} and must be neglected in the present approximation. Hence, we find that the only relevant contributions to the collision term (besides $V\mathcal{L}V$) are contained in the term:

$$\int_0^\infty d\tau \, V\mathcal{G}(\tau)V\tilde{\mathcal{U}}(-\tau)V$$

just as in Section 20.2. However, we must now retain only diagrams

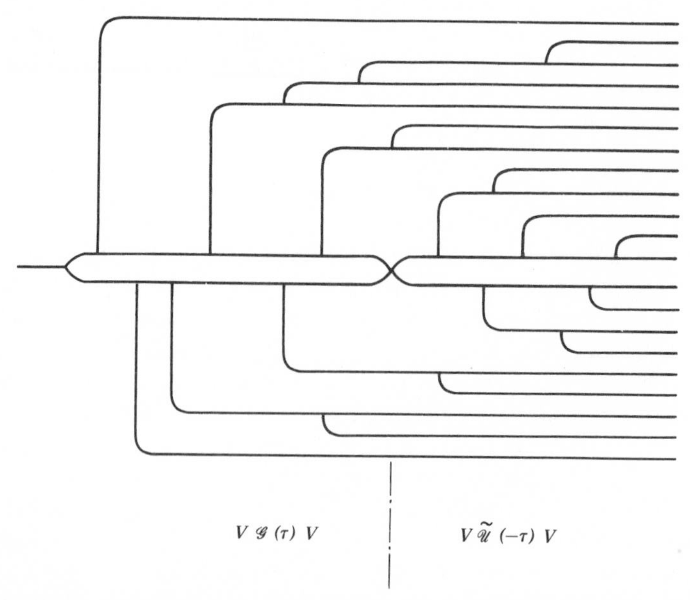

Figure 20.5.1. A typical diagram contributing to the plasma collision term.

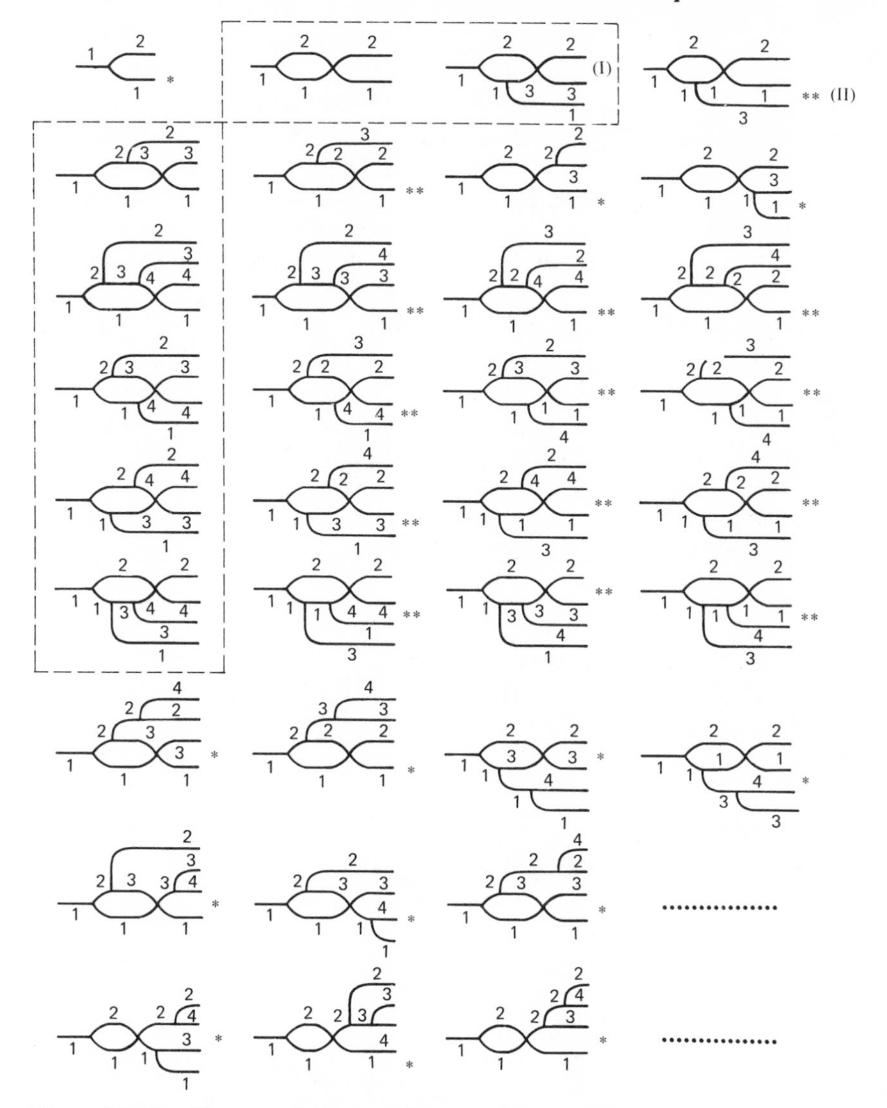

Figure 20.5.2. The set of plasma diagrams through order e^8.

constructed with the maximum number of B vertices and a single A vertex. Hence the approximation is at the opposite extreme as compared to Section 20.2. A typical diagram is shown in Fig. 20.5.1. It displays vividly an important characteristic of the plasma problem: The dominant diagrams involve a very large number of particles. Actually the contributions of order (20.5.2) are those that, in a given order in e^2, involve the maximum number of particles. This feature expresses the *collective character of the Coulomb collisions:* because of the long range of the forces, a large number of particles interact simultaneously.

To sum these diagrams systematically, we drew in Fig. 20.5.2 the first members of the class, through order e^8. The general case is not easily manageable. We shall therefore specialize once more to the case of spatially homogeneous systems, characterized by the conditions (20.3.1), (20.3.2).It is then easily checked that many of the diagrams shown in Fig. 20.5.2 vanish identically. In particular, all the diagrams involving the Vlassov vertex B acting directly on a vacuum pattern vanish because of the second equation (20.3.2): this excludes the diagrams marked by a single star. The remaining diagrams consist of a closed structure (ring) with simple external lines grafted along this ring, and with particle labels distributed in all possible ways. Just as the equilibrium graphs, the Coulomb expressions are most easily evaluated by going over to the Fourier representation. Using Eqs. (2.4.8), (2.4.9) we easily find the expression of the perturbation L'_{12}:

$$L'_{12} = (\nabla_1 V_{12}) \cdot \partial_{12} = \int d\mathbf{l} \exp[i\mathbf{l} \cdot (\mathbf{q}_1 - \mathbf{q}_2)] \tilde{V}_l \, i\mathbf{l} \cdot \partial_{12} \qquad (20.5.3)$$

We now evaluate the first two diagrams of order e^6 drawn in Fig. 20.5.2 to show that one of them vanishes identically. The first one (I) yields*

$$\mathrm{I} = \int_0^\infty d\tau \int_0^\tau d\tau_1 \int d\mathbf{p}_2\, d\mathbf{p}_3\, d\mathbf{q}_2\, d\mathbf{q}_3\, L'_{12} U^0_{12}(\tau - \tau_1) L'_{13} U^0_{123}(\tau_1) L'_{23} \bar{\varphi}(1)\bar{\varphi}(2)\bar{\varphi}(3)$$

$$= \int_0^\infty d\tau \int_0^\tau d\tau_1 \int d\mathbf{p}_2\, d\mathbf{p}_3\, d\mathbf{q}_2\, d\mathbf{q}_3 \int d\mathbf{l}_1\, d\mathbf{l}_2\, d\mathbf{l}_3 \exp[i\mathbf{l}_1 \cdot (\mathbf{q}_1 - \mathbf{q}_2)] \tilde{V}_{l_1}\, i\mathbf{l}_1 \cdot \partial_{12}$$

$$\times \exp[-(\mathbf{v}_1 \cdot \nabla_1 + \mathbf{v}_2 \cdot \nabla_2)(\tau - \tau_1)] \exp[i\mathbf{l}_2 \cdot (\mathbf{q}_1 - \mathbf{q}_3)] \tilde{V}_{l_2}\, i\mathbf{l}_2 \cdot \partial_{13}$$

$$\times \exp[-(\mathbf{v}_1 \cdot \nabla_1 + \mathbf{v}_2 \cdot \nabla_2 + \mathbf{v}_3 \cdot \nabla_3)\tau_1]$$

$$\times \exp[i\mathbf{l}_3 \cdot (\mathbf{q}_2 - \mathbf{q}_3)] \tilde{V}_{l_3}\, i\mathbf{l}_3 \cdot \partial_{23} \bar{\varphi}(1)\bar{\varphi}(2)\bar{\varphi}(3)$$

$$= \int_0^\infty d\tau \int_0^\tau d\tau_1 \int d\mathbf{p}_2\, d\mathbf{p}_3\, d\mathbf{q}_2\, d\mathbf{q}_3 \int d\mathbf{l}_1\, d\mathbf{l}_2\, d\mathbf{l}_3$$

$$\times \exp[i\mathbf{l}_1 \cdot (\mathbf{q}_1 - \mathbf{q}_2) + i\mathbf{l}_2 \cdot (\mathbf{q}_1 - \mathbf{q}_3) + i\mathbf{l}_3 \cdot (\mathbf{q}_2 - \mathbf{q}_3)] \tilde{V}_{l_1}\, i\mathbf{l}_1 \cdot \partial_{12}$$

$$\times \exp\{[-i\mathbf{l}_2 \cdot \mathbf{v}_1 - i\mathbf{l}_3 \cdot \mathbf{v}_2 + i(\mathbf{l}_2 + \mathbf{l}_3) \cdot \mathbf{v}_3](\tau - \tau_1)\} \tilde{V}_{l_2}\, i\mathbf{l}_2 \cdot \partial_{13}$$

$$\times \exp[-i\mathbf{l}_3 \cdot (\mathbf{v}_2 - \mathbf{v}_3)\tau_1] \tilde{V}_{l_3}\, i\mathbf{l}_3 \cdot \partial_{23} \bar{\varphi}(1)\bar{\varphi}(2)\bar{\varphi}(3)$$

$$= (8\pi^3)^2 \int_0^\infty d\tau \int_0^\tau d\tau_1 \int d\mathbf{p}_2\, d\mathbf{p}_3 \int d\mathbf{l}_1\, d\mathbf{l}_2\, d\mathbf{l}_3\, \delta(\mathbf{l}_3 - \mathbf{l}_1)\delta(-\mathbf{l}_2 - \mathbf{l}_3) \tilde{V}_{l_1}\, i\mathbf{l}_1 \cdot \partial_{12}$$

$$\times \exp\{[\cdots](\tau - \tau_1)\} \tilde{V}_{l_2}\, i\mathbf{l}_2 \cdot \partial_{13} \exp[(\cdots)\tau_1] \tilde{V}_{l_3}\, i\mathbf{l}_3 \cdot \partial_{23} \bar{\varphi}(1)\bar{\varphi}(2)\bar{\varphi}(3)$$

* In the intermediate calculations we do not write the factor $e^6 n^2$.

$$= (8\pi^3)^2 \int_0^\infty d\tau \int_0^\tau d\tau_1 \int d\mathbf{p}_2\, d\mathbf{p}_3 \int d\mathbf{l}\; \tilde{V}_l\,(-i\mathbf{l}\cdot\partial_{12})$$
$$\times \exp[-i\mathbf{l}\cdot\mathbf{g}_{12}(\tau-\tau_1)]\tilde{V}_l\,(i\mathbf{l}\cdot\partial_{13})$$
$$\times \exp(i\mathbf{l}\cdot\mathbf{g}_{23}\tau_1)\tilde{V}_l\,(-i\mathbf{l}\cdot\partial_{23})\bar{\varphi}(1)\bar{\varphi}(2)\bar{\varphi}(3)$$

We now go over from the unperturbed propagator to the resolvent; by using the convolution theorem for the Laplace transforms (16.1.21) we find

$$\mathrm{I} = (8\pi^3)^2 \int_0^\infty d\tau (2\pi)^{-1} \int dz$$
$$\times \exp(-iz\tau) \int d\mathbf{p}_2\, d\mathbf{p}_3 \int d\mathbf{l}\; \tilde{V}_l(-i\mathbf{l}\cdot\partial_{12})(i\mathbf{l}\cdot\mathbf{g}_{12}-iz)^{-1}$$
$$\times \tilde{V}_l(i\mathbf{l}\cdot\partial_{13})\,(-i\mathbf{l}\cdot\mathbf{g}_{23}-iz)^{-1}\tilde{V}_l\,(-i\mathbf{l}\cdot\partial_{23})\bar{\varphi}(1)\bar{\varphi}(2)\bar{\varphi}(3)$$
$$= \lim_{\varepsilon\to 0}(8\pi^3)^2 \int d\mathbf{p}_2\, d\mathbf{p}_3 \int d\mathbf{l}\; \tilde{V}_l(-i\mathbf{l}\cdot\partial_{12})\,(i\mathbf{l}\cdot\mathbf{g}_{12}+\varepsilon)^{-1}$$
$$\times \tilde{V}_l(i\mathbf{l}\cdot\partial_{13})(-i\mathbf{l}\cdot\mathbf{g}_{23}+\varepsilon)^{-1}\tilde{V}_l\,(-i\mathbf{l}\cdot\partial_{23})\bar{\varphi}(1)\bar{\varphi}(2)\bar{\varphi}(3)$$

where we used the identity (17.1.23). We, moreover, use the well-known representation of the singular δ distribution, which follows from Eq. (18.3.9):

$$\lim_{\varepsilon\to 0+}\frac{1}{x-i\varepsilon} = \pi i\delta_-(x) \tag{20.5.4}$$

We then write our expression as follows (restoring the factors e^6n^2):

$$\mathrm{I} = e^6n^2(8\pi^3)^2 \int d\mathbf{p}_2\, d\mathbf{p}_3 \int d\mathbf{l}\; \tilde{V}_l(-i\mathbf{l}\cdot\partial_{12})\;\pi\delta_-(\mathbf{l}\cdot\mathbf{g}_{12})$$
$$\times\, \tilde{V}_l\,(i\mathbf{l}\cdot\partial_{13})\;\pi\delta_-(-\mathbf{l}\cdot\mathbf{g}_{23})\;\tilde{V}_l\,(-i\mathbf{l}\cdot\partial_{23})\bar{\varphi}(1)\bar{\varphi}(2)\bar{\varphi}(3)$$

We finally note, by the argument of Eq. (11.6.19), that we may replace the difference of derivatives ∂_{12} by ∂_1, and ∂_{13} by ∂_1; only the last factor ∂_{23} involves two particles that occur in the functions to the left of this operator. Hence the equation is finally written as

$$\mathrm{I} = \int d\mathbf{p}_2\, d\mathbf{p}_3 \int d\mathbf{l}(-8\pi^3e^2ni\tilde{V}_l\,\mathbf{l}\cdot\partial_1)\;\pi\delta_-(\mathbf{l}\cdot\mathbf{g}_{12})$$
$$\times(8\pi^3e^2ni\tilde{V}_l\,\mathbf{l}\cdot\partial_1)\;\pi\delta_-(-\mathbf{l}\cdot\mathbf{g}_{23})(-e^2i\tilde{V}_l\,\mathbf{l}\cdot\partial_{23})\bar{\varphi}(1)\bar{\varphi}(2)\bar{\varphi}(3) \tag{20.5.5}$$

If we now go through exactly the same steps with the second diagram of order e^6, which differs from the first only by a permutation of two labels, we find

$$\mathrm{II} = \int_0^\infty d\tau \int_0^\tau d\tau_1 \int dx_2\, dx_3\; L'_{12}U^0_{12}(\tau-\tau_1)L'_{13}U^0_{123}(\tau_1)L'_{12}\bar{\varphi}(1)\bar{\varphi}(2)\bar{\varphi}(3) = 0$$

The vanishing appears at the fourth step in the pevious set of transformations: the $\mathbf{q}_3$ integration produces a function $\delta(\mathbf{l}_2)$, which is multiplied by a factor $\mathbf{l}_2$ from the operator L'_{13}. This particular feature occurs every time a label appears only once, on an external line. Because of this property all the diagrams marked by two stars in Fig. 20.5.2 vanish identically for homogeneous systems.

The diagrams remaining after this elimination are much simpler: we may take this opportunity and draw them in a simpler and more transparent way. Indeed, we note that in these diagrams, the "skeleton" obtained by removing all the external lines uniquely characterizes the diagram: hence we shall no longer draw these cumbersome lines (Fig. 20.5.3). From our previous detailed calculations we may infer rules allowing us to write down immediately the contributions to the kinetic equation.

The set of remaining diagrams is defined as the set of all *ring diagrams* shown in Fig. 20.5.3. We agree to label the fixed particle [i.e., unintegrated particle, labeled 1 in Eq. (20.5.5)] by the special index α and to always draw it on the lower line. We can generate these rings quite systematically as follows. Start from the set of all rings with n "loops" on the lines. The set of all rings with $n+1$ loops is obtained by adding to each member a loop on the upper line, to the left of all existing ones, and separately by adding a loop on the lower line to the left of the existing ones. Thus each n-loop ring generates two $(n+1)$-loop rings.

To translate the diagrams into equations we introduce the following abbreviations:

$$
\begin{aligned}
d_j &\equiv -8\pi^3 e^2 n \tilde{V}_l \, i\mathbf{l} \cdot \partial_1 \\
d_{jm} &\equiv -e^2 \tilde{V}_l \, i\mathbf{l} \cdot \partial_{jm} \\
\delta_-^{jm} &\equiv \pi \delta_-(\mathbf{l} \cdot \mathbf{g}_{jm}) \\
\int_{123\ldots} &\equiv \int d\mathbf{p}_1 \, d\mathbf{p}_2 \, d\mathbf{p}_3 \cdots
\end{aligned}
\tag{20.5.6}
$$

We note the following symmetries:

$$
\begin{aligned}
-d_j &= d_j^* \\
d_{mj} &= -d_{jm} = d_{jm}^* \\
\delta_-^{mj} &= (\delta_-^{jm})^*
\end{aligned}
\tag{20.5.7}
$$

The star denotes the complex conjugate. The rules are now as follows:

1. The expression $d_\alpha \delta_-^{\alpha 1}$ corresponds to the extreme left vertex.
2. The symbol d_j is written for each loop, j being the label of the line to the left of the loop. These factors are given a sign + or − according to

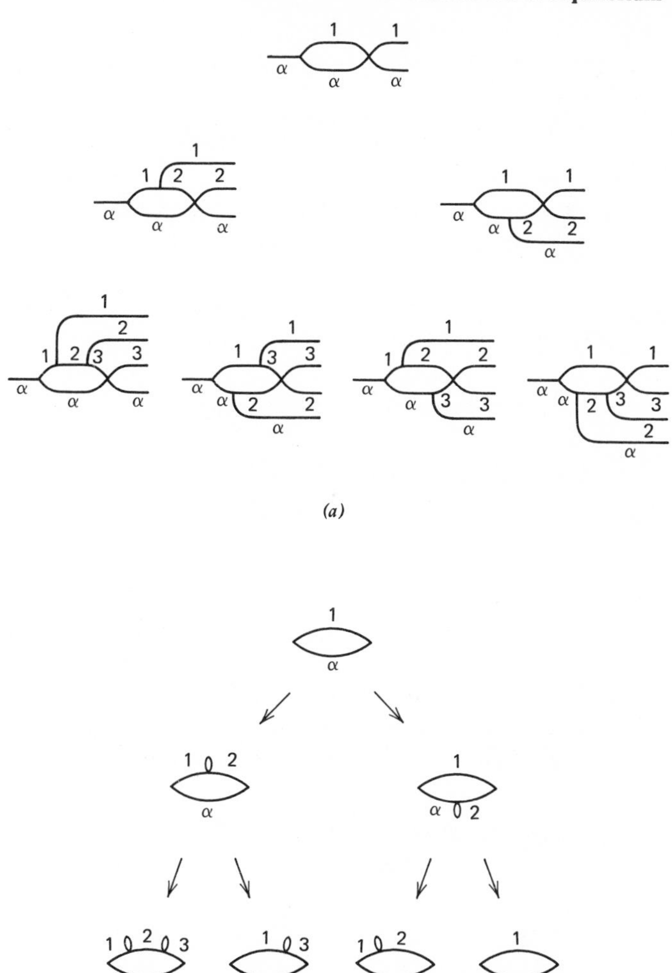

Figure 20.5.3. Simplified notation for the ring diagrams.

whether the loop lies, respectively, on the upper or on the lower side of the ring.

3. Factors δ_-^{is} are inserted between the factors d_j, i being the label of the lower line and s the index of the upper line in the state situated to the right of the loop d_j.

4. A factor d_{si} corresponds to the last vertex on the right, s being the label of the upper line and i that of the lower line in the preceding state.

5. The operator thus obtained acts on a product of kinetic-momentum distribution functions: $\bar{\varphi}(\alpha)\bar{\varphi}(1)\cdots\bar{\varphi}(n)$, $\alpha, 1, \ldots, n$ being the labels of all the particles involved in the ring.

6. The expression is integrated over the momenta $\mathbf{p}_1, \ldots, \mathbf{p}_n$ and over the wave vector $\mathbf{l}$.

We are now ready for the summation of the rings. We write down the contributions of the first classes of rings, through order $e^8 n^3$ (i.e., those of Fig. 20.5.3):

$$\begin{aligned}
\partial_t\bar{\varphi}(\alpha) &= \int d\mathbf{l} \int_1 d_\alpha \delta_-^{\alpha 1}\, d_{1\alpha}\bar{\varphi}(\alpha)\bar{\varphi}(1) \\
&+ \int d\mathbf{l} \int_{12} d_\alpha \delta_-^{\alpha 1}\, d_1 \delta_-^{\alpha 2}\, d_{2\alpha}\bar{\varphi}(\alpha)\bar{\varphi}(1)\bar{\varphi}(2) \\
&\qquad + \int d\mathbf{l} \int_{12} d_\alpha \delta_-^{\alpha 1}(-d_\alpha)\delta_-^{21}\, d_{12}\bar{\varphi}(\alpha)\bar{\varphi}(1)\bar{\varphi}(2) \\
&+ \int d\mathbf{l} \int_{123} d_\alpha \delta_-^{\alpha 1}\, d_1 \delta_-^{\alpha 2}\, d_2 \delta_-^{\alpha 3}\, d_{3\alpha}\bar{\varphi}(\alpha)\bar{\varphi}(1)\bar{\varphi}(2)\bar{\varphi}(3) \\
&\qquad + \int d\mathbf{l} \int_{123} d_\alpha \delta_-^{\alpha 1}(-d_\alpha)\delta_-^{21}\, d_1 \delta_-^{23}\, d_{32}\bar{\varphi}(\alpha)\bar{\varphi}(1)\bar{\varphi}(2)\bar{\varphi}(3) \\
&+ \int d\mathbf{l} \int_{123} d_\alpha \delta_-^{\alpha 1}\, d_1 \delta_-^{\alpha 2}(-d_\alpha)\delta_-^{32}\, d_{23}\bar{\varphi}(\alpha)\bar{\varphi}(1)\bar{\varphi}(2)\bar{\varphi}(3) \\
&\qquad + \int d\mathbf{l} \int_{123} d_\alpha \delta_-^{\alpha 1}(-d_\alpha)\delta_-^{21}(-d_2)\delta_-^{31}\, d_{13}\bar{\varphi}(\alpha)\bar{\varphi}(1)\bar{\varphi}(2)\bar{\varphi}(3) + \cdots
\end{aligned} \tag{20.5.8}$$

All the terms begin by a common feature: we can therefore write

$$\partial_t\bar{\varphi}(\alpha) = \int d\mathbf{l}\; d_\alpha F(\alpha) \tag{20.5.9}$$

where

$$\begin{aligned}
F(\alpha) = \int_1 \delta_-^{\alpha 1}\, d_{1\alpha}\bar{\varphi}(\alpha)\bar{\varphi}(1) &+ \int_1 \delta_-^{\alpha 1}\, d_1\bar{\varphi}(1) \int_2 \delta_-^{\alpha 2}\, d_{2\alpha}\bar{\varphi}(\alpha)\bar{\varphi}(2) \\
&+ \int_1 \delta_-^{\alpha 1}\, d_{1\alpha}^*\bar{\varphi}(\alpha) \int_2 (\delta_-^{12})^*\, d_{21}^*\bar{\varphi}(1)\bar{\varphi}(2) + \cdots
\end{aligned} \tag{20.5.10}$$

[We used the symmetry properties (20.5.7).] It is easily seen, by writing down further terms, that the two last terms in the rhs are each the first one in a series reproducing the functions F and F^*, respectively. Hence we obtain

$$F(\alpha) = \int_1 \delta_-^{\alpha 1}\, d_{1\alpha}\bar{\varphi}(\alpha)\bar{\varphi}(1) + \int_1 \delta_-^{\alpha 1}\, d_1\bar{\varphi}(1)F(\alpha) + \int_1 \delta_-^{\alpha 1}\, d_{1\alpha}^*\bar{\varphi}(\alpha)F^*(1) \tag{20.5.11}$$

We now introduce the following functions*:

$$\varepsilon(\alpha) \equiv 1 - \int_1 \delta_-^{\alpha 1}\, d_1\bar{\varphi}(1) = 1 + 8\pi^4 e^2 n i \tilde{V}_l m^{-1} \int d\mathbf{v}_1\, \delta_-(\mathbf{l}\cdot\mathbf{g}_{\alpha 1})\, \mathbf{l}\cdot\partial_1\bar{\varphi}(1)$$

$$q(\alpha) \equiv \int_1 \delta^{\alpha 1}\, d_{1\alpha}\bar{\varphi}(\alpha)\bar{\varphi}(1)$$

$$= \pi i e^2 m^{-1} \tilde{V}_l \int d\mathbf{v}_1\, \delta_-(\mathbf{l}\cdot\mathbf{g}_{\alpha 1})\, \mathbf{l}\cdot\partial_{\alpha 1}\bar{\varphi}(\alpha)\bar{\varphi}(1) \tag{20.5.12}$$

$$d(\alpha) \equiv 8\pi^3 e^2 n m^{-1} \tilde{V}_l\, \mathbf{l}\cdot\partial_\alpha\bar{\varphi}(\alpha)$$

With these new notations we now write our two basic equations, (20.5.9), (20.5.11) as follows:

$$\partial_t\bar{\varphi}(\alpha) = -8\pi^3 i e^2 n m^{-1} \int d\mathbf{l}\, \tilde{V}_l\, \mathbf{l}\cdot\partial_\alpha F(\alpha) \tag{20.5.13}$$

where the function $F(\alpha)$ obeys the integral equation:

$$\varepsilon(\alpha)\mathrm{F}(\alpha) = q(\alpha) + i\pi d(\alpha) \int d\mathbf{v}_1\, \delta_-(\mathbf{l}\cdot\mathbf{v}_\alpha - \mathbf{l}\cdot\mathbf{v}_1) F^*(1) \tag{20.5.14}$$

The result of the summation of the diagrams will be quite explicit, if we are able to solve Eq. (20.5.14). This is done in the next section.

20.6. KINETIC EQUATION FOR CLASSICAL PLASMAS

The function $F(\alpha)$ appearing in Eqs. (20.5.13), (20.5.14) depends on the velocity $\mathbf{v}_\alpha$, on the wave vector $\mathbf{l}$, and is also a functional of the distribution function $\bar{\varphi}(\alpha; t)$. Being a complex number, it can be split into a real and an imaginary part:

$$F(\alpha) = F_1(\alpha) + iF_2(\alpha) \tag{20.6.1}$$

The coefficients appearing in Eq. (20.5.12) are also complex and can be split in the same way. Equation (20.5.14) is thus equivalent to a set of two coupled equations for $F_1(\alpha)$ and $F_2(\alpha)$.

We first note that *only the imaginary part F_2 can contribute to the kinetic equation (20.5.13)*. This can be checked *a posteriori*; however, it can be easily understood that a real F_1 would yield an imaginary contribution to $\bar{\varphi}$, which would be unphysical (because $\bar{\varphi}$ is essentially a probability). We therefore only calculate $F_2(\alpha)$.†

* Note that we switch again from the momenta $\mathbf{p}_j$ to the velocities $\mathbf{v}_j$, whence the factors m^{-1} in the forthcoming formulas.

† $F_1(\alpha)$ can also be found exactly, although the procedure is more involved.

Next, we note that $F(\alpha)$, as well as $q(\alpha)$ and $d(\alpha)$, depend on the vector variable $\mathbf{v}_\alpha$, that is, on three scalar variables. But the kernel of the integral equation, $\delta_-(\mathbf{l}\cdot\mathbf{v}_\alpha - \mathbf{l}\cdot\mathbf{v}_1)$ depends on $\mathbf{v}_\alpha$ and $\mathbf{v}_1$ only through the scalar products $\mathbf{l}\cdot\mathbf{v}_\alpha$ and $\mathbf{l}\cdot\mathbf{v}_1$, that is, on the projection of the velocity on the wave vector $\mathbf{l}$. Similarly, $\varepsilon(\alpha)$ only depends on $\mathbf{l}\cdot\mathbf{v}_\alpha$. We introduce therefore the symbol:

$$\nu_\alpha \equiv \frac{\mathbf{l}\cdot\mathbf{v}_\alpha}{l} \tag{20.6.2}$$

Moreover, with every function $f(\mathbf{v})$ of $\mathbf{v}$, we associate a function $\tilde{f}(\nu)$ defined by integration over the transverse components of $\mathbf{v}$, that is:

$$\tilde{f}(\nu) = \int d\mathbf{v}\;\delta\left(\nu - \frac{\mathbf{l}\cdot\mathbf{v}}{l}\right) f(\mathbf{v}) \tag{20.6.3}$$

We now multiply both sides of Eq. (20.5.14) by $\delta(\nu_\alpha - \mathbf{l}\cdot\mathbf{v}_\alpha/l)$ and integrate over $\mathbf{v}_\alpha$. We note that

$$\int d\mathbf{v}_\alpha\;\delta\left(\nu_\alpha - \frac{\mathbf{l}\cdot\mathbf{v}_\alpha}{l}\right)\varepsilon\left(\frac{\mathbf{l}\cdot\mathbf{v}_\alpha}{l}\right)F(\alpha) = \varepsilon(\nu_\alpha)\tilde{F}(\nu_\alpha)$$

Also:

$$\begin{aligned}
&\int d\mathbf{v}_\alpha\;\delta\left(\nu_\alpha - \frac{\mathbf{l}\cdot\mathbf{v}_\alpha}{l}\right) d(\mathbf{v}_\alpha)\int d\mathbf{v}_1\;\delta_-(\mathbf{l}\cdot\mathbf{v}_\alpha - \mathbf{l}\cdot\mathbf{v}_1)F^*(\mathbf{v}_1)\\
&= \tilde{d}(\nu_\alpha)\int d\mathbf{v}_1\;\delta_-(l\nu_\alpha - \mathbf{l}\cdot\mathbf{v}_1)F^*(\mathbf{v}_1)\\
&= \tilde{d}(\nu_\alpha)\int d\nu_1\;\delta_-(l\nu_\alpha - l\nu_1)\tilde{F}^*(\nu_1)\\
&= l^{-1}\,\tilde{d}(\nu_\alpha)\int d\nu_1\;\delta_-(\nu_\alpha - \nu_1)\tilde{F}^*(\nu_1)
\end{aligned}$$

We, moreover, note the relation:

$$\left(\frac{\pi}{l}\right)\tilde{d}(\nu) = \varepsilon_2(\nu) \tag{20.6.4}$$

Putting these results together we find an equation for $\tilde{F}(\nu)$:

$$\varepsilon(\nu_\alpha)\tilde{F}(\nu_\alpha) = \tilde{q}(\nu_\alpha) + i\varepsilon_2(\nu_\alpha)\int d\nu_1\;\delta_-(\nu_\alpha - \nu_1)\tilde{F}^*(\nu_1) \tag{20.6.5}$$

This is a much simpler equation than (20.5.14) because it is one dimensional. But it is important to note that if (20.6.5) is solved, we automatically obtain the solution of (20.5.14) as well. Indeed, comparing the two equations, and using the obvious relation:

$$\int d\mathbf{v}_1\;\delta_-(\mathbf{l}\cdot\mathbf{v}_\alpha - \mathbf{l}\cdot\mathbf{v}_1)F^*(\mathbf{v}_1) = l^{-1}\int d\nu_1\;\delta_-(\nu_\alpha - \nu_1)\tilde{F}^*(\nu_1) \tag{20.6.6}$$

we can eliminate the integral between the two and obtain a simple algebraic relation between F and $\tilde{F}$:

$$F = \pi \frac{d}{l\varepsilon_2} \tilde{F} + \frac{1}{\varepsilon}\left[q - \frac{\pi d}{l\varepsilon_2} \tilde{q}\right] \tag{20.6.7}$$

In particular, the imaginary part is

$$F_2 = \frac{\pi d}{l\varepsilon_2} \tilde{F}_2 + \frac{1}{\varepsilon\varepsilon^*}\left[\varepsilon_1 q_2 - \varepsilon_2 q_1 + \frac{\pi d}{l} \tilde{q}_1\right] \tag{20.6.8}$$

We have used here the property:

$$\tilde{q}_2(\nu) = 0 \tag{20.6.9}$$

which follows from the definition (20.5.12). We now split Eq. (20.6.5) into real and imaginary components, using all the previous results:

$$\varepsilon_1 \tilde{F}_2 + \varepsilon_2 \mathscr{P} \tilde{F}_2 = 0 \tag{20.6.10}$$

$$\varepsilon_1 \tilde{F}_1 - \varepsilon_2 \mathscr{P} \tilde{F}_1 = \tilde{q}_1 + 2\varepsilon_2 \tilde{F}_2 \tag{20.6.11}$$

We introduced here a simple notation for the principal part:

$$\mathscr{P} f \equiv \frac{1}{\pi} \mathscr{P} \int_{-\infty}^{\infty} d\nu_1 \frac{f(\nu_1)}{\nu_\alpha - \nu_1} \tag{20.6.12}$$

Let us concentrate on Eq. (20.6.10): it has a number of simple features. First, we note that it does not involve $\tilde{F}_1$: This decoupling is extremely helpful: it is due to the property (20.6.4). Next, it is a homogeneous equation [because of Eq. (20.6.9)]. Moreover, it can be shown that the equation possesses a unique solution, provided a certain condition on $\bar{\varphi}(\alpha)$ is satisfied.* Here the argument is rather subtle, and involves the properties of singular integral equations. It would be leading us too far to go into these matters here. Let us accept the theorem and note that the extra condition is automatically satisfied if we are not too far from equilibrium.

From these remarks the solution follows without any further calculation. Equation (20.6.10) can be solved separately; being homogeneous it possesses a trivial solution, and by the third property, this solution is unique. Hence

$$\tilde{F}_2(\nu) = 0 \tag{20.6.13}$$

and from (20.6.8)

$$F_2(\mathbf{v}) = (\varepsilon\varepsilon^*)^{-1}\left[\varepsilon_1 q_2 - \varepsilon_2 q_1 + \left(\frac{\pi d}{l}\right)\tilde{q}_1\right] \tag{20.6.14}$$

* The condition is the *stability of the plasma,* which is expressed by the requirement that the function $\varepsilon_1 - i\varepsilon_2$ has no zeros in the upper half-plane. A typical example of an unstable plasma is one in which $\bar{\varphi}(\nu)$ has two maxima separated by more than a critical distance.

By using the definitions (20.5.12) and the abbreviations: $\tilde{f}'(\nu) = d\tilde{f}(\nu)/d\nu$ and $f'(\mathbf{v}) = l^{-1}\mathbf{l} \cdot \partial f(\mathbf{v})/\partial \mathbf{v}$, we obtain

$$F_2 = (\varepsilon\varepsilon^*)^{-1}\left(\frac{\pi e^2}{m}\right)\tilde{V}_l[\bar{\varphi}'(\alpha)\bar{\bar{\varphi}}(\alpha) - \bar{\varphi}(\alpha)\bar{\bar{\varphi}}'(\alpha)]$$

$$+ (\varepsilon\varepsilon^*)^{-1}\left(\frac{8\pi^4 e^4 n}{m^2}\right)\tilde{V}_l^2\{\mathcal{P}\bar{\bar{\varphi}}'(\alpha)[\bar{\varphi}'(\alpha)\bar{\bar{\varphi}}(\alpha) - \bar{\varphi}(\alpha)\bar{\bar{\varphi}}'(\alpha)]$$

$$- \bar{\bar{\varphi}}'(\alpha)[\bar{\varphi}'(\alpha)\mathcal{P}\bar{\bar{\varphi}}(\alpha) - \bar{\varphi}(\alpha)\mathcal{P}\bar{\bar{\varphi}}'(\alpha)]$$

$$+ \bar{\varphi}'(\alpha)[\bar{\bar{\varphi}}'(\alpha)\mathcal{P}\bar{\bar{\varphi}}(\alpha) - \bar{\bar{\varphi}}(\alpha)\mathcal{P}\bar{\bar{\varphi}}'(\alpha)]\}$$

The terms enclosed in curly brackets cancel each other and we are left with

$$F_2 = \frac{q_2}{\varepsilon\varepsilon^*} \tag{20.6.15}$$

Substituting this result into Eq. (20.5.13) we obtain the explicit kinetic equation:

$$\partial_t\bar{\varphi}(\mathbf{v}_\alpha; t) = 8\pi^4 e^4 n m^{-2}\int d\mathbf{l}\int d\mathbf{v}_1\, \mathbf{l} \cdot \partial_\alpha\left\{\frac{\tilde{V}_l^2}{|\varepsilon(\alpha)|^2}\right\}\delta(\mathbf{l} \cdot \mathbf{g}_{\alpha 1})\mathbf{l} \cdot \partial_{\alpha 1}\bar{\varphi}(\mathbf{v}_\alpha; t)\bar{\varphi}(\mathbf{v}_1; t) \tag{20.6.18}$$

and we recall that

$$\varepsilon(\alpha) = 1 + 8\pi^4 e^2 n m^{-1}\tilde{V}_l\int d\mathbf{v}_1\, i\delta_-(\mathbf{l} \cdot \mathbf{v}_\alpha - \mathbf{l} \cdot \mathbf{v}_1)\mathbf{l} \cdot \partial_1\bar{\varphi}(\mathbf{v}_1; t) \tag{20.6.17}$$

Equation (20.6.16) was derived independently in 1960 by Lenard, by Guernsey, and by the author. At first sight it looks very much like the Landau equation. This is satisfactory because a plasma, in zeroth approximation, does look like a weakly coupled gas (as was pointed out already in Section 11.6). The difference with the ordinary Landau equation is that the potential $\tilde{V}_l$ is replaced by an effective potential:

$$\tilde{V}_l^{\text{eff}} = \frac{\tilde{V}_l}{\varepsilon} \tag{20.6.18}$$

Hence ε plays the role of a *dielectric constant*. If it is evaluated in equilibrium, we find a function of $\mathbf{l}$ and $\mathbf{v}$, which for $\mathbf{v}=0$ reduces the effective potential to

$$V_l^{\text{eff}}(\mathbf{v}=0) = \frac{2\pi^2}{l^2}\frac{1}{1+\kappa_D^2/l^2} = \frac{2\pi^2}{l^2+\kappa_D^2} \tag{20.6.19}$$

This is the Fourier transform of the Debye potential, which we know from equilibrium theory (Section 6.5). Thus we understand that the summation of the ring diagrams produces a *screening* of the potential that has the effect of providing an effective cutoff at long distances (or short wave vectors). The screening is a collective effect involving a large number of particles.

Out of equilibrium, however, the screening mechanism is more complex. We note that the effective potential depends on the velocity of the particles; it actually is fixed by the instantaneous velocity distribution $\bar{\varphi}(\mathbf{v}; t)$. This feature introduces an important conceptual difference with the Landau equation. Equation (20.6.16) is much more strongly nonlinear: the distribution function appears in the denominator through $\varepsilon(\alpha)$. This expresses the fact that Coulomb "collisions" are effectively many-body collisions, involving a large number of particles (as is clear from the very nature of the ring diagrams). As a result, the effective potential changes in time as the distribution itself changes. The evolution process is much richer than in the previously studied cases in which only two-body collisions were involved. The effects induced by the peculiar nonlinearity still await a deeper study.

It is remarkable, however, that these complicated interaction mechanisms still lead to the same gross features as before. All the properties studied in Chapters 12 and 13 can be immediately extended to Eq. (20.6.16). The collision invariants are the same as before and, more important, the H theorem holds as for the Landau equation. Hence, we find again a truly irreversible evolution for the kinetic component of the distribution function.

We have no space here for going into a detailed analysis of the picture of a plasma emerging from the ring equation, and even less for discussing the refinements of the theory that were developed in recent years. We refer the reader to the literature, and stop at this point the illustration of a nontrivial summation technique in which the diagrams proved quite useful.

20.7. OTHER SOLUBLE PROBLEMS

Interacting many-body systems are obviously quite complicated objects, and we are not able to write down exact, explicit, and compact kinetic equations describing their time evolution. In the previous sections we met, however, two types of systems that, in certain limiting cases, can be treated exactly. More precisely, an infinite class of diagrams can be summed exactly in each case, and yields an explicit kinetic equation. The class of relevant diagrams is defined in each case by the requirement that a certain parameter μ is small and we are only interested in the limit $\mu \to 0$. In the first case, the smallness parameter is proportional to the density, in the second it is the plasma parameter. There are very few other problems of this kind imagined and solved to date, especially in non-equilibrium theory. We shall briefly describe in this section two such problems.

A. The Brownian Motion of a Heavy Particle

The problem of Brownian motion is a "classic" of statistical physics. Its study started at the beginning of this century, and the field is still very active today. The traditional approach to Brownian motion is through the theory of stochastic processes: this approach was described in Chapter 11. We stressed there its phenomenological character: a number of crucial assumptions have to be made in a completely *ad hoc* way in order to ensure the irreversible approach to equilibrium. One of the purposes of statistical mechanics is to provide a mechanical description of the Brownian motion. This would yield a definition of the phenomenological parameters (such as the friction constant) in terms of molecular quantities.

We saw in Section 11.8 that a weakly coupled gas provides us with a mechanical model of such a system. The approach to equilibrium of a "tagged" particle moving in a medium that is already in equilibrium is described by laws of the same kind as in the Brownian motion. In particular, we saw how the Fokker–Planck equation can be derived from the Landau equation, which in turn was derived from mechanics in Chapter 18. However, the physical mechanism of approach to equilibrium is quite different here and in the true Brownian motion. We first note that the distinguished particle is no different from the particles of the medium: it is "tagged" only by its initial condition. On the other hand, in the problem described in Sections 11.2 and 11.3 the interactions between particles are by no means weak: we may have arbitrarily strong collisions.

The system considered in the Brownian-motion problem can be described as a fluid of N particles of mass $m = 1$ and coordinates $(\mathbf{q}_1 \mathbf{v}_1)$ to $(\mathbf{q}_N \mathbf{v}_N)$ plus an extra particle of mass M and coordinates $(\mathbf{Q}, \mathbf{V})$. The Hamiltonian is

$$H = \tfrac{1}{2}MV^2 + \tfrac{1}{2}\sum_{j=1}^{N} v_j^2 + \sum_{i>j=1}^{N}\sum V_{ij}(\mathbf{q}_i - \mathbf{q}_j) + \sum_{i=1}^{N} U_i(\mathbf{q}_i - \mathbf{Q}) \qquad (20.7.1)$$

where V_{ij} describes the interaction between fluid particles and U_i the interaction between fluid and Brownian particles.

With the abbreviations

$$\partial_j = \frac{\partial}{\partial \mathbf{v}_j}, \qquad \partial = \frac{\partial}{\partial \mathbf{V}}$$

$$\nabla_j = \frac{\partial}{\partial \mathbf{q}_j}, \qquad \nabla = \frac{\partial}{\partial \mathbf{Q}}$$

the corresponding Liouvillian is

$$L_N = -\mathbf{V}\cdot\boldsymbol{\nabla} - \sum_j \mathbf{v}_j\cdot\boldsymbol{\nabla}_j + \sum_{i<j}\sum (\boldsymbol{\nabla}_i V_{ij})\cdot\partial_{ij} + \sum_i (\boldsymbol{\nabla} U_i)\cdot(\partial_i - M^{-1}\partial) \quad (20.7.2)$$

The typically Brownian case is obtained when $M \gg 1$. This assumption implies that the velocity $\mathbf{V}$ is predominantly small; as in equilibrium $\langle V\rangle \sim (k_B T/M)^{1/2}$, we assume in general that $V \sim M^{-1/2}$. We then introduce the smallness parameter:

$$\mu = M^{-1/2} \quad (20.7.3)$$

If we split the Liouvillian in the usual way:

$$L_{N+1} = L^0_{N+1} + L'_{N+1} \quad (20.7.4)$$

we find that *both* the free-motion Liouvillian and the interaction part contain a term proportional to μ:

$$\begin{gathered} L^0_{N+1} = L^0_N + \mu L^0_B \\ L^0_N = -\sum_j \mathbf{v}_j\cdot\boldsymbol{\nabla}_j, \qquad \mu L^0_B = -\mathbf{V}\cdot\boldsymbol{\nabla} \end{gathered} \quad (20.7.5)$$

and

$$\begin{aligned} L'_{N+1} &= L'_N + \mu L'_B \\ L'_N &= \sum_{j<n}\sum (\boldsymbol{\nabla}_j V_{jn})\cdot\partial_{jn} + \sum_j [\boldsymbol{\nabla}_j U_j(\mathbf{q}_j - \mathbf{Q})]\cdot\partial_j \\ \mu L'_B &= M^{-1}\sum_j [\boldsymbol{\nabla}_j U_j(\mathbf{q}_j - \mathbf{Q})]\cdot\partial \end{aligned} \quad (20.7.6)$$

We may note that the second term in L'_N amounts to describing the motion of the fluid particles under the action of an external potential due to the Brownian particle fixed at point $\mathbf{Q}$.

The program is now straightforward. We proceed exactly as in Chapters 18 and 17, defining formally the relevant operators $\mathcal{R}(z)$, $\tilde{\mathscr{E}}(z)$, $\mathbf{\Gamma}$, $\mathbf{C}$, and so on for the Liouvillian $\mathscr{L} = \mathscr{L}^0 + \mathscr{L}'$ derived from Eq. (20.7.4). The general expressions depend on μ through $\mathscr{L}^0$ and $\mathscr{L}'$. These expressions can be expanded, and only the lowest-order term (which is proportional to μ^2) is kept in the limit $\mu \to 0$.

This program was carried out by Lebowitz, Rubin, Résibois, and Davis. We shall not go into the details here but only quote the final result. The kinetic equation for the Brownian-particle distribution function $\bar{f}(\mathbf{Q}, \mathbf{V}; t)$ is

$$\partial_t \bar{f} + \mathbf{V}\cdot\boldsymbol{\nabla}\bar{f} = \zeta\partial\cdot\{\mathbf{V} + (\beta M)^{-1}\partial\}\bar{f} \quad (20.7.7)$$

This equation is identical to the *Fokker–Planck equation* (11.3.21), which has now been derived from mechanics in the natural limit $\mu \to 0$, with no approximation on the coupling strength, density, and so on. But now ζ is no longer a phenomenological parameter, but is expressed in terms of the dynamics:

$$\zeta = \tfrac{1}{3}\beta M^{-1} \lim_{\varepsilon\to 0} \int_0^\infty d\tau\, e^{-\varepsilon\tau} \langle \mathscr{F}(\tau) \cdot \mathscr{F}(0) \rangle \tag{20.7.8}$$

where $\mathscr{F}(\tau)$ is the force exerted on the Brownian particle by the fluid particles at time τ (in the Heisenberg representation), with the constraint that the former is fixed at position $\mathbf{Q}$:

$$\mathscr{F}(\tau) = \sum_j \nabla U_j[\mathbf{q}_j(\tau) - \mathbf{Q}]$$

The average $\langle \cdots \rangle$ is taken over an equilibrium ensemble. *The friction constant thus appears as the time integral of the force autocorrelation function.* Such quantities are very important in statistical physics and will be discussed more fully in Chapter 21. We already saw in Section 13.4 that the transport coefficients are quantities of this same type. This autocorrelation function is a complicated object: it cannot, in general, be calculated explicitly. But the important point in the mechanical derivation is to provide us with a link between a macroscopic parameter describing the dissipation ζ and the microscopic correlation function of the intermolecular forces.

B. The van der Waals Fluid Out of Equilibrium

We saw in Section 9.4 that an important system whose equilibrium properties are exactly soluble is provided by a fluid of particles interacting through a Kac potential:

$$V^{\text{tot}}(r) = V^{\text{HS}}(r) + \gamma^3 \varphi(\gamma r) \tag{20.7.9}$$

The notations and the properties of this long-range potential were explained in detail in Section 9.4. It is, of course, tempting to investigate the nonequilibrium properties of this system as well, in an attempt at an exact solution. This problem was successfully attacked recently by Résibois, Piasecki, and Pomeau, thus adding a new member to the class of exactly summable problems. Their work is technically rather difficult, but the ideas are clear and simple: We summarize them here.

If one tries to treat the problem by a straightforward perturbation expansion in powers of the small parameter γ (the inverse range of the attractive force), one finds that the individual terms diverge in the limit $\gamma \to 0$. Since the physical origin of the difficulty lies in the long range of the interactions, one would be tempted to apply the same techniques as for the case of plasmas (Sections 20.5 and 20.6). One realizes, however, that the problem is quite different.

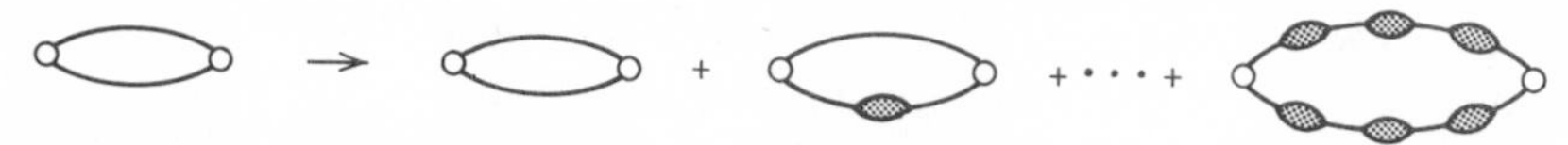

Figure 20.7.1. Renormalization of the propagators for a van der Waals fluid External lines are not represented in these diagrams (see Fig. 20.5.3). The shaded bubbles represent the sum of all possible collision operators. The open circle represents a vertex involving the long-range Kac potential.

(a) In the case of the Coulomb interactions, the potential energy $V(r)=e^2/r$ is a nonintegrable function: the integral of this function diverges at both short and long distances. This is not so for the Kac potential. It is constructed in such a way that its integral over all space is finite and independent of γ [see Eq. (9.4.4)].

(b) The existence of two kinds of charges, together with the global electroneutrality, provides a screening mechanism in the Coulomb case. Nothing analogous exists in the van der Waals problem, so that one should not select the diagrams on the basis of the same arguments.

In order to find a way out, we may argue as follows. A "collision process" involving the weak, long-range Kac potential has a very long effective duration. During this process the two (or more) partners cannot be considered to move freely as in the Landau approximation: rather, they feel the influence of the interaction with all the other particles in the medium.

Consider a simple Landau diagram as in Fig. 20.7.1, involving the long-range Kac potential. To express the previous idea we consider together with it all the diagrams involving collision operators on the two propagating lines. Analytically, it is not difficult to show that this summation involves the replacement of the free propagator* $(i\mathbf{k}\cdot\mathbf{v}-iz)^{-1}$ by a "dressed" propagator:

$$\tilde{X}_{\mathbf{k}}(\mathbf{v};z)=(i\mathbf{k}\cdot\mathbf{v}-K_{\mathbf{k}}(\mathbf{v};z;\gamma)-iz)^{-1} \tag{20.7.10}$$

$K_{\mathbf{k}}(\mathbf{v};z;\gamma)$ is a linearized collision operator, which, in the Landau approximation would be identical to the operator $K_{\mathbf{k}}$ of Section 13.3; however, it is now formally retained to all orders in the coupling parameter. Because of the occurrence of this collision operator, the renormalized propagator is a very complicated object. We may, however, represent it formally in the limit $z\to\varepsilon\to 0$, in terms of its own eigenfunctions $|\psi_\nu\rangle$ and eigenvalues $\lambda_{\mathbf{k}}^{\nu}(\gamma)$ in the form:

$$\tilde{X}_{\mathbf{k}}(\mathbf{v})=\sum_{\nu}|\psi_\nu\rangle[-\lambda_{\mathbf{k}}^{\nu}(\gamma)-i\varepsilon]^{-1}\langle\psi_\nu| \tag{20.7.11}$$

* In the Fourier–Laplace representation; see Section 20.5.

At this point there enters the simplification introduced by the Kac potential. Because of its long range, the only relevant wave vectors in the propagators are those of very small length, $k \leq \gamma$: all the others are cut down by the factor $\tilde{V}_k$ occurring in the vertex. But we know from the discussion of Section 13.3* that the dominant eigenvalues for small values of k are the *five hydrodynamical modes:* these are of order k and k^2, all others being at least of order k^3 or higher. Hence, the propagator (20.7.11) can be approximated, in the limit: $\gamma \to 0$, $k \to 0$, $k\gamma^{-1} \equiv y = \text{finite}$:

$$\tilde{X}_k(\mathbf{v}) \approx \sum_{\alpha=1}^{5} |\psi_\alpha\rangle [-\lambda_k^\alpha(\gamma) - i\varepsilon]^{-1} \langle \psi_\nu| \tag{20.7.12}$$

Moreover, these eigenvalues are expressed solely in terms of macroscopic parameters, such as the transport coefficients, the specific heats, and the like. Hence we arrive at the conclusion that the renormalized collision operator of the van der Waals gas is also expressed in terms of these parameters.

The previous argument refers to one simple class of diagrams. Résibois, Piasecki, and Pomeau extended the analysis to all the possible diagrams and selected on this basis all the dominant contributions in the limit $\gamma \to 0$. All these dominant diagrams can be expressed essentially in terms of hydrodynamic propagation during the long-range collision process: therefore they only depend on macroscopic parameters. The authors did not write down a kinetic equation but, rather, calculated directly the transport coefficients from the autocorrelation formulas (see Chapter 21). We refer the reader to the original papers for the details of this lengthy analysis and just quote a typical result, the expression for the thermal conductivity:

$$\kappa = \frac{k_B}{6\pi^2} \int_0^\infty dy \Biggl\{ \Biggl[\left(1 + \frac{n}{2U_s^2(y)}\, y\, \frac{\partial \tilde{\varphi}(y)}{\partial y}\right)^2 \frac{U_s^2(y)}{2\Gamma(y)} - \frac{U_s^{2\,\mathrm{HS}}}{2\Gamma^{\mathrm{HS}}} \Biggr] + 2T \Biggl[\frac{c_P(y)}{\eta^{\mathrm{HS}}/n + \kappa^{\mathrm{HS}}/n c_P(y)} - \frac{c_P^{\mathrm{HS}}}{\eta^{\mathrm{HS}}/n + \kappa^{\mathrm{HS}}/n c_P^{\mathrm{HS}}} \Biggr] \Biggr\} \tag{20.7.13}$$

Here $\tilde{\varphi}(y)$ is the Fourier transform of the Kac potential, as a function of $y = k/\gamma$; U_s^{HS}, c_P^{HS}, Γ^{HS}, η^{HS}, κ^{HS} are the sound velocity, specific heat, sound absorption, viscosity, and thermal conductivity, respectively, of the *reference system* of hard spheres; $U_s(y)$, $c_P(y)$, and $\Gamma(y)$ are generalized wave-vector dependent sound velocity, . . . , involving the Kac potential:

$$U_s^{2\,\mathrm{HS}} = \frac{c_P(y)}{c_V^{\mathrm{HS}}} \frac{1}{[(\partial P/\partial n)^{\mathrm{HS}} + n\tilde{\varphi}(y)]}$$

* The discussion of Section 13.3 refers to the Landau approximation; however, all those statements are quite general, in all orders of perturbation theory, as shown by Résibois.

We do not give the other definitions, which are complicated, and simply note that, if $\tilde{\varphi}(y)=0$, $U_s(y) \to U_s^{HS}$. Hence, the transport coefficients of the van der Waals fluid are expressed as quadratures involving the thermodynamic and transport coefficients of the reference hard-sphere fluid and the long-range Kac potential. From this point of view the result is of the same type as in equilibrium, but the expressions are, of course, much more involved. For instance, the Kac potential enters the equilibrium pressure, Eq. (9.4.31), only through the constant $\alpha \equiv \tilde{\varphi}(0)$. In both cases, the properties of the reference system are not known analytically; however, the relation between van der Waals properties and reference properties is exact: it does not depend on any specific model or approximation. We also note that the relation between real and reference fluid is by no means trivial for nonequilibrium properties and could hardly have been guessed beforehand, without doing the detailed analysis of diagrams. This remark remains true in spite of the fact that Résibois, Piasecki, and Pomeau showed *a posteriori*, at the end of their third paper, that a qualitative simpler argument could be used for deriving the gross features of their result.

BIBLIOGRAPHICAL NOTES

The Boltzmann Equation

Most of the general papers quoted in Chapter 17 deal with the classical problem of the foundation of the Boltzmann equation. We may add the following, more specific references:

G. E. Uhlenbeck and G. Ford, *Lectures in Statistical Mechanics*, American Math. Soc., Providence, 1963 (contains an elementary but clear introduction to the subject).

M. Kac, in *Proceedings of the Third Berkeley Symposium on Mathematical Statistics and Probability*, Univ. California Press, Berkeley, 1956.

M. S. Green, *J. Chem. Phys.* **25,** 836 (1956); *Physica* **24,** 393 (1958).

R. Brout, *Physica* **22,** 509 (1956).

H. Grad, *Handbuch. Phys.* **12,** 205 (1958).

Dense Gases

The *Enskog equation* was derived in D. Enskog, *Kungl. Svenska Vet. Akad. Handl.* **63,** No. 4 (1921).

It is discussed in detail in the books by Chapman and Cowling (quoted in Chapter 13) and by Hirschfelder *et al.* (quoted in Chapter 6).

The *Choh–Uhlenbeck equation* was derived in the report quoted in Chapter 17. See also the papers by Cohen quoted there, and P. Résibois, *Phys. Lett.* **9,** 139 (1964).

The *binary collision expansion* was developed by R. Zwanzig, *Phys. Rev.* **129,** 486 (1963).

J. Weinstock, *Phys. Rev.* **126,** 341 (1962); **132,** 454, 470 (1963).

K. Kawasaki and I. Oppenheim, *Phys. Rev.* **136,** A1519 (1964).

The *divergence of the density expansion* was found in:

J. Weinstock, *Phys. Rev.* **132,** 454 (1963).

J. R. Dorfman and E. G. D. Cohen, *Phys. Lett.* **16,** 124 (1965); *J. Math. Phys.* **8,** 282 (1967).

Goldman and E. Frieman, *Bull. Am. Phys. Soc.* **10,** 531 (1965); *J. Math. Phys.* **8,** 1410 (1967).

The summation of diagrams suppressing this divergence was done in:

K. Kawasaki and I. Oppenheim, *Phys. Rev.* **139,** A1763 (1965).

J. Weinstock, *Phys. Rev.* **140,** A460 (1965).

The *heat conductivity and the viscosity of a two-dimensional system of hard disks* were calculated by J. V. Sengers, *Phys. Fluids* **9,** 1333, 1685 (1966); see also *The Boltzmann Equation, Theory and Applications* (quoted in Chapter 11).

These matters are very clearly reviewed in E. G. D. Cohen, in *Statistical Mechanics at the Turn of the Decade* (E. G. D. Cohen, ed.), Dekker, New York, 1971.

The *quantum problem corresponding to the Choh–Uhlenbeck* equation was treated in:

P. Résibois, *Physica* **31,** 645 (1965).

J. T. Lowry and W. C. Schieve, *Transp. Theory Stat. Phys.* **1,** 225 (1971).

Plasmas

The integral equation (20.5.14) was first obtained by N. N. Bogoliubov in the book quoted in Chapter 17. This equation was solved and the kinetic equation was obtained independently by:

R. Balescu, *Phys. Fluids* **3,** 52 (1960).

A. Lenard, *Ann. Phys.* (N.Y.) **3,** 390 (1960).

R. L. Guernsey, *The Kinetic Theory of Fully Ionized Gases*, Off. Nav. Res. Contract No. Nonr. 1224 (15), July 1960.

This equation and other matters relative to plasma kinetic theory are discussed in the books of R. Balescu and of Iu. Klimontovich (quoted in Chapter 17). See also D. C. Montgomery and D. Tidman, *Plasma Kinetic Theory*, McGraw-Hill, New York, 1964.

See also the more elementary but excellent books:

N. G. van Kampen and B. V. Felderhof, *Theoretical Methods in Plasma Physics*, North Holland, Amsterdam, 1967.

D. C. Montgomery, *Theory of the Unmagnetized Plasma*, Gordon and Breach, New York, 1971.

The extension of the ring equation to an *unstable plasma* was done in:

R. Balescu, *J. Math. Phys.* **4,** 1009 (1963).

E. A. Frieman and P. Rutherford, *Phys. Fluids* **6,** 67 (1963); *Ann. Phys.* **28,** 134 (1964).

The kinetic equation for a *quantum-statistical plasma* in the ring approximation was derived by R. Balescu, *Phys. Fluids* **4,** 94 (1961). It is also extensively discussed in Balescu's book (quoted in Chapter 17).

Very comprehensive calculations of transport coefficients of classical and quantum plasmas in a very wide range of temperatures and densities were performed by:

W. B. Hubbard, *Astrophys. J.* **146,** 858 (1966).

M. Lampe, *Phys. Rev.* **170,** 306 (1968); **174,** 276 (1968).

W. B. Hubbard and M. Lampe, *Astrophys. J.* Suppl. 163 **18,** 297 (1969).

Among the more recent developments in plasma kinetic theory, we must mention the two papers that initiated the so-called "*quasilinear*" *theory* of plasmas:

A. A. Vedenov, E. P. Velikhov, and R. Z. Sagdeev, *Nucl. Fusion Suppl.* **2,** 465 (1962).

W. E. Drummond and D. Pines, *Nucl. Fusion Suppl.* **3,** 1049 (1962).

In these papers a kinetic equation for *weakly turbulent plasmas* is derived from the Vlassov equation. These ideas were applied to many problems. Among the most refined versions of this theory we quote:

B. B. Kadomtsev, *Plasma Turbulence,* Academic Press, New York, 1965.

T. H. Dupree, *Phys. Fluids* **9,** 1773 (1966).

J. Weinstock, *Phys. Fluids* **12,** 1045 (1969).

G. Benford and J. J. Thomson, *Phys. Fluids* **15,** 1496 (1972).

Among the recent extensions and the derivation of the kinetic equations beyond the ring approximation we quote:

A. Rogister and C. Obermann, *J. Plasma Phys.* **2,** 33 (1968); **3,** 119 (1969).

M. Baus, *Ann. Phys.* (N.Y.) **62,** 135 (1971).

M. Baus, *Phys. Lett.* **A40,** 213 (1972).

M. Baus, *Physica* **66,** 421 (1973).

An excellent book reviewing the present status of plasma physics is R. C. Davidson, *Methods in Non-Linear Plasma Theory*, Academic Press, New York, 1972.

Brownian Motion

Some of the important papers on the statistical mechanical foundations of this problem are:

J. L. Lebowitz and E. Rubin, *Phys. Rev.* **131,** 2381 (1963).

P. Résibois and H. T. Davis, *Physica* **30,** 1077 (1964).

P. Résibois and J. L. Lebowitz, *Phys. Rev.* **139,** A1101 (1965).

P. Mazur, in *Statistical Mechanics of Equilibrium and Non-Equilibrium* (J. Meixner, ed.), North Holland, Amsterdam, 1965.

The van der Waals Fluid

P. Résibois, J. Piasecki, and Y. Pomeau, *Phys. Rev. Lett.* **28,** 882 (1972).

P. Résibois, in *Irreversibility in the Many-Body Problem* (J. Biel and J. Rae, ed.), Plenum Press, New York, 1972.

Liquids

Let us quote the following papers in which kinetic equations were derived for the description of a liquid:

S. A. Rice and A. R. Allnatt, *J. Chem. Phys.* **34,** 2144, 2156 (1961).

I. Prigogine, G. Nicolis, and J. Misguich, *J. Chem. Phys.* **43,** 4516 (1965).

P. M. Allen and G. H. A. Cole, *Mol. Phys.* **15,** 549, 557 (1968).

P. M. Allen, *Physica* **52,** 237 (1971).

See an extensive discussion in the book by Rice and Gray quoted in Chapter 8.

Owing to lack of space we cannot go into a detailed discussion of the kinetic theories of liquids in this book. We should, however, make the following remark. The reader of the papers quoted above (and of many other papers in this field) will be struck by the fact that the kinetic equations for liquids are usually equations for the *pair distribution function* $\bar{g}_2(x_1x_2; t)$. This seems to contradict our definition of a kinetic equation as a closed equation for $\bar{f}_1$. The contradiction is, however, only illusory. In the kinetic regime $\bar{g}_2$ is a functional of $\bar{f}_1$. The basic kinetic equation is one for $\bar{f}_1$; from its solution $\bar{g}_2$ can be calculated through a creation operator evaluated within the same approximation. From a *practical* point of view it appears, however, simpler to write down directly an equation for $\bar{g}_2$ because we have a good qualitative picture of the behavior of this function, and this physical picture suggests the mathematical approximations to be used.

A very clear discussion of the Rice–Allnatt equation from this point of view (as well as an extension of that equation) appears in J. H. Misguich and G. Nicolis, *Mol. Phys.* **24,** 309 (1972).

The situation is the same in equilibrium theory (see Chapter 8). We know that in equilibrium the distribution function is in the $\mathbf{\Pi}$ subspace. Hence $\bar{g}_2^0$ can be calculated from $\bar{f}_1^0$ by applying the appropriate creation operator. However, rather than trying to construct this complicated operator, it turns out more economical to guess directly the approximate functional form of $\bar{g}_2^0$.

Solids

The kinetic equation for the transport processes due to phonon–phonon interactions was derived by R. Peierls, *Ann. Phys.* **3,** 1055 (1929).

See also R. Peierls, *Quantum Theory of Solids*, Clarendon Press, Oxford, 1955.

The application of the general formalism to this problem appeared in I. Prigogine and F. Henin, *J. Math. Phys.* **1,** 349 (1960).

Magnetic Systems

A very important class of problems studied recently with great success is the problem of *Heisenberg spins* describing magnetic crystals:

P. Résibois and M. de Leener, *Phys. Rev.* **152,** 305, 318 (1966); **178,** 806 (1969).

P. Résibois, *Bull. Cl. Sci. Acad. Roy. Belg.* **54,** 1493 (1968).

CHAPTER 21
DYNAMICS OF FLUCTUATIONS AND OF CORRELATIONS

21.1. DEFINITION AND CLASSIFICATION OF CORRELATION FUNCTIONS

We arrived now at a pretty comprehensive description of the behavior of many-body systems, both in equilibrium and out of equilibrium. It is, however, a well-known feature of any statistical theory that a description of the overall (or average) state of the system does not provide us with all the available information. The characteristic property of such systems is the existence of *fluctuations*. To study the phenomena related to fluctuations we need a refinement of the theory, which we consider in this chapter. To be sure, we have already studied fluctuations and correlations throughout this book; however, these were not yet of the most general type. In order to understand the problem, we review the various questions one might ask in increasing order of complexity.

Consider first a classical system in equilibrium, and let $Y \equiv Y(q, p)$ be some dynamical function. Its average value is denoted by $\langle Y \rangle$: it is a number, independent of time because the system is in a steady state. If we could however measure the actual value of the observable Y at a given time t, we would in general obtain a value $Y(t)$, different from $\langle Y \rangle$. Let us introduce the fluctuation $y(t)$ defined as

$$y(t) = Y(t) - \langle Y \rangle \qquad (21.1.1)$$

By definition, $\langle y(t) \rangle = 0$. On the other hand, fluctuations are spontaneous random departures from the average; they cannot be assigned a precise value. Rather, they must be considered as *stochastic variables* in the sense of Section 11.3. It must be clearly recalled, however, that in statistical mechanics the law of evolution in time is perfectly deterministic. In other words, to any initial value $y(0)$ corresponds, through the laws of Hamiltonian dynamics, a well defined value $y(t)$ at time t. It is only the stochastic distribution of $y(0)$ which induces the stochastic character of $y(t)$. This feature is opposed to the type of evolution described in Section

11.3, where the law of motion is given only in terms of transition probabilities.

The only accessible information about fluctuations is in the moments, that is, the averages of products of the stochastic variables. There is, however, a variety of possible moments, of which we consider a few simple and characteristic examples. We may consider first the average of the square of $y(t)$ for a *classical system* in equilibrium:

$$G^0_{yy} = \langle y(t)y(t)\rangle^0 \tag{21.1.2}$$

This is a number independent of time. This property can be inferred from the stationary character of the equilibrium, by a little calculation involving the properties studied in Sections 1.2 and 2.2:

$$\begin{aligned} G^0_{yy} &= \int dq\, dp\; y^2(q, p; t)F^0(q, p) \\ &= \int dq\, dp\; \{e^{t[H]}y^2(q, p)\}F^0(q, p) \\ &= \int dq\, dp\; y^2(q, p)\{e^{-t[H]}F^0(q, p)\} \\ &= \int dq\, dp\; y^2(q, p)F^0(q, p) \end{aligned} \tag{21.1.3}$$

The last step follows from the stationarity property $LF^0 = -[H]F^0 = 0$. Moments of this type have been studied in Section 4.6 where it was shown that they can be expressed in terms of macroscopic thermodynamic variables. The derivation of similar relationships in more general situations will be one of the aims of this chapter.

More general is the case when we consider several fluctuating quantities; we may then define correlations between $y(t)$ and $z(t)$:

$$G^0_{yz} = \langle y(t)z(t)\rangle^0 \tag{21.1.4}$$

These quantities are also time independent. An important special case is obtained when we consider classes of fluctuating quantities depending on a parameter. In particular, we consider *local dynamical quantities* defined in every point $\mathbf{x}$ of physical space. Such quantities have been studied before: They are of the general form [see also Eq. (12.1.2)]:

$$y(t; \mathbf{x}) = \sum_{j=1}^{N} \gamma[\mathbf{p}_j(t), \mathbf{q}_j(t)]\, \delta[\mathbf{x} - \mathbf{q}_j(t)] \tag{21.1.5}$$

If we consider products of the form $y(t; \mathbf{x})\; z(t; \mathbf{x}+\mathbf{r})$, these involve two different functions of q and p (even if $y \equiv z$ but $r \neq 0$); hence they give rise to a correlation function of the type (21.1.4), which we denote by

$$G^0_{yz}(\mathbf{r}) = \langle y(t; \mathbf{x})z(t; \mathbf{x}+\mathbf{r})\rangle^0 \tag{21.1.6}$$

These functions are independent of t and independent of $\mathbf{x}$, because the equilibrium state is homogeneous; they do depend, however, on the distance $\mathbf{r}$. Clearly, the pair correlation function studied in Chapter 7 is a particularly important example of a function of type (21.1.6). It should be clear that for $\mathbf{r}=0$, the correlation function $G^0_{yz}(\mathbf{r})$ reduces to the function of Eq. (21.1.4):

$$G^0_{yz}(\mathbf{r}=0)=G^0_{yz} \tag{21.1.7}$$

We may introduce the same kind of correlation functions but evaluate them in a nonequilibrium state. Now the correlation functions will depend on time:

$$G_{yz}(t)=\langle y(t)z(t)\rangle \tag{21.1.8}$$

Indeed, as the initial distribution function $F(q, p)$ is no longer stationary, the passage from the Heisenberg to the Schrödinger picture, analogous to (21.1.3) gives the two definitions:

$$\begin{aligned} G_{yz}(t) &= \int dq\, dp\; y(q, p; t)z(q, p; t)F(q, p) \\ &= \int dq\, dp\; y(q, p)z(q, p)F(q, p; t) \end{aligned} \tag{21.1.9}$$

Similarly, we may introduce time-dependent and space-dependent non-equilibrium correlation functions:

$$G_{yz}(t; \mathbf{r}, \mathbf{x})=\langle y(t; \mathbf{x})z(t; \mathbf{x}+\mathbf{r})\rangle \tag{21.1.10}$$

The correlation functions $g_2(\mathbf{r}; t)$ studied in Chapters 14–20 are precisely of this type, y and z being the local microscopic density.

The next step in generality is now clear. We may consider the correlation between quantity y measured at point $\mathbf{x}$ and at time t, and z measured at point $\mathbf{x}+\mathbf{r}$ and at a *different* time $t+\tau$:

$$G_{yz}(\tau, t; \mathbf{r}, \mathbf{x})=\langle y(t; \mathbf{x})z(t+\tau; \mathbf{x}+\mathbf{r})\rangle \tag{21.1.11}$$

These *two-time correlation functions* define a new important class that we did not yet study systematically,* and that will be the main object of this chapter. For $\tau=0$, these functions reduce to those of Eq. (21.1.10):

$$G_{yz}(\tau=0, t; \mathbf{r}, \mathbf{x})=G_{yz}(t; \mathbf{r}, \mathbf{x}) \tag{21.1.12}$$

A rather serious difference with the one-time correlation functions appears now. It is not possible to express these functions in the Schrödinger picture, that is, as averages of time-independent dynamical

* See, however, Sections 13.4 and 20.7.

variables taken with a time-dependent distribution function. Indeed, repeating the previous calculations we find

$$G_{yz}(\tau, t; \mathbf{r}, \mathbf{x}) = \int dq\, dp\, \{e^{t[H]}y(q, p; \mathbf{x})\}\{e^{(t+\tau)[H]}z(q, p; \mathbf{x}+\mathbf{r})\}\, F(q, p)$$

$$= \int dq\, dp\, [e^{t[H]}\{y(q, p; \mathbf{x})(e^{\tau[H]}z(q, p; \mathbf{x}+\mathbf{r}))\}]\, F(q, p)$$

$$= \int dq\, dp\, y(q, p; \mathbf{x})\{e^{\tau[H]}z(q, p; \mathbf{x}+\mathbf{r})\}\{e^{-t[H]}F(q, p)\}$$

$$= \int dq\, dp\, y(q, p; \mathbf{x})\{e^{\tau[H]}z(q, p; \mathbf{x}+\mathbf{r})\}\, F(q, p; t) \qquad (21.1.13)$$

Hence, $G_{yz}(\tau, t; \mathbf{r}, \mathbf{x})$ is expressed either in a pure Heisenberg representation or in a "mixed" representation as an average of a τ-dependent dynamical function taken with a t-dependent distribution. We shall see, however, that the Schrödinger picture can be extended by a suitable definition of two-time distribution functions in terms of which the two-time correlation functions can be evaluated just like ordinary averages. This matter (which has no direct bearing on the formal properties to be developed now) will be postponed to Section 21.6.

If we now specialize the definition (21.1.11) to an average taken with the equilibrium distribution function we find that the resulting correlation only depends on τ and on $\mathbf{r}$, not on t, nor on $\mathbf{x}$, as a result of the stationary and homogeneous character of equilibrium [see Eq. (21.1.9)]:

$$G^0_{yz}(\tau; \mathbf{r}) = \langle y(t; \mathbf{x})z(t+\tau; \mathbf{x}+\mathbf{r})\rangle^0 = \langle yz(\tau; \mathbf{r})\rangle^0 \qquad (21.1.14)$$

We may therefore just as well put $t=0$, $\mathbf{x}=0$ in the general definition. Equilibrium correlation functions of the type (21.1.14) will play a major role in this chapter.

When we consider *quantum-mechanical systems*, the definitions introduced above are very simply extended. As we know from Section 2.3, the averages are now defined in terms of traces involving the density matrix $\hat{\rho}$. The only subtler difference is in the fact that the products of operators $\hat{y}\hat{z}$ are ambiguous because $\hat{y}$ and $\hat{z}$ do not, in general, commute. We therefore define the correlation functions as averages of the symmetrized products, for instance:

$$G_{yz}(\tau, t; \mathbf{r}, \mathbf{x}) = \tfrac{1}{2}\langle\{\hat{y}(t; \mathbf{x})\hat{z}(t+\tau, \mathbf{x}+\mathbf{r}) + \hat{z}(t+\tau, \mathbf{x}+\mathbf{r})\hat{y}(t; \mathbf{x})\}\rangle$$
$$\equiv \tfrac{1}{2}\operatorname{Tr}\hat{\rho}\{\hat{y}(t; \mathbf{x})\hat{z}(t+\tau; \mathbf{x}+\mathbf{r}) + \hat{z}(t+\tau; \mathbf{x}+\mathbf{r})\hat{y}(t; \mathbf{x})\} \qquad (21.1.15)$$

An important concept related to the correlation functions is the *spectral density*. For nonlocal correlations it is defined as the Fourier transform of the correlation function with respect to the time delay τ:

$$\tilde{G}^0_{yz}(\omega) = \int_{-\infty}^{\infty} d\tau\, G^0_{yz}(\tau) e^{i\omega\tau} \tag{21.1.16}$$

In the case of local functions one may also introduce a Fourier transformation with respect to the distance $\mathbf{r}$:

$$\tilde{\tilde{G}}^0_{yz}(\omega; \mathbf{k}) = \int_{-\infty}^{\infty} d\tau \int d\mathbf{r}\, e^{i\omega\tau + i\mathbf{k}\cdot\mathbf{r}} G^0_{tz}(\tau; \mathbf{r}) \tag{21.1.17}$$

The spectral density corresponding to the equilibrium density–density correlation is directly observable. We saw in Section 8.1 that the Fourier transform of the pair correlation function is directly related to the structure factor $a_\mathbf{k}$ [see Eq. (8.1.5)]. The latter is measured from the intensity of elastic scattering of electromagnetic waves or of neutrons from a fluid. If we consider *inelastic scattering*, involving not only a transfer of momentum $\hbar\mathbf{k}$ but also a transfer of energy $\hbar\omega$, we can similarly define a form factor $a_\mathbf{k}(\omega)$ depending both on the wave vector $\mathbf{k}$ and on the frequency ω of the scattered radiation. It was shown by Van Hove that this form factor coincides with the spectral density (21.1.17). Since this work, inelastic neutron scattering became a quite important experimental tool for the study of dynamical, time-dependent phenomena in fluids.

Among the important topics of study is the investigation of the dynamical behavior of fluids near the critical point. Very active developments have been taking place in this field in the past few years. These theories are extensions of the idea of scaling laws to the dynamical phenomena. The concepts and arguments used in this field are very analogous to those of the corresponding equilibrium theory, which we described in detail in Chapters 8–10. The reader who mastered those chapters should have no difficulty in studying the original papers; we therefore do not treat these problems here but give some bibliographical references at the end of the chapter.

We shall, rather, concentrate here on some very characteristic and important nonequilibrium relations, which introduce new methods and ideas, besides being of great importance in the general framework of nonequilibrium statistical mechanics.

21.2. THE GREEN–KUBO FORMULAS FOR THE TRANSPORT COEFFICIENTS

We now show that the transport coefficients, that is, the crucial quantities of linear nonequilibrium thermodynamics, can be very naturally expressed in terms of time-correlation functions. The relation between transport coefficients and time-correlation functions was first discovered by Green in 1951, and independently developed quite systematically by Kubo since 1957.

The situation is simplest and clearest in the case of the so-called *mechanical transport coefficients*, which describe the reaction of the system to a perturbation of its Hamiltonian. Typical among these is the electrical conductivity. These coefficients must be contrasted with the *thermal coefficients* that were considered in Chapter 13 and that are due to the inhomogeneity of the state of the system. We consider here in detail the first class, and show how the electrical conductivity can be calculated in full generality.

Consider a system consisting of several species of charged particles (particle j bearing the charge $z_j e$), the total charge of the system being zero. The "internal" Liouvillian of the system will be denoted by L and is of the usual type:

$$L = L^0 + L' = \sum_j L_j^0 + \sum_{j<n}\sum L'_{jn} \tag{21.2.1}$$

At this stage, we prefer to use the phase-space formalism rather than the correlation patterns representation, because the formulas are more concise; eventually, however, the results must be expressed in terms of reduced distributions, as shown in Section 21.6.

We now consider that the system has been left to itself for a long time, so that at time $t=0$ it has reached the thermal equilibrium:

$$F(x_1, x_2, \ldots, x_N; 0) = F^0(x_1, \ldots, x_N) = (N!\, h^{3N} Z)^{-1} e^{-\beta H} \tag{21.2.2}$$

At time zero, we switch on an external electric field $\mathbf{E}$, which we assume for simplicity to be constant in space and time for $t>0$. Hence the Liouvillian is now [see Eqs. (2.4.21)–(2.4.23)]

$$\Lambda = L + L^F = L + \sum_{j=1}^{N} L_j^F \tag{21.2.3}$$

The external-field Liouvillian is a sum of one-particle operators:

$$L_j^F = -z_j e \mathbf{E} \cdot \partial_j \tag{21.2.4}$$

We now wish to solve the Liouville equation:

$$\partial_t F = (L + L^F) F \tag{21.2.5}$$

by assuming that L^F is a small perturbation. This is because we are interested in defining the electrical conductivity, which is a characteristic of the *linear* response of the system to the external field.*

* See also the discussion of Section 17.8.

The methods of Chapter 16 are readily adapted to the present problem. If we call $W(t)$ the propagator, it obeys an integral equation similar to (16.1.9), except that the role of the unperturbed propagator is now played by the complete internal propagator $U(t)$ (including the interactions), that is, precisely the object of the left-hand side of (16.1.9):

$$W(t)=U(t)+\int_0^t d\tau\, U(\tau)L^F W(t-\tau) \tag{21.2.6}$$

As we are interested in the linear response, we need only the first iteration solution of (21.2.6):

$$W^{(1)}(t)=U(t)+\int_0^t d\tau\, U(\tau)L^F U(t-\tau) \tag{21.2.7}$$

We now apply this operator to the initial condition (21.2.2) in order to obtain the solution, to first order in the external field:

$$F^{(1)}(t)=U(t)F^0+\int_0^t d\tau\, U(\tau)L^F U(t-\tau)F^0 \tag{21.2.8}$$

We note that the equilibrium distribution is a stationary solution of the internal Liouville equation; hence

$$U(t)F^0=F^0 \tag{21.2.9}$$

and Eq. (21.2.8) reduces to

$$F^{(1)}(t)=F^0+\int_0^t d\tau\, U(\tau)L^F F^0 \tag{21.2.10}$$

We now calculate the average value of the electric current at time t. Noting that the microscopic current is

$$\hat{\mathbf{j}}^e=\sum_{n=1}^{N}\left(\frac{z_n e}{m_n}\right)\mathbf{p}_n \tag{21.2.11}$$

we have

$$\mathbf{J}^e(t)=\int d^N x\, \hat{\mathbf{j}}^e F^{(1)}(t) \tag{21.2.12}$$

We note that in equilibrium, the average of the electric current is zero ($\hat{\mathbf{j}}^e$ is an odd function of $\mathbf{p}$, whereas F^0 is even). Hence, as expected, the presence of an average nonzero current is a measure of the deviation from thermal equilibrium:

$$\mathbf{J}^e(t)=\int_0^t d\tau \int d^N x\, \hat{\mathbf{j}}^e U(\tau)L^F F^0 \tag{21.2.13}$$

From Eqs. (21.2.2) and (21.2.4) we obtain

$$\begin{aligned} L^F F^0 &= -\sum_j z_j e\mathbf{E}\cdot\partial_j F^0 \\ &= \beta\sum_j\left(\frac{z_j e}{m_j}\right)\mathbf{p}_j\cdot\mathbf{E}F^0 \\ &= \beta\mathbf{E}\cdot\hat{\mathbf{j}}^e F^0 \end{aligned} \tag{21.2.14}$$

Hence,

$$\mathbf{J}^e(t) = \beta\mathbf{E}\cdot\int_0^t d\tau\int d^N x\,\hat{\mathbf{j}}^e e^{\tau L}\{\hat{\mathbf{j}}^e F^0\}$$

Going over now to the Heisenberg representation, as in Section 2.2:

$$\begin{aligned} \mathbf{J}^e(t) &= \beta\mathbf{E}\cdot\int_0^t d\tau\int d^N x\,F^0\,\hat{\mathbf{j}}^e e^{-\tau L}\hat{\mathbf{j}}^e \\ &= \beta\mathbf{E}\cdot\int_0^t d\tau\int d^N x\,F^0\,\hat{\mathbf{j}}^e\hat{\mathbf{j}}^e(\tau) \end{aligned} \tag{21.2.15}$$

or even more simply:

$$\mathbf{J}^e(t) = \beta\mathbf{E}\cdot\int_0^t d\tau\,\langle\hat{\mathbf{j}}^e\hat{\mathbf{j}}^e(\tau)\rangle^0 \tag{21.2.16}$$

We now consider what is expected to happen physically in this situation (Fig. 21.2.1). When the field is abruptly switched on, the system reacts and a current appears. This reaction, however, has some delay with respect to the field: the current varies in time until it reaches a steady value $\mathbf{J}^e_{st}$. This steady current is proportional to the field $\mathbf{E}$, and the constant of proportionality is, by definition, the *electrical conductivity*. If such a

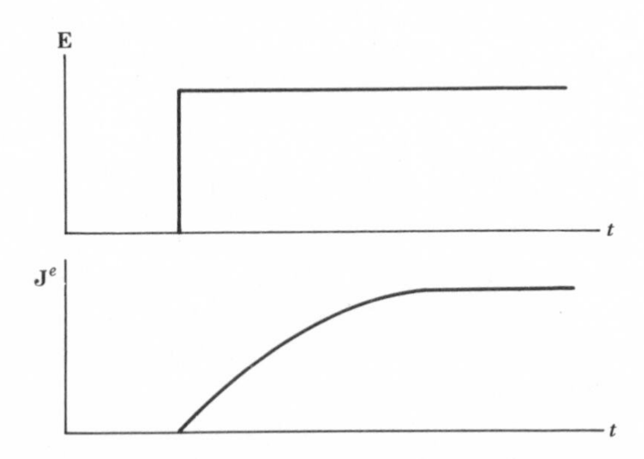

Figure 21.2.1. Establishment of a steady nonequilibrium current.

steady state exists at all, it will be reached by our system as $t \to \infty$. Thus:

$$\mathbf{J}^e_{st} = \lim_{t\to\infty} \mathbf{J}^e(t) = \mathbf{E} \cdot \beta \int_0^\infty d\tau \, \langle \hat{\mathbf{j}}^e \hat{\mathbf{j}}^e(\tau) \rangle^0 \tag{21.2.17}$$

Hence the electrical conductivity tensor is given by

$$\sigma = \beta \int_0^\infty d\tau \, \langle \hat{\mathbf{j}}^e \hat{\mathbf{j}}^e(\tau) \rangle^0 \tag{21.2.18}$$

(If the system is isotropic, and no magnetic field is present, the tensor reduces to a scalar.) We have obtained a remarkably beautiful and general result due to Kubo. It tells us that the *electrical conductivity is the time integral, from* 0 *to* ∞, *of the microscopic current auto-correlation in equilibrium.* The importance of this result is in its generality: no assumption has been made about any kind of particular model or approximation. It will be shown in Section 21.4 that formulas analogous to (21.2.18) can also be derived for the thermal transport coefficients. We shall see in Section 21.7 how these expressions can be further studied and evaluated.

21.3. THE FLUCTUATION–DISSIPATION THEOREM

The derivation of the Green–Kubo formulas for the transport coefficients is only a particular consequence of a very wide class of relations, generally assembled under the name of the *fluctuation–dissipation theorem.* (This name is misleading: the single theorem has many faces, so one should use a plural; moreover, it also applies to nondissipative systems as well). The first theorem of this type was derived by Nyquist in 1928, in connection with the theory of noisy circuits. Its general form is due to Callen and Welton (1951), who stressed its major role in statistical mechanics. Its importance comes first from its great generality: very few assumptions are necessary for its derivation. Moreover, it exhibits a beautiful link between equilibrium and nonequilibrium statistical mechanics. Finally, it provides us with simple expressions for microscopic quantities in terms of macroscopic observables. The aim of the fluctuation–dissipation theorem is to relate two apparently very different types of phenomena, which we now describe.

The first of these are the fluctuation phenomena, which were already discussed in some detail in Section 21.1. As was shown there, these random deviations from equilibrium are described by a collection of moments of various types, the most commonly known being the two-time equilibrium correlation functions $\langle A(t_1)B(t_2)\rangle^0$. The dynamical theory of fluctuations consists of the study of these moments as a function of time.

The second class of phenomena can be described as follows. We start with a system in equilibrium: such a system is characterized by a great symmetry. At some time we switch on an external field, possibly time dependent. The equilibrium is disturbed and the system tries to adjust to the new environment; some of the original symmetries are broken. This symmetry breaking manifests itself macroscopically by the appearance of new measurable quantities; the observables that existed already in equilibrium are also modified and become time dependent. Examples of such phenomena are very common: the appearance of an electric current as a result of an external electric field, the magnetization of a sample under the influence of an external magnetic field, and the like. The appearance of these new quantities (or deviations) are a very tangible measure of the response of the system to an external stimulus. By definition, they must vanish when the stimulus does not exist; therefore, if the latter is small, the response must be proportional to it, or more generally, must be a linear functional of the external forces. The proportionality coefficients, or the kernels of these linear functionals are important intrinsic properties of the undisturbed system, deeply related to its molecular background. The transport coefficients are just particular examples of such *response functions.*

The two classes of phenomena may seem very different, however one feels that they must be related. The idea of such a relationship goes back to Onsager's classical work of 1931. His postulate may be stated as follows. If a system is at time t_0 in some nonequilibrium state, it "does not know" whether it was brought into that state by the action of an external force or as a result of a random fluctuation. Its subsequent evolution toward equilibrium will therefore be the same in both cases (at least if the deviation is sufficiently small). This kind of relationship is made more precise by the fluctuation–dissipation theorem, as will now be shown. In order to avoid a too-strong analogy with the previous section, we now study a quantum-mechanical system.

Consider a system described by a Hamiltonian $\hat{H}^T$ consisting of an internal part, describing all the molecular interactions, plus an external part, which generally depends on time:

$$\hat{H}^T = \hat{H} + \hat{H}^F(t) \tag{21.3.1}$$

We assume that the external part is of the form:

$$\hat{H}^F(t) = -\hat{A}g(t) \tag{21.3.2}$$

where $\hat{A}$ is an operator and $g(t)$ a given function of time (not an operator).

The von Neumann equation (2.3.22) can be written compactly as

$$\partial_t \hat{\rho}(t) = -[\hat{H}]\hat{\rho}(t) - [H^F(t)]\hat{\rho}(t) \tag{21.3.3}$$

where

$$[\hat{H}]\hat{a} = (i\hbar)^{-1}[\hat{a}, \hat{H}]_-$$

We now assume that at time $t = -\infty$ the system was in equilibrium in the absence of the external field. The latter is then switched on, and the density matrix is expressed in terms of its initial value by means of a propagator $W(t)$:

$$\hat{\rho}(t) = W(t)\hat{\rho}(-\infty) \tag{21.3.4}$$

This propagator obeys the same formal equation as in (21.2.6), where $L^F \to -[\hat{H}^F(t)]$; and the lower limit of integration is $-\infty$ instead of zero. To first order in the field, the solution is therefore

$$\hat{\rho}(t) = \hat{\rho}^0 - (i\hbar)^{-1}\int_{-\infty}^{t} d\tau\, U(t-\tau)g(\tau)[\hat{A}, \hat{\rho}^0]_- \tag{21.3.5}$$

Let $\hat{B}$ be some dynamical operator whose average in equilibrium is $\langle\hat{B}\rangle^0$. The deviation from equilibrium of its average at time t is

$$\langle\Delta\hat{B}(t)\rangle = \langle\hat{B}\rangle - \langle\hat{B}\rangle^0 \tag{21.3.6}$$

$\langle\Delta\hat{B}(t)\rangle$ is typically a measure of the linear response of the system to the external stimulus, as discussed at the beginning of this section. It is easily calculated:

$$\begin{aligned}\langle\Delta\hat{B}(t)\rangle &= -\mathrm{Tr}\,(i\hbar)^{-1}\int_{-\infty}^{t} d\tau\, \hat{B}U(t-\tau)g(\tau)[\hat{A}, \hat{\rho}^0]_- \\ &= (i\hbar)^{-1}\int_{-\infty}^{t} d\tau\, g(\tau)\,\mathrm{Tr}\,\hat{B}(t-\tau)[\hat{\rho}^0, \hat{A}]_-\end{aligned}$$

where we went over to the Heisenberg picture operator $\hat{B}(t)$ in the usual way.

We now obtain $\langle\Delta\hat{B}(t)\rangle$ in the form of a linear functional:

$$\langle\Delta\hat{B}(t)\rangle = \int_{-\infty}^{t} d\tau\, \phi_{BA}(t-\tau)g(\tau) \tag{21.3.7}$$

This relation defines the *response function*, the explicit expression of which is

$$\phi_{BA}(t) = (i\hbar)^{-1}\,\mathrm{Tr}\,\hat{B}(t)[\hat{\rho}^0, \hat{A}]_- \tag{21.3.8}$$

Using the invariance of the trace with respect to permutations, we may write this as

$$\begin{aligned}\phi_{BA}(t) &= (i\hbar)^{-1}\,\mathrm{Tr}\,\hat{\rho}^0[\hat{A}, \hat{B}(t)]_- \\ &= (i\hbar)^{-1}\langle[\hat{A}, \hat{B}(t)]_-\rangle^0\end{aligned} \tag{21.3.9}$$

This expression can be further transformed in the following way. There exists an interesting identity valid when $\hat{\rho}^0$ is the canonical density matrix:

$$[\hat{\rho}^0, \hat{A}]_- = \int_0^\beta d\mu \, \hat{\rho}^0 e^{\mu \hat{H}} [\hat{A}, \hat{H}]_- e^{-\mu \hat{H}} \tag{21.3.10}$$

It is easily proved by substituting the explicit form (4.3.19) and differentiating both sides with respect to β: the identity becomes obvious. Moreover, it trivially holds for $\beta = 0$. We further note that $(\exp \mu \hat{H})$ has the same form as a propagator for an imaginary time $t = -i\hbar\mu$: we may therefore write

$$\begin{aligned}(i\hbar)^{-1} e^{\mu \hat{H}} [\hat{A}, \hat{H}]_- e^{-\mu \hat{H}} &= (i\hbar)^{-1} [e^{\mu \hat{H}} \hat{A} e^{-\mu \hat{H}}, \hat{H}]_- \\ &= (i\hbar)^{-1} [\hat{A}(-i\hbar\mu), \hat{H}]_- \\ &= \dot{\hat{A}}(-i\hbar\mu)\end{aligned}$$

Hence:

$$\phi_{BA}(t) = \int_0^\beta d\mu \, \langle \dot{\hat{A}}(-i\hbar\mu) \hat{B}(t) \rangle^0 \tag{21.3.11}$$

This provides a very compact expression for the response function.

We now specialize our study to the response to a periodic stimulus*:

$$g(t) = g e^{-i\omega t} = \lim_{\varepsilon \to +0} g e^{-i\omega t + \varepsilon t} \tag{21.3.12}$$

where g is a constant. The addition of a small imaginary part $-i\varepsilon$ to the frequency is necessary for ensuring the convergence of the integrals ("adiabatic switching"): we do not insist on this aspect here. The response formula (21.3.7) becomes in this case

$$\langle \Delta B(t) \rangle = \lim_{\varepsilon \to +0} \int_{-\infty}^{t} d\tau \, e^{-i\omega t + \varepsilon \tau} \phi_{BA}(t - \tau) g$$

or, changing the integration variable to $\theta = t - \tau$

$$\langle \Delta B(t) \rangle = \lim_{\varepsilon \to +0} \int_0^\infty d\theta \, \phi_{BA}(\theta) e^{i(\omega + i\varepsilon)\theta} g e^{-i(\omega + i\varepsilon)t}$$

Finally, the response formula appears in the form:

$$\langle \Delta B(t) \rangle = \chi_{BA}(\omega) g e^{-i\omega t} \tag{21.3.13}$$

where we introduce the important *generalized susceptibility* $\chi_{BA}(\omega)$:

$$\chi_{BA}(\omega) = \lim_{\varepsilon \to +0} \int_0^\infty d\theta \, e^{i(\omega + i\varepsilon)\theta} \phi_{BA}(\theta) \tag{21.3.14}$$

* These results can, of course, be used in the general case as well, by means of a Fourier transformation.

A characteristic feature is the fact that the response of the system appears at the same frequency as the stimulus. This is only true in the linear regime: beyond it there appear couplings of frequencies, generation of harmonics, and so on. All these phenomena are, of course, quite important in many fields (nonlinear optics, turbulence, ...) but are outside of our present problem.* The susceptibility $\chi_{BA}(\omega)$ has many fascinating properties: we have unfortunately not enough space for their detailed study here (see the references at the end of this chapter).

In order to see the connection with fluctuation theory, we rewrite an explicit expression for $\phi_{BA}(t)$, Eq. (21.3.9), in terms of matrix elements in a representation in which the internal Hamiltonian $\hat{H}$ is diagonal:

$$\begin{aligned}\phi_{BA}(t) &= (i\hbar)^{-1} \sum_n \sum_m e^{\beta(F-E_n)}\{A_{nm} e^{it(E_m-E_n)/\hbar} B_{mn} - e^{it(E_n-E_m)/\hbar} B_{nm} A_{mn}\} \\ &= (i\hbar)^{-1} \sum_n \sum_m e^{\beta(F-E_n)}(1-e^{\beta\hbar\omega_{nm}}) e^{-it\omega_{nm}} A_{nm} B_{mn} \end{aligned} \tag{21.3.15}$$

where we introduced the usual notation

$$\hbar\omega_{nm} = E_n - E_m \tag{21.3.16}$$

The advantage of this representation is that the whole time dependence appears in a single exponential. The susceptibility is therefore immediately obtained from Eq. (21.3.14):

$$\chi_{BA}(\omega) = (i\hbar)^{-1} \sum_n \sum_m e^{\beta(F-E_n)}(1-e^{\beta\hbar\omega_{nm}}) \pi\delta_+(\omega-\omega_{nm}) A_{nm} B_{mn} \tag{21.3.17}$$

The susceptibility is thus a complex number:

$$\chi_{BA}(\omega) = \chi'_{BA}(\omega) + i\chi''_{BA}(\omega) \tag{21.3.18}$$

and we obtain the following expressions for its real and imaginary parts:

$$\chi'_{BA}(\omega) = \hbar^{-1} \sum_n \sum_m e^{\beta(F-E_n)}(1-e^{\beta\hbar\omega_{nm}})[\mathcal{P}(\omega-\omega_{nm})^{-1}] A_{nm} B_{mn} \tag{21.3.19}$$

$$\chi''_{BA}(\omega) = -\pi\hbar^{-1}(1-e^{\beta\hbar\omega}) \sum_n \sum_m e^{\beta(F-E_n)}\, \delta(\omega-\omega_{nm}) A_{nm} B_{mn} \tag{21.3.20}$$

Consider now the equilibrium time-correlation function of $\hat{A}$ and $\hat{B}$ defined by Eq. (21.1.15). By a quite analogous calculation we write it in the form:

$$\begin{aligned} G^0_{BA}(\tau) &= \tfrac{1}{2}\mathrm{Tr}\, \hat{\rho}^0\{\hat{A}\hat{B}(\tau) + \hat{B}(\tau)\hat{A}\} \\ &= \frac{1}{2} \sum_n \sum_m e^{\beta(F-E_n)}(1+e^{\beta\hbar\omega_{nm}}) e^{-it\omega_{nm}} A_{nm} B_{mn} \end{aligned} \tag{21.3.21}$$

* They can, however, be studied by a natural extension of the present formalism.

The similarity with the response function is obvious. We now consider the associated spectral density (21.1.16). It should be noted that the latter is a *Fourier* transform, in contrast with the susceptibility which is a (one-sided) *Laplace* transform. Hence

$$\tilde{G}^0_{BA}(\omega) = \tfrac{1}{2}(1+e^{\beta\hbar\omega})2\pi \sum_n \sum_m e^{\beta(F-E_n)}\,\delta(\omega-\omega_{nm})A_{nm}B_{mn} \qquad (21.3.22)$$

Comparing this equation to (21.3.20) we obtain

$$\chi''_{BA}(\omega) = \hbar^{-1}\frac{1-e^{\beta\hbar\omega}}{1+e^{\beta\hbar\omega}}\,\tilde{G}^0_{BA}(\omega)$$

or

$$\chi''_{BA}(\omega) = \hbar^{-1}(\tanh\tfrac{1}{2}\beta\hbar\omega)\,\tilde{G}^0_{BA}(\omega) \qquad (21.3.23)$$

This is the celebrated *fluctuation–dissipation theorem*, which states that the imaginary part of the generalized susceptibility is proportional to the corresponding spectral density. It provides the link we were looking for between fluctuations and linear response.

Equation (21.3.23) can be inverted; however, as noted by Case, one must be careful because $\tanh 0 = 0$. Therefore the correctly inverted formula is

$$\tilde{G}^0_{BA}(\omega) = \hbar\,\coth(\tfrac{1}{2}\beta\hbar\omega)\,\chi''_{BA}(\omega) + C\,\delta(\omega) \qquad (21.3.24)$$

Here C is an arbitrary constant that can only be determined if we possess more information about the system.

We also note that, in the classical limit, as $\hbar \to 0$ the theorem reduces to

$$\chi''_{BA}(\omega) = \frac{\omega}{2k_B T}\,\tilde{G}^0_{BA}(\omega) \qquad (21.3.25)$$

The fluctuation–dissipation theorem is obviously a very general statement. It provides a variety of interesting relations, according to the choice of the quantities $\hat{A}$ and $\hat{B}$. We have no space here for treating many applications: these can be found in the literature. Let us simply check here that the formula (21.2.18) is indeed a special case of the theorem.

In the case of a plasma in an external time-dependent field $\mathbf{E}(t)$, the external Hamiltonian is

$$\hat{H}^F(t) = -\sum_n z_n e\hat{\mathbf{q}}_n \cdot \mathbf{E}(t) \qquad (21.3.26)$$

We choose for ΔB the rth component of the current:

$$\Delta B \equiv \hat{j}^{er} = \sum_n z_n e\dot{\hat{q}}^r_n \qquad (21.3.27)$$

Then the current response to the sth component of the field is given by the response function (21.3.11):

$$\phi_{BA}(t) = \int_0^\beta d\mu \left\langle \sum_m z_m e \dot{\hat{q}}_m^r(-i\hbar\mu) \sum_n z_n e \dot{\hat{q}}_n^s(t) \right\rangle^0$$

$$= \int_0^\beta d\mu \, \langle \hat{j}^{er}(-i\hbar\mu)\hat{j}^{es}(t)\rangle^0 \qquad (21.3.28)$$

The generalized susceptibility of this problem is taken from (21.3.14):

$$\chi_{BA}(\omega) = \lim_{\varepsilon \to +0} \int_0^\beta d\mu \int_0^\infty d\tau \, e^{i(\omega+i\varepsilon)\tau} \langle \hat{j}^{er}(-i\hbar\mu)\hat{j}^{es}(t)\rangle^0 \qquad (21.3.29)$$

In the classical limit $\hbar \to 0$, this equation reduces to

$$\chi_{BA}(\omega) = \lim_{\varepsilon \to +0} \beta \int_0^\infty d\tau \, e^{i(\omega+i\varepsilon)\tau} \langle \hat{j}^{er}\hat{j}^{es}(\tau)\rangle^0 \qquad (21.3.30)$$

Finally, we see that in the limit of zero frequency (i.e., the response to a constant external field) we recover the Kubo formula (21.2.18) for the electrical conductivity.

21.4. THE THERMAL TRANSPORT COEFFICIENTS

Kubo's theory and the fluctuation–dissipation theorem provide us with very general formulas for the transport coefficients describing the linear response of a system to an external field. However, we know that an important class of transport coefficients, such as viscosity, heat conduction, and diffusion coefficient, are not of this type. These describe the reaction of a system to a spatial inhomogeneity (see Chapter 13). The presence of the latter induces the appearance of flows of matter, momentum, or energy, which tend to homogenize the medium. However, the "driving forces" of these flows cannot be represented naturally as a term in the microscopic Hamiltonian. A single molecule does not behave differently in a homogeneous or in an inhomogeneous system; its law of motion does, however, depend on the presence or absence of an external field. Hence, we see that the "mechanical" and "thermal" processes are basically different at the microscopic level. On the other hand, macroscopically, the two types of phenomena are very similar, as shown for instance by the well-known close relationship between electrical conductivity and diffusion coefficient in an electrolyte solution. One is therefore motivated in looking for an extension of the fluctuation–dissipation methods to cover the thermal coefficients as well.

Some authors tried to imagine strange mechanical forces to which the system would react by setting up a momentum flow or a heat flow, or the like. If such forces are accepted, at least as a "Gedanken experiment," the fluctuation–dissipation theorem can be applied directly as before. This procedure is, however, rather artificial as it does not describe, for example, a real heat-conduction experiment.

A rather different approach was initiated by Mori in 1958. It also has its weakness, which we shall point out in the discussion. The final result is, however, confirmed by all the other existing theories and is therefore certainly correct. The crux of Mori's theory is the concept of *local equilibrium*, which was already discussed in Section 13.2. It was shown there that, if the macroscopic quantities vary very slowly over a mean-free path, the collisions quickly produce a state that is as close as possible to the equilibrium state, but only locally. In this state the one-particle distribution is of Maxwellian type, but its five characteristic parameters, density $n(\mathbf{x})$, local velocity $\mathbf{u}(\mathbf{x})$ and inverse temperature $\beta(\mathbf{x}) = [k_B T(\mathbf{x})]^{-1}$ depend slowly on space and time. The instantaneous state of the system always deviates but slightly from local equilibrium.

In a first step, Mori extends the local equilibrium concept from the one-particle distribution to the complete N-body distribution $F(t)$. This is a rather daring step. Indeed, the one-particle distribution is very directly related to local macroscopic quantities, and the arguments leading to the local equilibrium concept are a direct translation of our physical experience with hydrodynamics and kinetic theory. On the contrary, the local equilibrium assumption for $F(t)$ involves many more very detailed assumptions about the form of the correlations; such assumptions cannot be inferred from qualitative macroscopic arguments and therefore appear as being arbitrary. We may only hope that they will not affect the final result (and this hope is confirmed).

We now consider our "usual" classical system of N interacting particles whose Liouvillian L is defined by Eq. (2.4.15) with $L^F = 0$. A local equilibrium state of this system is *defined* by the distribution function F^{LE}*:

$$F^{LE} = Z^{-1} \exp\left\{-\beta \int d\mathbf{x} \left[\frac{T}{T(\mathbf{x})}\right] E(q, p; \mathbf{x})\right\} \tag{21.4.1}$$

This is a nice way of writing the distribution in terms of a local energy density $E(q, p; \mathbf{x})$ and a local temperature $T(\mathbf{x})$; $\beta = (k_B T)^{-1}$ relates to the space-averaged temperature T. The local energy density is defined as

$$E(q, p; \mathbf{x}) = H(q, p; \mathbf{x}) - \mathbf{j}(q, p; \mathbf{x}) \cdot \mathbf{u}(\mathbf{x}) + \tfrac{1}{2}\rho(q, p; \mathbf{x}) u^2(\mathbf{x}) \tag{21.4.2}$$

It appears as a superposition of five (scalar) dynamical functions that are

* In his paper, Mori uses quantum mechanics and the grand canonical ensemble; the difference is not important.

simply the five collisional invariants (Sections 12.3 and 12.4): the mass density:

$$\rho(q, p; \mathbf{x}) = m \sum_j \delta(\mathbf{q}_j - \mathbf{x}) \tag{21.4.3}$$

the three components of the momentum density:

$$\mathbf{j}(q, p; \mathbf{x}) = \sum_j \mathbf{p}_j \, \delta(\mathbf{q}_j - \mathbf{x}) \tag{21.4.4}$$

and the proper energy density:

$$H(q, p; \mathbf{x}) = \sum_j \delta(\mathbf{q}_j - \mathbf{x}) \left\{ \left(\frac{p_j^2}{2m} \right) + \frac{1}{2} \sum_{\substack{n \\ (\neq j)}} V_{jn} \right\} \tag{21.4.5}$$

The local velocity $\mathbf{u}(\mathbf{x})$ is defined as usual by

$$\mathbf{u}(\mathbf{x}) = \frac{\langle \mathbf{j}(q, p; \mathbf{x}) \rangle}{\langle \rho(q, p; \mathbf{x}) \rangle} \tag{21.4.6}$$

We now assume that at time zero, the system is in the state of local equilibrium: $F(0) = F^{\mathrm{LE}}$. The formal solution, at time t, can then be represented as follows:

$$F(t) = e^{tL} F(0) = F(0) + [F(t) - F(0)]$$
$$= F(0) + \int_0^t d\tau \frac{\partial}{\partial \tau} F(\tau) = F(0) + \int_0^t d\tau \, LF(\tau)$$

or

$$F(t) = F^{\mathrm{LE}} + \int_0^t d\tau \, U(\tau) L F^{\mathrm{LE}} \tag{21.4.7}$$

The evaluation of the rhs is pretty straightforward:

$$LF^{\mathrm{LE}} = -\beta F^{\mathrm{LE}} \int d\mathbf{x} \left[\frac{T}{T(\mathbf{x})} \right] LE(q, p; \mathbf{x}) \tag{21.4.8}$$

The calculation reduces to the action of the Liouvillian on the five collisional invariants. We first have [using Eq. (2.4.17)]*

$$L\rho(\mathbf{x}) = -m \sum_n \sum_j \mathbf{v}_n \cdot \nabla_n \, \delta(\mathbf{q}_j - \mathbf{x}) = -m \sum_j \mathbf{v}_j \cdot \nabla_j \, \delta(\mathbf{q}_j - \mathbf{x})$$
$$= \nabla \cdot \left\{ \sum_j \mathbf{p}_j \, \delta(\mathbf{q}_j - \mathbf{x}) \right\}$$

where, as usual, $\nabla = \partial / \partial \mathbf{x}$. Recalling that, in the Heisenberg picture, $LA = -\partial A / \partial t$ for every dynamical variable A, we obtained

$$-L\rho(\mathbf{x}) \equiv \dot{\rho}(\mathbf{x}) = -\nabla^\beta j^\beta(\mathbf{x}) \tag{21.4.9}$$

* We no longer write explicitly the arguments q, p.

where $j^{\beta}(\mathbf{x})$ is the mass flow, which is identical to the momentum density of Eq. (21.4.4). This equation is the *microscopic mass-balance equation* corresponding to the macroscopic continuity equation (12.4.3). It expresses the local conservation of mass.

A similar calculation provides us with

$$Lj^{\alpha}(\mathbf{x}) = \nabla^{\beta} \sum_{j} v_j^{\beta} p_j^{\alpha}\, \delta(\mathbf{q}_j - \mathbf{x}) + \sum_{j<n}\sum (\nabla_j^{\alpha} V_{jn})[\delta(\mathbf{q}_j - \mathbf{x}) - \delta(\mathbf{q}_n - \mathbf{x})] \quad (21.4.10)$$

If we invoke the *hydrodynamical approximation* as was done in Section 18.4, we may expand the second δ function around $\mathbf{q}_j = \mathbf{q}_n$ and keep only the leading term. We then find rather easily

$$-Lj^{\alpha}(\mathbf{x}) \equiv \left(\frac{d}{dt}\right) j^{\alpha}(\mathbf{x}) = -\nabla^{\beta} J_j^{\alpha\beta}(\mathbf{x}) \quad (21.4.11)$$

With the momentum flow (in the hydrodynamical approximation):

$$J_j^{\alpha\beta}(\mathbf{x}) = \sum_{n} \delta(\mathbf{q}_n - \mathbf{x}) \left\{ m^{-1} p_n^{\alpha} p_n^{\beta} - \frac{1}{2} \sum_{\substack{m \\ (\neq n)}} \left(\frac{r_{nm}^{\alpha} r_{nm}^{\beta}}{r_{nm}}\right)\left(\frac{\partial V_{nm}}{\partial r_{nm}}\right)\right\} \quad (21.4.12)$$

For the energy balance we finally obtain

$$-LH(\mathbf{x}) \equiv \dot{H}(\mathbf{x}) = -\nabla^{\beta} J_H^{\beta}(\mathbf{x}) \quad (21.4.13)$$

with the energy flow

$$J_H^{\beta}(\mathbf{x}) = \sum_{j} \delta(\mathbf{q}_j - \mathbf{x}) \left\{ \left(\frac{p_j^2}{2m}\right)\left(\frac{p_j^{\beta}}{m}\right) + \frac{1}{2} \sum_{\substack{n \\ (\neq j)}} m^{-1} p_j^{\beta} V_{jn} \right.$$

$$\left. - \frac{1}{4} \sum_{\substack{n \\ (\neq j)}} m^{-1} r_{jn}^{\beta} (\mathbf{p}_j + \mathbf{p}_n) \cdot \left(\frac{\mathbf{r}_{jn}}{r_{jn}}\right)\left(\frac{\partial V_{jn}}{\partial r_{jn}}\right)\right\} \quad (21.4.14)$$

We now substitute these results into Eq. (21.4.8) and perform an integration by parts:

$$\int d\mathbf{x} \left[\frac{T}{T(\mathbf{x})}\right] LE(q, p; \mathbf{x})$$

$$= \int d\mathbf{x} \left[\frac{T}{T(\mathbf{x})}\right] \{\nabla^{\beta} J_H^{\beta}(q, p; \mathbf{x}) - u^{\alpha}(\mathbf{x}) \nabla^{\beta} J_j^{\alpha\beta}(q, p; \mathbf{x}) + \tfrac{1}{2} u^2(\mathbf{x}) \nabla^{\beta} j^{\beta}(q, p; \mathbf{x})\}$$

$$= \int d\mathbf{x} \left\{ [-J_H^{\beta}(q, p; \mathbf{x}) + u^{\alpha}(\mathbf{x}) J_j^{\alpha\beta}(q, p; \mathbf{x}) - \tfrac{1}{2} u^2(\mathbf{x}) j^{\beta}(q, p; \mathbf{x})] \left[\nabla^{\beta}\left(\frac{T}{T(\mathbf{x})}\right)\right] \right.$$

$$\left. + \left[\frac{T}{T(\mathbf{x})}\right] [J_j^{\alpha\beta}(q, p; \mathbf{x}) - u^{\alpha}(\mathbf{x}) j^{\beta}(q, p; \mathbf{x})] \nabla^{\beta} u^{\alpha}(\mathbf{x}) \right\} \quad (21.4.15)$$

In Section 12.4 we had to take care in separating the flows into convective and dissipative parts: we must do the same here. We define the "dissipative" parts of the microscopic pressure tensor $J_V^{\alpha\beta}(q, p; \mathbf{x})$ and of the energy flow $J_{th}^{\beta}(q, p; \mathbf{x})$ by the equations

$$
\begin{aligned}
J_j^{\alpha\beta} = &\tfrac{1}{2}[j^\alpha(q, p; \mathbf{x})u^\beta(\mathbf{x}) + u^\alpha(\mathbf{x})j^\beta(q, p; \mathbf{x})] \\
&+\tfrac{1}{2}[j^\alpha(q, p; \mathbf{x})u^\beta(\mathbf{x}) + u^\alpha(\mathbf{x})j^\beta(q, p; \mathbf{x}) - 2\rho(q, p; \mathbf{x})u^\alpha(\mathbf{x})u^\beta(\mathbf{x})] \\
&+[J_V^{\alpha\beta}(q, p; \mathbf{x}) + P(q, p; \mathbf{x})\,\delta^{\alpha\beta}]
\end{aligned}
\tag{21.4.16}
$$

$$
\begin{aligned}
J_H^\alpha(q, p; \mathbf{x}) = &H(q, p; \mathbf{x})u^\alpha(\mathbf{x}) + J_{th}^\alpha(q, p; \mathbf{x}) \\
&+[J_V^{\alpha\beta}(q, p; \mathbf{x}) + P(q, p; \mathbf{x})\,\delta^{\alpha\beta}]u^\beta(\mathbf{x}) \\
&+\tfrac{1}{2}[j^\alpha(q, p; \mathbf{x}) - \rho(q, p; \mathbf{x})u^\alpha(\mathbf{x})]
\end{aligned}
\tag{21.4.17}
$$

where

$$P(q, p; \mathbf{x}) = \tfrac{1}{3}\sum_\alpha J_j^{\alpha\alpha}(q, p; \mathbf{x}) \tag{21.4.18}$$

is the local microscopic pressure. The "dissipative" flows $J_v^{\alpha\beta}$ and J_{th}^α have the important property of having zero average value in the local equilibrium state, as can be easily checked [see also Eqs. (13.2.9)]:

$$\langle J_V^{\alpha\beta}\rangle^{LE} = 0, \qquad \langle J_{th}^\alpha\rangle^{LE} = 0 \tag{21.4.19}$$

Hence these flows are typical measures of the response of the system to the presence of gradients, playing the same role as $\langle\Delta\hat{B}(t)\rangle$ in the previous section. We now calculate their average value in the nonequilibrium state $F(t)$ of Eq. (21.4.7). We thus obtain for the dissipative energy flow

$$
\begin{aligned}
\langle J_{th}^\alpha(\mathbf{x})\rangle = &\int dq\,dp\,J_{th}^\alpha(q, p; \mathbf{x})\beta Z^{-1}\int_0^t d\tau\,U(\tau) \\
&\times\exp\left\{-\beta\int d\mathbf{y}\left[\frac{T}{T(\mathbf{y})}\right]E(q, p; \mathbf{y})\right\}\cdot\int d\mathbf{x}'\,\{[J_H^\beta(q, p; \mathbf{x}') \\
&-u^\gamma(\mathbf{x}')J_j^{\gamma\beta}(q, p; \mathbf{x}') + \tfrac{1}{2}u^2(\mathbf{x}')j^\beta(q, p; \mathbf{x}')][-\nabla'_\beta TT^{-1}(\mathbf{x}')] \\
&+[J_j^{\gamma\beta}(\mathbf{x}') - u^\gamma(\mathbf{x}')j^\beta(q, p; \mathbf{x}')]\nabla'^\beta u^\gamma(\mathbf{x}')\}
\end{aligned}
\tag{21.4.20}
$$

We must now express the idea that we are only interested in the *linear response* of the system to the gradients. We therefore assume that the gradients are small and *constant:*

$$
\begin{aligned}
-\nabla^\beta\left[\frac{T}{T(\mathbf{x})}\right] &\approx T^{-1}(\nabla^\beta T) \\
\nabla^\beta u^\gamma(\mathbf{x}) &\approx \nabla^\beta u^\gamma
\end{aligned}
\tag{21.4.21}
$$

As all the terms in (21.4.20) are already proportional to one of these factors, we may replace everywhere else $T(\mathbf{x})$ by its constant average value T and $\mathbf{u}(\mathbf{x})$ by its constant average value $\mathbf{u} = 0$. The local-equilibrium function then reduces to the true equilibrium distribution and, using

(21.4.16), (12.4.17), Eq. (21.4.20) becomes*

$$\langle J^{\alpha}_{th}(\mathbf{x})\rangle = \int dq\, dp\, J^{\alpha}_{th}(q,p;\mathbf{x})\beta Z^{-1} \times \int_0^t d\tau\, U(\tau) e^{-\beta H(q,p)} \int d\mathbf{x}'\, J^{\beta}_{th}(q,p;\mathbf{x}')T^{-1}\nabla^{\beta}T \quad (21.4.22)$$

We finally average the result over the space, introducing the overall flows:

$$\begin{aligned}\mathcal{J}^{\alpha}_{th}(q,p) &= \mathcal{V}^{-1}\int d\mathbf{x}\, J^{\alpha}_{th}(q,p;\mathbf{x}) \\ &= \sum_{j=1}^{N}\Bigg\{\left(\frac{p_j^2}{2m}\right)\left(\frac{p_j^{\alpha}}{m}\right) + \frac{1}{2}\sum_{\substack{n\\(\neq j)}}\left(\frac{p_j^{\alpha}}{m}\right)V_{jn} \\ &\quad -\frac{1}{4}\sum_{\substack{n\\(\neq j)}}(\mathbf{p}_j+\mathbf{p}_n)\cdot\left(\frac{\mathbf{r}_{jn}}{r_{jn}}\right)\left(\frac{dV_{jn}}{dr_{jn}}\right)r^{\alpha}_{jn}\Bigg\}\end{aligned} \quad (21.4.23)$$

and similarly for the dissipative momentum flow:

$$\begin{aligned}\mathcal{J}^{\alpha\beta}_{v}(q,p) &= \mathcal{V}^{-1}\int d\mathbf{x}\, J^{\alpha\beta}_{v}(q,p;\mathbf{x}) \\ &= \sum_{j}\Bigg\{m^{-1}p^{\alpha}_j p^{\beta}_j - \frac{1}{2}\sum_{\substack{n\\(\neq j)}}\left(\frac{r^{\alpha}_{jn}r^{\beta}_{jn}}{r_{jn}}\right)\left(\frac{dV_{jn}}{dr_{jn}}\right)\Bigg\}\end{aligned} \quad (21.4.24)$$

Equation (21.4.22) then implies

$$\langle \mathcal{J}^{\alpha}_{th}\rangle = \kappa^{\alpha\beta}\nabla^{\beta}T \quad (21.4.25)$$

with

$$\begin{aligned}\kappa^{\alpha\beta} &= (k_B T^2\mathcal{V})^{-1}\int_0^t d\tau \int dq\, dp\, \mathcal{J}^{\alpha}_{th}\, e^{\tau L}[\mathcal{J}^{\beta}_{th}F^0] \\ &= (k_B T^2\mathcal{V})^{-1}\int_0^t d\tau \int dq\, dp\, F^0\mathcal{J}^{\beta}_{th}\, e^{-\tau L}\mathcal{J}^{\alpha}_{th}\end{aligned}$$

If we, moreover, assume that the time integral reaches a plateau value in a time of the order of the relaxation time, we may replace for such times the upper limit of integration by ∞, with the result:

$$\kappa^{\alpha\beta} = (k_B T^2\mathcal{V})^{-1}\int_0^{\infty} d\tau\, \langle \mathcal{J}^{\alpha}_{th}(\tau)\, \mathcal{J}^{\beta}_{th}\rangle^0 \quad (21.4.26)$$

Hence we finally obtained a formula for the *thermal-conductivity tensor*† that is completely analogous to the Kubo formula for the electrical

* We use the fact that the vector flow J^{α}_{th} cannot be coupled to the tensor flow $J^{\alpha\beta}_{v}$.
† Which reduces to a scalar for isotropic systems.

conductivity, (21.2.18). It is expressed as *the time integral of the heat-flow autocorrelation function in equilibrium.*

A similar calculation provides us with an expression of the viscosity (in an isotropic system):

$$\eta = (k_B T \mathscr{V})^{-1} \int_0^\infty d\tau \, \langle \mathscr{J}_v^{xy}(\tau) \, \mathscr{J}_v^{xy} \rangle^0 \tag{21.4.27}$$

The formulas (21.4.26) and (21.4.27) are remarkably compact. They give us the intellectual satisfaction that all the transport coefficients can be expressed in a unified form as integrals of autocorrelation functions of microscopic flows. They are extremely general, in the sense that no specific approximation has been made on the type of interactions. However, the local-equilibrium assumption is a very strong one that is difficult to justify in a N-body theory. We have seen in Section 13.4 that the expression of the transport coefficients obtained from the lowest-order approximation of the kinetic equation can indeed be cast in the form (21.4.27) [see Eq. (13.4.16)]. Actually, Résibois showed in 1964 that, to all orders of approximation, the expressions for the transport coefficients obtained by the method of the kinetic equations are the same as those given by the autocorrelation functions.

It must be noted, moreover, that, precisely because of the generality of the autocorrelation formulas, their explicit evaluation is very difficult: one is faced with the full N-body problem. One must first devise an algorithm by which the problem is properly reduced to a simpler problem, to which the techniques of kinetic theory can be applied. Such a formalism is outlined in Sections 21.6 and 21.7.

A further remark can be made here. We see from Eqs. (21.4.23) and (21.4.24) that both the heat flow and the momentum flow involve, in general, besides the "kinetic part" that appeared already in Eq. (13.4.14), a "potential part" related to the interactions. It would therefore seem that we need not only expressions of the one-particle function but also of the correlations in order to evaluate the transport coefficients. This, however, should be an illusion. If our considerations of Section 17.8 can be extended to the thermal-transport coefficients, these should be expressible as functionals of $V\bar{\mathfrak{f}}$ alone, that is, in terms of $\bar{f}_1(\mathbf{q}, \mathbf{p})$. Although a direct proof of the relation between the thermal transport coefficients and the Π subdynamics has not yet been given, there exists very convincing evidence in this sense. Indeed, in his recent work (which we reviewed in the simplest case in Chapter 13), Résibois showed that, to all orders of approximation, the transport coefficients—including their potential part—can be calculated in terms of $\bar{f}_1$ alone.

21.5. "LONG TAILS" IN THE CORRELATION FUNCTIONS

We devote the present section to a subject that has been vividly discussed in recent years. There was a kind of consensus among specialists of autocorrelation functions that these objects should behave like decaying exponentials, at least asymptotically, for long enough times. This belief was only based on the simple soluble models (which we treated in Chapter 11) such as the Brownian motion, the theory of Markovian stochastic processes, and the Boltzmann equation. A typical result of this kind appears in Eq. (11.2.15). There is, of course, no rational basis, let alone a proof of the general validity of this property. On the contrary, as early as 1960, Guernsey showed that in a plasma the correlations relax as t^{-1} for small wave vectors. His result, however, went unnoticed (maybe people believed this was one of those pathological effects related to long-range forces, which it actually is!). In 1968, Alder and Wainwright performed a molecular-dynamics computer calculation of the velocity autocorrelation function of systems of hard disks and of hard spheres, and found that this function has a definitely nonexponential behavior. Actually, for long times, it seemed to behave as $t^{-d/2}$, d being the dimensionality of the system. This result came as a big surprise and a shock to orthodoxy. However, the authors offered a simple hydrodynamical model for its explanation. A little later, Dorfman and Cohen showed that the result also comes out of the kinetic equation for moderately dense systems. We recall that, beyond the Boltzmann approximation, we had divergence difficulties in a density expansion (see Section 20.4). It turns out that precisely the terms responsible of those difficulties also lead to the nonexponential tail of the correlation functions. Finally, Ernst, Hauge, and van Leeuwen showed by a semiphenomenological calculation that the $t^{-d/2}$ behavior should be expected to be a quite general phenomenon, related to the hydrodynamical behavior of long-wavelength phenomena in a fluid. We summarize here the main features of their work.

Consider the viscosity of a fluid. From Eqs. (21.4.24), (21.4.27) its kinetic part can be written in the form:

$$\eta_K = \beta \int_0^\infty d\tau \, C_\eta(\tau) \tag{21.5.1}$$

with

$$\begin{aligned} C_\eta(\tau) &= \mathcal{V}^{-1} \left\langle (Nm v_1^x v_1^y) \left[\sum_j m v_j^x(\tau) v_j^y(\tau) \right] \right\rangle^0 \\ &= \mathcal{V}^{-1} \langle j^{xy} j^{xy}(\tau) \rangle^0 \end{aligned} \tag{21.5.2}$$

In order to approximately evaluate this quantity, the authors introduce a trick. They first average over the equilibrium ensemble with the extra constraint that particle 1 has at time zero the given velocity $\mathbf{v}_0$ and has positions distributed in a neighborhood of $\mathbf{x}_0$, with probability density $W(\mathbf{q}_1 - \mathbf{x}_0)$. The exact form of this function turns out to be irrelevant: it must only be normalized to 1. In a second step the result is averaged over $\mathbf{x}_0$, $\mathbf{v}_0$ with the one particle Maxwell distribution $\varphi^0(\mathbf{v}_0)$:

$$\begin{aligned} C_\eta(\tau) &= \mathcal{V}^{-1} \int d\mathbf{x}_0 \, d\mathbf{v}_0 \, \varphi^0(v_0) m v_0^x v_0^y \\ &\quad \times N \int d\mathbf{x} \int dq \, dv \, F^0(q, v) \sum_j m_j v_j^x(\tau) v_j^y(\tau) \\ &\quad \times \delta[\mathbf{q}(\tau) - \mathbf{x}] \, \delta(\mathbf{v}_1 - \mathbf{v}_0) W(\mathbf{q}_1 - \mathbf{x}_0) \\ &= \mathcal{V}^{-1} \int d\mathbf{x}_0 \, d\mathbf{v}_0 \, \varphi^0(v_0) m v_0^x v_0^y \int d\mathbf{x} \, j^{xy}(\mathbf{x}; \tau) \end{aligned} \tag{21.5.3}$$

where the current density $j^{xy}(\mathbf{x}; t)$ is defined by this equation. Its explicit evaluation is extremely difficult, because it involves the full N-body dynamical problem. However, one may assume (as usual) that if the spatial distribution $W(\mathbf{x})$ is sufficiently smooth and the gradients sufficiently small, the system will first quickly relax towards a local-equilibrium distribution with parameters $n(\mathbf{x}; \tau)$, $\mathbf{u}(\mathbf{x}; \tau)$, $T(\mathbf{x}; \tau)$ slowly varying in space and time. In a second step, this distribution relaxes towards total equilibrium over a much longer time scale, and according to the macroscopic hydrodynamic laws. With this assumption, one may approximate the complicated current density $j^{xy}(\mathbf{x}; \tau)$ by its local-equilibrium value:

$$j^{xy}(\mathbf{x}; \tau) \approx \int d\mathbf{v} \, j^{xy} F^{\mathrm{LE}}(\mathbf{x}, \mathbf{v}; t) \tag{21.5.4}$$

where the one-particle local-equilibrium function is

$$f^{\mathrm{LE}}(\mathbf{x}, \mathbf{v}; \tau) = n(\mathbf{x}; \tau) \left(\frac{m}{2\pi k_B T(\mathbf{x}; \tau)} \right)^{d/2} \exp\left(\frac{-(1/2) m [\mathbf{v} - \mathbf{u}(\mathbf{x}; \tau)]^2}{k_B T(\mathbf{x}; \tau)} \right) \tag{21.5.5}$$

The dimensionality of the space d appears here explicitly in the normalization factor. We now have

$$j^{xy}(\mathbf{x}; \tau) \approx m n(\mathbf{x}; \tau) u^x(\mathbf{x}; \tau) u^y(\mathbf{x}; \tau) \tag{21.5.6}$$

The functions $mn(\mathbf{x}; \tau) \equiv \rho(\mathbf{x}; \tau)$, $\mathbf{u}(\mathbf{x}; \tau)$, $T(\mathbf{x}; \tau)$ appearing in these formulas are solutions of the linearized hydrodynamic equations (12.5.29)–(12.5.31). To evaluate $j^{xy}(\mathbf{x}; \tau)$ we therefore solve these equations as an initial-value problem (rather than looking for their eigenvalues as in Section 12.6). Going over to the Fourier picture, we get for the velocity an equation similar to (12.6.2):

$$\partial_t u_{\mathbf{k}}^r + iCk^r \nu_{\mathbf{k}} + iAk^r \theta_{\mathbf{k}} + \eta' k^2 u_{\mathbf{k}}^r + \mu k^r k^s u_{\mathbf{k}}^s = 0 \tag{21.5.7}$$

and two analogous equations for $\nu_{\mathbf{k}}$ and $\theta_{\mathbf{k}}$. We now split the vector $\mathbf{u}_{\mathbf{k}}$ into a transverse and a longitudinal part as follows:

$$\mathbf{u}_{\mathbf{k}} = \mathbf{u}_{\mathbf{k}T} + \mathbf{u}_{\mathbf{k}L} \tag{21.5.8}$$

with

$$\begin{aligned} u_{\mathbf{k}T}^r &= u_{\mathbf{k}}^r - \hat{k}^r(\hat{\mathbf{k}} \cdot \mathbf{u}_{\mathbf{k}}) \\ u_{\mathbf{k}L}^r &= \hat{k}^r(\hat{\mathbf{k}} \cdot \mathbf{u}_{\mathbf{k}}) \\ \hat{\mathbf{k}} &= \frac{\mathbf{k}}{k} \end{aligned} \tag{21.5.9}$$

Note that in an incompressible fluid, the longitudinal part vanishes. The transverse part is related to shear flow, while the longitudinal part is related to sound propagation. We now find from Eqs. (21.5.7), (21.5.9) that the transverse part obeys a very simple closed equation:

$$\partial_\tau u_{\mathbf{k}T}^r + \eta'^2 k^2 u_{\mathbf{k}T}^r = 0 \tag{21.5.10}$$

The longitudinal part obeys an equation that is coupled to the density and temperature equations. For simplicity we only give here the calculation for the contribution of $\mathbf{u}_{\mathbf{k}T}$. The solution of Eq. (21.5.10) is

$$u_{\mathbf{k}T}^r(\tau) = u_{\mathbf{k}T}^r(0) e^{-\eta' k^2 \tau} \tag{21.5.11}$$

Hence, only the shear modes contribute to the transverse velocity (see Table 12.6.1). The current is therefore given, up to second order in the deviation from equilibrium, by

$$\begin{aligned} \int d\mathbf{x}\, j_{TT}^{xy}(\mathbf{x};\tau) &\approx \rho \int d\mathbf{x}\, u_T^x(\mathbf{x};\tau) u_T^y(\mathbf{x};\tau) \\ &= \rho(2\pi)^{-d} \int d^d\mathbf{k}\, u_{\mathbf{k}T}^x(0) u_{-\mathbf{k}T}^y(0) e^{-2\eta' k^2 \tau} \end{aligned} \tag{21.5.12}$$

In agreement with the spirit of the hydrodynamic approximation, we assume that the initial value of the field $\mathbf{u}_T(\mathbf{x};\tau)$ is determined by the distribution $W(\mathbf{x})$:

$$\mathbf{u}_T(\mathbf{x};0) = n^{-1} \mathbf{v}_{0T} W(\mathbf{x} - \mathbf{x}_0) \tag{21.5.13}$$

Substituting (21.5.12), (21.5.13) into Eq. (21.5.3) we find

$$\begin{aligned} C_{\eta T} \approx{}& m^2 n^{-1} (2\pi)^{-d} \int d\mathbf{v}_0\, \varphi^0(v_0) v_0^x v_0^y \\ &\times \int d^d\mathbf{k}\, [v_0^x - \hat{k}^x(\mathbf{k}\cdot\mathbf{v}_0)][v_0^y - \hat{k}^y(\mathbf{k}\cdot\mathbf{v}_0)]\, |\tilde{W}_{\mathbf{k}}|^2 \exp(-2\eta' k^2 \tau) \end{aligned} \tag{21.5.14}$$

We now note that, by our assumption of smoothness in space, the Fourier transform $\tilde{W}_{\mathbf{k}}$ is sharply peaked near $\mathbf{k}=0$; it can therefore be approximated by

$$\tilde{W}_{\mathbf{k}} \approx \tilde{W}_0 = \int d\mathbf{x}\, W(\mathbf{x}) = 1$$

Hence, going over to spherical coordinates, that is, expressing $d^d\mathbf{k}$ as $k^{d-1}\,dk\,d\Omega$, where $d\Omega$ is a solid-angle element, we obtain

$$C_{\eta T} \approx m^2 n^{-1} \int d\mathbf{v}_0 \varphi^0(v_0) v_0^{x2} v_0^{y2} (2\pi)^{-d}$$
$$\times \int d\Omega\, [1 - \hat{k}^{x2} - \hat{k}^{y2} + 2\hat{k}^{x2}\hat{k}^{y2}] \int_0^\infty dk\, k^{d-1} e^{-2\eta' k^2 \tau} \quad (21.5.15)$$

The form of the time dependence comes entirely from the integration over the absolute value of $\mathbf{k}$: the remaining integrations merely provide a numerical factor. It can be seen by a simple dimensional argument that $C_{\eta T}(\tau) \sim \tau^{-d/2}$: this is what we set out to show. A complete integration (which is elementary) gives

$$C_{\eta T}(\tau) = \frac{d^2-2}{d(d+2)\beta^2} (8\pi\eta'\tau)^{-d/2} \quad (21.5.16)$$

The contribution of the longitudinal part $\mathbf{u}_L(\mathbf{x};\tau)$ is calculated in similar fashion. Not surprisingly, it is determined by the sound modes and involves the sound absorption coefficient Γ [Eq. (12.6.17)]:

$$C_\eta(\tau) = \frac{1}{d(d+2)\beta^2} \left(\frac{d^2-2}{(8\pi\eta')^{d/2}} + \frac{1}{(4\pi\Gamma)^{d/2}} \right) \tau^{-d/2} \quad (21.5.17)$$

The nonexponential behavior is quite clear. However, it is suggestive to note its mathematical origin. We see from Eq. (21.5.15) that it is really a result of a *superposition of exponentials*. There is an infinite number of these, each being labeled by the continuous index k: they correspond to the superposition (or "coupling") of two hydrodynamic modes, in this case of two shear modes. The sum (which becomes an integral for continuous k) of these exponential modes is no longer an exponential.*

Various questions can be raised about this calculation. First, we may be surprised to see that we used the expressions of Table 12.6.1 for the hydrodynamic modes under an integral over k from 0 to ∞; we know that those expressions are only valid for *small* k. This objection can however

* One may invoke a weak analogy with a wave packet, which is a superposition of oscillating exponentials: $\exp(ickt)$, with k varying continuously. The individual waves interfere with each other, and as a result their superposition no longer has an exponential from (in general).

be easily answered. Indeed, the large values of k in the integral (21.5.15) are actually cut off by the exponential integrand. Moreover, any correction (say, of order k^3) to the hydrodynamic eigenvalue would lead to terms decaying faster in time: thus, the asymptotic behavior (within the present model) is correct.

The model itself may be open to questions. However, it has the advantage of clearly and simply showing the physical origin of the nonexponential tail of the correlation functions. It comes out of the propagation of disturbances over long distances by a mechanism dominated by the macroscopic hydrodynamic laws. This is a slow process that therefore lasts long after the decay of the faster exponential transients. Because of this feature, it is natural enough that the correlation functions are expressed, as in Eq. (21.5.17), in terms of the macroscopic transport coefficients $\eta, \Gamma, \ldots$. There is, however, a case in which the model appears equivocal. In the case of two dimensions, the time dependence of $C_\eta(\tau)$ is τ^{-1}. But the macroscopic viscosity η is expressed by Eq. (21.5.1) as the time integral of $C_\eta(\tau)$. If the integral is evaluated with the expression (21.5.17) it diverges: hence the coefficients η and Γ of that equation do not exist and therefore the model is self-contradictory. The argument is, however, not quite convincing, because Eq. (21.5.17) only provides an asymptotic approximation to $C_\eta(\tau)$; it would be a very brutal procedure to use it in an integral from 0 to ∞ over τ: it does not even provide a natural cutoff as in the problem discussed just above. Nevertheless, some doubt remains attached to this case.

In view of this discussion it appears quite necessary to investigate this problem from a more basic microscopic view point. This program was started by Résibois and Pomeau, who already made important contributions to the problem. Without going into details, it suffices to say that the idea of their approach is very similar to the one they used in the treatment of the van der Waals fluid (Section 20.7). It involves the study of the propagation during a collision process; we already know that, for small $\mathbf{k}$, this process is dominated by the hydrodynamic modes. These authors confirm the $\tau^{-d/2}$ term but can also go into much richer details. We refer the reader to their original work for further study.

21.6. MANY-TIMES REDUCED DISTRIBUTION FUNCTIONS

We now set up a general formalism allowing us to evaluate many-time correlations as easily as instantaneous ones. For this purpose we introduce a very natural generalization of the reduced distribution functions.

Referring to Eq. (3.1.20) we recall that the average of a dynamical function depending on the positions and velocities of two particles, evaluated at the same time t is expressed in terms of the function $f_2(y_1 y_2; t)$:

$$f_2(y_1 y_2; t) = \int d^N x \sum_{j_1} \delta(y_1 - x_{j_1}) \sum_{\substack{j_2 \\ (\neq j_1)}} \delta(y_2 - x_{j_2}) F(x_1 \cdots x_N; t) \quad (21.6.1)$$

We now consider that the position and the velocity of the second particle are displaced in time by an amount τ with respect to the first, and introduce the function:

$$f_{1|1}(y_1; \tau \mid z_1; t) = \int d^N x \sum_{j_1} \delta(y_1 - x_{j_1}[\tau]) \sum_{n_1} \delta(z_1 - x_{n_1}) F(x_1 \cdots x_N; t) \quad (21.6.2)$$

Here $x_j[\tau]$ represents the value of the dynamical variable x_j at time τ, in the Heisenberg picture, just as in (21.1.13):

$$x_j[\tau] = \exp(-\tau L_N) x_j \quad (21.6.3)$$

We recall that $x_j[\tau]$ is, in general, a complicated function of *all* the variables x_n $(n = 1, 2, \ldots, N)$: the latter are the "initial values" in the dynamical problem involving $x_j[\tau]$. More generally:

$$\sum_j \delta(y - x_j[\tau]) = \exp(-\tau L_N) \left\{ \sum_j \delta(y - x_j) \right\} \quad (21.6.4)$$

where it is clearly understood that the operator L_N acts only on the dynamical variables x_j, not on the *parameter* y.

We also note the difference in the summation prescription in Eq. (21.6.2) as compared to (21.6.1): in the former the term $j = n$ is *not* excluded. In the instantaneous case this term could only contribute when $y_1 = y_2$, otherwise it vanishes because of the product $\delta(y_1 - x_j)\,\delta(y_2 - x_j)$. Moreover, when $y_1 = y_2$, it reduces to the one-particle function. More precisely, comparing Eqs. (21.6.1), (21.6.2), and taking account of (21.6.3) we find

$$f_{1|1}(y_1; 0 \mid z_1; t) = \delta(y_1 - z_1) f_1(y_1; t) + f_2(y_1 z_1; t) \quad (21.6.5)$$

When $\tau \neq 0$, however, the "self-term" $j = n$ is no longer singular. Indeed, it is a nontrivial problem to consider the correlation between the successive positions of a given particle at different times. Mathematically, this absence of singularity comes from the fact that $x_j[\tau] \neq x_j$ (in general), hence the product of δ functions does not imply that $y_1 = z_1$. Actually, it is sometimes desirable to separate the function $f_{1|1}$ into a "self-distribution" $f_{1|1}^s$ and a "distinct-distribution" $f_{1|1}^d$:

$$f_{1|1}(y_1; \tau \mid z_1; t) = f_{1|1}^s(y_1; \tau \mid z_1; t) + f_{1|1}^d(y_1; \tau \mid z_1; t) \quad (21.6.6)$$

with

$$f^s(y_1;\tau\,|\,z_1;t)=\int d^Nx\sum_j\delta(y_1-x_j[\tau])\,\delta(z_1-x_j)F(x_1,\ldots,x_N;t) \tag{21.6.7}$$

$$f^d_{1|1}(y_1;\tau\,|\,z_1;t)=\int d^Nx\left\{\sum_{j_1}\delta(y_1-x_{j_1}[\tau])\right\}\left\{\sum_{\substack{n_1\\(\neq j_1)}}\delta(z_1-x_{n_1})\right\}F(x_1,\ldots,x_N;t) \tag{21.6.8}$$

We now consider the averaging process. If we want to calculate the correlation between a dynamical function b at time t and b' at time $t+\tau$, and if both these functions are of the form:

$$b=\sum_{j=1}^N b_1(x_j),\qquad b'=\sum_{j=1}^N b_1'(x_j)$$

we find

$$\langle b'(\tau)\,b\rangle_t=\int dy\,dz\,b_1'(y)b_1(z)f_{1|1}(y;\tau\,|\,z;t) \tag{21.6.9}$$

Indeed, by the definition (21.6.2):

$$\int dy\,dz\,b_1'(y)b_1(z)\int dx_1\cdots dx_N\sum_j\delta(y-x_j[\tau])\sum_n\delta(z-x_n)F(x_1,\ldots,x_N;t)$$
$$=\int dx_1\cdots dx_N\left\{\sum_j b_1'(x_j[\tau])\right\}\left\{\sum_n b_1(x_n)\right\}F(x_1,\ldots,x_N;t)$$

which is equivalent to the general definition (21.1.13). The main advantage of this formulation is in the fact that the whole time dependence, in both τ and t, lies in the reduced distribution function $f_{1|1}$. Hence the *correlations can be calculated by the same procedure as any ordinary average.*

The distribution function $f_{1|1}$ allows us to calculate correlations of any dynamical functions depending additively on single-particle functions. The formalism is, however, easily extended. We define the two-time function $f_{s|r}(y_1\cdots y_s;\tau\,|\,z_1\cdots z_r;t)$ as follows:

$$f_{s|r}(y_1\ldots y_s;\tau\,|\,z_1\ldots z_r;t)=\sum_{j_1\neq\cdots\neq j_s}\cdots\sum\;\sum_{n_1\neq\cdots\neq n_r}\cdots\sum\int dx_1\cdots dx_N$$
$$\times\,\delta(y_1-x_{j_1}[\tau])\,\delta(y_2-x_{j_2}[\tau])\ldots\delta(y_s-x_{j_s}[\tau])$$
$$\times\,\delta(z_1-x_{n_1})\ldots\delta(z_r-x_{n_r})F(x_1,\ldots,x_N;t) \tag{21.6.10}$$

We focus attention again on the summation prescriptions. The subscripts j_i must be all different; the same holds for the n_i's. But the contributions in which a j subscript equals an n subscript are included.

In terms of these functions, any two-time correlation of dynamical functions $b(x_1,\ldots,x_N)$, $c(x_1,\ldots,x_N)$, represented in the form (3.1.8) can be calculated by a formula similar to (3.1.18):

$$\langle b(\tau)c\rangle_t = \sum_{r=0}^{N}\sum_{s=0}^{N}\frac{1}{s!\,r!}\int dy_1\cdots dy_s\,dz_1\cdots dz_r$$

$$\times\, b_s(y_1,\ldots,y_s)c_r(z_1,\ldots,z_r)f_{s|r}(y_1,\ldots,y_s;\tau\,|\,z_1,\ldots,z_r;t) \quad (21.6.11)$$

We now show that the two-time distribution functions can be determined intrinsically by solving a succession of BBGKY hierarchies, to which the general methods used in previous chapters can be applied.

21.7. DYNAMICS OF TWO-TIME DISTRIBUTION FUNCTIONS

As the new distribution functions depend on two time variables, the problem of determining their time evolution is, in principle, more complicated. However, we will see that the problem can be reduced to a succession of two problems of the same kind as in Chapter 3.

Let us first consider the simple case of $f_{1|1}(y_1\tau\,|\,z_1;t)$. From Eqs. (21.6.2), (21.6.4) and using the theorem (2.2.10) we find

$$f_{1|1}(y;\tau\,|\,z;t) = \int d^N x\left\{e^{-\tau L_N}\sum_j \delta(y-x_j)\right\}\sum_n \delta(z-x_n)F(x_1,\ldots,x_N;t)$$

$$= \int d^N x \sum_j \delta(y-x_j)\,e^{\tau L_N}\left\{\sum_n \delta(z-x_n)F(x_1,\ldots,x_N;t)\right\} \quad (21.7.1)$$

Let us now consider the following function:

$$K_{(1)}(x_1,\ldots,x_N;\tau\,|\,z;t) = e^{\tau L_N}\left\{\sum_n \delta(z-x_n)F(x_1,\ldots,x_N;t)\right\} \quad (21.7.2)$$

As a function of $x_1,\ldots,x_N$ and τ, $K_{(1)}$ obeys the same Liouville equation as the ordinary distribution function F:

$$\partial_\tau K_{(1)}(x_1,\ldots,x_N;\tau\,|\,z;t) = L_N K_{(1)}(x_1,\ldots,x_N;\tau\,|\,z;t) \quad (21.7.3)$$

Hence $K_{(1)}(x_1,\ldots,x_N;\tau\,|\,z;t)$ can be considered as a generalized distribution function* depending parametrically on the variables z of one particle [whence the subscript (1)] and on the time t. To obtain $K_{(1)}$, the Liouville equation (21.7.3) must be solved with the initial condition:

$$K_{(1)}(x_1,\ldots,x_N;0\,|\,z;t) = \sum_{n=1}^{N}\delta(z-x_n)F(x_1,\ldots,x_N;t) \quad (21.7.4)$$

* The fact that $K_{(1)}$ is not normalized to 1 is of no importance here.

We also see that $f_{1|1}(y;\tau\mid z;t)$ is simply the one-particle reduced distribution corresponding to the N-particle distribution $K_{(1)}$ through the general formula (3.1.19):

$$f_{1|1}(y;\tau\mid z;t)=\int dx_1\cdots dx_N\sum_{j=1}^{N}\delta(y-x_j)K_{(1)}(x_1,\ldots,x_N;\tau\mid z;t)\quad(21.7.5)$$

These relations can easily be generalized. The function $f_{s|r}$ defined by Eq. (21.6.10) can be transformed as in Eq. (21.7.1) to yield

$$f_{s|r}(y_1\cdots y_s;\tau\mid z_1\cdots z_r;t)=\sum_{j_1\neq\cdots\neq j_s}\cdots\sum\int dx_1\cdots dx_N$$
$$\times\,\delta(y_1-x_{j_1})\cdots\delta(y_s-x_{j_s})K_{(r)}(x_1,\ldots,x_N;\tau\mid z_1\cdots z_r;t)\quad(21.7.6)$$

where

$$K_{(r)}(x_1,\ldots,x_N;\tau\mid z_1,\ldots,z_r;t)$$
$$=e^{\tau L_N}\sum_{n_1\neq\cdots\neq n_r}\cdots\sum\delta(z_1-x_{n_1})\cdots\delta(z_r-x_{n_r})F(x_1,\ldots,x_N;t)\quad(21.7.7)$$

For every value of r $(=0,1,2,\ldots)$, $K_{(r)}$ is a solution of the Liouville equation:

$$\partial_\tau K_{(r)}(x_1,\ldots,x_N;\tau\mid z_1,\ldots,z_r;t)=L_N K_{(r)}(x_1,\ldots,x_N;\tau\mid z_1,\ldots,z_r;t)\quad(21.7.8)$$

where the dependence on r particle variables $z_1,\ldots,z_r$ and on t is introduced through the initial value in $\tau=0$, which is easily read from Eq. (21.7.7).

We now note that each $K_{(r)}$, for fixed r, generates through (21.7.6) a set of reduced distribution functions. In other words, for each integral value of r we associate a generalized distribution vector $\mathfrak{k}_{(r)}(\tau\mid t)$ with the distribution function $K_{(r)}$ in just the same way as $\mathfrak{f}(t)$ was associated with F in Section 3.1. The components of this vector $\mathfrak{k}_{(r)}$ are simply the functions $f_{s|r}$ for fixed r and $s=0,1,2,\ldots$:

$$\mathfrak{k}_{(r)}(\tau\mid t)=\{f_{s|r}(y_1\cdots y_s;\tau\mid z_1\cdots z_r;t):s=0,1,2,\ldots\}\quad(21.7.9)$$

From Eq. (21.7.8) follows that the distribution vector $\mathfrak{k}_{(r)}(\tau\mid t)$ obeys, with respect to τ, the *same* matrix Liouville equation (3.3.12) as the ordinary distribution vector $\mathfrak{f}(t)$:

$$\partial_\tau\mathfrak{k}_{(r)}(\tau\mid t)=\mathscr{L}\,\mathfrak{k}_{(r)}(\tau\mid t)\quad(21.7.10)$$

The components of this equation constitute a BBGKY hierarchy:

$$\partial_\tau f_{s|r}(y_1 \cdots y_s; \tau \mid z_1 \cdots z_r; t)$$
$$= \left\{ \sum_{j=1}^{s} L_j^0 + \sum_{j<n=1}^{s} \sum L'_{jn} \right\} f_{s|r}(y_1 \cdots y_s; \tau \mid z_1 \cdots z_r; t)$$
$$+ \sum_{j=1}^{s} \int dy_{s+1} L'_{j,s+1} f_{s+1|r}(y_1 \cdots y_{s+1}; \tau \mid z_1 \cdots z_r; t) \quad (21.7.11)$$

We note that the operators L_j^0, L'_{jn} act only on the variables y_j, whereas z_n and t are mere parameters in this equation. It remains now to be noted that, in order for $f_{s|r}$ to be properly interpreted as a two-time correlation function, it must obey Eq. (21.7.11) together with a well-defined initial condition, which can be deduced from Eqs. (21.7.6) and (21.7.7).

Consider first the case $r = 1$. We then obtain

$$f_{0|1}(0 \mid z_1; t) = \int dx_1 \cdots dx_N \sum_{n=1}^{N} \delta(z_1 - x_n) F(x_1, \ldots, x_N; t)$$
$$= f_1(z_1; t) \quad (21.7.12)$$

Hence, from Eq. (3.1.19) we see that $f_{0|1}(0 \mid z_1; t)$ is just the *ordinary* reduced one-particle distribution, at time t. Considering now the case $f_{1|1}$, some care must be taken with the summation prescriptions:

$$f_{1|1}(y_1; 0 \mid z_1; t) = \int d^N x \sum_{j=1}^{N} \delta(y_1 - x_j) \sum_{n=1}^{N} \delta(z_1 - x_n) F$$
$$= \int d^N x \sum_{j=1}^{N} \delta(y_1 - x_j)\, \delta(z_1 - x_j) F$$
$$+ \int d^N x \sum_{j \neq n=1}^{N} \sum \delta(y_1 - x_j)\, \delta(z_1 - x_n) F$$
$$= \delta(y_1 - z_1) f_1(y_1; t) + f_2(y_1, z_1; t) \quad (21.7.13)$$

Similarly, one finds

$$f_{2|1}(y_1, y_2; 0 \mid z_1; t) = \delta(y_1 - z_1)\, \delta(y_2 - z_1) f_1(z_1; t)$$
$$+ \delta(y_1 - z_1) f_2(y_2, z_1; t) + \delta(y_2 - z_1) f_2(y_1, z_1; t)$$
$$+ \delta(y_1 - y_2) f_2(y_1, z_1; t) + f_3(y_1, y_2, z_1; t) \quad (21.7.14)$$

Also,

$$f_{0|2}(0 \mid z_1, z_2; t) = f_2(z_1, z_2; t) \quad (21.7.15)$$

$$f_{1|2}(y_1; 0 \mid z_1, z_2; t) = \delta(y_1 - z_1) f_2(z_1, z_2; t)$$
$$+ \delta(y_1 - z_2) f_2(z_1, z_2; t) + f_3(y_1, z_1, z_2; t) \quad (21.7.16)$$

and so on. We may infer from these examples that the initial value (in $\tau=0$) for $f_{s|r}(y_1\cdots y_s; 0 \mid z_1\cdots z_r; t)$ is a linear combination of the ordinary reduced distribution functions $f_r, f_{r+1}, \ldots, f_{r+s}$, at time t, in the combined phase space $y_1, \ldots, y_s, z_1, \ldots, z_r$. This relationship can be formally expressed as a relation between the distribution vectors $\mathfrak{k}_{(r)}(\tau; t)$ and $\mathfrak{f}(t)$:

$$\mathfrak{k}_{(r)}(0 \mid t) = \mathcal{P}_{(r)}\mathfrak{f}(t) \tag{21.7.17}$$

which can be resolved into components:

$$f_{s|r}(y_1\cdots y_s; 0 \mid z_1\cdots z_r; t) = \sum_{n=r}^{s+r}\int dw_1\cdots dw_n \\ \times P_{sn}^{(r)}(y_1\cdots y_s; z_1\cdots z_r \mid w_1\cdots w_n) f_n(w_1\cdots w_n; t) \tag{21.7.18}$$

The matrix elements $P_{sn}^{(r)}$ can be determined by comparison with Eqs. (21.7.12)–(21.7.16). For instance:

$$\begin{aligned} P_{0n}^{(1)} &= \delta_{n,1}\,\delta(w_1 - z_1) \\ P_{1n}^{(1)} &= \delta_{n,1}\,\delta(w_1 - z_1)\,\delta(y_1 - w_1) + \delta_{n,2}\,\delta(w_1 - y_1)\,\delta(w_2 - z_1) \end{aligned} \tag{21.7.19}$$

and so on. From this discussion, it is now easy to infer the following method of solution for the two-time distribution functions.

A. The system is defined at time $t=0$ as usual, by specifying all its reduced distribution functions, that is, the vector $\mathfrak{f}(0)$.

B. The distribution vector at time t is obtained by the action of the propagator $\mathcal{U}(t)$.

C. Applying the operators $\mathcal{P}_{(r)}$, we construct from $\mathfrak{f}(t)$ a set of derived distribution vectors $\mathfrak{k}_{(r)}(0 \mid t)$, $r = 1, 2, 3, \ldots$.

D. Each of these vectors, taken as an initial condition for $\tau = 0$, generates, by means of the propagator $\mathcal{U}(\tau)$, the final solution $\mathfrak{k}_{(r)}(\tau \mid t)$, whose components are the two-time distribution functions

$$f_{s|r}(y_1\cdots y_s; \tau \mid z_1\cdots z_r; t)$$

These operations are expressed by the following compact formula:

$$\mathfrak{k}_{(r)}(\tau|t) = \mathcal{U}(\tau)\mathcal{P}_{(r)}\mathcal{U}(t)\mathfrak{f}(0) \tag{21.7.20}$$

We thus defined a clear mathematical procedure for the solution of the initial-value problem for the two-time functions. This procedure is illustrated in Fig. 21.7.1. In the application of the steps B and D, all the methods described in the previous chapters can, of course, be applied quite directly.

The method of calculation is, of course, very much simplified in the important case when $\mathfrak{f}(0)$ is the *equilibrium distribution* $\mathfrak{f}^0$. The first two

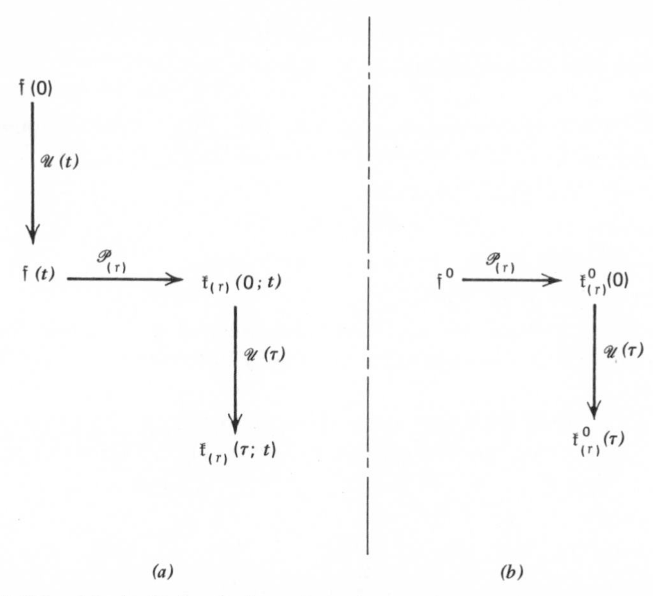

Figure 21.7.1. Method of solution of the initial-value problem for the two-time distribution vector; (*a*) the general case; (*b*) equilibrium correlations.

steps are then trivial, because $\mathfrak{f}(t)=\mathfrak{f}(0)=\mathfrak{f}^0$. There is only one step involving dynamics in this case, hence the calculation of $\mathfrak{f}_{(r)}(\tau)$ is by no means more difficult than the calculation of $\mathfrak{f}(t)$.

BIBLIOGRAPHICAL NOTES

The relation between the *time correlation and the inelastic scattering cross section* was derived by L. van Hove, *Phys. Rev.* **95,** 249 (1954).

The expression of the *transport coefficients in terms of autocorrelation functions* was derived in:

M. S. Green, *J. Chem. Phys.* **19,** 1036 (1951).

R. Kubo, *J. Phys. Soc. Japan* **12,** 570 (1957); *Lectures in Theoretical Physics,* Vol. 1 (Summer Inst. Theor. Phys., U. Colorado), Wiley-Interscience, New York, 1959.

The *fluctuation–dissipation theorem* was first derived by H. Nyquist, *Phys. Rev.* **32,** 110 (1928).

Onsager's principle appeared in L. Onsager, *Phys. Rev.* **37,** 405 (1931); **38,** 2265 (1931).

The general form of the fluctuation–dissipation theorem was given in H. B. Callen and T. A. Welton, *Phys. Rev.* **83,** 34 (1951).

An excellent recent review of the fluctuation–dissipation theorem is K. M. Case, *Transp. Th. Stat. Phys.* **2,** 129 (1972).

A detailed discussion of the analytic properties of the *generalized susceptibility* can be found in L. D. Landau and E. M. Lifshitz, *Statistical Physics* (quoted in Chapter 4).

The theory of *thermal transport coefficients* along this line can be found, among many papers, in:

H. Mori, *Phys. Rev.* **112,** 1829 (1958).

J. A. McLennan, *Phys. Fluids* **3,** 493 (1960).

J. L. Jackson and P. Mazur, *Physica* **30,** 2295 (1964).

L. P. Kadanoff and P. C. Martin, *Ann. Phys.* **24,** 419 (1963).

The *equivalence between the autocorrelation function method and the kinetic equation method* was proved in full generality by P. Résibois, *J. Chem. Phys.* **41,** 2979 (1964).

An important problem, very active in recent years, is the behavior of the *transport coefficients in the critical region.* The problem was introduced by M. Fixman, *J. Chem. Phys.* **36,** 310 (1961).

Recent developments were made by:

K. Kawasaki, *Phys. Rev.* **150,** 291 (1966).

L. P. Kadanoff and J. Swift, *Phys. Rev.* **166,** 89 (1968).

P. C. Hohenberg, M. de Leener, and P. Résibois, *Physica* **65,** 505 (1973).

Excellent reviews of the present state of the problem are found in the book of Stanley (quoted in Chapter 9) and in P. Résibois, in *Irreversibility in the Many Body Problem* (J. Biel and J. Rae, eds.), Plenum Press, New York, 1972.

The first mention of a *long tail in a correlation function* seems to have been made by R. L. Guernsey, *Relaxation Time for Two-Particle Correlation Functions in a Plasma,* Boeing Sc. Res. Lab. Report D1-82-0083, Dec. 1960; *Phys. Fluids* **5,** 322 (1962).

The recent interest in this problem was started by B. J. Alder and T. E. Wainwright, *J. Phys. Soc. Japan* (Suppl.) **26,** 267 (1968); *Phys. Rev. Lett.* **18,** 988 (1969).

The theoretical calculations referring to this problem appeared in:

J. R. Dorfmann and E. G. D. Cohen, *Phys. Rev. Lett.* **25,** 1257 (1970).

M. H. Ernst, E. H. Hauge, and J. M. van Leeuwen, *Phys. Rev. Lett.* **25,** 1254 (1970); *Phys. Rev. A* **4,** 2055 (1971).

Y. Pomeau, *Phys. Lett.* **27A,** 601 (1968).

See also Résibois's review paper quoted above.

A *reduced distribution function formalism for the study of time-correlation functions* seems to have been first used by J. L. Lebowitz, J. K. Percus, and J. Sykes, *Phys. Rev.* **188,** 487 (1969).

CHAPTER 22
CONCLUSION

We now wish to briefly sum up some of the features that appeared in our survey of statistical mechanics. The presentation given in the book was mainly motivated by pedagogical reasons. The best logical order is actually the reverse, in which one studies how a system starts from some arbitrary initial condition, undergoes a series of transformations, and ends up in equilibrium.

The first question one may ask is the following. Why is there such a science as nonequilibrium statistical mechanics? After all, we know from the very beginning what must be done to get the complete exact solution to the problem of evolution: we must solve the Liouville equation of the problem (Chapter 2).

This, however, is more readily said than done. Nobody ever solved a nontrivial Liouville equation for interacting systems. We must therefore proceed to simplifying the problem in a clever way. Here resides the problem of nonequilibrium statistical mechanics.

The guiding idea is based on the enormous disparity between the amount of information provided by the exact molecular description and the relatively very small amount of detail needed in the macroscopic level. Even if we succeeded in solving the Liouville equation for a real fluid (say) most of the information would be irrelevant for the calculation of its thermodynamic and hydrodynamic parameters. A good strategy would be to eliminate the irrelevant features of the problem at the very beginning, so that we do not spend time and effort on useless problems. This is why nonequilibrium statistical mechanics basically consists in a series of *successive contractions of the description* of a many-body system. This process looks like pruning off all the irrelevant branches of a tree, keeping only those that may be expected to bear fruit.

In performing these contractions, due account is taken of the characteristic properties of the systems studied in statistical mechanics. In particular, their *large size* is a fundamental ingredient of the theory, expressed by the famous *thermodynamic limit* (the physical meaning of which we discussed at length in Chapter 3). In the T-limit a number of qualitatively new properties appear, both in and out of equilibrium. This

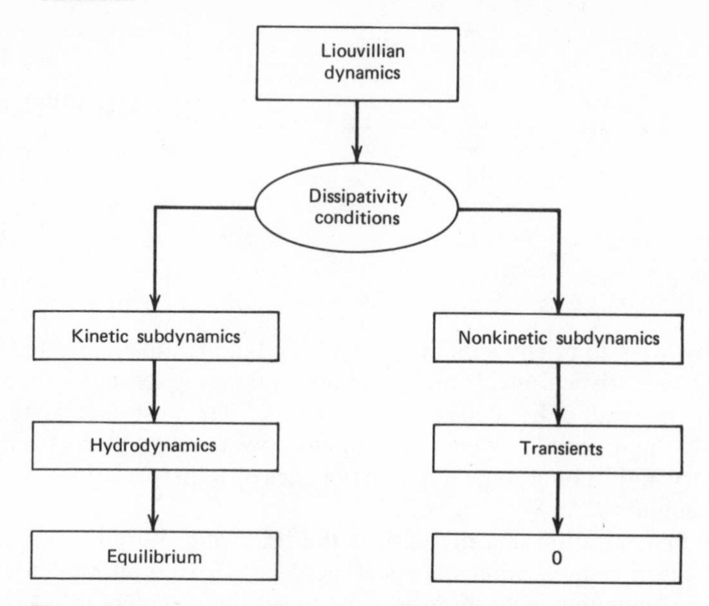

Figure 22.1. Levels of description in statistical mechanics.

symmetry breaking is essential for our understanding of macroscopic physics.*

Let us now recall the various stages in the contraction process, as they appear from our analysis. A first, essential reduction occurs for an important class of systems, when a set of "auxiliary conditions" are satisfied, ensuring that the system is a *dissipative* one. The dynamical description then splits into two completely separate and independent subdynamics (Chapters 16 and 17). The time evolution of the kinetic components is governed by a dynamical law that is by an order of magnitude simpler than the original dynamics. The simplicity is due to the reduction of the N-body (or ∞-body) problem to a one-body problem. The nonkinetic subdynamics is much more complex: it requires the specification of all the truly correlated correlation patterns.

The great progress introduced by the subdynamics picture (see Fig. 22.1) can be understood as follows. In the more traditional approaches to the problem, one attempted to show that the kinetic description provides us with a good *approximation* to the law of evolution of the systems. This goal cannot be general: it is only achieved for simple systems in which the time scales of collision and of relaxation are widely separated. The

* These questions are discussed in detail in a recent paper by I. Prigogine, C. George, F. Henin, and L. Rosenfeld, *Chem. Scripta* **4**, 5 (1973).

complex transients then decay to zero very fast and, over a time scale of the order of the relaxation time, the kinetic equation represents indeed a good approximation of the "last stage" of the evolution. When one studies, however, dense fluids or strongly coupled systems, the characteristic time scales are of the same order of magnitude. The former "transients" now interfere with the simple evolution to equilibrium. The mathematical description of this state of affairs is summarized in the Prigogine–Résibois master equation (Section 16.3). However, its source term demands the specification of all the initial *correlations*, and these data are not usually available. Therefore, the master equation can only be concretely used in simple limiting cases.

In the *subdynamics approach*, the whole philosophy changes. The kinetic part of the distribution function no longer appears as an approximation to the total distribution. The latter can be very different from the kinetic distribution and its overall time behavior could depart significantly from the kinetic evolution. The important facts are, however, the following:

A. *The kinetic part can be determined, without any reference to the remainder.*

B. *A certain class of important observables are completely* (or at least predominantly) *determined by the kinetic part alone.*

The kinetic behavior thus presents a curious double face. On one hand, it appears as a "universal" feature, which can be studied *per se*, without any assumptions on time scales, coarse graining, and the like. On the other hand, it manifests itself in a pure form only for a certain class of observables. In all other cases the macroscopic effects are more or less contaminated by complicated transient correlations.

Even the kinetic equation is, however, still a very complex object. A further important step in the simplification of the description sets in when we study *slowly varying processes*. In that case only the low-lying part of the spectrum of the kinetic equation plays a role; all other contributions have decayed to practically nothing. We have seen, in Chapter 13, that this part of the spectrum coincides with the spectrum of the macroscopic hydrodynamic equations. Hence, the distribution function in this regime is entirely determined by five macroscopic functions (or fields) describing the density, the velocity, and the internal energy. For practical purposes, this is the most important class of problems in nonequilibrium statistical mechanics. In this case the equations become sufficiently simple in order to be effectively soluble; in particular, the effects of the strong nonlinearities are not important in this domain. Powerful approximation techniques have been developed that allow us to effectively calculate concrete numbers, compare the results to experiments, and, conversely, infer molecular properties from macroscopic measurements.

The last step in the contraction process occurs when *equilibrium* is reached. Now the system is wholly described in terms of a single thermodynamic potential depending on a very small number of macroscopic variables, such as the density and the temperature. At this point, the dynamical problem becomes trivial, or rather nonexistent: equilibrium statistical mechanics is now coming onto the stage.

There is a considerable difference in methodology between equilibrium and nonequilibrium thermodynamics. The former is devoted to the study of the properties of *one known function*, namely, *the partition function* $Z(T, \mathcal{V}, N)$ (or of the derived concepts, such as the grand partition function). It is, of course, an extremely complicated function, and its study requires most elaborate mathematical techniques. In nonequilibrium theory, on the other hand, we are faced with *an infinite set of unknown functions*, describing all possible initial conditions. It is obvious that we cannot demand the same degree of detail in the two cases. In the nonequilibrium theory we must find out the common features of all the members of the infinite set. This is why the main stress is laid on the study of the *law of evolution in time*, that is, on the derivation of a differential equation. Such an equation is precisely the mathematical tool that includes in a single object the whole infinite set mentioned above. Once this equation is known, any particular state of the system can be studied by solving an initial-value problem.

In spite of the considerable difference in final goals, the reader may be struck by the *similarity of the mathematical techniques used in the two fields*. We have actually tried to stress this similarity in our presentation, because it is a unifying feature giving to statistical mechanics as a whole its unique flavor. To quote a few of these similarities, let us mention the series expansion techniques, the "graphology," and the resummation and renormalization methods. Although they apply to different objects, they possess a strong structural analogy. For this reason, most of the problems that have been solved (exactly or approximately) in equilibrium theory, have also been solved (to a corresponding degree of approximation) out of equilibrium. This is, of course, no accident. Basically, both problems go back to the study of the *Hamiltonian* of the system. This function plays simply a different role in the two theories: it *determines the distribution function in equilibrium*, and *it generates the motion out of equilibrium.*

In concluding this book, we do not wish to convey to the reader the impression that statistical mechanics in its present-day form is a perfect body of knowledge and that every one of its concepts is well defined and well understood. We are (fortunately!) still far from this ideal situation. The reader has certainly noticed the many open strings in various places of the development. Each one of these points toward an avenue for future research. There are, of course, an enormous number of possible applications of the general theory to specific systems, which still await elaboration. At the other extreme there is still much work for mathematicians in putting true mathematical rigor into the formulation of the concepts of statistical mechanics. Between these two extremes there remain a number of intriguing and difficult questions for which much work was and is being done and much progress achieved, but which are still not definitively settled. Among many questions of this type we may mention the following. What is the precise nature and origin of the singularities connected with the phase transitions and the critical phenomena? Can one formulate explicit conditions on the Hamiltonian of a system allowing one to decide whether or not it satisfies the dissipativity conditions for the existence of subdynamics?

These sample questions show that statistical mechanics is, more than ever, in a period of active growth. The review presented in this book is only a snapshot of what we consider to be the present state of the field. Even this statement is prudently formulated as a subjective one. Much of the material discussed here is by now considered as classic. Other topics are more recent and did not yet pass the test of survival. We personally consider them to be the more fascinating ones, even though it may turn out that in a few years they will have to be more or less thoroughly modified. Our hope is that, even under possibly different technical dressing, the basic ideas discussed here will remain valid, at least as a first approximation to the asymptotic solution.

APPENDIX
THE ERGODIC PROBLEM

As stated in the main text, we wish to give here an account of the recent developments of this very classical subject, which has been traditionally connected with statistical mechanics since the earliest days of this science. There has been an enormous progress in this field in the last decade. This progress resulted in a very paradoxical situation. On one hand, much light has been shed on the huge complexity of the behavior of dynamical systems. It now appears that even a single, small dynamical system displays in its evolution many features that were previously described as "statistical." On the other hand, it appears that ergodic theory detaches itself progressively from statistical mechanics. In the present state of development, it seems very difficult to invoke ergodic theory for a justification of the basis of statistical mechanics.

We shall try to give an idea of this development. The reader should be warned that ergodic theory is a province of mathematics. The original papers are hard to read for the average physicist. Here we tried to give a digest that involves practically no mathematics. It is intended to make the reader familiar with what is going on in this field. Many results are given without any proof, and are formulated in approximate terms. We hope, however, that the loss in precision might be balanced by a gain in physical insight.

A.1. TRAJECTORIES AND INTEGRALS OF THE MOTION

The ultimate question asked in classical dynamics is the following: Given a system of N degrees of freedom, characterized by a Hamiltonian $H(q, p)$ and by the equations of motion:

$$\dot{q}_i = \frac{\partial H(q, p)}{\partial p_i}$$
$$\dot{p}_i = -\frac{\partial H(q, p)}{\partial q_i} \tag{A.1.1}$$

and given the values of q and p at time $t=0$:

$$q_i(0)=q_i^0$$
$$p_i(0)=p_i^0 \tag{A.1.2}$$

determine the values of q and p at any other time t:

$$q_i=\mathfrak{q}_i(t;q^0,p^0)$$
$$p_i=\mathfrak{p}_i(t;q^0,p^0) \tag{A.1.3}$$

We know that this solution is unique (under very weak conditions) and can be expressed in the form (1.2.24). It is now an easy matter to invert these equations. Indeed, if we take q, p as a new initial condition and solve Eqs. (A.1.1) for time $(-t)$ we come back to the point q^0, p^0: this is a consequence of the *reversibility* of dynamics [see also Eq. (1.2.28)]:

$$q_i^0=\mathfrak{q}_i(-t;q,p)$$
$$p_i^0=\mathfrak{p}_i(-t;q,p) \tag{A.1.4}$$

We may look at these equations in the following way: We have found $2N$ functions, $\mathfrak{q}_i$, $\mathfrak{p}_i$ of the phase space variables q, p *and of time*, which have the property of being constant along any trajectory of a dynamical system. We get an even clearer picture if we eliminate the time among these $2N$ equations. We are then left with a set of $2N-1$ functions of the *phase-space variables alone*, which have the property of being constant along any trajectory; we may denote them by the symbols: $\Phi_j(q,p)$, $j=1,2,\ldots,2N-1$; these functions are called *conservative integrals*, or simply *integrals* or *constants* of the motion. We thus established quite generally the existence of $2N-1$ such integrals of the motion. Attributing a set of numerical values ϕ_j to these constants:

$$\Phi_j(q,p)=\phi_j, \qquad j=1,2,\ldots,2N-1 \tag{A.1.5}$$

is equivalent to determining completely the trajectory of the system in phase space.

We may look at the situation from a geometrical point of view. Each of the $2N-1$ equations (A.1.5), for given ϕ_j, defines a $(2N-1)$-dimensional "hypersurface" in the $2N$-dimensional phase space. The trajectory of the system must lie entirely on each of these surfaces, and therefore is simply the intersection of the $2N-1$ surfaces (see Fig. A.1.1).

We thus arrive at the following (provisional) picture. The knowledge of an integral for a given system allows us to put a restriction on the region of phase space where the trajectory is located: we "get rid of one dimension." Hence, determining more integrals reduces the problem further until we arrive at a one-dimensional line that is precisely the trajectory.

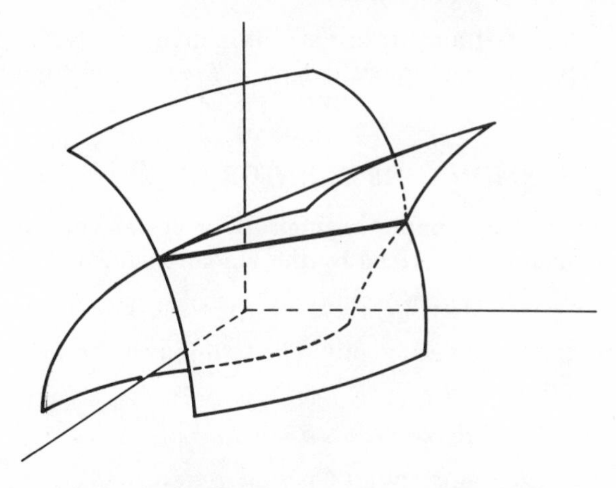

Figure A.1.1. In a three-dimensional phase space, the knowledge of two integrals completely determines the trajectory.

How do we characterize this problem analytically? It is easily seen from Eq. (1.2.7) that a necessary and sufficient condition for a dynamical function $\Phi_j(q, p)$ to be constant under the motion is the vanishing of its Poisson bracket with the Hamiltonian:

$$[H, \Phi_j]_P = 0, \qquad j = 1, 2, \ldots, 2N-1 \tag{A.1.6}$$

This equation has at least one obvious solution:

$$\Phi_1 = H(q, p) \tag{A.1.7}$$

Hence, for every conservative system,* the Hamiltonian is an integral of the motion: for given initial condition, the constant value of the Hamiltonian is simply the total energy of that system. Hence, the trajectory is bound to lie on the "energy surface":

$$H(q, p) = E \tag{A.1.8}$$

and therefore every dynamical problem can immediately be reduced to a $(2N-1)$-dimensional problem on the energy surface.

Our initial problem is now reduced to finding $2N-2$ other functions: $\Phi_2, \ldots, \Phi_{2N-1}$, which, together with the Hamiltonian $H \equiv \Phi_1$, form a complete set of functionally independent integrals. The knowledge of these functions would then restrict the trajectory to a single, well-defined line in phase space.

* A system whose Hamiltonian does not explicitly depend on time.

Although this program looks very attractive, we shall see that the situation is actually much more complex, even in the simplest possible cases.

A.2. TWO HARMONIC OSCILLATORS

Consider the trivially simple example of a system of two uncoupled harmonic oscillators, described by the Hamiltonian:

$$H=\tfrac{1}{2}(p_1^2+\omega_1^2q_1^2)+\tfrac{1}{2}(p_2^2+\omega_2^2q_2^2) \tag{A.2.1}$$

This problem can be solved exactly. The solution can be written in a form analogous to Eq. (A.1.4):

$$p_i \cos \omega_i t + \omega_i q_i \sin \omega_i t = p_i^0 \tag{A.2.2}$$

$$\omega_i q_i \cos \omega_i t - p_i \sin \omega_i t = \omega_i q_i^0 \tag{A.2.3}$$

Let us investigate in some detail the nature of the trajectories in phase space. We can eliminate the time among Eqs. (A.2.2) and (A.2.3) (for the same value of i) with the result:

$$p_i^2+\omega_i^2q_i^2=(p_i^{02}+\omega_i^2q_i^{02})=\text{const}, \qquad i=1,2 \tag{A.2.4}$$

Hence, we have identified two well-known integrals of the motion, which simply represent the separate energies of the two uncoupled oscillators:

$$\Phi_i(q,p)=\tfrac{1}{2}(p_i^2+\omega_i^2q_i^2), \qquad i=1,2 \tag{A.2.5}$$

In order to find the third integral, we have to eliminate t between the two equations (A.2.2) (for $i=1$, 2). Here we are immediately faced with a difficulty: the nature of this integral depends crucially on the relative values of the frequencies ω_1 and ω_2.

A. The Ratio of Frequencies ω_1/ω_2 Is Rational

We first consider the special case $\omega_2=2\omega_1$. Rather than doing analytical calculations, we use a graphical method. Equation (A.2.4) shows that in any section ($q_2=\text{const}$, $p_2=\text{const}$) of the phase space, the trajectory of the system is an ellipse; a similar statement holds for a section ($q_1=\text{const}$, $p_1=\text{const}$) (see Fig. A.2.1). The points marked $t=0$ on these graphs correspond to a particular choice of the initial condition. From these graphs it is easy to construct a graph of the trajectory projected on the plane (q_1, q_2). We simply let t vary, and plot corresponding abscissae taken from the two graphs on the single graph of Fig. A.2.2. We note that, as $\omega_2=2\omega_1$, by the time the representative point in Fig. A.2.1*a* circles the ellipse once, the corresponding point on Fig. 1.4.1*b* circles its ellipse twice. After a time $T=2(2\pi/\omega_1)$ both points come back to their initial position, hence the curve of Fig. A.2.2 is *closed*. This important feature

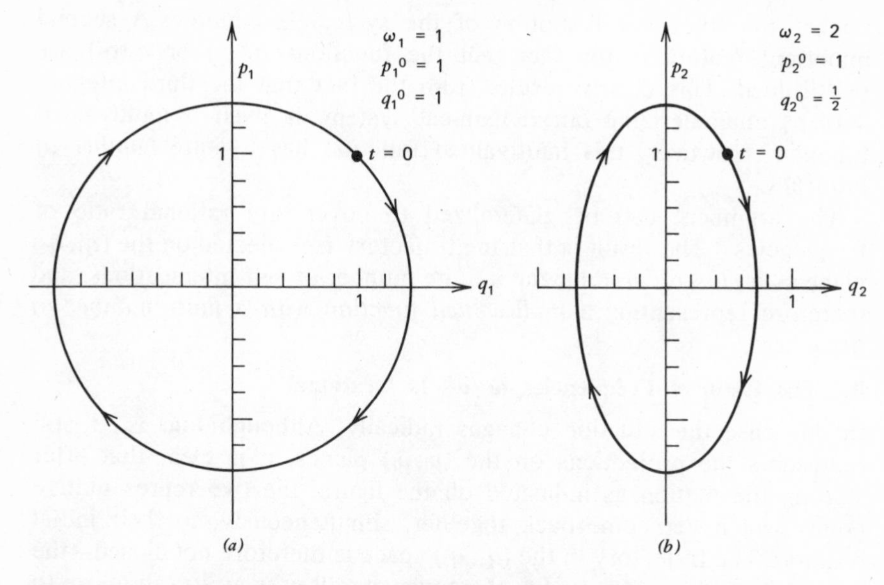

Figure A.2.1. Projections of the trajectory on the subspaces (q_1, p_1) and (q_2, p_2).

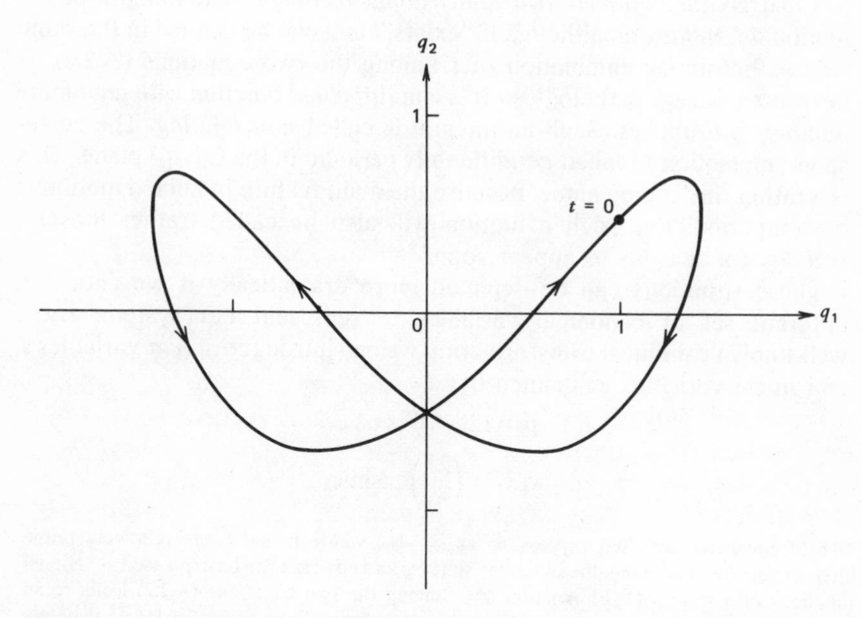

Figure A.2.2. Projection of the trajectory on the subspaces (q_1, q_2) in the case $\omega_2 = 2\omega_1$.

implies that the overall motion of the system is *periodic.* A second important feature is the fact that the functions $q_1(q_2)$ or $q_2(q_1)$ are *multivalued.* This clearly results from the fact that the third integral, $\Phi_3(q, p)$ characterizing our dynamical system, is itself a multivalued function. However, this multivalued function has a *finite number of branches.*

The argument can be generalized to cover any rational ratio of frequencies.* The result is that the trajectory is projected on the (q_1, q_2) plane as a *closed curve* having a finite number of self-intersections, and therefore representing a *multivalued function with a finite number of branches.*

B. The Ratio of Frequencies ω_1/ω_2 Is Irrational

In this case the situation changes radically. Although Fig. A.2.1 still represents the projections on the (p_j, q_j) planes, it is clear that after starting the motion as indicated on the figure, the two representative points will never come back together, simultaneously, to their initial position. The trajectory in the (q_1, q_2) space is therefore not closed—the *overall motion is not periodic.* Moreover, it will pass arbitrarily close to any point in the rectangle defined in the space (q_1, q_2) by the maximum amplitudes. The trajectory cannot be drawn as a (one-dimensional) line: *it fills densely the complete two-dimensional rectangle.* The integral of the motion Φ_3, therefore, although it "exists," and can be defined in the same way as before by elimination of t among the two equations (A.2.2), is nevertheless very pathological. It is a *multivalued function with an infinite number of branches.* Such an integral is called *nonisolating.* The corresponding motion is called *conditionally periodic* in the (q_1, q_2) plane. This is a rather misleading name, because the main feature of such a motion is its nonperiodicity. Such a motion will also be called (rather loosely) *ergodic,* for reasons to appear soon.

These situations can be depicted more dramatically if we choose a different set of canonical variables to represent our system. By a well-known canonical transformation we can introduce *action* variables J_i and *angle* variables φ_i defined by

$$p_i = (2\omega_i J_i)^{1/2} \cos \varphi_i$$
$$q_i = -\left(\frac{2J_i}{\omega_i}\right)^{1/2} \sin \varphi_i \qquad \text{(A.2.6)}$$

* The frequencies are then expressible as $\omega_j = l_j\omega$, where l_1 and l_2 are relatively prime integers. On the other hand, the functions $\sin(l_j\omega t)$ and $\cos(l_j\omega t)$ can be expressed as rational functions of $\tau = \tan(\frac{1}{2}\omega t)$. Elimination of τ among the two equations (A.2.2) leads to an integral Φ_3 such that $\Phi_3 = \text{const}$ represents an algebraic surface: such a surface has only a finite number of branches.

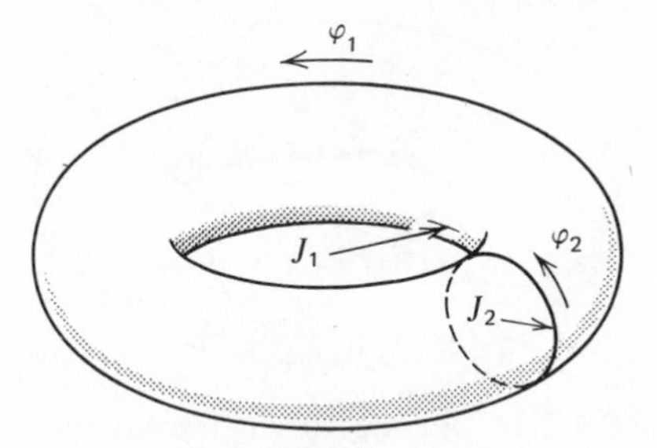

Figure A.2.3. The torus representing the phase space of the two-oscillator system.

In these variables, the Hamiltonian is simply

$$H = \omega_1 J_1 + \omega_2 J_2 \tag{A.2.7}$$

As H is independent of the angle variables, it immediately follows that

$$\dot{J}_i = -\frac{\partial H}{\partial \varphi_i} \equiv 0 \tag{A.2.8}$$

Hence *the actions are integrals of the motion* [they are simply proportional to the integrals Φ_1, Φ_2 defined in Eq. (A.2.5)]. The angles also have a simple law of motion: as the Hamiltonian is linear in the actions, it follows that

$$\dot{\varphi}_i = \frac{\partial H}{\partial J_i} = \omega_i$$

and hence

$$\varphi_i = \omega_i t + \varphi_{0i} \tag{A.2.9}$$

One must not forget, however, that an angle is always defined modulo 2π. Therefore a convenient representation of the four-dimensional phase space is obtained by introducing a *torus* (Fig. A.2.3). We agree to measure J_1 along the interior radius and J_2 along the radius of a transverse cross-section. The angle φ is measured along the "long" circle and the angle φ_2 along the "short" circle ("geographical coordinates").

If the frequencies ω_1, ω_2 are commensurate, the trajectory is a simple, closed, not self-intersecting line on the torus (see Fig. A.2.4). To every initial condition, with the same value of J_1, J_2 corresponds a closed curve of the same type, lying on the same torus. If J_1 and J_2 are different, the trajectory lies on another, concentric torus. If, however, the frequencies, are incommensurate (or "rationally independent"), the trajectory fills the surface of the torus densely: it never closes on itself.

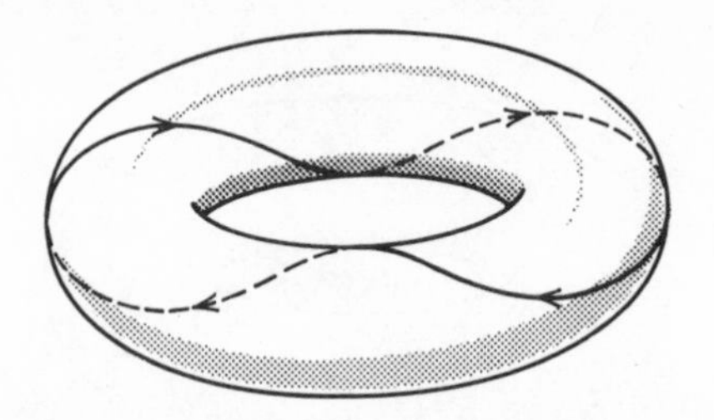

Figure A.2.4. A trajectory in the case $\omega_2 = 2\omega_1$.

A.3. THE KOLMOGOROV–ARNOLD–MOSER THEOREM

The discussion of the almost trivial example of Section A.2 is very illuminating in pointing out the fantastic complexity of the trajectories of even the simplest dynamical systems. It shows that the simple program set out in Section A.1 is illusory. It is not true in general that finding $2N-1$ integrals of the motion will confine the motion of the system to a nice regular curve in phase space. For a system of N uncoupled oscillators, only N integrals are isolating; the remaining $N-1$ are, in general,* nonisolating. Hence the trajectory of the system fills densely an N-dimensional region of phase space. In action-angle variables this region has the shape of an N-dimensional torus.

With these facts in mind, we may now restrict our ambition and ask a more modest question. *Is it possible to determine* N (*instead of* $2N-1$) *integrals that would restrict the trajectory to a* $(N-1)$*-dimensional region of phase space?*

The possible answer to this question is provided by the very classical *Hamilton–Jacobi theory of dynamics.* According to this theory, whenever the Hamiltonian $H(q, p)$ is independent of time, it is possible to define a canonical transformation to a new set of variables Q, P such that the transformed Hamiltonian $K = H(P)$ is a function of the new momenta alone. The equations of motion are then

$$\dot{Q}_i = \frac{\partial K}{\partial P_i} = \omega_i(P) \tag{A.3.1}$$

$$\dot{P}_i = -\frac{\partial K}{\partial Q_i} = 0 \tag{A.3.2}$$

Hence, just as in the case of the action-angle variables, the P_i's are integrals of the motion and the Q_i's are linear functions of time. If the

* We should not forget that rational numbers form a subset of measure zero within the set of real numbers.

system is confined in space, one can actually construct suitable combinations of the P_i's such that the N integrals are action variables, J_i and the conjugate variables are angles, φ_i. The canonical transformation defining the J's and the φ's is generated by a *characteristic function* $W(q)$ satisfying the Hamilton–Jacobi equation:

$$H\left(q, \frac{\partial W}{\partial q}\right) - \alpha_1 = 0 \tag{A.3.3}$$

A complete solution to this equation contains $(N-1)$ arbitrary constants that, together with α_1, form a set of N independent constants. One may now identify the J's with suitable combinations of these constants:

$$J_i = J_i(\alpha_1 \cdots \alpha_N) \tag{A.3.4}$$

Then W appears as a function of the q's and the J's:

$$W = W(q, \alpha) \equiv W(q, J)$$

and it is taken as a generating function for the canonical transformation, from which we obtain, by the usual rules

$$p_i = \frac{\partial W(q, J)}{\partial q_i}$$

and (A.3.5)

$$\varphi_i = \frac{\partial W(q, J)}{\partial J_i} = \omega_i(J)t + \beta_i$$

We thus seem to be led back to the simple picture of the oscillator system discussed before. The phase space is structured into a set of imbedded N-dimensional tori. Any possible trajectory lies on one of these tori. There is however a difference with the oscillator system. In that case the frequencies are absolute constants, given once and for all in the Hamiltonian. Hence, *all* the tori are covered either by closed curves or by dense ergodic trajectories, according to whether the frequencies are or are not commensurate. In the general case the frequencies depend on the actions, and therefore on the radii of the tori. It follows that, for a given system, some tori contain closed curves [if the actions are such that $\sum n_i\omega_i(J) = 0$ can be satisfied by nonzero integers n_i], while others are covered by ergodic trajectories. As however the rational numbers are a set of measure zero, *almost all the tori contain ergodic trajectories;* periodic motion is an exception.

Again, the picture is very attractive; but, beware of too much optimism! The Hamilton–Jacobi equation is, in general, an extraordinarily complicated nonlinear equation. Only in a very few cases can it be actually solved (these are the so-called *integrable* dynamical systems). In these cases we can, indeed, find a set of N nice action variables and conclude at the existence of tori.

But, in the general case, nothing tells us what the general nature of the integrals is. In particular, we do not know whether the action variables for an arbitrary system are isolating or not. If they are not, the trajectory is not restricted to a $(N-1)$-dimensional region, but fills densely a higher-dimensional domain.

In recent years some important new results appeared, which seriously activated this field. The first type of result is purely analytical. It is contained in a theorem suggested by Kolmogorov in 1954 and proved by Arnold and, independently, by Moser in 1963; it is therefore usually abbreviated as the *KAM theorem*. It is a perturbation-theory result, concerning the following problem. Consider an integrable system described by a Hamiltonian $H_0(J)$. It is characterized by a set of tori covered with ergodic trajectories. What happens if we introduce a small perturbation, thus considering a system with the Hamiltonian:

$$H(J, \varphi) = H_0(J) + \lambda H_1(J, \varphi) \tag{A.3.6}$$

The first guess is that the initial structure is completely destroyed even by the smallest perturbation. Indeed, the perturbation will couple the various oscillators, therefore causing the trajectories to leave the tori. Moreover, and more seriously, it is well known to everybody who had a course in celestial mechanics or (more likely) in elementary quantum mechanics, that special difficulties appear when the perturbation couples oscillators with frequencies such that $\sum n_i\omega_i = 0$. In that case there appears a *resonance phenomenon:* the effect of the perturbation is enormous, however small the coupling constant λ.

The KAM theorem now provides us with a surprise: it tells us that there is "much more order" than could be suspected from this discussion. Roughly formulated,* the theorem says that, *provided λ is sufficiently small* and $H_1(J, \varphi)$ is analytic in J and φ in a given domain, *the phase space can be separated into two regions of nonvanishing volume,† one of which is small compared to the other and shrinks to zero as $\lambda \to 0$. The larger region has the familiar structure of imbedded tori covered with dense trajectories.* In other words, for the majority of initial conditions, the trajectories of the system have the same character of a Lissajous curve restricted to $(N-1)$ dimensions as in the simple oscillator case. The

* The precise formulation and the proof of the KAM theorem are extremely difficult, involving advanced arguments from topology, analysis, and number theory. It is out of the question to reproduce them here.

† "Measure" in the Lebesgue sense is a more appropriate term.

initial tori are only slightly deformed. There is however a small region (i.e., a certain set of initial conditions) in which the trajectories are wildly erratic and may run away quite far from the nearby confined trajectories. This region that—we insist—is *not* of measure zero, is therefore a *region of instability*.

A.4. NUMERICAL EXPERIMENTS ON SIMPLE DYNAMICAL SYSTEMS

Celestial-mechanics experts have good reasons of being very happy with the KAM theorem. It is a great step forward toward proving the stability of planetary motion (a still unsolved problem!). All the conditions of the theorem are nicely satisfied in this problem: the system consists of the massive sun and of a small number of light planets that perturb each other's motion very slightly.

People doing statistical mechanics, on the other hand, feel frustrated by the KAM theorem. The thing they like most is maximum disorder. Therefore, a sizable amount of work has started in recent years in order to determine how the conditions of the KAM theorem may be violated in the systems of interest in statistical mechanics. The difficulties of an exact mathematical study are prohibitive. However, the advent of high-speed computers opened a new era of "experimental" investigation of these difficult problems; more and more important results are expected to come out of this study.

We first discuss a very simple problem that, although having no direct connection with statistical-mechanical systems, shows very strikingly the fantastic complexity of systems that at first sight seem trivial. The problem considered in the pioneering work of Henon and Heiles in 1963 concerns the motion of a single point in space under the influence of a cylindrically symmetric potential. (This problem simulates the motion of a star in the average field of a galaxy.) After taking account of the trivial integrals, that is, the total energy and the angular momentum, the problem is reduced to the motion of a particle in a plane, that is, to a four-dimensional phase space. For this reduced problem there exists an additional isolating integral:

$$E = \tfrac{1}{2}(p_1^2 + p_2^2) + U(q_1, q_2) \tag{A.4.1}$$

The trajectory is thus confined to three dimensions (see Fig. A.4.1).

A computer is now programmed to solve the equations of motion, for given energy and given initial condition, and to plot the intersection points of the trajectory with the q_2, p_2 plane; to simplify matters further, it only plots half of these intersections, namely, those where the trajectory goes

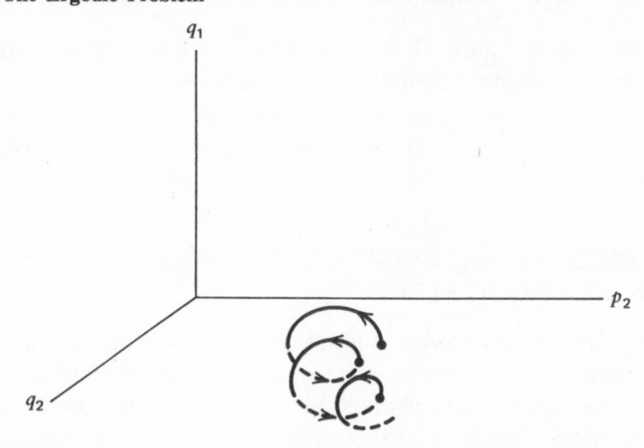

Figure A.4.1. A three-dimensional trajectory in the Henon–Heiles problem.

"up," that is, $p_1 > 0$. What can be expected to show up in these plots? The type of trajectory will be vividly displayed (see Fig. A.4.2). If the motion is periodic, the intersection is a single point. If the trajectory is conditionally periodic on a surface, that is, if it is restricted to a torus, the successive intersections will line up along a closed curve in the p_2, q_2 plane. This curve is simply the intersection of the "torus" with the plane. If, finally, the trajectory is ergodic in three dimensions; that is, it is no longer confined to a torus, the intersection point will wander erratically through the plane, eventually covering densely a finite area.

Henon and Heiles chose a simple potential, corresponding to the Hamiltonian:

$$H = \tfrac{1}{2}(p_1^2 + p_2^2) + \tfrac{1}{2}(q_1^2 + q_2^2) + q_1^2 q_2 - \tfrac{1}{3}q_2^3 \qquad \text{(A.4.2)}$$

Their results are shown in Figs. A.4.3. The first of these corresponds to low energy. We see that all trajectories form closed curves. This is a perfect illustration of the KAM theorem: practically (i.e., within computer accuracy) all the phase space consists of tori. If we now increase the energy (which effectively amounts to the possibility of larger perturbations) the picture changes drastically (Fig. A.4.3*b*). The remnant of the previous structure is still apparent; but in between the stable regions there appears a zone of ergodic behavior. It should be clearly understood that the scattered dots in this figure represent a *single* trajectory. When the energy is further increased, we reach the situation of Fig. A.4.3*c*. The torus regions have shrunk to almost nothing. Any trajectory outside these small domains travels erratically through almost the whole accessible phase space.

The techniques of Henon and Heiles have been applied later to

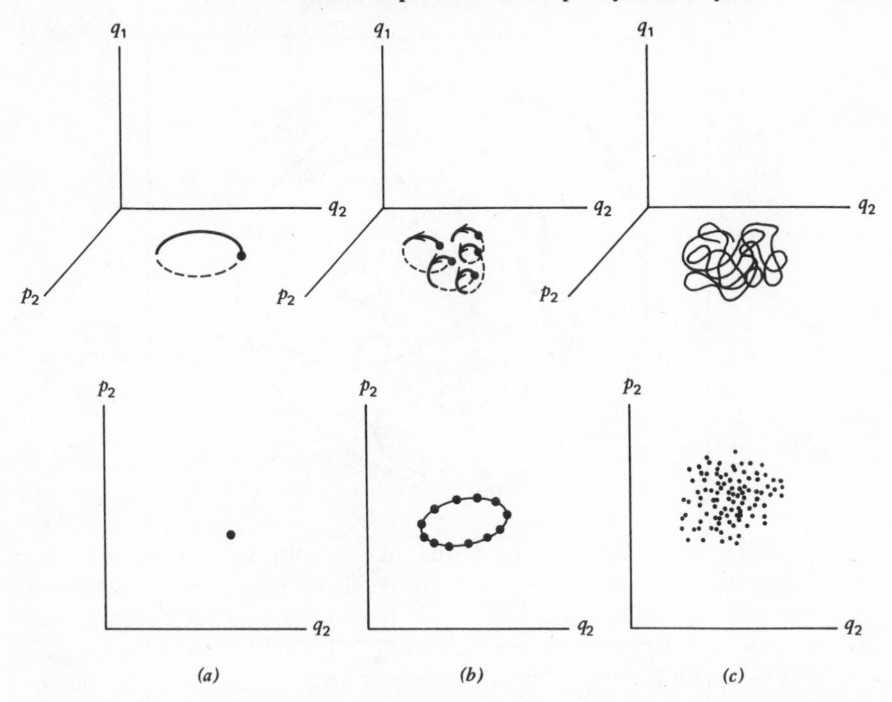

Figure A.4.2. Various types of trajectories: (*a*) periodic, (*b*) conditionally periodic, (*c*) ergodic in three dimensions.

problems of more direct statistical-mechanical interest. In particular the work of Ford and his co-workers is beginning to throw new light on the possible mechanisms of breakdown of the KAM theorem. The systems considered here consist of a small number of anharmonic oscillators (2 or 3) coupled together by small nonlinear terms. The latter are purposely chosen as a sum of potentials, each leading to resonant coupling. Typically,

$$H = H_0(J_1, J_2) + \alpha J_1 J_2 \cos(2\varphi_1 - 2\varphi_2) + \beta J_1 J_2^{3/2} \cos(2\varphi_1 - 3\varphi_2) \quad \text{(A.4.3)}$$

with

$$H_0(J_1, J_2) = J_1 + J_2 - J_1^2 - 3J_1J_2 + J_2^2 \quad \text{(A.4.4)}$$

and hence

$$\omega_1 \equiv \frac{\partial H_0}{\partial J_1} = 1 - 2J_1 - 3J_2$$
$$\omega_2 \equiv \frac{\partial H_0}{\partial J_2} = 1 - 3J_1 + 2J_2 \quad \text{(A.4.5)}$$

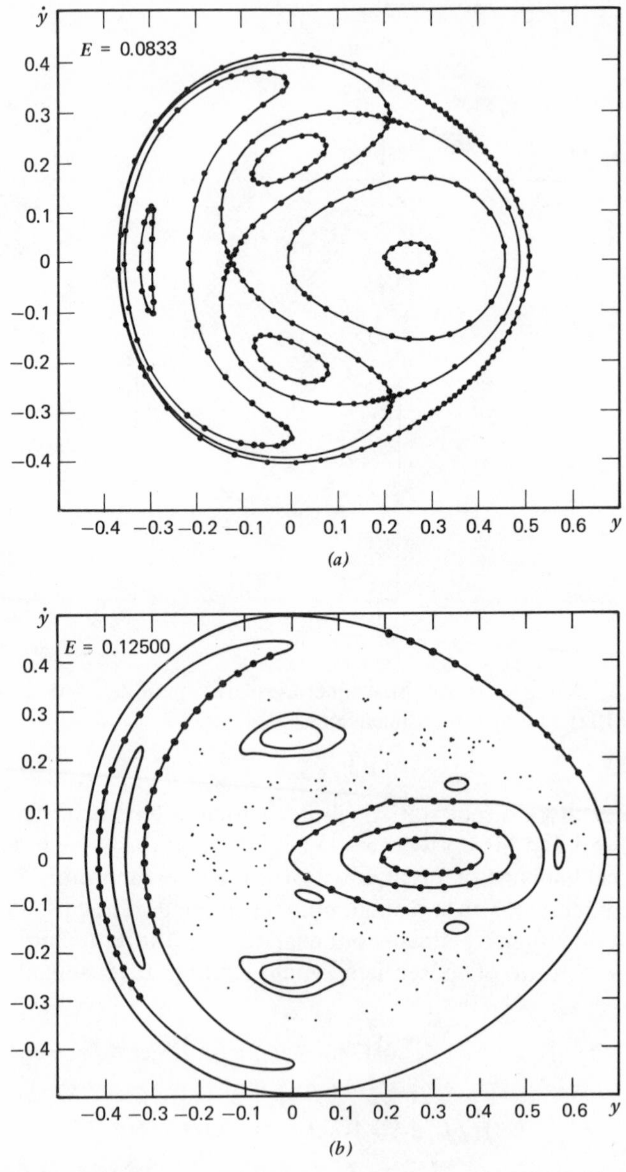

Figure A.4.3. Trajectories of the Henon–Heiles system at various energies. (From Henon and Heiles, 1964.)

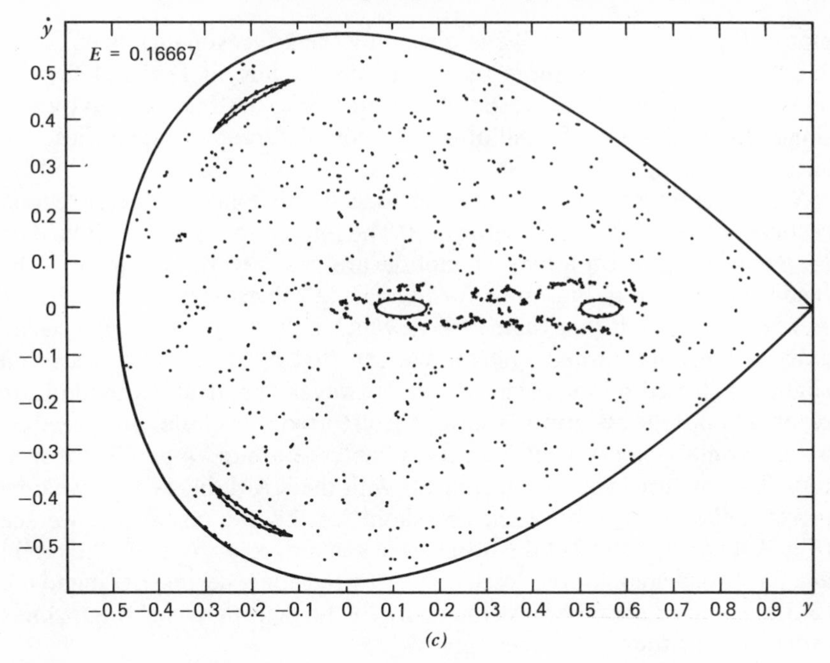

(c)

Figure A.4.3. (Cont'd).

Let us have a closer look at the resonances in this system. The relevant resonances, which will lead to "dangerous" behavior in presence of the coupling, are $2\omega_1 = 2\omega_2$ and $2\omega_1 = 3\omega_2$ (briefly, the 2–2 and the 2–3 resonances). From Eqs. (A.4.5) we see that the 2–2 resonance appears when

$$J_1 = 5J_2$$

Equation (A.4.4) then gives the values of J_1, J_2 corresponding to the resonant (unperturbed) torus as a function of the energy:

$$\begin{aligned} J_1 &= \tfrac{5}{13}[1-(1-\tfrac{13}{3}E)^{1/2}] \\ J_2 &= \tfrac{1}{13}[1-(1-\tfrac{13}{3}E)^{1/2}] \qquad \text{(2–2 resonance)} \end{aligned} \tag{A.4.6}$$

Similar calculations yield the following result for the 2–3 resonance:

$$\begin{aligned} J_1 &= \tfrac{1}{13} \\ J_2 &= \tfrac{5}{13} - (\tfrac{3}{13} - E)^{1/2} \qquad \text{(2–3 resonance)} \end{aligned} \tag{A.4.7}$$

We must further require that J_1, J_2 be positive. This is possible for all values of the energy, $E \geqslant 0$ for the (2–2) case, but it is only possible for $E \geqslant 14/169$ for the (2–3) case. Hence, the structure of the phase space for the *unperturbed* Hamiltonian, is represented by a Henon–Heiles type of plot in Fig. A.4.4. For small energies, only the 2–2 resonance is possible. On the corresponding torus (drawn in dotted line in Fig. A.4.4*a*) the motion is periodic: the corresponding trajectories are represented by any single point of the circle. All other tori contain ergodic trajectories: the latter are mapped by continuous circles.

When the energy becomes greater than the threshold value, a second resonant torus appears inside the first. The interaction is now switched on ($\alpha, \beta \neq 0$) and the equations of motion are fed into the computer. The results are shown in Fig. A.4.5 for increasing values of the enegy. Figure A.4.5*a* corresponds to a value of E at which only the (2–2) resonance can exist in the unperturbed system. We see that most of the tori are but slightly distorted by the perturbation. However, the neighborhood of the resonant unperturbed torus is seriously distorted: the circle is replaced by a zone containing crescentlike curves centered around a periodic trajectory. This picture is in nice agreement with the KAM theorem. If we now increase the energy above the threshold for the 2–3 resonance, we see (Fig. A.4.5*b*) that the 2–2 distorted zone has moved away from the origin and has broadened. Moreover, a new distorted zone, corresponding to the (2–3) resonance appeared. As the energy is further increased, both zones broaden until they start overlapping.

At this precise point, a violent instability sets in, and an important fraction of the phase space gets covered by ergodic trajectories (Fig. A.4.5*d*).

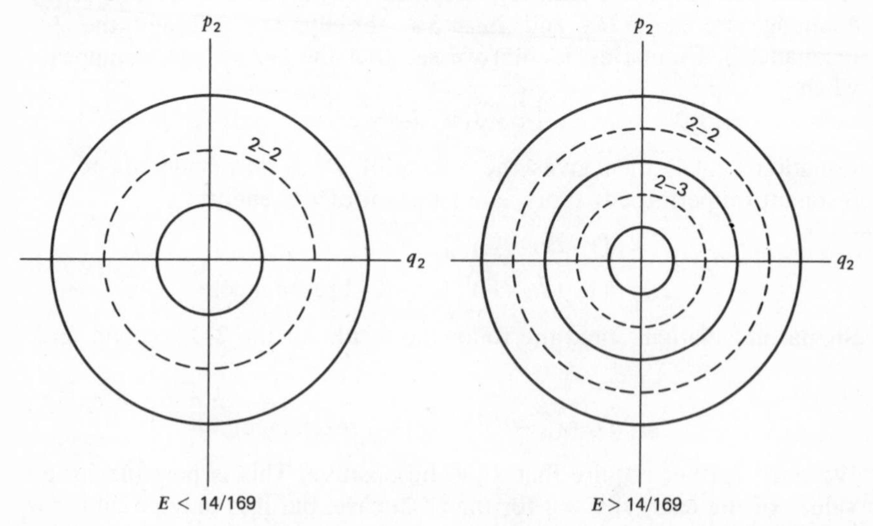

Figure A.4.4. Trajectories for the unperturbed Walker–Ford Hamiltonian (A.4.4).

This discovery, which has been further developed by Ford and his co-workers, is very important. It shows that *isolated resonances do not seriously affect the motion. But as soon as resonant zones overlap, instability sets in and a sizable fraction of the trajectories leave their tori and become ergodic.*

The significance of this result is the following. The systems discussed so far have very few degrees of freedom. It is amazing to see the degree of irregularity displayed already by such simple systems. A few years ago, nobody would have thought of possible ergodicity in two-oscillator systems. It was generally believed at that time that ergodicity is a property only of very large systems. Ford's results actually give us a hint toward the role of the number of degrees of freedom. As this number increases, an increasing number of resonances become possible, and more important, an increasing chance of overlap appears. Therefore it is likely that instability and ergodicity come in sooner. As the number of degrees of freedom becomes of the order of 10^{23}, as in realistic systems dealt with in statistical mechanics, it becomes almost certain that ergodicity becomes the rule rather than the exception: the KAM theorem then becomes irrelevant. The mathematical reason for this failure might be the following. For any finite system, and for sufficiently small perturbations, the KAM theorem tells us that the distorted region of phase space is small. However, in a many-body system, we may well have an accumulation of small effects that build up into a catastrophy as $N \to \infty$. This aspect is not at all covered by the KAM theorem.

Of course, none of the statements of this section is proved: they are simply conjectures, representing reasonable extrapolations from the results of numerical experiments. We shall see, however, in the next section that these ideas can be well connected to results obtained by diametrically opposite methods, and representing another extreme type of dynamical systems.

A.5. MEASURE IN PHASE SPACE

In parallel with the developments discussed in previous sections, another approach to the study of dynamical systems has been developed in the 1930s by a group of mathematicians led by von Neumann, Birkhoff, and Hopf. This approach is completely different in style and philosophy from the one described in previous paragraphs. The motivation of the first type of studies came from astronomical problems: Typically, the foundations of the theory were firmly laid by Poincaré in his classic *Traité de Mécanique Céleste.* The incentive for the work to be discussed now came directly from statistical mechanics: the mathematicians try hard to justify

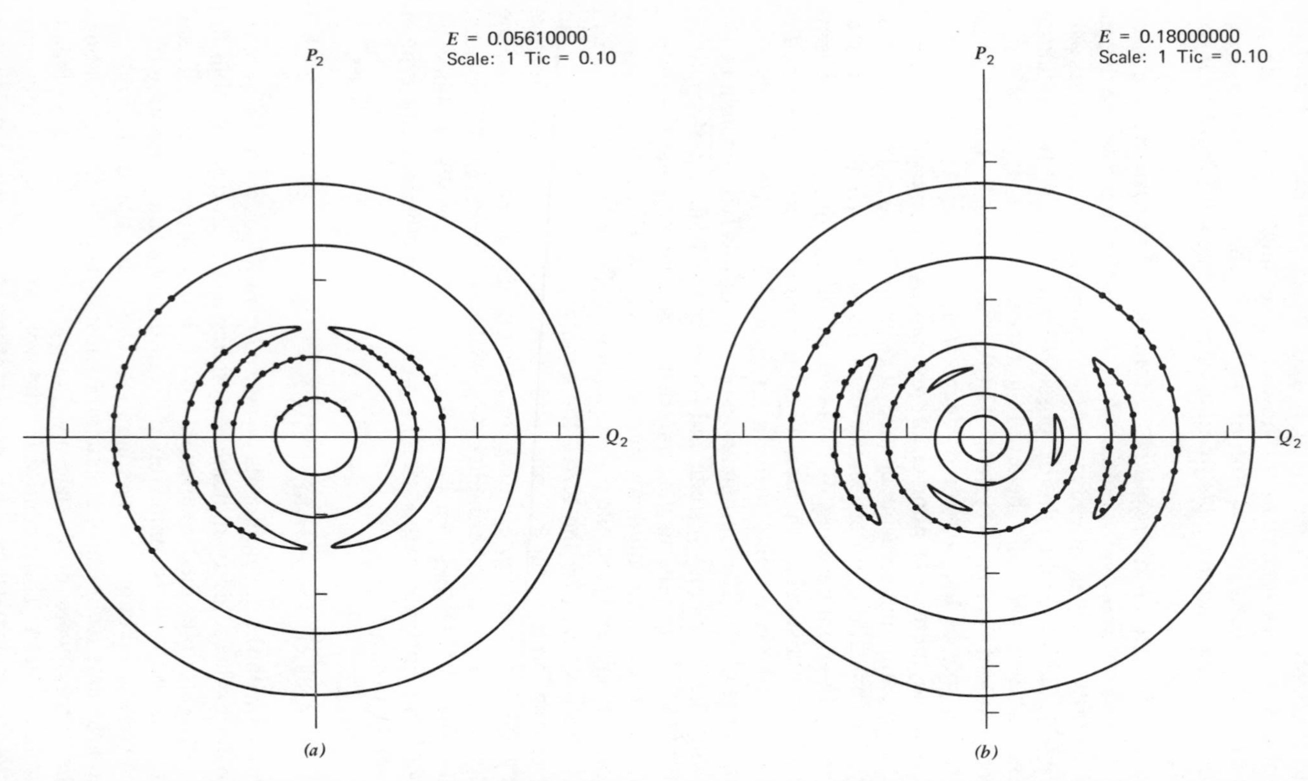

Figure A.4.5. Trajectories of the Walker–Ford system, for increasing energies. (From Walker and Ford, 1969.)

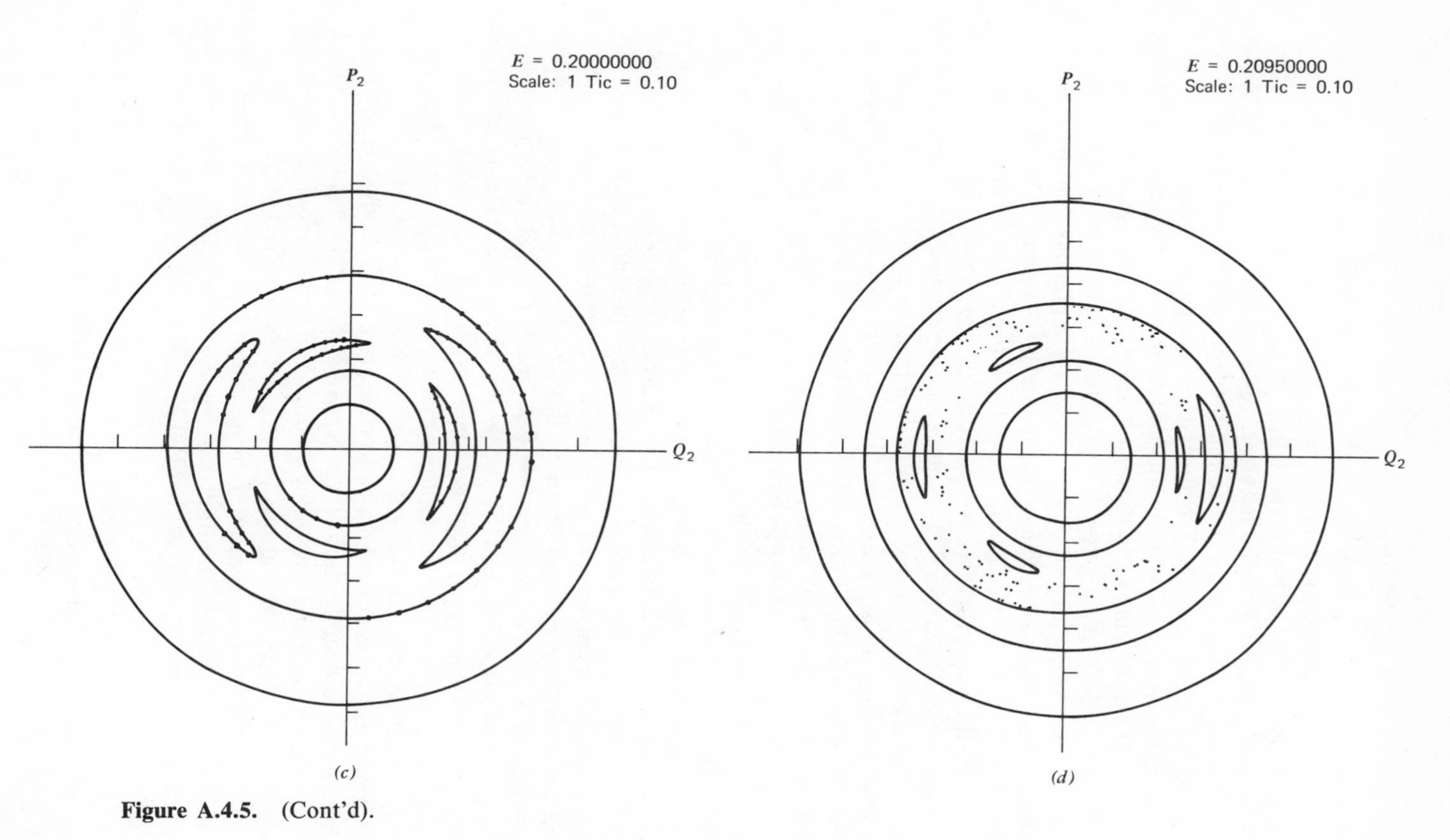

Figure A.4.5. (Cont'd).

some assumptions formulated by Boltzmann and Gibbs, which were considered basic to statistical mechanics, but which could not make sense in a rigorous mathematical formulation.

To clearly understand the nature of these problems, we must realize that, in going from Section 1.2 to the present one, we adopt a more and more global point of view. We started in Section 1.2 with the following question: "given the initial positions and momenta of a dynamical system at time $t = 0$, what will be the value of these quantities a short time later?" This is a strictly local question. Starting from a point x_0 in phase space, we construct the trajectory of x_t piecewise, by a gradual "unfolding" of a canonical transformation generated by the Hamiltonian.* The next step in globality is to ask: "For given initial condition, what is the topological nature of the trajectory? Is it a smooth line, or does it fill a domain of more than one dimension?" This question was studied in connection with the problem of the nature of the integrals. Next, we may ask: "Consider a system starting at an initial condition corresponding to periodic or conditionally periodic motion; what is the behavior of a system that starts at time zero in the neighborhood of the first? Does its trajectory remain close to the first or will it systematically deviate from it?" This is the type of problem studied in Poincaré's work and in the KAM theory, as well as in the numerical experiments of Section A.4. It is global in the sense that it studies trajectories as a whole; still it is local in the sense that it is concerned with the neighborhood of a given trajectory in phase space. One imagines an infinitesimally narrow "tube" around the axial trajectory (or a narrow sheet around an ergodic trajectory), and asks whether another trajectory remains or not in the tube for all times. We may call this a *local-stability* problem. We now reach the last, natural stage of study, by considering now the behavior of all the systems contained initially in a *finite* region of phase space. How will this region move as a result of the motion of its constituent points? How is its shape changed in time? What are the laws of motion of such *sets* of points?

Clearly, the problem at hand is to formulate a "hydrodynamics" in the phase space and study the various types of "flows" in this space. This is precisely what we are now going to do, after taking a look at Fig. A.5.1, which summarizes the various types of problems discussed.

Before being able to start such investigations, we must agree on a method of measuring quantitatively the size of a set of points in phase space. The simplest way of doing this is to consider its usual volume. Defining

$$dx = dq_1\, dq_2 \cdots dq_N\, dp_1\, dp_2 \cdots dp_N \qquad \text{(A.5.1)}$$

* In this section, we systematically use the shorthand x for the set of phase-space coordinates $\{q, p\}$ and x_t for $\{q(t), p(t)\}$.

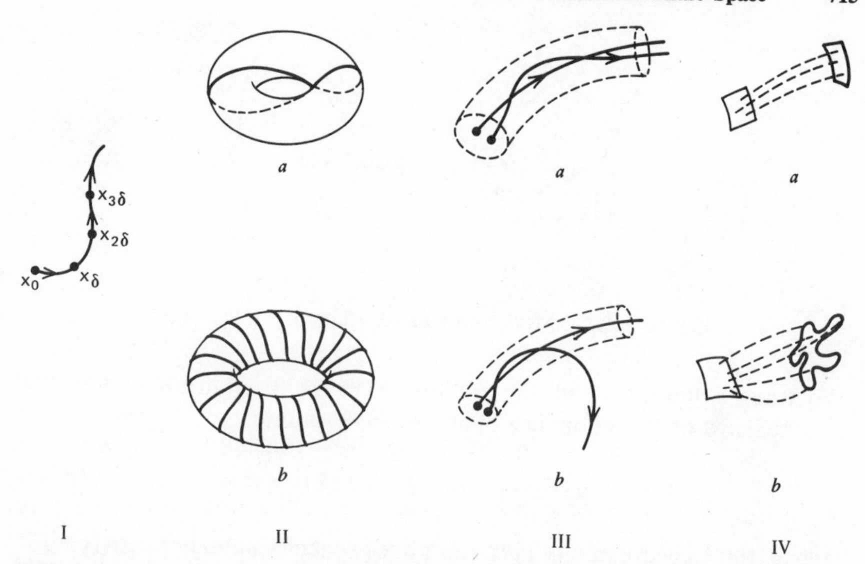

Figure A.5.1. Problems of increasing globality. I. Step-by-step integration of the equations of motion. II. Topological nature of complete trajectories: (*a*) periodic motion; (*b*) conditionally periodic motion on a torus. III. (*a*) Local stability; (*b*) local instability. IV. Types of flow in phase space.

we calculate the volume of a certain domain γ of phase space as

$$\mathcal{V}(\gamma)=\int_\gamma dx \tag{A.5.2}$$

Although this is a perfectly acceptable definition, it has a rather serious disadvantage for what we are going to do later. Indeed, if we extend the integration over the complete phase space Γ, the integral will be in general infinite. As we are interested in the global behavior of the fluid, this feature is clearly unpleasant. In order to avoid this inconvenience, we shall weight the various regions of phase space in a different way, by choosing once and for all, a certain nonnegative function:

$$F(x)\geqslant 0 \tag{A.5.3}$$

and defining the *measure* of an infinitesimal domain of phase space as

$$d\mu = F(x)\,dx \tag{A.5.4}$$

and, correspondingly, the measure of a finite domain γ as

$$\mu(\gamma)=\int_\gamma d\mu = \int_\gamma dx\,F(x) \tag{A.5.5}$$

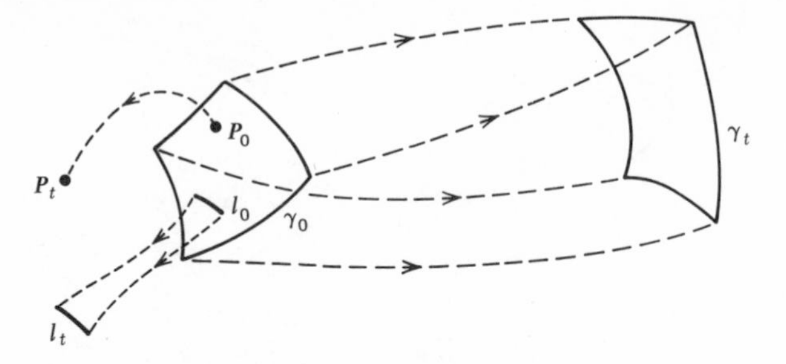

Figure A.5.2. Motion of a phase element.

We assume that the weighting function $F(x)$ is chosen in such a way that the measure of the complete phase space Γ is one*:

$$\mu(\Gamma) = \int_\Gamma d\mu = \int_\Gamma dx\, F(x) = 1 \tag{A.5.6}$$

Hence, the measure of any part γ of phase space is a number satisfying

$$0 \leq \mu(\gamma) \leq 1 \tag{A.5.7}$$

We shall be interested in the transformations of a domain γ as it moves in time. The important fact about this motion is the existence of certain invariants. We first show that the volume of every phase element γ, as defined by (A.5.2), remains constant under the motion. Let us denote by γ_0 a certain phase element at time 0, and by γ_t the element containing at time t at all the points that were initially in γ_0 (see Fig. A.5.2). The coordinates of individual points at time zero and at time t are correspondingly denoted by x_0 and x_t. By (A.5.2) we have

$$\mathcal{V}(\gamma_t) = \int_{\gamma_t} dx_t = \int_{\gamma_0} \frac{\partial(x_t)}{\partial(x_0)}\, dx_0$$

In the second step we changed the integration variables from x_t to the initial coordinates x_0. But we know that these two sets of variables are related by a canonical transformation. One of the important properties of such a transformation is that the corresponding Jacobian equals 1:

$$\frac{\partial(x_t)}{\partial(x_0)} = 1 \tag{A.5.8}$$

Hence

$$\mathcal{V}(\gamma_t) = \int_{\gamma_0} dx_0 = \mathcal{V}(\gamma_0) \tag{A.5.9}$$

* The weighting function $F(x)$ is, of course, identical to the phase-space distribution function introduced in Section 2.2. It is, however, introduced here in a different context.

The volume of any domain of phase space is preserved by the motion: this is the celebrated *Liouville theorem*. At present we note that it can be generalized to include the case of different measures. Let us further restrict the definition of the function $F(x)$ of Eq. (A.5.3). We choose it to be a function of x only through one or more isolating integrals of the motion:

$$F_0(x) = F_0(\Phi(x)) \tag{A.5.10}$$

It follows that $F_0(x)$ is itself a constant of the motion. It is then a trivial matter to show that not only the volume, but also *any invariant measure of a phase element is conserved under the motion:*

$$\mu_0(\gamma_t) = \mu_0(\gamma_0) \tag{A.5.11}$$

Before closing this paragraph, let us make the following remark. It is characteristic of all arguments based on the concept of measure that *individual erratic points or even infinite sets of points of measure zero are completely disregarded.* Thus, in our primitive two-dimensional figure A.5.2, it appears that the point P_0 and all the points on the line l_0 are "going astray" as compared to the bulk motion. Such points, however, form a subset of measure zero in the original set and are therefore disregarded: they do not contribute to the total measure. This important fact must always be kept in mind in all discussions of this type. It helps understanding the power, but also the limitations of such methods. On one hand, it allows us to describe the general, global properties of the motion in phase space, disregarding any "pathological" initial conditions. On the other hand, it may well happen in a physical problem that precisely these pathological systems are really the interesting ones.

A.6. ERGODICITY AND MIXING: SINAI'S THEOREM

We are now ready for the global study of evolution in phase space. The problem can be formulated as follows. We define a *flow* in phase space as a transformation, or a mapping of phase space onto itself, whereby every point x is transformed into some other point x_t by means of an operator $T(t)$:

$$x \to x_t = T(t)x \tag{A.6.1}$$

More generally, a function $b(x)$ on the phase space, that is, a dynamical function in our terminology of Section 1.2, transforms as

$$b(x) \to b(x_t) = b(T(t)x) \tag{A.6.2}$$

These transformations $T(t)$, considered for all values of the parameter

t, must, moreover form a *group*. The propagator $\bar{U}(t)$ introduced in Section 1.2 has precisely all these properties. However, the mathematicians who studied these problems have considered much more general flows, which are not necessarily generated by a Hamiltonian. They do, however, have one property in common with the Hamiltonian flows: they are assumed to be *measure preserving*, that is, to obey Eq. (A.5.11). Since the pioneering work of von Neumann, Birkhoff, and Hopf in the 1930s, this problem has developed into an enormous mathematical field, combining the methods of measure theory, spectral theory, functional analysis, topology, and so on. It is loosely known as the *ergodic theory* or *general dynamics*. It is clearly impossible, in the framework of this book, to give any detail, let alone any proof of the results of this highly abstract field. Our purpose is, rather, to discuss very qualitatively (and superficially) some of the concepts which evolve from this theory, and to disclose their physical relevance. In doing this, we shall not even follow the historical development of the problem.

The present trend of research in ergodic theory appears to be in a classification of the various types of flows, together with the study of the properties of the various classes. The Russian school, pioneered by Khinchine, and later led by Kolmogorov, Anosov, Arnold, and Sinai, is particularly active in this field.

To grasp this problem we must realize that in a measure-preserving flow, an initial element of phase space moves through the space, keeping its measure constant at all times. But besides its measure, we do not know anything else about the motion, the trajectory or the shape of this element: the wildest possibilities are open for consideration. We drew three typical examples in Fig. A.6.1.

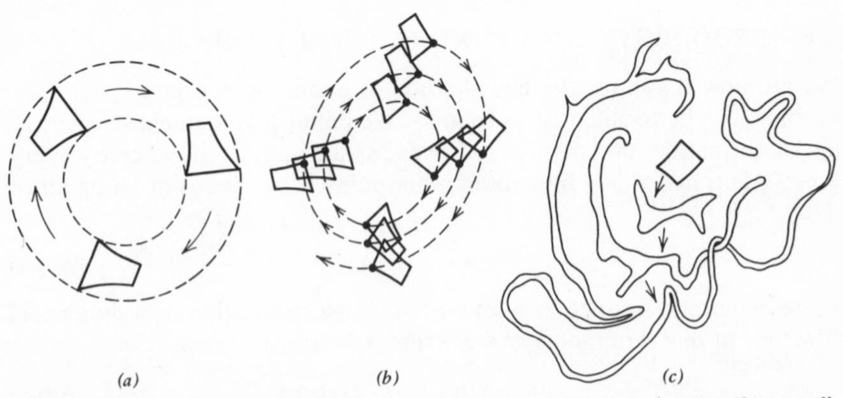

Figure A.6.1. Various types of flow in phase space: (a) nonergodic; (b) ergodic but not mixing; (c) mixing.

In the case (a), the phase element moves without distortion, returning to its initial location every T sec. It all looks like a periodic translation of a solid body. During its motion, the "drop of phase fluid" sweeps a finite fraction of the available phase space. Such a situation can arise in real mechanical systems. Consider, for example, a set of harmonic oscillators with commensurate frequencies: the trajectories of their representative points in phase space are closed curves on a torus (see Section A.2). If we consider a patch on the surface of one of these tori, its motion will look precisely as in Fig. A.6.1a. Another type of motion is depicted in Fig. A.6.1b. Here, the shape of the volume element is only slightly altered during the motion. But it never returns to its initial location. As one watches it for a long enough time, it sweeps out a large fraction of the phase space, possibly even the whole phase space. Moreover, as one waits for an infinite time, the element crosses every domain of phase space an infinite number of times. Such a flow is called *ergodic*.

But even more complicated types of flow are possible as shown in Fig. A.6.1c. Here, the *shape* of the initial element is very violently distorted. An initial square will send out amoebalike arms, then look like an octopus with highly contorted arms, and eventually it grows out into extremely fine filaments (the measure must be preserved!) that spread out uniformly over the whole space. The classical image for depicting this situation is that of a glass containing 45 cm^3 of vodka, into which one carefully deposits a drop of 5 cm^3 of schnaps. After stirring the fluid, and sampling a spoonful of liquid, you will find everywhere in the glass a delicious cocktail containing 10% schnaps and 90% vodka.* For obvious reasons, such a flow is called *mixing*. It can occur only if the individual points starting out close together in space have widely diverging trajectories; typically, one can prove the mixing property for a system in which initially neighboring points diverge exponentially in time.

How can one characterize such complex motions mathematically? Consider first the *ergodic motion*. In this case, the trajectory of any individual point crosses an arbitrarily chosen domain in phase space an infinite number of times (as $t \to \infty$). This is true wherever you pick up the test domain, and however small its measure. You cannot, however, let it shrink to a single point. It is a known topological theorem that a one-dimensional trajectory, even if it fills densely a higher-dimensional

* The history of this image is quite interesting. It was first introduced by Gibbs in a gloomy Puritan way using ink and water. Later, Halmos pepped it up by using gin and vermouth. Arnold and Avez seem (strangely enough) to have a more American taste: rum and coca cola. It seems fair enough to us to give the experiment a German and Russian flavor, considering the contributions of Hopf, Khinchine, Kolmogorov, and Sinai to this important problem.

domain, cannot go through every point of that domain.* In spite of this, a very interesting property can be proven. Consider a dynamical function $b(x)$, which is integrable over the phase space:

$$\int_{\Gamma} d\mu_0 \, |b(x)| < \infty \tag{A.6.3}$$

Birkhoff proved that the following time integral has a finite value, independent of the lower limit of integration:

$$\lim_{T\to\infty} \frac{1}{T-t_0} \int_{t_0}^{T} dt \, b(x_t) = \bar{b}(x) \tag{A.6.4}$$

This function $\bar{b}(x)$ is adequately called the *time average* of the dynamical function $b(x)$ along a trajectory of the system. It is a function of the coordinates x at time t_0. Birkhoff's theorem asserts its existence for almost all initial conditions x. ("Almost all" means "all, except possibly for a set of measure zero.")

Birkhoff then proved a second theorem (which is really a consequence of the first), which for a long time was thought to be of utmost importance to statistical mechanics. He showed that, *under a certain condition*, to be discussed below:

(a) $\bar{b}(x)$ *is constant almost everywhere in space* (i.e., *it is independent of* x);

(b) *the constant time average* $\bar{b}(x)$ *equals the space average* $\langle b \rangle$:

$$\bar{b}(x) = \langle b \rangle \equiv \int_{\Gamma} d\mu_0 \, b(x) \tag{A.6.5}$$

This theorem is known by a name proposed by Boltzmann: the *ergodic theorem*.† We now *define* an *ergodic flow* as a flow for which Eq. (A.6.5) is valid. Intuitively, Eq. (A.6.5) expresses the fact that almost every trajectory of the system spends equal times in equal regions everywhere in phase space.

The importance of the theorem stems from the following considerations. *If* one is interested in calculating time averages of dynamical function, one is confronted with the exceedingly difficult problem of evaluating the function x_t, that is, of solving the equations of motion.

* A more familiar example is the set of rational points on the segment [0, 1]. These points fill densely the segment, that is, in every arbitrarily small interval there is an infinite number of rational points. However, the set is of measure zero: there are still "many more" irrational numbers.

† Various forms of the ergodic theorem have been proved: they differ in the type of regularity conditions required for $b(x)$ and in the type of convergence assumed in Eqs. (A.6.4), (A.6.5). We do not discuss these technical matters here.

However, under certain conditions, Birkhoff's theorem tells us that the time average is equal to the space average, the computation of which is by orders of magnitude simpler.

We now come to the condition under which (A.6.5) is valid. Birkhoff showed that the equality holds if and only if the phase space is *metrically indecomposable* (equivalently, one says that a flow admitting no metrical decomposition of the space is *metrically transitive*). Metrical indecomposability means that the phase space cannot be separated into two invariant regions, say, γ_i, $i = 1, 2$, whose measures are different from zero or from 1. Intuitively, no trajectory can be confined to a finite portion of phase space but, rather, must wander through the whole space. Another way of expressing this is to state that every invariant subset of the phase space has either measure zero or measure one (in which case it comprises almost all of the phase space).*

Although this concept is pretty clear intuitively, it is extremely difficult to prove in full rigor that a given flow is metrically transitive.

We immediately note that for a Hamiltonian system, the flow can certainly not be ergodic in phase space. Indeed, there exists at least one isolating integral, which is the energy. We know therefore that every trajectory is confined to a hypersurface of constant energy. It then follows that the part of phase space defined by the inequality $H(x) < \alpha$ where α is a given number, is an invariant region of positive measure. Hence the phase space is metrically decomposable.

This is, however, not a serious difficulty. It is more interesting to study the motion on a surface of constant energy. Provided we can properly define an invariant measure on the surface (which is almost always possible), we can now study the ergodicity of the motion on this $(2N-1)$-dimensional surface. If there are other isolating integrals, the process of reduction is correspondingly continued. Nonisolating integrals, on the other hand, do not prevent ergodic motion, as appears from our discussions of Sections A.2–A.4. However, it is not proven rigorously that if all integrals except the "trivial" ones are nonisolating, the motion is necessarily ergodic.

We now come to the characterization of a *mixing flow*. To understand the mathematical definition, we consider at time zero two regions of phase space, A and B, both of finite measure $\mu_0(A)$ and $\mu_0(B)$ (see Fig. A.6.2). As time goes on, we hold the region B fixed, but let A evolve according to the laws of the flow. It is therefore transformed at time t into a new region A_t. If the flow is mixing, the initial region A will spread out through the whole phase space, eventually intersecting B. Let $A_t \cap B$ denote the

* Strictly, there is a third possibility: the subset is not measurable.

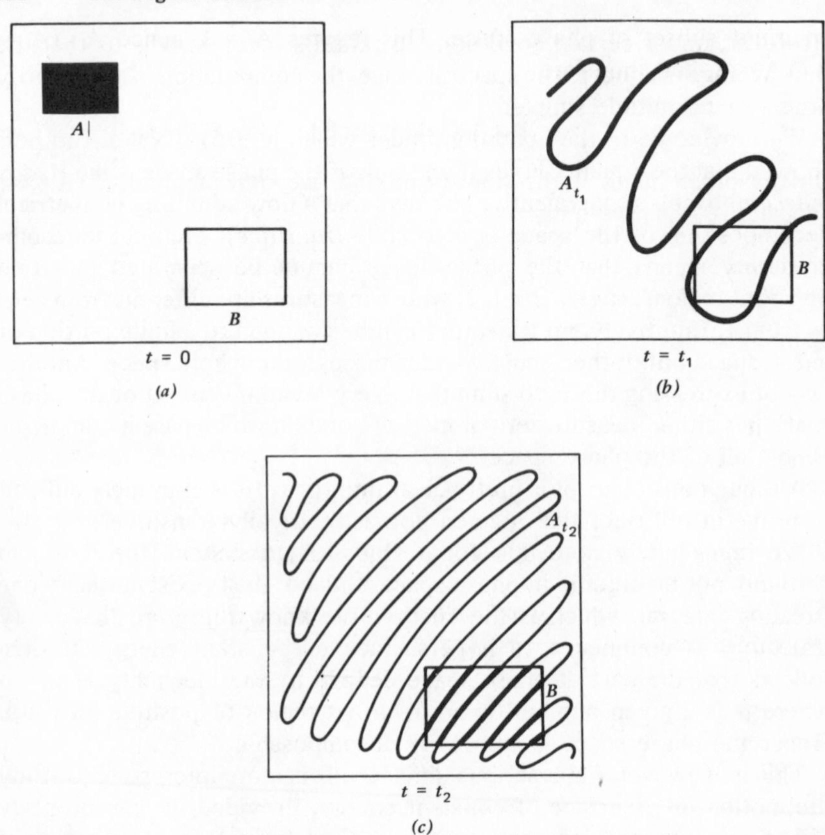

Figure A.6.2. Characterization of a mixing flow.

region of intersection of A_t and B, and $\mu_0(A_t \cap B)$ be its measure. As the spreading becomes more and more uniform, any part B will look locally as the complete space. In particular, the measure of the intersection, $\mu_0(A_t \cap B)$ will be to the measure $\mu_0(B)$ in the same ratio as the total measure $\mu_0(A_t)$ is to the measure of the complete space, which is 1 by definition. But the flow is measure preserving, hence $\mu_0(A_t) = \mu_0(A)$. We may therefore summarize this discussion by *defining a mixing flow* through the property:

$$\lim_{t\to\infty} \frac{\mu_0(A_t \cap B)}{\mu_0(B)} = \mu_0(A) \tag{A.6.6}$$

It is easily shown that *mixing implies ergodicity*. Indeed, let A be an *invariant* subset of phase space. This implies $A_t = A$ hence $A_t \cap A = A \cap A = A$, and therefore

$$\mu_0(A_t \cap A) = \mu_0(A)$$

On the other hand, if the flow is mixing, we may apply Eq. (A.6.6), choosing $B = A$; we obtain

$$\lim_{t \to \infty} \mu_0(A_t \cap A) = [\mu_0(A)]^2$$

Combining this with the previous result we have:

$$\mu_0(A) = [\mu_0(A)]^2$$

so that $\mu_0(A)$ equals 0 or 1. Therefore, the flow is metrically transitive and hence ergodic.

The converse is, of course, not true: Ergodicity does not imply mixing. In a flow of the type depicted in Fig. A.6.1*b* (realized, for instance, by the motion of harmonic oscillators on a torus), the intersection of A_t with a fixed domain is alternatively empty and nonempty; its measure tends toward no limit at all as $t \to \infty$. Hence there exist flows that are ergodic but not mixing.

We shall not further develop these concepts here. The important question to ask next is the following. Given a dynamical system, whose Hamiltonian is a known function $H(x)$, is it ergodic? Is it mixing? For a very long time, this question remained unanswered. People were beginning to wonder whether abstract ergodic theory had anything to do with real mechanical systems. The situation however changed dramatically in 1962, when Sinai announced his proof that a *system of N hard spheres enclosed in a box with hard walls is a mixing system.* Such a system, although not yet very realistic, is already quite close to the kind of systems studied in statistical mechanics. The most startling feature of this result is, however, the fact that the mixing property holds whenever $N \geq 2$. This destroys an old belief that ergodicity can only be a property of very large systems. The fact that two billiard balls on a table have not only ergodic, but mixing, and even "exponentiating" behavior is of utmost importance in pointing out the extraordinary complexity of the motion of the simplest systems. Although the difficulty of deriving such results is beyond the possibilities of today's average theoretical physicist, there is no doubt that Sinai's theorem will be considerably developed and extended in coming years. More important even will be the effort in understanding the consequences of these results for the foundations of statistical mechanics. This effort is only at its very beginning.

A.7. ERGODIC THEORY AND STATISTICAL MECHANICS

The first question that comes up in statistical mechanics (see Section 2.2) is the following: How do we establish a correspondence between microscopic dynamical quantities b and macroscopic dynamical quantities B? At this stage a *postulate* of a physical nature is needed.

Boltzmann argued that the measurement of a macroscopic quantity always requires a finite amount of time, due to the inertia of the measuring apparatus. Hence the result of a measurement provides us with an average behavior of the system over a certain time interval. Extrapolating this argument, Boltzmann finally came with the suggestion of the following definition:

$$B = \lim_{T \to \infty} \frac{1}{T} \int_0^T dt\, b(x_t) = \bar{b} \tag{A.7.1}$$

In other words, a macroscopic quantity is the time average of the corresponding microscopic quantity evaluated over an infinite time.

If this view is accepted, the problem of the evaluation of such an average would be enormously simplified by the *ergodic theorem*. Indeed, if this theorem holds, the practically insoluble dynamical problem of calculating an average of b along the trajectory (to be determined) of a single system, is replaced by the much simpler problem of calculating the average of this quantity over the energy surface. The latter method leads to a very appealing physical interpretation. The concept of measure, which is so crucial in ergodic theory, is just as crucial in probability theory. We are thus led to look at the dynamical quantity b as a *random variable*. Instead of considering a single system we consider an infinite number of identical copies of the system, spread continuously throughout the phase space. The set of all these systems is called an *ensemble*. The density of the representative points $F(x)$ is interpreted as the *probability density* for finding the system of interest at a particular point in phase space. (In other words, the measure of a domain in phase space is interpreted as the probability of finding the system in that region.) The total measure of phase space being equal to 1, the system is certainly somewhere in the available phase space. The macroscopic dynamical quantity B is now identified with the usual *expectation value* of the random variable b, corresponding to the given probability distribution:

$$B = \int d\mu_0\, b = \int dx\, F_0(x) b(x) \equiv \langle b \rangle \tag{A.7.2}$$

Many authors have adopted this point of view, which we summarize again as follows:

A macroscopic quantity is the time average of a microscopic quantity over an infinite time. Through the ergodic theorem, this time average can be replaced by a statistical average taken over an ensemble uniformly distributed over an energy surface.

There are, however, serious objections to this argument, which made a number of authors adopt a different point of view. These objections became even more strongly justified in recent times. Let us analyze them here.

If Boltzmann's view is to hold, the ergodic theorem becomes the cornerstone of statistical mechanics. Disregarding topological subtleties that were opposed to Boltzmann's original formulation of the ergodic hypothesis, we may stick to Birkhoff's rigorous formulation. For the theorem to hold, it must be shown that the energy surface is metrically indecomposable (see Section A.6). For a long time, this statement had the status of a working assumption. But in recent years we do have a rigorous mathematical statement in Sinai's theorem, which we mentioned in Section A.6. This theorem tells us that a system of N hard spheres ($N \geqslant 2$) is ergodic. However, Sinai's result clearly shows that the *ergodic theorem might be necessary for the foundation of statistical mechanics, but cannot possibly be sufficient.* Indeed, it has always been thought before Sinai's result, that only a very large system can be ergodic. On the other hand, the results of statistical mechanics have shown that some of the most important properties of matter, such as the existence of a specific heat per unit mass, independent of the size and shape of the sample (see Section 4.7), or the existence and properties of phase transitions such as condensation or crystallization (see Section 9.4), can *only* be understood as *properties of very large systems.* For instance, our considerations of Section 4.7 show that for a foundation of statistical thermodynamics, the thermodynamic limit (properly intrepreted as in Sections 3.2 and 3.3) is a crucial ingredient for proving the existence of intensive quantities and the thermodynamic stability of a macroscopic system. Hence, Sinai's result definitely shows that there exist ergodic systems that are of no interest to statistical mechanics.

There is, however, a much more serious objection against founding statistical mechanics on the ergodic theorem. Boltzmann's definition necessarily provides only macroscopic dynamical quantities that are *independent of time.* But such quantities are rather exceptional. Only in the state of equilibrium are the dynamical quantities stationary. Hence Boltzmann's definition (A.7.1) may provide good definitions of static thermodynamical quantities, but cannot possibly be taken as a foundation of hydrodynamics, electrodynamics, or any other branch of macroscopic physics in which the process of time evolution is of basic importance.

We thus definitely join the group of physicists (including, among others, Tolman and Landau) holding that the ergodic theorem is an interesting property of dynamical systems, but is irrelevant as a foundation of statistical mechanics. The way out of the difficulties mentioned above is *to consider the ensemble average* (A.7.2) *as the primary definition of a macroscopic dynamical function,* not to be related to any other concept considered as more fundamental. The ergodic theorem is thus side-stepped. Moreover, the main difficulty mentioned above no longer subsists. The macroscopic quantity B in Eq. (A.7.2) can now be a function of time. Indeed, the corresponding b can be taken to be a function of time, to be averaged over the ensemble: the expectation value clearly will depend on time. We do not need any nonmechanical assumption to define the law of evolution in time: the latter is provided by the equations of mechanics: $b(t) = \bar{U}(t)b$ [see Eq. (1.2.24)]. Through Eq. (A.7.2) this mechanical law of evolution *induces* a law of evolution for macroscopic quantities, $B(t)$ [see Eq. (2.2.9)].

Although, in the present view, the concept of an ensemble is considered as a postulate not needing further justification, it is worthwhile discussing for a while the reasonableness of such a postulate. The main argument here seems to be in the violent instability of mechanical motion, which clearly appeared in this Appendix. If we were able to determine exactly the initial condition of a dynamical system, we would make some prediction of its behavior at time t. But if we make a tiny error, picking up a slightly different initial condition, very close to the former, we would very quickly be off by an appreciable amount. This is a result of the *mixing* property that seems to be widespread among systems of interest to us. In macroscopic physics, on the other hand, we do not generally find such violent sensitivity on the choice of initial conditions. Hence a macroscopic observation involves some kind of smoothing. Any such observation allows us to determine a relatively small number of parameters (density, velocity, temperature, etc.) of the system; this number is always less by an enormous order of magnitude than the 10^{23} numbers required for the specification of a mechanical initial condition. Hence we cannot make a definite prediction about the system, as we do not know its initial condition. The only features we can predict are those that are common to all, or at least to most systems whose mechanical initial condition is compatible with the few things we do specify about the system. Hence *we cannot predict definitely the outcome of a given experiment. We may, however, predict the most probable or the average result of this experiment if we imagine it repeated a large number of times under the same conditions.* The modern physicists need not to be shocked by this statistical nature of the prediction process. This formulation of the basic postulate is the same as in the interpretation of quantum mechanics.

We thus arrive at the following practical method of statistical mechanics. Having specified all that we know of the system under study, we construct at time zero an ensemble including all the systems whose phase-space coordinates are compatible with the known information. We then study the evolution of the initial ensemble in the course of time, agreeing to define macroscopic quantities at all times by Eq. (A.7.2). If the system is of mixing type, the initial ensemble will eventually spread out uniformly over the entire energy surface, with the macroscopic consequence that the system reaches thermal equilibrium.

It therefore appears from the present discussion that the *mixing property of a mechanical system is much more important for the understanding of statistical mechanics than the mere ergodicity*. However, it should be stressed again that *mixing alone does not provide a necessary and sufficient justification of the methods of statistical mechanics.* It does not tell us, for instance, that the definition (A.7.2) gives the value of B observed in most experiments and that any deviation or fluctuation from the average is a negligibly unfrequent event. It is shown in Section 4.6 that this important property is only ensured if the number of particles in the system is very large. Again, none of the known mixing theorems tell us anything about the rate at which equilibrium is approached or about the mechanism of this process.

A very significant insufficiency of the mixing property as a foundation of nonequilibrium statistical mechanics was pointed out by Lebowitz (reference at the end of this Appendix). He considered a system of a *finite* number of particles and calculated the integral, from $t=0$ to $t=T$ of the velocity autocorrelation function. In the limit $T\to\infty$, this quantity should be related to a transport coefficient, namely, the diffusion coefficient (see Section 21.4). Lebowitz showed that, if one assumes the system to be mixing, the limit of this integral as $T\to\infty$ is strictly *zero*. This example clearly shows that dissipativity, an essential property of macroscopic systems, cannot be derived from mixing.

A detailed rigorous study of the way in which the concepts of mixing and the concept of large numbers of degrees of freedom influence the macroscopic laws of motion is still lacking. However, at the present time, statistical mechanics can be developed on the basis of the postulate (A.7.2) together with the assumption of large numbers of particles and leads to a well organized body of knowledge, allowing predictions that in most cases are remarkably well verified by experiments. This fact is the ultimate justification of the methods of statistical mechanics.

BIBLIOGRAPHICAL NOTES

The *ergodic hypothesis* was introduced by L. Boltzmann, *J. f. Math.* **100,** 201 (1887).

It was also extensively discussed in Ehrenfests' article quoted in Chapter 11. It was proved to be mathematically untenable, in its first form, by:

A. Rosenthal, *Ann. Phys.* **42,** 796 (1913).

M. Plancherel, *Ann. Phys.* **42,** 1061 (1913).

It was weakened in the form of the *quasiergodic hypothesis* by E. Fermi, *Phys. Z.* **24,** 261 (1923).

The *Hamilton–Jacobi theory* is explained in detail in the book by Goldstein (quoted in Chapter 1).

The problem of the *analyticity of the integrals of the motion* was first studied by H. Poincaré. *Méthodes Nouvelles de Mécanique Céleste,* Vol. I (reprinted by Dover, New York, 1957).

A clear account can be found in A. Wintner, *The Analytical Foundations of Celestial Mechanics,* Princeton Univ. Press, 1947.

Recent studies of this problem, with reference to statistical mechanics, are:

P. Résibois and I. Prigogine, *Bull. Cl. Sci. Acad. Roy. Belg.* **46,** 53 (1960).

I. Prigogine and A. Grecos, *Physica* **59,** 77 (1972).

A detailed study of the ergodic properties of an impurity in a harmonic oscillator chain is given by R. I. Cukier and P. Mazur, *Physica* **53,** 157 (1971).

The *KAM theorem* was announced by A. N. Kolmogorov, *Dokl. Akad. Nauk USSR* **98,** 527 (1954).

The theorem was proved independently by

V. I. Arnold, *Uspekhi Mat. Nauk* **18,** 13 (1963) [English transl.: *Russian Math. Surv.* **18,** 9 (1963)].

J. Moser, *Nachr. Akad. Wiss. Gottingen,* No. 1, 1962.

A very extensive review (but mathematically very difficult) is given in V. I. Arnold, *Uspekhi Mat. Nauk* **18,** 91 (1963) [English transl.: *Russian Math. Surv.* **18,** 85 (1963)].

A clear review paper, more specifically addressed to physicists, is G. M. Zaslavsky and B. V. Chirikov, *Usp. Fiz. Nauk* **105,** 3 (1971).

In the field of *numerical experiments*, a classic is the paper by E. Fermi, J. Pasta, and S. Ulam, *Studies of Non-Linear Problems,* Los Alamos Scient. Lab. Report LA-1940 (1955).

The papers discussed in the text are:

M. Henon and C. Heiles, *Astron. J.* **69,** 73 (1964).

G. Walker and J. Ford, *Phys. Rev.* **188,** 416 (1969).

J. Ford and G. H. Lunsford, *Phys. Rev.* A **1,** 59 (1970).

The role of *metrical indecomposability* appeared clearly in:

G. D. Birkhoff, *Proc. Nat. Acad. Sci.* (*USA*) **17,** 650, 656 (1931).

J. von Neumann, *Proc. Nat. Sci.* (*USA*) **18,** 70, 263 (1932).

This problem is extensively discussed in

E. Hopf, *Ergodentheorie,* Ergebn. d. Math., Vol. 5, no. 2, Springer, Berlin, 1937.

A. I. Khinchine, *Mathematical Foundations of Statistical Mechanics*, Dover, New York, 1949.

I. E. Farquhar, *Ergodic Theory in Statistical Mechanics*, Wiley-Interscience, New York, 1965.

The concept of *mixing flows* was introduced physically by J. W. Gibbs, *Elementary Principles in Statistical Mechanics*, Yale Univ. Press, New Haven, 1902 (reprinted by Dover, New York, 1960).

It was formulated in a mathematically precise way by Hopf (see the book quoted above).

A modern review of ergodic and mixing systems is found in V. I. Arnold and A. Avez, *Ergodic Problems of Classical Mechanics*, Benjamin, New York, 1968.

The proof that a *system of hard spheres is ergodic and mixing* was obtained by

Ia. Sinai, in *Statistical Mechanics, Foundations and Applications*, (T. Bak, ed.), IUPAP meeting, Copenhagen, 1966.

Ia. Sinai, *Russian Math. Surv.* **25**, 137 (1970).

Recently, a number of excellent *reviews* appeared, specifically addressed to physicists, in which mathematical technicalities are omitted:

A. S. Wightman, in *Statistical Mechanics at the turn of the Decade*, (E. G. D. Cohen, ed.), Dekker, New York, 1971.

J. L. Lebowitz, in *Statistical Mechanics, New Concepts, New Problems, New Applications* (S. A. Rice, K. F. Freed, and J. C. Light, eds.), Univ. of Chicago Press, 1972.

I. E. Farquhar, in *Irreversibility in the Many-Body Problem* (J. Biel and J. Rae, eds.), Plenum Press, New York, 1972.

J. L. Lebowitz and O. Penrose, *Physics Today*, Feb. 1973, p. 23.

AUTHOR INDEX

SUBJECT INDEX